THE COMPLETE BOOK OF PESTICIDE MANAGEMENT

THE COMPLETE BOOK OF PESTICIDE MANAGEMENT

Science, Regulation, Stewardship, and Communication

FRED WHITFORD, Ph.D.

A JOHN WILEY & SONS, INC., PUBLICATION

DISCLAIMER

The content of this publication is for educational purposes only. Specific chapters have been reviewed in accordance with U.S. Environmental Protection Agency policy and approved for publication. Mention of trade names or commercial products does not constitute endorsement or recommendation. The authors' views have not been approved by any governmental agency, business, or university. The book is distributed with the understanding that the authors are not engaged in rendering legal or other professional advice, and that the information contained herein should not be regarded or relied upon as a substitute for professional consultation. This book is not all-inclusive; it is not necessarily a complete compilation of information pertinent to legal compliance. The use of information contained herein, by any person, constitutes an agreement to hold the authors harmless for any liability claims, damages, or expenses incurred as a result of reference to or reliance on the information provided.

This book is printed on acid-free paper. ♾

Published simultaneously in Canada.

For ordering and customer service, call 1-800-CALL-WILEY.

Library of Congress Cataloging-in-Publication Data is available.

ISBN 0-471-40728-3

Printed in the United States of America

10 9 8 7 6 5 4 3 2 1

CONTENTS

CONTRIBUTORS

John Acquavella, Senior Fellow, Epidemiology, Agricultural Sector, Monsanto Company

Butch Ambler, MSDS & Trademark Specialist, DuPont Agricultural Products

Gail Arce, Regulatory Toxicologist, Griffin

Bob Avenius, Regional Technical Manager, TruGreen-ChemLawn

Roy Ballard, Extension Educator, Purdue University Cooperative Extension Service

Lynn Ballentine, Specialist in Poison Information, Indiana Poison Center, Methodist Hospital

Dan Barber, Consultant, Product Research and Development, OSKER, LLC

Michael Barrett, Ground Water Section, United States Environmental Protection Agency

Joe Becovitz, Pesticide Program Specialist, Office of the Indiana State Chemist

Richard Bennett, Wildlife Toxicologist, United States Environmental Protection Agency

Larry Bledsoe, Research and Extension Entomologist, Department of Entomology, Purdue University

Arlene Blessing, Developmental Editor and Designer, Purdue Pesticide Programs

Amy Breedlove, Program Analyst, Office of Pesticide Programs, United States Environmental Protection Agency

Sarah Brichford, Extension Water Quality Specialist, Department of Agronomy, Purdue University

AMY BROWN, Coordinator, Pesticide Education and Assessment Program, University of Maryland

JEFF BURBRINK, Extension Educator, Purdue Cooperative Extension Service, Purdue University

CAROL BURNS, Epidemiologist, The Dow Chemical Company

JAY CASTLEMAN, Health and Safety Coordinator, E.M.P. Co-op, Inc.

BOBBY CORRIGAN, President, R.M.C. Consulting

OTTO DOERING, Professor of Agricultural Economics, Department of Agricultural Economics, Purdue University

JEFFREY DRIVER, Principal and Director, Health Sciences, infoscientific.com, Inc.

ROBERT EARL, Vice President, Public Health, International Food Information Council

C. RICHARD EDWARDS, Professor and Integrated Pest Management Coordinator, Department of Entomology, Purdue University

RICHARD FEINBERG, Professor and Director, Center for Customer Driven Quality, Purdue University

BILL FIELD, Professor, Department of Agricultural and Biological Engineering, Purdue University

JANE FRANKENBERGER, Assistant Professor, Department of Agricultural and Biological Engineering, Purdue University

TIMOTHY GIBB, Extension Entomologist, Department of Entomology, Purdue University

DAVID GUNTER, Attorney, Dean Mead

LARRY HAMBY, Hazmat Specialist, Indiana State Fire Marshal

E. MARK HANNA, Labor and Employment Attorney, the Law Officer of E. Mark Hanna

GERALD HARRISON, Professor, Agricultural Economics Department, Purdue University

DAVE HUTH, Manager, Bennett's Greenhouses

MARGARET JONES, Regional Pesticide Expert, United States Environmental Protection Agency, Region 5

JAMES KLAUNIG, Professor and Director of Toxicology, Indiana University School of Medicine

JOEL KRONENBERG, Manager, Toxicology, Monsanto Company

LARRY LEJEUNE, Assistant Pesticide Director, Louisiana Department of Agriculture and Forestry

ROSIE LERNER, Extension Consumer Horticulturist, Department of Horticulture, Purdue University

CURT LUNCHICK, Human Exposure Assessment Specialist, Aventis CropScience

TRACY MACMILLIAN, Director of Marketing, North Safety Products

DONNA MARRON, Attorney, Plews Shadley Racher & Braun

ANDREW MARTIN, Training Specialist, Purdue Pesticide Programs, Purdue University Cooperative Extension Service

NICOLE MASON, IPM Technician, Mark M. Holeman, Inc.

MONTE-MAYES, Advisor, Global Toxicology, Dow AgroSciences

SCOTT MCGINNESS, Vice President, Connolly, Ford, Bower & Leppert

BRIAN K. MILLER, Extension Wildlife Specialist, Department of Forestry and Natural Resources, Purdue University

BRIAN R. MILLER, Manager, Environmental/Auditing, Agrium Retail

AMY MYSZ, Environmental Health Scientist, United States Environmental Protection Agency, Region 5

JONATHAN NEAL, Professor of Entomology, Department of Entomology, Purdue University

HENRY NELSON, Surface Water Branch, Office of Pesticide Programs, United States Environmental Protection Agency

THOMAS NELTNER, President, Improving Kids' Environment

MICHAEL OLEXA, Professor and Director, Agricultural Law Center, University of Florida

LARRY OLSEN, Director, North Central Region Pest Management Center, Department of Entomology, Michigan State University

KEVIN PASS, President, Action Pest Control

KELLY PEARSON, Extension Educator, Purdue Cooperative Extension Service, Purdue University

JEAN SEAWRIGHT PILEGGI, Certified Management Consultant, Seawright & Associates

LAWRENCE PINTO, Entomologist and Pest Control Consultant, Pinto & Associates, Inc.

K. S. RAO, Global Product Registration Manager (retired), Dow AgroSciences

DANIEL REARDON, President, S.E.C.U.R.E Insurance

ZAC REICHER, Extension Turfgrass Specialist, Department of Agronomy, Purdue University

KEN ROGERS, Consultant, Title III Support Services

KATHERINE ROWAN, Associate Professor of Communication, George Mason University

GAIL RUHL, Senior Extension Plant Disease Diagnostician, Department of Botany and Plant Pathology, Purdue University

CLIFF SADOF, Professor of Entomology and Extension Ornamental Specialist, Purdue University

DAVID SCOTT, Pesticide Administrator, Office of the Indiana State Chemist, Purdue University

PAUL SOMMERVILLE, Management Consultant, Professional Labor Relations Services

HENRY SPENCER, Pharmacologist (retired), United States Environmental Protection Agency

JANIS STONE, Professor, Extension Textiles and Clothing Specialist, Iowa State University

ALLEN SUMMERS, President, Asmark, Inc.

MARK THORNBURG, Attorney, Indiana Farm Bureau, Inc.

RICK TINSWORTH, Senior Vice President, Novigen Sciences

MIKE TITUS, Assistant Chief, Flora Fire Department

ROBERT TOMERLIN, Exposure Assessment Specialist, Senior Vice President, Novigen Sciences

RONALD TURCO, Professor in Soil Microbiology, Department of Agronomy, Purdue University

DOUGLAS URBAN, Senior Scientist, Environmental Fate and Effects Division, United States Environmental Protection Agency

IAN VAN WESENBEECK, Environmental Fate Scientist, Dow AgroSciences

JOHN WARD, Chief, Pesticide Programs Section, United States Environmental Protection Agency, Region 5

DAN WEISENBERGER, Research Agronomist, Department of Agronomy, Purdue University

FRED WHITFORD, Coordinator, Purdue Pesticide Programs, Purdue Cooperative Extension Service, Purdue University

ARTHUR-JEAN WILLIAMS, Chief, Environmental Field Branch, United States Environmental Protection Agency

CARL WINTER, Extension Food Toxicologist, University of California at Davis

BOB WOLF, Extension Specialist Application Technology, Kansas State University

JEFF WOLT, Risk Assessment Leader, Dow AgroSciences

PREFACE

The genesis of *The Complete Book of Pesticide Management: Science, Regulation, Stewardship, and Communication* occurred 10 years ago when I became coordinator of Purdue Pesticide Programs. As part of the Cooperative Extension Service, I was asked to develop a publication series on pesticides, one that would address all aspects of pesticides and pesticide use: current issues, regulations, inventory management, use on Indiana specialty crops, and use in and around the home. Early on, I solicited assistance from people with diverse skills, expertise, and experience. The knowledge they shared and the assistance they so willingly provided throughout the years helped form the framework for this book.

Publication Authors The authors who helped on each chapter of this book are representative of the many professionals who have contributed to Purdue Pesticide Programs' publications over the years. They helped in spite of being inundated with their own work. They are scientists and professionals and friends. They come from government, manufacturing, retail industries, universities, trade associations, and not-for-profit organizations. So, to all who have helped me over the years, I thank you for your time, expertise, and encouragement. This book is a tribute to your hard work.

Reviewers Not enough can be said for the hundreds of reviewers who have given unselfishly of their time and expertise to assure technical and scientific accuracy. Your suggestions on individual publications have contributed to each and every chapter in this book. Although you are too numerous to list, you are nevertheless deserving of credit for your contributions to this book.

Editor Over the years, I have taken much credit for the articulate, easy-to-read prose of our publications; but it is our professional editor, Arlene Blessing, who consistently turns rough drafts into polished publications. Under her direction, every paragraph is examined, every sentence reviewed, every word scrutinized. Her tireless, patient editing demonstrates that words do matter and that every word should speak to the reader. I am extremely grateful for her efforts.

Illustrators I learned very early that well-edited technical material is greatly enhanced by quality illustration; the advantages of helping the reader to visualize what the text imparts are immeasurable. I am privileged to work with Stephen and Paula Adduci who consistently develop illustrations to complement the written word.

Funders It would be negligent on my part not to mention the government agencies—especially the Office of the Indiana State Chemist and EPA Region 5—private companies, and individuals who have contributed funds toward the development of many of our extension publications. Without them, it would be very difficult to bring the science and management of pesticides to the public.

ACKNOWLEDGMENTS

My journey to Purdue University involved a lot of "life learning." In retrospect, I realize that much of what I've learned about life came not from the classroom, but from working in factories, restaurants, and auction houses; on farms and off-shore rigs; and in door-to-door sales. Along the way, some very special people gave me irreplaceable gifts, and it is to these "gift givers" that this book is dedicated. Without them, I would not be who I am today, and you would not be reading this book.

My mother, Irene Péchard Whitford, gave me a gift called "work ethics." Early in my life she instilled the value of hard work and taught me that every job should be done to the best of my ability, no matter how mundane the task.

I spent countless hours with my French grandmother, Jéanne Péchard, during my early childhood in France; and she instilled in me a special love for animals, flowers, and the outdoors—a love that will remain with me always.

My childhood friend Ronald Williams, along with his late mother, Mary Ann, created some of the fondest memories of my youth. Ronald's gift of friendship remains with me today, and Mrs. Williams taught me the only Bible verses I've ever known.

My wife, JoAnn Jimmerson Whitford, queen of my heart and a true Southern belle, has given me gifts too numerous to mention. Her special love and support have never waned, even when money was nonexistent and times were tough. She also taught me to love reading—a quite remarkable and invaluable gift.

James White, Bill Davis, and Larry Sellers, three professors at Louisiana Tech University where I earned my bachelor's degree, inspired me with encouragement and excitement. They unlocked my imagination to the world of biology, whether it was catching frogs, snakes, and turtles, identifying plants, or just learning about science in general. From the onset, they encouraged me to finish my undergraduate degree, even when working a full-time job made it very difficult.

Bill Showers, my major professor at Iowa State University during my master's and Ph.D. studies, was truly my mentor. He gave me the gifts of critical thinking and self-confidence, and he taught me how to think analytically. To him, all problems have solutions: one has only to think about them long enough. It is to him that I owe special gratitude for helping to develop my professional career.

I also owe thanks to various groups within the state of Indiana: the county Extension educators who gave me the opportunity to speak to their groups; the trade, environmental, and health associations who allowed me to address their members; and the business owners and farmers who allowed me to visit their facilities and farms. I want to thank David Petritz for giving me the opportunity to work at Purdue University and for his vote of confidence throughout the years. It was through these groups and individuals that I learned practical application. Their efforts and their unselfish gifts of time helped shape my ability to understand pesticide issues, people, Extension, and teaching. In return, the royalties from this book will be donated to the Indiana Make-A-Wish Foundation, to help others in need.

A special thanks to Stephanie Lentz, Manish Gupta, and the production team at TechBooks for all of their efforts in helping me work through the publication process.

Lastly, I am indebted to my children Scott and Robyn, who loved and accepted me even when I couldn't spend time with them because I was too busy studying or working. I hope that, in some way, I have given back to them that same love and acceptance.

I am truly grateful for the gifts I have received, and I look forward every day to meeting new people who may offer another "special gift." It is my wish that in some way, I, too, will have a positive impact on the lives of others. Hopefully, others will be able to look back on their lives and count me as one of their gift givers.

FRED WHITFORD

West Layfayette, Indiana
March 2002

INTRODUCTION

During the 1950s, pesticide production escalated in the United States as companies began to commercialize their products. Technology spurred the federal government to amend the Federal Insecticide, Fungicide, and Rodenticide Act (FIFRA 1947) to ensure adequate oversight of pesticides. The USDA was required to register products and to establish standards for labeling (Fig. 1).

Figure 1 Making a pesticide application around 1900.

PESTICIDES PROVIDE BENEFITS

Pesticides manage pests in homes, schools, restaurants, museums, and hospitals; in orchards and landscapes; at industrial sites and on farms. They target mosquitoes, fleas, ticks, and rodents that carry disease to people, pets, and livestock; they disinfect our drinking water; they protect our buildings from wood-destroying termites and carpenter ants.

The motoring public and transportation industries benefit from herbicides that keep plants from obscuring road signs and encroaching on rights-of-way such as roadsides, railroad tracks, and power lines. Pesticides protect indigenous flora from invasive plant species and are used in conservation tillage programs aimed at reducing soil erosion.

Pesticides play a role in farm production by managing pests of fruits, vegetables, grains, fibers, and livestock. Thus, they contribute to an abundant, economical U.S. food supply, casting a positive impact on our balance of trade. Thanks in part to the use of pesticides, our farm production needs are met by a mere 2% of the total U.S. population, which frees the rest of us to pursue other vocations. With the world population projected to reach eight billion by the year 2025, and with limits on the amount of "new" land that can be converted for farm production, pesticides are expected to remain a major contributor in meeting the worldwide demand for food and fiber.

PESTICIDES CARRY RISKS

A number of issues surfaced in the 1960s, posing alarming questions about environmental risks associated with pesticides. The release of Rachel Carson's *Silent Spring* in 1962 engaged and united the public, and government leaders subsequently addressed environmental issues as a priority in assessing pesticides for registration.

The creation of the U.S. Environmental Protection Agency (EPA) in 1970 was the dawning of dramatic change in the federal regulation of pesticides. Emphasis during the registration process shifted from product efficacy to human health and environmental risk potential. Registration hinged on the manufacturer's ability to demonstrate that benefits outweighed associated risks. It also was dependent upon comprehensive testing and research conducted by the manufacturer in support of specific, critical, human health and environmental policy objectives: to assess risk potential and, through proper labeling, ensure safe use.

Clarity and openness in the risk assessment process facilitate informed debate on pesticide use; and ultimately the registration of a pesticide must withstand scientific inquiry and public scrutiny—and sometimes legal review.

In ecological and human health risk assessment, scientific information is used to identify potential risks associated with the use of a pesticide. Good regulatory decisions depend on documented scientific research, an understanding of the data (strengths and weaknesses), and sound professional judgment in drawing conclusions from the data. Solid risk assessments clearly distinguish fact from assumption, and thorough evaluations that yield clear and concise conclusions add a vital dimension to the EPA's decision-making process.

The pesticide label links research data and consumer decisions on safe pesticide use. It is the primary source of general and technical information passed down from regulatory agencies and pesticide manufacturers to consumers: the agricultural community, the commercial pest management industry, and the general public. The label is the one document where scientific review, regulatory oversight, and public policy are interwoven to achieve a common objective: the conveyance of information on storing, handling, applying, and disposing of pesticides and their containers in a manner consistent with good health and environmental stewardship.

THE ONGOING DEBATE

Risks associated with pesticide use draw public attention. Obviously, a pesticide has to be effective against the target pest; yet it is difficult to develop a chemical that will affect *only* the target pest. The goal is to pose *no* risk to humans, nontarget plants, domestic animals, wildlife, and the environment; however, no pesticide is risk-free, and certainly no pesticide is totally "safe" in all situations. All pesticides carry the potential for harm.

The pesticide debate often is polarized, contentious, and speculative. People are concerned about how "safe" or "dangerous" pesticides truly are. They want to know if the benefits justify the risks, if their food is safe, and if their family and pets might suffer adverse effects from pesticide applications in the home. The public wants answers to these questions and more—but they are easier asked than answered.

Questions on pesticide risk are at the heart of public debate, and agenda-laden interest groups have an abundance of answers. But often their focus is narrow, their opinions based on inadequate scientific information. Pesticide issues are complex, and we must draw from many disciplines (anthropology, psychology, sociology, chemistry, history, toxicology, economics, and others) to address the whole picture.

Figure 2 Pesticide safety is a consequence of multiple, interactive strategies.

Government faces increasing responsibilities and ever-changing agendas, its regulatory policies being steered by the questions we pose. Meanwhile, science, by nature, addresses a set agenda within a marked time frame, yielding definitive answers. But science is not *exact* and, in fact, can be highly uncertain. So it is no surprise that the pesticide debate is not clear-cut. Two common words answer most questions regarding pesticides: "It depends."

Likewise, proper discussion of risk assessment, risk management, and risk communication requires the integration of many disciplines. Although the pesticide debate has been defined as oversimplistic, the new and improved pesticide education will integrate economic, social, political, psychological, educational, and scientific analysis.

The quest to ensure that the benefits of pesticides outweigh associated risk potential is ongoing. New scientific knowledge and changing public expectations cause local, state, and federal agencies to evaluate pesticide use continually. Resulting changes in public policy and legislative mandates are intended to ensure that the benefits of pesticides outweigh their potential risk to human health and environmental quality. Risk assessment, pesticide registration, product labeling, government enforcement, applicator education, and open communication lay the foundation of a comprehensive framework to regulate pesticides: the manufacturing process as well as distribution, use, and disposal (Fig. 2).

Do pesticides pose risks? Are they beneficial? The answer to both questions is yes. With definitive policies and requirements for pesticide review before a product enters the marketplace, with clear and precise labeling, and with good consumer education, pesticides can play an important role in maintaining the quality of life we enjoy.

THE COMPLETE BOOK OF PESTICIDE MANAGEMENT

CHAPTER 1

THE EVOLUTION OF PESTICIDE REGULATIONS: THE SHIFT FROM BENEFITS TO RISKS

Pesticides are perhaps the most highly regulated item of commerce in the United States today. The risks and benefits associated with pesticide use are largely reflected in the evolution of a complex regulatory framework that leads to product registration and use. Captured within this framework is a formal process whereby the effects of pesticide use on man and the environment are evaluated.

The U.S. Congress has enacted two major federal laws to manage health and environmental risks from pesticides: the Federal Insecticide, Fungicide, and Rodenticide Act (FIFRA) and the Federal Food, Drug, and Cosmetic Act (FFDCA). FIFRA gives the U.S. Environmental Protection Agency (EPA) the authority to register pesticides; to require appropriate supporting chemical, toxicological, environmental, and residue studies; and to develop labeling requirements based on these studies. Pesticides that come into contact with food or animal feed are regulated under the FFDCA, which gives the EPA the authority to establish legal limits for pesticide residues in or on food and feed.

Regulations for pesticide registration specify data requirements, methods for conducting studies, procedures for risk assessment, and makeup of the product label. The EPA uses these tools to determine whether a pesticide can be used without unreasonable harm to human and environmental health. In addition to addressing specific risks to human health and the environment, the EPA appraises potential economic, social, and environmental impact associated with use of the pesticide. In effect, the decision-making process balances potential risk to humans and the environment against social, environmental, and projected economic benefits (Fig. 1.1).

There have been many changes in pesticide registration requirements and the understanding of toxicology during the past decade: what was acceptable yesterday may not be acceptable today or tomorrow. Policies and decisions on acceptable risk change over time as science and public policies change. And as public awareness and concerns over potential risk change, so do registration requirements.

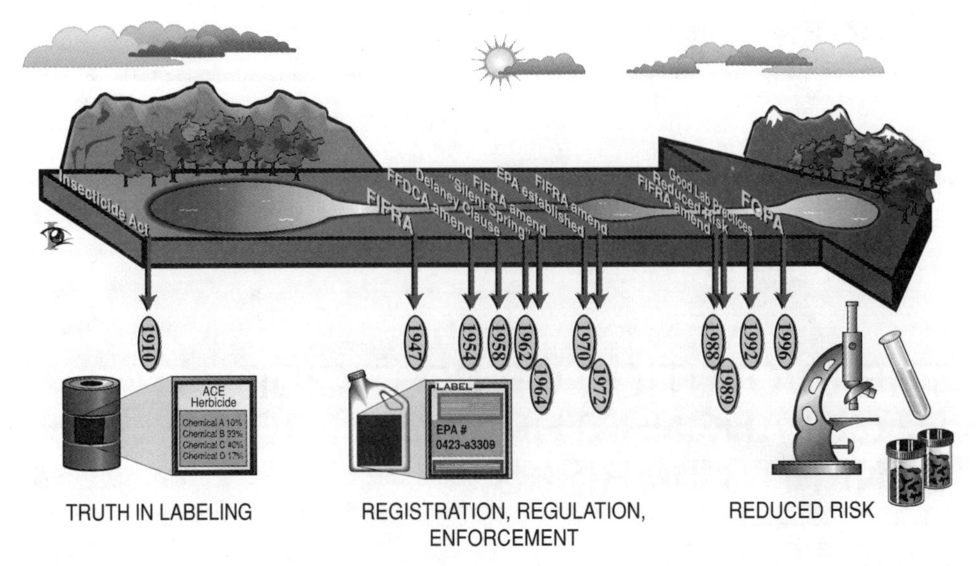

Figure 1.1 Pesticide regulations have evolved from performance-based standards to risk-based standards.

EARLY FEDERAL LAWS FOCUSED PRIMARILY ON BENEFITS

The 1906 Pure Food and Drug Act prohibited unsafe substances in foods. It was followed by the Insecticide Act (1910) which prohibited the interstate sale or transport of impure or improperly labeled insecticides and fungicides. Its primary focus was to ensure that products were labeled adequately and that container contents were stated precisely on the label. The Insecticide Act contained no registration requirements and did not set safety standards.

The Insecticide Act was replaced in 1947 by a more comprehensive law: the Federal Insecticide, Fungicide, and Rodenticide Act. FIFRA was the first law to require pesticide manufacturers to register their products with the U.S. Department of Agriculture (USDA), which was responsible for registering all pesticides prior to sale or movement via interstate or foreign commerce. The first USDA registration number (USDA Reg# 1-1) was issued to *PRATT'S Dip and Disinfectant* on November 14, 1947. The entire FIFRA text can be found at http://www.epa.gov/pesticides/fifra.htm.

Pesticide regulations were expanded again in 1954 with the Miller amendment to the Federal Food, Drug, and Cosmetic Act. The amendment required the establishment of tolerances for pesticide residues in or on agricultural commodities. A tolerance was defined as the legal limit (amount) of pesticide residue that could remain in or on a harvested food crop after application; it was established primarily on the basis of good agricultural practices.

In 1958, an amendment to FFDCA, commonly referred to as the Delaney clause, prohibited the use of any food additive shown to cause cancer in man or experimental animals. Pesticide residue concentrations in processed foods (e.g., tomato paste and tomato sauce) at levels higher than those found in the raw agricultural commodity (e.g., whole tomatoes) were considered food additives and were thereby subject to the provisions of the Delaney clause. But pesticides that did not concentrate in processed foods were not considered additives and thus were not subject to the Delaney clause.

Figure 1.2 *Silent Spring* was an important milestone in the recognition of ecological consequences of pesticide use.

ENVIRONMENTAL MOVEMENT CHANGES PUBLIC PERCEPTION OF PESTICIDES

Increasing environmental concerns in the 1960s, exemplified by Rachel Carson's *Silent Spring* (1962), changed forever how pesticides will be viewed by the American public (Fig. 1.2). The most commonly used insecticides at that time were a class of chemicals called chlorinated hydrocarbons that included such insecticides as DDT, aldrin, and dieldrin. Environmental groups and the news media portrayed these pesticides as chemicals that bioaccumulate in the environment, disrupt links in the food chain, and poison wildlife. *Silent Spring* captured the public's attention, rallied greater public awareness of environmental issues, and called for a ban on numerous pesticides.

GOVERNMENT POLICIES SHIFT TOWARD RISK REDUCTION STRATEGIES

In 1970, Congress created the U.S. Environmental Protection Agency and shifted the regulation of pesticides from the U.S. Department of Agriculture to the EPA. In 1972, FIFRA was dramatically strengthened to give the EPA more regulatory authority.

Changes in FIFRA since 1970 have resulted in a major philosophical shift in pesticide regulation: from benefits to risks. Originally, FIFRA required regulators to review and register pesticide products. But in 1972, Congress changed FIFRA from a labeling law to a comprehensive statute designed not just for the distribution of pesticides but also for the use of pesticides. Pesticide manufacturers then were required to demonstrate that use of the product would not cause "unreasonable adverse effects on human health or the environment," thereby including a safety requirement.

The 1972 amendment to FIFRA also created a distinction between lower risk, unclassified pesticides (commonly called general-use products) and the higher risk pesticides classified for restricted use. The general-use pesticides could be purchased and used by the general public, whereas the higher risk, restricted-use pesticides could be purchased only by certified pesticide applicators and used only by certified applicators or persons under their direct supervision. By establishing the need for standards of competency for applicators of restricted-use pesticides, Congress clearly and specifically acknowledged that proper training is fundamental to their proper use.

In addition to requiring scientific data in support of pesticide registration, FIFRA was modified to prohibit any use of a pesticide inconsistent with its labeling. In other words, the label became the law, and violations for not following the label could result in label enforcement via license revocation, fines, and imprisonment.

RISK–BENEFIT CONSIDERATIONS

Because Congress did not intend the 1972 amendment to FIFRA to be solely an environmental bill, an industry bill, or a farm bill, efforts were made to balance the needs of all stakeholders. Regulatory decisions were to be based on the balancing of potential health and/or environmental risks against potential benefits stemming from use of the pesticide; that is, decisions would depend on risk–benefit analysis.

STUDY REQUIREMENTS AND SCIENTIFIC TESTING GUIDELINES

Under FIFRA, the EPA has issued requirements since 1975 specifying the types of toxicological, ecotoxicological, residue, and environmental-fate studies that must be conducted to support pesticide registration. The EPA also has issued scientific testing guidelines specifying the methodologies that should be used in conducting these studies. The lists of required studies and recommended methodologies are updated periodically as the science advances and as new health and environmental concerns are raised.

PESTICIDE MANUFACTURERS AND GOOD LABORATORY PRACTICES

Some years ago, fraudulent practices surfaced in a contract toxicology laboratory that triggered concern within the Food and Drug Administration (FDA) and EPA regarding the integrity of data being submitted to support pesticide registration. As a result, Good Laboratory Practices (GLPs) were established and implemented in 1989 (40 CFR Part 160).

GLP regulations require that each testing facility develop written procedures that describe all aspects of testing, from start to finish, including data development, collection, and security. These procedures are viewed as the codes of experimental conduct and are commonly referred to as Standard Operating Procedures (SOPs). GLP requirements also include the following:

- Identify a study director who has overall responsibility for the conduct of the study.
- Develop study protocols prior to testing.
- Ensure that the test pesticide is appropriately characterized as to its composition, purity, and stability.
- Train, allocate, and schedule personnel.
- Provide adequate resources and facilities, instrumentation and equipment, animal care, and archives.
- Establish an independent quality assurance unit to inspect and ensure the quality and reliability of all study data.
- Ensure high-quality documentation of all aspects of study conduct.
- Review reports for accuracy and for consistency with raw data.

GLPs describe a process within which laboratory and field studies must be planned, monitored, recorded, and reported. GLP regulations require laboratories to accurately document every step of an experiment. All data and observations are recorded in study notebooks or electronically recorded by computers interfaced with data-generating instruments such as weight scales and blood-analysis instruments. Changes in procedures or corrections of errors must be recorded, dated, and initialed and reasons for changes in data or text must be provided.

The EPA's Office of Enforcement and Compliance Assurance (Agriculture and Ecosystem Compliance Division) inspectors audit the accuracy and integrity of data generated by pesticide research facilities. EPA inspectors look at employee education and training records, check calibration of analytical equipment, and review compliance with written SOPs. Inspectors also examine the integrity of the data; the housing, feeding, handling, and care of test animals; the handling of test, control, and reference substances; and the accuracy of the study reports. Collectively, the procedures established by the manufacturer and the inspections by the EPA assure that testing is accurate, scientifically sound, and properly documented and that the experiments are generating meaningful, reliable results. They afford EPA reviewers a degree of confidence in the validity of data compiled during human and environmental risk assessment.

EPA MOVES TOWARD A MORE COMPREHENSIVE REVIEW OF RISKS

Comprehensive risk assessments were rare prior to 1980 because of insufficient scientific knowledge and tools to interpret the data consistently. But in the 1980s the emphasis in regulatory evaluations shifted from toxicity assessment alone to include an exposure assessment, a measure of uncertainty analysis, and an assessment of potential risk. With the advent of this more complete information, a shift from hazard-based assessment to risk assessment was possible. Implementation of these additional considerations, coupled with improved scientific assessments, has improved the regulatory decision-making process.

POLICY SHIFTS TO REDUCED-RISK PESTICIDES

The EPA developed a policy (Pesticide Registration Notice 97-3, September 4, 1997) that focuses on reduced-risk pesticides and offers manufacturers the incentive of quicker registration decisions for "low-risk" products. The policy favors pesticides that have less potential to cause adverse health and environmental effects than those currently registered. Registration applications documenting low-risk characteristics are granted priority in the review process; and accelerated reviews allow low-risk pesticides to move more quickly into the marketplace, ideally in 1–2 years compared to the usual 4–6 years.

The first pesticide to be registered under the EPA's reduced-risk pesticide initiative was hexaflumuron, an insect growth regulator that controls termites by inhibiting the molting process. Structures are monitored for termite activity by placing a number of wood stakes in the surrounding soil. Pest management companies check the stakes periodically for signs of termite feeding, and those stakes found to be infested are replaced with stakes impregnated with 0.1 g of hexaflumuron bait. Worker termites ingest the bait and return to feed other members of the colony. In this manner, colony numbers diminish to a point where the termites have difficulty sustaining themselves. When termite activity at a bait stake ceases, it is replaced with a monitoring stake containing no hexaflumuron.

The EPA agrees that hexaflumuron meets the definition of a reduced-risk pesticide, due largely to its negligible potential for environmental exposure, its low human toxicity, and its environmentally friendly application rate. The EPA believes that hexaflumuron substantially reduces quantities of liquid pesticide used to control subterranean termites in and around structures.

AGGREGATE AND CUMULATIVE RISKS

On August 3, 1996, President Clinton amended FIFRA by signing into law the Food Quality Protection Act (FQPA). FQPA passed unanimously in Congress and was heralded by many as a new beginning for food safety.

Probably the most important aspect of FQPA is that it amended both FFDCA and FIFRA to create a single, health-based standard for all pesticides on all foods. FQPA mandates that tolerances for foods must be "safe," which is now defined as ". . . a reasonable certainty that no harm will result from aggregate exposure to the pesticide chemical residue, including all anticipated dietary exposures and all other exposures for which there is reliable information." The use of a single standard for all foods eliminates inconsistencies between allowable residues on processed foods and those on raw agricultural commodities.

FQPA has also required changes to the EPA's pesticide risk assessment process. Risk assessment for pesticides now addresses aggregate exposure to a given chemical from nonoccupational sources and the combined risk for groups of chemicals with common mechanisms of toxicity (cumulative risk assessment). The mandates of FQPA also provide additional protection for infants and children and limit the consideration of benefits when registering or reregistering a pesticide.

SUMMARY

Pesticide regulation in the United States has evolved from a focus on efficacy and labeling to a process that takes a holistic view of pesticide manufacture, use, and disposal. This framework

Figure 1.3 Risk assessment considerations are reflected in the development of pesticide labels to ensure responsible use.

has evolved over time as scientists, regulators, and the general public have become more aware of the risks which pesticides may pose to human health and the environment. As our base of scientific understanding has increased, so has the body of data and the complexity of assessments needed to properly balance benefits and risks (Fig. 1.3).

Regulatory assessment has evolved from a hazard-based assessment (e.g., is it toxic?) to one where the risks (the integration of hazard and exposure) and benefits (such as improved efficacy and the potential for use in integrated pest management programs) are integrated into registration decisions. The potential risk to subgroups (e.g., children and the elderly) and the risk from exposure to pesticides with similar modes of toxicity are now being considered. Overall, the regulation of pesticides has become more rigorous and focused. It has forced both regulators and manufacturers to employ the best science and to conduct thorough pesticide assessment.

CHAPTER 2

HUMAN HEALTH RISK ASSESSMENT: EVALUATING POTENTIAL EFFECTS OF PESTICIDES ON HUMANS

Health issues are the subject of lively public debate concerning pesticide use. We may be exposed to pesticides in the food and water we consume and in the air we breathe; we may be exposed at home, at work, and at play. Questions continually arise as to how much risk pesticides pose. The general public and government policymakers want clear, definitive answers. Answers to questions on the relationship of pesticides and public health are based largely on risk assessment (Fig. 2.1).

INITIAL FOCUS ON DIETARY RISKS

Nearly all Americans are exposed to some level of pesticides in their diet. Thus, understanding the risk potential of pesticide residues in food is critical not only for consumers but for producers, food processors, pesticide manufacturers, and government agencies as well; their efforts must interlink to ensure a stable and wholesome food supply.

ATTENTION SHIFTS TO OCCUPATIONAL RISKS

Risk assessors in the 1950s began to question the risk posed to workers handling concentrated pesticide products (e.g., pesticide applicators) and to field-workers exposed to residues on foliage (e.g., workers picking apples). This focus on worker exposure was driven by physicians' and industrial hygienists' attempts to determine how workers became exposed. The advent of occupational risk assessment necessitated new methods for calculating risk: avenues previously unexplored in dietary risk assessment. Exposure from various routes at varying frequencies and durations had to be considered.

PUBLIC SCRUTINY OVER RESIDENTIAL RISKS

Risk assessors in the late 1980s began to focus on risks from pesticides used in and around the home and workplace. Previously, risk assessors and risk managers had thought that

Figure 2.1 A regulatory review of product data is integral to understanding the risks that pesticides may pose.

demonstration of minimal risk to applicators and field-workers provided (by default) adequate safeguards for residential use of a given pesticide. This was based on the belief that the pesticide exposure is many times lower with residential use than with occupational use; however, in recent years that assumption has been questioned.

Occupational risk assessment conducted for pesticide handlers and field-workers may not reflect risk to the young, the elderly, the sick, or other potentially more susceptible segments of society. Safety standards for occupational settings are calculated for healthy workers, typically males aged 20–50 years. So the assumptions made, information used, and conclusions drawn via the occupational risk assessment process may not pertain to those more sensitive to pesticides.

THE MULTISTEP RISK ASSESSMENT PROCESS

Human risk assessment is best described as a three-step process:

- Toxicity assessment: an evaluation of intrinsic toxicity or hazard potential of the chemical to mammalian species
- Exposure assessment: an estimation of potential human exposure to the chemical
- Risk characterization: an evaluation of potential risk to humans

What Are the Effects from the Chemical?

The purpose of assessing the toxicological properties of a pesticide is to determine whether it has the potential to produce adverse effects on human health. Carefully controlled experimental studies on animals form the basis for distinguishing the toxicological properties of a pesticide. Animal studies employ a wide range of pesticide doses, including levels far above those to which humans are ever exposed.

What Are the Routes and Levels of Human Exposure?

Human exposure to pesticides usually occurs via ingestion of residues in food and water. However, dermal and inhalation exposure and the incidental ingestion of residues stemming from residential or occupational pesticide use also are recognized as potential routes of exposure. The extent of exposure depends on the type of use (e.g., crop, lawn, or garden applications; mosquito control; indoor pest management), the application rate, the method and frequency of application, and the breakdown and movement of the chemical in the environment.

What Is the Relationship between Exposure and Toxicity?

Risk is a function of both toxicity and exposure, and risk characterization is the integration of pesticide toxicity and exposure data to predict the likelihood of adverse human health effects. Though toxicity data and exposure data are evaluated separately, the resulting assessments are used together to characterize risk. A highly toxic chemical may not pose significant risk if exposure is minimal, but, on the other hand, a slightly toxic chemical may pose unacceptable risk at high doses or prolonged exposure.

GENERAL PRINCIPLES OF TOXICOLOGY

Toxicological testing evaluates whether exposure to a pesticide will produce acute effects (e.g., eye and skin irritation, neurotoxicity) or chronic effects (e.g., impaired liver function, reproductive abnormalities, cancer).

Toxicological evaluations are conducted with experimental animals exposed to various levels of the pesticide for various lengths of time, from hours to years. Results often lead investigators to additional research on the interaction of the pesticide with biological systems. Understanding the biological mechanisms that underlie effects observed in animals allows toxicologists and risk assessors to predict the chances of harm to human populations exposed to the pesticide.

Consideration of exposure levels and effects produced at specific doses is essential in determining toxicity. Exposure, in and of itself, does not necessarily produce harmful effects; for instance, people who are exposed to low levels of pesticides in their food or drinking water, or through contact at the workplace, normally suffer no harm. But harm may occur when people are exposed, accidentally or otherwise, to higher levels that have been shown to produce adverse health effects in laboratory animals.

The duration and magnitude of exposure determine the nature and severity of the effect. In other words, the length of time during which exposure to the chemical occurs (duration) plus the size (magnitude) and number (frequency) of doses combine to determine the severity of the effect.

Scientific inquiry into the toxic properties of a pesticide requires studying how an organism reacts to the pesticide and what internal changes it triggers. Toxicology is an interdisciplinary science; that is, it requires input from numerous disciplines, including pathology, biochemistry, hematology, genetics, endocrinology, and physiology, in order to deduce cause-and-effect relationships. No single study provides all of the information necessary to identify the toxicological properties of a pesticide; rather, a series of studies generally classified as either phenomenological (descriptive) or mechanistic must be conducted.

Pesticide and Animal Interaction

It is important when investigating the toxicology of a pesticide to understand the effects of the chemical on the animals, and vice versa. The genetic, physiological, anatomical, and biochemical variability among and within animal species helps scientists understand why a pesticide may be highly toxic to rats but nontoxic to dogs or people, for example, or why another is toxic to rats and mice but not to fish or birds.

Pesticides cannot be categorized as "safe" or "dangerous" to humans merely because they are classified as substances that kill pests. Each active ingredient has its own unique chemical structure and toxicological characteristics. Pesticides with very similar chemical structures in many instances produce dramatically different effects. One chemical may generate a highly toxic effect whereas another may exhibit no toxicity whatsoever to the same animal at the same dose.

Effect of the Chemical on the Animal Laboratory studies are useful in predicting and explaining pesticide toxicity. Effects on animals are determined by the chemical structure of the pesticide, its mode of action, and the fate of the chemical within the animal. Not all animals react to all pesticides in the same manner and the response can be species- or individual-specific.

Species-Specific Pesticidal effects often vary with the species of animals studied; for instance, one species may exhibit kidney disease, and another, liver disease. The degree of sensitivity may vary as well; for example, one species may exhibit severe toxicity compared to mild toxicity in another.

Individual-Specific Individual animals within a species can exhibit dissimilar responses to the same pesticide. Toxic effects can vary with the size, sex, age, and general health of the test animals.

The Relationship between Dose and Response

The Swiss physician Paracelsus (1493–1541), the father of toxicology, believed the relationship between dose and response to be inseparable. Paracelsus asked, "What is it that is not poison? All things are poison and nothing is without poison. The right dose differentiates a poison and a remedy" (Fig. 2.2).

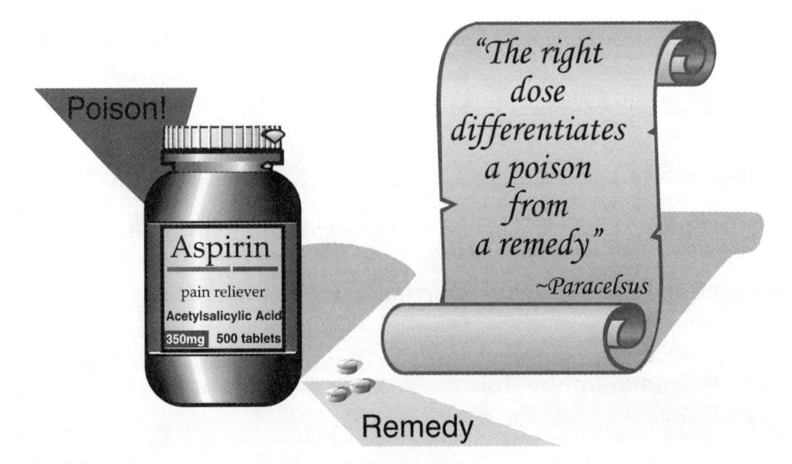

Figure 2.2 The dose makes the poison.

Dose–response is a familiar and critical concept for toxicological testing. High doses are likely to produce detectable injury, whereas low doses may produce little or no injury. For example, ingesting one or two sleeping tablets might have a beneficial effect, but consuming a bottle of them could be lethal. One glass of beer may not affect an adult, but the same amount could easily intoxicate a child. A little salt improves the taste of food, but a lot can cause serious health consequences—even death. Therefore, the question of how much exposure (dose amount) can be tolerated becomes a critical factor in evaluating safety.

Toxicologists follow the basic premise that all chemicals, both natural and synthetic, are toxic at some dose. Certain chemicals, such as table salt, cause adverse effects only at high doses; others, such as cyanide, exhibit toxicity at a very low dose. When toxicologists determine the gradation of effects resulting from increasing doses of a chemical, they have established a dose–response for that chemical.

The Bell Shaped Dose–Response Curve It is important to discuss sensitivity differences among individuals of a single species. Just as people do not look alike, they also do not respond in exactly the same manner to medicines, pesticides, or other chemicals. We know, for example, that penicillin is a lifesaving antibiotic for most people; however, some hypersensitive individuals exhibit serious side effects from it. Certain individuals become intoxicated after a few alcoholic drinks, whereas others can consume considerable quantities of the same drink without adverse effects.

People and all living organisms exhibit a broad spectrum of reaction and sensitivity to chemicals. Some people are very resistant to high doses whereas others are very susceptible even to low doses; most individuals are somewhere in-between. In fact, when the sensitivities of a large number of people (or rats) are graphed against increasing doses of a drug, a pesticide, or some other chemical, the resulting distribution curve is usually shaped like a bell. The far left and right sides of the bell lip represent the small number of people who are either very susceptible or very resistant to the chemical, and the sensitivities of the remaining majority of people fall between those extremes. If a certain dose of a toxic chemical is given to a large number of animals, some show no effect, some get sick, and

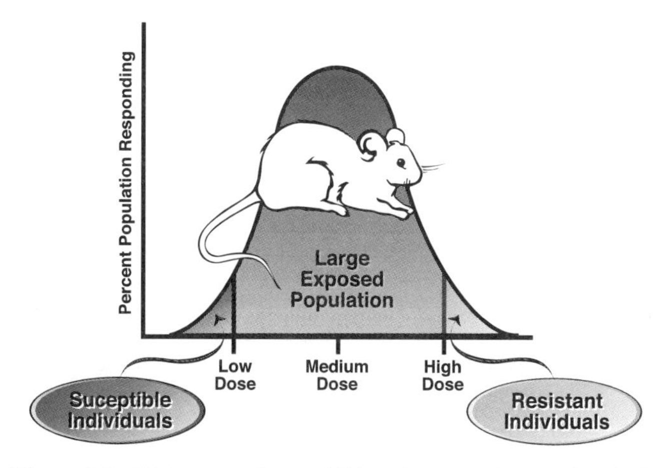

Figure 2.3 Living organisms exhibit various reactions to chemicals.

some die. Thus, in any toxicological test, we do not expect all animals to be affected, nor do we expect all affected animals to exhibit the same degree of severity (Fig. 2.3).

Understanding the Concept of Dose–Response The toxicity of a chemical is determined by quantifying the response of laboratory animals to a series of increasing doses. The relationship between administered dose and animal response is graphically depicted as the dose–response curve. The graph includes the measured response (e.g., number or percent of animals affected, or the severity of response) on the vertical axis and increasing doses of the test chemical on the horizontal axis. For a measured response such as death, the percentage of animals that die increases proportionally as the dose increases. The LD_{50} is a common measure used to define acute toxicity; that is, the lethal dose for 50% of the animals tested (Fig. 2.4). More than 50% of the animals die at doses higher than the LD_{50}, whereas fewer (or no animals) die at lower doses. Thus, the higher the LD_{50}, the less acutely toxic the pesticide.

A generalized dose–response curve has three distinct regions:

- No detectable response
- Increasing linear response
- Plateau (maximum) response

Animals exposed to low doses exhibit no signs of toxicity. The specific point on the dose–response curve where the more susceptible animals are first affected is termed the threshold level: the lowest dose that produces a measurable response in the most sensitive animals. The threshold level is the beginning of the linear response region of the curve and is the demarcation between the "no observed adverse effect level" (NOAEL) and the "lowest observed adverse effect level" (LOAEL). Increasing the dose beyond the threshold level usually increases not only the proportion of animals that show response, but the severity of the effect as well. It is this linear region of the dose–response curve—increased dose, increased response—that is used to measure, describe, and predict the toxicological properties of a pesticide.

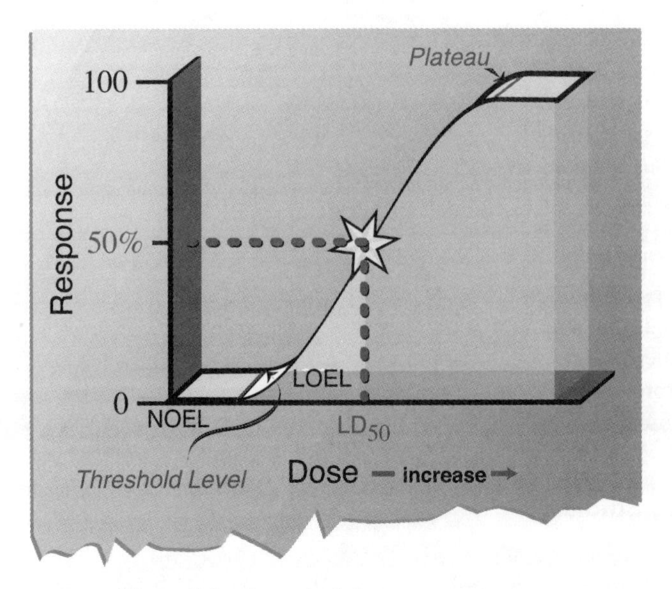

Figure 2.4 A standard dose–response curve.

The third part of the curve, the plateau, begins at the point where an increase in dose no longer produces an increase in response. The point at which the response levels off (at the upper end of the dose range) is known as the maximum effect level. The response may plateau at 100% because all of the animals tested are affected, or it may plateau at a lower response level because the animals are resistant to even the highest dose tested.

Using Dose–Response Curves The following should be considered when reviewing toxicology studies and interpreting dose–response relationships.

- No single dose–response curve can describe the entire range of toxicological responses exhibited by an experimental animal. Each response (death, cell injury, etc.) is a separate end point and can have a different dose–response curve.
- End points may be actual observations, such as changes in animal behavior or food consumption patterns; they also may be an indirect indicator of toxicity. Toxicologists can use measurements of blood and urine constituents as indicators of toxicity, disease, or deterioration in test animals without having to resort to surgery, X rays, or whole body scanning. For instance, a toxicologist might use changes in blood enzymes as an indicator of damaged liver cells. A chemical measurement such as a change in blood chemistry may indicate exposure but have no direct relation to the toxicological effect.
- The dose required to produce a given effect (end point) may vary, depending on the pesticide.
- The data used to develop each dose–response curve are unique to the organism (test animal).
- Dose–response curves may differ dramatically, depending on the route of exposure to the pesticide: oral, dermal, or inhalation.

Dose–response curves provide the toxicologist with valuable information. Examination of these curves and their supporting data provides a basis for comparing pesticide toxicity threshold levels as well as median (LD_{50}) doses. For example, regulatory agencies categorize pesticides according to LD_{50} values which, in turn, are useful in defining label language for precautionary statements, first aid directions, packaging restrictions, and transport recommendations.

The slope of the dose–response curve is of critical importance. A steep curve indicates only a slight difference between a nontoxic dose and a toxic dose: even a small increase in dose produces a significantly different effect. Conversely, a somewhat flat dose–response curve indicates that a relatively large increase in dose has little effect. A flat dose–response curve indicates a larger margin of safety between a nontoxic dose and a toxic dose. Clearly, correct interpretation of toxicological data and valid conclusions on the toxicity of a pesticide require an in-depth understanding of the dose–response curve.

Descriptive Studies

Descriptive (phenomenological) studies form the basis of toxicology where "dose makes the poison." The most important aspect of toxicological evaluation is the determination of the dose–response relationship between amount of exposure and incidence or severity of observed effects (Fig. 2.5). Effects may be observed from studies using isolated cells or tissue cultures or from studies in which small mammals, such as rodents, rabbits, and dogs, are used. The design of descriptive studies varies according to length of exposure (days, months, years), route of exposure (dermal, oral, inhalation), and toxicological measurements (e.g., reproductive toxicity, cancer, organ toxicity, developmental toxicity, neurotoxicity, and immunotoxicity).

Threshold Effects With the possible exception of some types of cancer, most toxicological phenomena occur at or above specific dose levels. These dose levels are referred to as *threshold doses*, and the observed effects are referred to as *threshold effects*. Within a full suite of studies, there may be a different threshold dose for each adverse effect observed, but the precise threshold dose for each effect is rarely determined. One of the most important aspects of toxicological studies is the identification of the NOAEL, which is the highest dose that does not cause any observable adverse effect. The lowest dose level that results in an adverse response is the LOAEL. The threshold dose, although not precisely determined, lies somewhere between the NOAEL and the LOAEL (Fig. 2.6). Table 2.1 briefly outlines U.S. phenomenological and mechanistic toxicology studies on pesticides.

Describing Adverse Toxicological Effects

Effects Depend on Exposure Duration The effects of a pesticide vary with duration of exposure:

- Acute: short-term exposure; a single exposure or multiple exposures within a very short period of time
- Subchronic: intermediate-term exposure; repeated exposure over a longer period of time
- Chronic: long-term exposure; repeated exposure over a very long time

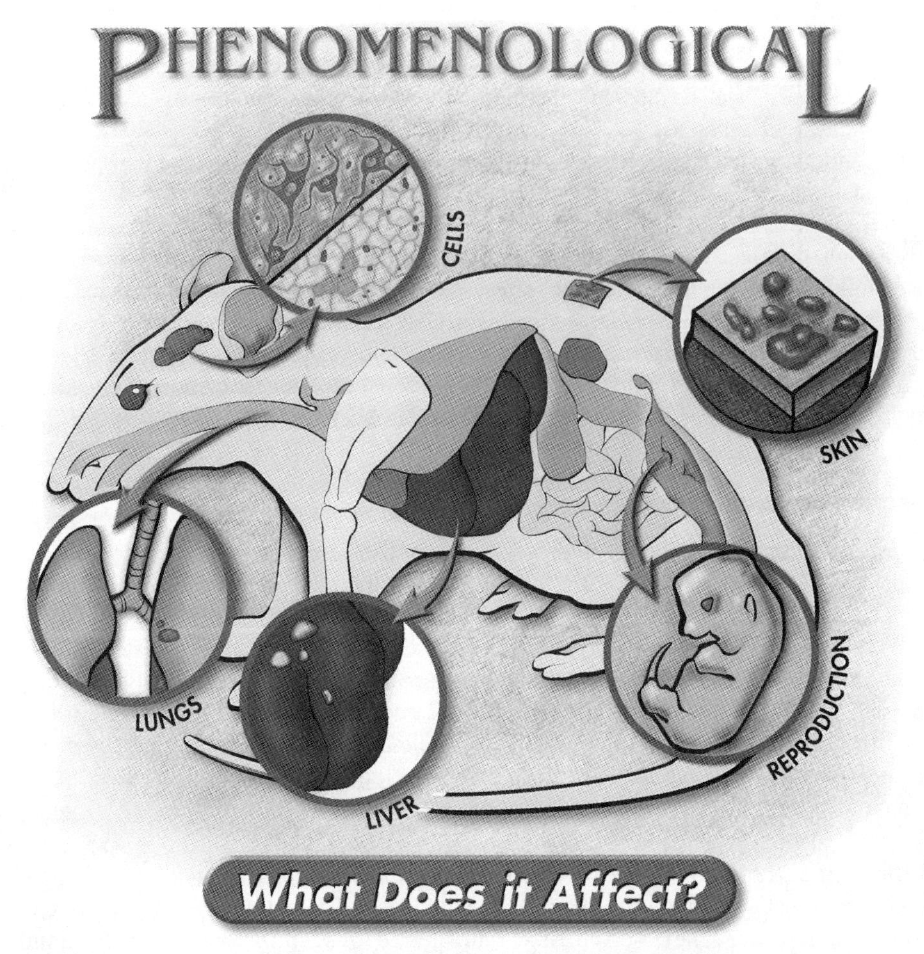

Figure 2.5 Phenomenological studies are used to measure the relationship between the amount of the exposure and the observed effects.

For most pesticides, the response to acute exposure is very different from the response to subchronic or chronic exposure; that is, a dose administered once may evoke little or no response, whereas multiple exposures (at the same dose) over an intermediate or long period of time might generate a significant response.

A pesticide is said to be acutely toxic when adverse effects result from a single exposure, usually at a relatively high dose. But it should be noted that exposure to the same or smaller doses multiple times within a very short period of time (e.g., 24 hr) also is termed acute. Acute effects in humans often result from accidents, such as a child ingesting a pesticide or an applicator not taking proper precautions during mixing, loading, and application. Suicide attempts or accidents, and in some cases the blatant misuse of pesticide products, may constitute acute exposure. Any time a pesticide causes adverse effects following acute exposure, it is said to exhibit acute toxicity.

Subchronic toxic effects manifest after frequently repeated exposure, over weeks or months, to pesticide doses which might produce only minimal or no response to a single acute

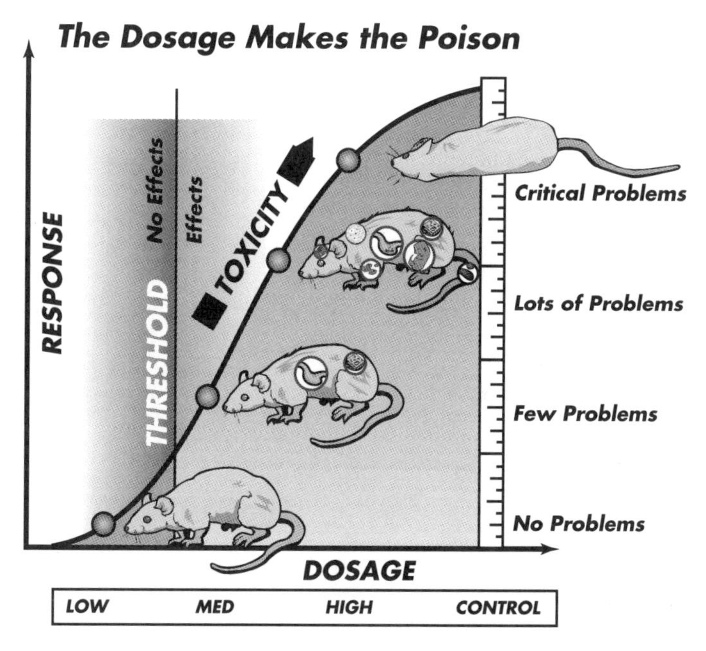

Figure 2.6 Effects are determined by the dose.

exposure. The body may not be allowed time to eliminate the pesticide before successive exposure, thus resulting in a buildup or constant exposure that triggers adverse subchronic effects. Daily exposure can cause repeated insult and long-term effects even if each dose is rapidly eliminated.

Chronic effects result from continual exposure over a long period of time—a lifetime, for example. Pesticides can have cumulative effects on the body, even at doses so low that no immediate or short-term effects are apparent. Although the body might be able to recover from minimal effects that a single dose or a few low doses might cause, it may not be able to recoup totally between repeated exposures over a long period of time (Fig. 2.7).

Effects Can Be Reversible or Irreversible The toxicity of a pesticide is described as reversible if its effects subside or disappear when exposure ends or shortly thereafter. But in situations where adverse pesticidal effects persist even when exposure is eliminated, the toxicity is considered irreversible.

The nonlethal toxic effects of some pesticides are reversible when exposure is eliminated, almost regardless of the dose; however, the effects of other pesticides may be reversible at low-dose exposures but irreversible at high doses. Toxic effects sometimes are reversible, initially, but with continued exposure become irreversible, the dose notwithstanding.

Toxicity Characterized by Effect Toxicity often can be described according to the observable or measurable effect it causes.

TABLE 2.1 Toxicology Studies Generally Conducted for Pesticides in the United States

- Phenomenological studies
  - Acute toxicity
    - Acute oral toxicity (rat)
    - Acute dermal toxicity (rat or rabbit)
    - Acute inhalation toxicity (rat)
    - Eye irritation (rabbit)
    - Skin irritation (rabbit)
    - Skin sensitization (guinea pig)
  - Subchronic toxicity
    - 28-day feeding[a] studies in rats, mice, and dogs
    - 90-day feeding studies in rats, mice, and dogs
    - 21-day and/or 90-day dermal studies in rats or rabbits
  - Chronic toxicity and carcinogenicity
    - 1-year dog feeding study
    - 18-month mouse feeding study
    - 2-year rat feeding study
  - Reproduction/developmental toxicity
    - Rat and rabbit developmental toxicity (teratology) studies
    - Two-generation rat reproduction studies
  - Neurotoxicity
    - Acute rat neurotoxicity
    - 90-day rat neurotoxicity
    - Acute and subchronic hen delayed neurotoxicity
    - Rat developmental neurotoxicity
  - Genetic toxicology
    - Ames *Salmonella* bacterial point-mutation assay
    - Mouse micronucleus assay
    - *In vitro* mammalian point-mutation assay (mouse lymphoma)
    - *In vitro* and/or *in vivo* chromosomal aberration assays
    - *In vitro* and/or *in vivo* unscheduled DNA synthesis assays
- Mechanistic studies
  - Absorption
  - Distribution
  - Metabolism
  - Excretion
  - Pharmacokinetics

[a] In some cases, the pesticide may be administered via drinking water, gavage (stomach tube), or capsule (for dogs) instead of being mixed into the animal's diet.

- Death is the ultimate toxic effect; it occurs when critical bodily functions are altered or inhibited.
- Irritation is observed when a pesticide affects cells of the skin, eye, or respiratory tract; corrosion occurs when the integrity of the outer layer of cells is destroyed. The effect frequently is referred to as a "burn." Less severe irritation might appear as redness, swelling, or inflammation of the skin. Irritation/corrosion can result from single or cumulative exposures.
- Skin sensitization is an allergic reaction; sensitization requires multiple exposures over a period of time. The initial exposure "sensitizes" the person, and subsequent exposures cause the individual to react to the chemical by developing a rash.

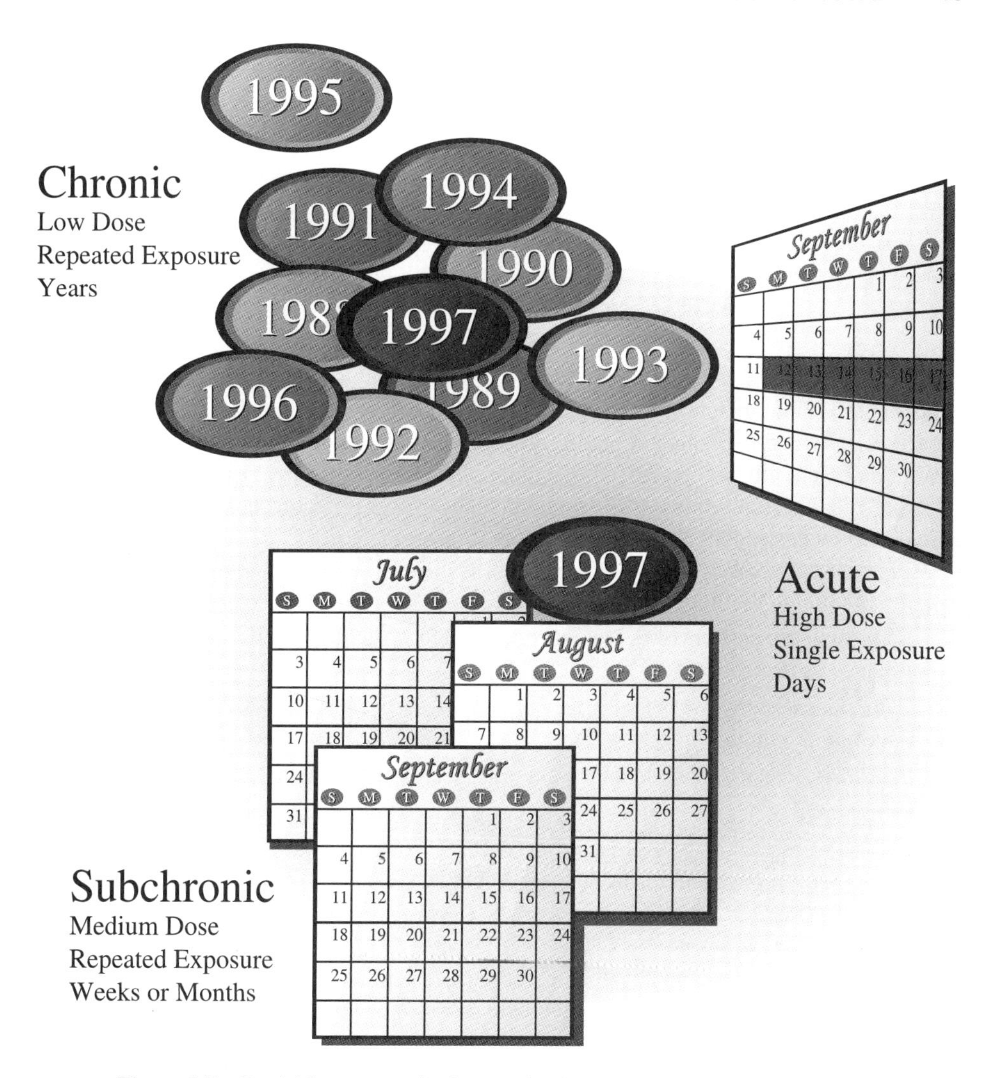

Figure 2.7 Pesticide exposure is characterized as acute, subchronic, or chronic.

- Mutagenicity (also called genotoxicity) results from a change in the genetic material of a cell. There are two general types: a gene mutation that changes the genetic code of the DNA, and a structural mutation that causes structural chromosome damage. A mutagenic compound may produce chromosomal aberrations by modifying the physical structure or number of chromosomes; the result is chromosomes that are fragmented or mismatched, or chromosomes that fail to undergo cell division.

Gene mutations include the deletion, addition, or substitution of the chemical components of DNA, which contain all the coded information that allows organisms to function. Disruptions in genes or chromosomes can lead to diseases (including cancer) and birth defects. A mutagen is of most concern when it damages egg or sperm cells, enabling the defect to be passed on to successive generations.

Toxicity Described by Target Organ/System Effects Toxicological effects often are described according to the organ or system that they impact: cardiovascular, respiratory, gastrointestinal, urinary, muscular, skeletal, and dermal effects; central or peripheral nervous system and sense organ effects; immune system effects; endocrine gland effects; and reproductive system effects. Organ effects in experimental animals often, though not always, are predictive of the effects expected from excessive human exposure to the same pesticide. A toxicological insult to one organ may have indirect repercussions on other parts of the same system, or on multiple systems, due to the complex interaction and coordination of various systems of the body.

TUMORS Tumors—also called neoplasms—are abnormal growths of tissue; they can be either benign or malignant. Most benign tumors are not life threatening because cell division usually is slow and the cells are noninvasive; that is, they will not spread to surrounding tissue. Malignant tumors (cancer) divide rapidly, in an uncontrolled fashion, and spread to other body tissues; this, coupled with their tendency to intercept nutrients needed by healthy tissue, thereby destroying it, renders them life threatening.

Malignant tumors may be one of four cancer types:

- Leukemias are cancers of red blood cells, certain white blood cells, and the tissues that produce these cells.
- Lymphomas are cancers that affect organs of the lymphatic system, such as lymph nodes.
- Sarcomas are cancers of connective tissues such as bone, muscle, and cartilage.
- Carcinomas are cancers of the internal or external epithelial tissues.

Alterations at the Subcellular Level Following are examples of pesticide interaction at the subcellular level:

- Enzymes are proteins that speed up chemical reactions of specific molecules. A pesticide that interferes with an enzymatic process can prevent, slow down, or speed up a chemical reaction within a cell. Enzymatic interference can lead to a toxic response by the cell, tissue, organ, or system. For example, acetylcholinesterase is an enzyme essential to the proper function of the nervous system; it can be inhibited by organophosphorus insecticides, leading to nervous system toxicity.
- Pesticides can interact with critical cellular components (e.g., DNA, hormone receptors, energy-producing chemicals, nerve-impulse-transmitting chemicals, and cell membrane transport proteins) to produce harmful effects. Pesticides may interfere with molecules that serve specific purposes. For example, hemoglobin is a special molecule whose primary function is to transport oxygen in red blood cells. Interfering with hemoglobin so that it does not perform effectively can result in injury stemming from changes in oxygen transport.

Mechanistic Studies

Mechanistic studies detail the processes by which an adverse effect is manifested. Some are conducted to determine how a pesticide is absorbed, distributed, metabolized, and eliminated (Fig. 2.8). Others attempt to identify the underlying physiological processes

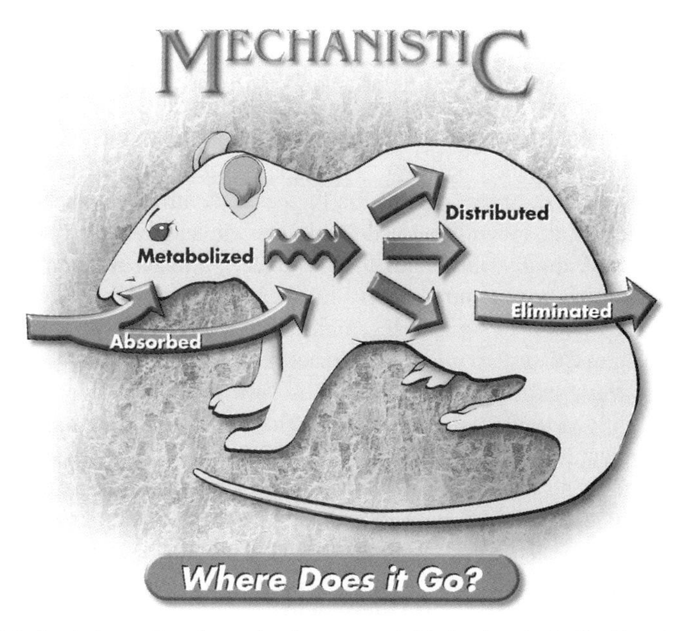

Figure 2.8 Mechanistic studies determine how a pesticide is absorbed, distributed, metabolized, and eliminated.

and/or biochemical pathways that are affected by the pesticide, that is, to determine the mechanism responsible for producing adverse effects.

Scientists need to determine if tumors identified in animal studies actually result from pesticide–DNA interaction or if they are secondary to other toxicity. Many pesticides induce cancer in rodents at high dose levels, but not all doses induce genetic changes. Although the precise mechanism for carcinogenic response at high dose levels cannot always be determined, many tumors are thought to be secondary responses to some other toxic effect, such as an attempt to replace dead cells via enhanced cell proliferation. The secondary response of cell replacement leads to more opportunities for genetic mistakes that may lead to cancer. An increased incidence of cancers of this type presumably would occur only at or above a threshold dose for cell proliferation; increased occurrence would not be expected at a lower dose.

For instance, bladder tumors were reported in animals exposed to high doses of a chemical in a chronic study. Without additional detailed information, it would be assumed that the cancers resulted from a nonthreshold effect. However, in this case the pesticide was shown to cause cell proliferation in the bladder only at high dose levels; thus, a threshold for the tumors can be assumed. The type of tumor and mode of action of the carcinogenic response noted in animals is very important for other reasons as well. In some cases, the tumors observed in animal studies may not occur at all in humans. For example, some chemicals produce kidney tumors in male rats through a process involving a protein that is found in male rats but not in humans. Similarly, due to physiological and biochemical differences, rats (particularly males) are far more susceptible to thyroid follicular tumors than are humans. Thus, the development of data to understand the mechanism by which the chemical induces a carcinogenic effect in animals is extremely valuable in determining the potential of the chemical to cause cancer in humans.

An animal's response to a pesticide may hinge on how its internal systems "process" the chemical. The system of one animal species may metabolize (alter or convert its structure) a pesticide to a nontoxic metabolite, whereas that of another species may not (species-specific response). Furthermore, individual animals of a species also can metabolize the chemical differently (individual-specific response).

Evaluation of the effect of test animals on the pesticide involves consideration of how readily the chemical is absorbed into the animals' system and how the system distributes, breaks down, and eventually eliminates the pesticide. Animals whose systems retain a pesticide for a long time may exhibit effects not seen in animals whose systems eliminate the chemical more rapidly. In some cases, a chemical may appear toxic only after being converted by an animal's system into a more reactive form.

Test animals whose skin does not readily absorb a given pesticide may exhibit only minimal effects following dermal exposure, but the same animals might react seriously when the contaminant is administered orally or through inhalation.

The disposition of a pesticide from its point of contact with and elimination from the body of laboratory animals is studied to gain an understanding of how humans might respond to the same chemical. A pesticide may gain access to the circulatory system through dermal, inhalation, or oral exposure and subsequently be carried in blood from the point of entry to various organs and tissues. The body's physiological processes may facilitate the storage or excretion of the toxin or cause it to metabolize, thus modifying its toxicological effects. Body tissue might absorb the pesticide from the blood and store it, or it might release the contaminant back into the bloodstream for elimination in urine, in feces, or through exhalation.

These processes are complex and interactive. The mechanisms by which biological systems handle pesticides help determine whether toxic effects will be produced. Each pesticide is unique; pesticide molecules differ in their chemical structure, size, shape, stability, and electronic charge. These factors determine how pesticides are absorbed, metabolized, distributed, and eliminated in the human body. For example, the chemical structures of some pesticides make them highly fat soluble, whereas others are not soluble in fat but are highly soluble in water. The significance is that the former may be stored for lengthy periods of time in fatty tissue, whereas the latter might be rapidly eliminated through urination. The shape and electronic charge of some pesticide molecules allow binding to critical site receptors such as enzymes that mediate nerve function in cells or on cell membranes. These and other variables often lead to significant differences in biological response to pesticides.

Absorption into the Body The rate and extent of absorption is highly dependent on the physical and chemical properties of the pesticide and the site of exposure (Fig. 2.9). Pesticides may enter the human body through:

- the skin (dermal exposure),
- the mouth (oral exposure), or
- the lungs (respiratory or inhalation exposure).

Skin is a natural barrier to many pesticides, but penetration can occur, especially if the skin is breached by cuts and abrasions.

Each region of the gastrointestinal tract—mouth, esophagus, stomach, small and large intestines, colon, and rectum—has its own internal environment that dramatically affects

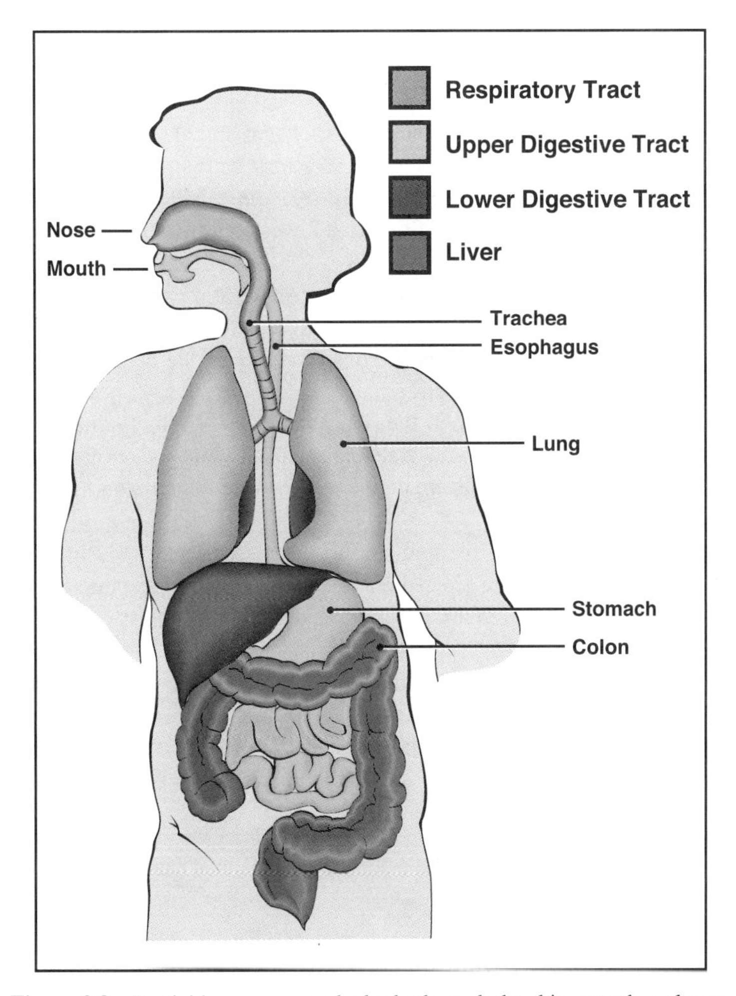

Figure 2.9 Pesticides may enter the body through the skin, mouth, or lungs.

the absorption of a pesticide. Some compounds readily absorb in the mouth, whereas others are absorbed only in the intestines. Some pesticides simply pass through the organism (man or animal) without being absorbed at all.

Pesticides inhaled into the lungs need only to cross the thin barrier separating lung tissue from the blood supply to gain rapid access into the bloodstream.

Distribution and Uptake Blood is the medium through which pesticides are transported to organs, tissues, and cells. The path of transport of a pesticide within the body depends on whether it is absorbed through the skin, the lungs, or the gastrointestinal (GI) tract. Pesticides absorbed in the GI tract enter the bloodstream flowing directly to the liver—the major site of pesticide metabolism—where they are usually broken down soon after absorption. Those that are absorbed into the bloodstream through the skin or through inhalation actually circulate throughout the body before reaching the liver to be broken down. Therefore, the same pesticide dose may be more toxic through inhalation than ingestion.

However, some organophosphate pesticides are oxidized to more toxic metabolites by the liver and are more toxic through ingestion than the inhalation route. This illustrates the importance of understanding uptake.

Uptake by cells is dependent on a pesticide's physical and chemical properties and the type of cell involved. Therefore, different pesticides may be distributed to different tissues in the body. Pesticides can enter a tissue passively, by simple diffusion; that is, the pesticide can move from a high concentration in blood to a lower concentration in body tissue.

Metabolism Pesticides are subjected to chemical alterations by enzymes in the body via a process known as metabolism. Metabolism refers to chemical reactions that alter the structure and the physicochemical properties of the pesticide by adding, removing, or substituting various chemical components.

Metabolism takes place primarily in the liver, where cells usually change the original pesticide molecule to a less toxic, more water soluble form which makes the chemical easier to excrete. However, some molecules are converted to more toxic forms.

Storage Sites Pesticides may accumulate in body tissues, particularly in fat. It is fat-soluble pesticides, primarily, that are stored in the body for long periods of time. The depletion of body fat can release them into the bloodstream.

Excretion and Elimination Metabolized chemicals in the body ultimately are eliminated by the kidneys, in urine, or are carried to the intestine in bile from the liver. Pesticides may be reabsorbed into the bloodstream from the intestinal tract or excreted in fecal material. Pesticides also can be eliminated through the lungs in expired air or in body secretions such as tears, saliva, and milk.

GENERAL TESTING METHODOLOGY

Animal Testing Crucial to Safety Evaluation

Toxicological studies are used by pesticide manufacturers to judge whether to proceed with the development of a pesticide.

Use of Animals Biomedical research has for decades relied on animals as human surrogates. The development of new medicines and the characterization of safe and effective doses are impossible without the use of laboratory animals specifically bred for this purpose. The use of animal models for describing the toxicological properties of pesticides was prescribed by the EPA at its inception in the 1970s and by the Food and Drug Administration which preceded it. Regulatory entities around the world require similar testing of pesticides.

The use of specially bred research animals in toxicological research is not without controversy. Views on the ethics of animal use and welfare issues are polarized, as is scientific dialogue on the relevance of data derived from animal models. Animal rights organizations contend that experimental use of laboratory animals is cruel, unethical, and indefensible. Although, at present, animals cannot be totally eliminated from the testing process, pesticide manufacturers and government agencies worldwide have made great strides in reducing the number of animals used, in eliminating unnecessary experimentation, and in ensuring that animals are housed properly and treated humanely.

The reliability of predicting human hazards from animal data has long been debated. Many argue that drawing conclusions about human safety from animal models is fraught with uncertainty. For example, experimental results can be influenced both by the choice of species and by the strain. Data from an animal species that absorbs, metabolizes, or eliminates a pesticide differently from humans makes extrapolation to humans less relevant. It is essential to reliable conclusions that toxicologists use all available knowledge, data, and expertise in selecting the appropriate animal species for experimentation and evaluation.

Although physiological differences exist between humans and experimental animals, most scientists agree that similarities outweigh the differences and that animals are the only alternative to direct human testing. There is overall agreement that animal models generally provide reliable information that helps to safeguard public health.

The inclusion of laboratory animals as models in pesticide testing programs provides many advantages to scientists and regulators, including the following:

- Animals are the only alternative to human testing.
- Disease-free animals bred for uniformity are available commercially at a reasonable cost.
- Laboratory animals can be produced in numbers sufficient for toxicological investigations. They can be housed in a relatively small space, have reasonable food requirements, and are amenable to frequent physical examinations.
- The historical information background on a particular animal strain or species—normal responses, disease rates, and tumor frequencies—provides accuracy in toxicological data interpretation.
- The use of genetically similar individuals (inbred strains) can lead to more consistent results.
- The relatively short life span of laboratory animals facilitates the observation of effects and diseases associated with lifetime exposures.
- Certain diseases such as cancer can be observed through the use of animals bred specifically for their susceptibility to a disease or for an anticipated response.
- Experimental animals reach sexual maturity at an early age, have relatively short pregnancies, and produce large litters; these characteristics facilitate the study and evaluation of pesticidal impact on reproduction over multiple generations.
- Routes of entry in test animals—oral, dermal, or inhalation—simulate human exposure, yielding a relevant understanding of the chemical fate and properties of a pesticide inside the body.
- Experiments can be standardized by monitoring and controlling uncertainties. For instance, food and water quality can be monitored and adjusted as required to achieve standardization; environmental factors such as temperature and light can be manipulated as well.
- Historically, regulatory agencies around the world make pesticide registration decisions based on their established familiarity with certain test animal species. This standardization of test species, along with experimental methodology, forms a database by which to judge the toxicity of the pesticide being tested.

Species Commonly Used in Pesticide Testing Programs Researchers and regulators do not rely on any one animal species in conducting safety assessments. Human responses to a pesticide cannot be mimicked exactly or modeled by a single animal species; therefore, toxicologists must use multiple species—rats, mice, rabbits, guinea pigs, dogs—to predict the entire range of pesticide toxicity to humans. Hamsters, monkeys, pigs, chickens, and cats are used less frequently.

Toxicologists repeatedly test the same strain of animals to facilitate toxicity comparisons between new and existing pesticides. Animals are purchased from sources that document the history and purity of the genetic strain and guarantee the animals to be healthy and disease-free.

Mouse The mouse (*Mus musculus*) is commonly used in pesticide and pharmaceutical testing. Mice are used predominantly for pesticide carcinogenicity tests, offering these advantages: They are small; they are easy to maintain; and they have relatively short life spans.

Rat Strains derived from the Norway rat (*Rattus norvegicus*, commonly called the laboratory rat) have been used in agricultural and pharmaceutical research since the 1850s. The rat offers many of the same advantages as mice in toxicological testing.

Albino Rabbit Albino rabbits (*Lepus cuniculus*) are used to evaluate skin and eye irritation as well as birth defects. They breed readily, produce large litters, and are easily reared in quantity. Their large bodies and eyes facilitate skin and eye exposure studies.

Guinea Pig The guinea pig (*Cavia porcellus*) through decades of testing has been a reliable human surrogate in identifying pesticides that induce skin sensitization, that is, allergies.

Domestic Hen The nervous system of the domestic hen (*Gallus domesticus*) is sensitive to organophosphorus insecticides; thus, it is used to evaluate nervous system toxicity for this class of pesticides.

Dog The beagle dog (*Canis familiarus*) is commonly used as the nonrodent species of choice. Dogs share many physiological properties with humans and fully complement rodent studies. Their size facilitates difficult surgical procedures, and their ample blood supply allows larger and more frequent samples to be taken without affecting the animals' health.

Table 2.2 compares the biological and physiological parameters of animals with those of humans.

Animal Husbandry Newly purchased animals brought into the testing facility are quarantined from animals already housed there. They are acclimated to their new quarters, food, and environment for up to 2 weeks prior to testing. The animals' diet and water are monitored and analyzed for impurities, and their bedding is changed regularly. Cages, racks, and other equipment are cleaned thoroughly on a regular basis. Animals showing signs of illness or disease during the quarantine period are not used for study.

TABLE 2.2 Comparison of Biological and Physiological Parameters

Parameter	Rat	Mouse	Rabbit	Guinea Pig	Hen	Dog	Human
Life span (years)	2–3	2	6	6	3–5	15	70
Adult weight (kg)	0.2–1	0.02–0.04	2.5	0.5–0.8	1.5–3.5	8–12	70
Estrus cycle (days)	4–5	4–5	Induced	15–18	Daily	21–28	21
Age at maturity (weeks)	13	7–9	24–35	12–16	22	40–72	15–18 years
Gestation period (days)	21	19–21	29–35	58–70	21	56–58	270
Litter size	12–16	10–12	5–10	2–4	270/year	4–8	1–2
Birth weight (g)	4–6	1–2	30–70	80	50	300–500	3200
Eye opening (days)	12–14	10–12	7	0	0	10	0
Weaning age (days)	21	21	28	18–24	Hatched	42–56	250
Weaning weight (g)	40–50	10–12	1800	250	50	1500–2500	8000
Body temperature (°F)	99.5	99.0	103.0	103.0	103.0	101.3	98.6
Heart rate (beats/min)	330–480	320–780	205–220	230–380	280	130–150	72
Blood volume (percent body weight)	6–7	5	5.5	7–7.5	9	8	8
Respiratory rate (breath/min)	100	163	35–65	84	12–30	20–25	10–13

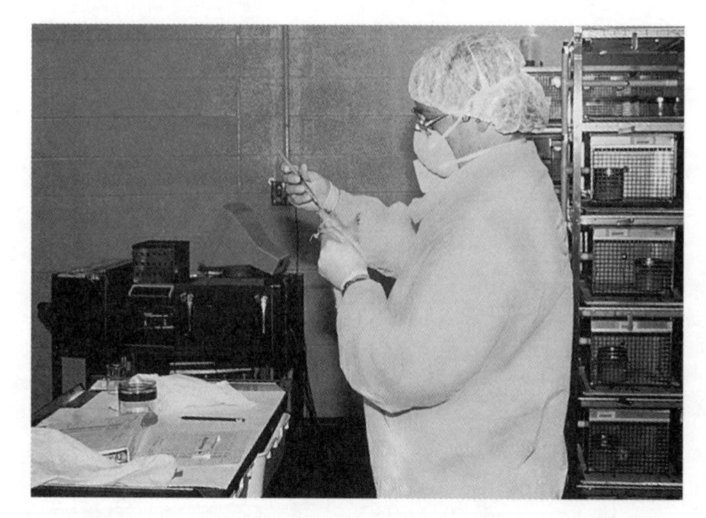

Figure 2.10 A tube syringe is used to insert a pesticide into the stomach of a laboratory rat.

Individual studies are isolated from all others. Animals are housed in well-ventilated rooms with a constant environment: lighting, temperature, and humidity are preset and monitored to prevent significant fluctuation. All instruments used during each experiment are calibrated routinely to ensure accurate measurement.

Administration of Pesticides to Animals

Typically, the pesticide is administered in the animals' food or water or in the air they breathe; dermal effects are studied by placing the chemical on the test animals' skin. Animals typically are administered a pesticide via the anticipated, predominant route of human exposure: oral, dermal, or inhalation. The dose is usually based on each individual animal's weight and is expressed in milligrams of administered chemical per kilogram of body weight (mg/kg). As an example, two rabbits weighing 4 and 6 kg are to be administered the same pesticide dose of 5 mg/kg. The smaller rabbit should be administered 20 mg, whereas the larger should receive 30 mg. The dose also can be administered as mg/l (air or water) or ppm (air, water, diet). To assure proper physiological function in the test animals, a maximum dose is prescribed for administration by various routes of exposure; for example, the maximum dose administered by oral gavage is a function of the animal's normal stomach capacity.

Oral Administration The method chosen for administering an oral dose often depends on the chemical, the animal species, and the duration of the study. In a short-term study, the pesticide might be administered to dogs as a gelatin capsule or to rodents through a stomach tube; these methods place the entire dose directly into the stomach (Fig. 2.10). In longer term studies, the pesticide usually is incorporated into the animals' feed or water, allowing access to small amounts each time they eat or drink.

Dermal Administration The animals' fur is clipped prior to placing a pesticide dose directly on the skin. Solid materials are crushed and mixed with a liquid to form a paste, slurry, or solution. The site of application is bandaged to keep the animals from licking the treated area and ingesting the chemical. Another method employed to deter licking is the placement of a large "collar" around the neck to restrict the animals' access to the application site.

Inhalation Administration Animals are confined in airtight chambers into which pesticide vapor, aerosol mists, or dusts are introduced. It is critical that the test substance be uniformly distributed throughout the chamber for the time period during which animals are obliged to breathe the treated air. The pesticide concentration and particle size in the air is monitored regularly. If dust particles are too large, they are ground to assure accessibility to the lungs. Placing animals in chambers exposes not only the respiratory system but all external body surfaces as well. Alternative testing systems are available for exposure of the animals' nose or face.

Establishing Dose Levels Most tests require four groups of animals, each receiving a different dose level of the pesticide:

- None (control group: no pesticide whatsoever)
- Low (an amount of pesticide estimated to produce no toxic response)
- Medium (enough pesticide to evoke a moderate response)
- High (in acute studies, enough pesticide to cause death; in chronic studies, enough to produce significant signs of toxicity but not death)

Making Comparisons between Treated and Untreated Animals

An essential part of any toxicological program is the response comparison between animals exposed to a pesticide and those not exposed. Animal groups exposed to a pesticide are called "treatment groups," and those left untreated are referred to as "controls." Control groups are handled exactly like treatment groups, except that they are not administered a pesticide. Toxicologists use controls to demonstrate normal growth and development and to provide valuable information on the occurrence, type, and frequency of background disease in untreated animals. Data from both groups are evaluated to differentiate an abnormal from a normal response.

The type of control group used in a toxicological study may be specified by regulatory protocol, or the manufacturer may choose control groups known to ensure levels of performance and reliability of the testing program.

Untreated Control Group The untreated control group is also called the negative control group. For dietary studies, the untreated control group is fed an identical diet as the test group—minus the pesticide. Otherwise, they are treated identically.

Vehicle Control Group Some pesticides are in capsule form or dissolved into a solvent such as corn oil prior to administration to the test animals. Animals assigned to a vehicle control group receive the capsule or solvent without the pesticide, which helps differentiate effects caused by the vehicle.

Positive Control Group The positive control group is not treated with the test chemical, but with a substance known to produce a specific effect; thus, treated control groups ensure that the test system is appropriately sensitive to the end point of interest. For instance, tri-ortho-tolylphosphate (TOTP) is a chemical known to produce certain neurotoxic symptoms; so neurotoxic effects observed in test groups treated with TOTP demonstrate those which might be expected in groups treated with a similarly neurotoxic pesticide. Positive control groups using known mutagens are most commonly used for *in vitro* genotoxic studies.

Historical Control Group Facilities that supply animals to toxicology test labs typically maintain historical control records, and information from those records is used to assess changes in diseases over time. Testing laboratories also maintain historical control group records of animal parameters and disease rates for the testing facility. The large numbers of animals represented in historical records yield better estimates of normal disease rates than do the relatively small numbers of control animals in individual studies.

Self Control Group An animal, under some circumstances, can serve as both a treated and an untreated control. Self control groups are especially useful in eye and skin irritation studies; a response to a pesticide in one eye or on one area of skin on an albino rabbit, for instance, is compared to the untreated (control) eye or skin of the same animal.

Evaluating Toxicological Effects

Many observations in toxicological studies are used to determine if pesticides impact animal health. In general, effects can be observed as

- changes in appearance and behavior,
- findings from routine physical examinations,
- changes in body weight,
- shifts in food consumption,
- alterations in blood chemistry and hematology,
- physical and chemical changes in urine,
- changes in appearance and weight of internal organs, and
- microscopic changes of internal organs and tissues.

Observations of General Behavior Experimental animals are typically observed twice daily for mortality and signs of toxicity. The observer needs extensive, on-the-job training and experience to become proficient in recognizing differences between normal and abnormal behavior and symptoms.

Routine Physical Examinations Animals are handled and their reactions and reflexes observed during detailed physical examinations. Each animal also is examined for unusual growths or lumps; abnormalities are recorded each time they are observed (Figs. 2.11 and 2.12).

Figure 2.11 A scientist examining a laboratory rat, looking for abnormalities.

Observable End Points

Activity
- Hyperactivity
- Hypoactivity
- Nonresponsiveness
- Prostrate position
- Aggressiveness

Ears
- Discharge
- Tears/lacerations
- Pallor
- Redness
- Swelling
- Scabs
- Encrustation

Eyes
- Red discharge
- Bloodlike discharge
- Puslike discharge

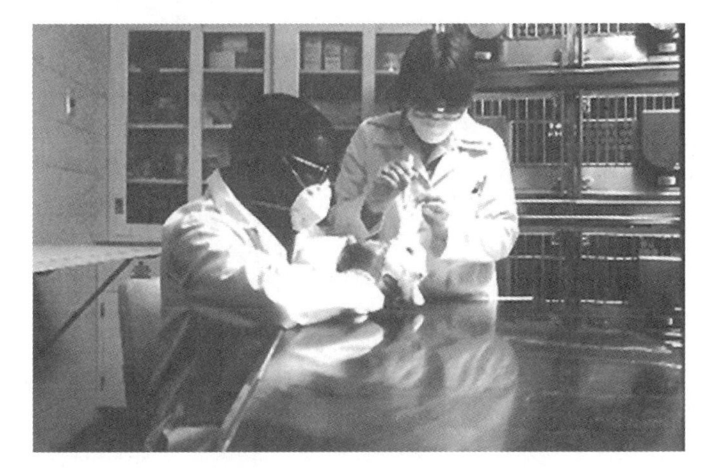

Figure 2.12 Detailed observations are made on all test animals.

Eyes
- Lacrimation (tearing)
- Wetness around the eyes
- Excessive blinking
- Partially/completely closed eyelids
- Opacity (cloudy eyes)
- Pupil contraction
- Pupil dilation (expansion)
- Pallor
- Yellow/brown conjunctival discoloration
- Diffuse conjunctival redness
- Dilation of conjunctival blood vessels
- Pitted/raised corneal surface
- Protrusion
- Necrosis/rupture of globe
- Loss of eye
- Periorbital encrustation

Excreta
- Unusual urine color
- Bloodlike urine color
- Decreased urination
- Excessive urination
- Unusual fecal color
- Bloodlike fecal color
- Black stool
- Soft stool
- Diarrhea
- Decreased defecation
- Hard stool
- Bright yellow urine

Feet and limbs
- Focal loss of hair
- Swelling
- Abrasions
- Sores on feet
- Torn toenails
- Loss of limb/paw/toes
- Cysts

General appearance
- Emaciation
- Dehydration
- Distended abdomen
- Self-mutilation
- Intra-abdominal swellings
- Protrusion of tissue from rectum

Genitalia
- Urogenital discharge
- Vaginal discharge
- Red discharge
- Yellow/brown discharge
- Clear discharge
- Bloodlike discharge
- Swelling
- Protrusion of tissue
- Greenish discharge
- Abnormal penile erection

Mouth
- Emesis (vomiting)
- Salivation
- Wetness around the mouth
- Swollen mouth
- Bleeding gums
- Ulceration of lips
- Ulceration of oral mucosa
- Bloodlike discharge
- Broken teeth
- Missing teeth
- Mismatched teeth
- Overgrown teeth
- Pale mucosa

Movement/posture
- Head tilt
- Ataxia (incoordination)
- Convulsions
- Twitching
- Tremors
- Misuse of limbs
- Arching of back
- Circling movements
- Somersaulting
- Lateral rolling movements
- Side-to-side head motion

Movement/posture

- Up-and-down head motion
- Backward walking

Nose

- Red/pink discharge
- Clear discharge
- Yellow/brown discharge
- Bloodlike discharge
- Frothy discharge
- Perinasal encrustation

Respiration

- Labored breathing
- Gasping
- Rapid breathing
- Slow breathing
- Shallow breathing
- Sneezing
- High-pitched sounds
- Rattling sounds
- Intermittent apnea (stopped breathing)

Skin

- Hair loss
- Abrasions
- Scabs
- Scars
- Swelling
- Flaking/peeling
- Sloughing
- Necrosis
- Ulceration
- Discoloration
- Pallor
- Bloodstained appearance
- Red/brown encrustation
- Urine-stained hair
- Feces-stained hair
- Anogenital staining
- Mammary gland staining
- Mammary gland secretion
- Cysts

Changes in Body Weight Body weight is an important end point in assessing the toxicological effects of a pesticide (Fig. 2.13). Animals in short-term studies are weighed at least weekly. Those in longer studies are weighed weekly for 12 weeks and, thereafter,

Figure 2.13 A rabbit is weighed on an automatic scale system that simultaneously enters the weight into a computer.

monthly, as the rate of weight change slows. Decreased body weight may correlate to toxicity, in that an animal that does not feel well likely will not eat well; another correlation might be that the toxin inhibits the animal's ability to utilize the food consumed.

Shifts in Food Consumption A measured amount of food is offered to each animal. At specified intervals, uneaten food is weighed. The difference between the amount offered and the amount uneaten represents food consumption for that time interval. Comparison of the amount consumed by controls and that consumed by treated animals reveals early indications of toxicity. Water consumption also may be recorded.

Changes in Hematology Hematology is the study of cellular components in whole blood. Small amounts of blood are taken from experimental animals for hematological studies. The following cellular constituents are recognized as key measurements in describing toxicity: hemoglobin concentration and platelet, erythrocyte (red cell), and leukocyte (white cell) counts.

Alterations in Blood Chemistry When cells are removed from blood, what remains is the liquid portion known as plasma. Blood chemistry measurements are important because cells, tissues, and organs are in close contact with circulating blood. Damaged organs often release enzymes and other substances into the bloodstream that are recognizable as indicators of toxicity. The measurement of enzymes, electrolytes, and biochemical changes in plasma facilitates indirect detection of organ damage.

The following chemicals are typically measured in blood:

Acid/base balance
Alanine aminotransferase
Albumen
Alkaline phosphatase
Aspartate aminotransferase
Bilirubin
Blood creatinine
Calcium
Chloride
Cholesterol
Cholinesterase
Glucose (blood sugar)
Hormones
Lipids
Phosphorus
Potassium
Protein
Sodium
Total bilirubin
Urea nitrogen

Alterations in Urine Measurements of hematology, blood chemistry, and urinalysis are similar to analyses performed during human medical examinations. The following parameters of urine provide clues to toxic effects from pesticides:

Appearance
Bilirubin
Cast kidney cells
Glucose
Ketones
Occult blood
pH
Protein
Specific gravity
Urobilinogen
Volume
White blood cells

Gross Observations of Internal Tissues Autopsies are performed on animals that have died or been euthanized during or at the conclusion of studies. Internal body cavities

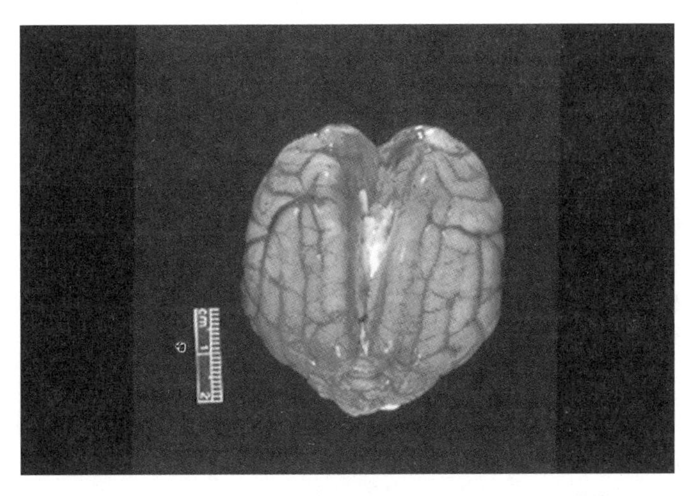

Figure 2.14 Some studies require the examination of internal organs for abnormalities.

are examined. The testes, thyroid, thymus, spleen, heart, lungs, ovaries, brain, liver, and kidneys may be weighed and examined for changes in size and color and for visible lesions (Fig. 2.14). Organs are observed to verify presence and proper location within the body cavity. Once all body tissues have been evaluated and removed, the focus turns to muscles and bones.

Microscopic Evaluation of Tissues As many as 50 tissues from each animal are examined microscopically. Each is sliced into thin sections—a process which can take as long as 6 months—for microscopic observation to detect the presence of lesions and abnormal growth. Pathologists look at each tissue from each animal and examine cell structure, organization, size, and shape; some chronic studies require examination of as many as 20,000 tissue slices. Organs and tissues are subject to routine microscopic evaluation; there is no limitation on the types or numbers of tissues examined.

Typical Tissues Examined Microscopically

Adrenals	Jejunum
Aorta	Kidneys
Bone marrow	Liver
Brain	Lungs
Cecum	Lymph nodes
Colon	Mammary glands
Duodenum	Muscle
Esophagus	Nose
Eye	Ovaries
Femur (bone)	Pancreas
Gall bladder	Peripheral nerve
Heart	Pituitary
Ileum	Rectum

Salivary glands	Thymus
Skin	Thyroid
Spinal cord	Tongue
Spleen	Trachea
Sternum	Urinary bladder
Stomach	Uterus
Testes	

STUDY DESIGN FOR TOXICOLOGICAL TESTING OF PESTICIDES IN THE UNITED STATES

Describing the acute and chronic effects of a pesticide on experimental animals requires a meticulously planned program. The tests are designed to determine dose–response.

Acute tests are conducted first—to measure observable effects at relatively high doses administered over short periods of time. Longer term testing generally necessitates the use of lower doses administered regularly for extended periods, allowing the animals to survive for the duration. This is especially important for chemicals that accumulate in the body. The identification of subtle effects on animals exposed to pesticides in longer term studies requires sophisticated analysis.

Toxicological conclusions are dependent on the complete compilation of data and the integration of all studies. This is critical because no single species of laboratory animal or any single study is necessarily predictive of human toxicity.

Acute Tests: Single Exposure, High Dose, Short Duration

Six tests are used for acute testing. Three measure lethal and sublethal effects (oral LD_{50}, dermal LD_{50}, inhalation LD_{50}), and the remaining three examine eye irritation, skin irritation, and dermal sensitization (allergy). Not only do all active ingredients of pesticides undergo this battery of testing, but so does each product formulation containing the active ingredient. These tests are critical in assessing the potential impact of accidental exposure to pesticides.

Mortality Studies Single-dose mortality studies assess the consequences of short-term, high dose pesticide exposure on animal health; subsequent autopsies allow valuable observation of organ damage.

Young adult males and nonpregnant females are randomly assigned to treatment groups for acute toxicity studies, with treatment of the groups differing primarily by route of exposure and dose.

Rats in acute oral studies are given a specific pesticide dose administered through a stomach tube. Dermal studies use rabbits or rats, with the pesticide applied to a shaven area of skin on the back (roughly 10% of the body). The pesticide is held next to the skin with a gauze dressing for 24 hr. Inhalation studies expose groups of rats to various pesticide air concentrations in tightly sealed chambers (Table 2.3).

Animals in all three lethality studies are observed for 14 days after exposure; signs of illness or changes in behavior are noted. Changes in weight, length of recovery period, and time of death are also recorded. Finally, each animal is autopsied, allowing careful visual inspection of external body parts, internal organs, and body cavities.

The primary objective of the three lethality studies is to estimate the relative toxicity of the pesticide. The number of deaths that occur in each group is noted, as is the dose or concentration level which is lethal to 50% of the treated animals: the LD_{50} (oral or dermal

TABLE 2.3 Acute Mortality Protocols

Parameter	Oral	Dermal	Inhalation
Preferred test species	Rat	Rabbit or rat	Rat
Age of animals	Young adult	Young adult	Young adult
Pregnancy status	Not pregnant	Not pregnant	Not pregnant
Substance tested	Active ingredient, formulation	Active ingredient, formulation	Active ingredient, formulation
Number of dose levels	1–6	1–4	1–4
Number of animals/dose	10	10	10
Male : female ratio	50 : 50	50 : 50	50 : 50
Type of control group	None	None	None
Route of administration	Stomach	Skin	Air
Exposure	Single dose	Single dose	Single dose
Duration of exposure	Single dose	24 hr	4 hr
Observation period (days)	14	14	14

exposure) or LC_{50} (inhalation) is calculated. Information collected from single-dose acute toxicity studies is useful in developing dosage regimens for subchronic studies. Acute lethality data is useful in comparing the acute toxicity of one pesticide to that of another, in placing the pesticide into relative toxicity categories, and in developing label language, warnings, and precautions for the proper handling and use of the product.

Irritation and Sensitization Testing Data gathered from eye irritation, dermal irritation, and dermal sensitization studies are used to estimate the ability of a pesticide to produce temporary irritation or permanent damage to eyes and skin. The type of damage, its duration, and the severity of injury are recorded for each pesticide (Table 2.4).

TABLE 2.4 Acute Irritation and Sensitization Protocols

Parameter	Eye Irritation	Dermal Irritation	Dermal Sensitization
Preferred test species	Rabbit	Rabbit	Guinea pig
Age of animals	Adult	Adult	Adult
Pregnancy status	Not pregnant	Not pregnant	Not pregnant
Substance tested	Active ingredient, formulation	Active ingredient, formulation	Active ingredient, formulation
Dose	0.1 ml	0.5 ml (liquid)	(depends on skin reaction)
	—	0.5 g (solid)	(depends on skin reaction)
Number of animals/dose	3–6	3–6	10–20
Male : female ratio	50 : 50	50 : 50	50 : 50
Type of control group	Self	Self	Positive, negative
Route of administration	Eye	Skin	Skin
Exposure	Single dose	Single dose	Multiple doses
Duration of exposure	24 hr	4 hr	6 hr × 4 weeks
Observation period (days)	3–21	3–14	35

Eye Irritation Study Rabbits are used for eye irritation studies. The lower eyelid of one eye is pulled down and the pesticide is applied. The eyelids are held together for a few seconds to ensure that the pesticide remains in place, and the other eye is left untreated to serve as a control. The pesticide must remain in the treated eye for 24 hr before it is washed out.

Effects on the treated eye are recorded 1 hr after treatment and again at 24, 48, and 72 hr. If irritation is not observed at 72 hr, recording ceases; if irritation persists, however, daily recording may continue for up to 21 days. Eyes are scored relative to the opacity of the cornea and any redness, swelling, or discharge observed. The pesticide is classified as corrosive if the eye injury is permanent.

Skin Irritation Study A section of fur on the back of each rabbit is shaved one day prior to the start of skin irritation studies. A single dose of pesticide is applied to a 1-inch square of skin that is then covered with a gauze patch. The patch is removed after 4 hr, and the treated area is washed. Observations are recorded after 1 hr and again at 24, 48, and 72 hr, then daily for up to 14 days. Observations on skin damage, redness, and swelling are scored.

Dermal Sensitization Study Dermal sensitization studies are used to evaluate the potential of a pesticide to produce allergic contact dermatitis after repeated skin contact. Guinea pigs' fur is clipped or shaved from their backs, then a minimally irritating pesticide concentration is applied once a week for 6 hr, for three consecutive weeks (induction phase). At that point, the animals are given a 2-week rest period after which a challenge dose—at a concentration previously determined to be nonirritating—is applied to a previously untreated area of skin. The challenge dose site is examined for redness and swelling 24 and 48 hr after exposure. The pesticide is considered a sensitizer if the challenge dose produces a skin reaction.

Subchronic Tests: Repeated Exposure, Intermediate Dose, Moderate Duration

Subchronic exposure occurs repeatedly over weeks or months and subchronic effects are those resulting from such exposure. Subchronic studies assess sublethal toxic effects, generating a portfolio of toxicological information distinctly different from that of acute testing. Acute studies involve high doses administered during a very short period of time, the result of which often is death of the test subjects. Subchronic testing, on the other hand, employs lower doses to facilitate keeping test animals alive for long periods of time; this is essential in identifying any subtle effects resulting from repeated doses (Table 2.5).

Subchronic testing programs must utilize toxic and nontoxic dose levels. High doses must elicit sublethal effects, middle doses must evoke only minimal adverse effects, and low doses should trigger no toxic effects whatsoever. Generally, 3–5 dose levels are tested. Typical subchronic studies expose animals for 1–3 months. Exposure routes are identical to those of acute testing programs: oral, dermal, inhalation.

Each animal is observed twice daily for signs of toxic effects such as abnormal appearance and altered behavior. Thorough weekly examinations are conducted to detect signs of toxicity and to track growth (including weight and development). Blood is examined periodically for hematological and chemical parameters.

Animals that die are autopsied, and those that survive are euthanized and autopsied at termination of the study. Tissues are examined, then evaluated microscopically for signs of disease or toxic effects; major organs are weighed. In some studies, the persistence or reversibility of pesticidal effects may be tested upon cessation of exposure.

TABLE 2.5 Subchronic Protocols

Parameter	Oral	Dermal	Inhalation
Preferred test species	Rat, mouse, dog	Rat, rabbit, or guinea pig	Rat
Age of animals	Mouse/rat, 6–8 weeks; dog, 4–6 months	Adult	6–8 weeks
Pregnancy status	Not pregnant	Not pregnant	Not pregnant
Substance tested	Active ingredient	Active ingredient	Active ingredient
Number of dose levels	3	3	3
Number of animals per dose	Rat, 20; Beagle, 8	10	10
Male : female ratio	50 : 50	50 : 50	50 : 50
Type of control group	Untreated	Untreated	Untreated
Route of administration	Diet/stomach tube	Skin	Air
Exposure (day)	continuous (diet)	6 hr	6 hr
Duration of exposure (days)	90	21 or 90	90
Observation period (days)	90	21 or 90	90

Subchronic Oral Studies Young rats and mice (6–8 weeks old) are commonly fed diets containing a pesticide for 90 days. Beagle dogs (4–6 months old) may be tested likewise, although they may receive the pesticide in a gelatin capsule rather than in their food.

Subchronic Dermal Studies The pesticide is applied to the shaved skin of rats or rabbits and immediately covered with gauze to hold the material next to the skin. The dose is administered for 6 hr/day for at least 5 days a week over the 21-, 28-, or 90-day duration of each study.

Subchronic Inhalation Studies Groups of rats are exposed to various air concentrations of the pesticide in tightly sealed chambers 6 hr each day, 5 days per week, for 1–3 months.

Chronic Tests and Carcinogenicity: Multiple Exposure, Low Dose, Long Duration

Chronic studies measure the effects of daily exposure to a pesticide over a 1–2 year period. Chronic toxicity research (typically performed on rats and dogs) examines cumulative effects of a pesticide on body organs: lungs, kidneys, liver, etc. Results are interpreted as indicators of the potential for chronic exposure to result in illnesses such as kidney or liver disease, or cancer. The chronic rodent toxicity studies generally involve 400 or more animals and require 3 years to complete, that is, from the start of the experiment to compilation of the data into a final report. The total cost for these studies may exceed $1,000,000 each.

Chronic studies assess specific toxic effects on body organs and identify cumulative effects resulting from repeated exposure. For instance:

- Animals used in chronic studies are exposed over a longer period of time—typically 1–2 years, depending on the species.

- Longer exposure extends observation periods, allowing latent symptoms to develop.
- Daily dose levels generally are lower in chronic studies; however, the total cumulative dose is greater due to the longer duration.
- Microscopic tissue and organ evaluations are key measurements in determining long-term effects of pesticide exposure; in a cancer study, about 50 tissues per animal—15,000 to 20,000 tissue sections—are examined.

Carcinogenicity studies are conducted specifically to examine the potential of a pesticide to cause cancer. Scientists look for growths of new or abnormal tissues and determine whether they are benign or malignant.

In determining what routes of exposure should be studied, preference is given to those through which human exposure is most likely. Dietary administration is generally preferred for chronic toxicity and carcinogenicity studies because it is more representative of potential human exposure from small amounts of the pesticide over time.

Testing Procedures, Dose Administration, and Measured Responses Alternative methods of exposure may be used when animals refuse to eat a pesticide-treated diet or when exposure is negated by rapid degradation of the pesticide in the diet. Options might include dissolving the pesticide in drinking water, administering the pesticide in capsules, or using a stomach tube.

A study may be designed to maintain a constant exposure level in the animals' diet, or exposure might be based on the weight of the animals (i.e., milligrams of pesticide per kilogram of body weight). The latter approach requires periodic adjustment of the individual dose to account for the growth of each animal. Regardless, it is preferable that animals being administered the low dose of pesticide exhibit no adverse physical or behavioral effects. A midlevel dose regimen should produce only minimal signs of toxicity, whereas animals on the high-dose regimen should display significant signs of toxicity (but not death).

The EPA defines the Maximum Tolerated Dose (MTD) as the highest dose that causes no more than a 10% decrease in body weight gain of the test population (compared to control groups) without imposing life-threatening diseases (other than a cancer response) or producing mortality. The definition and interpretation of the MTD differ among international regulatory agencies, and there is considerable debate among toxicologists on the ramifications of testing at doses so high that they alter the physiology of the test animals.

Chronic studies require intense observation and time-consuming measurements of each animal. Test animals are examined weekly, using visual inspection and observation to assess the general condition and overall behavior of each animal (Fig. 2.15). The animals are weighed weekly for the first 13 weeks and monthly thereafter. Food consumption is monitored on a similar schedule.

Blood composition (hematology and chemistry parameters) and urine composition (pH, protein levels) are indicators of the general health of the test animals. Each is collected and analyzed at 6-month intervals and at the conclusion of the study. The analysis is much the same as for humans during physical checkups and for the diagnosis of disease. Blood may be drawn and analyzed more frequently if subchronic testing of the pesticide has indicated toxic tendencies.

All animals that die during the course of study are autopsied, and those remaining are euthanized and autopsied at the conclusion of the study. All organs and tissues within the body cavity are carefully examined, and specific organs such as the liver, kidneys, brain, and testes are removed and weighed. Approximately 50 organs and tissues are removed, cut

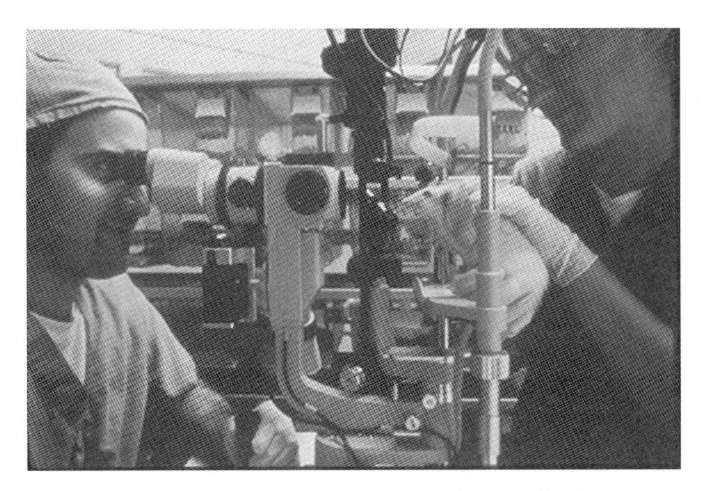

Figure 2.15 Examining the eyes of a test animal.

into thin slices, and mounted onto glass slides; the samples are then stained for microscopic examination to determine whether the pesticide caused changes in cell numbers, size, type, or structure. Microscopic abnormalities are documented for each organ and each animal. The findings are summarized by sex and dose and compared to control groups to ascertain whether the pesticide has influenced the incidence or severity of effects (Table 2.6).

Reproductive and Developmental Toxicology

The toxicological evaluation is incomplete without a thorough assessment of the chemical's potential to interfere with reproductive processes. Adverse effects on reproduction include infertility, the tendency to abort, reduced weight of offspring, and birth defects. Toxic "insults" to reproduction processes can cause diseases or behavioral abnormalities as well as learning and functional deficits in offspring. Toxicological studies on reproduction address

TABLE 2.6 Chronic Protocols

Parameter	Chronic Toxicity	Oncogenicity
Preferred test species	Rat, mouse, and beagle dog	Rat and mouse
Age of animals	Mouse and rat, 6–8 weeks; dog, 4–6 months	6–8 weeks
Pregnancy status	Not pregnant	Not pregnant
Substance tested	Active ingredient	Active ingredient
Number of dose levels	3	3
Number of animals/dose	Rat, 100 animals; beagle, 8 animals	100
Male : female ratio	50 : 50	50 : 50
Type of control group	Untreated or vehicle	Untreated or vehicle
Route of administration	Diet, inhalation, stomach tube	Diet, inhalation, stomach tube
Exposure (day)	Daily	Daily
Duration of exposure (days)	12–24 months	18–30 months
Observation period	Rodents, 12–24 months; dog, 12 months	Rats, 24–30 months; mice, 18–24 months

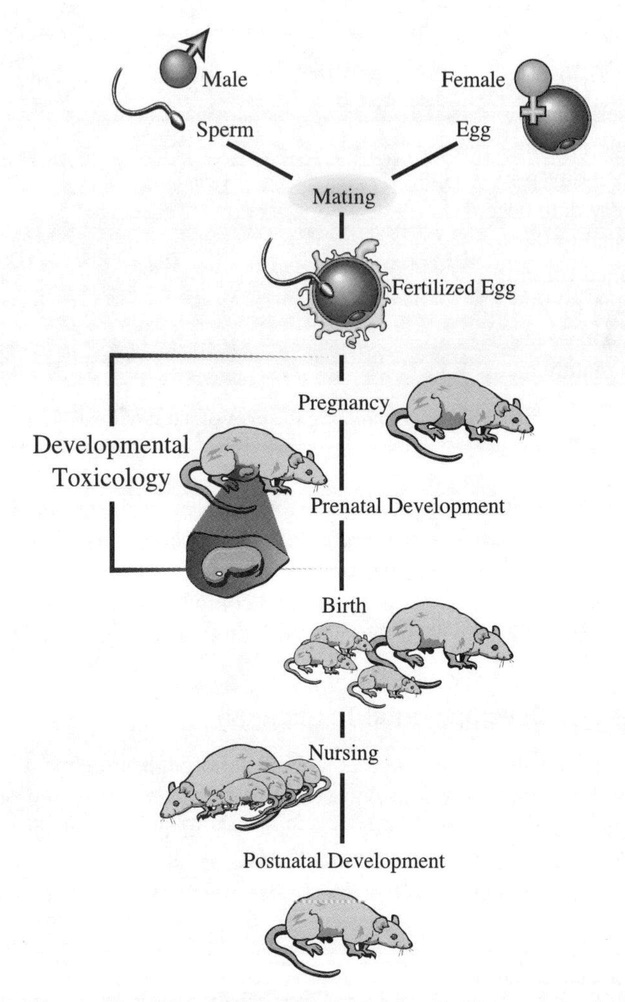

Figure 2.16 Developmental toxicity studies evaluate the chemical's impact on the mother and fetus from the time of egg implantation through birth.

the pesticide's influence on the fetus and the chemical's ability to interfere with normal reproduction processes. Two studies are routine: developmental/teratological (e.g., birth defects) and reproduction/fertility.

Developmental Toxicology These studies assess pesticidal toxicity to pregnant females, fetal growth and development, and physical birth defects. Abnormalities include spontaneous abortion (miscarriage), embryo death, reduced birth weights, extra ribs, and birth defects. The latter may manifest as obvious physical deformities such as cleft palate or missing appendages, or they may occur more subtly as internal organ or skeletal damage.

Developmental toxicity studies evaluate chemical effects on the mother and fetus from the time of implantation of the fertilized egg into the wall of the uterus through birth (Fig. 2.16). The highest dose should produce some toxic effects to the mother, such as weight loss, but it should not be lethal; this ensures the opportunity to evaluate fetal and birth defects, where present. Ideally, a medium dose should not produce maternal toxicity but should stress the developing fetus. A low dose should have no adverse effect on the mother or the developing fetus.

Generally, one rodent species (usually rats) and one nonrodent species (usually rabbits) are used to assess developmental toxicity. Females are mated or artificially inseminated and the pesticide is administered daily, by stomach tube, from the time of fertilized egg implantation to 1-day prior to delivery. The mothers are euthanized 1-day prior to their projected delivery date because they sometimes will eat their abnormal young—which, of course, are critical to the study. The uterus is removed from the euthanized female and its contents examined for embryonic and fetal death and live fetuses. Mean litter weights are recorded for each mother. Uterine implantation sites are counted and correlated with the number of live and dead fetuses to determine if any of the fertilized eggs implanted in the uterus failed to mature and were reabsorbed.

Each fetus is weighed and measured and its sex determined. Each is examined for external and internal anomalies. The soft tissue is removed from some or all of the fetuses to facilitate skeletal examination, and all bones—particularly the vertebrae, long and short bones, and head bones—are checked for size, shape, position, and hardening.

Reproduction/Fertility Studies Reproduction studies address the effects of the pesticide on male and female reproductive processes, from egg and sperm production and mating through pregnancy, birth, nursing, growth and development, and maturation. The studies are conducted through two generations of offspring, that is, three generations including the parents (Fig. 2.17). Groups of young adult rats, both male and female, are fed diets containing various concentrations of the pesticide (Table 2.7).

The animals are fed the treated diet continually for approximately 10 weeks prior to mating and through pregnancy, birth, and nursing of the young. Potentially, the fetuses are exposed to the pesticide in the womb, through the mother's blood, and later when nursing. At about 21 days of age, the pups are weaned and thereafter fed the same treated diet as the mother. At sexual maturity, they are mated with other pups from the same treatment groups; for example, females on the high-dose treatment regimen are mated to males also on the high dose. The males are then euthanized and the females are fed the high-dose diet throughout pregnancy. Thus, the cycle is repeated as the females deliver second-generation pups, which are allowed to nurse and grow for 21 days. The adult females and pups are then euthanized.

All animals are observed daily: behavioral changes, signs of toxicity, food consumption, body weight, and mortality are recorded for each animal. Litters are examined for number of pups, stillbirths, and live births. Live pups are weighed at birth and thereafter on days 4, 7, 14, and 21. All animals are autopsied, with special attention directed to the organs of the reproductive system.

Records are kept on

- egg and sperm formation and viability (Figs. 2.18 and 2.19);
- estrus cycles;
- mating, conception, and pregnancy;
- fetal development and survival;
- birthing and nurturing;
- growth, development, and survival of offspring through two generations; and
- toxic effects in mothers or offspring.

Two Generation Reproduction Study
With Daily Dosing of Adults and Offspring

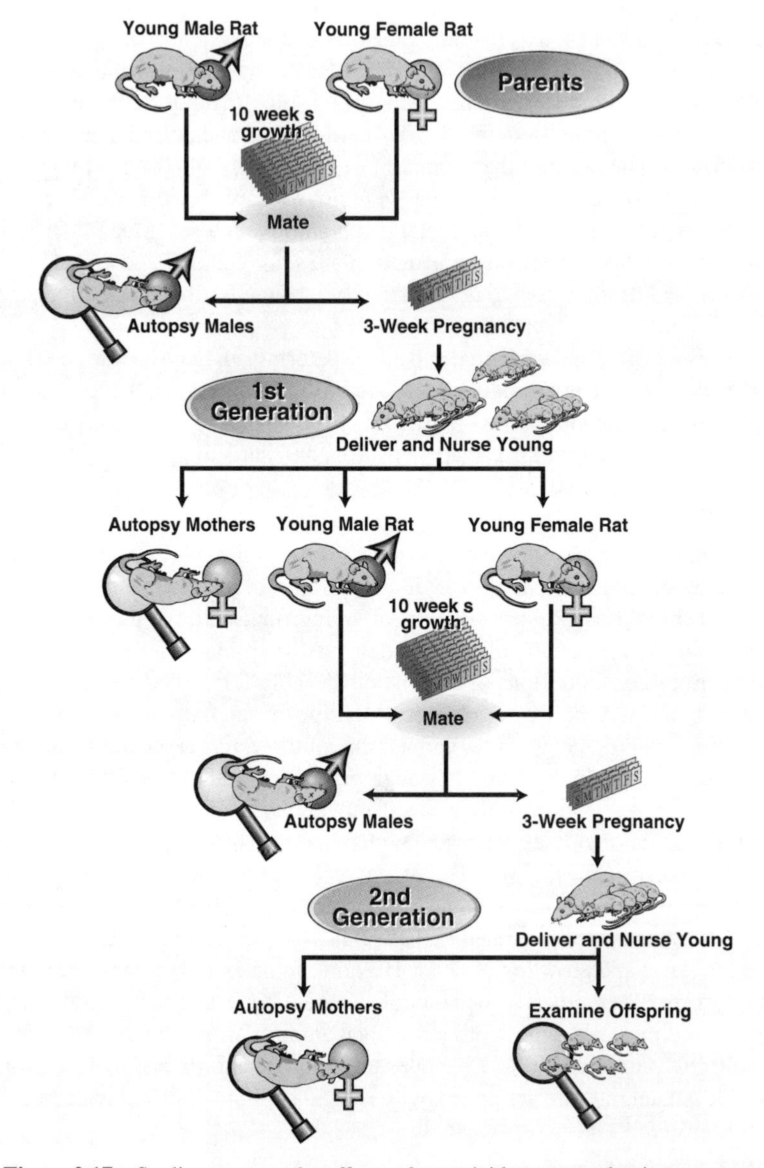

Figure 2.17 Studies measure the effects of a pesticide on reproductive processes.

Neurotoxicity

Increased attention has been given to evaluating the potential effect of pesticides on the structure and function of the nervous system. Acute and subchronic studies in rats are conducted frequently as part of the pesticide development process. These studies are intended to detect gross functional changes in behavior, motor activity, simple reflexes, and

TABLE 2.7 Reproductive Protocols

Parameter	Development	Reproduction
Preferred test species	Rat and rabbit	Rat
Age of animals	Young adults	6–8 weeks old at start of dosing
Pregnancy status	Pregnant	Not pregnant at start of dosing
Substance tested	Active ingredient	Active ingredient
Number of dose levels	3	3
Number of animals/dose	Rat, 20; rabbit, 20	60
Male : female ratio	Females only	50 : 50
Type of control group	Untreated or vehicle	Untreated
Route of administration	Stomach	Oral (in diet or water)
Exposure (days)	Daily through pregnancy	Continuous through two generations

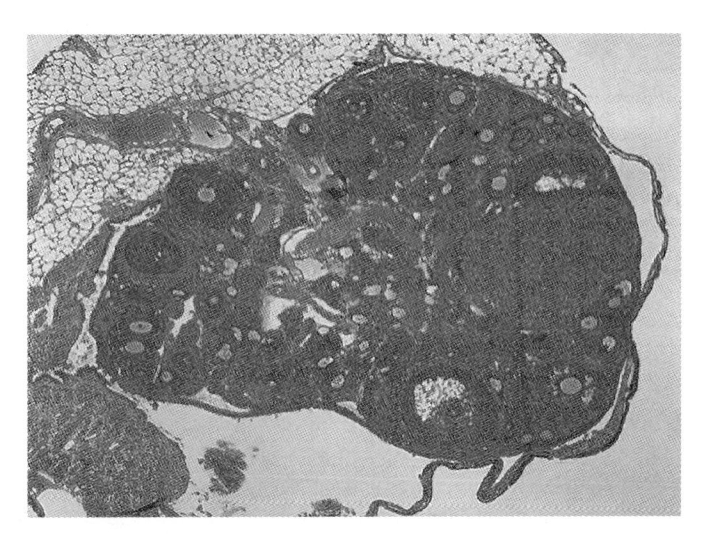

Figure 2.18 Normal ovary from a mouse (magnified) showing numerous eggs at various stages of development.

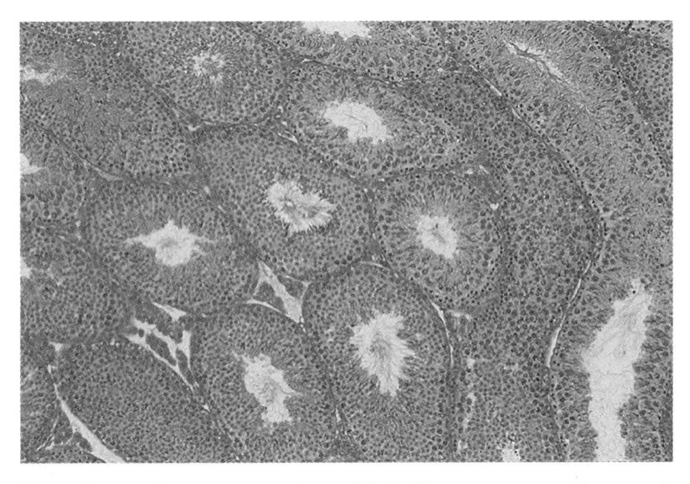

Figure 2.19 Normal testes from a mouse (magnified). Spermatozoa can be seen in the middle of the tubules.

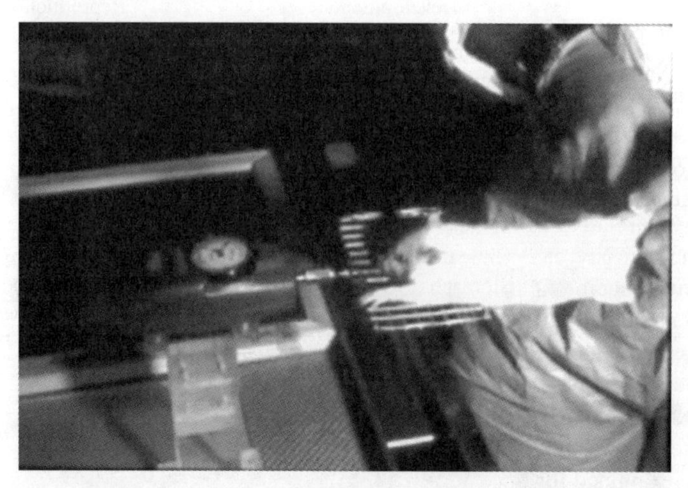

Figure 2.20 Grip strength is one of many evaluations used to assess the potential of the chemical to impact the nervous system.

any histopathological changes in the central or peripheral nervous systems. The studies generally consist of a Functional Observational Battery (FOB), a motor activity test, and a microscopic neuropathological examination (Fig. 2.20). The FOB uses predefined scoring scales to quantify and describe gross functional changes observed in the home cage, during handling, and while the animal is allowed to move freely in an open space. This could include changes in gait, reflexes, ease of handling, grip strength, right ability, etc. The motor activity test involves an automated device that measures the movement or activity of individual animals. The neuropathological assessment is achieved by a thorough microscopic examination of tissues from the central and peripheral nervous systems (Table 2.8).

In the acute study, young adult rats are administered a single high dose of the test substance, generally by gavage, then evaluated over a 2-week period. In a subchronic study, young adult rats are fed lower levels of the test chemical for 90 days and evaluated during that period. In some cases, a third developmental neurotoxicity study is also performed

TABLE 2.8 Neurotoxicity Protocols

Parameter	Acute	Subchronic	Delayed
Preferred test species	Rat	Rat	Domestic hen
Age of animals	7 weeks	6–8 weeks	Adult
Substance tested	Active ingredient	Active ingredient	Active ingredient
Number of dose levels	3	3	3
Number of animals per dose	20	20	10
Male : female ratio	50 : 50	50 : 50	0 : 100
Type of control	Positive, untreated	Positive, untreated	Positive, untreated
Route of administration	Oral	Diet/stomach	Oral
Exposure	Single dose	Single dose	Single dose
Observation period (days)	21	90	21 to 42
Microscopic pathology	Central and peripheral nervous system	Central and peripheral nervous system	Central and peripheral nervous system

to evaluate the potential neurotoxic effects in the offspring. In this case, the test chemical is administered to pregnant female rats during gestation and lactation. The offspring are then evaluated over a 60-day period. The evaluations generally include detailed clinical observations, motor activity tests, response to auditory startle, assessment of memory and learning, and a microscopic neuropathological examination.

Some organophosphate insecticides are known to produce a form of neurotoxicity that is delayed relative to initial exposure. Scientific studies have determined that this delayed neurotoxicity can be readily detected in hen chickens but not in mice or rats. Acute delayed neurotoxicity testing is required for organophosphate and carbamate insecticides and other pesticides suspected of causing delayed neurotoxicity. The pesticide is administered to domestic hens orally, as a single dose. Behavioral and locomotor abnormalities (e.g., splayed, waddling gait; circling; somersaults; paralysis) are observed for 21 days. If no neurotoxic responses are observed, the hens are redosed after 21 days and evaluated for 21 additional days. Nervous system tissues—brain, spinal cord, and nerves—are examined for microscopic changes.

Genetic Toxicology

Genetic toxicology is the study of how a chemical interferes with the genetic material of a cell (Fig. 2.21). It is studied in two ways:

- One is by visual inspection, through a microscope, of the damage to whole chromosomes. Damage at this level is referred to as *chromosome damage,* or *aberrations.*
- The second is by determination of any damage to genes (DNA sequences) that are too small to view under a microscope. Damage to genetic material inside a chromosome is referred to as *gene mutation.*

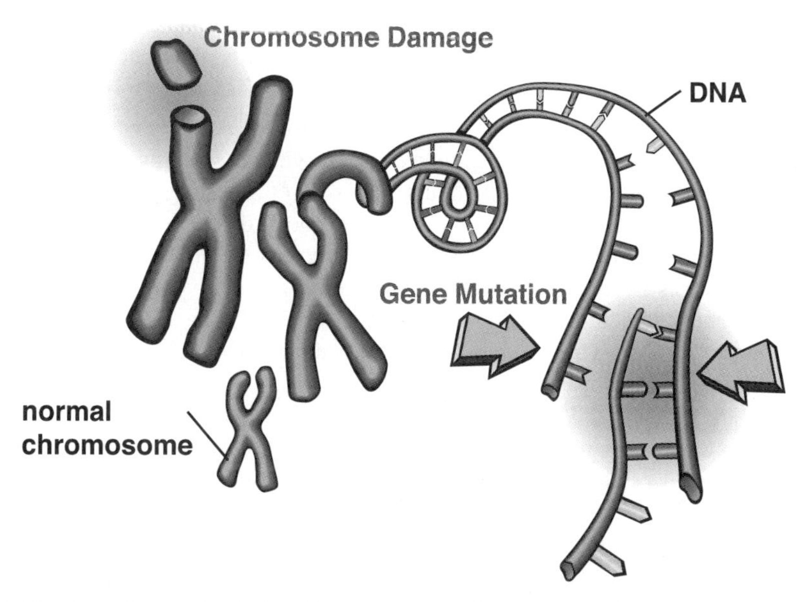

Figure 2.21 Genetic toxicological studies are used to evaluate the potential of a chemical to interfere with the genetic material of a cell.

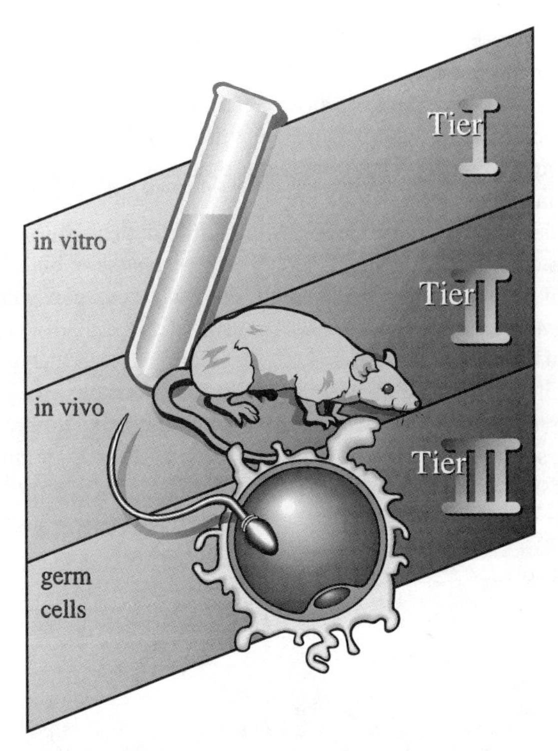

Figure 2.22 A tiered approach to assessing the impact of a chemical on genes and chromosomes is used in the pesticide registration process.

The body often can repair genetic damage, but unrepaired damage can result in mutation. In general, a mutagen is defined as a substance that permanently changes the genetic material of a cell. Mutations in germ cells (eggs or sperm) can cause heritable effects (those that can be passed on to future generations). Mutations in somatic cells (i.e., all cells but germ cells) may result in health problems, including cancer.

To determine if a pesticide is a mutagen, a battery of genotoxicity assays is conducted to evaluate the potential of the chemical to cause damage to both chromosomes and genes, both in cell cultures and in animals. Although the goal of this testing program is to identify mutagens, especially those that could affect future generations, the tests also are valuable as screens for potential carcinogens, since most mutagens are also carcinogens. The various individual tests which comprise the battery detect different kinds of genetic damage (Fig. 2.22). Genetic toxicology uses *in vitro* (cell culture tests in bacteria or yeast and mammalian cells) and *in vivo* (whole animal) tests. A tiered approach to assessing each end point (gene or chromosome) is used. A positive response in a lower tier test may lead to more informative and complicated higher tier testing.

An *in vitro* test battery provides the foundation of genotoxicity screening. It is the first tier of testing (*in vivo* studies are required as part of second-tier studies) and often the most sensitive; because these tests are so sensitive, many positive responses that need further characterization are obtained.

The second tier generally is performed in the whole animal. These tests are not as sensitive but they are more relevant. As in other toxicological tests in animals, the chemical must be absorbed and metabolized and must reach the target organ. These tests allow complex physiological factors to be examined.

The third tier focuses on germ cells. A positive result in this higher order test is a serious indication that the chemical can cause effects in future generations. A positive genotoxic response in somatic cell assays requires investigation into the potential of the chemical to cause germ cell mutation. In almost all cases, a germ cell mutagen is also a somatic cell mutagen; but most somatic cell mutagens are not germ cell mutagens. Most chemicals do not reach the germ cell; and of those that do, most tend not to interact with DNA.

Interpretation of test results can be challenging; because a battery of tests is used, not all tests will be positive, nor will all be negative. A chemical is assessed as a mutagen by first evaluating individual test end points. A single positive response in a test does not necessarily determine that a chemical is a mutagen. The overall assessment takes into consideration whether the test is *in vitro* or *in vivo,* whether it involves mammalian or nonmammalian cells, and whether somatic or germ cells are used.

Questionable results generally are resolved by advancing to a higher level in the hierarchy of testing. For example, a positive bacterial cell test may be confirmed by an *in vitro* test or in the whole animal, and a negative mammalian test normally supersedes a positive response in a bacterial test. A negative germ cell test, however, does not negate mutagenicity in somatic cells.

Regulatory Testing for Mutagenicity Numerous tests have been developed to detect chemical induction of both gene and chromosome damage. A battery of tests is performed initially to assess a chemical's genotoxic potential. The EPA requires that the battery include a bacterial gene mutation test, an *in vitro* mammalian gene mutation test, an *in vitro* mammalian chromosomal aberration test, and an *in vivo* mammalian chromosomal aberration test.

Bacterial Gene Mutation Test The *Salmonella typhimurium* bacterial gene mutation assay was developed in 1973 by Bruce Ames and is commonly referred to as the Ames test. It has proven to be an efficient and reliable method for testing large numbers of chemicals and has enhanced the detection of potentially mutagenic and/or carcinogenic compounds.

There are a number of *Salmonella* strains used in the gene mutation assay. Each is engineered to detect different types of gene mutations such as a frameshift or a base pair. To make the system sensitive for the detection of potential mutagens, the bacterial cell walls are altered to be more permeable to chemicals otherwise excluded, and the bacteria are modified to assure that they will not repair some types of DNA damage. Because bacteria do not metabolize chemicals as do animals, rat liver enzymes are added to the bacterial cultures to approximate the metabolism of mammalian systems and allow the detection of mutagenic chemicals that are mutagenic only in their metabolized form.

Procedures for conducting the Ames test are quite simple. Bacteria are treated with the chemical, with and without metabolic activation, and grown on agar plates in such a way that only the mutated bacteria will grow. Bacterial colonies are counted after 2 days and compared to the number of untreated colonies. A positive mutagenic response occurs when the number of treated colonies is 2–3 times higher than the number of colonies from untreated bacteria cultures.

In Vitro *Mammalian Gene Mutation* This group of tests is very similar in concept to the *Salmonella*/Ames bacterial test. Mammalian cell lines that are sensitive to gene mutation

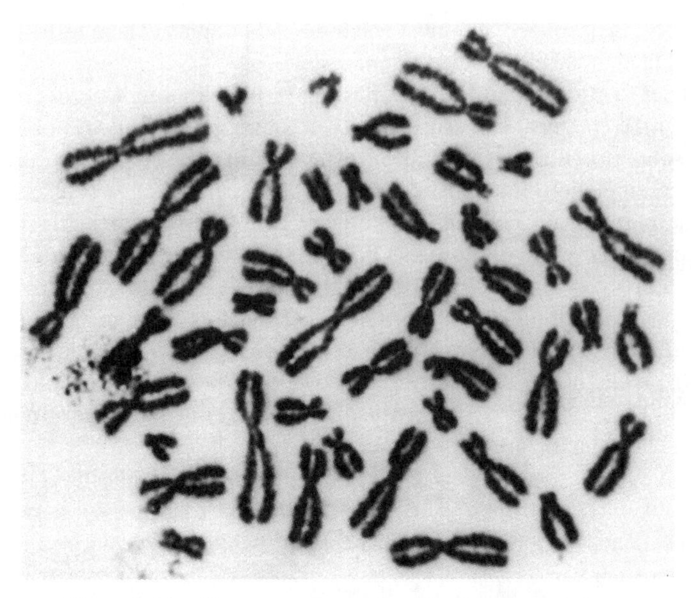

Figure 2.23 Chromosomes are identified by size and shape.

are selected. The most commonly used mammalian cells are cultured from Chinese hamster ovary (CHO), Syrian hamster embryo (SHE), V79 hamster, and mouse lymphoma cells. One of the most common mammalian *in vitro* gene mutation tests is the CHO/HGPRT assay, so named because it uses CHO cells and monitors gene mutation in the HGPRT gene. Metabolic activity is achieved by introduction of rat liver enzymes. Cells are treated for 4–18 hr, after which the chemical is removed and the cells are allowed to grow for a short time. A selective agent, 6-thioguanine, is then added to kill all normal cells so that only mutant cells will grow. Growing cell colonies are counted and mutant frequencies are determined.

In Vitro *Mammalian Chromosomal Aberration* In this type of assay, CHO cell lines or cultured human cells (lymphocytes) are most commonly used. Cells are exposed to the chemical in the presence and absence of rat liver enzymes. The cells are stained to facilitate microscopic examination to count the number of chromosomes present and to identify chromosome abnormalities. Scientists look for deformed and extra chromosomes and for loss of chromosomes (Fig. 2.23). The findings are compared to those of untreated control cells.

In Vivo *Mammalian Chromosomal Aberration* The two most commonly used tests for this end point are the mouse micronucleus test and the rat bone marrow chromosomal aberration assay. Animals are administered the chemical by stomach tube or injection. After 24 hr, they are sacrificed and their bone marrow cells are removed. In the *in vivo* chromosomal assay, as with an *in vitro* chromosome assay, scientists look for deformed and extra chromosomes and for loss of chromosomes. The findings are compared to those of untreated animals.

In the micronucleus test, bone marrow is removed from treated and untreated mice and stained for the presence of DNA. Normal red blood cells (RBCs) do not have a nucleus and therefore do not have chromosomes or DNA, that is, normal red blood cells do not stain. The DNA of mutated cells, on the other hand, does stain. Evaluations are conducted under the microscope. The presence of small bodies of DNA called micronuclei indicates that the chemical damaged the chromosomes.

Germ Cell Mutations There are two basic tests for germ cell effects: a *Drosophila* (fruit fly) sex-linked recessive lethal assay, and a rat dominant lethal test. In rare cases, a mouse spot (specific locus) test or a mouse heritable gene translocation assay is required. The mouse spot and the heritable gene translocation tests require a large number of animals and are costly and cumbersome to perform.

Pharmacokinetics: Absorption, Distribution, Metabolism, and Excretion

Pharmacokinetic studies determine how a pesticide moves into, gets distributed within, and finally leaves the body. The studies are designed to address several major areas of interest:

- The quantity of pesticide absorbed;
- The distribution of the pesticide in tissues, organs, blood, and urine;
- The identity, quantity, and location of major metabolites;
- The ability of the pesticide to be stored in tissues and organs;
- The routes of excretion; and
- The differences in absorption, metabolism, excretion, and distribution of a pesticide when animals are administered single doses versus repeated doses, or small doses versus large doses.

Special studies may be required to better define the pharmacokinetic properties of a pesticide. These studies might address tissue residues, placental transfer, occurrence in breast milk, and metabolism by specific organs, tissues, and cells. Comparative studies with different animal species are useful, also, in explaining species differences in toxicity.

Testing Protocols A radioactive tracer (carbon-14, sulfur-35, tritium) is incorporated into the pesticide molecule, making it possible to track the pesticide or its by-products as they move within and are expelled from the body. Young adult rats are assigned to several groups for which the dose differs in level, method of administration, and/or frequency. The adult rats are housed in metabolism cages equipped to collect urine and feces and, if necessary, carbon dioxide (Fig. 2.24). Animals are observed for 7 days after the administration of the dose or until 90% of the dose has been excreted in urine and feces, whichever comes first. Four typical groups are summarized below.

- Group A: A low, single dose of the tracer pesticide is administered by intravenous injection.
- Group B: A low, single dose of the tracer pesticide is administered orally, by capsule or by stomach tube.
- Group C: A low dose of the pesticide, without the tracer, is administered by capsule or by stomach tube daily for 14 days. The animals then receive a low, single dose of the tracer pesticide 24 hr after the last dose of the unlabeled pesticide.
- Group D: A high, single dose of the pesticide is administered orally, by capsule or by stomach tube.

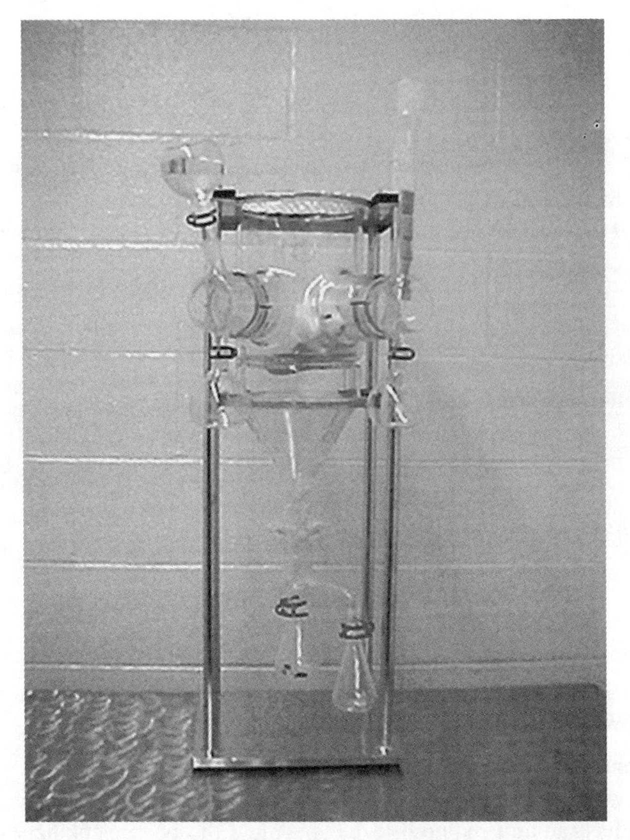

Figure 2.24 The urine, feces, and exhaled air is collected and evaluated for pesticide residue.

Urine, feces, blood, and various animal tissues are analyzed at various points in time following pesticide administration to determine levels of pesticide and metabolites.

Pharmacokinetic studies can be subdivided conceptually into three distinct phases:

- The *distribution* phase determines where the pesticide goes in the body.
- The *metabolism* phase chemically analyzes the identity and amount of metabolites in urine, feces, and tissues.
- The *excretion* phase determines the extent to which the pesticide and its metabolites are eliminated from the body and how long it takes.

EXTRAPOLATION FROM ANIMALS TO HUMANS

The descriptive and mechanistic studies are used in two important extrapolations (Fig. 2.25):

- Animal to human
- High dose to low dose

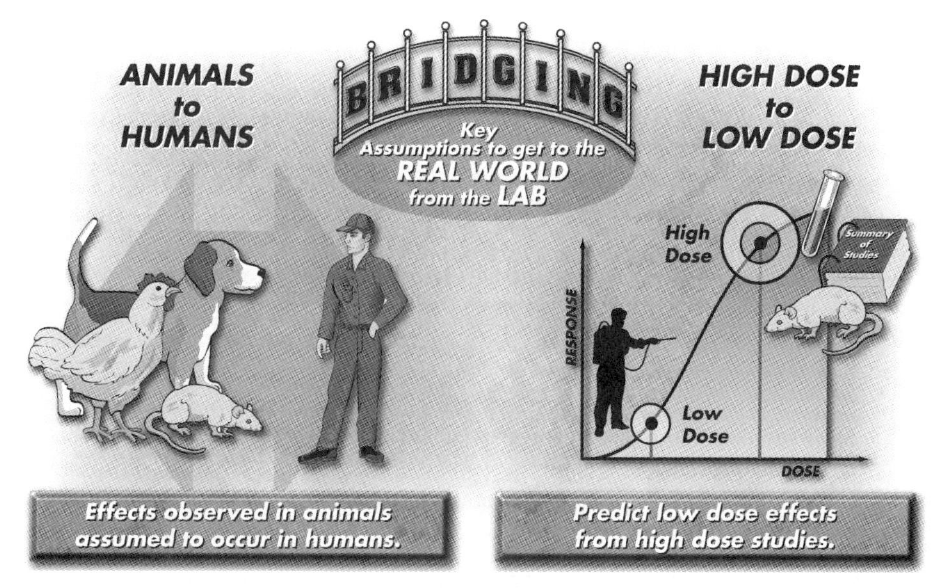

Figure 2.25 Laboratory animals are key predictive models for effects on humans.

Animal to Human Extrapolations

Risk assessment has traditionally relied on laboratory animals as predictive models for humans since we share many biological characteristics. Risk assessors generally assume that adverse effects in animals may be replicated in humans and that humans may be up to 10 times more sensitive than the most sensitive animal species tested. This is assumed unless there is sufficient information to indicate otherwise.

High Dose to Low Dose Extrapolations

In general, pesticide levels to which most humans might be exposed are far lower than those used in animal toxicity studies. Higher pesticide levels are used in animal testing to maximize detection of potential adverse effects from overexposure. Because of the limited number of animals that can be tested, animal studies at lower doses may not detect a subtle effect that may occur in very large human populations exposed to the chemical. However, high doses used in animal studies may overload the metabolic and/or physiological processes of the animals and thus lead to adverse effects that are not predictive of those expected at lower exposure levels. This dilemma leads to one of the major challenges for toxicologists and risk assessors today: determining whether or not the effect is relevant to humans.

Studying the Human Experience

As previously indicated, most of the toxicology data used in human risk assessment is derived from animals, and questions are sometimes raised as to the relevance of these data. But in some cases there are several ways to evaluate the likelihood of similar effects in humans.

Human Cell Research New approaches to cell and tissue culture studies allow the use of isolated human cells and tissues in evaluating pesticide toxicity. Isolated human cells and tissue can be obtained (from persons who have died in accidents or undergone surgery) and placed in a nutrient solution that allows the cells to continue their normal metabolic processes. Once tissue and cell cultures are established, a pesticide can be introduced and its effect studied for varied lengths of exposure. If metabolism of the chemical is similar between human cells and animal cells, validity for using the animal model is assumed.

Clinical Studies Although rare, human volunteer studies are conducted with some pesticides, but only after thorough review and approval by an ethics review board. Such studies are conducted under carefully controlled conditions, with multiple safeguards to protect human health. Volunteers are monitored carefully by physicians before, during, and after the studies to verify that there are no adverse effects. The comparison of data generated from human volunteer studies to those generated in animal studies leads to more accurate human risk assessment. In some studies, absorption, metabolism, and excretion are studied by exposing volunteers to a small, nontoxic dose. More information on the EPA's policy regarding human testing can be found at http://www.epa.gov/science1/ec0017.pdf.

Epidemiological Investigations Epidemiological studies can be used to corroborate predictions and extrapolations from animals to humans. Individuals working in pesticide manufacturing facilities are ideal subjects for epidemiological studies. In these particular work situations, much is known about the workers' medical history, work history, and exposure levels. Thus, an association between chemical exposure and a particular chemical can be evaluated by comparing the health of an individual exposed to a pesticide to that of someone not exposed. See Chapter 3 for more information on epidemiological investigations.

DIETARY EXPOSURE ASSESSMENT

Pesticide residues in the diet probably represent the most common source of pesticide exposure for the general public. Dietary exposure is a function of the type and amount of food consumed and the pesticide residues in or on that food. The total dietary intake of a single pesticide for any population is calculated by summing the potential pesticide intake from all food items that potentially contain its residues. The basic model for estimating dietary exposure to chemical residues in food is very simple:

$$\text{Pesticide Ingested} = \text{Residue Concentration} \times \text{Foods Consumed}$$

There are numerous levels of dietary exposure modeling, ranging from single (point) exposure residue estimates to complex simulation analyses using probabilistic approaches. But all models, however complex, are based on the basic relationship: Exposure depends on the residue concentration in the food and on the amount of food consumed (Fig. 2.26).

Two types of dietary exposure are generally considered: chronic and acute. Chronic exposure occurs over a long period of time. It is calculated for typical exposure levels and therefore uses mean consumption and mean residue values.

In contrast, acute exposure considers short-term exposure to the pesticide residues in food, usually over the course of one day. Acute dietary exposure is calculated using individual consumption data. The residue values range from tolerances (maximum values from studies) to the entire range of residue values observed in surveys. Although dietary

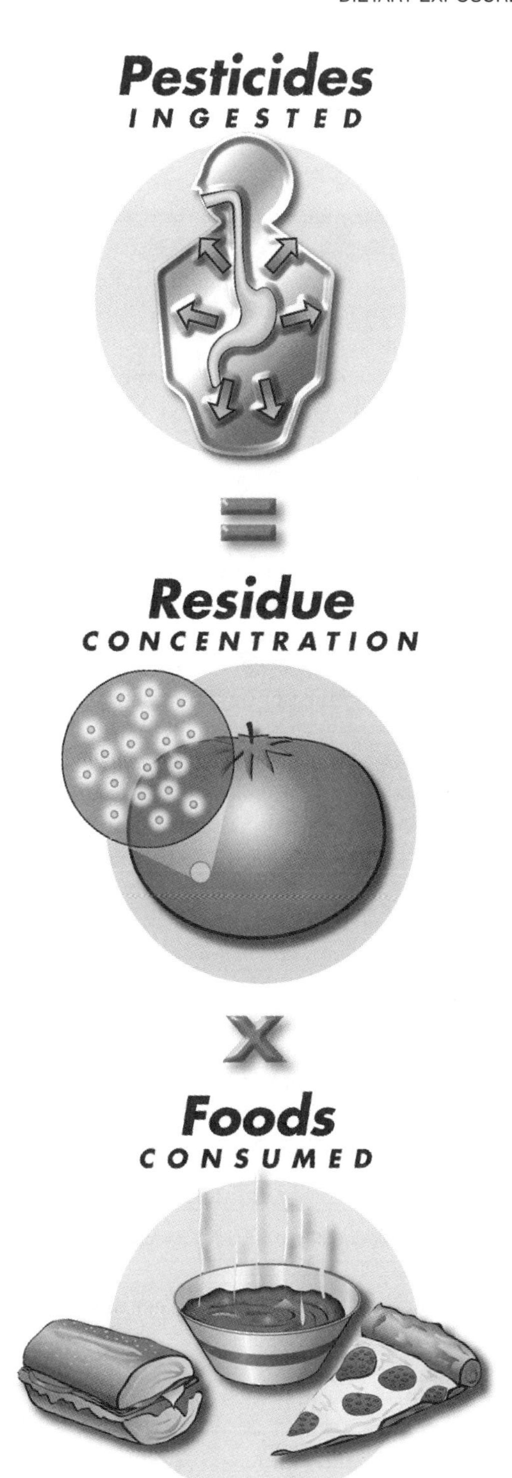

Figure 2.26 Pesticide exposure is a function of residue concentrations on food and the amount of food consumed.

risk assessment is simple in principle, it can be very complicated in practice depending on which assumptions are made and which models are used.

How Much Food Do We Consume?

USDA Estimates Food Consumption The U.S. Department of Agriculture is the primary agency that collects information on food consumption by the American public. Nationwide Food Consumption Surveys were conducted by the USDA in 1977–1978 and in 1987–1988. In 1989, the USDA began a new cycle of surveys: the Continuing Survey of Food Intake by Individuals (CSFII).

CSFII surveys are important because food consumption patterns change with time and have an impact on estimates of pesticide exposure. For example, overall fruit consumption has remained unchanged, but children are drinking more fruit juices. People are eating leaner cuts of meat, more chicken and fish, and less beef than they did 10 years ago. We are eating more restaurant meals, both dine-in and carry-out, and more ready-to-heat microwave meals.

USDA food consumption surveys are intended to measure daily consumption patterns for households across the United States at various times throughout the year (Fig. 2.27). The surveys ask participants to complete questionnaires that deal with the total household food intake over 2 or 3 consecutive days. Each participant is asked to describe the type and quantity of each food eaten, the time of day it was consumed, and its origin (home or restaurant).

Completed questionnaires are submitted to USDA nutritionists who convert the foods eaten to corresponding raw agricultural commodity ingredients. This determination is based on generic or product-specific recipes and food label ingredient statements. For instance, if a person eats two slices of a supreme pizza, they actually consume tomato paste, bell peppers, onions, wheat, olives, sugar, milk products, pork, and oil.

The total amount of each raw agricultural ingredient consumed is calculated by adding the contribution from each food eaten. For instance, the daily dietary intake of wheat is calculated by adding the total amount of wheat consumed in bread, bakery goods, cereals, pasta, and other food items that contain wheat.

The final calculation is to divide the weight of each agricultural ingredient eaten by the weight of the individual. The food consumption estimate is expressed as grams of raw agricultural commodity per kilogram of body weight per day.

If a 69-kg (150-lb) woman consumes 100 g of wheat per day, her consumption is expressed as 1.45 g of wheat per kg of body weight:

$$100 \text{ g} \div 69 \text{ kg} = 1.45 \text{ g/kg}$$

A 113-kg (250-lb) man who eats 100 g of wheat per day consumes approximately 1 g of wheat per kg of body weight:

$$100 \text{ g} \div 113 \text{ kg} = 0.88 \text{ g/kg}$$

A 27-kg (60-lb) child who eats 100 g of wheat per day consumes approximately 4 g of wheat per kg of body weight:

$$100 \text{ g} \div 27 \text{ kg} = 3.7 \text{ g/kg}$$

Figure 2.27 Food consumption patterns are based on individual responses which are then used to establish the food intake pattern in a population.

Understanding Consumption Patterns by Subpopulations USDA dietary consumption surveys are designed to represent the entire U.S. population as well as specific subpopulations. Each person in the survey is classified according to demographic characteristics: age, sex, ethnic group, pregnancy and lactation status, and household income. Based on individual responses, food consumption patterns are established for populations and subpopulations as listed in Table 2.24.

How Much Pesticide Is on Food?

The amount of a food commodity consumed is only part of the equation for estimating total pesticide ingestion via the diet. The other critical measurement is the amount of residue in or on those foods. Pesticide residues can be estimated by many methods, each with its own strengths, limitations, and assumptions.

Tolerances A tolerance is a legally enforceable maximum level, generally expressed in parts per million (ppm), of a pesticide and/or its metabolites that can be legally present in or on a commodity such as fresh or processed foods, animal feed, meat, milk, and eggs. International tolerances are referred to as Maximum Residue Levels (MRLs). Pesticide tolerances for specific crops can be found at 40 CFR Part 180 at http://www.epa.gov/pesticides/cfr.htm.

Tolerances for Crops and Products Derived from Crops Crop tolerances are based on results from controlled field trials conducted in various geographical locations. The trials are designed to identify the highest concentrations expected on a crop, often referred to as a Raw Agricultural Commodity (RAC), using good agricultural practices, maximum application rates, maximum number of applications, and the shortest application-to-harvest interval. Because they are conducted under maximum conditions, they yield maximum residue levels—or what are termed worst-case estimates.

The registrant petitions the EPA to set a pesticide tolerance for each crop that appears on a product label. As a general rule, the EPA requires a slightly higher tolerance than the highest residue found in field tests. The higher tolerance allows for the occurrence of slightly higher residues that may occur under environmental conditions not tested or under differing agricultural practices (Fig. 2.28).

Table 2.9 shows residue levels in samples taken from three crops for the purpose of setting a tolerance for Insecticide X. The registrant conducted 16 residue studies, as mandated by EPA residue chemistry guidelines. Since the maximum residue observed in the apple field trials was 0.27 ppm, the registrant might petition EPA for a pesticide tolerance of 0.3 ppm, which is slightly above the highest residue detected on the apples in any of the 16 trials (Fig. 2.29).

Tolerances for Animal Products Direct application of pesticides to livestock can leave residues in meat, as can livestock consumption of treated feed. If the results of animal metabolism studies indicate that pesticide residues are likely to be found in animal products from livestock that is fed crops treated with pesticides, tolerances for products such as meat, milk, and eggs must be established. To establish tolerances, chickens and ruminant animals (goats and cattle) are fed diets containing various levels of the pesticide for 28 days. Eggs, meat, and milk from these animals are then analyzed for pesticide residues. The registrant would petition the EPA for individual tolerances for beef, poultry, eggs, and milk, depending on the residues detected (Fig. 2.30).

Anticipated Residues Since the tolerance represents the maximum residue concentrations possible on a commodity, its use for dietary risk assessment represents a conservative or worst-case scenario for dietary exposure (Fig. 2.31). More realistic estimates for dietary exposure result from pesticide data generated under normal use patterns. For example,

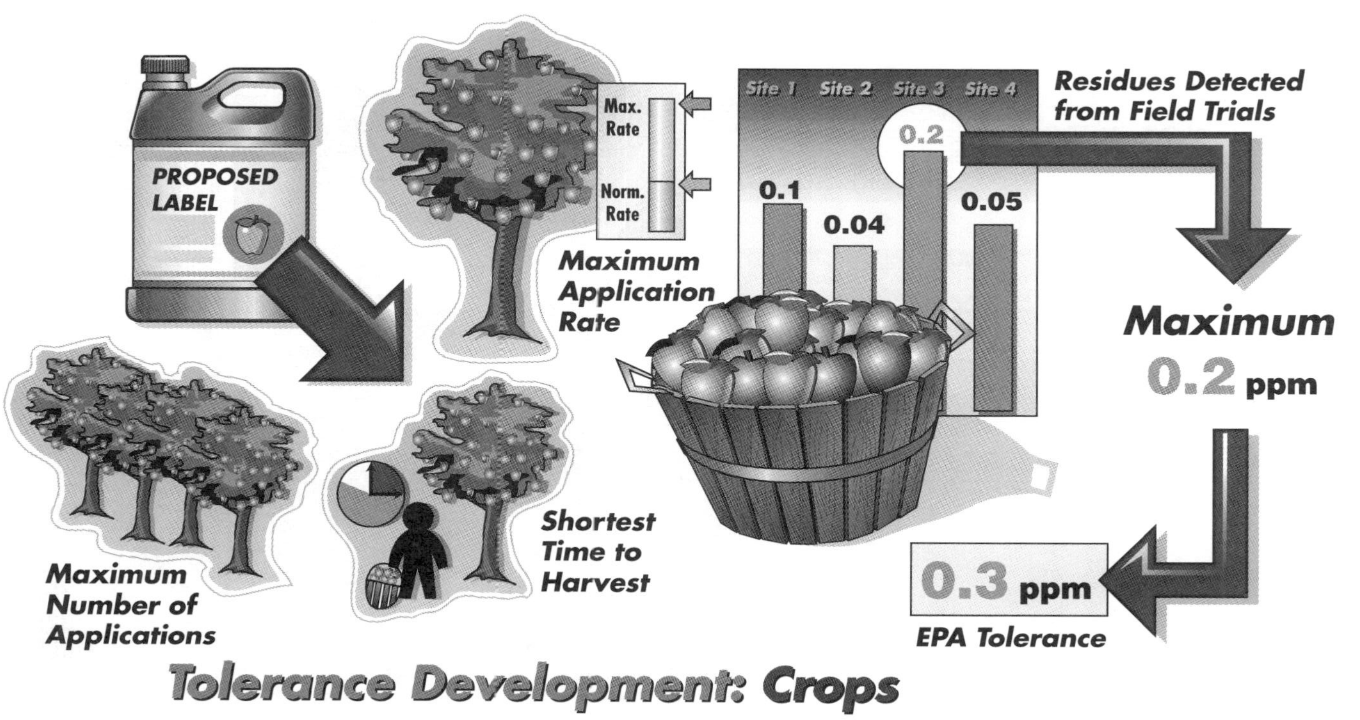

Figure 2.28 Testing under worst-case field conditions is used to determine pesticide tolerances.

TABLE 2.9 Residue Levels (ppm) from Field Trials for Insecticide X

Sample	Apples	Oranges	Tomatoes
1	0.27	1.20	0.44
2	0.24	1.10	0.42
3	0.21	1.00	0.39
4	0.19	0.94	0.33
5	0.18	0.93	0.31
6	0.14	0.91	0.27
7	0.13	0.83	0.27
8	0.13	0.81	0.24
9	0.11	0.80	0.20
10	0.09	0.77	0.19
11	0.09	0.75	0.18
12	0.08	0.73	0.17
13	0.07	0.66	0.16
14	0.06	0.64	0.14
15	0.04	0.63	0.11
16	0.04	0.54	0.09
Mean	0.13	0.83	0.24

pesticides are not always applied at the maximum rate and frequency permitted by the label, and crops are not always harvested as soon as legally allowed following pesticide application. Also, residue levels may decrease over time as a result of storage, washing, trimming, and cooking.

The assumptions and data used to calculate anticipated residue estimates generally depend on the crop and/or whether acute or chronic risks are being evaluated. Depending on

Figure 2.29 Applying a pesticide by air blast sprayer in an orchard to determine residue levels.

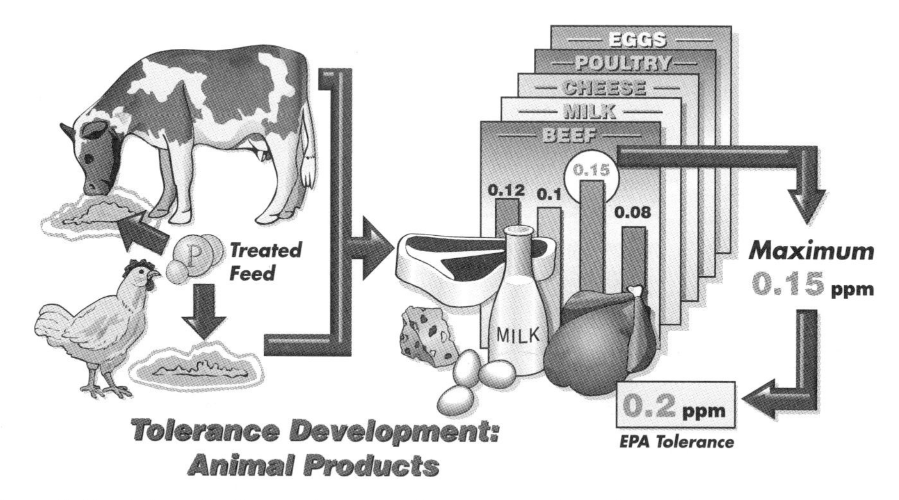

Figure 2.30 Animal feeding studies are used to determine food chain effects on the development of tolerance.

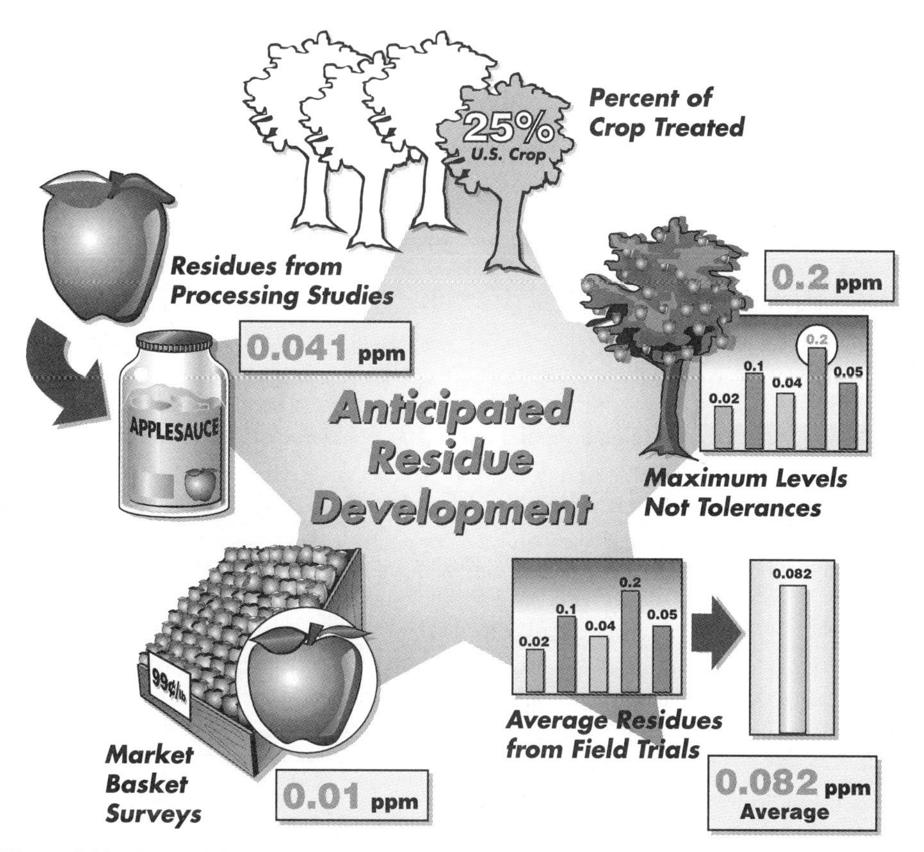

Figure 2.31 Determining anticipated residue levels requires comprehensive date on crop management and commodity distribution.

the exposure scenario and the degree of refinement desired, anticipated residues may be derived as follows:

- By taking into account the percentage of a crop treated with the pesticide. Only a portion of any crop in the United States is likely to be treated with a given pesticide, and only the treated portion is expected to yield pesticide residues. Dietary exposure based on tolerances may often, as a worst case, assume that 100% of the crop is treated. Information on the percentage of crops treated can be found at http://www.ncfap.org/ncfap/index.html and http://ipmwww.ncsu.edu/opmppiap/.

  Note: Adjusting for the fact that only 15% (0.15) of the apple crop is treated (Table 2.12), the average residue would be 0.02 ppm (0.15 × 0.13 ppm); with oranges, adjusting for treatment of 20% of the crop, the average residue would be 0.17 ppm (0.20 × 0.83 ppm); and with tomatoes, adjusting for treatment of only 10% of the crop, the average residue would be 0.02 ppm (0.10 × 0.24 ppm).
- By using the maximum residue from field trials, rather than the tolerance. This is often used when assessing potential acute risks from consuming whole foods such as apples, potatoes, or tomatoes. The maximum residue generally is slightly lower than the tolerance and is more representative of the highest residues expected. For instance, in Table 2.9 the highest pesticide residue found on apples is 0.27 ppm, while the tolerance is 0.3 ppm.
- By using the average residue from field trials when considering dietary exposure for blended commodities. A blended commodity generally is not eaten intact; instead, it is mixed with like crops from other farms: for example, wheat that is ground into flour and apples that are made into juice or pie filling. In the example shown in Table 2.9, the mean residue value of 0.13 ppm for apples could be used instead of the tolerance (0.3 ppm) or maximum residue (0.27 ppm) in assessing potential chronic exposure from residues in processed apples.
- By using factors from processing studies. Registrants often conduct studies to evaluate the effect of processing on residue levels. Depending on the physicochemical properties of the pesticide (e.g., solubility in fat and water) and the nature of the crop, residues in processed fractions (e.g., corn oil, tomato paste) may be higher or lower than in the raw agricultural commodity.
- By using residues based on monitoring data, the FDA (http://vm.cfsan.fda.gov/~dms/pesrpts.html), the USDA (http://www.ams.usda.gov/science/pdp/index.htm), and the states routinely collect and analyze foods such as fresh produce, meat, milk, and eggs to determine the levels of pesticide residue present.

  Food may be monitored by collecting samples at or near the farm, at the point of entry into the United States (for imported foods), and at close-to-consumer locations (e.g., at produce markets or grocery distribution centers). Such monitoring programs typically show much lower residues than those from field residue studies (Tables 2.10 and 2.11).
- By using residues based on market-basket surveys. On occasion, registrants measure actual residues present in food offered for purchase by the consumer. These studies are conducted by sampling and analyzing fresh and/or processed products at retail locations throughout the country.

TABLE 2.10 Examples from the USDA Pesticide Data Program

Commodity/Product	Pesticide Type	Percent Detection	Mean Residue (ppm)	90th Percentile[a] Residue (ppm)	Ratio of 90th Percentile to Tolerance
Apple juice/thiabendazole	Fungicide	38	0.12	0.28	0.03
Apples/diphenylamine	Fungicide	86	0.72	1.60	0.16
Carrots/linuron	Herbicide	69	0.03	0.08	0.08
Grapes/captan	Fungicide	42	0.07	0.02	0.0004
Green beans/acephate	Insecticide	33	0.03	0.07	0.02
Oranges/imazalil	Fungicide	58	0.14	0.30	0.03
Peaches/iprodione	Fungicide	79	0.72	1.70	0.09
Spinach/permethrin	Insecticide	60	0.97	3.30	0.17
Sweet potatoes/dicloran	Fungicide	64	0.24	0.65	0.07
Tomatoes/methamidophos	Insecticide	37	0.01	0.03	0.03
Wheat/chlorpyrifos methyl	Insecticide	73	0.07	0.18	0.03

[a]The 90th percentile means that at least 90% of that commodity had residues (ppm) of at least that much or less.
Source: Pesticide Data Program Annual Summary, Calendar Year 1996. USDA Agricultural Marketing Service, 1998.

In addition, the FDA conducts a Total Diet Study that uses a market-basket approach to analyze pesticide residues in food prepared for eating. Over 200 food items are selected and purchased in grocery stores in four geographic areas, four times a year; the food is then prepared in institutional kitchens, where it is analyzed for pesticide residues after it is table-ready or in final form for consumption. These studies provide more realistic estimates of pesticide residue concentrations actually consumed because they take into

TABLE 2.11 Examples from the Food and Drug Administration Pesticide Testing Programs

		How Much Residue Was Found (%)		
	Sample Numbers	None	Within Tolerance	Above Tolerance
Domestic production				
Grains/grain products	363	53	46	1
Milk/dairy products/eggs	781	97	3	0
Fish/shellfish	520	62	38	0.5
Fruits	1194	46	53	1
Vegetables	1958	64	35	1
Animal feeds	506	61	39	1
Import production				
Grains/grain products	230	80	19	1
Milk/dairy products/eggs	129	95	5	0
Fish/shellfish	124	88	12	0
Fruits	1735	57	4	3
Vegetables	2378	63	34	3
Animal feeds	76	75	20	5

Source: Residue Monitoring 1996. Food and Drug Administration Pesticide Program, 1998.

Figure 2.32 Selecting fruits and vegetables for pesticide analysis.

account changes that result during storage, cleaning, processing (e.g., apples into applesauce), and cooking. More information on the FDA's Total Diet Study can be found at http://vm.cfsan.fda.gov/~comm/tds-toc.html.

One of the most useful residue databases for exposure assessment is the USDA's Pesticide Data Program (PDP), which is designed to provide residue data for risk assessment. Several features distinguish it from typical monitoring databases such as those compiled by the FDA, whose main objective is enforcement. Some PDP samples are washed or peeled before analysis which is not done in standard FDA monitoring. In general, the PDP typically uses methods 5–10 times more sensitive than those used in enforcement program studies. The PDP monitoring program is designed to provide data for use in dietary risk assessment.

Tolerances are established on the RAC (i.e., the harvested crop), and when food samples are collected and analyzed for FDA enforcement programs, the RACs are tested. For example, the FDA analyzes pesticides in whole oranges, including the peel. The USDA's PDP, on the other hand, analyzes residues in the edible pulp of the fruit only, excluding the peel. Therefore, the PDP provides residue information on foods "as eaten." Such data are more suitable for risk assessment than those collected during enforcement program studies (Fig. 2.32).

PDP sampling is based on a rigorous statistical design. Samples are collected from large distribution centers that account for approximately 60% of the nation's food supply. Often, regulatory agencies such as the EPA want to evaluate potential dietary exposure to foods consumed in relatively large quantities. The PDP focuses on fresh fruits and vegetables, although milk and processed foods such as canned green beans, grape juice, and corn syrup sometimes are sampled as well. The EPA and USDA collaboratively select specific foods to be sampled and analyzed in each study.

How Much Pesticide Do We Consume, Long Term?

There are two basic techniques for estimating long-term exposure to pesticides in food:

- Use of tolerance levels to calculate the Theoretical Maximum Residue Contribution (TMRC), also referred to as the Theoretical Maximum Daily Intake
- Estimation of the Anticipated Residue Contribution (ARC), also called the Estimated Daily Intake

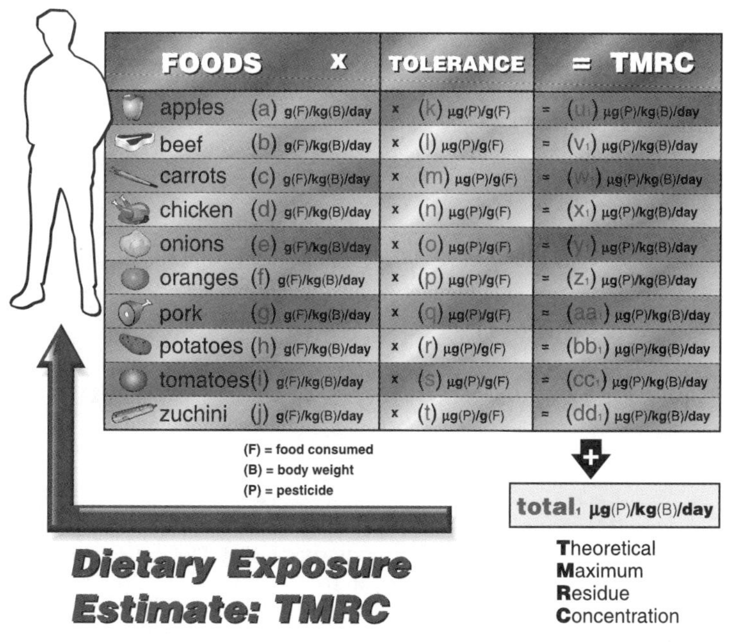

Figure 2.33 Theoretical maximum residue contribution estimates dietary exposure based on worst-case consumption of foods, and assuming pesticide residues are at tolerance levels.

Theoretical Maximum Residue Contribution TMRCs are calculated individually for each crop on a pesticide label. They are calculated for each agricultural commodity by multiplying the amount consumed by the corresponding tolerance level. The theoretical maximum amount of pesticide consumed is then calculated by summing the TMRCs from each individual commodity. The estimate of total exposure developed via this method represents the theoretical, worst-case, maximum legal amount that an individual might consume (Fig. 2.33).

TMRCs are calculated on the assumptions that 100% of the crops for which the pesticide is registered are treated and that pesticide residues are present at tolerance levels. TMRC analysis further assumes that postharvest storage, handling, processing, or cooking does not reduce residues. For example, Insecticide X is registered and has tolerances as listed in Table 2.12.

Tolerances established for raw agricultural commodities are valid for all forms of the crop consumed, unless scientific experiments indicate the need for different tolerances for different forms of the food. For example, if a tolerance of 0.3 ppm is established for apples, that is the legal pesticide limit on fresh apples, apple juice, apple juice concentrate, dried apples, etc. But if an experiment shows that the tolerance on dried apples should be 1 ppm, that tolerance would be assigned to dried apples only, while the tolerance for all other apple products would be 0.3 ppm. A sample TMRC calculation using the tolerances shown in Table 2.12 and consumption data from the 1994–1996 CSFII is given in Table 2.13.

Anticipated Residue Contribution More realistic estimates of dietary exposure can be obtained by considering pesticide use patterns and/or residue levels anticipated rather than the TMRC. The methodology and assumptions used to estimate residues vary somewhat, depending on whether short- or long-term risk is being evaluated (Fig. 2.34).

TABLE 2.12 Established Tolerances for Insecticide X

Commodity	Tolerance (ppm)	Percentage of Crop Treated
Apples	0.30	15
Corn	0.02	5
Oranges	1.50	20
Tomatoes	0.50	10
Wheat	0.02	100
Meat	0.02	100
Milk	0.02	100
Poultry	0.02	100
Eggs	0.02	100

TABLE 2.13 Estimates of Chronic Dietary Exposure Using Theoretical Maximum Residue Contributions and Anticipated Residue Contributions

	TMRC Exposure (mg/kg/day)	ARC Exposure (mg/kg/day)
U.S. population (total)	0.005408	0.000152
U.S. population (spring)	0.005299	0.000150
U.S. population (summer)	0.005122	0.000146
U.S. population (autumn)	0.005544	0.000157
U.S. population (winter)	0.005677	0.000156
Northeast region	0.006741	0.000185
Midwest region	0.005245	0.000152
Southern region	0.004740	0.000135
Western region	0.005453	0.000151
Hispanics	0.006858	0.000180
Non-Hispanic whites	0.005011	0.000145
Non-Hispanic blacks	0.005957	0.000162
Non-Hispanic/nonwhite/nonblack	0.006957	0.000185
All infants (< 1 year)	0.003911	0.000103
Nursing infants (< 1 year)	0.001463	0.000041
Nonnursing infants (< 1 year)	0.004627	0.000122
Children 1–6 years	0.015889	0.000418
Children 7–12 years	0.008507	0.000239
Females 13–19 years (not pregnant or nursing)	0.005341	0.000147
Females 20+ years (not pregnant or nursing)	0.003555	0.000104
Females 13–50 years	0.003976	0.000115
Females 13+ years (pregnant/not nursing)	0.005251	0.000144
Females 13+ years (nursing)	0.004984	0.000143
Males 13–19 years	0.005864	0.000165
Males 20+ years	0.003638	0.000108
Seniors 55+ years	0.003329	0.000098

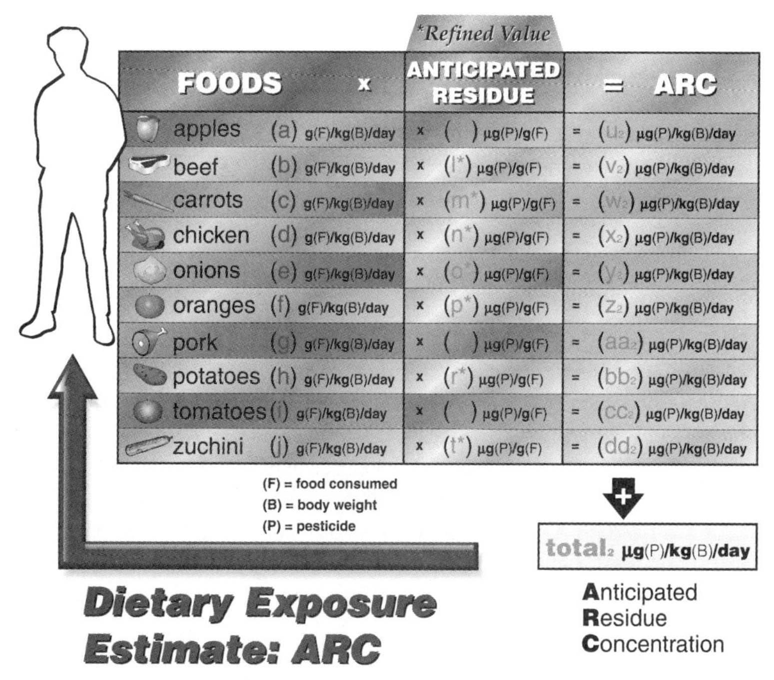

Figure 2.34 Anticipated residue contributions use refined values of pesticide residues as they actually occur on food and food products.

Anticipated residue estimates are further refined by considering residue data from field trials, food processing studies, and monitoring. Reduced residues are expected when data from processing and monitoring studies are incorporated. Such studies analyze residues in crops not immediately after harvest, as in field trials, but after storage, handling, and processing. They provide a more realistic estimate of potential human exposure.

The ARC values in Table 2.13 were calculated using more detailed information:

- Mean residue values from field trials (instead of tolerance)
- Percentage of crop treated (instead of assuming that 100% was treated)
- Adjustment for the effects of processing

In addition, theoretical residues in edible animal tissues were calculated by using data from animal metabolism and livestock feeding studies (instead of simply using tolerance values). As shown, even relatively simple refinements have tremendous impact on the results of exposure estimation. It is expected that the procurement of monitoring data would reduce exposure estimates even further.

How Much Pesticide Do We Consume, Short Term?

Acute analysis does not calculate a single estimate of exposure as with TMRCs and ARCs. As with TMRC and ARC calculations for chronic exposure, however, acute distributional analysis may be worst case or refined. The first level of acute analysis assumes that 100% of

TABLE 2.14 Acute Dietary Exposure Estimates (mg/kg/day) for the Upper 95th, 99th, and 99.9th Percentile for Selected Subgroups

	95th Percentile	99th Percentile	99.9th Percentile
U.S. Population (all seasons)	0.022468	0.049183	0.103411
Nonnursing infants (<1 year)	0.015741	0.035950	0.061152
Children (1–6 years)	0.055713	0.101169	0.201126
Children (7–12 years)	0.032254	0.058032	0.088462

all registered crops contain tolerance level residues, and distribution of exposure is calculated from individual daily consumption data. Since there is no variation in residue data, variation in consumption forms the basis for distribution. So in the initial acute analysis we can evaluate safety at the extreme end of the exposure distribution because all consumption data are included in the analysis—even those for individuals who consume large quantities of food. In contrast, the chronic analysis uses mean consumption data to estimate typical exposure over a relatively long period of time.

Table 2.14 summarizes the results of a Tier 1 acute dietary exposure analysis for Insecticide X, using tolerance levels shown in Table 2.12. In this worst-case approach, the entire U.S. apple supply was assumed to contain 0.3 ppm, all milk was assumed to contain 0.02 ppm, and so on for all commodities considered (100% treated at residue tolerance levels).

In Table 2.13, the TMRC estimate for the U.S. Population is a simple point estimate of 0.005408 mg/kg/day. In contrast, exposures for the Tier 1 acute analysis shown in Table 2.14 are summarized for the 95th, 99th, and 99.9th percentiles of the exposure distribution. In fact, the acute analysis yields a complete exposure distribution, whereas Table 2.14 reports only the high-end exposure estimates. Comparison of the two tables clearly demonstrates the difference between chronic and acute assessment.

The estimated exposure in the upper 95th percentile is 0.055713 mg/kg/day for children 1–6 years old; the other 5% of children in that age group would experience exposures above 0.055713 mg/kg/day. Ninety-nine percent of children 1–6 years old would be exposed to 0.101169 mg/kg/day or less, and 1% would be exposed to more than 0.101169 mg/kg/day.

Monte Carlo Analysis The Monte Carlo approach is a well-established statistical technique employed to refine exposure estimates for acute dietary risk assessment. Repeated sampling from the complete distribution of food consumption and the entire spectrum of pesticide residue data (from field trials, monitoring data, and market-basket surveys) is used to predict the amount of pesticide likely to be consumed by one individual on a given day. Percentage of crop treated also can be included in the Monte Carlo analysis, but differently than for chronic exposure (Fig. 2.35).

Some individuals do not consume fresh apples, whereas others may consume several apples per day. Insecticide X residues on fresh apples may vary from no residue (85% of apples not treated) up to 0.27 ppm, as shown in Table 2.9. The residue distribution on apples (from which the Monte Carlo analysis would sample randomly) initially would consist of the 16 residues from field trials (Table 2.9). But since only 15% of the apples were treated with Insecticide X, 90 zero-residue samples would be added to the distribution so that the

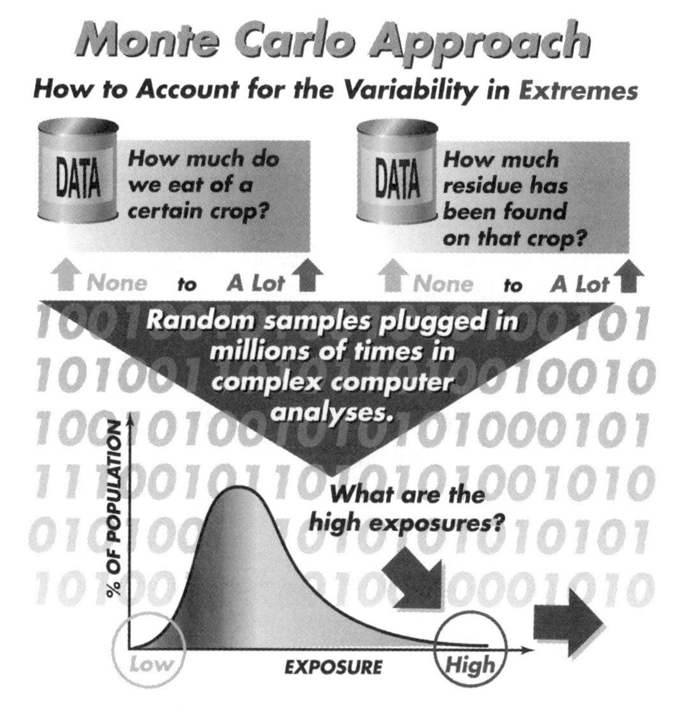

Figure 2.35 Statistical approaches are used to better understand the probability of residue exposure.

16 Insecticide X residues would comprise 15% of the total (106). Thus, the entire residue distribution would be 16 residues plus 90 zeros. In the Monte Carlo sampling, a zero would be selected approximately 85% of the time, and a residue from the 16 residue values would be selected at random the other 15% of the time.

The Monte Carlo analysis in Table 2.15 uses residue samples from actual field trial data on apples, oranges, and tomatoes. Only selected population groups are displayed because of the complexity of the acute distribution.

The acute exposure estimates in Table 2.15 are much lower than those in Table 2.14. This is because Table 2.14 is based on tolerance level residues assumed to be present in all crops that could be treated with Insecticide X, whereas the refined analysis shown in Table 2.15 uses actual residue values (only a portion of the crop was treated).

TABLE 2.15 Monte Carlo Acute Dietary Exposure Estimates (mg/kg/day) for the Upper 95th, 99th, and 99.9th Percentiles for Selected Subgroups

	95th Percentile	99th Percentile	99.9th Percentile
U.S. Population (all seasons)	0.002682	0.005916	0.012706
Nonnursing infants (<1 year)	0.001727	0.004583	0.011140
Children (1–6 years)	0.006711	0.012376	0.023540
Children (7–12 years)	0.003913	0.006959	0.010900

OCCUPATIONAL EXPOSURE ASSESSMENT

Workers who formulate or package pesticide products in factories, those who apply pesticides for commercial businesses, and those who farm come into contact with pesticides in the course of their work. In addition, workers who enter treated fields or greenhouse facilities also may be exposed to pesticide residues. Although pesticide exposure in the work environment cannot be totally eliminated, worker contact with pesticides can be minimized by following product label directions, using appropriate protective clothing and equipment, and practicing good industrial hygiene.

Worker Exposure Related to Work Practices

Exposure assessments are most precise when worker exposure is described clearly and accurately. Variables that influence exposure are

- duration and frequency of exposure,
- protective gear used,
- product formulation,
- route of exposure,
- quantity of pesticide handled,
- type of mixing/loading operations,
- type of application equipment,
- environmental conditions, and
- nature of work tasks following entry into a treated field.

The worker exposure scenario and individual work practices determine estimated worker exposure. For example, a pesticide applicator with one company may take 30 min each day to dilute and mix the pesticides that he will use that day, whereas the rest of the day is spent driving to and from job sites and making applications. Another company may assign one worker the responsibility of handling, mixing, and loading all pesticides for applicators whose sole job is to operate application equipment from within as the products are applied. Workers may be exposed at the job site as they walk through treated areas (e.g., turf) or as they work in crawl spaces beneath homes treated for termites. Reentry exposure may involve a field-worker exposed to pesticide residues when harvesting crops by hand (Fig. 2.36).

Job activities have a direct bearing on how much and when a worker is exposed to pesticides. A person who mixes and loads concentrated pesticides throughout the workday is exposed differently than a person who applies dilute solutions all day but does no mixing, and differently than workers in a field or greenhouse where pesticides were applied several days earlier.

Work-related activities that bring a worker into contact with a pesticide—storing, mixing, loading, rinsing containers, application, and harvesting—should be identified. The use pattern and label information for the pesticide can be used to predict situations in which a worker could potentially be exposed to pesticides. Work regimens can be determined by considering use rates, how long the worker is exposed during each use, how often applications are made, the method of application, the crop/target being treated, the time of day the application takes place, and protective clothing and equipment needed.

Figure 2.36 Worker exposure studies are conducted under actual use conditions.

Techniques for Measuring Worker Exposure

A work-activity pattern describes the tasks that bring a worker into contact with pesticides. The next step in quantifying exposure is to estimate the amount of pesticide to which the worker is potentially exposed during each specific work task. Accurate estimations of total daily exposure require quantification of the amount of exposure from each activity, such as

- handling the container,
- opening the container,
- removing the product from the container,
- loading the product into water and mixing,
- rinsing the container,
- handling safety clothing, and
- application.

Exposure monitoring studies have been conducted for a variety of pesticides, using commercial applicators, farmers, and field-workers. These studies usually are conducted by the pesticide registrant to fulfill federal and state data requirements for the registration or reregistration of pesticide products. Workers are informed about personal protective equipment and its importance during mixing and application. Then they are monitored for exposure as they carry out the various aspects of their job: mixing and loading, application, reentry for harvest, etc. The amount of pesticide found on and under clothing and the amount found in the breathing zone are monitored and quantified during work activity (Fig. 2.37).

Approaches to Quantifying Exposure

Dosimetry The more common of the two methods for quantifying exposure is dosimetry, which estimates the amount of pesticide in contact with clothing, skin, and/or the breathing zone of the worker (Fig. 2.38). There are several passive dosimetry methods, but patches typically are placed under or attached to the outside of clothing on the chest, back, upper arm,

Figure 2.37 Conducting a worker exposure study using a granular pesticide product.

forearm, thigh, and lower leg. Patches also can be attached to the front and sides of caps when estimations of exposure to the face and neck are needed. The patches trap residues that would otherwise come into contact with the clothing or skin. At the end of the exposure period, the patches are collected and the trapped residues are removed with solvent and analyzed to determine the quantity of pesticides contacting the worker. The findings are approximate because the matrix used for the trap will have different characteristics than

Figure 2.38 Dosimetry is used to accurately determine residue levels in worker exposure studies.

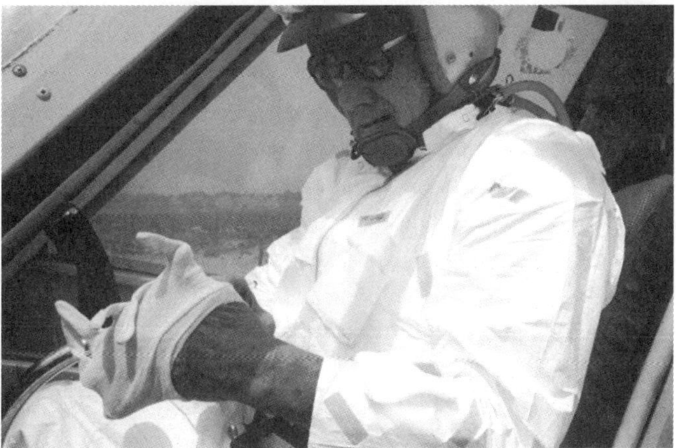

Figure 2.39 Conducting worker exposure studies using lawn care technicians and an aerial applicator. Note the patches used for dosimetry.

skin. Perspiration, acidity, hairiness, etc., may influence how much compound would really be intercepted by a worker.

The amount of pesticide recovered from a patch generally is reported as micrograms of pesticide per square centimeter (μg/cm^2). It can then be standardized per pound of active ingredient handled or per unit of time. The assumption is that the concentration of pesticide on the patch is indicative of the amount deposited over the entire corresponding body region. The amount of residue (μg) per square centimeter (cm^2) indicated by the patch is multiplied by the total surface area of the body region (Fig. 2.39).

As an example, a researcher placed patches on the outer clothing on the chest/stomach region of workers picking apples in an orchard. The amount of pesticide retrieved from the patches averaged 0.10 μg/cm^2. The chest/stomach area for an average adult male is 3454 cm^2, so the calculation for total exposure of the chest/stomach region was as follows:

$$0.10\ \mu\text{g/cm}^2 \times 3454\ \text{cm}^2 = 345\ \mu\text{g}$$

Figure 2.40 All facets of mixing, loading, and applying are evaluated in worker exposure studies.

Dosimeter measurements for all body regions are summed to derive a value for total pesticide exposure to the outside of the body.

Whole-body dosimetry uses clothing actually worn by workers instead of patches as dosimeters: T-shirts, long-sleeved shirts, socks, trousers, long underwear, etc. The type of work clothing varies with geographic region, time of year, and product label. Pesticide applicators perform their jobs as usual: mixing and loading, making applications, etc. Their clothing is collected after the completion of each task or at the end of the workday (Fig. 2.40). A single amount of pesticide representative of exposure to the entire body, or an amount for each body region, can be determined by analyzing the intact garments or specific sections, respectively.

Whole-body dosimetry studies occasionally utilize techniques to estimate the penetration of pesticides through outer clothing to underclothing. Clothing penetration is derived by dividing the concentration detected on undergarments (inside measurement) by the sum of concentrations found on outer garments (outside measurement) and the inside measurement (Fig. 2.41).

In the absence of data, an estimate of possible penetration may be used. The EPA assumes that 50% of pesticide deposited on outer clothing can penetrate to or be deposited on underclothing unless study data demonstrates otherwise. The California Environmental Protection Agency (Cal-EPA) assumes this value to be 10%. For example, if 20 $\mu g/cm^2$ of pesticide were collected from the back and front portions of coveralls assumed to have a clothing penetration of 10%, it is assumed that 2 $\mu g/cm^2$ would reach the underclothing. And underclothes would provide additional protection by preventing a portion of the chemical that reaches them from penetrating to the skin.

Rinses and wipes can be used to measure pesticide residues on the hands, face, and neck. Historically, pesticide residues on the hands have been measured by analyzing the pesticide content of rinsate after washing the hands, by measuring the amount found on hand wipes used after exposure, or by calculating the amount left on cotton glove dosimeters. For instance, field-workers picking strawberries might be asked to wash their hands at specific

Figure 2.41 Whole-body dosimetry uses exposure on outer clothing to determine how much residue a worker contacts.

times, with the rinsate collected for analysis. Pesticide residue from the face and neck can be collected by swabbing the skin (Fig. 2.42).

Personal air samplers are used to estimate the amount of pesticide in the breathing zone of workers. A battery-powered monitoring pump is clipped to the belt and, typically, a flexible tube is run up the back and over the shoulder where it is clipped to the collar of the worker. Each monitoring pump pulls the air through an absorbent filter such as polyurethane foam or organic resin, which removes the pesticide from the air and traps it. The pesticide is then extracted with a solvent for residue analysis (Fig. 2.43).

These two techniques used to estimate external exposure can also be used to estimate the amount of pesticide residue actually absorbed by the worker. In the absence of specific data pertaining to dermal or inhalation absorption, 100% of the amount inhaled and 10% (Cal-EPA) to 100% (EPA) of residues on the skin are assumed to be absorbed.

Biological Monitoring The second approach to quantifying occupational exposure—biological monitoring—provides a measure of the total amount of pesticide actually absorbed by the worker via all routes (oral, dermal, and inhalation). The technique estimates

Figure 2.42 Hand rinses and face wipes are used to determine the amount of exposure of unprotected skin.

the actual absorbed dose via analysis of urine or blood, and/or exhalation of the pesticide and/or its metabolites. This technique generally provides a more accurate estimate of the total absorbed dose than do external dosimetry techniques, but it does not differentiate between routes of exposure (Fig. 2.44).

Most biological monitoring studies use urine as the sampling medium because it can be collected easily in a quantitative and noninvasive manner. Workers are monitored 1–2 days prior to exposure to the pesticide to confirm that they have not been exposed to the pesticide previously. Total urine output generally is collected and assayed for pesticide residues for 48–96 hr after the pesticide is handled or until prescreening levels are regained.

Estimating Occupational Exposure

When direct measurements of worker exposure by biomonitoring are not available, occupational exposure is estimated. The absorbed daily dose is the estimated total amount of a

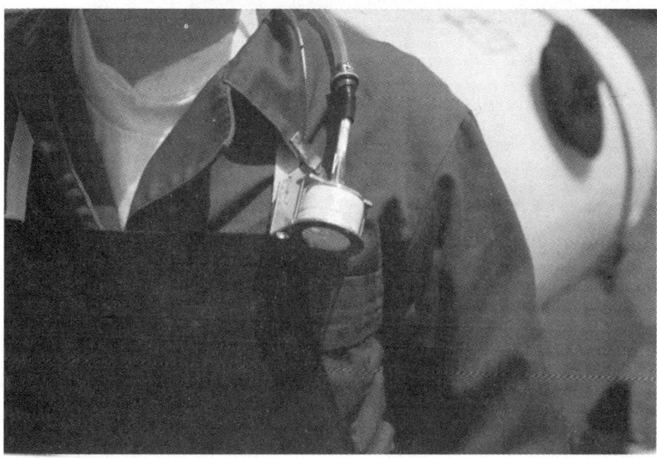

Figure 2.43 A personal monitoring air sampler is used to estimate the amount of pesticide a worker could contact by inhalation.

pesticide that a person absorbs systemically for each day that they are exposed. All routes of exposure are considered: oral, dermal, and inhalation.

Absorbed daily dose also may be expressed as route-specific, such as the dermal absorbed daily dose, and it can be determined using estimates of external exposure (from passive dosimetry studies) in conjunction with estimated absorption percentages for each route of exposure.

Exposure Calculated by Various Approaches Exposure estimates are obtained from three general sources: generic data, generic data with chemical-specific attributes, and chemical-specific exposure monitoring.

Generic Data The magnitude of exposure to pesticides generally is not chemical-specific; it depends mostly on the type of formulation, method of application, use rate, and protective clothing used. Therefore, occupational exposure often can be estimated by using surrogate

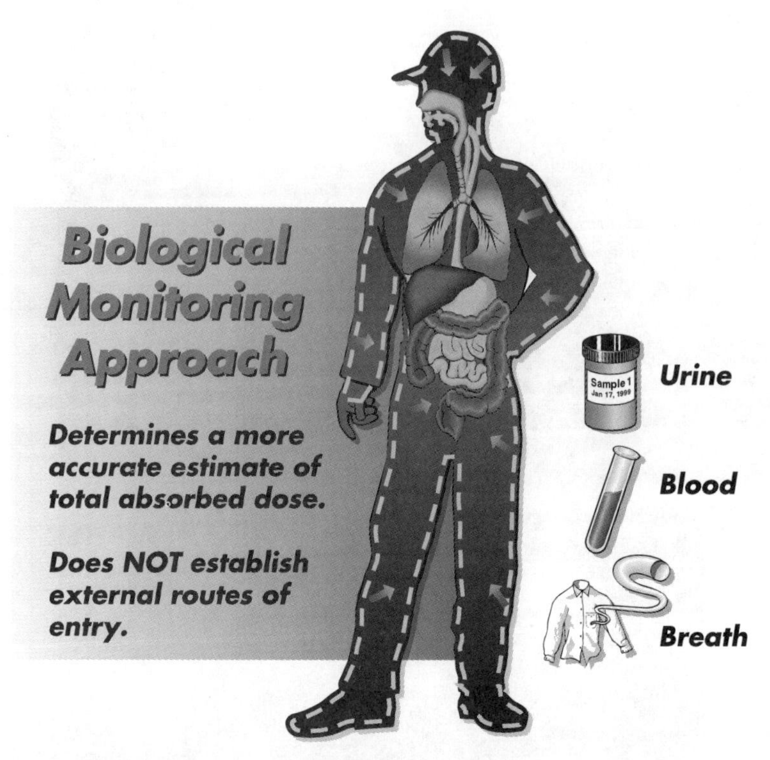

Figure 2.44 Biological monitoring provides direct measurement of residue exposure to workers.

data developed previously for other chemicals. Pesticide companies have voluntarily pooled a large amount of exposure data into a single generic database called the Pesticide Handlers' Exposure Database (PHED); it is available to the public through the EPA.

PHED data may be segregated based on type of formulation (granular, liquid, wettable powder), method of application (aerial, ground boom, air blast, backpack), use rates, and protective equipment required. PHED contains data to estimate exposure to a mixer/loader, an applicator, a combined mixer/loader/applicator, and a flagger. Thus, PHED may be used for a surrogate estimate of inhalation and dermal exposure for many exposure scenarios. Exposure calculations in a preliminary assessment typically use the following assumptions to predict the absorbed daily dose:

- Maximum use rates for assessing short-term exposure, and average use rates for assessing intermediate and long-term exposure.
- Pesticide penetration through outer clothing is assumed to be 10–50%. However, PHED typically provides actual measurements under a single layer of clothing, which negates the need to estimate clothing penetration.
- Dermal absorption is assumed to be 100% when pesticide-specific data are not available, although regulatory agencies other than the EPA may assume 10%.
- Maximum acres treated per day when assessing short-term risk, and average acres treated per day or year when assessing intermediate or long-term risk.

TABLE 2.16 PHED Output in Micrograms per Pound Active Ingredient Sprayed

Patch Location	Distribution Type	Median	Mean	Geometric Mean	Observations
Head	Lognormal	2.0	9.0	3.0	20
Neck (front)	Lognormal	0.6	0.9	0.7	20
Neck (back)	Lognormal	0.1	0.4	0.2	20
Upper arms	Other	0.3	0.7	0.5	30
Chest	Other	0.4	1.3	0.7	30
Back	Other	0.4	0.9	0.6	30
Forearms	Other	0.2	0.7	0.3	30
Thighs	Other	0.4	2.5	0.8	30
Lower legs	Other	0.2	0.7	0.4	30
Feet	—	—	—	—	0
Hands	Normal	12	16	7	22

Total dermal exposure: 21.8 μg
Total inhalation exposure: 0.449 μg

An example of a PHED output using an enclosed tractor cab, air blast application of liquid sprays is provided in Table 2.16. In this scenario, the applicator is wearing long pants, a long-sleeved shirt, and protective gloves. The dermal exposure is 21.8 μg and the inhalation exposure is 0.449 μg per pound of active ingredient applied.

Generic Data with Chemical-Specific Attributes This approach to assessment uses product-specific data. For example, a dermal absorption study may indicate that 10% of the pesticide that reaches the skin may actually penetrate the skin; so 10% would replace the value of 100% dermal absorption. Another study may show only 10% penetration through clothing to the skin, in which case 10% would replace the value of 50% clothing penetration.

Chemical-Specific Exposure Monitoring Chemical-specific exposure monitoring relies on field studies that provide actual exposure data on the pesticide, relative to specific tasks. It is common for these studies to include measurements of pesticides on skin surfaces as well as actual biological monitoring of the applicator.

Predicting Exposure for a Mixer/Loader/Applicator by External Methods The following example illustrates the absorbed daily dose calculation when using chemical-specific measurements of external exposure.

Total External Deposition on Clothes and Exposed Skin Pesticides were extracted from patches, hand washes, and face and neck wipes for quantifying exposure of adult males during mixing/loading and application activities; residue levels were reported for each body region. The total amount of pesticide present on outer clothing or exposed skin was determined to be 6958 μg per person per workday (Table 2.17).

With patches, this was calculated by multiplying the surface area for each region by the amount of pesticide per centimeter squared. (The surface area and dosimeter residue values per cm^2 are not used for wipes and washes, as these techniques collect the total residue from exposed skin.) The results for all body regions were summed to yield the total external deposition, often referred to as the potential dermal exposure (Fig. 2.45).

TABLE 2.17 External Pesticide Deposition on Clothes and Skin

Location on Body	Sample Type	Surface Area of Region (cm^2)	Mean Patch Residue ($\mu g/cm^2$)	Total Residue (μg)
Head (face excluded)	Patch	630	0.02	13
Face	Wipe	N/A	N/A	19
Back of neck	Wipe	N/A	N/A	4
Front of neck	Wipe	N/A	N/A	6
Chest/stomach	Patch	3454	0.87	3005
Back	Patch	3454	0.58	2003
Upper arms	Patch	1479	0.20	296
Forearms	Patch	1211	0.16	194
Hands	Wash	N/A	N/A	302
Thighs	Patch	3663	0.09	330
Lower legs	Patch	2455	0.32	786
				6958

Total Dermal Exposure (μg/person/day) The total estimated amount of pesticide deposited on the clothing and exposed skin (i.e., the potential dermal exposure) in the example is 6958 μg. However, clothing essentially intercepts a portion of the pesticide that reaches it, preventing it from contacting the skin beneath. An evaluation of concurrent exposure values on and under clothing, in studies reported in the Pesticide Handlers' Exposure Database, indicates a 10% default for protection afforded by one layer of clothing. In this example,

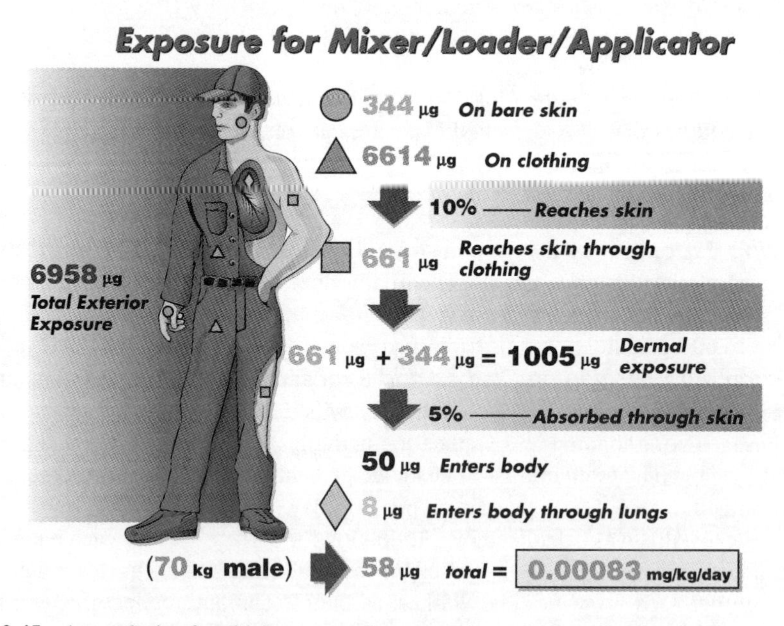

Figure 2.45 An analysis of multiple routes of exposure is used to predict the amount of pesticide that contacts a mixer/loader/applicator.

it is assumed that the worker wears a long-sleeved shirt and long pants; 6614 μg are deposited on clothing and 344 μg on uncovered skin (head, face, neck, and hands). Thus, 661 μg (6614 μg $\times$ 0.10) of pesticide would be expected to contact the skin after penetration through the clothing of the upper and lower arms, upper and lower legs, and front and back torso. Adding the 661 μg to the 344 μg that directly contacted the skin indicates a total estimated dermal exposure of 1005 μg per person per day.

Penetration through Skin Human dermal absorption studies indicate that 1–10% of many dermally applied pesticides actually is absorbed through the skin. *In vivo* and *in vitro* studies with rat skin, which is more permeable than human skin for many pesticides, indicate an absorption range of 1–30%.

In the previous example, 1005 μg of the pesticide reaches the skin surface. The next step is to estimate the percentage that penetrates through the skin and enters the circulatory system. A study of laboratory animals indicated 5% dermal absorption. Exposure for a full workday is calculated as follows:

$$1005\ \mu\text{g deposited on skin/day} \times 5\%\ \text{dermal absorption} = 50\ \mu\text{g absorbed/person/day}$$

Penetration to the Lungs There is very little data on the absorption percentage of pesticides through inhalation. But it is generally assumed that 100% of pesticides collected in air monitoring pumps would reach the lungs and penetrate the lung membranes if inhaled. Thus, if studies using monitoring pumps show that workers inhale 1 μg/hr, the resulting daily absorption (dose) through inhalation is calculated as follows:

$$1\ \mu\text{g/hr} \times 8\text{-hr workday} = 8\ \mu\text{g inhaled daily}$$

The calculation assumes continuous exposure and a constant breathing rate over the specified period of time, but exposure assessments typically have shown that less than 5% of total pesticide exposure is through inhalation.

Absorbed Daily Dose In the preceding example, the Absorbed Daily Dose (ADD) of 58 μg/day includes contributions from the skin (dermal absorbed daily dose of 50 μg/day) and the lungs (inhaled absorbed daily dose of 8 μg/day). The ADD of 58 μg/day typically is then converted to milligrams per kilogram per day (mg/kg/day) as shown in the following example.

The example converts the absorbed daily dose values (μg/day) to absorbed dose expressed as mg/kg/day for a 70-kg male. The 70-kg male is representative of an average adult male; a 60-kg adult female also could be used. Body weight values actually used vary with the regulatory agency (EPA, Cal-EPA, Canada, Japan, Europe).

$$\frac{58\ \mu\text{g/day}}{1000\ \mu\text{g/mg}} \times \frac{1}{70\ \text{kg}} = 0.00083\ \text{mg/kg/day}$$

Predicting External Exposure for a Reentry Worker Workers may reenter treated fields to perform tasks such as weeding, thinning, and harvesting. Estimates of reentry worker exposure typically are based on the amount of pesticide residue on crop foliage and the amount that transfers from the foliage onto the skin or clothing of the workers. The type of crop and the particular work activity are influencing factors. The remaining steps used

Figure 2.46 Measurements of actual residues on treated crops are used to determine the amount of total residue that can be dislodged from leaves.

to calculate the absorbed daily dose are similar to those previously discussed for mixers, loaders, and applicators.

Dislodgeable Foliar Residue Dislodgeable foliar residue (DFR) is the amount of residue that can be removed from plant foliage and that could serve as a source of exposure for workers who weed, harvest, or perform other work activities in treated fields (Fig. 2.46). DFR measurements are repeated over an extended period of time following application so that the rate of residue dissipation can be measured. DFRs are pesticide-specific and depend on the physical, chemical, and environmental fate properties of the pesticide formulation.

DFR studies are conducted by applying the end use product to various crops at the highest concentration and the shortest reapplication interval allowed by the pesticide label. The design of the study indicates when leaf samples are to be collected; for example, samples could be taken within 4–12 hr and at 1, 2, 5, 7, 14, 21, 28, and 35 days after application. A leaf punch is used to sample foliage from the top, middle, and lower part of the plant. The collected leaf tissue is placed into a container with aqueous surfactants (e.g., mild detergents), then shaken or agitated. The amount of pesticide measured in the aqueous solution is considered potentially dislodgeable. The relevant DFR is taken from the part of the foliage with which a worker typically comes into contact (Fig. 2.47).

The amount of residue that workers collect on their clothing or exposed skin obviously depends on the amount of foliage they contact. For example, a person harvesting strawberries is expected to come into contact with less foliage than a worker harvesting peaches or apples.

Figure 2.47 Foliar residue studies are important in establishing the amount of time that must elapse before workers are allowed to reenter fields following a pesticide application.

As a default, the EPA assumes that 20% of pesticides applied to agricultural crops are initially available as DFRs. This assumption is based on multiple DFR studies submitted to the EPA. Day 0 (day of application) values for agricultural crops commonly range from 0.5 $\mu g/cm^2$ to 10 $\mu g/cm^2$, depending on application rates. Dissipation of DFRs after Day 0 is chemical-specific.

Transfer Coefficient A Transfer Coefficient (TC), referenced in earlier literature as the transfer factor, is used to estimate the amount of pesticide transferred onto workers from a previously treated surface (Fig. 2.48). The TC is different from a dislodgeable residue in that it is not pesticide-specific; it is more dependent on the crop treated and the extent of foliar contact. Transfer coefficients are calculated as follows:

$$\mathrm{TC}\,(\mathrm{cm}^2/\mathrm{hr}) = \frac{\text{Measured Exposure } (\mu\text{g/hr})}{\text{Dislodgeable Foliar Residue } (\mu\text{g/cm}^2)}$$

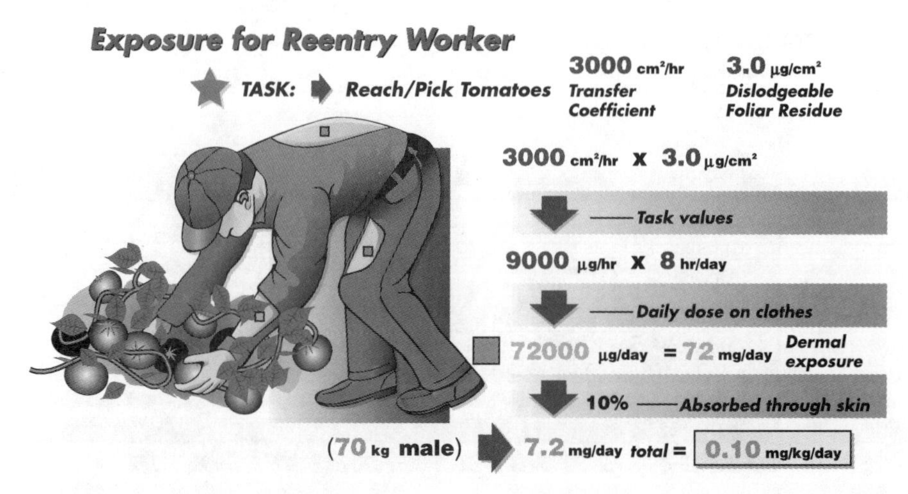

Figure 2.48 An analysis of multiple routes of exposure is used to predict actual residues for workers reentering pesticide-treated areas.

For example, a DFR of 1.9 μg/cm^2 was determined for cantaloupe vines for a fungicide where harvest exposure was 324 μg/hr. Thus, using the above formula, the TC for harvesting cantaloupes is 170 cm^2/hr (324 μg/hr $\div$ 1.9 μg/cm^2).

Table 2.18 presents generic transfer coefficients for five work tasks classified by type of crop, method of harvest, and body regions that come into contact with pesticide residues during harvest.

Dermal exposure can be estimated by using generic default transfer coefficients in conjunction with chemical-specific, dislodgeable foliar residues, as illustrated in the following examples. In the tomato reach-and-pick work task, the initial dislodgeable foliar residue for one pesticide was found to be 3.0 μg/cm^2, and a generic transfer coefficient of 3000 cm^2/hr was assumed. Thus:

$$\begin{aligned}\text{Total Dermal Exposure} &= 3.0\ \mu\text{g/cm}^2 \times 3000\ \text{cm}^2/\text{hr} = 9000\ \mu\text{g/hr}\\ &= 8\text{-hr/day} \times 9000\ \mu\text{g/hr} = 72{,}000\ \mu\text{g/day}\\ &= \frac{72{,}000\ \mu\text{g/day}}{1000\ \mu\text{g/mg}} = 72\ \text{mg/day}\end{aligned}$$

$$\text{Residue Penetrating the Skin} = 72\ \text{mg/day} \times 0.1(\text{dermal penetration}) = 7.2\ \text{mg/day}$$

$$\begin{aligned}\text{Dermal Absorbed Daily Dose for a 70-kg male} &= 7.2\ \text{mg/day} \times \tfrac{1}{70}\ \text{kg}\\ &= 0.10\ \text{mg/kg/day}\end{aligned}$$

TABLE 2.18 Generic Transfer Coefficients

Work Task	Body Contact Areas	Crop Type	TC (cm^2/hr)
Sort/select	Hand	Tomatoes (mechanical)	100
Reach/pick	Hand + arm	Lettuce	200–700
Reach/pick	Hand + arm + leg	Tomatoes (pole)	1000–3000
Search/reach/pick	Upper body	Tree fruit	3000–6000
Expose/search/reach/pick	Whole body	Grapes	8000–25000

TABLE 2.19 Example of Occupational Exposure Assessment Using Pesticide Handler Exposure Database

Work Scenario	Exposure Type	Exposure Per Pound Handled (μg/lb active ingredient)	Active Ingredient Handled (lb/day)	Exposure (μg/day)	Absorbed Dose (μg/day)
Mixer/loader	Inhalation	0.68	6.25	4	4
	Dermal	93.20	6.25	583	58
Applicator	Inhalation	1.80	6.25	11	11
	Dermal	16.60	6.25	104	10
			Total	702	83

$$\text{Daily Dermal Exposure (mixer/loader)} = \frac{583\ \mu\text{g/day}}{1000\ \mu\text{g/mg} \times 70\,\text{kg}} = 0.0083\,\text{mg/kg/day}$$

$$\text{Daily Dermal Exposure (applicator)} = \frac{104\ \mu\text{g/day}}{1000\ \mu\text{g/mg} \times 70\,\text{kg}} = 0.0015\,\text{mg/kg/day}$$

$$\text{Total Daily Dermal Exposure (mixer/loader/applicator)} = 0.0083 + 0.0015 = 0.0098\ \text{mg/kg/day}$$

$$\text{Absorbed Daily Dose (mixer/loader)} = \frac{62\ \mu\text{g/day}}{1000\ \mu\text{g/mg} \times 70\ \text{kg}} = 0.00089\ \text{mg/kg/day}$$

$$\text{Absorbed Daily Dose (applicator)} = \frac{21\ \mu\text{g/day}}{1000\ \mu\text{g/mg} \times 70\ \text{kg}} = 0.0003\ \text{mg/kg/day}$$

$$\text{Absorbed Daily Dose (mixer/loader/applicator)} = 0.0012\ \text{mg/kg/day}$$

Case Study Exposure Assessment Using PHED

An assessment of potential exposure and risk to workers associated with occupational use of Insecticide X, an insecticide applied to corn, was performed using the following information. This example is typical of actual assessment and combines default assumptions with reliable data extracted from the initial exposure assessment of the insecticide (Table 2.19).

- Pesticide type: foliar-applied corn insecticide
- Pesticide formulation: water-dispersible granules
- Pesticide use rate: maximum label rate of 0.0312 pounds active ingredient per acre
- Application timing allowed by the label: once every 14 days
- Application method: ground boom sprayer attached to a truck or tractor equipped with spray tank
- Water volume: 10–30 gallons per acre
- Based on agricultural census data (http://www.nass.usda.gov/census), it is assumed that a 400-acre cornfield is treated by a commercial applicator at the rate of 200 acres per day. The commercial applicator is assumed to handle the product for 30–60 days of a 90-day period during May/June/July.
- Exposure input values:
    a. Exposures are estimated for a mixer/loader and an applicator.
    b. One hundred percent of the estimated inhaled dose is absorbed.

c. Ten percent of the chemical that contacts the worker's skin is absorbed by the body.
d. The insecticide is applied to 200 acres of corn per day.
e. The individual will handle a maximum of 6.25 lb of the active ingredient in 1 workday: (0.0312 lb active ingredient/acre × 200 acres)

- Mixer/loader exposure values:
  a. Wears clothing required by the label: long pants, long-sleeved shirt, and gloves.
  b. PHED default inhalation rate is 25 liter/min for light work activity.
  c. Exposure rates were determined from PHED, as follows:
     — Inhalation Exposure = 0.68 μg/lb active ingredient
     — Dermal Exposure = 93.2 μg/lb active ingredient
- Applicator exposure input values:
  a. Wears clothing required by the label: long pants and long-sleeved shirt.
  b. Default inhalation rate of 25 liter/m for light work activity.
  c. Applicator exposure was estimated from PHED, as follows:
     — Inhalation Exposure Rate = 1.8 μg/lb active ingredient
     — Dermal Exposure Rate = 16.6 μg/lb active ingredient

RESIDENTIAL EXPOSURE ASSESSMENT

Increasingly, government and industry scientists are asked about human risk stemming from pesticide use in and around the home (e.g., indoor applications to carpet and pets and outdoor applications to turf, vegetable gardens, and ornamental plantings). A key aspect in evaluating residential exposure is recognition of the unique "properties" of the residential environment, including the recognition that routine home activities bring people into contact with treated areas. Residential exposure assessment involves consideration of multiple routes (oral, dermal, and inhalation) and pathways (e.g., contact with treated turf or pets) and the source of exposure (indoors or out).

Outdoor Exposure Studies

Turf applications are the major source of human exposure to pesticides used outdoors, although applications to ornamental and landscape plantings are also significant. Exposure from treated surfaces such as turf can be determined by measuring dislodgeable and transferable residues. Dislodgeable residues are those that potentially can be transferred from a given surface. Transferable residues are those that are actually transferred during normal human contact with the treated surface. In pesticide use studies on turf, the highest labeled rate is applied within the time frame of peak local use. Dislodgeable residues are measured for samples taken just prior to application, immediately after (once residues have dried), and at various intervals for several days thereafter. These measurements are used to determine how fast the pesticide residue dissipates after application and how much of it potentially could be transferred by human contact.

Indoor Exposure Studies

Indoor exposure assessments are complicated by the diversity of pesticide application methods. Treatments may entail crack and crevice, carpet, moth repellent, termiticide,

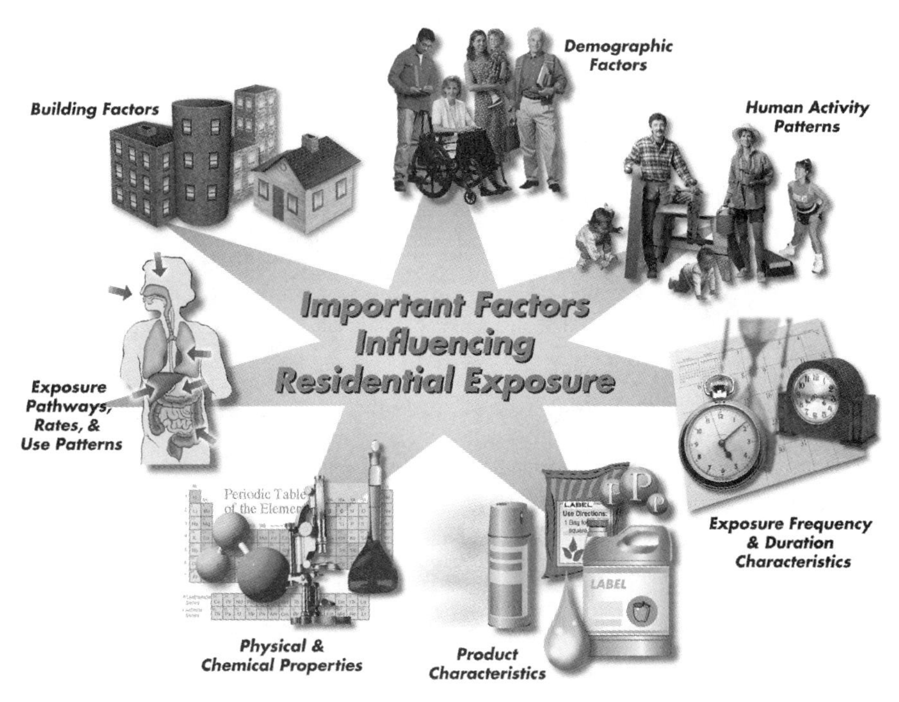

Figure 2.49 Predicting residential exposures takes into consideration numerous factors.

disinfectant, or pet product applications; room foggers also may be used. Human contact with indoor pesticide residues may vary significantly. Exposure may be highly unlikely when pesticides are applied behind cabinets, for example; but the likelihood of exposure increases significantly in the case of broadcast applications to carpets for flea control. Humans may experience dermal exposure to pesticide residues on carpets, vinyl tile, upholstery, countertops, and pets. Airborne residue and dust may cause inhalation exposure. Potential human residential exposure is influenced by the type of product used, the physical/chemical characteristics of the product, and the indoor environment: room size, air exchange rates, temperature, types of surfaces (e.g., carpet, upholstery, vinyl), and the nature of human activities within (Fig. 2.49).

Important Factors That Influence Residential Exposure

Residential Building Factors Room configuration, construction materials, and ventilation determine the probability of human exposure following indoor pesticide applications. The number of open windows and doors, the rate of mechanical ventilation and air mixing, and the rate of outside air infiltration influence the dilution of pesticide-contaminated indoor air. Climatic influences such as season and temperature also have an effect.

Demographic Factors Infants, toddlers, and the elderly are considered more sensitive to pesticide exposure than other age groups. Other important factors include body weight; inhalation rates, which vary primarily by age, gender, and activity level; activity patterns; and the relationship of these physiological and behavioral factors to demographics, geographic location, and time of application.

Figure 2.50 Exposure potential for the residential use of a pesticide is determined in experiments that simulate actual use conditions.

Human Activity Patterns The ways that people are exposed to pesticides in residential settings are remarkably different from those of workers exposed on the job. Most work-related tasks are routine and repetitive; therefore, work habits that lead to worker exposure are predictable. Home activities are less routine, less repetitive, and less predictable.

Infants and toddlers spend considerable time crawling and playing on floors and carpets and therefore breathe air that is nearer the floor; they wear relatively little clothing and spend more time indoors than adult members of the same family. A person exercising on a treated carpet and teenagers playing on a treated lawn are but two behaviors that can bring older family members into contact with pesticide residues.

Exposure Frequency and Duration Characteristics Frequency (days per year, years per lifetime) and duration (minutes or hours per day) are critical in estimating residential exposure. Both factors depend on how the product is used and the kinds activities that bring individuals into contact with treated areas.

Product Characteristics Perhaps the most important contributing factors to dermal and inhalation exposure are the nature of the pesticide product, the form in which it is released (e.g., fine respirable aerosol, nonrespirable coarse aerosol, vapor), the concentration of the active ingredient, the formulation, and the method of application (Fig. 2.50).

Physical and Chemical Properties Several factors are important with respect to the physical and chemical properties of the pesticide: molecular weight, vapor pressure (does it release as a vapor, and how quickly?), solubility in fat and water, and breakdown to other chemicals. These factors determine the chemical rate of evaporation into the air (after application) and how much of the pesticide actually transfers from carpet to hands—and from hands to mouth, in the case of young children (Fig. 2.51).

Exposure Pathways and Routes Key exposure routes and pathways routinely considered for adults, children, and infants following either indoor or outdoor residential

Figure 2.51 Testing is also conducted for end-of-hose sprayers that are commonly used around the home.

pesticide use include:

- Potential consumer exposure (dermal and inhalation) during application
- Potential postapplication dermal exposure
- Potential postapplication inhalation exposure
- Potential postapplication, nondietary, incidental oral exposure (e.g., from toys or hand-to-mouth transfer)

Incidental ingestion of soil, grass, and other environmental media also may be considered. Incidental ingestion of surface residues via the hands is based on the assumed transfer of residue from surface to hands to mouth. Inherent to this assumption is that children, through mouthing the hands (or contaminated objects), can incidentally ingest pesticide residues. The uncertainty associated with the frequency of hand- or object-to-mouth behavior must be acknowledged explicitly. Studies generally show that the conservative, screening-level exposure estimates of potential incidental ingestion are exaggerated as compared to estimates derived via hand rinse and wipe monitoring data.

It is important to acknowledge that current direct and indirect monitoring data do not suggest that hand- or object-to-mouth transfer and incidental ingestion are significant routes of exposure. Additional data are needed to better define the variability and uncertainty of residue transfer from treated surfaces to hands (wet or dry) to mouth.

Product Use Patterns Monitoring data must be considered with label and use information such as application rate, method of application, site of application, timing, and frequency of application to gain an understanding of residential exposure. Oral, dermal, and inhalation exposure may be calculated separately or combined as a single estimate of systemic exposure or absorbed dose.

Techniques Used in Monitoring Residential Exposure

There are several methods for measuring residential exposure using external dosimetry and/or biological monitoring.

Indoor and Outdoor Rollers The roller technique typically constitutes use of a polyurethane foam (PUF) pad placed over a metal roller which holds the PUF in place. Transferable residues are measured by pushing the roller in two directions over a portion of the treated area. Inside the roller is a premeasured weight that provides consistency in pressure as it rolls across a treated carpet or lawn. When sampling is complete, the foam is removed from the roller and the pesticide residue quantified (Fig. 2.52).

Figure 2.52 The use of rollers helps determine the maximum amount of residue that can be dislodged from the grass or tile.

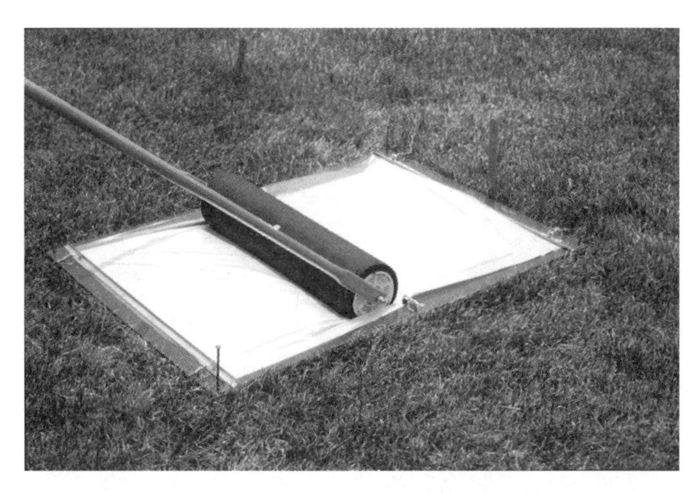

Figure 2.53 Another method of measuring the amount of dislodgeable residue from grass is placing a sheet over treated turf, passing a roller over the sheet, and measuring the amount of residue collected on the sheet.

An alternate method is to place a sheet of cloth over the carpet or lawn and push the metal roller across it. The cloth is then analyzed for pesticide residue (Fig. 2.53).

Drag Sleds A drag sled is a weighted block with a removable cotton (denim) dosimeter attached to the bottom surface. The sled is dragged across a predefined treated area and the denim analyzed for pesticide residue (Fig. 2.54).

Hand Presses Adult subjects press their hands with predetermined force against a treated surface. The hands are immediately wiped or washed in a solvent such as isopropanol and the solvent analyzed for pesticide residue (Fig. 2.55).

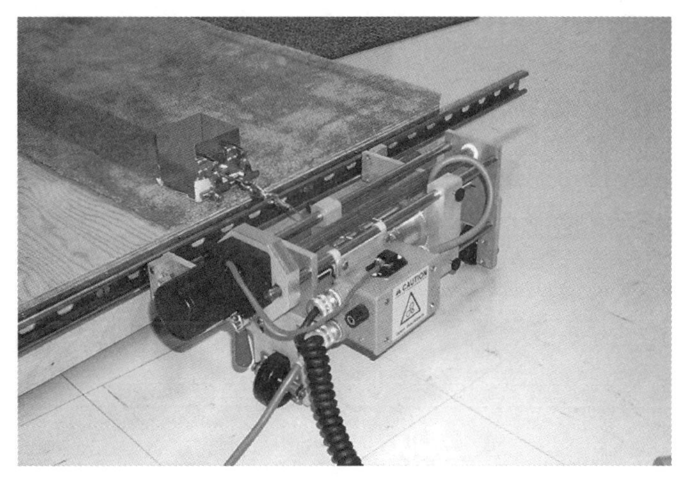

Figure 2.54 A drag sled is moved across the treated carpet, and the amount collected on the bottom surface is measured.

Figure 2.55 Hand pressing is used to determine how much residue can be removed from treated carpet.

Coupons Coupons made of cotton, aluminum foil, or glass are placed throughout the area to be treated, then collected for quantification after the pesticide application. Those collected immediately following application can be used to estimate the amount of pesticide deposited per unit of surface area treated. Those collected and analyzed later can provide information on the rate of degradation of the pesticide. Fresh coupons placed following application can be used to measure movement and repositioning of residues (Fig. 2.56).

Area and Personal Air Monitoring Stationary air sampling devices measure airborne contamination throughout a treated house. They are strategically placed in kitchens, basements, bedrooms, and family rooms to measure pesticide movement throughout the home. Each device has a pump that draws air through a pesticide-extracting charcoal or resin filter. Samples are taken near the floor; at heights representative of a child's breathing area; and

Figure 2.56 Coupons are used to measure the amount of residue available after the application of an aerosol.

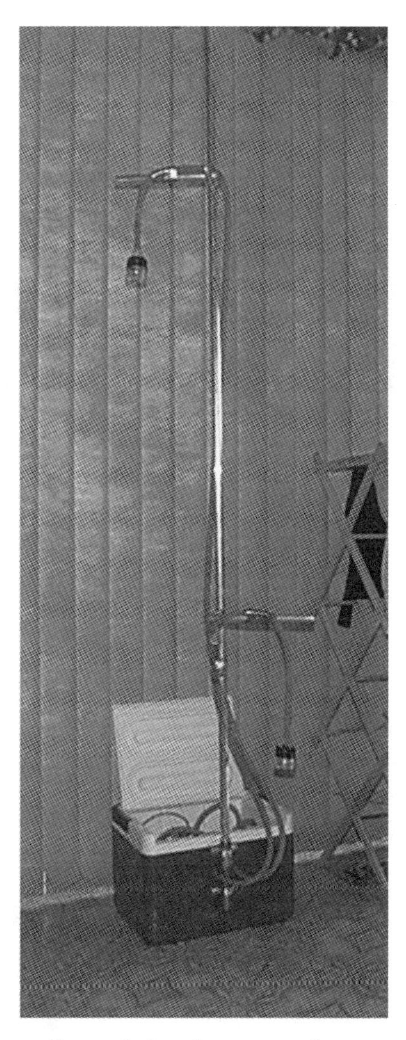

Figure 2.57 Air monitoring studies are designed to measure the amount of pesticide in the breathing zone of children and adults.

at heights representing adults' breathing space, seated and standing. Indoor air concentrations of the pesticide are measured during application and repeated several times during the first 24 hr then less frequently. Personal air samples measure contamination levels in the breathing zone of individual household members. Personal sampling pumps generally are clipped to the shirt collar to measure the amount of pesticide residue in air reaching the mouth and nose (Fig. 2.57).

Human Volunteer Monitoring Studies Human volunteer monitoring studies often involve the use of whole-body dosimeters, air sampling, or biological monitoring methods. Although study designs vary, volunteers' activities are documented. Choreographed activities such as crawling across a treated carpet facilitate researchers' ability to relate environmental measurements to actual human exposure.

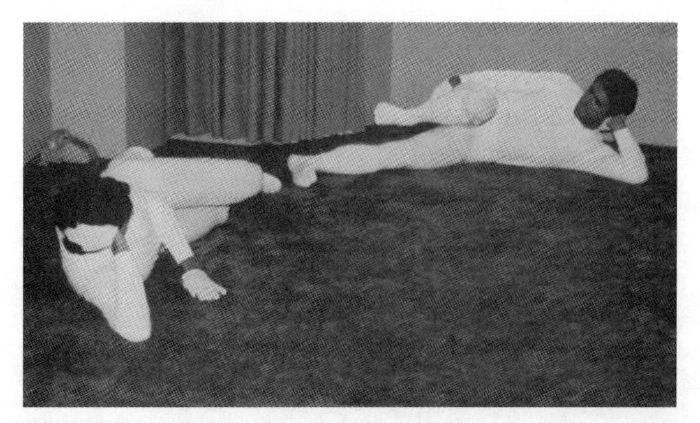

Figure 2.58 Exercise routines on treated surfaces can be used to determine the amount of dislodgeable residue from a treated indoor surface.

Jazzercise routines have been used to measure inhalation and dermal exposure following pesticide treatment. Jazzercise is an exercise program consisting of a set number of 3-min routines led by certified instructors. The exercises selected are those that bring volunteers into repeated intensive contact with a pesticide-treated surfaces such as carpeting. Adult volunteers are provided a complete set of cotton underclothing and outer wear. They are assigned to specific areas within the treated room where they perform the exercise routines. At the conclusion of the program, volunteers place the clothing into separately marked plastic bags for chemical analysis (Fig. 2.58).

Case Study of Residential Exposure Assessment

The following sample calculations are representative of methods used to assess same-day, postapplication exposure of children 1–6 years old. Similar methods are used for assessing exposure of other population subgroups.

Label Directions and Product Use Information The product used in this example is Insecticide X, formulated as a 6-oz fogger. Instructions for this product indicate the following:

- The fogger will treat up to 6000 ft^3 of unobstructed space (i.e., a room with approximate dimensions of 26 × 30 × 8).
- Only one container should be used in rooms 12 × 12 × 8 or smaller.
- The fogger should not be used in areas less than 100 ft^3.
- Insecticide X should not be used in serving areas where food could be exposed.
- The user is instructed to open cabinets and doors in the treatment area; remove or cover exposed food, dishes, and food-handling equipment and surfaces; remove pets and plants; shut off fans and air conditioners; and close doors and windows.
- The user is instructed to keep the treated home closed for 2–3 hr before reentering.
- Prior to reoccupying the treated area, the user should ventilate the area by opening all doors and windows for 30 min.

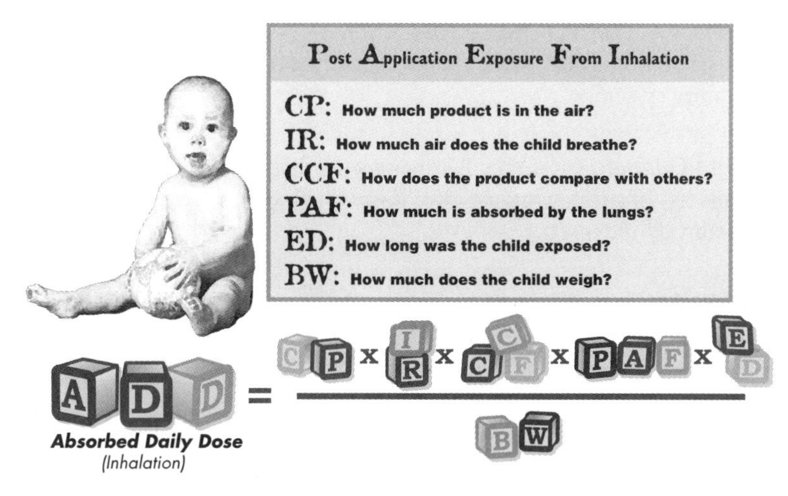

Figure 2.59 Factors necessary to estimate exposure from inhalation.

Potential Postapplication Exposure of Children 1–6 Years Old by Inhalation of Airborne Aerosols For the purpose of estimating potential "day of application" inhalation exposure to the insecticide, the airborne concentration estimate is based on the results of a total release fogger exposure monitoring study. In the current example, when the consumer is instructed to reenter the home 2–3 hr after application, the mean indoor air concentration equals the analytical detection limit of 0.000175 mg/m^3. Based on the fact that samples corresponding to the time of reentry were at or below the detection limit, the average aerosol air concentration value of 0.000175 mg/m^3 is used to represent the highest potential airborne exposure.

The equation for estimating potential inhalation exposure and absorbed dose is developed as follows (Fig. 2.59).

CP: Concentration of Product (active ingredient in mg/m^3) As noted previously, a conservative estimate of postapplication air concentration is 0.000175 mg/m^3.

IR: Inhalation Rates (m^3/hr) Inhalation rates are affected by numerous individual characteristics including age, gender, weight, health status, and level of activity (e.g., sleeping, walking, jogging). The EPA's Exposure Factors Handbook (www.epa.gov/ncea/expofac.htm) reviews studies that provided inhalation rates at various activity levels. The handbook summarizes the average number of hours per day spent resting and performing various levels of activity (light, moderate, and heavy). For this sample assessment, the inhalation rate for children is estimated at 0.47 m^3/hr.

CCF: Concentration Correction Factor (unitless) This factor adjusts air concentration based on a comparison of the amount of active ingredient released from the product being evaluated to the amount released in previous studies. This assumes that all other factors that affect air concentrations (temperature, air exchange, etc.) remain the same as those recorded during the monitoring study. For this example, the 6-oz fogger releases 1.87 times the amount used in a surrogate fogger study.

PAF: Pulmonary Absorption Factor (percentage) A default value of 100% absorption generally is used; that is, 100% of the chemical entering the lungs is assumed to be absorbed by the respiratory system.

ED: Exposure Duration (hr/day) Air monitoring data suggest that aerosols are airborne for approximately 2 hr following use of a fogger. This assessment assumes that 2 hr is a reasonable estimate of exposure.

BW: Body Weight (kg) The mean body weight of male and female children 2–7 years old (from the EPA's Exposure Factors Handbook) is 18.9 kg. The postapplication inhalation daily exposure and absorbed daily dose for children is calculated as follows.

$$\text{Inhalation Daily Exposure} = (\text{CP}) \times (\text{IR}) \times (\text{CCF}) \times (\text{ED})/(\text{BW})$$

$$\text{Inhalation Daily Exposure} = \frac{(0.000175\,\text{mg/m}^3 \times (0.47\,\text{m}^3/\text{hr}) \times (1.87) \times (2\,\text{hr/day})}{18.9\,\text{kg}}$$

$$\text{Inhalation Daily Exposure} = 0.000016\,\text{mg/kg/day}$$

$$\text{Inhalation Absorbed Daily Dose} = \text{Inhalation Daily Exposure} \times (\text{PAF})$$

$$\text{Inhalation Absorbed Daily Dose} = (0.000016\,\text{mg/kg/day}) \times (1.0)$$

$$\text{Inhalation Absorbed Daily Dose} = 0.000016\,\text{mg/kg/day}$$

Potential Postapplication Exposure of Children 1–6 Years Old from Dermal Contact with Treated Surfaces The Jazzercise study method was used for estimating potential day-of-application dermal contact with floor surfaces. The procedure for estimating potential dermal exposure is based on the use of transfer factors from indoor rollers. The general equation for estimating potential dermal exposure and absorbed dose is as follows (also Scc Fig. 2.60).

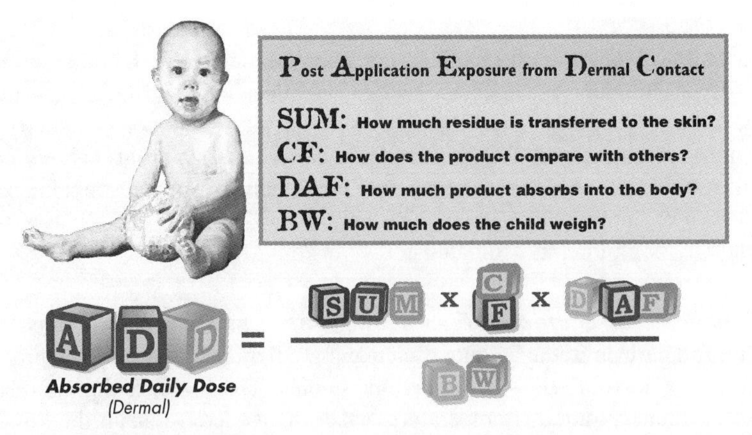

Figure 2.60 The factors associated with estimating exposure amounts from dermal contact with a pesticide.

TABLE 2.20 Surface Areas for Clothing Scenarios Used in Dermal Exposure Calculations for Children 1–6 Years Old

Body Part	Body Surface (cm^2)
Arms (uncovered)	1085
Upper body (uncovered)	1615
Legs (uncovered)	1650
Lower body (uncovered)	1220
Hands (uncovered)	452
Feet (uncovered)	553

Sum: Total Dermal Exposure This exposure summation represents the combination of body-specific transfer factors (TF), transferable residues (TR), and surface area (SA). A dislodgeable chemical residue is the portion deposited on a solid surface, which may be dislodged by direct contact to human skin or clothing: the maximum amount potentially available on a given day.

A transferable pesticide residue is the amount that can be removed from a treated surface onto other objects, including humans. The scenario for infants and children assumes nakedness; thus, the TFs and SAs used in these calculations are for uncovered body areas. The clothing scenario used for dermal exposure calculations resulted in the following estimations of body surfaces (Table 2.20).

As an example, the total dermal exposure (mg) summation calculation (summed across TF × TR × SA for each body part) for children is as shown in Table 2.21.

CF: Dermal Experimental Correction Factor The dermal experimental correction factor adjusts the milligrams of dermal exposure (derived from summation calculations previously described) based on the amount of active ingredient released from the product used in the surrogate dermal monitoring study versus the amount of active ingredient released from the product being evaluated. The product assessed in this example released 1.87 times the active ingredient released in the California reference study cited in Table 2.21.

DAF: Dermal Absorption Fraction for Active Ingredient (unitless) It is assumed for this example that 10% of the pesticide on the skin is absorbed into the body.

TABLE 2.21 Dermal Exposure Calculation for Children[a]

Body Area	TF (unitless)	TR (mg/cm^2)	SA (cm^2)	Dermal Exposure (mg)
Arms + upper body	2.4	0.0000064	2700	0.0415
Legs + lower body	2.4	0.0000064	2870	0.0440
Hands	12.6	0.0000064	452	0.0364
Feet	13.6	0.0000064	553	0.0481
			Total	0.1700

[a]The TF for each body part and the mean TR estimates used in this example were obtained from a study published by the California Department of Pesticide Regulation.

BW: Body Weights for Children (kg) The mean body weight for male and female children 2–7 years old is 18.9 kg (EPA Exposure Factors Handbook).

Postapplication Dermal Daily Exposure and Absorbed Daily Dosage is calculated as follows:

$$\text{Dermal Daily Exposure} = \frac{(\text{Sum}) \times (\text{CF})}{(\text{BW})}$$

$$\text{Dermal Daily Exposure} = \frac{0.17\,\text{mg} \times 1.87}{18.9\,\text{kg}} = 0.017\,\text{mg/kg/day}$$

$$\text{Dermal Absorbed Daily Dose} = \text{Dermal Daily Exposure} \times (\text{DAF})$$

$$\text{Dermal Absorbed Daily Dose} = (0.017\,\text{mg/kg/day}) \times (0.1) = 0.0017\,\text{mg/kg/day}$$

Potential Postapplication Exposure of Children Aged 1–6 Years by Hand-to-Mouth Transfer from Treated Surfaces Hand-to-mouth transfer residue data from 20 min of Jazzercise is used as a surrogate. It is substantiated by a study involving broadcast application to carpets, with attention to postapplication inhalation as well as to dermal exposure monitoring and biomonitoring. Usual child activities (playing with blocks, crawling, walking, etc.) were performed by adult volunteers over a 4-hr period. Comparison of hand-rinse data to residues transferred onto glove dosimeters revealed remarkably similar totals (Fig. 2.61).

The general equation for estimating potential dermal exposure to hands and subsequent incidental oral exposure or absorbed dose is as follows.

DH: Daily Dermal Exposure to Hands (mg/day) Hand exposure estimates represent a combination of the hand transfer factor (TF), transferable residue (TR), and hand surface area (HSA): DH = TF × TR × HSA. Daily dermal exposure to hands is calculated as 0.0364 mg.

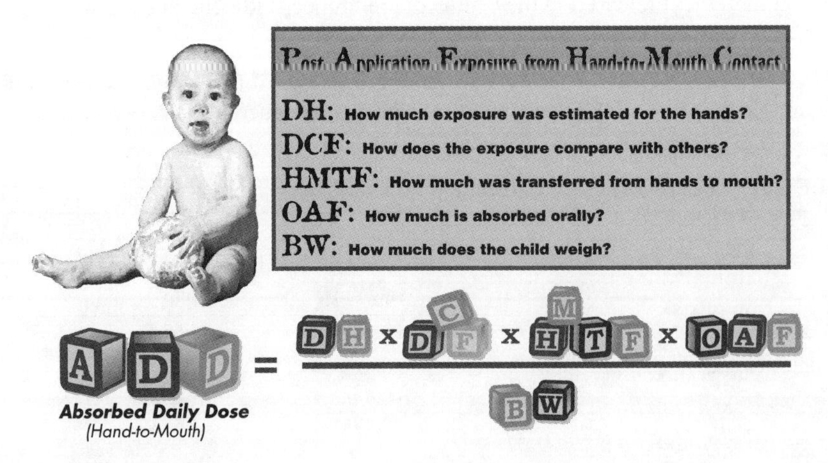

Figure 2.61 The factors associated with the calculation of exposure amounts for hand-to-mouth transfer by children.

DCF: Dermal Experimental Correction Factor (unitless) The DCF adjusts milligrams of dermal exposure (derived from summation calculations as previously described) based on the amount of active ingredient released from the surrogate dermal monitoring study versus the amount of active ingredient released from the product being evaluated. The product assessed in this example released 1.87 times the active ingredient released in the California reference study cited in Table 2.21.

HMTF: Hand-to-Mouth Transfer Factor (unitless) Hand-to-mouth transfer estimation for children and infants is based on data from hand-wash removal efficiency studies. The data for lipophilic compounds suggest that water-only rinsing of hands results in less than 5% removal. In contrast, more rigorous, solvent-based rinsing removes 20–40% of the pesticide. So incidental oral exposures are based on the assumption that approximately 10% of residues on children's hands are ultimately ingested as a result of hand-to-mouth transfer.

OAF: Oral Absorption Fraction for Active Ingredient (unitless) An oral (gastrointestinal) absorption factor of 100% is used as a default assumption.

BW: Body Weight (kg) The mean body weight across male and female children aged 2–7 is 18.9 kg (from the EPA's Exposure Factors Handbook). Postapplication Oral Daily Exposure and Absorbed Daily Dose is calculated as follows.

$$\text{Oral Daily Exposure} = \frac{(\text{DH}) \times (\text{DCF}) \times (\text{HMTF})}{(\text{BW})}$$

$$\text{Oral Daily Exposurc} = \frac{(0.0364\ \text{mg/day}) \times (1.87) \times (0.1)}{18.9\ \text{kg}}$$

$$= 0.00036\ \text{mg/kg/day}$$

$$\text{Oral Absorbed Daily Dose} = \text{Oral Daily Exposure} \times \text{OAF}$$

$$\text{Oral Absorbed Daily Dose} = 0.00036\ \text{mg/kg/day} \times 1 = 0.00036\ \text{mg/kg/day}$$

RISK CHARACTERIZATION

The final step in risk assessment is risk characterization, which involves integration of toxicological data with exposure data to estimate the level of human risk. Risk characterization also includes a description of assumptions and uncertainties that go into the evaluation of risk.

Approaches to risk characterization differ, depending on whether the toxicity end point of concern has a threshold. It is generally assumed that most types of toxic effects have thresholds below which adverse effects will not occur. Other types, such as genotoxic carcinogens, often are assumed to have no threshold (i.e., there is some probability of harm at any level of exposure), and the nature of the risk characterized is different from that described for threshold effects (Fig. 2.62).

Figure 2.62 Risk characterization requires information on exposure and toxicity.

Threshold Effects

For threshold effects, risk assessments normally are conducted by utilizing a Reference Dose (RfD) or Margin of Exposure (MOE) approach. In the MOE approach (known outside the United States as the Margin of Safety), the anticipated human exposure level is compared to the lowest NOAEL from an appropriate toxicology study.

$$\text{Margin of Exposure} = \frac{\text{No Observed Adverse Effect Level}}{\text{Estimated Human Exposure}}$$

For example, if the NOAEL is 30 mg/kg/day and the estimated human exposure is 0.5 mg/kg/day, the MOE is 60:

$$\text{Margin of Exposure} = \frac{30\,\text{mg/kg/day}}{0.5\,\text{mg/kg/day}} = 60$$

This indicates the estimated human exposure is 60 times lower than the NOAEL.

Important considerations in selecting a study from which the NOAEL is derived are as follows:

- Animal model used
- Type of study
- Study design
- Route of administration
- Study duration

TABLE 2.22 Critical Toxicological End Points (NOAELs) Identified for Use in Risk Assessment for Insecticide X

Exposure Scenario	Appropriate Toxicological Study	NOAEL
Acute dietary	Acute rat neurotoxicity	25 mg/kg/day
Chronic dietary	2-year rat feeding	1 mg/kg/day
Short-term (1–7 days) and intermediate (1 week to several months) inhalation	28-day rat inhalation	3 mg/m^3 (0.003 mg/l, or ca. 0.235 mg/kg/day)
Short-term and intermediate dermal	21-day rat dermal	250 mg/kg/day
Chronic dermal and inhalation	2-year rat feeding (assuming 100% inhalation and 10% dermal absorption)	1 mg/kg/day
Short-term and intermediate (multiroute, systemic)	90-day dog feeding study (assuming 100% oral absorption)	15 mg/kg/day

Ideally, the route of administration and study duration should be comparable to those of the human exposure scenario being evaluated. Depending on the pesticide's uses, NOAELs from several different studies may be utilized in a comprehensive risk assessment (Table 2.22).

The greater the MOE, the greater the degree of safety. In general, an MOE should be at least 100 (current EPA policy) if the NOAEL is derived from an animal study; it should be at least 10 if the NOAEL is derived from human data. An MOE of 100 means that the estimated level of human exposure is 100 times lower than the highest dose tested that produced no adverse effects in the toxicology study. Larger MOEs may be required under certain conditions, for instance, if there are concerns about the quality or completeness of the database or if there is concern about possible increased sensitivity of infants or children (Fig. 2.63).

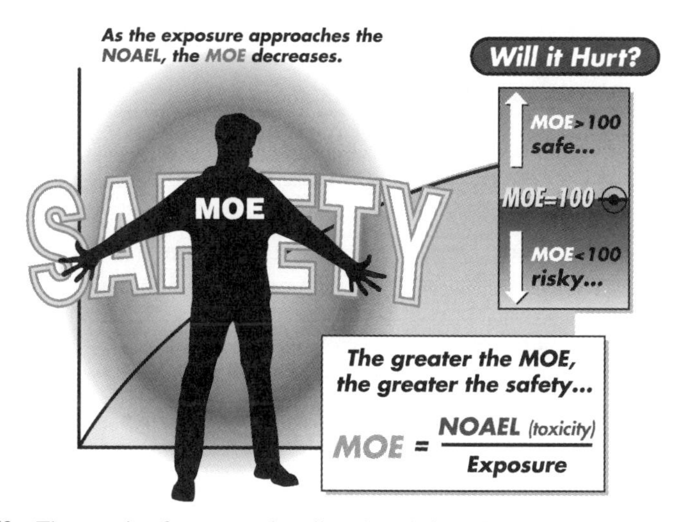

Figure 2.63 The margin of exposure describes the relationship between exposure and effects.

The RfD approach is similar to the MOE approach except that the anticipated human exposure level is compared to the appropriate RfD instead of the NOAEL. RfD is defined as the estimated human exposure level believed to have no adverse impact on human health. A chronic RfD (also called the Acceptable Daily Intake) is defined as the level to which a human can be exposed every day for a lifetime without experiencing adverse effects. Acute RfDs—that is, estimates of the amount of pesticide to which an individual can be exposed in one day without experiencing adverse health effects—also have been established.

RfDs are calculated by dividing the lowest NOAEL from an appropriate toxicology study using the most sensitive animal species (or humans) by the appropriate uncertainty factors (also referred to as safety factors).

$$\text{RfD} = \frac{\text{No Observed Adverse Effect Level}}{\text{Uncertainty Factors}}$$

Uncertainty factors are established by EPA policies. Most commonly, uncertainty factors of 10× each are applied to account for interspecies extrapolation (animals to humans) and intraspecies variation (differences among humans), for a total uncertainty factor of 100. Additional uncertainty factors of 3–10× each also may be applied to account for lack of an appropriate NOAEL or an incomplete toxicity database, or, as a result of FQPA, to provide additional safety margins for infants and children. The total uncertainty factor can range from 10× (if the NOAEL is derived from a human study) to 10,000×, although it rarely exceeds 1000×. The division of the NOAEL by these uncertainty factors is thought to provide reasonable assurance that exposure to the chemical at a dose less than or equal to the RfD will not pose significant human risk (Fig. 2.64).

Consider the following examples of calculating an RfD for pesticides for which the following toxicological data are available:

- NOAEL of 200 mg/kg/day from a rat developmental toxicity study
- NOAEL of 50 mg/kg/day from a rabbit developmental toxicity study
- NOAEL of 100 mg/kg/day from a two-generation rat reproduction study
- NOAEL of 75 mg/kg/day from a 1-year dog feeding study
- NOAEL of 25 mg/kg/day from an 18-month mouse feeding study
- NOAEL of 10 mg/kg/day from a 2-year rat feeding study

In this case, the chronic RfD generally would be calculated as 0.1 mg/kg/day, utilizing the lowest NOAEL (10 mg/kg/day from the 2-year rat feeding study) and a 100-fold safety factor.

In the risk characterization process, calculation of unacceptable MOEs or estimated exposures greater than the RfD indicate that

- a more refined exposure assessment needs to be completed,
- additional data may be required,
- mitigation measures need to be used (e.g., use of a different formulation, protective clothing, enclosed tractor cabs, longer reentry intervals),
- the use should not be registered, or
- the product should be taken off the market (if previously registered).

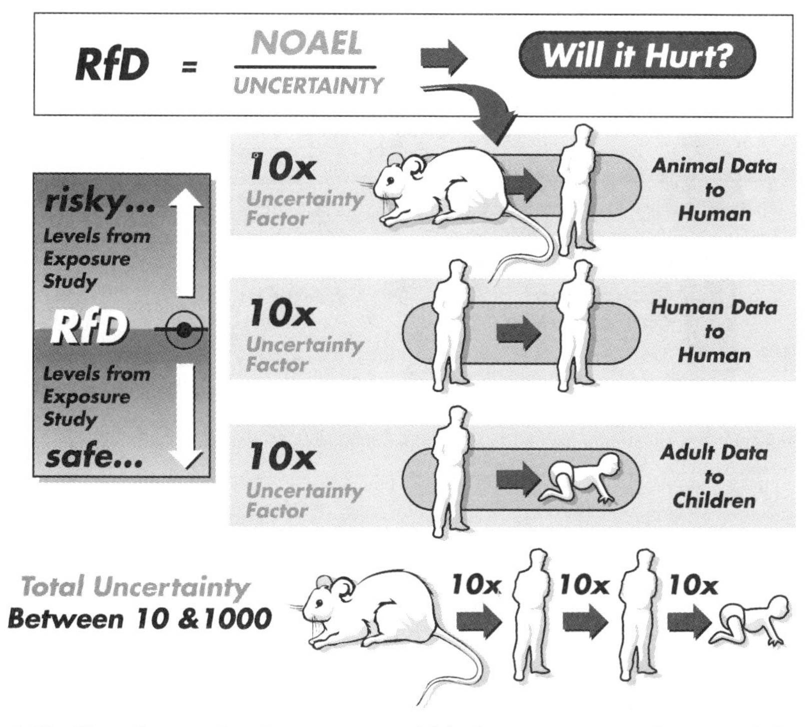

Figure 2.64 The reference dose incorporates multiple factors to account for uncertainties in the extrapolation of effects.

Nonthreshold Effects

The EPA generally considers cancer to be a nonthreshold effect. Therefore, cancer risk assessment in the United States usually does not compare anticipated human exposure levels to an RfD, nor is an MOE determined. Instead, a cancer assessment provides an estimate (expressed as a probability) of the excess risk of cancer resulting from exposure to the pesticide. For instance, a calculated risk of 1×10^{-6} (1 in 1,000,000) means that a person would have no more than a one-in-a-million chance of developing cancer in excess of the background incidence in the general population. This level of excess cancer risk generally is considered acceptable for the general public, whereas higher estimated levels such as 1×10^{-5}(1 in 100,000) or even 1×10^{-4}(1 in 10,000) may be considered acceptable for some occupational exposures (Fig. 2.65).

Mathematical Models In the United States, the potential risk to humans from exposure to carcinogens is most often estimated using mathematical models. All mathematical models used for cancer risk assessment extrapolate from high-dose levels used in animal studies to much lower human exposure levels. But results of extrapolation can differ substantially, depending on the model used. The slope of the dose–response curve calculated by models that assume linearity at low doses often is used to describe cancer potency: the steeper the response curve, the more potent the carcinogen. Since there is always some uncertainty associated with the calculated dose–response curve, there is always a chance that the slope of the true dose–response curve could be higher or lower than calculated. Statisticians have

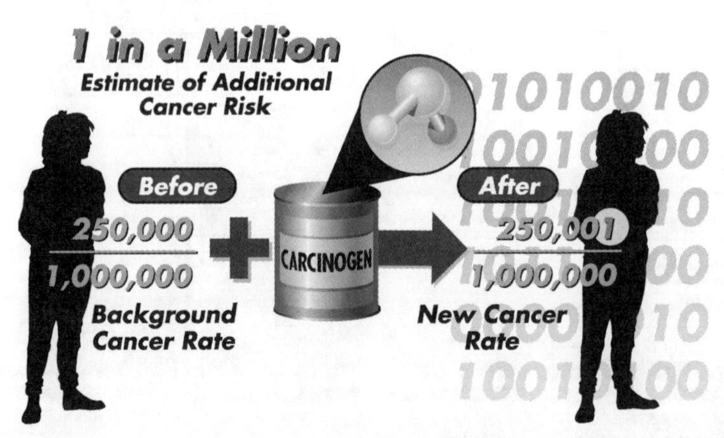

Figure 2.65 Cancer risk assessment is concerned with the occurrence of an additional cancer per 1,000,000 individuals.

developed methods that allow estimation of both the upper and lower limits of the calculated dose–response curve; thus, they say that the true dose–response curve will fall somewhere between the lower and upper estimates 95% of the time.

The upper estimate of tumor potency (often referred to as Q*, pronounced "q-star") developed by mathematical models is most frequently reported as the most conservative: it produces the highest estimation of potential risk from standard models. However, the lower estimation of risk, which can be zero, has the same chance as the upper estimate of being the true estimate of risk. So to provide an unbiased assessment and an indication of uncertainty of the derived estimate, the risk assessment must yield the most likely estimate of risk as well as the upper and lower estimates.

For example, the best estimate of cancer risk from lifetime exposure to a 1 mg/kg/day dose of an animal carcinogen might be 1×10^{-8} (1 in 100,000,000), with upper and lower estimates of 1×10^{-6} and zero, respectively. In other words, 95 times out of 100 the true risk of cancer from such exposure will fall somewhere between zero and 1×10^{-6}. The EPA focuses on and conservatively regulates the upper estimate. In this example, risk is calculated as one in a million (the upper-bound risk) instead of one in 100 million (the mean risk).

Route-Specific and Systemic

Dietary risk assessment is more straightforward than occupational or residential risk assessment. It involves evaluating potential risk from a single-route exposure similar to that used to generate most toxicology data. Dietary risk assessment generally is conducted by comparing estimated dietary exposure to results from toxicology studies in which the pesticide was administered orally (diet, stomach tube, or capsule). On the other hand, occupational and residential risk assessment usually involves evaluating multiple routes of exposure: dermal; inhalation; and, particularly for residential exposure to children, incidental oral ingestion (Fig. 2.66).

Two different approaches can be used to assess potential risk from multiple sources of exposure: a route-specific approach, as in dietary risk assessment; and a systemic (oral equivalent) dose approach. Each has its advantages and disadvantages.

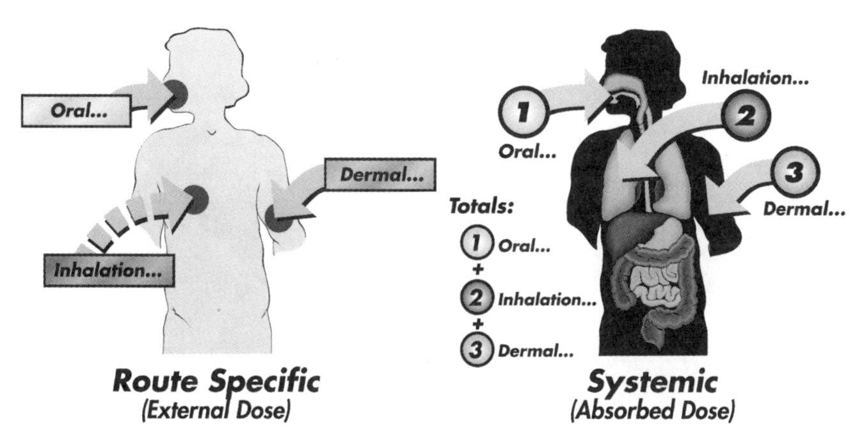

Figure 2.66 Risk assessments for humans consider external absorbed doses.

In a route-specific risk assessment, the estimated dermal or inhalation exposure is compared to the appropriate end point from a toxicology study in which a comparable route and duration of exposure were used. For example, consider that a farmworker who uses pesticides for a few weeks each year is at risk of exposure; the potential for dermal exposure during application or reentry can be evaluated by comparing his estimated exposure level to the NOAEL from a 21- or 90-day dermal toxicity study.

Similarly, potential risk from repeated inhalation exposure can be compared to the NOAEL from a 28- or 90-day inhalation study. If the toxic effect is the same regardless of the route of exposure, potential risk from occupational and residential exposure also can be assessed on a systemic basis.

In systemic risk assessment, the total amount of pesticide absorbed into the body via combined dermal, inhalation, and/or oral exposure is calculated and compared to a systemic (or oral equivalent) NOAEL probably derived from a subchronic feeding study.

The advantages of route-specific risk assessment are that

- it accounts for possible differences in the way chemicals behave (pharmacokinetics) among various routes of exposure, and
- it can be used even if the most sensitive toxic end point differs depending on the route of exposure.

Most toxicity studies are based on oral exposure; few are based on dermal or inhalation exposure. Thus, route-specific toxicity data of appropriate duration may not always be available for the occupational or residential exposure scenario in question. Conversion of all exposures to systemically absorbed or oral equivalents offers two advantages: considering all exposures simultaneously and comparing them to a more comprehensive toxicity database. Ideally, the systemic method requires knowing the rate or percentage of dermal and inhalation pesticide absorption. As a default, dermal and inhalation absorption often are assumed to be 100% and to occur at the same rate as oral absorption. The decision whether to use route-specific or systemic risk assessment methodology generally depends on the proposed exposure scenario and the toxicity data available.

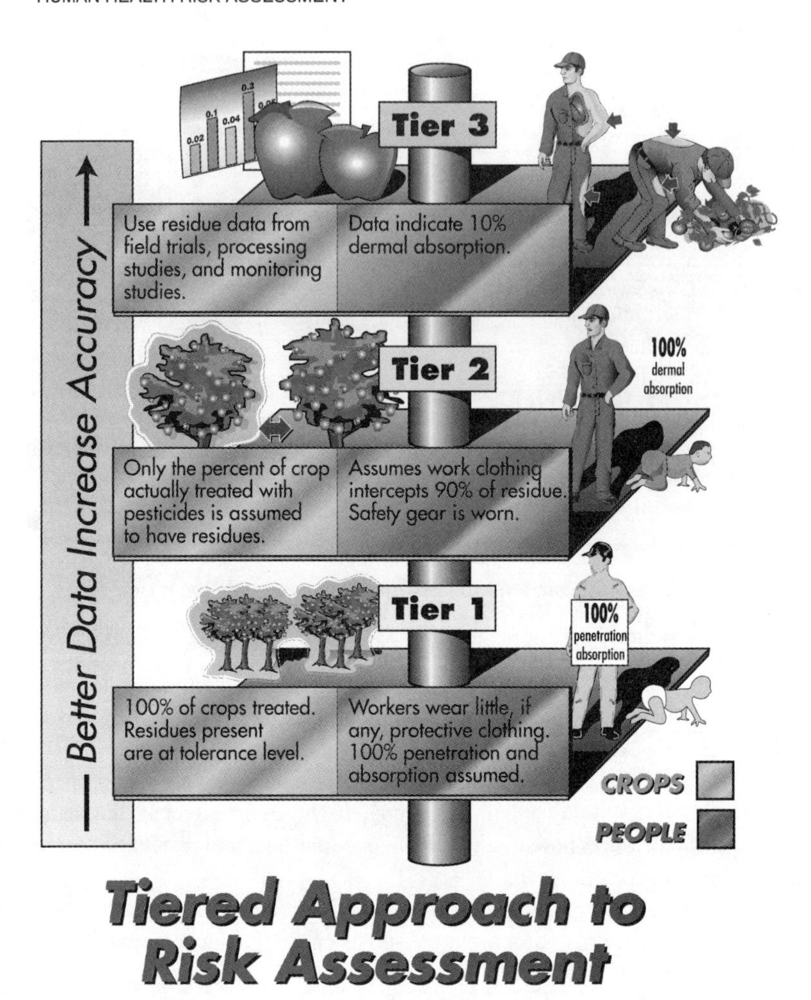

Figure 2.67 Human health risk assessment follows an approach that begins with conservative, worst-case assumptions and leads to more realistic assessments of exposure.

The Tiered Approach to Risk Assessment

It is common for regulatory authorities to screen pesticide risk by using conservative, worst-case estimates of exposure. This expedites screening and facilitates the budgeting of scarce resources. Worst-case assessments exaggerate exposure, yielding higher estimates than would actual exposure. They are considered "conservative" in that they represent the worst case scenario, thereby affording a wide margin of safety (Fig. 2.67).

Risk assessment is generally a multitiered process. The initial or Tier 1 risk assessment uses very conservative "default" assumptions in the absence of more specific, reliable exposure data. For example, Tier 1 assessments assume that all crops are treated; that all residues reach tolerance levels; that workers wear little, if any, protective clothing; and that 100% of the pesticide that contacts the skin is absorbed. Dietary, residential, and/or occupational exposure estimates calculated under these conditions may be hundreds or thousands of times higher than actual exposure. However, if the risk estimates from these conservative assumptions are acceptable (below the Acceptable Daily Intake), no further evaluation is necessary.

Initial risk calculations that do not yield acceptable risk estimates using default assumptions do not necessarily indicate excessive risk. Rather, they indicate the need to incorporate more detailed and reliable data for key parameters: frequency, duration, and magnitude of exposure. It is important to remember that refinements to the exposure assessment do not change actual exposure; they simply modify estimates of exposure. The resulting, more realistic exposure estimates form the basis for higher tier analysis.

The greater the concern for risk posed by a pesticide, the greater the need to replace default assumptions with more reliable data. Only when the refined, upper tier risk assessments yield unacceptable risk is there true cause for concern and, possibly, the need for action.

Following is an example of a multitiered approach to dietary risk assessment; a similar approach can be used in occupational and residential assessment. This example considers five hypothetical pesticides that differ in toxicity and exposure.

Tier 1 The initial tier of dietary risk assessment uses conservative default assumptions of 100% crop treated and all residues present at the tolerance level. The TMRCs for five pesticides are calculated to be

- 0.001 mg/kg/day for Pesticide A,
- 0.01 mg/kg/day for Pesticide B,
- 0.1 mg/kg/day for Pesticide C,
- 1.0 mg/kg/day for Pesticide D, and
- 3.0 mg/kg/day for Pesticide E

Although most pesticides have different chronic RfDs, for simplicity it is assumed that the RfD for each of the five chemicals is the same: 1.0 mg/kg/day. The 1.0 mg/kg/day chronic reference dose is compared to the total amount of each pesticide consumed, based on tolerances. Individual dietary consumption of Pesticides A, B, and C is substantially below the chronic RfD; thus, it is assumed that dietary consumption of these pesticide residues will not cause adverse human health effects.

Pesticide D results are borderline, so it does not pass Tier 1 risk assessment. And the possibility that Pesticide E may pose dietary risk to humans cannot be excluded since the Tier 1 estimated consumption exceeds the chronic RfD. Therefore, both Pesticides D and E are candidates for Tier 2 risk assessment.

Tier 2 In the Tier 2 dietary risk assessment, only the percentage of crops actually treated is assumed to contain residues. In this example, TMRCs for Pesticides D and E are 1.0 and 3.0 mg/kg/day, respectively. However, if data were available to indicate that not more than 50% of labeled crops are in fact treated, the assumed ARCs would be 0.5 and 1.5 mg/kg/day, respectively. Thus, the anticipated dietary exposure to Pesticide D (0.5 mg/kg/day) is clearly below the chronic RfD of 1.0 mg/kg/day, and no further refinement of the risk assessment process is needed. Pesticide E, however, is a candidate for a Tier 3 assessment since the Tier 2 exposure estimate (1.5 mg/kg/day) exceeds the RfD.

Tier 3 In Tier 3 risk assessment, using residue data from field trials, processing studies, and/or monitoring studies refines even further the anticipated residues. In this example, data indicate that Pesticide E readily degrades during storage and that much of the residue is removed during handling or washing of fruits and vegetables prior to distribution to grocery stores. Based on these data, the ARCs from Pesticide E are further reduced to 0.14 mg/kg/day, which is well below the RfD. Thus, the use of ARCs that incorporate better data allows risk assessors to conclude that there is no unreasonable risk from consuming foods from crops treated with this pesticide.

Dietary Risk Assessment

Organization of the food consumption data may vary, depending on the purpose of the dietary exposure analysis, and the purpose of analysis may vary according to whether the toxicological effect under consideration is chronic (long-term) or acute (short-term).

In assessing dietary exposure to the chronic toxicological effects of pesticides, most regulatory authorities consider some measure of typical food intake, such as mean or median food consumption values. But for compounds that might be acutely toxic, it is important to know if the dietary intake over a relatively short period of time (such as a day) is safe. By examining extreme exposure, acute assessment protects the safety of people who ingest more pesticide residues than virtually anyone else in the population.

Acute Dietary Risk Acute dietary risk is evaluated with the MOE or the acute RfD (aRfD) approach. The toxic effect used for acute dietary risk assessment may yield the toxic effect after only one or two exposures. In most cases, end points are derived from acute neurotoxicity studies, developmental toxicity studies, or studies on cholinesterase inhibition.

The exposure estimate in acute risk assessment is intended to represent the highest total amount of residue that an individual might ingest in a single day. In a Tier 1 acute risk assessment, it is assumed that all commodities for which a pesticide tolerance is established contain residues at the tolerance level. In subsequent tiers, the highest residue level observed in field trials, monitoring data, or market-basket surveys is used to estimate residues for commodities consumed as a single item (e.g., apple, orange, banana, potato). Adjustments for percentage of crop treated and the use of average rather than maximum residue levels help in estimating residues in blended commodities such as corn oil, flour, juice, and milk; since these items are derived from multiple crops, it is unlikely that all were treated with the pesticide.

A distribution of single-day exposures is calculated for acute dietary assessment, based on the distribution of individual consumption values within the population. The MOE is then calculated for each of those exposure values, yielding a distribution of MOEs for the population. The results of the acute dietary risk assessment are presented as the MOE for specified percentiles (95th, 99th, 99.9th) of the population subgroup. Alternatively, the results can be expressed as a percentage of the RfD.

Table 2.23 shows the output from a Tier 1 acute dietary risk assessment of Insecticide X for the U.S. population. The estimated exposures summarize a complete distribution as in Table 2.14. The acute oral toxicity end point for Insecticide X is a NOAEL of 25 mg/kg/day, from an acute rat neurotoxicity study utilizing a 100-fold uncertainty factor (Table 2.22). The aRfD is then calculated as 0.25 mg/kg. Exposure at the 90th percentile of the population is 0.014371 mg/kg/day, or less, and accounts for 6.7% of the aRfD

TABLE 2.23 Tier 1 Acute Dietary Risk Assessment for Insecticide X

U.S. Population Percentile	Exposure (TMRC, mg/kg/day)	Percent Acute RfD (aRfD = 0.25)	Margin of Exposure (NOAEL = 25 mg/kg/day)
10th	0.000186	0.1	134,409
20th	0.000381	0.2	65,617
30th	0.000629	0.3	39,746
40th	0.000992	0.4	25,202
50th	0.001583	0.6	15,793
60th	0.002644	1.1	9,455
70th	0.004591	1.8	5,445
80th	0.008122	3.2	3,078
90th	0.014371	6.7	1,740
95th	0.022468	9.0	1,113
97.5th	0.032846	13.1	761
99th	0.049183	19.7	508
99.5th	0.064399	25.8	388
99.75th	0.079281	31.7	315
99.9th	0.103411	41.4	242

and an MOE of 1740. Exposure of the remaining 10% of the population, then, is at least 0.014371 mg/kg/day. If an MOE were less than 100 or the exposure greater than the aRfD, further refinements would be necessary; that is, more precise and/or reliable estimates of exposure would be needed to demonstrate adequate margins of safety, and a revised risk assessment at a higher tier would be conducted. If the MOE or the RfD percentage remained unacceptable after refinement, various mitigation steps would have to be taken to reduce exposure. For example, the application rate could be reduced or the preharvest interval could be extended. If the compound was still efficacious, these modifications could reduce residue sufficiently. An alternative would be to omit specific crops from the product label.

Chronic Dietary Risks Potential risk from chronic dietary exposure to pesticide residues is estimated by comparing average residue consumption to the chronic RfD. Risk is considered acceptable as long as the estimated exposure level is less than or equal to the chronic RfD. If the pesticide is a nonthreshold carcinogen, an estimated cancer risk potential is calculated using mathematical models.

Table 2.24 represents a chronic dietary risk assessment for Insecticide X. Exposure estimates are based on mean residues from field trials, and a chronic RfD of 0.01 mg/kg/day is assumed, based on a NOAEL of 1 mg/kg/day in the chronic rat study and a total uncertainty factor of 100. It also is assumed that the database for Insecticide X is complete and that no evidence of increased sensitivity in infants or children is noted. Thus, in this case, no additional uncertainty factor is needed to protect children.

In this example, the average dietary intake of Insecticide X for the total U.S. population (0.000152 mg/kg/day) represents 1.5% (0.000152/0.01) of the chronic reference dose. Exposures are relatively constant throughout the year and do not appear to be affected by geography or race. Children aged 1–6 years comprise the group with the highest exposure potential, ingesting an average of 0.000418 mg/kg/day of Insecticide X, or 4.2%

TABLE 2.24 Chronic Dietary Risk Assessment for Insecticide X

Population Subgroup	Total Exposure (ARC, mg/kg/day)	Percent of Reference Dose
General U.S. population		
48 States, all seasons	0.000152	1.5
General U.S. population by season		
Spring	0.000150	1.5
Summer	0.000146	1.5
Autumn	0.000157	1.6
Winter	0.000156	1.6
U.S. population by region		
Northeast	0.000185	1.9
North Central	0.000152	1.5
Southern	0.000135	1.4
Western	0.000151	1.5
U.S. population by race		
Hispanics	0.000180	1.8
Non-Hispanic Whites	0.000145	1.5
Non-Hispanic Blacks	0.000162	1.6
Non-Hispanic other than Black or White	0.000185	1.9
U.S. population by pregnancy status		
13+, pregnant/not nursing	0.000144	1.4
13+, nursing	0.000143	1.4
U.S. population by age and gender		
Nursing infants < 1 year old	0.000041	0.4
Nonnursing infants < 1 year old	0.000122	1.2
Children 1–6 years old	0.000418	4.2
Children 7–12 years old	0.000239	2.4
Males 13–19 years old	0.000165	1.7
Females 13–19 years old,	0.000104	1.0
not pregnant or nursing	0.000147	1.5
Males 20+ years old	0.000098	1.0
Females 20+ years old,	0.000115	1.2
not pregnant or nursing	0.000115	1.2

(0.000418/0.01) of the chronic RfD. Thus, in all cases—even with conservative assumptions regarding anticipated exposure—the total consumption of Insecticide X is estimated at well below the chronic RfD level assumed to cause no adverse human health effects. Chronic dietary exposure to Insecticide X is judged acceptable.

Dietary Cancer Risks Potential cancer risk from dietary exposure to nonthreshold carcinogens generally is estimated by multiplying the average consumption of pesticide residue by Q*, that is, by the upper potency estimate associated with that chemical. This calculation provides a 95th percentile upper estimate of excess cancer risk resulting from ingestion. In other words, the true excess risk of cancer is equal to or less than the calculated value 95% of the time.

In this example, it is assumed that Insecticide X is a nonthreshold carcinogen, and that Q* is $3 \times 10^{-3}(\text{mg/kg/day})^{-1}$. Estimates of dietary consumption used to evaluate chronic dietary risk are used to estimate potential dietary cancer risk. In this case, the average dietary consumption by the overall U.S. population is 0.000152 mg/kg/day. Based on these exposure estimates, the 95th percentile upper estimate of excess cancer risk for the overall population is calculated as 4.0×10^{-7}[0.000152 mg/kg/day $\times$ 0.003 (mg/kg/day)$^{-1}$]. The upper estimate of potential cancer risk for the overall U.S. population is less than 1×10^{-6} (one in a million), which is considered acceptable.

Occupational Risk Assessment

Route-Specific Dermal Risk Assessment for Mixer/Loader/Applicator The total dermal exposure to Insecticide X during mixing, loading, and application to corn was previously estimated to be 687 μg/day. In this example, it is assumed that a 70-kg worker mixes, loads, and applies Insecticide X on 10–15 days over a 3-month period. If the NOAEL for Insecticide X in a 21-day dermal toxicity study is 250 mg/kg/day, the short-term or intermediate route-specific dermal MOE for this worker is calculated as follows:

$$\text{MOE dermal} = \frac{\text{NOAEL}}{\text{Daily Dermal Exposure}}$$

$$\text{Daily Dermal Exposure} = \frac{687\ \mu\text{g/day}}{1000\ \mu\text{g/mg} \times 70\ \text{kg}}$$

$$\text{Daily Dermal Exposure} = 0.0098\ \text{mg/kg/day}$$

$$\text{MOE dermal} = \frac{250\ \text{mg/kg/day}}{0.0098\ \text{mg/kg/day}} = 25{,}510$$

Example of a Systemic Risk Assessment for a Mixer/Loader/Applicator In the same example, the total absorbed daily dose for a worker mixing, loading, and applying Insecticide X to corn was estimated at 0.0012 mg/kg/day (Table 2.19). If the lowest NOAEL for Insecticide X in a 90-day feeding study is 15 mg/kg/day, the MOE for intermediate exposure for this worker is calculated as follows.

$$\text{MOE}_{\text{systemic}} = \frac{\text{NOAEL}}{\text{Absorbed Daily Dose Intermediate Term}}$$

$$\text{Absorbed Average Daily Dose} = \frac{0.0012\ \text{mg/kg/day applied} \times 15\ \text{days applied}}{90\ \text{days}}$$

$$\text{Absorbed Daily Dose} = 0.0002\ \text{mg/kg/day}$$

$$\text{MOE}_{\text{systemic}} = \frac{15\ \text{mg/kg/day}}{0.0002\ \text{mg/kg/day}} = 75{,}000$$

To assess the potential risk posed by chronic exposure or the risk for cancer produced by a threshold carcinogen, the MOE is calculated by comparing the average absorbed daily dose (ADD) over 1 year (for chronic effects) or a lifetime (for threshold carcinogens) to the appropriate chronic NOAEL. The MOE can be refined by replacing the maximum application rate with average or typical rates; however, the large MOE indicates that further refinement is unnecessary.

The potential risk for any acute toxic effect generally is evaluated by comparing the highest ADD (i.e., without averaging, to the appropriate acute oral toxicity end point). Thus, for Insecticide X the MOE for acute exposure is calculated by comparing the total ADD derived from dermal plus inhalation exposure to the NOAEL from the acute neurotoxicity study.

$$\text{MOE} = \frac{\text{NOAEL}}{\text{ADD}} = \frac{5\ \text{mg/kg/day}}{0.0012\ \text{mg/kg/day}} = 4167$$

These MOEs are well above the value of 100 usually deemed acceptable for agricultural workers, even when conservative assumptions are used in the estimation of exposure potential. Therefore, the application of Insecticide X to corn poses no significant risk to agricultural workers.

Residential Risk Assessment

As in the case of occupational exposure, residential risk assessment for children is conducted using either the route-specific or the systemic approach. In this example, an acute potential toxicity is assumed; thus the focus is on day-of-application exposure and absorbed dose. It is assumed that Insecticide X is not used outdoors, so the residential risk assessment is conducted as follows.

Inhalation Daily Exposure = 0.000016 mg/kg/day

Inhalation Absorbed Daily Dose = 0.000016 mg/kg/day (assuming 100% inhalation absorption)

Dermal Daily Exposure = 0.017 mg/kg/day

Dermal Absorbed Daily Dose = 0.0017 mg/kg/day (assuming 10% dermal absorption)

Incidental Oral Exposure = 0.00036 mg/kg/day

Incidental Oral Absorbed Daily Dose = 0.00036 mg/kg/day (assuming 100% oral absorption)

Short-term to intermediate dermal NOAEL = 250 mg/kg/day

Short-term to intermediate inhalation NOAEL = 0.235 mg/kg/day

Short-term to intermediate (multiroute, systemic) NOAEL = 15 mg/kg/day

Route-Specific Residential Risk Assessment

$$\text{MOE inhalation} = \frac{\text{NOAEL}}{\text{Inhalation Daily Exposure}} = \frac{0.235\ \text{mg/kg/day}}{0.000016\ \text{mg/kg/day}} = 15{,}000$$

$$\text{MOE dermal} = \frac{\text{NOAEL}}{\text{Dermal Daily Exposure}} = \frac{250\ \text{mg/kg/day}}{0.017\ \text{mg/kg/day}} = 15{,}000$$

$$\text{MOE incidental oral} = \frac{\text{NOAEL}}{\text{Incidental Oral Absorbed Daily Dose}} = \frac{15\ \text{mg/kg/day}}{0.00036\ \text{mg/kg/day}}$$

$$= 42{,}000$$

Systemic Residential Risk Assessment

$$\text{MOE} = \frac{\text{NOAEL}}{\text{Average Absorbed Daily Dose}} = \frac{15\ \text{mg/kg/day}}{\underset{(\text{lungs} + \text{skin} + \text{hand-to-mouth})}{0.000016 + 0.0017 + 0.00036}} = 7225$$

These MOEs are well above the value of 100, which is usually deemed acceptable for potential indoor exposure. Therefore, indoor application of Insecticide X as a fogger (according to label directions and following reentry restrictions) should pose no significant risk to children.

Drinking Water Risk Assessment under FQPA

The EPA's Office of Pesticide Programs (OPP) has strengthened its consideration of pesticides consumed in drinking water under FQPA. Drinking water exposure and risk consideration may be approached from numerous perspectives, and the actual procedures used continue to evolve.

Within OPP, initial drinking water exposure estimates are obtained using Tier 1 (screening level) tools. Typically, models such as SCI_GROW and FIRST are used to obtain conservative estimates of pesticide levels in surface and ground water, respectively (See Chapter 5). These estimates of exposure concentration are very conservative since they represent raw water present in vulnerable settings. The model estimates of pesticide concentrations are therefore compared to actual monitoring data for both raw and finished drinking water to obtain an understanding of the model estimates as conservative predications of drinking water exposures.

Exposure estimates may be compared to HAL and MCL if they meet the general standards of the EPA's Office of Water. The significance of drinking water as a component of dietary ingestion of pesticides is determined by calculating drinking water consumption of pesticides as a percentage of the RfD. If this percentage is significant, the drinking water exposure is added to the overall exposure estimate for dietary risk. Finally, drinking water exposure is aggregated with other exposure data as described next.

AGGREGATE AND CUMULATIVE RISK ASSESSMENT

The general public may be exposed to pesticides via multiple routes (inhalation, oral, dermal) and sources (air, water, food, soil, and various surfaces in and outside the home). FQPA now requires the EPA to evaluate potential aggregate risk to an individual who may be exposed to pesticides from one or more sources simultaneously.

The Risk Cup Analogy

The "risk cup" symbolizes how the new safety standard—reasonable certainty that no harm will result from aggregate exposure to pesticide residues—will be evaluated and implemented by the EPA. It represents the total amount of pesticide residue to which a person might be exposed from all sources (diet, water, residential uses) without significant risk. The total allowable exposure (i.e., the size of the risk cup) is based on findings from toxicological studies including appropriate uncertainty and safety factors. For example, in

Figure 2.68 Modern health risk assessments entail the consideration of all potential exposure sources.

assessing chronic toxicity to the general population, the risk cup is based on the pesticide's chronic reference dose. A different size risk cup generally is used to assess potential risk from acute exposure and, depending on whether or not additional safety factors are imposed, to assess potential risk to infants and children (Fig. 2.68).

The assumption is that when the predicted exposure from pesticides or groups of pesticides exceeds risk cup capacity, the pesticide or group of pesticides fails to meet the "reasonable certainty of no harm" standard written into FQPA. Conversely, exposure levels that do not exceed total risk cup capacity are deemed to meet the standard.

The risk cup analogy can be applied to aggregate risk assessment conducted on a single chemical or to a combined or cumulative risk assessment conducted on multiple chemicals with a common mechanism of toxicity. Determining the best way to assess risk from multiple routes of exposure, and/or from multiple chemicals, will be a major risk assessment and scientific challenge over the next few years.

The development of a risk cup including aggregate and cumulative exposures—multiple sources and common mechanisms of toxicity—will require new methodologies. A more sophisticated risk assessment than those used previously will be needed. Although highly conservative assumptions often have been used to demonstrate negligible risk for single compounds, the inclusion of multiple routes of exposure and multiple chemicals in the same risk cup will require more comprehensive and accurate data to demonstrate reasonable certainty of no harm.

SUMMARY

The risk of pesticide exposure to human health is a function of exposure and toxicity. Since both measurements involve a degree of uncertainty, risk assessments generally use very conservative assumptions to assure adequate margins of safety. The risk assessment process

Figure 2.69 Risk assessments are refined as new and more comprehensive data becomes available.

generally proceeds in tiers from assessments based on very limited data, with very conservative assumptions, through assessments with extensive data and a solid understanding of the pesticide and its human exposure effects. In the tiered approach, low-risk pesticides with large margins of exposure are screened out of the risk assessment process at a very early stage; this facilitates the direction of resources to the assessment of risk posed by those pesticides of greatest concern to human health (Fig. 2.69).

State-of-the-art risk assessment methodologies are used to assess exposure and risk to special subpopulations. Therefore, risks to infants and children and to workers are evaluated separately from those posed to human populations in general. Risk assessments are increasingly concerned with the aggregate risk of pesticide exposure to humans where the combination of risk from multiple sources (air, food, water, playground, home, etc.) is considered.

Despite the public desire for zero risk, the world is not risk-free. Recognition of the risks associated with pesticide use leads to informed decision making in identifying levels of risk acceptable to society. Risk assessment, product labeling, governmental enforcement, and applicator and consumer education form the foundation of a comprehensive framework to regulate the manufacture, use, and disposal of pesticides and to ensure that adverse effects on human health and the environment are minimized. Responsible management of pesticide benefits and risks allows optimal benefits in terms of public health, safety, and prosperity.

CHAPTER 3

EPIDEMIOLOGY: VALIDATING HUMAN RISK ASSESSMENTS

THE SCIENCE OF EPIDEMIOLOGY

Epidemiology is the study of the distribution and causes of disease in human populations. Epidemiologists focus on determining which factors cause disease and which factors protect against disease. Although modern epidemiology is considered a relatively young science, its basic concepts have aided society for hundreds of years in understanding causes of diseases such as cholera and lung cancer. By identifying causes of disease and populations which may be at highest risk, steps can be taken to prevent disease.

Human risk assessment is estimated by measuring the suitable routes of exposure in controlled experiments on laboratory animals. This facilitates predictions of how pesticides may affect human health; data from such assessments are critical in setting acceptable human exposure limits. Nevertheless, it is important to continue to investigate potential human health risks, directly, through epidemiological studies that examine whether the rate of disease in an exposed population is higher or lower than in a similar, unexposed population (Fig. 3.1). Epidemiological studies rest on one key assumption: In the absence of exposure, two human study populations will exhibit similar or identical rates of disease.

Pesticides are developed to be toxic to specific living organisms; therefore, it is logical to be concerned about the potential of these chemicals to adversely affect exposed human populations. Epidemiological investigations increasingly address pesticides and their potential association with human disease. This increased concern for human toxicity potential addresses various levels (high, medium, low, absent) through various routes of exposure (food, air, water, soil).

The process of identifying causes of disease within populations exposed to pesticides is complex, primarily because pesticides are merely representative of environmental exposures that people may encounter. Therefore, determining that a pesticide is associated with increased adverse health effects—cancer, respiratory problems, immune disorders, birth defects—requires:

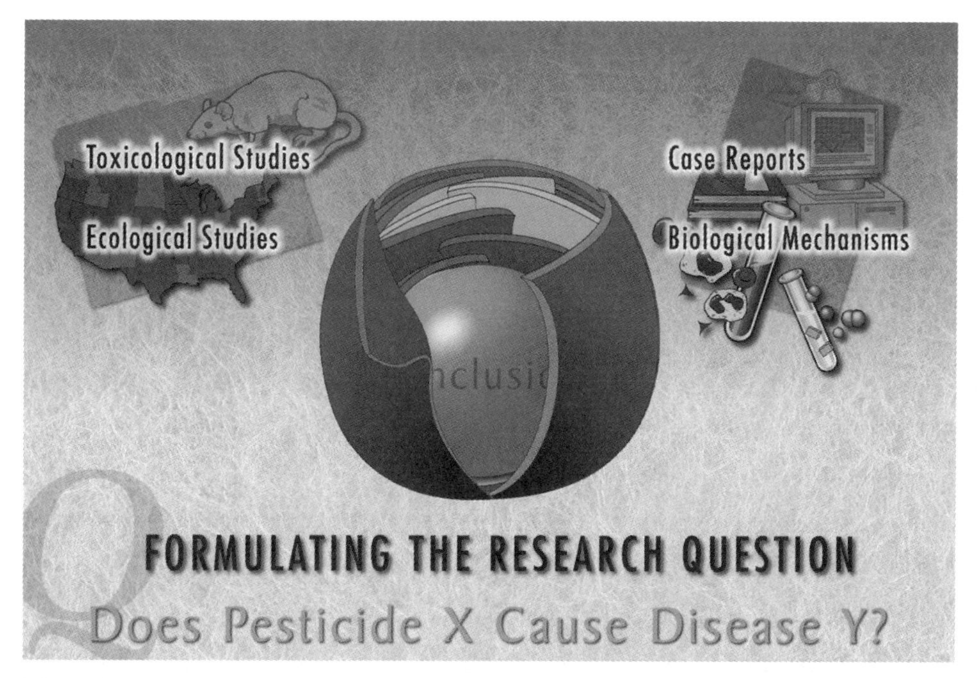

Figure 3.1 Defining the question is the first step in answering the question, does pesticide x cause disease y?

- evidence of a positive association between pesticide exposure and a specific disease, and
- a control for the effect of other causes of that specific disease.

For instance, farmers are exposed not only to pesticides but also to other potential risk factors such as fertilizers, nitrates, fuels and engine exhausts, solvents, organic and inorganic dusts, electromagnetic radiation, ultraviolet radiation, and animal pathogens. Behavioral, dietary, and genetic factors may impact their risk of disease as well.

PRINCIPLES OF EPIDEMIOLOGY

Person, Place, and Time

The main objectives of epidemiology are to

- describe the occurrence of disease, and
- explain the possible causes of disease by identifying and quantifying etiologic (risk) factors.

Describing a disease requires the gathering of information on the distribution of disease in human populations based on age, gender, race, and geographical area. When little is known about the cause and occurrence of a disease, epidemiologists study disease patterns as the first step in generating a hypothesis on causal factors.

Figure 3.2 Person, place, and time are key principles in epidemiology.

Characteristics used to describe patterns of disease fit into three general categories (Fig. 3.2):

- Person: Who is getting the disease (based on personal characteristics such as age, gender, race, religion, occupation, and socioeconomic status).
- Place: Where the disease is occurring, geographically.
- Time: Seasonal patterns, or whether the disease rate is increasing, decreasing, or staying the same.

Risk Factors

Risk factors (e.g., personal characteristics and environmental factors) are known to influence the distribution of disease within a population. Differences in disease patterns between two study populations often can be explained by one or more risk factors: age; gender; ethnicity; and personal factors such as cigarette smoke, diet, exercise, occupation, etc. (Fig. 3.3).

Exposure Relationships

Epidemiologists study acute and chronic diseases. Acute diseases develop soon after an exposure has occurred, whereas chronic diseases develop over months or years after exposure. The time that elapses between initial exposure and disease detection is divided into two periods: induction and latency.

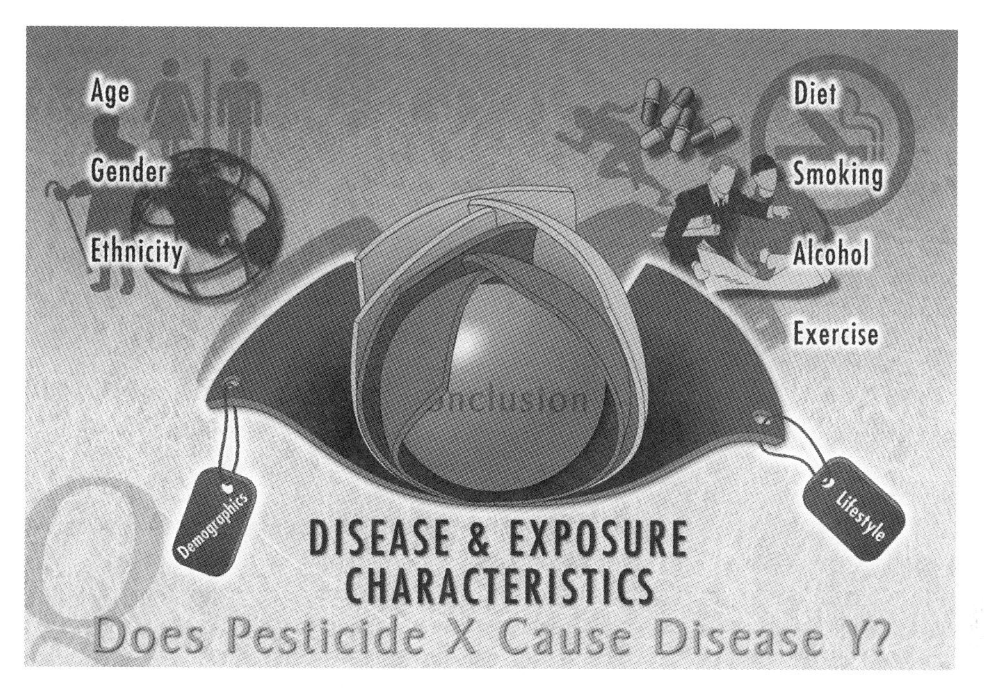

Figure 3.3 Risk factors are known to influence the distribution of disease within a population.

The induction period is the time between exposure and disease onset. The induction period may last hours (e.g., from the time a foodborne pathogen is ingested to the onset of food poisoning) or decades (as in the correlation between long-term smoking and the onset of lung disease).

Latency refers to the time that lapses between disease onset and detection. The total time lapse between exposure and disease diagnosis, therefore, is the sum of induction and latency. The longer the time lapse of each, the more difficult it is to link a specific exposure to a specific disease in an epidemiological study; imprecise memory of exposure, lack of records, and people moving out of the study may contribute to the uncertainty. Studies that correlate pesticides to chronic diseases (e.g., cancer) have to consider exposures that occurred decades before the disease diagnoses.

SOURCES OF INFORMATION

Disease Records

Personal Medical Records Personal medical records are the most reliable source of information to confirm disease diagnosis and the date of diagnosis, but gaining legal access to medical files requires concerted effort on the part of the investigator. A protocol must be in place to protect the confidentiality of the study participants, and participation must be voluntary to conform to ethical research guidelines.

Hospital Discharge Data Hospital discharge data include disease-related information such as the diagnosis, length of stay, and date of discharge; information regarding the patient's date of birth, race, gender, and area of residence is often available as well. For certain diseases, however, diagnosis can be merely preliminary and should be verified by medical record review. Hospital discharge data can be useful to identify a cohort (a group of people who share common characteristics); however, the identity of subjects is revealed only through informed patient consent, which can be cumbersome to acquire.

Disease Registries Physicians are required by law to report certain diseases to public health authorities. In addition, there are numerous population-based registries in the United States that arrange to get diagnoses from various medical sources. In addition, the Centers for Disease Control registry program interacts significantly with state cancer registries, and many states have their own state-funded disease registries.

While some registries compile information on the incidence of cancer, others track the occurrence of birth defects, communicable diseases such as AIDS, and kidney disease.

Death Certificates A death certificate is the official and primary source of information on the cause of death. Underlying and supplemental causes are recorded on the death certificate, then coded according to underlying cause of death and recorded by state vital statistics agencies. A death certificate also contains other information on the deceased: gender; age; race; marital status; occupation; education; place of residence; and the date, time, and place of death (Fig. 3.4).

Birth Certificates Birth certificates can be utilized in studies of etiologic (risk) factors that adversely affect reproduction; information on the newborn might include gender, birth

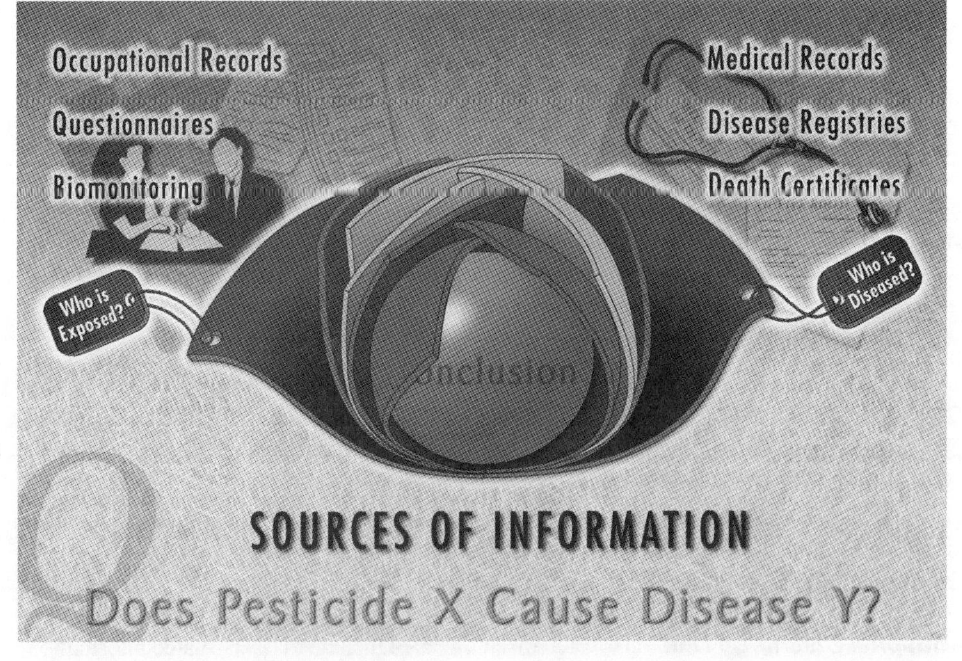

Figure 3.4 Epidemiologists utilize a variety of sources in determining exposures and diseases.

weight, and documentation of any malformations. The ages of the parents and the number of previous pregnancies of the mother often are noted as well.

Exposure Records

Under certain circumstances, epidemiologists can utilize actual measures of exposure as the basis for classifying the exposure status of study subjects. Exposure measurements in the workplace and personal exposure measurements for study subjects improve study validity and the overall interpretation of the study. For example, measurements of workplace surfaces frequently touched or an individual's breathing zone obviously are preferred bases for classifying workers' exposure.

Biomonitoring Scientists have begun estimating exposures through biomonitoring: measuring the pesticide or one of its components in blood, urine, or fat. Such measurements form a good basis for determining who is exposed and, more important, the degree of exposure. Compounds that are fat soluble remain measurable in the body for years, whereas compounds that are fat insoluble can be measured in blood or urine for only a short time after exposure.

Occupational Records Occupational records are another source of exposure data. The best occupational information comes from employment records maintained by individual companies. Sometimes these records are detailed and well documented, with a complete listing of an employee's work assignments, the beginning and ending dates of each, and even some personal exposure monitoring data. Frequent employment changes complicate the situation because the potential for exposure and the quality of employment records vary significantly among employers.

Other times, work records are quite poor or nonexistent. Epidemiological studies sometimes assume a decedent's occupation based on the death certificate, although it usually states only the decedent's *last* job. Therefore, the data entered do not necessarily represent objective documentation. Results based on occupational data from death certificates are not always valid.

Questionnaires Questionnaires may be distributed and completed in person, by telephone, or by mail. However, the resulting data must be interpreted cautiously because they are only as valid as the respondents' memory and state of mind at the time they complete the questionnaire. Undoubtedly, there are differences in circumstantial recall of various respondents; for example, diseased individuals may be more thorough in filling out a questionnaire and may have thought more about their exposure histories than those who are disease-free. It has been demonstrated that the care taken in questionnaire development and the circumstances of questionnaire administration affect the reliability of response interpretation. Data obtained from questionnaires can, however, provide valuable historical information.

Among the various types, person-to-person interviews provide the best response rate and quality of information, although they are more expensive to conduct and the results depend largely on the interviewer. Questionnaires completed by telephone or through the mail, while cheaper, often result in lower response rates—especially by mail. There are always questions about how respondents differ from those who fail to respond.

Another issue affecting questionnaire response is the sensitivity of questions asked. If questions deal with sensitive information, subjects may refuse to provide answers or may give answers that reflect their own bias. Sexual history, alcohol use, and drug abuse are examples of factors that might not be reported accurately on questionnaires unless special procedures are established to gain the respondents' trust.

EPIDEMIOLOGY AS AN OBSERVATIONAL SCIENCE

Epidemiological research often begins with a clearly formulated question and hypothesis:

- Question: Do commercial pesticide applicators who apply lawn care chemicals exhibit elevated rates of peripheral nerve damage?
- Hypothesis: Rates of peripheral nerve damage in pesticide applicators differ from rates in the general (unexposed) population.

The epidemiologist then must design a study to evaluate the hypothesis to determine whether the population-at-risk—in this case, lawn care applicators—has an elevated rate of peripheral nerve damage. The term *population-at-risk* is misleading because it is not meant to imply that a group of people is actually experiencing an increased risk of disease. Rather, population-at-risk defines the population in which a disease–exposure relationship can be studied. For example, the population-at-risk is commercial lawn care applicators, the condition of interest is peripheral nerve damage, and the contributing risk factor is hypothesized to be pesticide exposure.

Researchers using traditional scientific methods with experimental animals control the circumstances of a study; that is, they determine who will be exposed and who will not be exposed. Epidemiologists, however, cannot control exposure except when the exposure may have a beneficial effect; ethical considerations render it impossible to expose populations to potentially toxic substances to observe whether disease develops. Hence, detection of the potential toxic effects of pesticide exposure can only be accomplished through observational studies.

Although epidemiologists conduct observational studies under real-world conditions, which allow examination of a multitude of factors and interactions, often it is these real-world conditions which cloud the disease–exposure relationship. Results from an observational study are not necessarily evidence of a causal disease–exposure relationship, but they do indicate that exposure is associated with disease.

Causation and association are two distinctly different concepts of the relationship between exposure and effect. Causation indicates that there is sufficient, strong evidence for scientific consensus. Association means that a relationship has been reported, but that the evidence is not strong enough to effect consensus.

One challenge in observational research is the identification, within the study groups, of any important characteristics other than pesticide exposure that might contribute to disease. Epidemiologists must rule out confounding factors such as age, gender, and diet when attempting to link a disease with a specific exposure; without this accountability, association of the studied exposure cannot be verified. Epidemiologists must eliminate the role of all other factors in determining a valid association between test groups and incidence of disease.

STUDY DESIGNS IN EPIDEMIOLOGY

Cohort Design

Cohort studies begin with a group of people that share common characteristics—the cohort—and evaluate their health over an extended time period. A cohort might include Kansas wheat farmers, golf course superintendents, or certified pesticide applicators. The basic question addressed by a cohort design is this: Is the exposed population more or less likely to develop disease than the unexposed population?

A cohort design requires all subjects to be free of disease at the start of the study. All subjects are followed over time, and their individual exposures and diseases are documented. Ultimately, the cohort is separated, based on those who were exposed to a specific agent and those who were not. Disease occurrence is then analyzed to see if frequency varies between the exposed and unexposed groups (Fig. 3.5).

Cohort designs can be either retrospective or prospective. The difference is simply the timing of data collection: whether the study proceeds from a previous point in time (retrospective) or from the current time, forward (prospective).

Retrospective Cohort Design A retrospective cohort design focuses on a group exposed at some point in the past. The exposure point can be a documented historical event such as an explosion, a fire, a spill, or a date of employment. Once the retrospective date of initial exposure is determined, epidemiologists trace the study subjects' health status from that point in time to the end-of-study date. This type of study can be used on occupational groups such as employees of a pesticide manufacturing facility or those of a commercial pesticide application company.

Consider this example. An epidemiologist might decide, today, to study the mortality of all current and former workers at a pesticide manufacturing plant, who worked for at least 1 year between 1950 and 1995. The workers at the manufacturing facility comprise the retrospective cohort. The epidemiologist reviews work histories with knowledgeable experts in exposure information and with other available exposure information (e.g., air monitoring data at the worksite) on individuals in the cohort to determine the exposure status of each worker.

The epidemiologist conducts extensive research into each worker's history from 1950 (or from the date of first employment, if after 1950) through 1995 to determine whether any of the workers died and, if so, their cause of death. Death rates of exposed workers in the cohort are compared with those of unexposed workers; local, state, and/or national mortality rates are compared as well. Additionally, the epidemiologist may use rates of surrounding counties to ensure that the comparison population is most similar to the worker population.

Prospective Cohort Design A prospective cohort design focuses on a group of people from a current point in time through a future point in time. Consider the example of the Agricultural Health Study being conducted in Iowa and North Carolina by the National Cancer Institute and the National Institute of Environmental Health Sciences. The study involves farmers, commercial pesticide applicators, and their families; each individual filled out a baseline questionnaire, up front, and will complete subsequent questionnaires over the course of the study. Follow-up evaluations with each cooperating individual will be made every 5 years. Data on pesticide use and other risk factors is being collected in an attempt

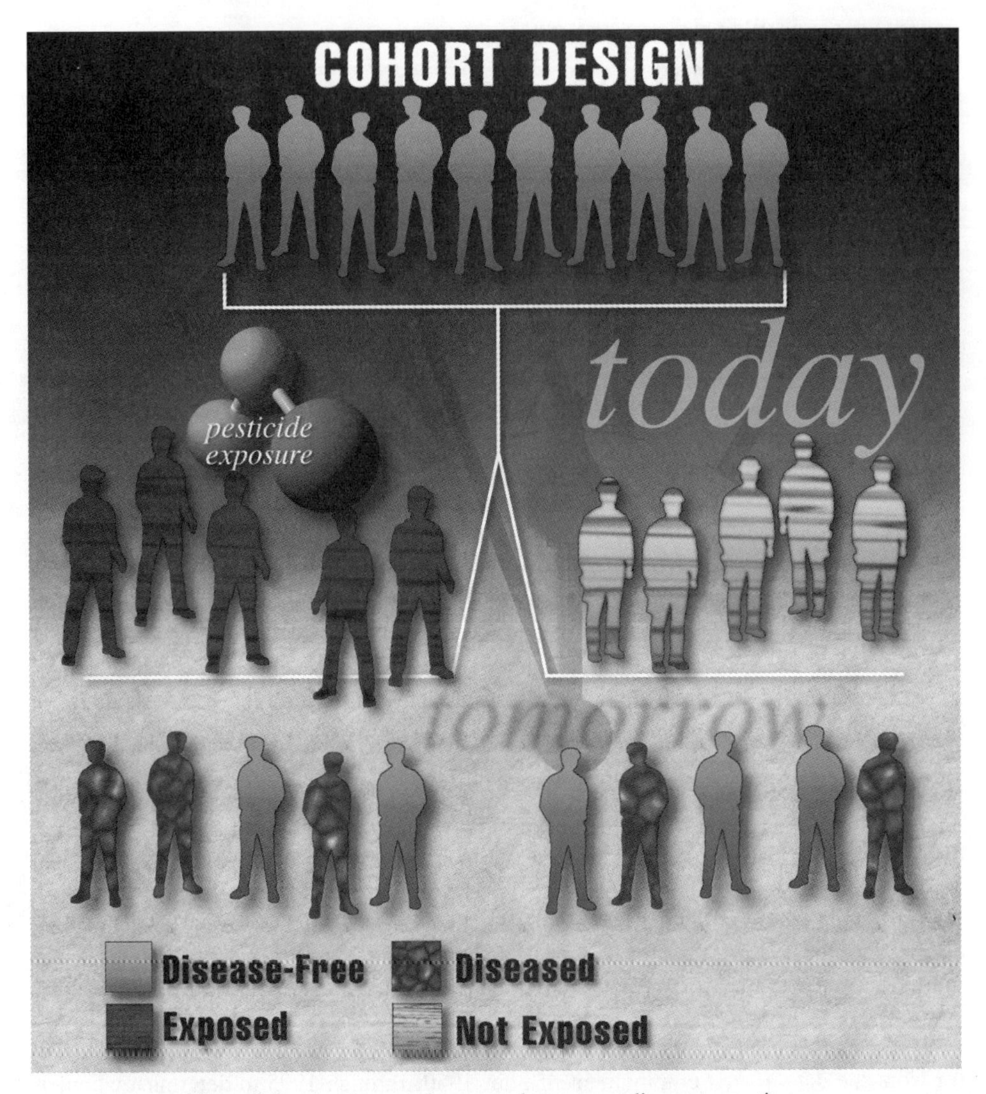

Figure 3.5 A cohort study can evaluate many diseases over time.

to relate exposure to disease. Cohort disease rates will be determined at regular intervals and compared to find those with and without exposure to specific pesticides.

Advantages A cohort study is the design of choice when studying what diseases may result from a specific exposure. The follow-up aspect of the cohort design provides very useful information on the interval of time between the first known exposure and disease detection: how long it takes for the disease to develop. Cohort studies are advantageous when the investigator wishes to evaluate a large number of diseases.

Limitations Cohort studies require the cooperation of large numbers of people for many years and are therefore expensive to conduct. A long-term commitment of resources and professional staff is required for the collection of accurate, useful information. Some diseases are so rare that a cohort of 100,000 people is needed to yield an adequate number of diagnosed cases to get precise information on exposure–disease correlations.

Case-Control Design

Case-control design often is used for studying a single disease. The basic question addressed by case-control design is this: Are individuals with diagnosed disease more (or less) likely to have been exposed than those without disease? The distinguishing characteristic of case-control studies is that subject selection is based on disease status. Cases are identified among disease reistries, hospitals' and physicians' records, and volunteers; ideally, disease-free members of the population that gave rise to the cases are selected as controls.

Exposure information is developed from existing records and/or from detailed questionnaires completed by the subjects. It is used to compare the frequency of exposure among cases and controls and to adjust, statistically, for other factors that may influence disease (Fig. 3.6).

An example of a case-control study is one that investigates the likelihood that children with brain cancers were exposed to pesticides used by their parents, inside or outside the home. Cancer registries are used for subject selection. Parents of cases and controls are interviewed on their use of pesticides in and around the home, and the frequency of exposure (based on parents' recall) is compared.

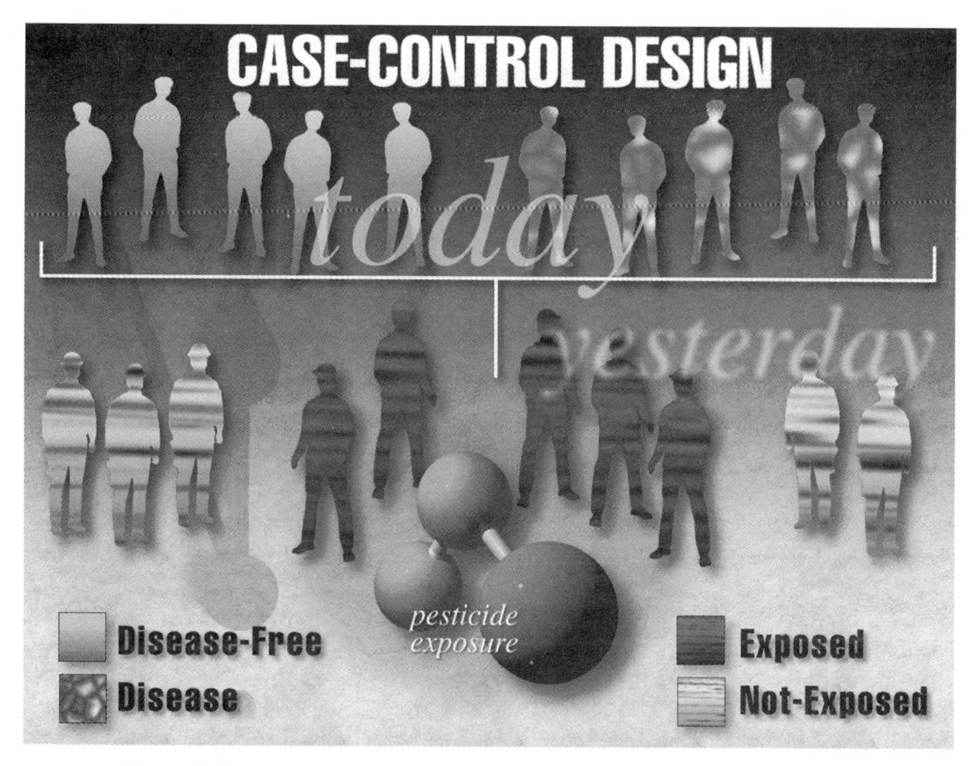

Figure 3.6 A case-control study can evaluate many exposures from the past.

Advantages Case-control designs are extremely useful in studying uncommon diseases and those that take many years to develop. Since they start with individuals who have the disease, fewer subjects (than in cohort studies) are involved, and since the studies can be completed in a relatively short period of time (months or years), expense is often much less than for the cohort study.

Limitations One major disadvantage to case-control design is that information on exposure is collected after disease diagnosis. Diseased individuals (or their next of kin) may remember exposures or events differently than those who remain healthy. They also may be more highly motivated to participate in a case-control study.

Additional Study Designs

Epidemiologists can generate disease and exposure information from study types other than the cohort and case-control designs. These study types—case reports, cross-sectional studies, and ecological designs—are best used to develop hypotheses for more rigorous testing via cohort and case-control studies (Fig. 3.7).

Case Reports A case report is simply a description of a patient's diagnosis and disease progression, often published in medical literature by physicians who sense a pattern or something unusual. Initial research on the correlation of pesticides to cancer stemmed from Swedish physicians' observance of a potential association between lymphoma (lymphoid tissue malignancy) and exposure to herbicides in patients diagnosed with the disease. Case reports provide no information on cause-and-effect, nor can they be extrapolated to larger populations. However, they are extremely useful in bringing forth observations, which alerts epidemiologists to the suspected relationship; this directs the focus of future studies toward specific disease–exposure relationships.

Cross-Sectional Study The cross-sectional study simultaneously examines exposure and disease; that is, the epidemiologist starts with a defined population and, for each member of the population, collects exposure and outcome information at (or from) a certain point in time. In an investigation of pet handlers and health complaints, for example, the handlers would be asked about their activities and products used. At the same time, they would be asked about a range of symptoms such as skin rashes and fatigue. The number of exposed workers with symptoms would be compared to the number of unexposed workers with the same symptoms. A critical problem with cross-sectional design is that the epidemiologist does not know whether the onset of disease (or symptoms) began before or after exposure. Also, symptoms may influence the subject to report exposure.

Ecological Design Unlike case reports where individuals are described, and unlike case-control and cohort studies in which data are collected on individuals, ecological studies examine exposure and disease patterns for groups or populations. Generally, they utilize data that have been collected for other purposes. Hypothetically, an ecological study might link data on mortality rates of non-Hodgkin's lymphoma (NHL) patients in certain Minnesota counties, as reported by the Minnesota Department of Health, to rates of herbicide use in the same counties as reported by the Minnesota Department of Agriculture. If high NHL death rates are recorded in counties where large quantities of herbicides have been applied, it is possible that exposure to herbicides is a causal factor. The problem with this design is that diseased individuals may not have been exposed to herbicides, a premise that cannot

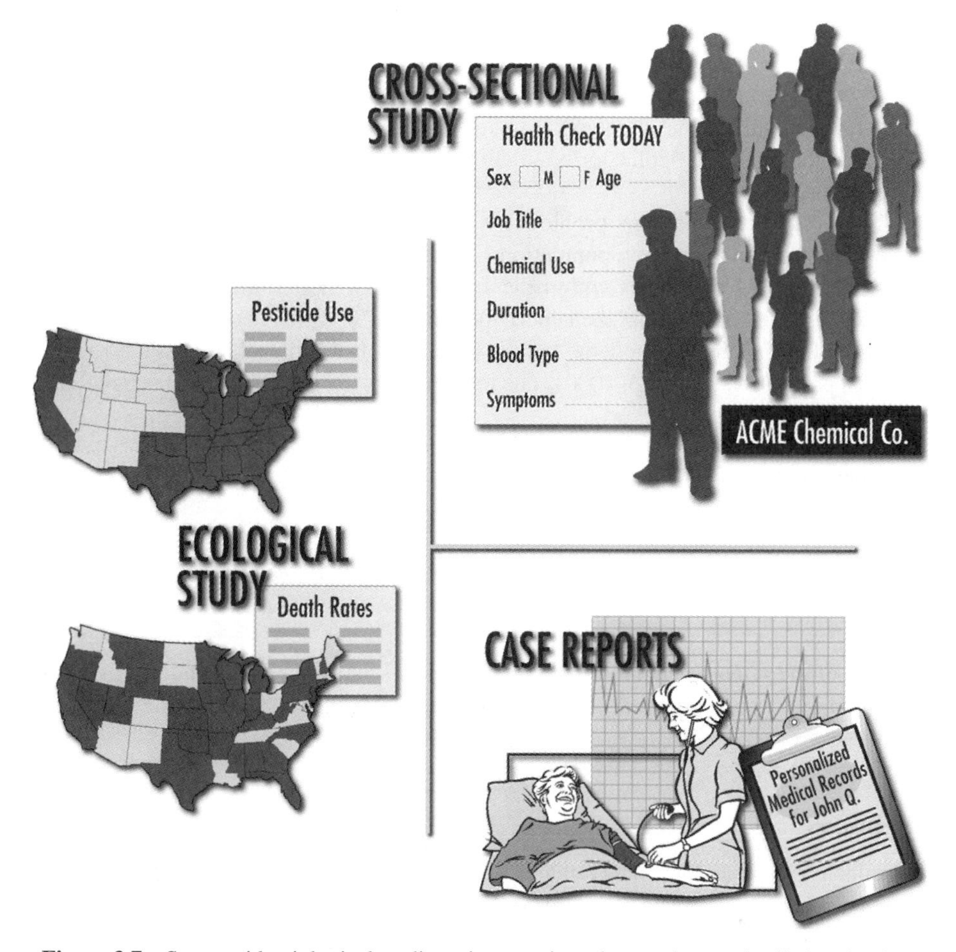

Figure 3.7 Some epidemiological studies raise questions that need more detailed evaluation.

be substantiated because the data are available only for the population, not for individuals within the population. Epidemiologists are reluctant to base conclusions on ecological data, but, nonetheless, such studies can foster research hypotheses for future case-control or cohort studies.

REPORTING EPIDEMIOLOGICAL DATA

The presentation of epidemiological research in scientific journals and research reports varies with study design and the types of information collected. Following are some common methods for summarizing and presenting data in scientific literature.

Disease Rates

A disease rate is a measurement of the frequency of a disease within a defined population, over a defined period of time. The frequency of disease is meaningless unless it can be defined with respect to the population involved and the time period of occurrence.

For instance, a 1-year study reports that 93 children with birth defects were among 3379 live births in a pesticide applicator population; in that same year, birth defects were detected in 1493 children out of 68,493 live births among the general population. Superficially, it might appear that the general population experiences more birth abnormalities (1493) than the pesticide applicator population (93), but raw number comparison distorts the proportion.

Epidemiologists address this problem by converting the number of cases found in a sample population to a common population size. For instance, the study recorded 93 birth defects among 3379 live births within the applicator population and 1493 birth defects among 68,493 live births within the general population. So it was necessary to convert the prevalence of birth defects in both the sample (applicator) and control populations to reflect a common population size (e.g., 1000). The raw numbers (93 and 1493) are used merely as factors in the equation.

$$\frac{\text{93 birth defects}}{\text{3379 live births}} \times 1000 = \begin{array}{l}\text{27.5 birth defects per 1000 live births for the}\\ \text{applicator group}\end{array}$$

$$\frac{\text{1493 birth defects}}{\text{68,493 live births}} \times 1000 = \begin{array}{l}\text{21.8 birth defects per 1000 live births among the}\\ \text{general population}\end{array}$$

The focus of the research should be to compare the 27.5 birth defects per 1000 live births among pesticide applicators with the 21.8 defects per 1000 live births among the general population.

Rate Ratios

Rate ratios are comparisons of two rates commonly used to measure disease associations between two populations. The two most commonly used rate ratios are the relative risk ratio (RR) and the odds ratio (OR).

Relative Risk Ratio A measure of association calculated for a cohort study is a ratio called the relative risk (RR). It is a comparison of disease rates among exposed versus unexposed persons. It is often written as follows:

$$\text{RR} = \frac{\begin{array}{l}\text{Number with disease in}\\ \text{exposed group}\end{array}}{\begin{array}{l}\text{Total number in exposed group,}\\ \text{times number of years followed}\end{array}} \div \frac{\begin{array}{l}\text{Number with disease in}\\ \text{unexposed group}\end{array}}{\begin{array}{l}\text{Total number in unexposed group,}\\ \text{times number of years followed}\end{array}}$$

An RR of 1.0 means that rates for a specific disease are the same for exposed and unexposed subjects. An RR greater than 1.0 indicates a higher disease rate for exposed versus unexposed subjects, thus implying a possible relationship between exposure and disease. An RR less than 1.0 indicates a reduced disease rate in exposed subjects, possibly indicating a protective (or beneficial) effect of that exposure.

Consider this example. An epidemiologist presents the results of a 20-year cohort study among 3500 company employees: applicators, business managers, and office staff. The incidence of lung cancer was of particular interest. The study identified 3000 employees from the cohort who were exposed to insecticides in their course of employment. The remaining 500 were assigned jobs that did not bring them into contact with insecticides. Of the 3000 exposed employees, 32 were diagnosed with lung cancer. Among the 500 employees who were not exposed to insecticides, 20 were diagnosed with lung cancer during the same 20-year period. An RR of 0.27 was calculated for the prospective study, which means that individuals in the group exposed to insecticides were less likely to develop lung cancer than those in the group not exposed.

$$\text{RR} = \frac{32}{3000 \times 20 \text{ years}} \div \frac{20}{500 \times 20 \text{ years}} = 0.27$$

Odds Ratio The measure of association in a case-control study is the odds ratio. It is the ratio of the odds of exposure in the diseased group (case) to the odds of exposure in the nondiseased group (control). The odds ratio is analogous to the relative risk under most circumstances.

Following is the formula for calculating the OR.

$$\text{Odds ratio} = \frac{\begin{array}{c}\text{Number of cases}\\ \text{with exposure}\end{array}}{\begin{array}{c}\text{Number of cases}\\ \text{without exposure}\end{array}} \div \frac{\begin{array}{c}\text{Number of controls}\\ \text{with exposure}\end{array}}{\begin{array}{c}\text{Number of controls}\\ \text{without exposure}\end{array}}$$

Consider this example. An epidemiologist conducts a case-control study of prostate cancer, identifying (from the state cancer registry) 500 men aged 65 or older who had been diagnosed during the preceding 2 years; 1500 cancer-free controls were selected from a local population registry. Willingness to participate in the study was confirmed with individuals in both groups. Interviews and questionnaires were used to collect pertinent background histories (e.g., socioeconomic information) and to document past exposures based on the recollection of each participant. A total of 420 cases and 315 controls provided pertinent information. Analyses of the data indicated that 45 cancer cases had been exposed, at some time, to herbicides; 20 of the controls likewise had been exposed, at some time, to herbicides. The epidemiologist calculated an odds ratio of 1.8.

$$\frac{45}{375} \div \frac{20}{295} = 1.8$$

What does an OR of 1.8 mean? As with the RR, the baseline for comparison is 1.0. An OR of 1.0, by analogy to the cohort study, implies that the rate of disease is equal in exposed and unexposed subjects. In this example, an OR of 1.8 is interpreted to mean that the prostate cancer rate was 80% higher for exposed subjects than for unexposed subjects. When the OR is greater than 1.0, it suggests an elevated disease rate among exposed subjects. Conversely, odds ratios less than 1.0 imply a reduced disease rate for exposed subjects; that is, the exposure may be protective.

Confidence Intervals

The confidence interval is a valuable statistic that communicates information on the preciseness of the odds ratio and the relative risk. The odds ratio and relative risk are single point estimates of the ratio of disease rates for exposed and unexposed populations, but they are vulnerable to statistical variability; that is, the true value could be higher or lower than the point estimate, due to sampling variability. The confidence interval is usually constructed to provide the theoretical upper and lower 95% probability limits for the calculated OR or RR.

Epidemiologists typically use a 95% confidence interval (CI). For example, a report indicates that the OR is 0.9 and the 95% CI is from 0.4 to 2.0. In this example, there is a 95% probability that the upper (2.0, an adverse effect) and lower (0.4, a protective effect) limits include the true estimate of risk in this population.

The confidence interval is very useful in judging the variability in the RR or OR. For instance, an epidemiologist reports an OR of 1.8 and a 95% confidence interval of 0.2 to 2.9 for the association between birth defect rates in the general population and commercial pesticide applicators. This is interpreted to mean that the rate of birth defects was elevated nearly twofold among exposed subjects. However, the 95% confidence interval also indicates that the estimate could lie between the lower interval (0.2) and the upper limit (2.9). Such a wide confidence interval would make the epidemiologist cautious about concluding that exposure is truly associated with a higher rate of disease (Fig. 3.8).

If all of the values of the 95% confidence interval are greater than 1.0 (the level of risk) it can be concluded that exposure is associated with an increased rate of disease. Similarly, if the upper and lower limits of the CI are less than 1.0, exposure is interpreted to be associated with a reduced rate of disease.

However, if the 95% confidence interval includes 1.0 (e.g., 0.2–2.9), the results probably are inconclusive. As such, even though the OR or RR may exceed 1.0, the confidence interval means that the data do not clearly support the conclusion of an disease–exposure association.

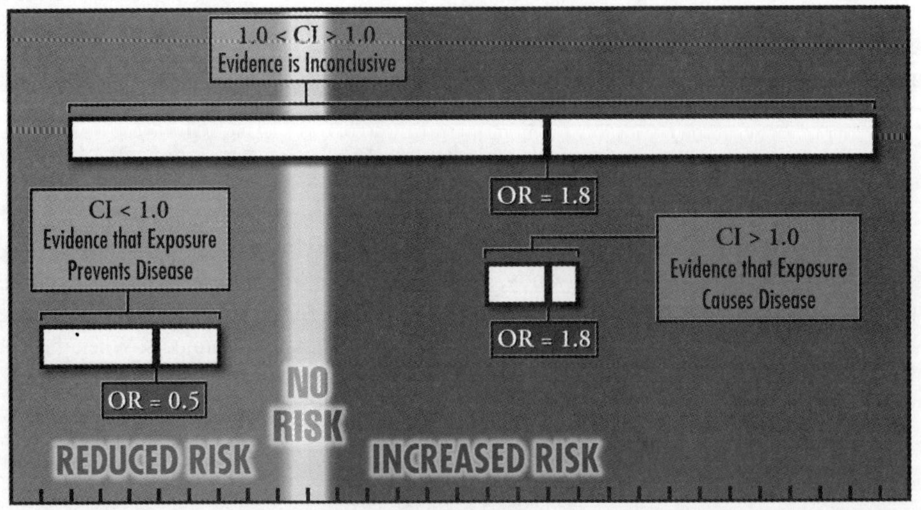

Figure 3.8 The confidence interval is very useful in judging the variability of risk estimates.

BIAS COMPLICATES STUDY RESULTS

The undoing of epidemiological research is bias, which is present to some extent in all human studies. Mistakes in planning, conduct, or analysis produce bias. Misclassification of individuals as exposed (or diseased) can result when subjects recall and report events differently based on their disease status. Bias also can occur when factors other than those being measured—so-called confounding factors—contribute to the disease and exposure. Bias can introduce error into information from which study conclusions are drawn. That is, the RR/OR and related 95% CI can be distorted by bias.

Possible sources of bias are noted in well-conducted epidemiological studies. Potential pitfalls must be addressed before the results of a study can be considered valid evidence of a causal relationship. Following are types of bias that need to be minimized in all epidemiological studies.

Selection Bias

Selection bias occurs when there are major differences between the characteristics of people selected for the study and the characteristics of those who are eligible but either not selected or not choosing to participate (Fig. 3.9). For instance:

- Since control subjects do not have the disease under study, they may be less motivated (than the cases) to participate. For instance, if only 50 out of 100 individuals who are eligible to participate as controls agree to do so, the 50 participants may not reflect the underlying population that gave rise to the selected cases.

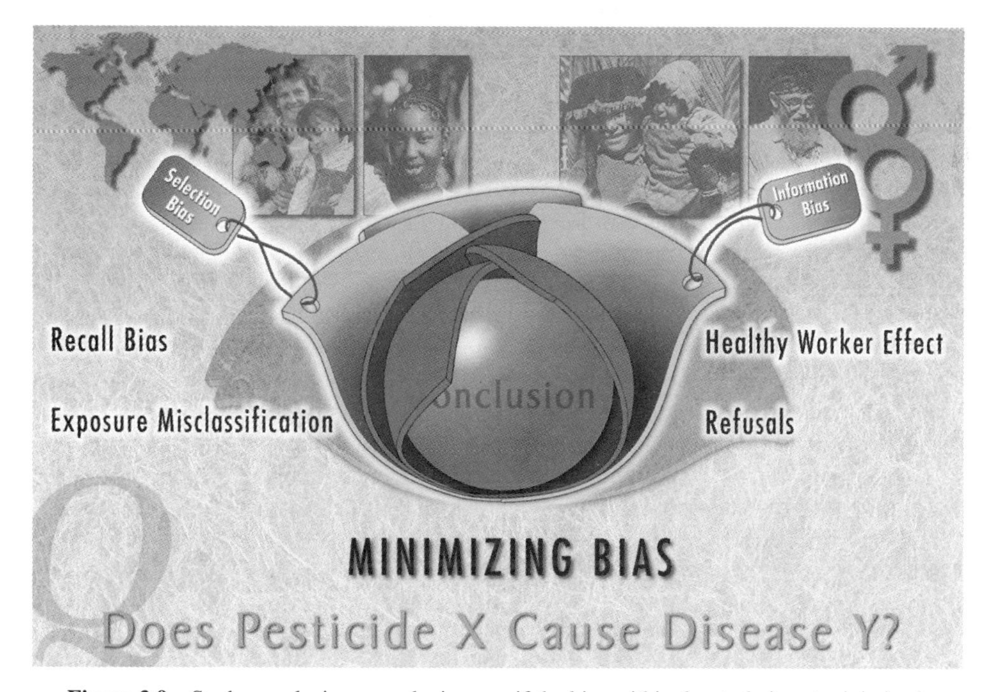

Figure 3.9 Study conclusions may be in error if the bias within the study is not minimized.

- Individuals with disease may be selected from a clinic where a high percentage of all patients treated have a significant characteristic or circumstance in common: migrant farmworkers, for example. Controls selected randomly by phone would represent a much broader socioeconomic range, thereby invalidating comparison for purposes of the study.
- Occupational studies exclude persons who are too sick to seek employment; therefore, mortality rates for workers are lower than rates for the general population—the so called "healthy worker" effect.
- A sample of right-of-way operators would not be representative of all operators if the sample contained only workers who were allowed to take time off to participate in the study.

Information Bias

Information bias refers to mistakes in obtaining the necessary information on study subjects, such as a person being classified incorrectly with respect to exposure or disease. Perception of symptoms can be highly variable among individuals; this is a problem when disease is self-reported, especially with subjective complaints such as headache, fatigue, and arthritis.

Self-reported exposure presents misclassification problems as well. Using herbicides as an example, certain subjects will claim exposure if herbicides were applied to their lawn by a lawn service company or if they walked on the sidewalk in a park where herbicide application signs were posted. Others would not consider either of these situations an actual exposure; in fact, some individuals report exposure only if they, personally, used the product. The truth may be that all, some, or none of the subjects were actually exposed (particularly if label-recommended protective clothing and equipment were used).

Recall Bias

Recall bias occurs when subjects remember past events differently; it is selective recall influenced by disease status. Recall bias is of particular concern because epidemiologists must depend on the accuracy of information provided by respondents. An example of such a problem is in the study of birth defects. It is common for mothers who deliver children with abnormalities (case mothers) to wonder about the cause of birth defects by reliving every aspect of their pregnancies. When interviewed, the case mothers often recall the type and amount of pesticides and the number of times they were used in their homes or gardens or on their pets. By contrast, mothers with healthy births (control mothers) generally do not remember such details because they have no incentive to remember such details. Mothers of unhealthy babies may report higher frequencies of all kinds of exposure, suggesting, in this hypothetical case, a false disease–exposure relationship.

Recall bias is likely to be greater if the case subject is not available to be interviewed, that is, when the epidemiologist must interview a proxy (e.g., a daughter answering questions on behalf of her deceased father) to access case information. An example of this is the reported use of the herbicide 2,4-D in a study of Nebraska farmers. Case subjects who responded for themselves did not indicate evidence of a relationship between 2,4-D use and the risk of non-Hodgkin's lymphoma, whereas analyses based on proxy responses showed evidence of a disease–exposure association. The quality of detailed information from proxies is variable, although their responses to the general questions is more reliable.

Confounding Bias

Confounding bias occurs when the association between exposure and disease is distorted due to related extraneous factors. Establishing true relationships between exposure and disease often requires the epidemiologist to consider personal factors such as the age, gender, ethnicity, education, marital status, occupation, social class, diet, smoking, and geographic location of each individual in the study. Age is the most important of the personal factors that influence disease. Incidence of chronic disease (e.g., cancer) generally increases with age. Accordingly, the effect of age must be accounted for when evaluating disease risks attributable to specific exposures; that is, the epidemiologist must remove the effect of the confounding personal factor to establish the true relationship between exposure and disease. Failure to account for personal factors can produce false associations ("finding" an association that does not exist) or obscure true causal associations (missing a relationship that actually exists).

PLACING SCIENTIFIC STUDIES IN PERSPECTIVE

Individual Study

Epidemiologists and other scientists communicate to the scientific community by exchanging project reports and by publishing their research in scientific journals. They frequently report on ongoing or recently completed research projects at professional conferences, government meetings, industry workshops, and public forums. Although all of these forms of communication are important, it is the publication of epidemiological research in scientific journals that is the most important, primarily because journal articles are peer-reviewed prior to publication. They are scrutinized by scientific experts on the subject matter, whose favorable opinions are required for acceptance into the journal. Scientists around the world accept published papers as contributions to their own research. Although publication in a scientific journal does not guarantee that the research results are valid, it usually does indicate that peer researchers judged the research methods to be sound or the topic to be interesting.

Understanding the research paper—the review process and the publication sections—and knowing how to judge it critically are of utmost importance in researching disease–exposure relationships.

Independent Peer Review Process The peer review process is similar for all researchers who present their findings for publication in scientific journals. The author submits a written manuscript to a journal editor; the editor, in turn, submits the publication to one or more scientists familiar with the subject matter. The reviewers examine the study: objectives, design, data acquisition, findings, and conclusions. These are blind reviews in that the author of the submitted publication does not know who is reviewing the manuscript. This process of independent and anonymous review is known as *independent peer review*.

The reviewers submit their recommendations to the editor, in writing. They can suggest that the manuscript be accepted for publication either as written or with minor or major revisions or that it be rejected. The editor sends to the author all written comments, remarks, and suggestions from each anonymous reviewer. A manuscript that is accepted with revisions must be resubmitted with suitable resolution of the changes requested by reviewers. Authors do have the prerogative to present arguments about why certain changes should not be made, and if the editor agrees with the author, he or she may overrule the reviewers.

This process, from the time the manuscript is submitted until it appears in press, often takes a year or more. Once published, the research paper is open to more scrutiny by a wider range of peers. It is then that the scientific community at large may review, criticize, and/or attempt to replicate the findings.

Layout of the Scientific Paper Author submission and journal publication of research findings follow specific guidelines established by journal editorial boards. Although there are obvious differences in style and format among journals, the basic information is quite similar. The following sections are found in most journal publications.

Title Titles of scientific papers are like titles of books: They need to convey the subject of the research.

Authors The individual who heads a study usually is listed first on the publication, as senior author. The authors' affiliations (e.g., university, foundation, government) and addresses are referenced as footnotes in the paper, generally indicating which author to contact for reprints or correspondence. Funding sources for the research usually are presented as footnotes as well.

Abstract The abstract is the summary of the purpose, methods, results, and conclusions of a study. The information in an abstract should never be used alone, without reading the article, because the abstract omits important details and qualifications that may be critical to proper interpretation of the study.

Introduction The introduction provides a brief synopsis of the more pertinent literature on the subject. The authors often cite scientific papers published in other journals, but on occasion they will make reference to other types of written materials: research theses and dissertations, manuscripts and reports, and personal communications. This information in the introduction is used to construct what is known and what questions have not previously been asked and to identify gaps in the research. The questions that have not yet been answered often form the rationale for the research undertaken.

Materials and Methods The materials and methods section is critical to any research paper. It tells how the subjects were selected, how the study was conducted, what measurements were taken, and how the data were analyzed.

Results The findings of the study are presented in the results section. The written text is normally augmented by data presented in tables, figures, graphs, and charts. Important information found in this section includes population characteristics and measures of disease association (OR and RR) and their statistical precision. The results section should contain actual findings only.

Discussion The discussion section in most journals is used by the scientist to interpret study results. Authors frequently use their data and those from published literature to offer their professional judgment relative to their findings. This is a very useful exercise because it provides a forum from which the scientific community advances new theories and new ideas. It offers suggestions for further research.

References Cited The references section provides the list of publications referenced in the text: author, title of the paper, journal and page number, and date. This is a very important part of any paper that merges past research with current; it directs readers to cited sources, facilitating their personal review of the references used by the authors to form conclusions.

Documenting a Study and Its Findings

The validity of study results requires accurate diagnosis of the disease, large enough populations of exposed and unexposed subjects so that meaningful differences can be isolated, assessment of the reported cause (etiologic agent), evidence of exposure (actual or estimated), and consideration of confounding factors and other biases.

It is important to understand that most epidemiological studies do not make causal inferences such as "this causes that." Instead, they find statistical associations that suggest causal relationships. A statistical association does not necessarily imply causation; rather, it means only that the study has found that one or more factors appear related to disease.

Conversely, the lack of a statistical association does not mean absolutely that a risk factor is not contributing to disease. Other factors not measured or accounted for may be masking or confounding researchers' ability to recognize a statistical association. A major question to ask about epidemiological studies is whether an appropriate study design was used for the questions, hypotheses, and conclusions drawn by the epidemiologist.

Consider the following points when reviewing an epidemiological study:

- What kind of study design was used? Look for the research plan and, if present, the rationale for choosing one method of study over another.
- Were the objectives stated clearly, and was the study true to its objectives? Clear objectives for conducting the study should be presented. Research discussions should focus on what the study attempted to measure.
- Were the underlying assumptions and limitations of study design presented?
- How were the subjects selected? The logic for selecting or rejecting individuals in any study should be articulated, describing the specified populations from which subjects were drawn and the methods used in their selection. It is the most important point that a study must address if the results are to have significant meaning.
- Were exposures and medical outcomes assessed using objective and reasonably accurate procedures?
- Did the study include an appropriate control or comparison group?
- Were the rationale and criteria for inclusion and exclusion of cases and controls presented?
- Were the rationale and criteria for disease ascertainment and exposure classification discussed?
- Did the study find differences between groups for the stated hypotheses? More confidence is placed in a study that finds differences in the prespecified set of questions or hypotheses proposed in the introduction, whereas less confidence is placed in unexpected findings. Confidence also is placed in findings that make clear biological sense and in those that have been replicated.

- May confounding variables explain the association? Scientists often ask whether the rise in one factor (exposure) which gives rise to an outcome (disease) is actually dependent on an unmeasured variable.
- Were some of the results inconsistent with the conclusions of the authors?

Weight of Evidence

A single study, except under extraordinary circumstances, cannot establish a cause-and-effect relationship. It is necessary to link all of the studies as pieces of a puzzle to see how they fit together. The ultimate causation judgment of the scientific community should be based on a weight-of-evidence approach (Fig. 3.10).

Hill's Criteria A widely used method for reviewing scientific evidence involves applying Bradford Hill's criteria for causality. Hill's criteria emphasize the necessary precedence of exposure to disease, the size of the risk estimate (RR or OR), and whether the disease rate or risks increase or decrease with increased exposure. Hill's criteria include discussions on the following points.

Biological Plausibility Biological plausibility is inherently judgmental and limited by our current knowledge of basic disease processes. Can a biological mechanism be shown to explain how a particular agent could have caused the disease? Are there biological explanations that link exposure pathways and disease pathology? Does the associated risk align with the study findings?

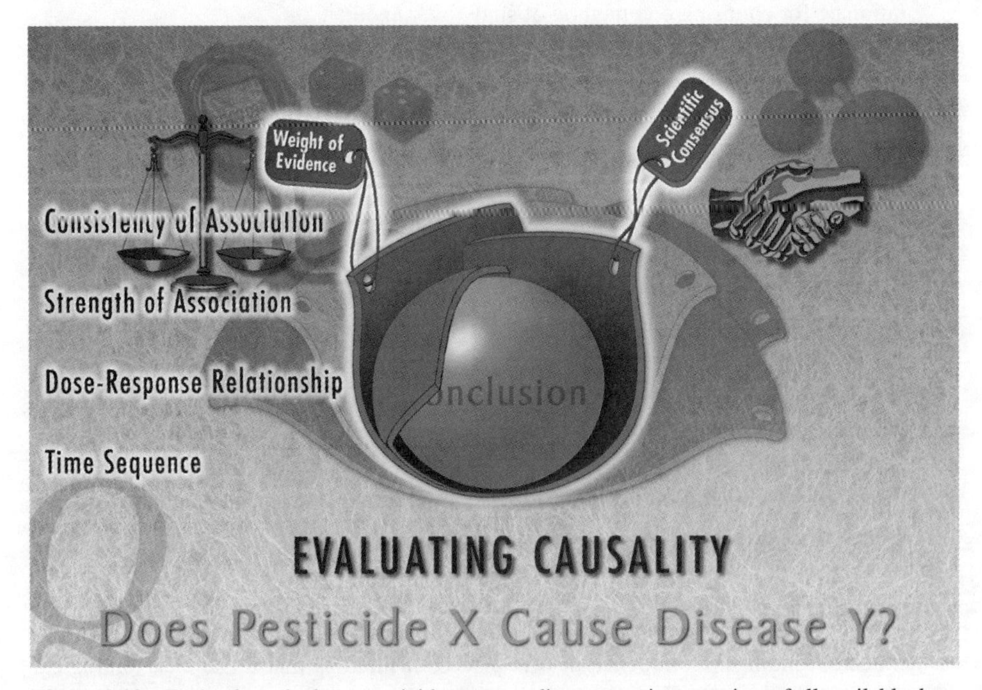

Figure 3.10 Evaluating whether a pesticide causes a disease requires a review of all available data.

Time Sequence The appropriate time relationship between first exposure and disease detection must be demonstrated. Each disease requires a certain length of time between environmental exposure and the manifestation of disease in humans. For example, associating a risk factor with cancer usually requires years or decades consistent with the length of time that most cancers require to become clinically evident.

Dose–Response Relationship As in toxicological assessment, a dose–response relationship is also an important component in Hill's criteria (See Chapter 2 for more information on dose–response). The RR should increase or decrease as the level of exposure increases to satisfy this criterion. Persons exposed at high levels should experience greater effects than those exposed at lower levels.

Strength of the Association The strength of association for each risk factor is a major consideration. For example, a relative risk of 4.0 is given more weight as a potential causal factor than a relative risk of 2.0 in a similar study. An odds ratio of 1.2 is less convincing than an odds ratio of 5.0, given equal statistical variability.

Consistency of the Association An association between exposure and disease generally needs to be demonstrated in several similar studies before epidemiologists begin to consider causality. Research errors and chance findings do happen, and history has shown the scientific community that it is easy to be misled by an apparently sound (but isolated) finding.

Statistical Association Statistical significance testing is a tool used to objectively evaluate the role of chance or sampling variation in the observed findings of the study. Conventionally, epidemiologists have considered results with a probability value (P value) of less than 0.05 to be statistically significant. In other words, under ideal circumstances, findings as extreme or more extreme than those observed in a study have less than 5 chances in 100 of occurring due to chance. Over the years, many scientists have used statistical significance as a decision rule for separating valid from invalid findings, but this practice has fallen into disfavor for two reasons. First, when there is bias in a study, significance probability calculations are misleading. Second, a disease–exposure relationship may be truly causal but not statistically significant, due merely to a small study population. Accordingly, the use of statistical significance should be viewed only in the context of the other strengths and weaknesses of the study.

Scientific Consensus

Consensus is often sought by governmental agencies, medical communities, and other public organizations that are considering reducing exposures or allocating funds for public education. Panels are assembled to discuss the available findings and to derive a consensus statement or conclusion. Such organizations frequently use Hill's criteria as a point of reference, but they may place more emphasis on the subjective opinions of committee members.

It is important to know that these opinions may or may not be reflective of the views of the broader scientific community. In many instances, the panel's findings may subsequently change scientific consensus. There are many examples where not all of the members of a consensus committee agree on major issues. This often leads to publishing not only a majority report but a minority report as well. Those in support of the minority report make their case concerning why they deem the majority report erroneous.

SUMMARY

Epidemiologists and the medical community are increasingly researching the potential human health effects of pesticides outside the laboratory setting. This is important because epidemiological studies can provide information that cannot be predicted from testing on nonhuman species. Studying the effects of human behavior, as well as multiple exposures under real-world conditions, adds to the toxicological evidence derived from laboratory studies.

The new information derived from these studies becomes available to the public in bits and pieces. It requires great care on the part of the media and others not to exaggerate preliminary findings from single studies. Epidemiological evidence needs to be viewed with a critical eye because of limitations inherent in any study. This holds true regardless of the completeness, accuracy, or perceived objectivity of the press or the investigators in the study.

Even experienced senior epidemiologists have difficulty interpreting some epidemiological findings. Those not trained in epidemiology face additional difficulty due to their incomplete understanding of the field, and that difficulty is compounded by the fact that news reports of epidemiological studies are abbreviated versions of scientific journal articles. Often, the news media misconstrue or overemphasize certain findings without mention of the authors' own scientific disclaimers. This is why importance is placed on obtaining the original published work to review for yourself the evidence presented in the publication (Fig. 3.11).

Figure 3.11 Does pesticide x cause disease y? can be answered only by addressing the many factors involved in disease formation.

The nature of epidemiology as a science lends itself to the difficulties of studying the health effects of pesticides. Indeed, each human is unique, and their behavior is often unpredictable. Studying groups of people to determine if exposure to pesticides causes disease is a challenging task. Nonetheless, the science of epidemiology has contributed significantly both to the understanding of human health risks from exposure to pesticides and to the validation of risk assessments based upon animal data and population assumptions. Studies that have shown a disease–exposure relationship have led to some products being replaced by safer ones. Others have been used to set exposure guidelines for manufacturers or professional applicators. Studies showing no adverse health relationships have increased our understanding of the potential benefit of pesticides and have directed research toward other possible causes of disease. Collectively, the goal is to reduce risk associated with human exposure to pesticides and to maximize benefits from their use. Epidemiology plays a major part in this goal.

CHAPTER 4

ECOLOGICAL RISK ASSESSMENTS: EVALUATING PESTICIDE RISKS TO NONTARGET SPECIES

The diversity of wildlife habitats throughout our country is surprising. In rural areas, fence rows, fields, pastures, ponds, wetlands, and woodlands host many species of wildlife. Wild species also inhabit urban landscapes, finding food and shelter in lawns, cemeteries, golf courses, parks, and homes. Collectively, these habitats—urban, rural, public forest, and pristine areas—provide resources to support plants (flora) and animals (fauna) that are ecologically important and that have a high social value (Fig. 4.1).

Many wildlife species share habitat with humans and are influenced by human activity. Wildlife living adjacent to farmland may benefit from the crops grown but also may be inadvertently exposed to pesticides used to control insects, weeds, and diseases. Urban expansion for new housing, manufacturing facilities, and other activities not only eliminates valuable habitat but also may bring wildlife into contact with pesticides used on turf, ornamental and landscape plantings, gardens, highway rights-of-way, parks, and rodent and mosquito abatement programs.

The EPA evaluates the potential for adverse effects on wildlife and other components of the environment through the ecological risk assessment process. The risk assessment process is primarily a tool for organizing information to answer assessment questions defined at the beginning of the process by risk assessors and risk managers. The risk assessment process produces an evaluation of the likelihood and/or magnitude of risk, which is used along with information on societal benefits in environmental decision making (in this case, pesticide registration decisions).

BENEFITS OF WILDLIFE

Watching wildlife in natural settings appeals to persons of all ages and all ethnic, educational, and social backgrounds. Eighty-five percent of Americans participate in some wildlife observation activity (e.g., whale, bird, or butterfly watching). Ecotourism has become a lucrative market; many businesses and communities actively advertise and attract tourists

Figure 4.1 The bobwhite quail is a upland game bird typically found in agricultural settings.

to observe wildlife. Wildlife viewing tours organized by professional wildlife biologists are in demand, and tourism is supported by a cadre of jobs related to lodging, meals, transportation, art, equipment, and publications. Taxes dedicated from the sale of hunting and fishing licenses and certain outdoor equipment provide revenue to purchase, maintain, and restore wildlife habitats, while monetary support from some government agencies is dwindling. Private organizations (e.g., The Nature Conservancy) and private businesses also purchase, donate, and manage wildlife areas. These user-based taxes also help finance scientific research on wildlife communities. It is important that we as a society do all we can to maintain and preserve the natural world and continue to benefit from its existence (Fig. 4.2).

Figure 4.2 Butterflies have important ecological and aesthetic value.

Figure 4.3 Wild rice is an important food source for Native Americans.

Fewer than 20 plant species are responsible for feeding most of the world's population. However, it is estimated that 80,000 species of edible plants (e.g., wild rice; Fig. 4.3) may have potential as new food sources, and wild plant and animal species may become new sources of genetic material with agricultural or medical applications. For example, the purple coneflower, a native prairie plant, has an oilseed content greater than that of commercially grown sunflowers; South American corn varieties have been used to breed resistance to northern corn leaf blight into North American corn hybrids. Plants such as the Pacific yew, which produces taxol, also have been shown to possess an array of pharmaceutical chemicals whose medicinal properties may be used to combat disease. Wildlife species also play a role as natural enemies of pest species (e.g., birds eat mosquitoes; snakes consume rodents).

Many of these relationships are known, but many are yet to be discovered. Therefore we must ensure that the vast array of plants and animals on our planet is maintained for future generations; that is, we must maintain biodiversity. It is likely that important connections exist among species diversity, environmental quality, and the long-term sustainability—and profitability—of farming operations.

Figure 4.4 The impact of a pesticide on wildlife requires a full understanding of animal and plant biology.

PESTICIDE IMPACT DEPENDS ON WILDLIFE ECOLOGY

The term *wildlife* as used here includes insects, spiders, mammals, birds, fish, amphibians, reptiles, and plants. Each species has specific food, cover, water, space, and breeding site preferences. The location where a species can meet all of its living requirements becomes that species' habitat. Wildlife habitats are not just the Grand Canyon, ancient forests of the Pacific Northwest, or rich coastal marshes on the eastern seaboard; habitats exist across the American landscape. Wildlife habitats—large and small, native and man-made—exist in urban settings, in agricultural fields, and in the wilderness.

Wildlife ecologists and natural resource managers study the needs and habits of wildlife. An important goal of wildlife research is to discover and understand the critical factors that affect survival and sustainability of viable populations. Most wildlife will adapt and flourish, given sufficient and suitable habitat, even in the presence of human activity. Although ecological studies may pinpoint very specific requirements for individual species, the lives of plants and animals can be integrated with their habitats, collectively, into a matrix (ecosystem) (Fig. 4.4).

Knowledge of the biological and ecological relationships of any given plant or animal and the role that that species plays in the ecosystem is required to evaluate the potential impact of a *specific* pesticide on a *specific* plant or animal species. The impact of a specific pesticide may be negative, neutral, or positive to a species or its habitat. The interaction of pesticides and wildlife in their habitat is evaluated by scientists trained in wildlife ecology, population dynamics, physiology, environmental chemistry, and environmental toxicology (Fig. 4.5).

Assessing and characterizing ecological risk, including a myriad of nontarget aquatic and terrestrial organisms, populations, communities, and ecosystems, is a much younger and more complex science than the assessment of risk to human health. Ecological risk assessment considers a greater range of complex issues and covers more species than does human health risk assessment.

Each species within an ecosystem fulfills specific ecological roles. Plants are the primary producers of chemical energy in any terrestrial or aquatic ecosystem. They capture

Figure 4.5 Studying aquatic vegetation in a wetland.

sunlight and convert it to energize new plant growth, forming the bottom of the food chain. Energy flows through the food chain when organisms consume plant tissues and are, in turn, consumed. For instance, green algae are one-celled microscopic organisms that are a staple food item for invertebrates such as water fleas and mysid shrimp; these invertebrates, in turn, become food for young fish and small fish species. The fish then are consumed by predators such as larger fish, amphibians, birds, and aquatic mammals. Because of the dynamics of the flow of energy, perturbations (disturbances) of the most seemingly minor species may lead to observable (measurable) changes in the entire ecosystem. However, because of the ability of organisms and populations to adapt to perturbations—because they are resilient—effects on one or more components of an ecosystem may result in minimal ecological change. That is, structures (species composition) may change, but the function (interaction) of the system remains stable. Alternately, if only resilient species survive the perturbation, species diversity can be reduced. As noted previously, maintaining biodiversity is key to maintaining the overall quality of the environment.

Adverse environmental effects at various levels may involve more than energy flow. For instance, adult mussels are nearly sedentary at the bottom of moderately flowing streams, but they filter algae and other small organisms from the water. Young mussels of some species attach to the gills or fins of certain fish, where they remain as harmless parasites (for weeks or months) until their internal organs develop; then they drop from their host into the stream bank substrate. Without the host species, some mussel populations cannot survive (Fig. 4.6).

PESTICIDE POISONING OF WILDLIFE

Pesticides are applied in many forms, via various delivery methods, to forests, rangeland, aquatic habitats, farmland, rights-of-way, urban turf, and gardens. Their widespread use makes contact with pesticide residues inevitable for some wildlife. Pesticide poisonings to wildlife may result from *acute* (short-term) or *chronic* (long-term) exposure. Additionally, pesticides may impact wildlife via *secondary exposure* or through *indirect effects* on its food source or habitat.

Figure 4.6 Many mussels living at the bottoms of streams are on various state and federal endangered species lists.

Not all pesticides have detrimental effects on all wildlife, nor do pesticide residues necessarily have negative consequences for wildlife. The potential impact must be evaluated by simultaneously considering the availability of the pesticide or its degradation product(s), the toxicological properties of the pesticide, and the ecological characteristics of the exposure. Due to the complexity of these issues, many scientific disciplines must play a role in both the studies and the interpretation of results. The results from scientific studies aid numerous federal and state natural resource agencies to assess and manage the effects of pesticides on wildlife, including endangered species.

The degree of direct impact that a pesticide has on wildlife is determined by the sensitivity of the species to the chemical and by the degree of exposure. The following questions help to summarize the complexity that biologists and toxicologists face when attempting to evaluate pesticidal effects on wildlife.

- What level of a pesticide residue or its degradation product (metabolite) is introduced into a wildlife habitat through direct application or via the transportation of residues in air, water, food, or soil?
- How long does the pesticide remain in the environment?
- Should acute or chronic exposure be considered?
- By what route is the animal or plant exposed to the pesticide (e.g., direct contact, inhalation, or consumption of contaminated food or water)?
- Is the pesticide capable of producing biochemical effects, illness, or death through either single or multiple exposures?

Acute Poisoning

Short exposure to high concentrations of some pesticides may kill or sicken wildlife. Examples of acute wildlife poisoning include fish kills caused by pesticide residues carried to ponds, streams, or rivers by surface runoff or spray drift; bird die-offs caused by foraging on pesticide-treated vegetation or insects or by consumption of pesticide-treated granules, baits, or seeds are another example. These types of poisonings generally can be substantiated by analyzing tissues of affected animals for the suspected pesticide or by investigating impacts on biochemical processes (e.g., cholinesterase levels in blood and brain tissue). In general, acute poisoning to wildlife takes place over a relatively short time, impacts a very localized geographical area, and is linked to a single pesticide.

Chronic Poisoning

Exposure of wildlife over an extended period of time to low levels pesticides, not immediately lethal, may result in chronic poisoning. The most well-known example of a chronic effect in wildlife is that of the organochlorine insecticide DDT (via the metabolite DDE) on reproduction in certain birds of prey. DDT and other organochlorine pesticides such as dieldrin, endrin, and chlordane have been implicated in bird mortality resulting from chronic exposure. The decrease in use of these compounds in the 1970s and early 1980s resulted in decreased organochlorine residues in most areas, and reproduction in birds such as the bald eagle has greatly improved. Organochlorine pesticides used in some foreign countries may pose risk to migratory birds which overwinter there.

Secondary Poisoning

Pesticides may impact wildlife through secondary poisoning when an animal consumes prey species that contain pesticide residues. Examples of secondary poisoning are (1) birds of prey becoming intoxicated after feeding on an animal that is dead or dying from acute exposure to a pesticide, and (2) the accumulation and movement of persistent chemicals through the wildlife food chain.

Indirect Effects

Residues from pesticides used in forests and aquatic habitats, and on rangeland or farmland, inevitably make contact with wildlife. Early research studied whether pesticides result in increased mortality, cause secondary poisoning, or impair wildlife breeding. But today's research is attempting to measure whether pesticides impact wildlife in more subtle, indirect ways. The following are today's questions:

- Do herbicides affect wildlife adversely by reducing their food supplies, cover, and nesting sites?
- Do insecticides diminish insect populations on which bird or fish species feed?
- Do insecticides reduce pollinator (bee) populations, resulting in reductions in food for seed-eating birds?

These are difficult questions. Researchers must have an in-depth understanding of the biology of the animal species in question and must know the extent to which pesticides interact with those organisms. Today there is an equal emphasis on ecology and toxicology in ecotoxicology.

A review of indirect pesticidal effects on birds breeding on English farms was presented by the Joint Nature Conservation Committee. The committee tried to determine whether or not there is published evidence of indirect, adverse pesticidal effects on bird populations on farms. To answer this question, the researchers had to glean from published literature a full review of the ecology of farmland birds (population trends, diets, food items) and correlate that information to pesticide use trends.

The population trends of 40 bird species were examined over three decades. Census data indicated that populations of two dozen species breeding on farms were declining—some by as much as 70%. Populations of the remaining 16 species either remained stable or increased.

The major sources of food for the birds under study were seeds and invertebrates. Population trends for invertebrate and plant species indicated that both diversity and abundance were declining on farmland. Other published studies have indicated that pesticide use can result in short-term reductions in food abundance; thus, some bird species may be impacted if reductions occur during a time when adult birds are feeding their chicks.

Sufficient documentation on when populations began to decline was available for only 12 bird species. The decline of 11 species coincided with increased use of pesticides on English farms, somewhere between 1974 and 1985. However, there was direct evidence only that gray partridge chick survival was impacted, indirectly, by herbicides and insecticides, and that reduced pesticide use resulted in higher survival of the chicks.

This report clearly demonstrates the difficulty in linking indirect pesticide use to adverse effects. The main difficulty is the lack of detailed, long-term biological studies on birds, insects, and plants. Many of the conclusions on indirect effects are based on circumstantial evidence; for instance, 11 bird species experienced population decline as pesticide use increased. But scientists would have to rule out other agricultural factors such as crop rotation, acreage, and tillage practices, which also impact bird species, in order to conclude that pesticide use was the contributing factor. Predation, competition, parasitism, disease, and loss of habitat also may impinge on bird survival. Thus, the process of linking adverse effects on wildlife indirectly with pesticide use is fraught with scientific complexity. To find concrete answers, wildlife biologists and agricultural researchers must work together to develop new research to quantify the indirect effects of pesticides on wildlife.

ECOLOGICAL RISK ASSESSMENT PROCESS

Before a risk assessment is conducted, there should be a clear understanding of the objective and the expected outcome. The initial stage of the process is called problem formulation, wherein the purpose of the assessment is articulated, the problem is defined, and a plan for analyzing and characterizing risk is determined. The problem formulation process begins with the initial stages of characterizing exposure and effects by examining the type of pesticide (What chemical class? How quickly does it degrade? What is its mode of action?), by asking where it is going to be used (What crops? What geographical regions? What species may be exposed?), and by noting the types of ecological effects expected or observed. Problem formulation results in three products: (1) assessment end points that adequately reflect management goals and the ecosystems they represent, (2) conceptual models that describe the key relationships between a pesticide and the assessment end points, and (3) an analysis plan for evaluating toxicity and exposure information.

The ecological risk assessor working with disciplinary experts, risk managers, and an interested public has the responsibility of properly framing the problem and translating it into a conceptual model for evaluation. For instance, a risk assessment designed to investigate acute lethal effects is of little use if the question is really where low-level, long-term exposure impacts population fecundity. A risk assessment can fail in the absence of adequately defined questions and a conceptualized approach to address those questions.

Ecological risk assessment is a process whereby toxicity (effects data) and exposure estimates (environmental concentrations) are evaluated for the likelihood that the intended use of a pesticide will adversely affect terrestrial and aquatic wildlife, plants, and other organisms. It is a process of organizing information toward characterization of the risks. Data required to conduct an ecological risk assessment include the following:

- Toxicity to wildlife, aquatic organisms, plants, and nontarget insects
- Environmental fate
- Environmental transport
- Estimated environmental concentrations
- Where and how the pesticide will be used
- What animals and plants may be exposed
- Climatological, meteorological, and soil information

Risk Assessment: The Multistep Process

Initial ecological risk assessments often are preliminary in nature and may be based on limited data and/or very conservative assumptions. As more research data are compiled and more accurate assumptions are considered, the more precise and comprehensive the risk assessment—and the greater the confidence in conclusions drawn. However, if initial risk assessments indicate no cause for concern, a more refined risk assessment may not be necessary.

Quantitative risk assessment is a function of toxicity and exposure. The risk assessment process involves multiple steps: a formulation of the problem to be addressed, followed by an appraisal of toxicity and exposure, and concluding with a characterization of risk.

Toxicity Assessment Toxicity testing identifies concentrations that, when administered to surrogate animals or plants, result in a measurable adverse biological response. These measured concentrations and associated toxicity end points basically describe what the chemical does to the environment—in this case, a single living organism (Fig. 4.7).

Toxicological assessment is commonly based on laboratory studies; that is, it reflects adverse effects observed when animals are intentionally administered a range of concentrations of the pesticide being studied. Toxicity can be characterized by mortality or by sublethal effects within the range of doses tested.

An important aspect of toxicological evaluation is determination of the relationship between magnitude of exposure and extent and severity of observed effects—commonly referred to as *dose–response*. The dose–response relationship identifies dose levels at which adverse effects occur as well as the no observed effect concentration (NOEC). For a preliminary risk assessment, the lowest NOEC, LD_{50}, etc., is used to estimate risk. See Chapter 2 for more information on dose–response.

Exposure Assessment Contact with a chemical in the environment constitutes exposure. Exposure concentrations may be either estimated or measured, based on the data from laboratory and field experiments. Nontarget exposure assessments ascertain the exposure

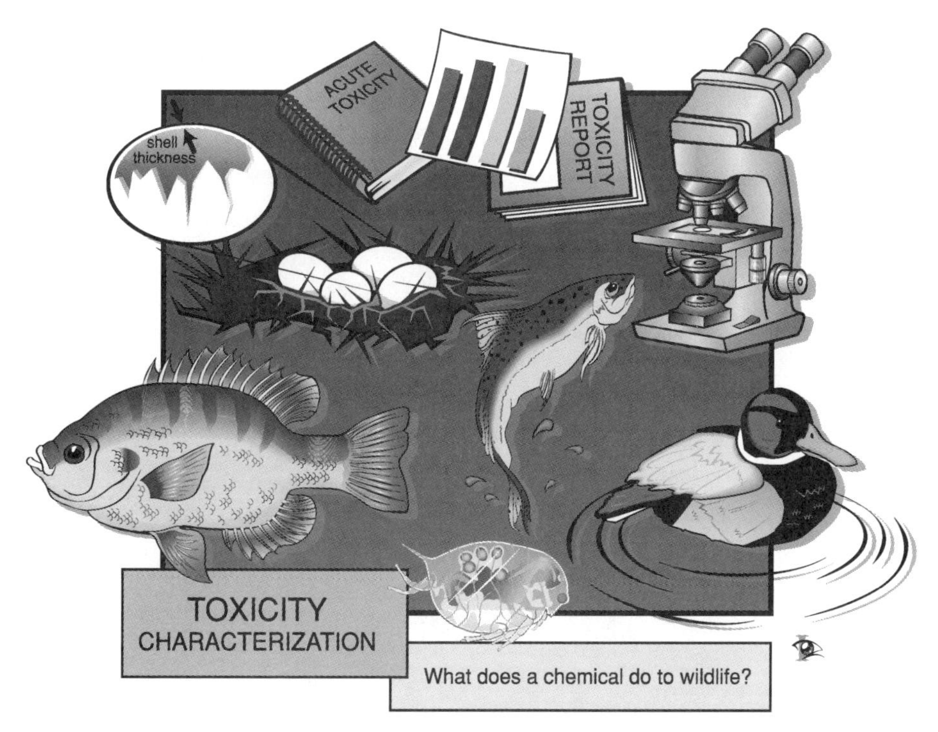

Figure 4.7 Ecological toxicity studies measure the effects of the chemical on wildlife.

of wildlife and plants to pesticides in the environment. The extent of exposure depends on the type of use (crop, lawn, and garden treatment; mosquito control; indoor pest management); application rate; method of application; frequency of application; and the breakdown, movement, and partitioning of the chemical in the environment (Fig. 4.8).

Risk Characterization Risk characterization defines the likelihood that wildlife will be exposed to hazardous concentrations. Thus, risk characterization describes the relationship between exposure and toxicity (Fig. 4.9). Risk assessors identify species likely to be exposed, the probability of exposure, and effects that might be expected. An adverse effect is predicted only if exposure approaches or exceeds dose levels that have resulted in adverse effects in previous toxicology studies.

Suppose that a sampling for a given pesticide in the environment yields an estimated exposure level of 3 parts per billion (ppb) in water, and that a short-term laboratory study shows that an exposure level of 100 ppb produces an adverse effect in bluegill sunfish. How could this information be integrated to predict the outcome of fish exposed to the chemical?

In this particular example, it is understood that fish may be exposed to 3 ppb, but adverse effects do occur in bluegill exposed, short-term, to 100 ppb. A risk assessor might express no concern for fish at an exposure level of 3 ppb since it is significantly below the 100 ppb threshold for injury. But risk characterization often is not that simple. The bluegill is merely a surrogate for "all" fish species; therefore, a risk assessment may apply a safety factor (for example, 10) to the bluegill toxicity value to protect species that may be more sensitive than the bluegill. In this case, 10 ppb is still greater than expected exposure, but it provides a margin of safety for untested species that may be more sensitive than bluegill.

Figure 4.8 Exposure characterization must address ways that a pesticide can be distributed in the environment.

The risk assessor may need to consider whether prolonged exposure of the fish, at a level of 3 ppb or lower, might trigger adverse effects, or whether another life stage (e.g., embryo) might be more vulnerable than the adult. Another consideration is that organisms (predators) higher in the food chain might be at risk if the pesticide accumulates in fish on which they prey.

Figure 4.9 The toxicity of the product and the amount of the exposure are considered in characterizing the risks that pesticides pose to wildlife.

TABLE 4.1 Estimated Time and Costs for Meeting the Toxicity Testi

	Time	
Acute Fresh Water Aquatic Toxicity (3 species)	5 months	
Acute Marine Aquatic Toxicity (3 species)	4 months	
Chronic/Reproductive Fish	6–12 mont[illegible]	
Avian Acute Oral LD_{50}	4 months	
Avian Dietary LC_{50} (2 species)	4 months	$24,[illegible]
Chronic/Reproductive Avian (2 species)	15 months	$190,000
Acute Honeybee	3 months	$4,000
Aquatic and Terrestrial Plants	9 months	$70,000

STUDY DESIGNS FOR TOXICOLOGICAL TESTING OF PESTICIDES IN THE UNITED STATES

Specific terrestrial and aquatic tests are mandated by federal law and described in 40 CFR Part 158 (http://www.epa.gov/pesticides/cfr.htm.). The wildlife tests required for registration of a pesticide (Table 4.1) are clearly described in 40 CFR:

- Part 158.490, wildlife and aquatic organisms data requirements
- Part 158.540, plant protection data requirements
- Part 158.590, nontarget insect data requirements

Some tests are not required for every pesticide product but are listed in the regulations as *conditionally required.* Conditionally required tests are not mandated unless the use pattern or information generated from required tests indicates a need for additional testing (See Chapter 6 for more details on the registration process). For instance, early life stage fish tests or invertebrate life cycle tests are conditionally required for most pesticide products. These tests become mandatory when scientific data indicate that the pesticide is relatively persistent (stable) in the aquatic environment and that concentrations at or below 1 part per million (ppm) produce significant acute mortality.

The technical grade of an active ingredient normally is the substance used in pesticide screening tests. Federal regulations specify the conditions under which the end use product (the formulation) must be tested. In all cases, effects observed and measurements recorded for animals and plants exposed to pesticides must be compared to control organisms (fish, invertebrates, birds, plants, or animals) held under identical conditions but not exposed to the chemical.

Testing surrogate species allows scientists to observe adverse effects resulting from acute (short-term) and chronic (long-term) exposure. Exposure for short periods at high concentrations is used to determine concentration levels necessary to produce mortality—lethal doses—and sublethal effects on one stage of an organism (e.g., juvenile or adult).

The more complex, chronic exposure tests are used to examine, in detail, how a pesticide impacts an animal at various stages of its life cycle (e.g., bird and fish embryonic development). As noted previously, chronic data are required when acute tests indicate that the toxicity of the pesticide exceeds a trigger limit and when the environmental fate or use of the pesticide indicates that nontarget organisms may be impacted. The data from chronic tests are essential to the accurate prediction of long-term effects on fish and wildlife.

Impacts Modeled by Indicator Species

Surrogate organisms used in toxicological testing are selected to represent various trophic levels within an ecosystem. For instance, adverse effect data on bobwhite quail exposed to a pesticide are used to generalize how that pesticide might adversely affect all upland game birds. *Daphnia* (the water flea) models fresh water crustaceans, and the Eastern oyster models fresh water and marine mollusks. An assumption key to ecological risk assessment is that data on surrogate species adequately predict how a pesticide will impact the broader spectrum of plants and animals.

It is impossible, inadvisable, and illegal to test every species—abundant, threatened, or endangered—with every pesticide. In the regulatory testing process, the test species selected are intended to broadly represent nontarget organisms. Chosen wildlife species typically satisfy the following criteria:

- Ecologically significant
- Abundant and broadly distributed, geographically
- Susceptible to chemical exposure
- Commercially available for testing
- Easy to handle in the laboratory
- A relatively short life span for life cycle tests
- Aesthetically, recreationally, or commercially important

These indicator species provide the research scientist and the regulatory decision maker with an information base for assessing the potential risks to a broad range of nontarget birds, mammals, fish, aquatic invertebrates, predatory insects, insect pollinators, and plants. Figure 4.10 lists typical indicator species.

A Tiered Approach to Testing

Toxic levels of a pesticide and descriptions of potential adverse effects are developed using a tiered (sequential) approach (Fig. 4.11). Tier 1 studies, the most fundamental, are primarily acute laboratory studies that examine concentrations necessary to cause mortality or acute sublethal effects. Tier 2 involves longer term, reproduction and life cycle studies that provide more complex data on the pesticide's impact on the entire life cycle of the test animal. Tier 3 studies examine the impact of a pesticide on animals in simulated environmental or actual field conditions.

The Science of Wildlife Testing

Toxicological testing and scientific measurements are conducted under recommended guidelines, approved methodologies, and specified reporting requirements. Exacting standards are necessary for consistency in evaluating pesticide safety and for comparison among chemicals. The EPA's pesticide assessment guidelines stipulate the following general practices when conducting the various tests required for registration:

- Toxicological or phytotoxicological testing is not performed on endangered or threatened species.
- Only EPA-recommended wildlife and aquatic organisms should be used for laboratory testing purposes. This provides consistency in analysis of all pesticides.

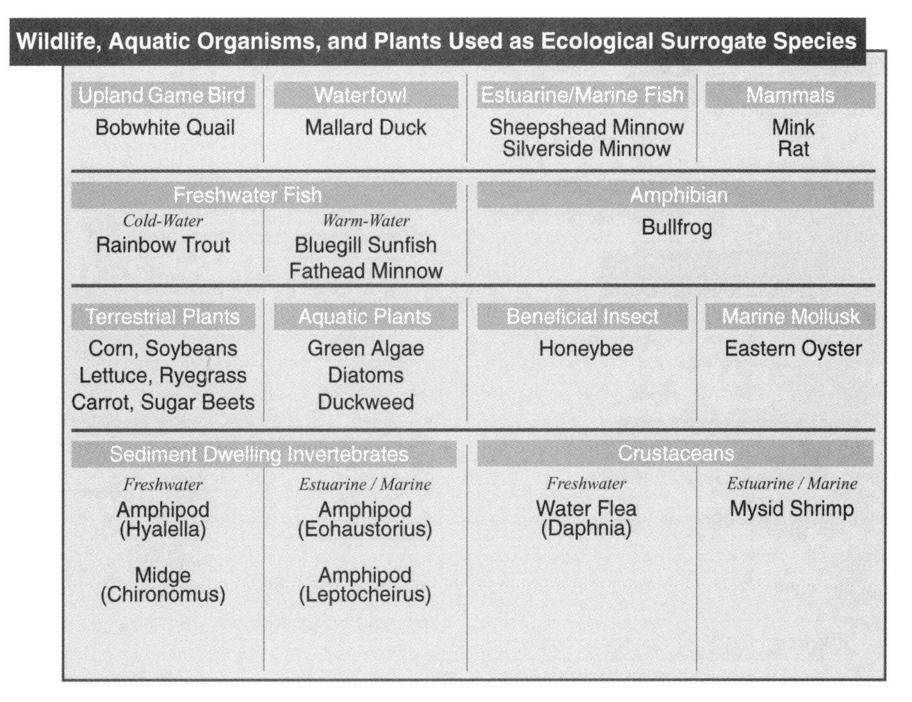

Figure 4.10 Surrogate species are used to determine the ecological impact on birds, fish, plant, and other species.

- The test organisms should be uniform in weight, size, and age.
- Control groups—those not exposed to pesticides—should be maintained in a manner similar to that of the test groups.
- The substance to be tested—the technical grade of the active ingredient, or the end use product—is clearly specified. If the test substance is diluted or dissolved for administration, the carrier should not interfere with absorption, distribution, or metabolism of the test material; alter the chemical properties of the substance; enhance or reduce the toxic characteristics of the test substance; affect food and water consumption; or impact the physiological processes of the test organism.
- Detailed descriptions of the nature, incidence, time of occurrence, severity, and duration of all observed toxic effects should be recorded.
- All data generated must be in accordance with established Good Laboratory Practices for handling and care of test organisms.
- Final reports should include all information necessary to provide a complete and accurate description of test procedures and evaluation of test results.
- Responsible parties must confirm by signature that appropriate quality assurance and quality control methods were followed.

Avian Species

Northern bobwhites (upland game species) and mallards (waterfowl) are used to conduct a series of toxicological tests to quantify the short- and long-term impacts of pesticides on

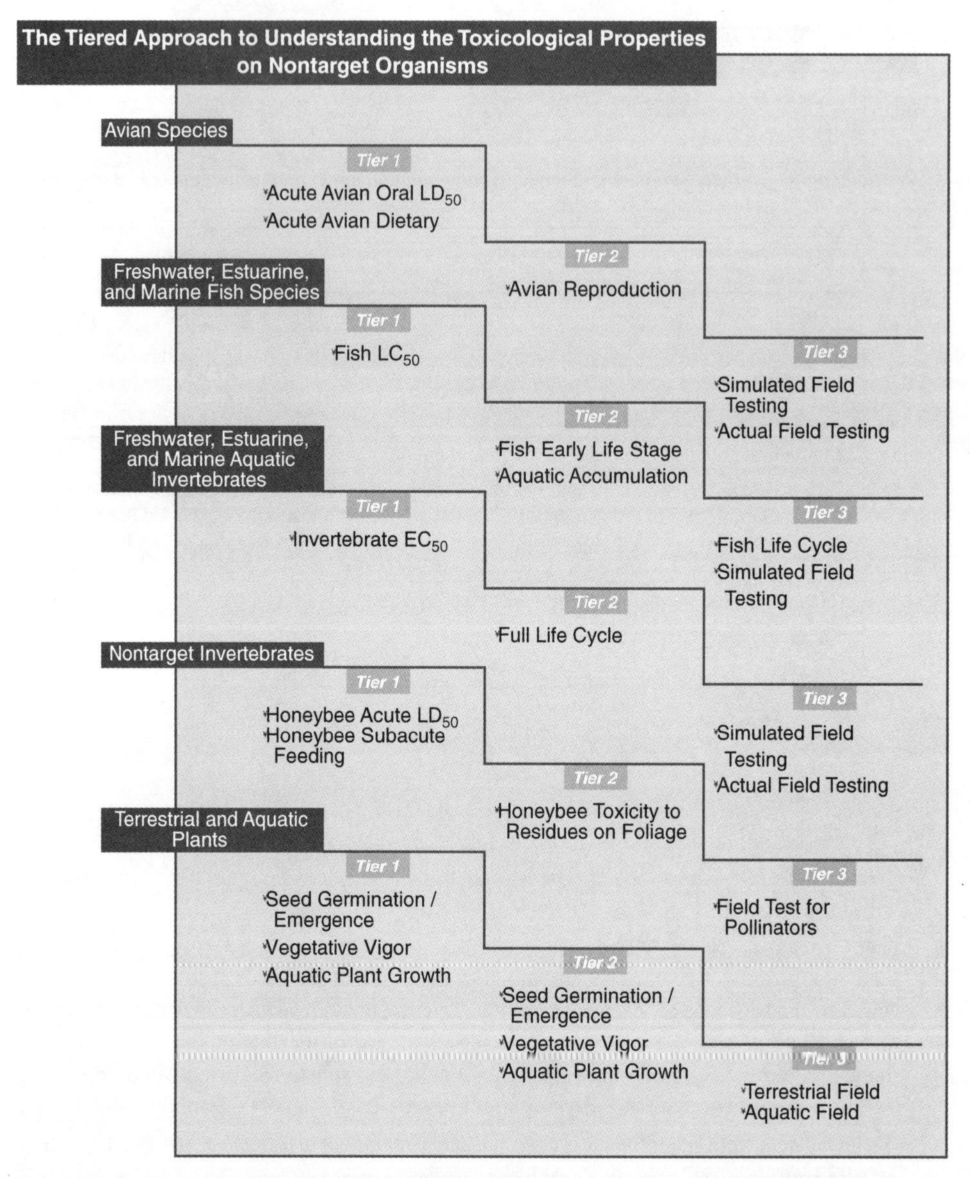

Figure 4.11 Toxicity testing proceeds through lower tiers that utilize worst-case information to higher tiers that use data derived from more complex studies.

avian wildlife (Fig. 4.12). These species generally are obtained from pen-reared stock and maintained under conditions of temperature, humidity, lighting, and pen size that conform to good husbandry practices and protocols established by the EPA.

There are three major laboratory tests for avian effects:

- Acute oral LD_{50}
- Acute dietary LC_{50}
- Reproduction

Figure 4.12 Paired mallards for an avian reproduction study.

In all three studies, the primary route of exposure is oral: either the pesticide is introduced directly into the subjects' crops (acute oral exposure), or it is incorporated into their diet (dietary exposure). For example, *acute oral exposure* occurs when a bird ingests a large single dose at one feeding, and *dietary exposure* may result from ingestion of pesticide residues on food items over a period of a few days.

The purpose of these acute oral LD_{50} studies is to determine the chemical's acute oral toxicity, expressed as a single dose of material (milligrams per kilogram of body weight) that will result in 50% mortality among test birds (LD_{50}). The test provides a measure of a species' sensitivity to a toxic substance. Birds tested must be in good health, from the same source, and preferably from the same hatch. Birds must be at least 16 weeks old at test initiation and must have been preconditioned to the test facilities for at least 15 days prior to experimentation. The standard study uses 10 birds for each of five dose levels. The test material is administered orally to each bird by direct injection into the stomach or crop or through the use of capsules. Birds are observed for a minimum of 14 days, and any mortality or signs of intoxication are recorded. In addition, an internal examination may be made to determine the condition of major organs on birds that die during the study.

The *acute avian oral* LD_{50} test assesses the effect of a single, oral dose of a pesticide administered to bobwhite quail or mallard ducks (Fig. 4.13). Birds that survive a dose are observed for 2 weeks for subsequent mortality and sublethal effects such as abnormal behavior and reduced food consumption. The test yields the LD_{50} level and the NOEC. The LD_{50} represents the dose which can be expected to kill 50% of the test population, and the NOEC reflects the maximum dose that produces no observed effect on the test population.

Deriving the *acute avian dietary* LC_{50} [concentration of pesticide in the diet, in ppm (mg/kg), that is lethal to 50% of the test population] involves feeding Northern bobwhite chicks and mallard ducklings a pesticide-treated diet for 5 days. A 3-day observation period follows, during which the birds are fed a control diet; abnormal behavior and food consumption are recorded (Fig. 4.14).

The most commonly used test for chronic pesticide effects on terrestrial wildlife is the avian reproduction test. The objectives of the study are to determine pesticidal effects

Figure 4.13 Administering a dose to a bobwhite quail to determine acute oral toxicity.

on the health and reproductive performance of egg-laying adults, on embryo viability, on eggshell thickness, on the number of eggs laid, on the number of normal hatchlings, and on the survival of hatchlings. Generally, one control and three pesticide-treated dietary concentrations (selected to bracket environmental exposure) are fed to four test groups of first year breeders. Exposures begin 10 weeks prior to egg laying and continue during 10 weeks of egg laying. Eggs are collected daily, incubated artificially, and checked periodically for embryonic development; hatchlings are grown on untreated feed for 2 weeks to monitor their viability and growth. This test is now conducted for virtually all pesticides (Fig. 4.15).

In addition to the three major tests, *simulated* and *actual field* tests are conditional; that is, they may be required by the EPA on a case-by-case basis as part of the registration process.

Figure 4.14 Evaluating the effects of a pesticide incorporated in the diet of bobwhite quail.

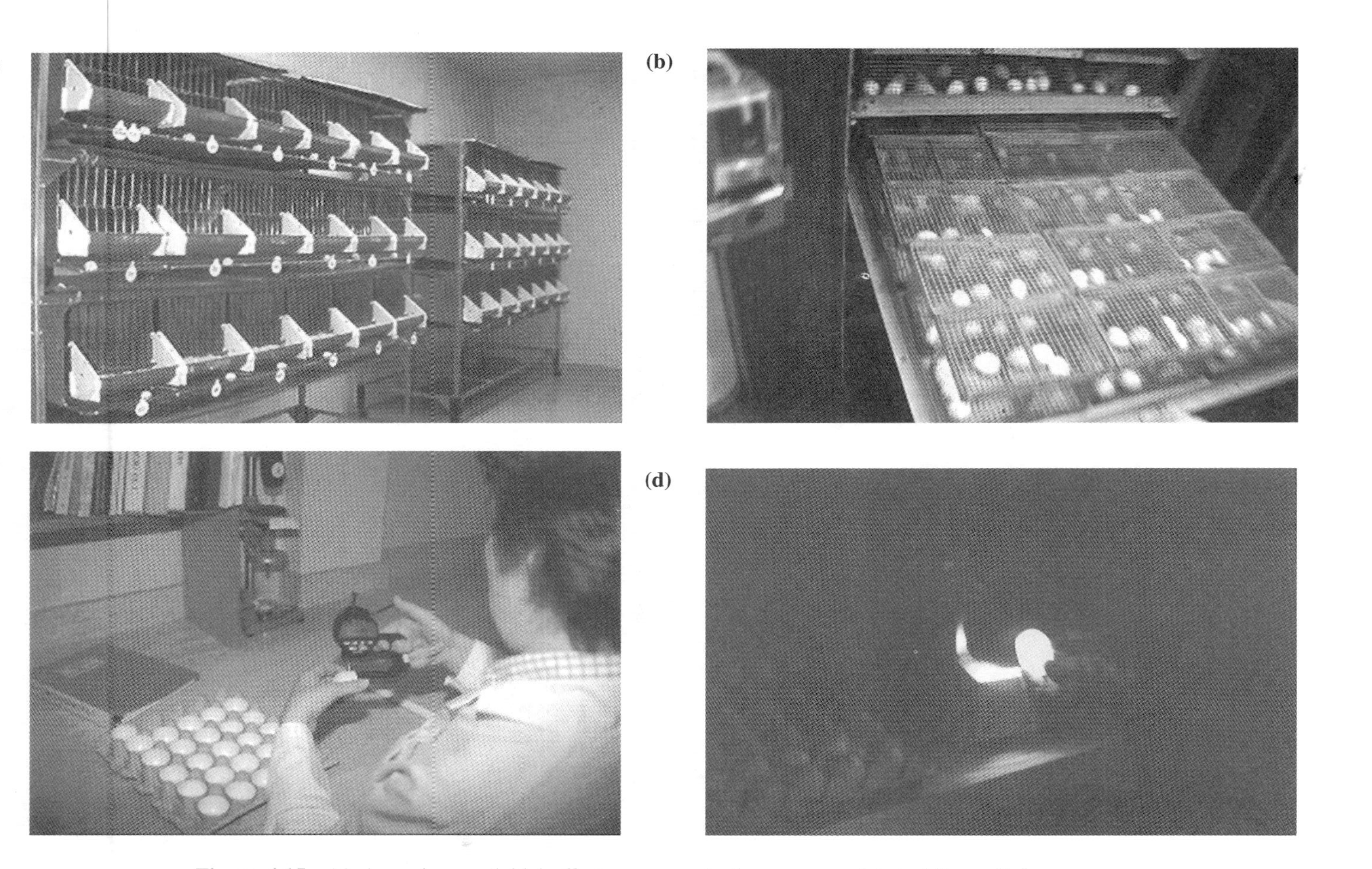

Figure 4.15 (a) Assessing pesticidal effects on reproductive success of bobwhite quail from a one-generation study. (b) Placing eggs in an incubator. (c) Measuring the effect of the pesticide on egg shell thickness. (d) Embryo viability is observed by candling the eggs.

Figure 4.16 Canada geese are placed on treated and untreated turf plots to measure their avoidance of the pesticide.

A *simulated field* test might include placing mallards, bobwhite quail, or geese in outdoor cages, under circumstances that mock exposure in the wild, and exposing them to the pesticide at a rate and frequency prescribed by the pesticide label (Fig. 4.16).

An *actual field* test might involve a pesticide application to an orchard, after which various data are recorded. Besides mortality, data on bird abundance, species composition, survival of dependent young, residues on wildlife food sources, and residues in animal tissues (as evidenced by autopsy and tissue sampling) may be collected.

From 1987 to 1992, the EPA's Office of Pesticide Programs increasingly required ecotoxicological field studies, both for new chemical registrations and as part of the ongoing reregistration process for older pesticides. Two major documents, *Guidance for Conducting Terrestrial Field Studies* and *Guidance for Conducting Aquatic Mesocosm Studies,* were issued to provide detailed recommendations on the implementation of these studies. The former document was concerned primarily with field monitoring of the pesticidal impact on birds under actual use conditions, the latter concerned the use of test systems composed of constructed ponds (0.1- to 0.25-acre surface area) treated with pesticides at rates approximating chemical contamination from runoff and spray drift following agricultural applications.

After approximately 5 years of ecotoxicological field testing under these guidance documents, more than 45 avian field studies and 10 aquatic mesocosm studies were conducted. Because of the complexity of the studies, and the magnitude of the data, interpretation was difficult. The EPA determined that the results did not add sufficient, interpretable information to agency risk assessments to justify the time and resources necessary to support the testing. These studies are no longer routinely requested by the EPA.

In the absence of such field testing, the EPA is evaluating the risks of pesticides based largely on laboratory test results, estimation of environmental exposures through the use of computer models and literature sources, measured pesticide residues, and pesticide incident data. When such analyses indicate a potential for adverse environmental effects, the EPA requires pesticide registrants to implement changes in product use recommendations to lessen exposure. Under this new regulatory approach, the EPA may still require field testing

Figure 4.17 The bluegill is an example of a surrogate species commonly used in ecological testing.

in special circumstances or may require field monitoring to determine if pesticide use changes have adequately reduced exposures or effects.

Fresh water and Estuarine/Marine Fish

Acute LC_{50} tests most often employ rainbow trout and bluegill sunfish (tested separately) that are actively feeding but have yet to spawn (Fig. 4.17). Fish are exposed to various concentrations of the pesticide to determine the dose–mortality response and also the sublethal responses over 96 hr (4 days).

The results generated from *acute* LC_{50} tests include the 96-hr LC_{50} and the NOEC. Behavioral changes and gross pathological observations, such as erratic movement and swimming at the surface, also are recorded. Gross pathological observations could include protrusion of the eyeball, sloughing of skin, and increased skin pigmentation (Fig. 4.18).

Figure 4.18 Typical aquatic toxicity laboratory.

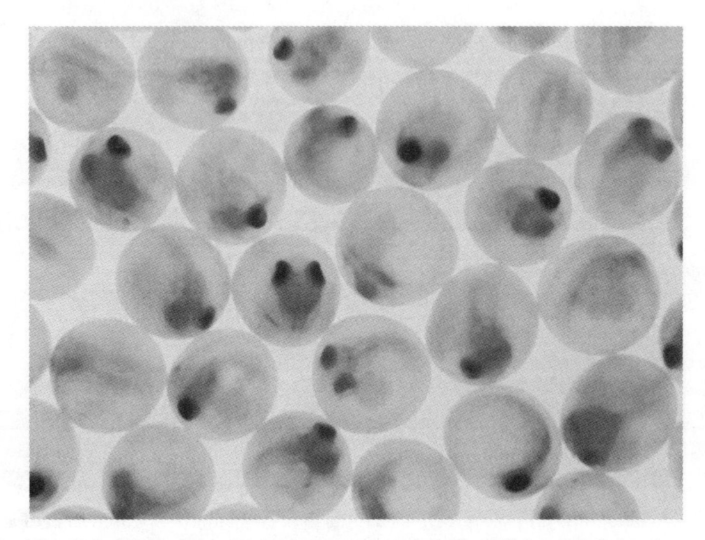

Figure 4.19 The early life stage test examines how a pesticide will impact the embryonic and larval stages of fish.

The *early life stage* test examines how a pesticide impacts the embryonic and larval stages of fish (Figs. 4.19 and 4.20). This test is initiated with newly fertilized eggs (embryos) and continues for at least 28 days (fathead minnows) or 70 days (rainbow trout). Data generated by early life stage tests include:

- Number of embryos hatched
- Amount of time embryos require to hatch
- Embryo and larval mortality
- Larval weight
- Larval length

The NOEC for these parameters is determined by statistically comparing treatment groups with controls.

Fish life cycle studies use fathead minnows to represent fresh water fish species and sheepshead minnows as surrogates for estuarine fish species. The time needed to complete a fish life cycle study is 260 to 300 days. Fish are exposed to a pesticide from one stage of their life cycle to the same stage in the subsequent generation (embryo to embryo). Fish embryos are placed into water containing known pesticide concentrations, and observations are made from the egg stage through spawning. The eggs laid by the first generation and the larvae that emerge are followed for an additional 28 days. NOEC determination is based on control response and treatment response (Fig. 4.21).

Fish bioaccumulation studies begin with fish exposed to a known pesticide concentration in water. During specific time periods, fish are sacrificed to determine the pesticide concentration in their tissues. Fish are separated into filet samples (edible portion) and samples with all other parts combined [visceral (inedible) portion]. Bioconcentration studies are conducted until concentrations identified in fish tissues remain fairly level or for at least 28 days. The fish are moved to clean water and are periodically analysed to determine

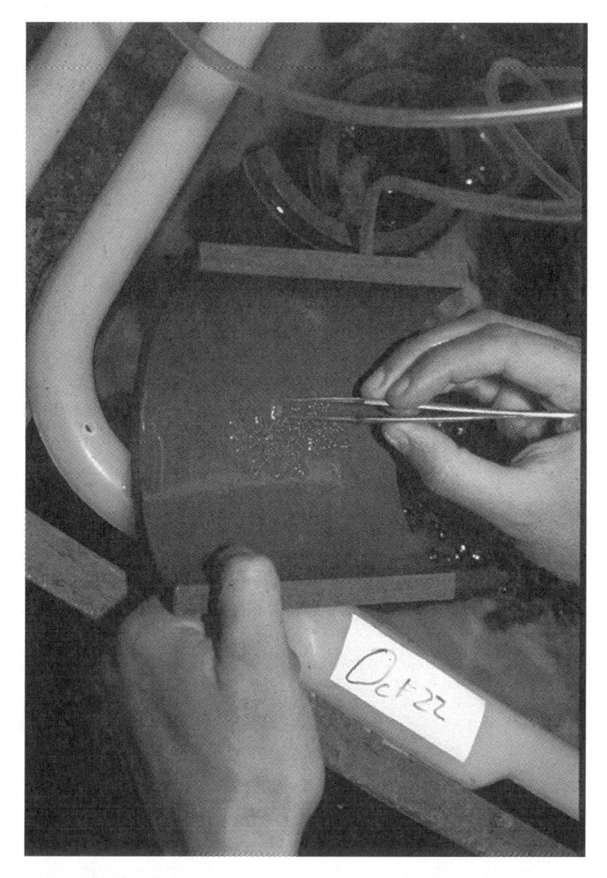

Figure 4.20 Removing fish embryo for use in toxicity tests.

the decline of residue concentrations in body tissues. This provides information regarding potential food chain effects and the persistence of the chemical in aquatic life.

Fresh water Aquatic Invertebrates

Two primary tests are included in assessment of pesticidal effects on invertebrate species: *acute* EC_{50}, and *aquatic invertebrate life cycle. Daphnia* (fresh water crustaceans in which females can self-fertilize) are typically used in both studies (Fig. 4.22). The LC_{50} is a measure of a concentration that is predicted to result in 50% mortality. The EC_{50} is the concentration that will result in 50% of the population exhibiting a predescribed effect such as immobility or reduction in growth.

Invertebrate acute EC_{50} tests provide information on how newly hatched daphnids respond to a pesticide over a 48-hr exposure period. The objective is to estimate the concentration of pesticide that will immobilize the daphnid. The concentration calculated to immobilize 50% of the daphnids is considered the EC_{50}. Immobilization is measured by gently shaking a vessel containing daphnids; those that swim for less than 15 sec are considered immobile.

Figure 4.21 Rainbow trout are used to illustrate growth response to varying pesticide concentrations.

Estuarine and Marine Organisms

The *acute toxicity* test is the primary study used to address the toxicity of pesticides to estuarine organisms. A crustacean (mysid shrimp), an estuarine/marine fish (sheepshead minnow), and a marine mollusk (Eastern oyster) are used to evaluate pesticides. Sheepshead minnows and mysid shrimp are placed into separate aquariums, containing specific

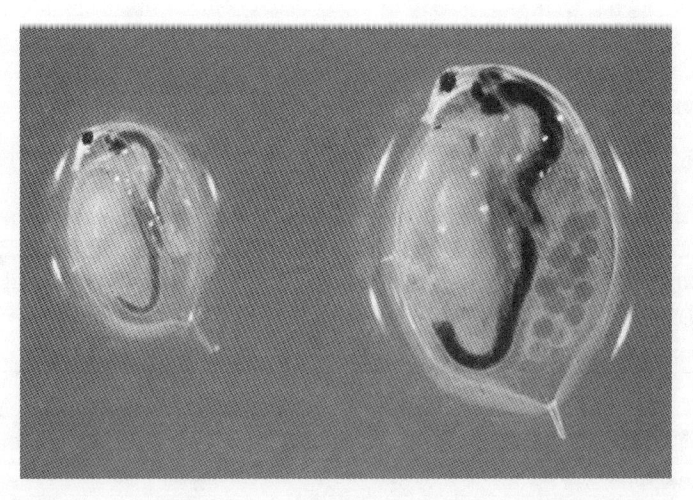

Figure 4.22 *Daphnia* are widely used to test for adverse effects on fresh water invertebrates.

Figure 4.23 An acute toxicity test using sheepshead minnows.

concentrations of the pesticide, for 96 hr (Fig. 4.23). The mortality data generated is used to calculate the LC_{50} (fish and mysid) and/or the EC_{50} (oyster).

Typically, one end point—the pesticide concentration that inhibits shell growth by 50%—is sought to estimate the impact of a pesticide on mollusk species. The shell growth of young Eastern oysters is assessed by placing them in a series of pesticide concentrations in water for 96 hr (Fig. 4.24). The concentration that inhibits shell growth by 50% is calculated and is reported as an EC_{50}. Such a study is regarded as an indirect measure of the impact of the pesticide on the nutritional status of the animal. A mollusk that closes its shell in the presence of a pesticide will not feed; thus, shell growth may be severely limited.

Figure 4.24 Eastern oysters placed in aquariums to evaluate the effect of a pesticide on shell growth.

Mammals

Testing on wild mammals normally is not required. Data from testing for human health risks (using rats, mice, dogs, and rabbits) are used to predict toxicity to wild mammals (See Chapter 2 for more details). A situation where wild mammalian toxicity testing may be required is when a highly toxic rodenticide or predacide is used as a broadcast bait; minks often are used as test subjects to determine the potential for secondary toxicity (toxic effects in predators that have consumed poisoned animals).

Nontarget Insect Pollinators

The honeybee *(Apis mellifera)* is used as the principal pollinator insect. Tests used to measure effects of a pesticide on nontarget pollinator insects are the *honeybee acute contact* LD_{50} and the *honeybee toxicity of residues on foliage.*

The *acute contact* LD_{50} test requires that honeybees be anesthetized to facilitate placement of the pesticide directly on the abdomen or thorax. Afterward, the bees are monitored for 48 hr; the LD_{50} is calculated and expressed in micrograms (μg) of active ingredient per bee (Fig. 4.25).

The *honeybee toxicity of residue on foliage* test determines if honeybees can receive a toxic dose from dislodgeable residues on foliage. The pesticide is applied to foliage which is then harvested at prescribed points in time (e.g., 3, 6, 24 hr). Harvested foliage is placed into cages containing worker bees, and the number of bees that die from contact with the foliage is recorded and compared with controls. This test provides information on the effect of weathered residues and helps determine best agricultural practices for the product.

Nontarget Plants

Nontarget Terrestrial Plant Toxicity Corn, soybeans, root crops, tomatoes, cucumbers, lettuce, cabbage, oats, ryegrass, and onions often are used to test effects on nontarget plants (Fig. 4.26). Either the soil is treated or the pesticide applied to the foliage at the maximum rate allowed by the label—or at a concentration three times the expected environmental concentration. Data collected in specific tests include

- root length,
- percent germination,
- percent emergence,
- time to emergence,
- plant height,
- dry plant weight, and
- percent of plants exhibiting phytotoxic (morphologic) changes.

The NOEC is determined for each end point, such as growth and root length. The most sensitive NOEC may be used in risk assessment.

Nontarget Aquatic Plant Toxicity

Tier 1 The two aquatic plant species most commonly tested for nontarget plant toxicity are *Selenastrum capricornutum* (a fresh water green alga) and *Lemna gibba* (a macrophyte

(a)

(b)

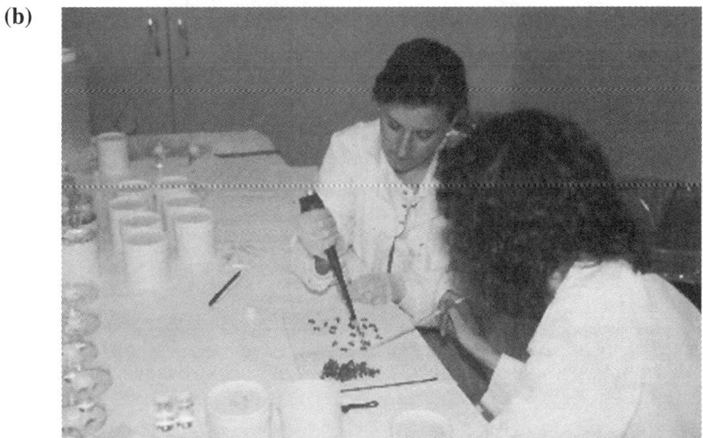

Figure 4.25 (a) Evaluating the effect of pesticides on honeybees. (b) Administering a dose to honeybees to determine contact toxicity.

duckweed) (Fig. 4.27). Water containing the two species is treated with either

- a single dose representing the maximum allowed rate, or
- a concentration three times the expected environmental concentration.

Tier 2 Five aquatic plant species are used for Tier 2 testing: *S. capricornutum*, *L. gibba*, *Anabaena flos-aquae* (a blue-green alga), *Skeletonema costatum* (a marine diatom), and

Figure 4.26 Nontarget plant studies measure the impact of a pesticide on seedling emergence and vegetative vigor.

Navicula pelliculosa (a fresh water diatom). The focus is on growth rate data measured either as increases in cells (algae and diatoms) or as fronds (duckweed). Effects are expressed as EC_{50} (the concentration of pesticide that reduces cell growth by 50%). Follow-up tests may determine if the effects are temporary, sublethal, or lethal.

EXPOSURE CHARACTERIZATION

Knowledge of pesticide environmental fate and transport characteristics is essential to accurately estimate the form and amount of the chemical that wildlife and aquatic organisms might encounter in the environment. Environmental processes result in transformation,

Figure 4.27 Algae species are used to predict pesticide toxicity to aquatic plants.

transfer, and transport of pesticides. The characterization of environmental fate assesses *what the environment does to the chemical* so that environmental exposure—that is, the concentrations an organism might actually encounter—can be estimated. Laboratory and field studies make it possible to predict the concentration of the pesticide and its metabolites available to nontarget organisms.

Understanding how a pesticide can be modified by the environment also is critical in judging how it will affect its intended target (i.e., its efficacy). For instance, a herbicide may offer great promise in the laboratory in suppressing hard-to-control perennial weeds found in cantaloupe and tomato fields. But if environmental factors in the field break down the herbicide before the weeds can absorb a sufficient dose, weed control is diminished. Likewise, photodegradation of an insecticide may reduce exposure to nontarget species but at the same time reduce its efficacy. If a pesticide is somewhat resistant to environmental breakdown, concerns are raised about its residual action—persistence—impacting water quality, wildlife, and rotational crops.

Complex Interactions

The environment plays a major role in determining

- how much pesticide residue remains.
- what form it might assume.
- how long it remains.
- where it ultimately goes.

How a pesticide reacts within a specific environment (the site of application) is dependent on its physical and chemical characteristics as well as environmental properties such as soil type, landscape position, and weather. Soil properties such as pH, temperature, moisture, and nutrient concentrations influence how chemicals are changed in the environment. Similarly, climatic factors such as temperature, humidity, and rainfall impact pesticide persistence and movement.

Site-specific differences that influence how a pesticide reacts over time are critical in estimating environmental exposure. For example, very different pesticide fate and behavior might be observed in a Louisiana sugarcane field versus an Indiana cornfield. The circumstances of use, the unique properties of each chemical, and the nature of the receiving environment make predicting environmental exposure across the United States very complex.

How Will the Pesticide Be Used? Application rates and techniques have direct bearing on how a pesticide enters the environment. Thus, a pesticide applied at a rate of *ounces or less* per acre has a lower potential for exposing fish and wildlife to toxic concentrations than the same chemical applied at a rate of *pounds* per acre.

A pesticide may be so reactive to sunlight that it decomposes soon after application. If the same product is protected from sunlight by incorporation into the soil, however, its persistence in the environment may be extended.

An incorporated product is less likely to impact wildlife than one which is left exposed on the soil surface. Thus, the formulation, the application methods, and the application intervals are considered critical factors in estimating environmental exposure.

How Does the Application Formulation Affect the Fate of the Pesticide? A pesticide product consists of one or more active ingredients and several inert ingredients. The product is mixed with adjuvants and/or water to make up the application mixture. Adjuvants can affect the behavior of the pesticide by prolonging its residence time in the environment, by slowing or enhancing its breakdown into by-products, or by facilitating the pesticide target interaction. The application method also must be considered when calculating the fate of a pesticide in the environment.

How Will the Pesticide Be Transformed by the Environment? Research has shown that the original (parent) molecule often is modified as it enters and interacts with the environment. Pesticides often are degraded in water (hydrolysis), by sunlight (photodegradation) and by soil and aquatic microorganisms (microbial degradation). Knowledge of these processes is key to assessing ecological risk.

Chemical properties of pesticides, such as volatility or water solubility, influence their transfer in the environment. For instance, the initial distribution of a pesticide in soil and water, or between soil particles and the air that surrounds them, might be 90% in one medium (e.g., soil) and 10% in another (e.g., water). The physical characteristics of a molecule, in combination with chemical properties of the environment, influence whether a pesticide molecule partitions in soil, air, or water.

How Long Will the Pesticide Persist in the Environment? A pesticide's continued presence in the environment—*persistence*—is a key factor in predicting potential exposure of wildlife. Persistence is generally described in terms of an environmental half-life, that is, the length of time it takes for the disappearance of one half of the applied pesticide from an environmental compartment. Biological and chemical processes that degrade or dissipate the pesticide influence its persistence.

How Will the Pesticide Be Transported from the Original Application Site to Off-Site Environments? Pesticides and their metabolites (breakdown products) move in a number of ways. They can

- be taken up by target pests and nontarget plants and animals,
- leach downward into ground water,
- attach to soil particles and be washed away in surface water runoff,
- volatilize into the atmosphere, or
- drift off-target during application.

Leaching, runoff, volatilization, and drift often are modeled to show how pesticides move in the environment. Compounds which are not absorbed into target plants and do not readily adsorb to soil particles often have a high potential for leaching to ground water or entering streams via surface water runoff. Highly volatile pesticides may escape the soil environment and dissipate into the atmosphere; when redeposited later, (e.g., via rainfall), they may interact with nontarget plants and animals.

What Is the Range of Pesticide Concentrations Expected to Come in Contact with Biological Systems? Knowledge of the application rate and an understanding of how a pesticide can be transformed, transferred, transported, and partitioned within

Figure 4.28 A quail equipped with a radio transmitter is used to evaluate the effects of a pesticide on survival and reproductive performances.

the environment facilitate prediction of the range of concentrations with which organisms interact.

Determination of expected environmental concentrations depends on information developed from a large number of environmental fate and residue chemistry studies that describe transformation, transfer, and transport. The descriptions of studies that follow demonstrate the extensive laboratory and field data required for estimation of environmental concentrations.

What Organisms Are Expected to Be Exposed to a Pesticide? Characterization of environmental exposure requires consideration of the inhabitants—wildlife, aquatic organisms, and nontarget plants. An understanding of the natural history (distribution, abundance, breeding habits, food sources) of nontarget species facilitates identification of the predominant route of exposure (Fig. 4.28).

Environmental Fate Studies

Studies That Measure Environmental Concentrations The interaction of pesticides with soils, surface water, and ground water is complex. Pesticide fate is controlled by numerous simultaneous biological, physical, and chemical reactions. Comprehending the fate of pesticides requires an understanding of certain processes: *transformation, transfer,* and *transport. Transformation* refers to biological and chemical processes that change the structure of a pesticide or completely degrade it. *Transfer* refers to the way in which a pesticide is distributed between solids and liquids (e.g., between soil and soil water), or between solids and gases (as between soil and the air it contains). *Transport* is the movement from one environmental compartment to another, such as the leaching of pesticides through soil to ground water, volatilization into the air, or runoff to surface water.

When a pesticide is applied to a field, certain processes result. Foliar-applied pesticides stick to leaves, where a portion is absorbed. Rainfall inevitably washes some of the chemical off the leaf surface onto the soil below, and some may be transformed by sunlight. Soil-applied pesticides generally interact first with moisture around and between soil particles, influencing how the chemical ultimately will react in the environment. Thus, a *soil solution* can be viewed as a chemical staging area for most reactions that control environmental fate. For instance, sorption processes (transfer), degradation by microbial and chemical reactions (transformation), volatilization to the atmosphere, leaching deeper into the soil profile, and overland flow (transport) all occur predominantly from soil solution.

Four basic questions on environmental fate must be answered in the data supporting the EPA registration: (1) How rapidly does the original pesticide degrade in the environment? (2) What are the breakdown products? (3) What is the persistence of the metabolites in the environment, and (4) Where will the parent molecule and metabolites partition in the environment (air, soil, water)? Environmental fate data generated from laboratory and field experiments are then used to assess the magnitude of exposure in different environmental compartments. Risk assessment involves comparing effects data from various toxicological studies with fate and exposure data to predict potential health and environmental impacts—and to protect natural resources in general.

A tiered approach to testing is used to simplify the process of investigating environmental fate data for pesticide registration. The first tier of environmental fate studies is conducted in the laboratory and addresses hydrolysis, photodegradation in water and soil, aerobic soil and anaerobic aquatic metabolism, mobility, and terrestrial field dissipation. The second tier consists of field soil studies and additional laboratory research triggered by results of first-tier work.

Finally, a third tier addresses key transport mechanisms indicated (by preliminary studies) as significant in assessing the ultimate environmental impact of the product—for example, small-scale ground water monitoring studies or surface water runoff evaluations. If a product's margin of safety is low in any environmental compartment, large-scale surface and ground water monitoring programs and ecosystem investigations may be conducted under conditional registration for the purpose of demonstrating in-use safety.

Laboratory and Field Data Required for Estimation of Environmental Concentrations of Pesticides

Environmental Fate Data Requirements

- Hydrolysis
- Photodegradation in water
- Photodegradation in air
- Aerobic soil metabolism
- Anaerobic soil metabolism
- Anaerobic aquatic metabolism
- Leaching and adsorption
- Laboratory volatility
- Field volatility
- Field dissipation, terrestrial uses
- Field dissipation, aquatic uses

- Forestry field dissipation
- Confined rotational crop
- Field rotational crop
- Accumulation in irrigated crop
- Accumulation in fish
- Accumulation in aquatic nontarget organisms

Residue Chemistry Data Requirements

- Chemical identification
- Nature of residue in plants
- Residue analytical methods, plants/animals
- Magnitude of the residue in potable water
- Magnitude of the residue in fish

Spray Drift Data Requirements

- Droplet size spectrum
- Drift field evaluation

Abiotic Degradation Studies Abiotic (chemical) degradation is the breakdown of pesticides by *nonbiological* reactions (i.e., without the involvement of living organisms) occurring in soil solution and on the soil surface. Factors which affect abiotic degradation include the chemical nature of the pesticide as well as environmental properties such as temperature, water content, and pH. Hydrolysis (reaction with water) is important for the degradation of many pesticides, as is photodegradation (reaction with sunlight); these two processes generally are the most important abiotic mechanisms involved. As a general rule, abiotic degradation is less consequential for transformation of a molecule than is biological degradation.

Hydrolysis is a common chemical reaction—a process by which a pesticide reacts with a water molecule. Hydrolysis reactions generally substitute the hydroxyl (OH) group from water (HOH or H_2O is the chemical structure of water) into the structure of the pesticide, displacing another group. Reaction with water breaks the chemical bonds holding the parent molecule together, and the extent of breakdown is pH dependent.

Photodegradation (photolysis) involves the breakdown of organic pesticides by direct or indirect energy from sunlight. Light energy can be absorbed by the pesticide or by secondary materials (e.g., organic matter) which become activated and, in turn, transfer energy to the pesticide. In either case, pesticides absorb energy from sunlight, become unstable or reactive, and degrade. Photolysis can occur in water, in air, or on surfaces such as soil or a plant leaf. Photolytic reactions occur near the surface of the ground (in the top few hundredths of an inch) or near water surfaces, where light can penetrate.

The hydrolytic and photolytic reactions of a pesticide are important in predicting the ultimate environmental fate of a chemical. When a pesticide reacts with water or absorbs solar energy, either directly or indirectly, the chemical bonds holding the parent molecule together are broken. Degradation studies are conducted to document the formation and decline of the parent compound as well as transformation products.

ABIOTIC TESTING PROCESS Controlled hydrolysis studies are conducted according to federal regulations to identify how a pesticide reacts with water (Fig. 4.29). The chemical is added to buffered water that has been sterilized to kill pesticide-degrading

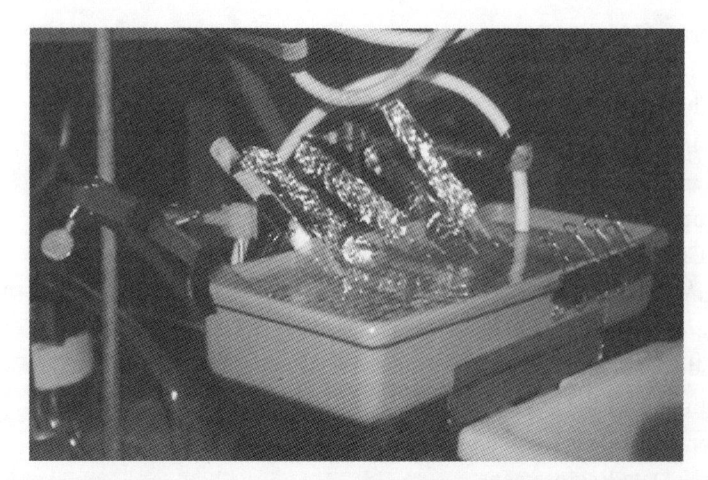

Figure 4.29 Experimental design for measuring the effects of hydrolysis on the pesticide.

microorganisms to test hydrolysis under acidic, alkaline, and neutral conditions. Studies are typically conducted at pH 5, 7, and 9. Pesticides specifically made to contain tagged radioactive carbon-14 can be used by scientists to help track the environmental fate of the chemical through its degraded products. Placement in dark incubators prevents the pesticide from reacting with light. Test samples are collected periodically for 30 days and analyzed to determine the amount of parent molecule remaining and the products generated and to account for all of the radioactive carbon-14.

Studies of photodegradation in water are conducted in a manner similar to hydrolysis experiments, except that they are conducted in the presence of simulated or actual sunlight (Fig. 4.30). Photodegradation on soil is studied by applying a radiolabeled pesticide to a thin layer of sterile soil. The treated water or soil is irradiated with simulated or actual sunlight and degradation is observed and measured. Samples are collected over a 30-day period and analyzed as in hydrolysis studies. Aqueous photodegradation studies, compared to hydrolysis studies, show how sunlight affects chemical breakdown in water. Soil photodegradation studies show how soil components (e.g., clay and organic matter) affect pesticide breakdown. These data can be compared and pathways for the two systems established.

Metabolism Studies Communities of soil microorganisms are very diverse. For example, researchers have estimated that between 5000 and 7000 different bacterial species may exist in a single gram of fertile soil. Populations of bacteria often can exceed 100 million individuals in 1 g of soil, and populations of fungal colonies can exceed 10,000.

Microbial degradation is a transformation process that results when soil microorganisms (bacteria and fungi) metabolize (break down) a pesticide, either partially or completely (Fig. 4.31). Microorganisms can cause changes in a pesticide when this activity occurs; in the presence of oxygen it is termed *aerobic* metabolism, and in the absence of oxygen, *anaerobic* metabolism. Most microorganisms inhabiting the soil profile, where oxygen degrade pesticides via aerobic metabolism. As a pesticide undergoes aerobic metabolism, it is normally transformed into carbon dioxide (CO_2) and water. Under anaerobic metabolic conditions, microorganism degradation may produce additional end products such as methane. Microorganisms that use anaerobic metabolism for breaking

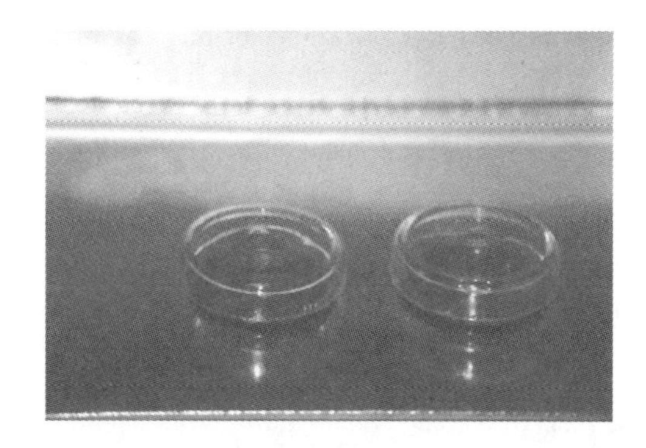

Figure 4.30 Samples prepared for exposure in on photolysis apparatus.

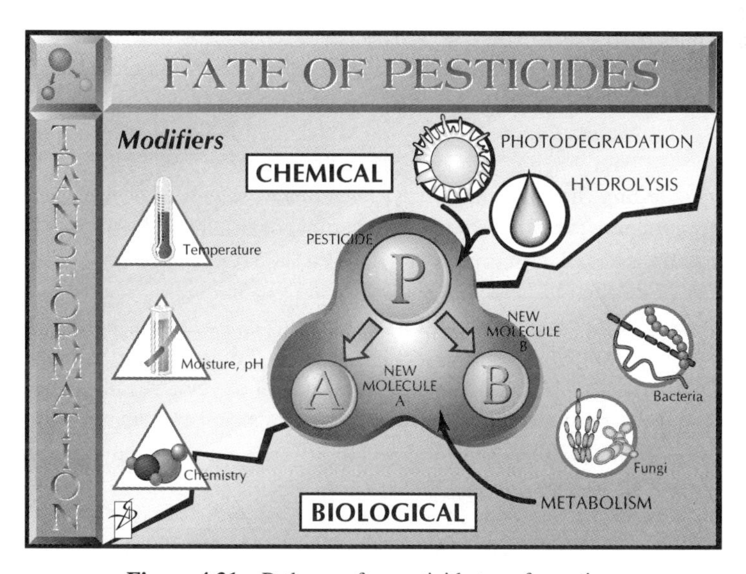

Figure 4.31 Pathways for pesticide transformation.

down pesticides are typical of the microbes that inhabit waterlogged soils in terrestrial systems or that live in the bottom sediments of ponds, lakes, and rivers. These organisms also are present in ground water and, to some extent, in the soil profile.

Pesticides, along with many other naturally occurring organic molecules, may serve as a source of food or energy for soil microbes. Because they occur at very low environmental concentrations, however, it is unlikely that their capacity to serve as a food source is adequate to sustain high numbers of microbes. Pesticides are more apt to serve as *incidental* food sources for microbes also drawing from other food sources.

Most soil microbes are associated in colonies on the soil surface, not free in soil solution. A pesticide in soil solution has to move to these microbial colonies and cross the microbial cell membrane into the cell to metabolize. Some microbes produce enzymes which are exported from the cell to predigest pesticides that are poorly transported. Once inside an organism, a pesticide can metabolize via internal enzyme systems. Any energy derived from the breakdown of the chemical can be used for growth and reproduction; any portion not fully degraded to CO_2 or incorporated into cells is released back into soil solution as intermediate chemical metabolites.

Studies have revealed that multiple organisms often are involved in the degradation phenomenon. Previous notions that single species are solely responsible for microbial degradation of a pesticide probably are not correct. Different species have different capabilities, and together they can form a pool of talent resulting in degradation of the pesticide. The likelihood that the chemical will be completely degraded is decreased if any of the microbes are missing from the pool. The ability of microbes to degrade a pesticide is related to their metabolic capacity and the complexity of the molecule and to environmental factors that regulate microbial activity (water content, temperature, aeration, nutrients).

METABOLISM TESTING PROCESS Plant metabolism is of more consequence to human health than to ecological risk—for example, in determining the nature and magnitude of the residue in food. In addition to plant metabolism and uptake studies, the biological degradation of pesticides by microorganisms in soil is examined. The term *metabolism* is used since most pesticides are degraded primarily by microorganisms in soil that metabolize all or part of the pesticide molecule.

Three kinds of environmental metabolism studies are required under FIFRA: a 1-year aerobic soil metabolism study using selected field soils; a 30-day aerobic aquatic metabolism study using sediment and natural water; and a 1-year (maximum) anaerobic aquatic metabolism study using sediment and natural water (Figs. 4.32 and 4.33). These studies seek to determine how fast the parent molecules are degraded by biological processes (mainly microorganisms) in different soils, in aerobic sediment and water, or in anaerobic sediment and water and to determine what metabolites are formed. To be valid, these studies also must account for the radioactive carbon-14 used as a tracer.

Field soils known to be previously unexposed to pesticides are used in the various metabolism studies. Oxygen is maintained for aerobic metabolism studies, whereas anaerobic studies require that the soils or sediments be purged of oxygen. A radiolabeled pesticide is applied to the soil or sediment and, at specified intervals, researchers remove and analyze samples for the parent pesticide and any metabolites. If volatile materials and CO_2 are released, they are trapped and analyzed.

Mobility Studies Sorption is a transfer process by which pesticides in soil are dispersed between solid matter and water; it is important in regulating the concentration of pesticides

Figure 4.32 Aerobic soil metabolism study showing connection of the flask to an oxygen manifold, which helps maintain aerobic conditions within the vessel.

in soil solution and surface or ground water. One important environmental sink (retention or storage site) for many pesticides is organic matter. The transfer—called *partitioning*—of a pesticide into organic matter in soil is a somewhat nonspecific mechanism and may often control sorption in soils and sediments.

Much organic matter (humus) is made up of a series of organic polymers (long chains or mats of molecules) and generally consists of two systems: a hydrophilic (water-loving) surface and a hydrophobic (water-hating) interior. The convention of "like dissolves like" holds for pesticide interactions with organic matter in soil. Nonionic (noncharged or neutral) pesticides escape from soil solution into the hydrophobic interior and, as a result, a pesticide equilibrium is set up between organic matter and soil solution. Pesticides move between organic matter and water in soil. Also, pesticides may undergo an aging process over time, whereby the chemical moves deeper into organic matter and becomes unavailable to move back into soil solution. Pesticides that are water soluble tend to remain at the surface of soil organic matter, whereas those that are insoluble penetrate to the hydrophobic interior.

The amount of pesticide sorbed is largely a function of the total amount of organic matter (sorption regions) in the soil. Sorption to clay mineral particles also occurs but usually is less significant than sorption to organic matter in determining environmental fate, unless the soil has very low organic matter content (Fig. 4.34).

Many pesticides develop a charge as the result of soil solution pH (a measure of acidity); that is, neutral pesticide molecules can become ionic (charged) and more reactive. If the

Figure 4.33 A biometer system used to maintain an anaerobic condition within the vessel.

pH-induced charge is positive, the pesticide can bind to negatively charged soil particles. If the induced charge is negative, the pesticide may actually be repelled from the negatively charged surfaces of soil solids.

Sorption to soil particles is also dependent on soil water content because water is necessary for chemical movement; water molecules compete with pesticide molecules for attachment sites on clay and organic matter. Therefore, pesticide sorption tends to be greater in dry soils than in wet soils. Decreased soil water content forces the pesticide to interact with soil surfaces; however, the amount of sorption also depends on the *type* of clay and organic matter content.

The bond between a pesticide molecule and a soil particle determines, to a large degree, the environmental fate of the pesticide. For instance, pesticides that are tightly sorbed to soil particles have decreased mobility and are less likely to contaminate ground water. The bond may decrease the rate at which the pesticide is degraded by soil microbes, leading to longer environmental persistence. Pesticides strongly sorbed to soil particles may travel primarily with eroded soil and enter surface water, whereas weakly sorbed pesticides that are more water soluble may be released into soil water solution and enter surface water as runoff.

The sorption process may be somewhat more complex for a pesticide that exhibits volatility. In these cases, sorption reflects competitive reactions that partition the molecule between soil solids, soil solution, and soil air. Depending on the balance of soil air and

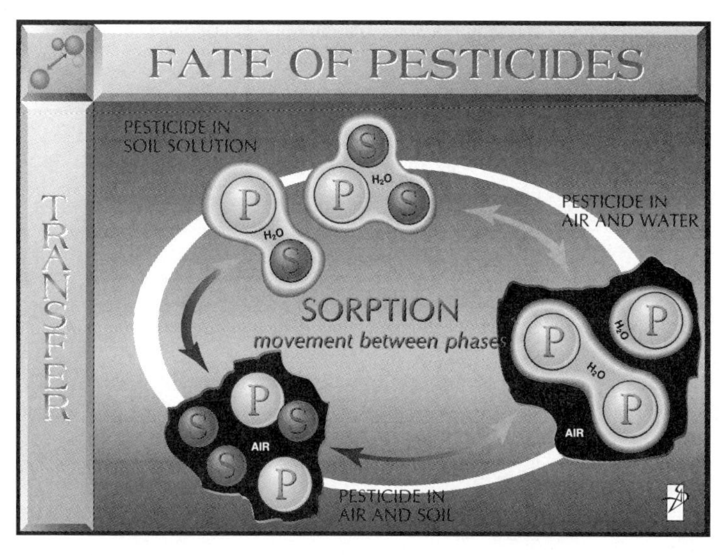

Figure 4.34 Pesticide transfers within the soil environment.

water and on environmental factors such as temperature, a weakly sorbed, volatile pesticide may move in soil air to deeper points in the soil profile or into the atmosphere.

MOBILITY TESTING STUDIES Scientists estimate potential mobility of a pesticide by first determining its sorption in soil. Soil and water are made into a slurry which is then treated with a range of pesticide concentrations. After a period of time, the slurry is centrifuged to separate the soil from the water, and the chemical concentration in each is determined (Fig. 4.35).

Pesticide retention is expressed as a sorption coefficient (K_d) which is expressed as a ratio: the *concentration of chemical sorbed to soil* over the *concentration of*

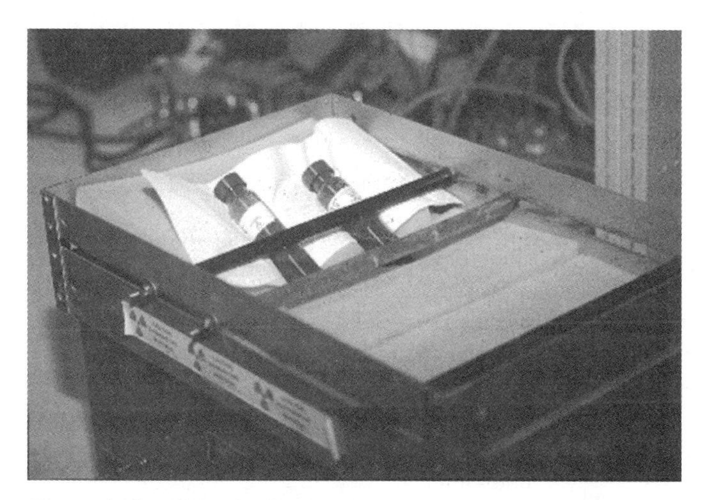

Figure 4.35 Vials of radiolabeled pesticide for a soil sorption study.

chemical remaining in water. The K_d equals concentration sorbed divided by concentration in solution.

The K_d is relevant to understanding pesticide transport since chemicals remaining in soil solution can leach or become available in the water of a pond or stream. Because pesticides in soil solution are subject to leaching, the extent of sorption as measured by the K_d serves as a predictor of mobility: the higher the K_d, the lower the tendency to move in soil. For example, if a K_d is lower than 2, the molecule is termed *highly mobile;* if it is between 2 and 5, the molecule is considered *mobile;* and if the K_d is greater than 5, it's deemed *immobile* with respect to leaching.

Frequently, the K_d is expressed as a K_{oc} by dividing the K_d by the fraction of organic carbon present in the soil: $K_{oc} = K_d$ divided by fraction organic carbon. This expression is useful because soil organic carbon usually is the major sink for pesticides in soil or sediment. The expression of sorption in terms of K_{oc} facilitates comparison of the potential mobility of a chemical to be compared with that of other chemicals, regardless of soil type. A K_{oc} value greater than 500 usually is associated with immobile pesticides. The K_{oc} is often used to describe sorption when organic carbon in the soil is the most important sink for the pesticide.

Field Dissipation Studies Terrestrial field studies are conducted to verify the integrated routes and rates of pesticide degradation and mobility demonstrated in the laboratory; the length of time required to complete the terrestrial studies is estimated from data generated in the lab. Pesticides that are persistent—those that have a soil metabolism half-life greater than 6 to 12 weeks under optimal conditions for degradation—generally require at least 18 months to conduct studies in the field. Field studies with less persistent pesticides often can be completed within a single field season.

When a pesticide is proposed for use over large areas and/or on multiple crops, several field test locations and cropping scenarios are required for field dissipation studies. The test sites must be established and maintained according to best management practices for the intended crop or noncrop use. Preapplication sampling and analysis of the soil to a depth of 3 ft are performed to confirm that no pesticide is present. An end use product is applied with typical application equipment at the highest rate stated on the proposed label. At timed intervals, representative samples are removed at prescribed depths and analyzed for the presence of the parent product and environmentally significant metabolites. Since terrestrial dissipation studies are most frequently conducted without radiolabeled chemicals, precise method development to assure sensitive analysis of soil residues is necessary. Pesticides that are active at low rates require sophisticated and highly sensitive analytical methods for extraction and analysis of the parent molecule and significant metabolites; measurements in *parts per billion* are necessary.

Dissipation studies also must be conducted to determine the environmental fate of pesticides designated for aquatic crop and noncrop uses. Protocols for conducting aquatic studies and the timetables involved are similar to those of terrestrial field studies. Water and sediment (and sometimes animal and plant) samples are collected for analysis to detect the parent molecule and significant metabolites.

Estimated Environmental Concentrations

Potentially, all pesticides pose some risk to sensitive nontarget organisms, and environmental concentration estimates are critical in estimating ecological risk. Data developed on the environmental fate of a pesticide, along with use information as stated in the proposed pesticide labeling, are used to generate a value known as an Estimated Environmental Concentration (EEC). The EEC is an estimate of how much of a pesticide might reach nontarget areas, potentially exposing wildlife, bees, worms, aquatic animals, and plants. EECs generally are predicted for ground and surface water, soil, and wildlife food items.

The key word in *Estimated Environmental Concentration* is *estimated*. Scientists cannot measure actual concentrations for every conceivable environmental situation. An actual concentration measured today most likely would not match measurements taken sometime earlier or later. Therefore, EECs should not be viewed as *hard*, *fixed*, or *absolute* values, but as *estimates* based on available data. Actual concentrations can and do fluctuate according to numerous variables: time of year, geographic location, weather patterns, soil conditions, application interval, cropping systems, etc.

Mathematical models that simulate the fate of pesticides in the environment are used for developing EECs. When used early in the development process, modeled EECs rely on preliminary environmental fate information and proposed use patterns. This initial lack of specific pesticide information—the unknown—leads to this uncertainty which is compensated by very conservative assumptions.

As researchers gain a better understanding of the pesticide molecule through refined laboratory data and in-depth analysis—and as the influence of factors such as weather conditions and soil characteristics on pesticide fate become known—input assumptions are modified and new EECs are determined.

The challenge in refining the EEC is to provide greater scientific certainty and improved interpretations of the available data. In this way, an improved understanding and approximation of the actual environmental concentration is achieved. Confidence in the risk assessment is increased with more precise predictions of exposure.

Estimated Environmental Concentrations for Aquatic Systems

There are four basic tiers used to estimate environmental exposure concentrations for aquatic systems. Each tier requires additional data or more refined data analysis. The higher tiers employ sophisticated models where principal parameters and site-specific scenarios have been developed.

Aquatic EEC Tier 1: Single Event EEC Based on a High-Exposure Scenario

The *Generic Estimated Environmental Concentration* (GENEEC) model was developed by the EPA to determine a generic EEC for aquatic environments. EECs derived from GENEEC models reflect the pesticide concentration expected under worst-case conditions: an application on a highly erosive and very steep upland slope, with heavy rainfall occurring immediately after application. The watershed model is essentially 10 acres of surface area. It is assumed that the entire area is treated and that the treated area has uniformly high slopes so that all runoff drains directly into a 6-ft-deep, 1-acre pond.

The GENEEC model utilizes environmental fate parameters identified by laboratory testing protocols as well as information obtained from the proposed pesticide labeling. It also includes fixed soil and weather parameters. The model estimates pesticide runoff to the pond on the basis of rate and method of application, water solubility, soil binding (adsorption) characteristics, and persistence of the pesticide as determined by hydrolysis, photolysis, anaerobic and aerobic metabolism; spray drift is also a factor.

An example of the inputs and outputs to a GENEEC model is presented for an insecticide used on corn.

Inputs

Rate (pounds per acre): 1.0
Number of applications and interval: 1
Soil K_d: 2
Solubility (ppm): 30
Application type (percent drift): 2.9
Incorporated (inches): 0
Metabolic field half-life (days): 14
Days until rain/runoff: 2
Hydrolysis half-life (days) in pond: n/a
Photolysis half-life (days) in pond: 0
Metabolic half-life in pond (days): 21
Combined metabolic half-life in pond (days): 21

Outputs

Peak Generic Estimated Environmental Concentration (ppb): 40
Maximum 4-day average Generic Estimated Environmental Concentration (ppb): 38
Maximum 21-day average Generic Estimated Environmental Concentration (ppb): 29
Maximum 60-day average Generic Estimated Environmental Concentration (ppb): 17
Maximum 90-day average Generic Estimated Environmental Concentration (ppb): 13

Aquatic EEC Tier 2: Single Site, Variable Weather Tier 2 assessments determine EECs based on geographic areas nationwide and on use sites (e.g., corn) in close proximity to ponds; many input variables are the same as those for GENEEC models, but additional parameters more descriptive of the use site may be factored as well. These data are used in more comprehensive models such as PRZM/EXAMS (See Chapter 5 for more information on models). Conditions typical of product use sites, including specific soils and weather information (a distribution of weather, including a one-in-ten-year, high-runoff incidence), are used. Single median values for chemical characteristics are selected from laboratory-derived environmental half-lives in the upper 10% of the statistical distribution that leads to estimates biased toward the upper end. Contributions from spray drift also are factored into the estimate. The goal of Tier 2 analysis is to better define the *range* of EEC—as compared to the single, worst-case Tier 1 assessment—that can be reasonably expected under variable weather conditions. Frequently, a case more typical of the intended site is analyzed, as well, for comparison against the worst-case scenario.

Aquatic EEC Tier 3: Multiple Sites, Multiple Weather Conditions Tier 3 differs from Tiers 1 and 2 in that both use-site and weather parameters are varied. Tier 3 assessments examine hypothetical circumstances representative of the regions in and conditions under which the pesticide is likely to be used. Tier 3 modeling results in development of a distribution of EECs that might be expected across use markets, recognizing that both soil properties and weather patterns vary significantly by market region and years of use. Tier 3 analysis is used by pesticide registrants to address environmental exposure concerns that arise during product reregistration processes. These assessments require very complete data describing pesticide environmental fate as well as comprehensive soil and weather data describing use sites. These assessments require thousands of computer simulations and are frequently summarized as regional maps using Geographical Information Systems (GIS).

Aquatic EEC Tier 4: Watershed Site Assessment Tier 4 assessments are complex analyses that investigate how pesticides are likely to interact with a landscape composed of hundreds of thousands of acres. The landscape has diverse soils and climates, varied proximities of treated fields to receiving waters, and randomly distributed bodies of water.

GISs are commonly used at Tier 4; they allow graphical evaluation of concurrent risk factors (within the regions of use) that heighten concerns. In other words, GISs distinguish high-risk versus low-risk areas of use on a regional basis.

Tier 4 procedures sometimes shift from modeling to actual environmental residue monitoring. Expected environmental concentrations are validated by actual measurements in the environment, called *Actual Environmental Concentrations* (AECs). Although sampling provides actual residue data, each set of data is valid only for the point in time when and locations where the samples are taken. Each sample yields only a hint of the scope of residue incidence. Modeling and monitoring often are combined within Tier 4 to provide a more comprehensive understanding of the distribution of exposure that occurs within treated watersheds.

Estimated Environmental Concentrations for Terrestrial Systems

Terrestrial (unlike aquatic) wildlife are exposed to pesticides primarily through the plant or animal material that they consume as food. Other routes of exposure—such as dermal, inhalation, and ocular—are considered less important, and effects are thought to at least resemble those associated with oral and dietary routes. For birds, the primary route of exposure is dietary exposure to residues on food items, although there is a potential to ingest granular pesticides. EECs are developed for these routes of exposure.

Exposure estimates for wildlife vary according to the amount of pesticide residue on food items and the actual consumption of those items. In determining EECs for terrestrial organisms, it is assumed that residue levels on food items increase as application rates per acre rise.

Predicting Concentrations on Food Items The EPA and pesticide manufacturers, in performing Tier 1 risk assessments, use a series of tables that establish guidelines on how much pesticide residue might be expected on various types of plants and insects (Fig. 4.36). The original tables were developed by Fred Hoerger and Gene Kenega and refined by John Fletcher, James Nellessen, and Thomas Pfleeger.

Predicting the total amount of residue available on vegetation, based on what is called the Kenega Nomagram, depends on two variables: application rate and plant type. Plant types are assigned to a simple plant characterization scheme:

- Short rangegrass
- Long grass

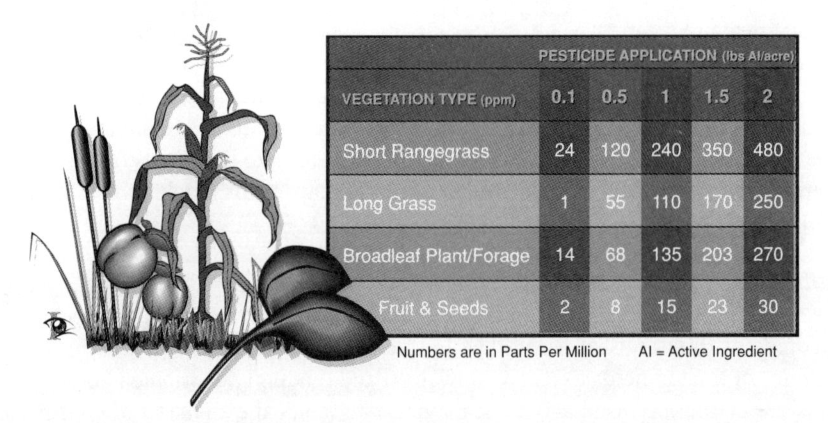

VEGETATION TYPE (ppm)	PESTICIDE APPLICATION (lbs AI/acre) 0.1	0.5	1	1.5	2
Short Rangegrass	24	120	240	350	480
Long Grass	1	55	110	170	250
Broadleaf Plant/Forage	14	68	135	203	270
Fruit & Seeds	2	8	15	23	30

Numbers are in Parts Per Million AI = Active Ingredient

Figure 4.36 Estimates of pesticide residue concentrations based on application rates are used to conservatively predict concentrations.

- Broadleaf plants/forage
- Fruits
- Seeds

A portion of the revised Kenega table in Fig. 4.36 illustrates maximum expected residues per plant species as a function of application rate. For instance, the table predicts a maximum EEC of 240 ppm on rangegrass immediately after a pesticide application at one pound per acre.

EECs also are proposed for birds and mammals that consume pesticide-treated insects, without the benefit of empirical data. It is important to note that these EECs are based on application rates without regard to the characteristics of the pesticide. When actual residue data are not available, EECs of 58 and 135 ppm (based on pounds applied per acre) can be used as estimates of residues on large and small insects, respectively.

Models Can Account for Residue Declines Residue levels from the Kenaga table can be refined. First, the most appropriate environmental fate half-life value is determined. This allows for degraded residue values to be calculated, which consider multiple applications and time intervals between repeat applications. These values can be used as more realistic estimates of terrestrial residues. Further refinements can factor in more specific feeding habits, food sources, body weights, and ingestion rates for sensitive species likely to inhabit or visit the treated area.

Food Consumption Patterns Dictate the Amount of Exposure Exposure of birds or mammals can be refined by incorporating the weight of the animal, the percentage of food consumed relative to body weight, and the amount of pesticide residue on the food item.

Consider this example. A 100 g (0.1 kg) bird is known to consume a quantity of seeds equivalent to 10% of its body weight, each day, following an application of 1.5 lb of active ingredient per acre. The estimated EEC is predicted to be 23 ppm by using the Kenega table for fruits and seeds. If it is assumed that this avian species feeds exclusively on the seeds, what is the EEC (total amount of pesticide per kilogram of seeds consumed daily)?

$$0.01 \text{ kg seeds/day} \times 23 \text{ mg pesticide/kg seeds} = 0.23 \text{ mg pesticide/day}$$

Therefore, the bird ingests 0.23 mg of pesticide per day. What is the estimated daily exposure?

$$\frac{0.23 \text{ mg pesticide/day}}{0.1 \text{ kg (weight of bird)}} = 2.3 \text{ mg pesticide/kg body weight/day}$$

Granular Product LD_{50} per Square Foot There is a potential for birds and mammals to ingest pesticide granules. The initial risk assessment of oral exposure of terrestrial vertebrates to a granular insecticide considers an estimated number of unincorporated granules per square foot in relation to the number per square foot that results in 50% mortality, that is, LD_{50}. This procedure is based on

- application rate,
- portion of granules unincorporated,
- concentration of the pesticide per granule,
- a laboratory-derived LD_{50}, and
- body weight of the bird or mammal in question.

The conservative procedure converts the amount of active ingredient in exposed pesticide granules in a given area into the number of toxic doses potentially available per unit area. Previous field studies have provided evidence that wildlife mortality may be observed when the number of LD_{50}/ft^2 exceeds established criteria. If a granular pesticide use results in LD_{50}/ft^2 values higher than established criteria, additional research may be conducted to either confirm or refute the presumption of adverse effects.

RISK CHARACTERIZATION

Risk characterization is the integration step in risk assessment. Once all available data on exposure (EEC) and toxicity are assembled, the overall ecological risk for a pesticide can be characterized. The exposure and toxicity characterizations are integrated into a comprehensive, scientifically defensible description of the potential risk to the environment from use of the pesticide.

Currently, the key components of risk characterization include:

- calculation of risk quotients,
- level-of-concern analysis,
- use of probabilistic techniques if there is a presumption of risk at the Tier 1 level, and
- weight-of-evidence analysis.

Risk characterization should yield clear, concise information on the scientific rationale applied during the assessment process.

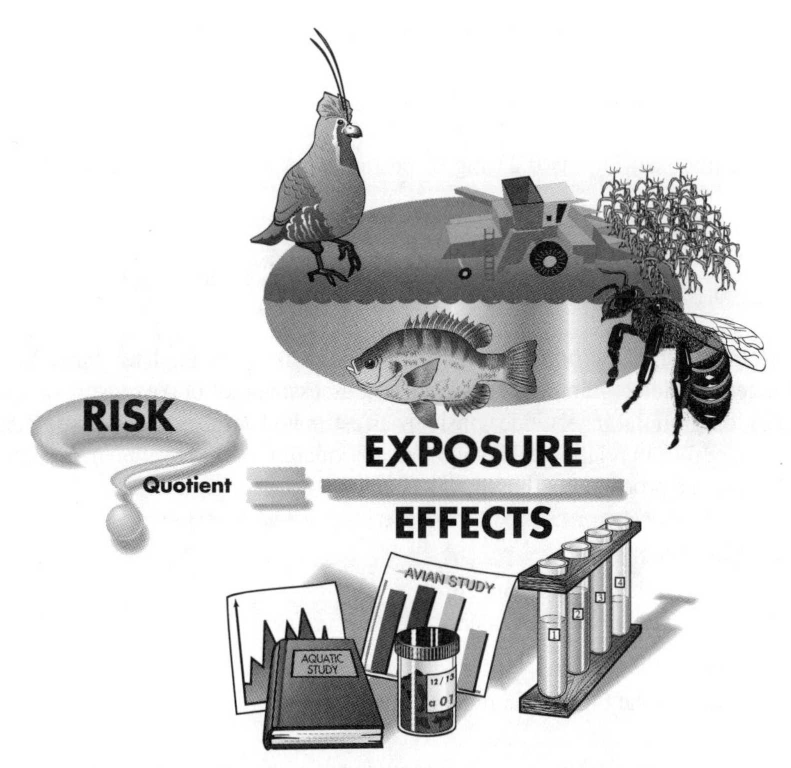

Figure 4.37 Risk is a function of toxicity (effects) and exposure.

Risk Quotients: The Integration of Toxicity and Exposure

The integration of toxicity and exposure is accomplished by developing an index called the risk quotient (RQ = EEC/Effects concentration). This begins with a conservative Tier 1 assessment that utilizes the highest EEC and the most sensitive effects end point to determine the quotient. An RQ provides general guidance on potential risk posed by a pesticide. It is derived by dividing the EEC for a particular environmental compartment (such as water) by a toxicological end point (such as LC_{50}) for an organism (e.g., fish) subject to exposure in that compartment.

In other words, an EEC for *water* may be divided by an LC_{50} for *fish* to determine the risk quotient. An RQ of less than one indicates an estimated exposure concentration below the toxicity end point. A risk quotient greater than one indicates that exposure may exceed levels shown in laboratory tests to produce adverse effects; it may lead to refined estimates of exposure and effects to gain a better understanding of the risks which are likely to occur in the environment (Fig. 4.37).

Levels of Concern

LOCs are trigger ratios used by regulatory authorities for comparison against calculated RQs. LOCs incorporate (into risk assessment) uncertainties due to possible exposure of sensitive populations and estimated environmental concentrations. It has long been recognized that the number of organisms used to examine adverse effects is limited; thus, there

Level of Concern (LOC)

End Point and Scenario	Risk Quotient	Nonendangered	Endangered
Mammalian acute (granular)	LD_{50} / FT^2	0.5	0.1
Mammalian acute (spray)	EEC / LC_{50}	0.5	0.1
Mammalian chronic (spray)	EEC / NOEC	1.0	1.0
Avian acute (granular)	LD_{50} / FT^2	0.5	0.1
Avian dietary (spray)	EEC / LC_{50}	0.5	0.1
Avian chronic (spray)	EEC / NOEC	1.0	1.0
Aquatic acute	EEC / LC_{50}	0.5	0.05
Aquatic chronic	EEC / NOEC	1.0	1.0
Terrestrial plants	EEC / EC_{25}	1.0	1.0
Aquatic plants	EEC / EC_{50}	1.0	1.0

NOEC = No Observed Effect Concentration

Figure 4.38 Risk quotients which compare exposure and effects are used as early indicators of ecological concern.

remains the possibility that untested organisms in the same environment may be more or less sensitive that those tested to a particular pesticide. The EPA established LOC trigger values to ensure adequate protection for more-sensitive, untested, and/or endangered species.

There are two general categories of LOCs—acute and chronic—for each nontarget fauna group; and there is one category (acute) for each nontarget flora group. To determine if an LOC has been exceeded, a risk quotient must be determined and compared to trigger values. The table in Fig. 4.38 provides risk quotients and LOCs.

LOCs are *regulatory triggers* used to categorize whether the potential risk is of low, medium, or high concern. For instance, a risk quotient less than 0.5 developed for an avian acute response is of minimal concern to nonendangered species, whereas a quotient of 0.5 or greater suggests potentially higher acute risk.

LOCs differ among biological indicators and types of tests as well as between nonendangered and endangered species. If the risk quotient is categorized as minimal for a chemical where no LOC is triggered, the use of the pesticide is predicted to cause no adverse effects when used in accordance with the label; and registration or reregistration is granted. The LOC is not a hard and fast target; it is something that is used to make initial evaluations. For instance, if an LOC is 1 and the RQ is 3, it would be of moderate concern. A moderate (for fish, LOC > 0.1 but < 0.5) risk quotient indicates that applicators should be educated on use of the pesticide to minimize the likelihood of adverse environmental effects; pesticides with moderate RQs usually are granted restricted-use registration.

Figure 4.39 Levels of concern allow for the categorization and prioritization of risks.

Registrants of pesticides with risk quotients that generally exceed the LOC face risk mitigation requirements prior to product registration. Optimally, mitigation efforts lower potential risk concerns below the LOC, frequently through a refined EEC. Reduced rates of application and number of applications, buffer strips, in-furrow application (vs. broadcast), and ground application (vs. aerial) exemplify measures that can be taken to minimize fish and wildlife exposure (Fig. 4.39).

A Case Study The following is an example of the use of toxicological data and EEC values to calculate a risk quotient; it is based on an application rate of one pound of active ingredient per acre.

The data and assessment show that the risk quotients for Pesticide A do not exceed LOCs of acute and chronic exposures for fish and birds; thus, no additional testing or mitigation is required. For Pesticide B, which is more toxic, the risk quotients are greater than the threshold for presumption of risk for both acute and chronic exposure. Additional testing, or more extensive evaluation of the EEC, may be required to demonstrate a reduced risk, or risk mitigation measures might be adopted. This example illustrates how two pesticides with different toxicological properties but similar application rates can have different presumptions of risk, based on their toxicity.

Application rates also influence risk assessment. In the second illustration (Fig. 4.40b), the toxicological data assumes that Pesticide A has an application rate three times that of Pesticide B and illustrates that, although Pesticide B is more toxic to fish and birds, its presumption of risk is similar to the less toxic Pesticide A. In the second illustration (Fig. 4.40b), neither pesticide's RQ value exceeds the triggers for presumption of risk, although pesticide B is acutely toxic to fish and birds.

(a)

Calculated Risk Quotients (application rate 1lb/acre)

Organism	Toxicity Test	Pesticide A	Pesticide B	Estimated Environmental Concentration (EEC) Water Residue	Estimated Environmental Concentration (EEC) Soil Surface or Food Residue	Risk Quotient A	Risk Quotient B
Fish	Acute LC_{50} (mg/L)	2	0.125	0.08 mg/L		0.04[a]	0.64
	Chronic NOEC (mg/L)	0.5	0.03	0.08 mg/L		0.16	2.67
Bird	Acute LD_{50}[c] (mg/kg)	1000	75		10.41 mg/ft^2	0.06[b]	0.78
	Dietary LC_{50} (ppm)	1000	75		120 ppm	0.12	1.60
	Reproduction NOEC (ppm)	350	25		50 ppm	0.14	2.0

[a]RQ=EEC/Toxicity; thus, 0.08/2=0.04

[b]$RQ = \dfrac{\text{Application rate (lb active ingredient per acre)} \times (453{,}590\ \text{mg/lb}/43{,}560\ \text{ft}^2/\text{acre})}{LD_{50}\ \text{mg/kg} \times \text{weight of bird (grams)}/1000\text{g/kg}}$

[c]Bobwhite quail with a mean weight of 178 grams

(b)

Calculated Risk Quotients (application rate A=1lb/acre, B=.33lb/acre)

Organism	Toxicity Test	Pesticide A	Pesticide B	Estimated Environmental Concentration (EEC) Water Residue	Estimated Environmental Concentration (EEC) Soil Surface or Food Residue	Risk Quotient A	Risk Quotient B
Fish	Acute LC_{50} (mg/L)	2	0.125	A=0.08 mg/L B=0.025		0.04[a]	0.2
	Chronic NOEC (mg/L)	0.5	0.03	A=0.08 mg/L B=0.025		0.16	0.83
Bird	Acute LD_{50}[c] (mg/kg)	1000	75		A=10.41 mg/ft^2 B=3.47	0.06[b]	0.26
	Dietary LC_{50} (ppm)	1000	75		A=100 ppm B=33	0.10	0.44
	Reproduction NOEC (ppm)	350	25		A=60 ppm B=20 ppm	0.17	0.80

[a]RQ=EEC/Toxicity; thus, 0.08/2=0.04; 0.025/0.125=0.2

[b]$RQ = \dfrac{\text{Application rate (lb active ingredient/acre)} \times (453{,}590\ \text{mg/lb}/43{,}560\ \text{ft}^2/\text{acre})}{LD_{50}\ \text{mg/kg} \times \text{weight of bird (grams)}/1000\text{g/kg}}$

[c]Bobwhite quail with a mean weight of 178 grams

Figure 4.40 Examples of how multiple factors (e.g., toxicity and application rate) influence the estimation of risk.

Probabilistic Risk Assessment

Ecological risk assessment is evolving into a more complex process than that previously described. The quotient method is the current Tier 1 approach, which is designed to be conservative. If a compound does not trigger a level of concern one can reasonably assume that there is a low potential for adverse environmental effects. However, if a compound fails this Tier 1 screen, it does not automatically mean that the compound will cause adverse effects to wildlife. Failure simply means that a more detailed examination of exposure and potential effects must be conducted.

Probabilistic methods for assessment of risk have been recommended for higher tier assessments. The approach is to use distributions of exposure (intensity and duration) as well as a distribution of species sensitivity. These data are combined to assess the likelihood that a given pesticide concentration in water, soil, or vegetation will exceed an effects threshold concentration, leading to an adverse environmental effect. The major advantages of using probabilistic distributions are that (1) they use all the available toxicity data and expand beyond point estimates of exposure, and (2) they provide a quantitative description of the magnitude and probability of the adverse effect(s). In many cases, toxicity tests beyond those routinely required will be needed to complete the assessment.

Weight-of-Evidence Analysis

Risk characterization requires that the complete data and analysis describing environmental fate and ecological effects of a pesticide be considered in a weight-of-evidence analysis. Strengths, limitations, and uncertainties, as well as magnitude, frequency, and spatial and temporal patterns of previously identified adverse effects, are discussed in this analysis. Monitoring data and reported incidents of wildlife mortality are included to help confirm risk potential as predicted from laboratory and field studies. Dissipation and application characteristics of the pesticide—distance from application site, duration of effects, and time of year at which wildlife and aquatic organisms may be most susceptible—are discussed in terms of likelihood of the pesticide to affect wildlife and aquatic organisms. Very often, the weight-of-evidence analysis suggests a number of potential mitigation measures that may be used to reduce risk while maintaining benefits from continued registration of the pesticide (Fig. 4.41).

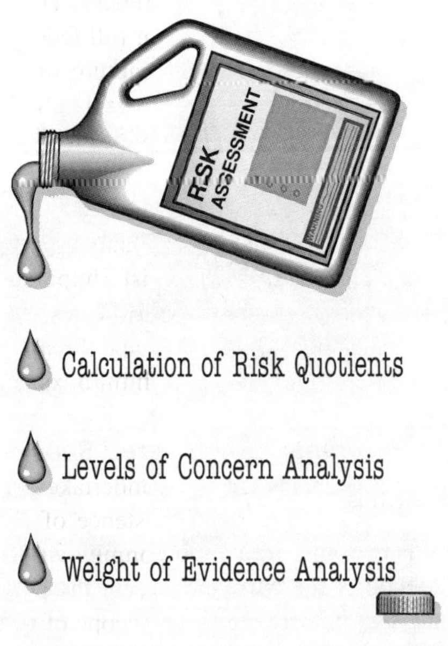

Figure 4.41 Assessing the risk to wildlife requires a number of approaches.

A further important step in characterizing the ecological risks posed by a pesticide is to compare potential risks from pesticides already used on the same site and, typically, for the same pests. This helps decision makers to view the overall picture of potential ecological risk while making registration and reregistration decisions. Thus, the risk conclusions drawn from ecological risk assessments consider the likelihood of adverse effects, based on evidence from extensive data analysis as well as the professional judgment of experienced regulatory authorities.

PUBLIC POLICY ON WILDLIFE

Endangered Species Act

The Endangered Species Act (ESA) of 1973 and its subsequent amendments comprise the major federal legislation that protects not only threatened and endangered wild plants and animals, but also critical habitats and ecosystems that support those and many other species. Referring to endangered and threatened plants and animals, ESA states in the preamble that "these species of fish, wildlife, and plants are of aesthetic, ecological, educational, historical, recreational, and scientific value to the nation and its people."

The U.S. Fish and Wildlife Service and the National Marine Fisheries Service are responsible for identifying candidate species for federal endangered species protection. Species proposed for listing are made public through a notice of review in the Federal Register. This notice of review is the process whereby these two federal agencies ask all interested persons and organizations for biological and ecological information on each species on the proposed list. Currently 3600 species or subspecies of plants and animals have been identified as candidates for listing.

In the United States, information sufficient to move approximately 1000 species from candidate status to the United States List of Endangered and Threatened Wildlife and Plants has been gathered. Federally listed species are given full federal protection: No one shall "harass, harm, pursue, hunt, shoot, wound, kill, trap, capture, or collect, or attempt to engage in any such conduct." All federal agencies (e.g., EPA and U.S. Army Corps of Engineers) must comply with the ESA by ensuring that their activities will not jeopardize the continued existence of a listed species.

The Federal List of Endangered Species Plants are in the majority on the U.S. Fish and Wildlife Service's endangered species list (http://endangered.fws.gov/listing/index.html). The term species means a species, subspecies, or distinct population. The total number of U.S. species considered endangered and threatened on June 30, 2000, was 961 and 273, respectively (http://ecos.fws. gov/tess/html/boxscore.html) (Table 4.2).

EPA Office of Pesticide Programs' Endangered Species Protection Program
The ESA mandates that federal agencies shall not undertake activities or make decisions whose consequences will adversely impact the existence of federally threatened or endangered species or their habitats. The EPA must comply with the provisions of the ESA in assuring that a pesticide registration does not create the potential for exposure of, or otherwise jeopardize, a federally listed species. The scope of this program covers all outdoor uses of pesticides, including home and garden uses.

TABLE 4.2 Total Number of U.S. Species Endangered and Threatened

Endangered	Category	Threatened
63	Mammals	9
78	Birds	15
14	Reptiles	22
10	Amphibians	8
69	Fishes	44
20	Snails	11
61	Clams	8
18	Crustaceans	3
30	Insects	9
6	Arachnids	0
564	Flowering plants	141
2	Conifers and cycads	1
24	Ferns and allies	2
2	Lichens	0

The Endangered Species Protection Program has two phases: consultation and implementation. The EPA's "may affect" determination takes place prior to formal consultation with the U.S. Fish and Wildlife Service (FWS) and is the key to initiating consultation. Following are the fundamental steps in the process.

- *Species which potentially could be affected by the use of pesticides are identified.* The EPA, the U.S. Department of Agriculture, and the FWS have collaboratively ranked approximately 93 species for pesticide vulnerability.
- *Pesticides that may impact any of these species are identified.* The EPA identifies the pesticides registered for use in areas within the range of a protected species and issues what is known as "may affect" determination.
- *The EPA may eliminate a "may affect" determination.* The EPA may remove a "may affect" determination through pesticide use limitations that are sufficient to achieve a "no effect" determination.
- *Environmental Protection Agency consults with the Fish and Wildlife Service on the remaining "may effect" determinations.* The EPA requests a formal consultation with the Fish and Wildlife Service. A thorough review of the species allows the FWS to develop a Biological Opinion, which indicates if harm is likely to result from pesticide exposure to a specific organism in a specific habitat. This Biological Opinion specifies reasonable and prudent measures, such as specific pesticide use limitations, that the EPA must implement to protect the species.
- *Habitat maps are developed.* Where there is potential for impact, the EPA develops species habitat maps within an Endangered Species Bulletin (http://www.epa.gov/espp/). The bulletin identifies pesticides that may harm the species and describes use limitations necessary to protect them.
- *Pesticide users must read labels.* Pesticide labels alert the pesticide user to refer to county Endangered Species Bulletins. If the area in which the user is making an application is included in the bulletin, the user must comply with all of the provisions.

The bulletin becomes part of the labeling and therefore carries the full force of law if not properly followed once the Endangered Species Protection Program becomes final. Currently, a voluntary approach to protection protects many species from exposures to pesticides.

Program implementation includes several components, depending on the approach to protection selected by state pesticide regulatory agencies. Currently, there is an EPA interim program in which some pilot states are conducting activities to protect endangered species from pesticides. The federal approach to protection is through labelings, bulletins, and fact sheets. The label refers the user to a bulletin and a toll-free endangered species hotline number (800-447-3813) to call for information about endangered species, such as whether there is a bulletin available for the county. The user must comply with use restrictions in the bulletin, which contains a map, a list of pesticides, and use limitations such as buffer zones or limitations on application methods. An example of a Endangered Species Bulletin for fresh water mollusks in Indiana is presented in (Fig. 4.42).

State "Protection from Pesticides" Plans

About one-fourth of the states manage or are developing their own programs to protect federally listed species from pesticide injury as an alternative to, or in addition to, the EPA labeling program. In many cases, protection is accomplished by providing information and education on endangered species and pesticides directly to affected landowners, land managers, operators, applicators, and dealers. Pesticide management plans for lands near these species are negotiated jointly with users. Some states are involved in mapping, developing protection guidelines, or in other ways providing protection from potential harm from pesticides. State plans need the approval of the FWS and the EPA and can substitute for EPA bulletins and fact sheets. Some excellent brochures have been developed by state programs.

State Wildlife Resource Management Programs

States have considerable responsibility to protect wildlife. State biologists are actively creating species inventories from which a better understanding of distribution and abundance can be obtained. Using the guidelines established by the Minnesota Department of Natural Resources, most states adopt similar strategies to help protect wildlife and habitat.

Identification of species and habitats and setting priorities for conservation. A comprehensive biological inventory of the state's endangered species and natural habitats is the first step toward their protection. Without this information, responsible management decisions cannot be made concerning the fauna and flora of an area.

Protection by saving the best and the rarest. After identification, ecologically significant lands are protected by acquisition, conservation easement, or landowner registry. Creative partnerships among public agencies, private conservation organizations, and private landowners are the key to protecting all the major types of wildlife habitat (Fig. 4.43).

Stewardship by managing endangered species and unique habitats. The protection of threatened natural lands is critical in the conservation process. Active management, including monitoring and restoration, often is required to maintain the ecological conditions necessary for long-term survival of endangered species and their habitats.

How To Use This Information

1) On the county map, find the specific shading pattern(s) in or near the area where you intend to apply pesticides.
2) Read the descriptor under the Shading Key for the pattern(s) to identify the specific area involved.
3) In the "Table of Pesticide Active Ingredients," locate the active ingredient in the pesticide you intend to apply.
4) Locate the code to the right of the active ingredient name and under the shading pattern(s) that apply to you.
5) When using the pesticide, find the code(s) described under "Limitations on Pesticide Use" and follow the limitation given.
6) If you are applying more than one listed active ingredient or applying a listed active ingredient in an area with more than one shading pattern (species), multiple codes may apply. If so, you should follow the most restrictive limitation.
7) Read the information on Reducing Runoff and Drift in this pamphlet.

Table of Pesticide Active Ingredients

Active Ingredient	Shading Pattern Code
BENOMYL	1m
CARBARYL	1m
CHLORPYRIFOS	
Alfalfa	43
All Other Uses Except as a Termiticide	1m
DIAZINON	1m
DICOFOL	1m
DIMETHOATE	1m
DISULFOTON	1m
MALATHION	1m
NALED	1m
PARATHION(ethyl)	1m
PHOSMET	1
PROPICONAZOLE	1m
PYRETHRINS	1m
TRICHLORFON	1m

Limitations on Pesticide Use

Codes/Limitations

1 Do not apply this pesticide within 20 yards from the edge of water within the shaded area shown on the map for **ground applications**, nor within 100 yards for **aerial applications**.

1m Within the shaded area shown on the map and ½ mile up all streams that join the shaded area do not apply this pesticide within 20 yards from the edge of water for **ground applications**, nor within 100 yards for **aerial applications**.

43 Do not apply this pesticide within 100 yards from the edge of water within the shaded area shown on the map for **ground applications**, nor within ¼ mile for **aerial applications**.

Figure 4-42 An endangered species bulletin for mussel species in Indiana.

DeKalb County, Indiana

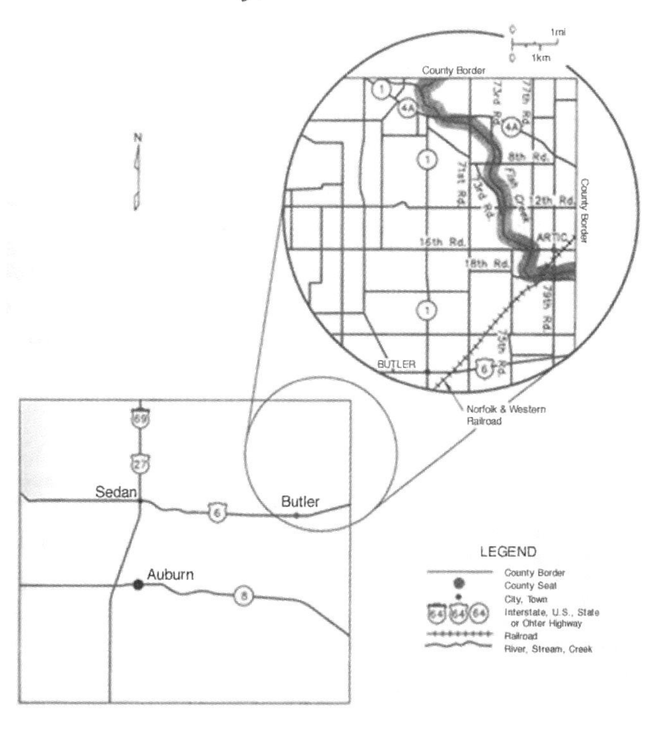

SHADING KEY

Freshwater mollusks [**Clubshell mussel,** *Pleurobema clava,* ***Northern riffleshell,*** *Epioblasma torulosa totulosa.* **White cat's paw pearly mussel,** *Epioblasma (=Dysnomia) suicata delicata*].

Figure 4.42 *(Continued)*

Figure 4.43 Protecting habitat is a key principal in wildlife conservation.

Promoting public awareness through education. Educational programs which enhance public awareness of the loss of natural habitat and the potential jeopardy to endangered species will result in increased public support for conservation initiatives.

SUMMARY

Among the goals of society is our ambition to provide citizens with meaningful employment, appropriate food, good health, decent housing, a safe environment, and quality education. Government, industry, and the general public must approach economic productivity in a manner that is ecologically and environmentally sound. Most wildlife species do not have the luxury of moving to new habitats when exposed to pesticides; they either adapt to changes in their habitat or cease to exist.

A diverse, healthy flora and fauna are indicators of a healthy ecosystem. It behooves us to take our environmental stewardship seriously and to take the reasonable steps necessary to protect wildlife from hazards posed by pesticides.

The responsibility for ensuring that wildlife is protected from potential adverse pesticidal effects belongs to *manufacturers, government,* and *the pesticide user* and can be viewed as a *triangle of wildlife protection* (Fig. 4.44). The manufacturer must develop products,

Figure 4.44 Protecting wildlife against harm from pesticides requires a partnership among government, industry, applicators, and the public.

supported by sound scientific studies, that allow for the maximum benefits of use with minimal risk to wildlife and its habitat. Local, state, and federal government must establish standards for pesticide use, promote research addressing wildlife contaminant issues, and educate and certify pesticide users. The pesticide user—farmer, homeowner, and professional applicator—must follow pesticide label instructions and strive to apply pesticides as carefully as possible, with wildlife protection in mind. Protection and sustainability of our environmental heritage is a task requiring support.

CHAPTER 5

WATER QUALITY RISK ASSESSMENT: PREDICTING COMPLEX INTERACTIONS BETWEEN PESTICIDES AND THE ENVIRONMENT

Today we face challenges concerning water as a *renewable* and *limited* resource, nationwide. But it was once considered an *unlimited* resource. Rivers and streams were viewed as cheap, dependable sources of water that supported the national surge in manufacturing, construction, irrigated agriculture, and employment; unfortunately, they were also viewed as prime avenues for the disposal of waste materials (Fig. 5.1).

During the 1960s, a significant degradation of the nation's surface water became evident. The dumping of sewage effluent and by-products of manufacturing and agriculture into surface water had become associated with *contamination* and *pollution*. Once-pristine rivers and streams were tainted from the repeated introduction of waste. Several contamination incidents (such as the Cuyahoga River catching on fire) received significant media attention, sparking public concern for water quality. These events evoked such emotional outrage that public coalitions demanded legislative attention. The resulting multitude of new laws, both state and federal, specified policies and goals for establishing water quality, placing the responsibility for compliance squarely with the cities and industries releasing pollutants into water.

Protecting water quality is now an environmental priority. However, pollution control is shifting from dependence on government regulation, alone, to cooperative activities among groups with diverse viewpoints. This cooperation evolved as various factions recognized community prosperity as a function of the development of water policies that blend economic goals and environmental incentives. But even though stakeholder interaction is common, major differences still exist among public, industrial, agriculture, regulatory, and environmental groups.

The issue of pesticides affecting water quality is complex technically, socially, and politically. But if community life, agriculture, industry, recreation, wildlife, and natural habitats are to flourish, input from an educated public is essential. Put simply, we must understand. We must make informed decisions.

Figure 5.1 Water quality has emerged as a national priority.

CYCLING OF WATER IN THE ENVIRONMENT

Oceans contain 97% of the world's water. The remaining 3% is fresh water, of which approximately 70% is stored as ice in glaciers. Nearly all of the unfrozen fresh water on the planet occurs in belowground aquifers; only 1% of unfrozen fresh water is stored in lakes, streams, and rivers (Fig. 5.2).

Water drawn from rivers and tapped from deep within the earth's aquifers is not "new." It has been continuously recycled between land and the atmosphere for thousands of years

Figure 5.2 Distribution of the world's water supply.

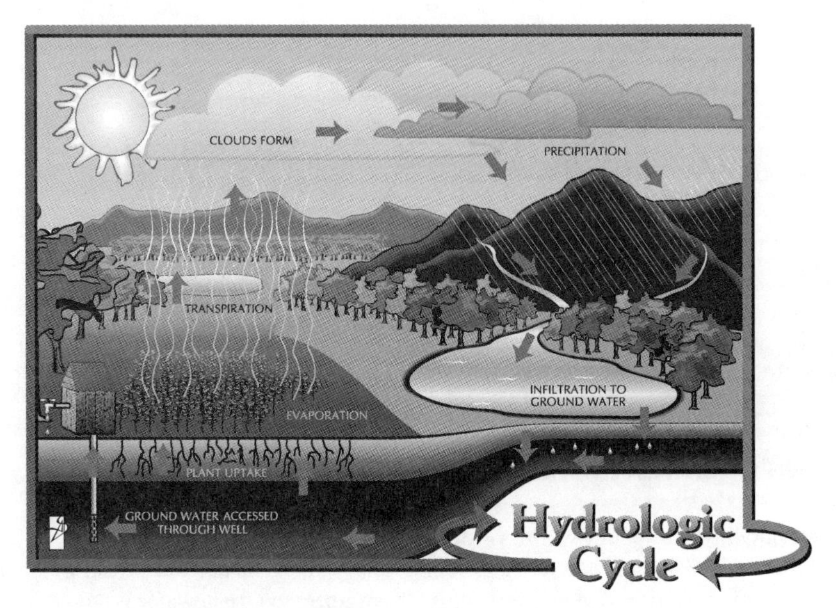

Figure 5.3 The cycling of water in the environment.

through intricate processes of evaporation, transpiration, precipitation, overland runoff, infiltration, and interactions between ground and surface water. Together these processes are linked as the hydrologic (water) cycle (Fig. 5.3).

The hydrologic cycle, energized by the sun, links surface water to both ground water and atmospheric water. Solar energy converts surface water to water vapor which moves into the atmosphere during evaporation. Plants absorb water from the soil and can release it into the atmosphere by transpiring (giving off) water vapor from their leaves. Water vapor rises, then condenses in the cooler atmosphere to form clouds; water stored in clouds is eventually returned as precipitation in the form of rain, hail, sleet, or snow. Water entering the soil can percolate deeper to reach ground water, which in turn can discharge to lakes and streams or arrive back at the surface through wells, marshes, and springs (Fig. 5.4). Ground water can contribute to the base flow of streams, lakes, wetlands, and other waterways. Once on the surface, water is again energized by the sun; the evaporation and transpiration processes are repeated, providing water vapor for cloud formation and continuation of the hydrologic cycle.

Water at the Surface

A surface water system is characterized by its watershed or drainage basin. A watershed is the area of land draining to a specific river; the watershed boundary is defined by a region's topography. Watersheds vary in size and can be nested within other watersheds of increasing size, similar to the branching of a tree. For example, the entire Mississippi River watershed that drains into the Gulf of Mexico encompasses most of the central United States; it consists of thousands of smaller subwatersheds, each contributing to the total water volume flowing into the Gulf (Fig. 5.5).

Within a watershed, surface water occurs in a network of storage and flow areas. Soil constitutes a large internal catchment (storage body) for water within a watershed; reservoirs

Figure 5.4 Ground water can appear on the surface as a spring.

serve as catchments of flowing water. These can include natural bodies of water such as rivers, lakes, and wetlands as well as constructed (artificial) water reservoirs such as canals, man-made lakes, and drainage ditches. Catchments can *cycle* their water; that is, a "new" volume of water can replace the "old" volume. The storage time of water—also known as hydrologic residence time—depends on the hydrologic characteristics of the catchment. The mean hydrologic residence time, defined as a ratio of the average volume to the average flow, represents how long it takes to replace an "old" volume of water with a "new" volume.

Rainfall and melting snow infiltrate the top layers of soil at a rate commensurate with the soil's porosity and initial water content. Infiltration is controlled by the intensity and duration of precipitation. The upper few inches of soil can become saturated, temporarily exceeding its capacity to hold water. The water's capacity to percolate downward can be likewise decreased, and water accumulating on the land surface flows overland to a lower elevation. This movement, termed *overland flow* or *surface runoff,* can occur across small or large areas, depending on the amount and intensity of precipitation and the local terrain, soil type, soil moisture, land slope, and vegetative cover. A gentle rain lasting all day may result in only moderate runoff, but an intense summer thunderstorm producing a large amount of rainfall in a short time may yield significant runoff. Runoff also can result from miscalculated or inappropriate timing, intensity, and duration of irrigation. Runoff flows down a slope or gradient until it reaches a storage area (e.g., a stream, pond, or low spot); when the storage/infiltration capacity of that area is exceeded, runoff flows even farther down gradient. Flooding occurs when precipitation exceeds the storage capacity of a given area.

Figure 5.5 The flow of water within a watershed.

Water in the Ground

Ground water is a widely distributed natural resource found beneath the earth's surface. However, most ground water occurs in tiny voids between grains of gravel, sand, silt, and clay or in cracks and fractures in the bedrock in geologic formations called aquifers.

The geology of a particular location dictates the depth and volume of ground water. Ground water for wells and springs comes from geologic formations called aquifers, which may be shallow (near the earth's surface) or very deep (hundreds of feet below the surface). As a general rule, the most commonly used fresh water aquifers lie 30–300 feet belowground.

The solid portions of aquifers are composed of rock, sand, or gravel, depending on local geology. Most aquifers are unconsolidated (loose) deposits of sand, clay, silt, or gravel and contain water in the voids between particles and rock fragments. Other aquifers occur as cracks in bedrock or in consolidated (solid) materials such as igneous rock (granite, basalt), sedimentary rock (limestone, siltstone, sandstone), or metamorphic rock (slate, shale).

Aquifers are characterized as either confined or unconfined. Confined aquifers lie below a confining layer of less permeable clay or rock which slows the vertical movement of water into them. The water in confined aquifers can be recharged from water that moves into the water-bearing zone at distant areas where there are no confining layers.

Unconfined aquifers do not have a confining layer and are "open" to water moving down from surfaces directly above. The water surface of unconfined aquifers—the water table—fluctuates with changes in atmospheric pressure, rainfall, withdrawal via pumping wells, and other factors. Unconfined, unconsolidated aquifers are particularly vulnerable to contamination because typically they are quite shallow, allowing surface water to reach the water table (ground water) very quickly.

Both confined and unconfined aquifers are saturated zones in which the voids between solids are completely filled with water. Between topsoil and water-saturated soils, voids of unconsolidated materials fill with water and air, forming the vadose (unsaturated) zone. The portion of the vadose zone near the soil surface is where plants root, vegetation decays, and animals burrow; it is in this area that most terrestrial plants and soil organisms reside. The lower portion of the vadose zone hosts less biological activity than the upper soil layer.

Precipitation either runs off sloping land or infiltrates into soil. All soils can store water in voids. The ability of a soil to store and transfer water downward in saturated or unsaturated conditions is a function of numerous soil features. For example, the nature of soil particles and the way they aggregate influence key features such as porosity and how water is absorbed by soils. Soil with small voids (such as clay) can hold more water than those with larger voids (such as sand).

Natural ground water movement within aquifers often is somewhat horizontal, whereas water movement in vadose zones is usually vertical and follows the direction of the topography (e.g., subsurface runoff). Horizontal flow of ground water generally is slow, depending on the porosity and the permeability of the materials that make up the aquifer; it is measured in centimeters per day. Conversely, water flow through the vadose zone can occur very rapidly (meters per day), especially in soils prone to preferential flow through cracks, worm holes, root channels, etc.

THE INTRODUCTION OF PESTICIDES INTO THE WATER CYCLE

The watershed concept is important because it links land area to bodies of surface water. Land use within a watershed impacts the quality of local surface water; the quality of water leaving a watershed can, in turn, affect the cumulative quality of water great distances downstream. For example, pesticides detected in a city's drinking water supply could come from lawn applications and other urban uses *or* from an upstream watershed where agriculture is predominant. Certainly the potential exists for compounds to move off-site (downstream) in surface water, but numerous biological, physical, and chemical processes can alter the fate of pesticides in the environment.

When water contamination is detected, the logical first response is to identify the contaminant and determine where and how it entered the water. *Point-source* contaminants such as discharges from industrial or municipal waste treatment plants can be traced to discrete sources of entry. But contaminants from agricultural runoff and housing developments, such as eroded soil, nutrients and pesticides, form *nonpoint-sources,* and there is not always a clear distinction between point- and nonpoint-source contamination.

The outfall (place of discharge) of a municipal storm water system may be a point source, but pesticide contaminants that it delivers into a river, stream, or lake may originate from thousands of lawn applications within a community (i.e., from nonpoint-sources).

In 1999, the U.S. Geological Survey released *The Quality of Our Nation's Waters,* a comprehensive, in-depth evaluation of the impact of pesticides and nutrients on water; it can be viewed on the Internet at <http://water.usgs.gov/pubs/FS/test313.html>. The 8-year study examined more than 50 major river basins in urban and rural areas—more than half the land acreage of the United States, where more than 60% of the U.S. population live and work. Some of the findings from this work include the following:

- Stream quality in urban settings is impacted more by insecticides than by herbicides.
- Agricultural application of herbicides has the greatest influence on streams and shallow ground water.

- Pesticide and nutrient concentrations in surface water vary from season to season and from watershed to watershed. Small streams experience rapid increases and decreases in pesticide residues during and after heavy rains. In contrast, pesticide concentrations in large rivers remain lower than those found in smaller streams because of increased dilution. However, pesticides are detectable in large rivers for longer periods of time.
- Shallow ground water (less than 100 ft) overlain with sand and gravel, karst, or other highly conductive materials is most vulnerable to contamination.
- The data suggest that public health is not generally threatened because levels of individual contaminants do not exceed drinking water standards.
- Pesticide concentrations in many urban and agricultural streams were found to exceed safety levels for some aquatic life.

Movement of Pesticides into Surface Water

Surface water usually flows in defined channels such as streams and rivers. The amount and rate of flow vary with precipitation, channel structure (substrate and geometry), and flow gradient.

Flowing water can carry dissolved and absorbed pollutants in suspended sediment from eroded soil (carried in runoff) and from the sides and bottom of the water channel. Sediments travel various distances, depending on the size and surface characteristics of the sediment and on the water flow rate. Generally, sediment is deposited, resuspended, and redeposited by flowing water. When flowing water meets stored water, such as when a stream enters a pond, the flow rate is decreased and the larger, heavier sediment settles to the bottom. Smaller, finer, lighter weight materials such as clay may remain in suspension for longer periods of time. Whereas flowing water tends to *carry* pollutants, storage water tends to act as a *repository* because water transports (flows) out of it very slowly, allowing sediment to settle to the bottom and remain there. However, because of dilution from waters that enter a storage area ahead of the pollutant, contamination levels in storage water can be lower than in a stream.

Runoff and erosion have the potential to move or carry more pesticide off-site than leaching because runoff is a surface event and many pesticides are surface applied. Generally it takes much less time for a pesticide to reach surface water via runoff than to reach ground water via leaching.

Volatilization is the process whereby a solid or liquid evaporates into the atmosphere as a gas, which provides a significant pathway of transfer for some pesticides. The tendency of a pesticide to volatilize from water is approximated by the ratio of its vapor pressure to its aqueous solubility (equivalent to a Henry's law constant). Compounds with high vapor pressure (an inherent chemical property) and low solubility have a strong tendency to volatilize from water.

Specific environmental factors—high soil temperature, low relative humidity, and increased air movement—tend to increase volatilization. Pesticides that sorb tightly to soil are less likely to volatilize. In dry soils, the absence of moisture allows pesticides to sorb tightly to soil particles; thus, less volatilization occurs. Volatile pesticides usually are incorporated (plowed into the soil) after application or subsurface injection to reduce loss into the atmosphere. However, it also has been shown that pesticide volatilization from soil is complex and highly dependent on the movement of water to and from the soil surface.

Once a pesticide enters the atmosphere as a gas, it can break down in water droplets and be highly susceptible to long-range transport from the application site. Within the atmosphere, the pesticide may react with light (photolysis) and water (hydrolysis) and sorb to suspended materials such as dust particles. Pesticides in a gaseous state may dissolve in atmospheric water and be transferred back to the soil surface or surface waters during rainfall.

Exposure Routes There are a number of routes by which pesticides can reach surface water following application. The primary routes are through runoff and erosion, but in some circumstances spray drift, subflow from field tile, and direct applications (herbicides applied to water to control aquatic plants) contribute to detectable levels of contamination. The mechanisms of various routes of surface transport suggest that, although areas directly adjacent to fields or streams may be the most vulnerable, distant pesticide applications also can have an impact.

Spray Drift Spray drift can occur when liquid pesticides are applied with a tractor-mounted or handheld sprayer. The amount of drift depends primarily on wind speed, droplet size, and application height. Studies have shown that small droplets can be carried hundreds of feet under windy conditions. If movement to bodies of water, nontarget plant species, or dwellings is a concern, an appropriate downwind buffer zone is required.

Subflow Soil-applied pesticides have the potential to move very rapidly through drainage tile to surface waters. Field tile is designed to remove excess water from cropland, especially after early spring thaws or large amounts of rainfall during the growing season (Fig. 5.6). But highly soluble, persistent, and weakly sorbed pesticides (not strongly attached to soil particles or organic matter) can be transported along with excess water and become part of the local surface water system, moving from drainage ditches to wetlands, streams, and rivers. Drainage systems must be managed carefully to reduce this movement of pesticides.

Direct Introduction Direct introduction of pesticides into surface water often results from accidental spills and, occasionally, from illegal use of a pesticide.

Figure 5.6 Drainage tiles can move pesticides directly from a field into a body of water.

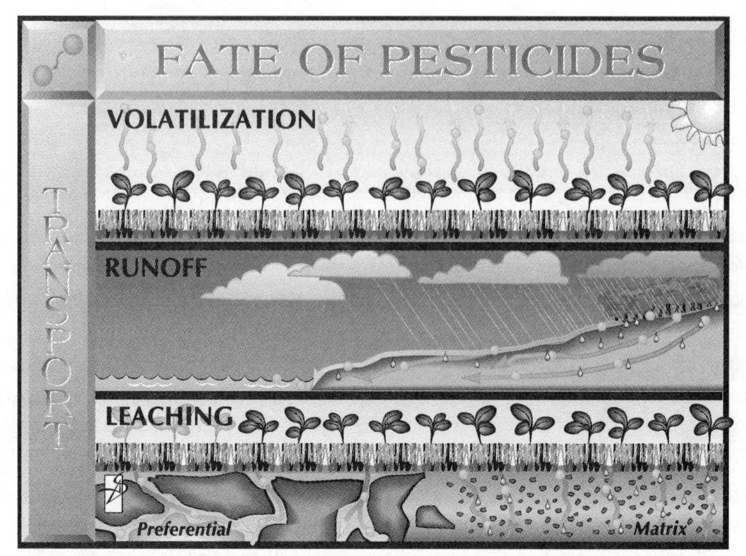

Figure 5.7 Transport mechanisms for pesticide movement in the environment.

Runoff and Erosion The runoff potential of a pesticide is influenced by five factors:

- Soil type
- Slope of terrain
- Intensity, amount, and timing of rainfall (with respect to pesticide application)
- Properties of the pesticide
- The cropping system

All five factors should be considered when estimating runoff potential. Weakly sorbed pesticides can be transported from an application site to surface water within minutes when heavy rain occurs shortly after application. Pesticides that exhibit strong sorption to soil usually have a lower runoff potential than those that exhibit weak sorption, but they can still reach surface water if sorbed to eroding soil particles. Herbicides are more likely transported in water than with eroded sediment. Foliar-applied herbicides that bind strongly to foliage and those that degrade rapidly on the leaf surface are less likely to move off-target in runoff than those that are soil applied.

The relationship between rainfall intensity and pesticide concentration in runoff is complex. Although significant, intense rainfall may result in a greater total amount of pesticide entering surface water, the resulting pesticide *concentration* may be lower than for a smaller amount of less intense rainfall, due to dilution.

Leaching of Pesticides into Ground Water

Leaching is the term for the process of downward movement of water through the soil profile. If the water contains pesticides, they will leach along with the water. In general, there are two kinds of phenomena associated with leaching: *preferential* and *matrix flow.* Although one may predominate, both can occur simultaneously.

Preferential flow occurs when water flows rapidly through wormholes, root channels, cracks, and large structural voids in soil. This allows pesticides to move rapidly through the soil profile with reduced likelihood that they will be retained by soil particles or undergo microbial degradation.

Matrix flow is the slow migration of water and chemicals into and through small pores in soil. The slow movement allows more time for pesticides to make contact with soil particles over a larger soil surface (Fig. 5.7).

The most important factors in determining whether a pesticide will leach are its degradation rate (persistence), its sorption (mobility) characteristics, and its propensity to release rapidly back into soil solution, once it is bound. Pesticides that are weakly sorbed by soil and resist degradation are more likely to leach to ground water than are those that degrade rapidly or remain tightly bound to soil. Factors such as soil type, topography, and rainfall also may impact the leaching potential of a pesticide, and factors such as application rate, frequency, and type (foliar, pre-and postemergence) also have an effect.

WATER QUALITY ASSESSMENT FOR PESTICIDE REGISTRATION

Water quality and environmental fate studies are conducted during the registration process to quantify the potential for the pesticide to contaminate surface or ground water; labeled use conditions and worst-case environmental conditions (e.g., heavy rainfall) are employed. Environmental concentration data from these studies are combined with toxicological data and used to determine the risk of exposure to humans as well as nontarget aquatic and terrestrial plant and animal species. Due to the expense and the high level of technical effort required to conduct field studies, sometimes computer simulation models are used to extrapolate results to a wider geographic region or to different climatic conditions and soil types. This probabilistic approach affords registrants and regulators a larger data set to work with in evaluating exposure and risk—without the expense and technical support needed to conduct additional field studies.

Tools Used to Assess Water Quality

There are a number of methods used to estimate the impact of pesticides on water quality. They range from simple calculations such as trigger values and chemical property comparisons to complex techniques such as computer modeling and laboratory and field experiments. Typically, the EPA requires laboratory and field dissipation studies, prior to registration, to determine the potential for water contamination. The EPA's evaluation of the entire data package, including toxicology, may be favorable, but if there are concerns about the potential of the molecule to reach ground or surface water, prospective or retrospective ground and surface water monitoring may be required as well.

Trigger Values Laboratory studies on environmental fate provide the basis for screening pesticidal effects on water quality prior to product registration (see Chapter 4 for more details on environmental fate studies). The scope of the regulatory screening process tends to focus on two aspects of pesticide behavior which affect leaching and runoff: (1) persistence, that is, how long it takes for the pesticide to be transformed or degraded, and (2) mobility, or how easily the pesticide and its degradation products can be transported to ground or surface water, based on the pesticide's sorption properties.

Information on the persistence of a pesticide is obtained primarily from laboratory soil metabolism, hydrolysis, and photolysis studies and from field dissipation studies. Information on a pesticide's mobility is obtained primarily from laboratory soil sorption experiments, column and lysimeter leaching experiments, and field dissipation studies. A complete set of environmental fate studies generally is required in identifying pesticides that have the *potential* to leach into ground water or to enter surface water as runoff.

Trigger values are based on laboratory data. More-refined water assessments consider additional field parameters such as application rate, soil, and target crops as well as the potential for exposure of nontarget species. Trigger values are determined from a group of reference pesticides for which use history *and* extensive water monitoring data are available.

Surface Water Triggers Data describing persistence and sorption are used to categorize a pesticide and its major degradation products into one of nine categories that qualitatively separate pesticides in terms of magnitude and anticipated duration, according to their relative potential to contaminate surface water. Pesticides also are distinguished according to their relative propensity to occur in the dissolved or sorbed phase.

In evaluating surface runoff potential (for labeling), pesticides are sometimes assigned to one of the nine categories, that is, nine possible combinations of three different ranges of their half-lives and three separate ranges of sorptive K_{oc}. (See Chapter 4 for more information on K_{oc}.)

The following criteria apply for surface water:

- Sorptive K_{oc}
  1. Low sorption: $K_{oc} \leq 500$
  2. Intermediate sorption: $K_{oc} > 500$ and ≤ 5000
  3. High sorption: $K_{oc} > 5000$
- Persistence (Half-Life) in Soil
  1. Short: half-life $\leq$ 2 weeks
  2. Intermediate: half-life $>$ 2 weeks but $\leq$ 2 months
  3. Long: half-life $>$ 2 months

Thus, using the prescribed criteria, pesticides can be grouped into nine categories representing each possible combination of low, intermediate, and high sorption relative to short-term, intermediate, and long-term persistence (e.g., low sorption/short persistence; or high sorption/intermediate persistence).

Persistence grouping helps predict how long a pesticide will remain in the soil and therefore be susceptible to runoff. Such considerations are important since some mitigation procedures that reduce soil erosion do not necessarily reduce runoff volume, and vice versa. Pesticides with a low sorption coefficient are more likely to move off-target in runoff. Pesticides with high sorption coefficients remain attached to soil and move only if the soil is eroded; therefore they are less likely to be found in lakes and streams.

Ground Water Triggers It may take a pesticide months or years to leach through soil to reach ground water. Therefore, a pesticide needs to be persistent *and* mobile to reach most aquifers. Federal and state regulators use several methods to assess a pesticide's leaching potential, but most estimates rely heavily on chemical half-life and soil sorption coefficients (K_{oc}) as the controlling parameters.

Pesticides are presumed by the EPA to have ground water contamination potential if environmental fate studies trigger a number of the following criteria for both persistence and mobility.

Trigger Values Related to Persistence

1. Aerobic soil metabolism half-life > 3 weeks
2. Field dissipation half-life > 3 weeks
3. Photolysis half-life > 1 week
4. Hydrolysis half-life > 60 days in sterile water

Trigger Values Related to Mobility

1. $K_{oc} < 300$
2. Weak- to moderate-acid pesticide which is not attracted to soil particles
3. Water solubility > 30 parts per million (ppm)

Models *Screening model* is the term used to (1) describe relatively simple computer techniques (ranking systems) based on chemical properties used to estimate a worst-case concentration in water which may occur at the intended pesticide application site, or (2) assign a relative rank compared to other pesticides.

Screening models are also called Tier 1 models because they constitute the first and simplest step in the assessment of a pesticide's potential impact on the environment (Fig. 5.8).

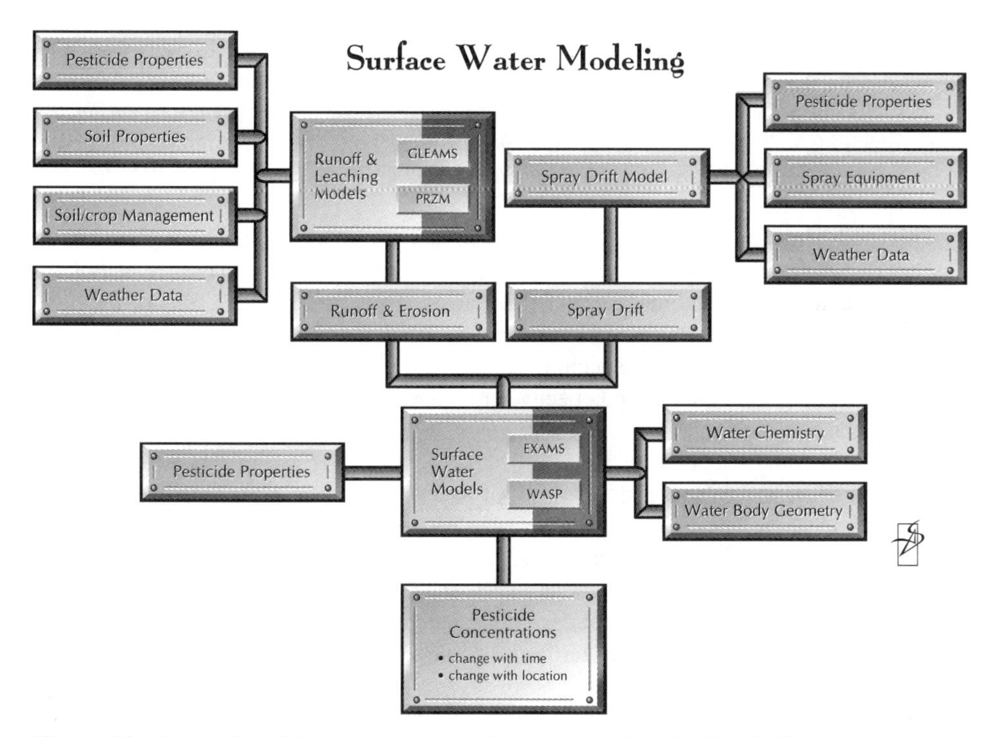

Figure 5.8 Data and modeling approaches used in asssessing the risk of pesticides to surface water.

Screening models are easy to use; require minimal data input; and provide quick estimates of pesticide levels that may occur in water, even if worst-case assumptions hold true and environmental conditions are conducive to leaching or runoff.

Modeling of pesticide leaching and runoff potential uses tools (with varying degrees of complexity, sophistication, and assumptions) that attempt to describe, mathematically, what happens in the real world. Computer models, using chemical and environmental parameters as inputs, can be used to predict pesticide contamination of surface or ground water in more locations, over longer periods of time, and under more diverse conditions than those feasible in field use or monitoring studies. Runoff and surface water monitoring studies are subject to unpredictable weather factors, but computer modeling uses historical weather data gathered over long timespans and wide geographic areas. Using probabilistic techniques or Monte Carlo analysis, regulators and scientists can easily evaluate the effects of varying soil or weather conditions or use practices. Water exposure models used by the Office of Pesticide Programs and described in this chapter can be found at http://www.epa.gov/oppefed1/models/water/models4.htm (Fig. 5.9).

No model consisting of sets of mathematical equations can cover every possible contingency in the natural environment. Therefore, models are validated to varying degrees by comparison with field studies of pesticide fate and behavior. This validation work yields "real-world" regulatory modeling results, although it probably never will be sophisticated enough to eliminate the need for collection of field data.

Field Studies Water quality field studies required for registration of a new pesticide range from relatively simple terrestrial field dissipation (lysimeter) studies to more complex and expensive prospective ground water monitoring; retrospective ground water monitoring; tap water monitoring; and surface water monitoring or runoff studies. The EPA requires (at a minimum) a terrestrial field dissipation study for assessing the persistence and potential environmental mobility of a new pesticide. These studies usually are conducted at two or more locations within the pesticide use region. The product is applied according to label directions to a small (typically < 0.1 ha) soil plot which is then sampled to a depth of 2–3 ft, in 3- to 6-inch increments and analyzed for the pesticide and metabolites of interest.

This study enables scientists to calculate the half-life of the compound (how fast it degrades) under field conditions and to determine whether the compound or its metabolites are significantly mobile and move out of the plowed layer. Finally, the results of the field study are used to verify laboratory findings. If the results of the field study suggest a potential for the molecule to contaminate ground or surface water, the EPA may require additional, more intensive studies such as the prospective ground water (PGW) study which is discussed in detail later in this chapter.

Water Monitoring When a sufficient set of data on surface and ground water concentrations of widely used pesticides is available, the EPA uses it as a basis for risk assessment. The EPA and other federal agencies, including the U.S. Geological Survey and the U.S. Department of Agriculture, work together to provide monitoring data for pesticides in surface and ground water. The process of designing such monitoring studies is complicated by the need to assess both ecological hazards and drinking water exposure. If monitoring data for a pesticide that has been on the market for several years is scarce, the EPA may require a retrospective ground water monitoring study or a tap water monitoring study; both are discussed later in this chapter.

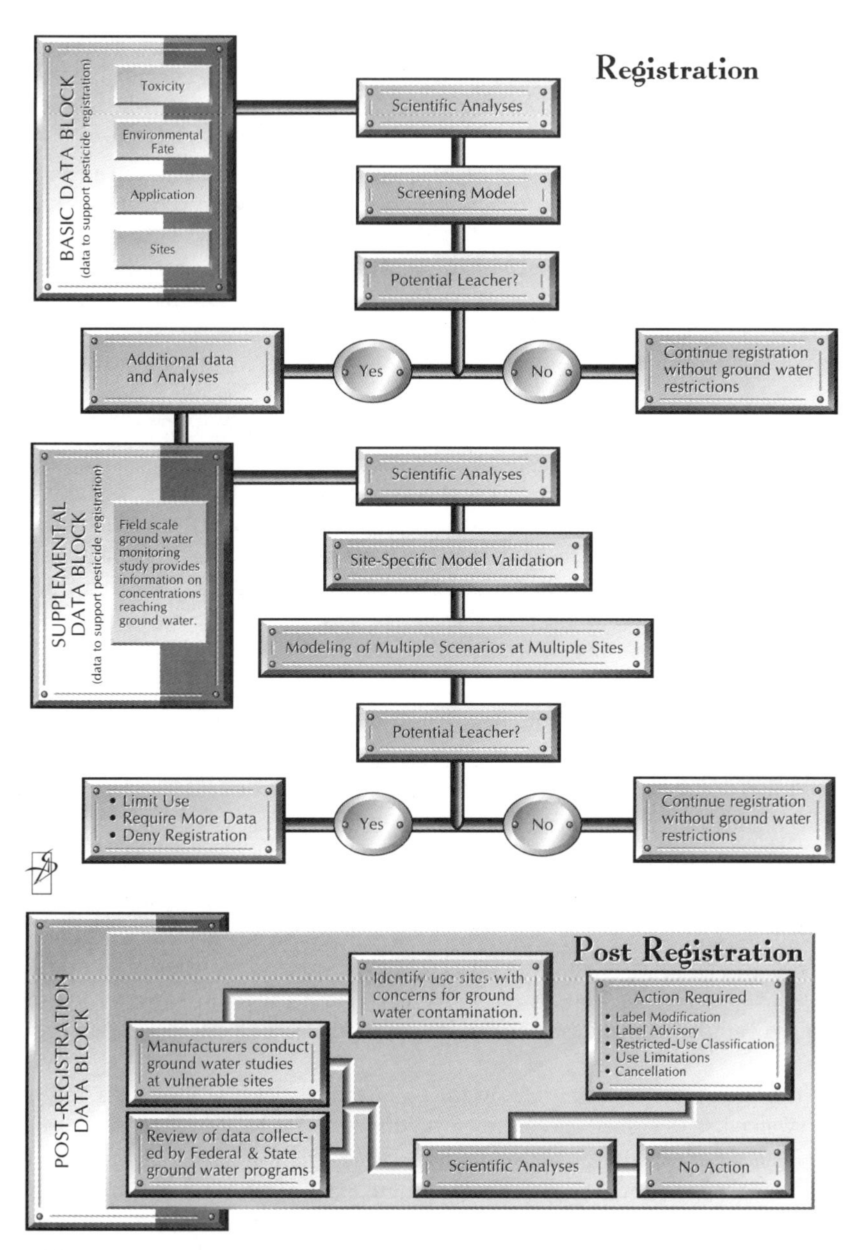

Figure 5.9 The registration process for determining the potential of a pesticide to leach to ground water.

Surface Water Exposure Assessments

Techniques to Predict Surface Water Pesticide Concentrations

Tier 1 Screening Models The Generic Estimated Environmental Concentration program (GENEEC) is used for Tier 1 estimation of pesticide concentrations in surface water and for subsequent evaluation of ecological effects. The first tier model is designed as a coarse screen; it estimates concentrations based on fundamental chemical parameters and on application information from the pesticide label. GENEEC uses a chemical's soil–water partition coefficient and degradation half-life values to estimate runoff from a 10-ha field into a 1-ha, 2-m-deep pond (see Chapter 4 for details on GENEEC).

GENEEC calculates both acute and chronic generic estimated environmental concentration (EEC) values. It considers reduction in dissolved pesticide concentration due to adsorption of the pesticide to soil or sediment, incorporation, degradation in soil before washoff to surface water, direct deposition of spray drift into surface water body, and degradation of the pesticide within a body of water. It is designed to mimic a high-exposure simulation obtained with more sophisticated models.

Tier 1 is used to screen chemicals to determine whether or not their risk potential warrants higher level modeling. Chemicals that fail in the screening process advance to the more sophisticated, parameter intensive Tier 2 modeling: PRZM, EXAMS, and AgDrift.

Drinking Water Estimates. The Tier 1 model FQPA Index Reservoir Screening Tool (FIRST) is used in estimating pesticide concentrations in surface water and in subsequent evaluation of drinking water exposure levels. It is a program to estimate potential acute and chronic pesticide concentrations in untreated surface water used as a source of drinking water.

The program considers

- reduction in dissolved concentration due to the percentage of the watershed that is cropland (percent cropped area),
- adsorption of the pesticide to field soil and to reservoir sediment,
- incorporation of the pesticide following application,
- degradation in soil before washoff to the reservoir, and
- degradation of the pesticide within the body of water.

Reservoir water concentrations may be increased due to the deposition of spray drift into a feeding stream or directly into the reservoir. The FIRST program requires less time and effort than other models but is designed to mimic complex simulations that use sophisticated models.

The FIRST program assumes that up to 6% of the pesticide applied to a 173-ha (ca. 427-acre) watershed is washed into the reservoir during one large storm. The actual amount, which appears as the dissolved concentration estimate, is a function of the equilibrium partition coefficient (K_d) or the organic carbon normalized equilibrium partition coefficient (K_{oc}). This parameter is used in Tier 1 to partition the pesticide in this field–reservoir system into two separate phases: a dissolved (in water) phase, and an adsorbed (to soil) phase. The pesticide in the dissolved-in-water phase is considered toxicologically available during consumption. This concentration is used to calculate the concentration of pesticides in reservoirs used for drinking water.

Tier 2 Mechanistic Models Both GENEEC and FIRST use scenarios which are likely to result in the highest exposure to pesticides in a vulnerable body of water: a small pond or reservoir with a long residence time, that is, where water is held for long periods. Higher tiers require a more specific assessment to determine the pesticide levels likely to occur under various actual use conditions.

A combination of three models generally is used for high-tier assessments:

- Pesticide Root Zone Model (PRZM), edge-of field runoff/leaching
- AgDrift, spray drift
- Exposure Analysis Modeling System (EXAMS), fate in surface water

PRZM, developed by the EPA in 1984, provides site-specific leaching and runoff estimates. As with other pesticide soil fate and transport models, PRZM incorporates soil characteristics and hydrology, weather, irrigation, and crop management practices into complex mathematical formulas used to estimate leaching potential. The EPA uses PRZM to make multiple site comparisons of a pesticide's leaching potential to that of older "reference" pesticides with histories of use and ground water monitoring. Models such as PRZM also are used to calculate estimates of the concentration of a pesticide that will leach. PRZM is a useful tool for comparing different pesticides at the same site and/or for comparing the same pesticide at different sites. Model estimates are most useful when laboratory and field data specific to the pesticide are available for comparison and interpretation.

Input data needed for PRZM includes pesticide fate properties, soil characteristics, management practices, and daily weather; output includes estimated runoff volumes, sediment yields, and associated pesticide concentrations at the edge of the field. Estimated pesticide runoff concentrations from PRZM and estimated concentrations from the spray drift model, AgDrift, are used in surface water models such as EXAMS; pesticide fate properties and receiving water characteristics are input, as well. Output, as a function of time and location, includes estimated peak and average pesticide concentrations present in water and sorbed to suspended and bottom sediments.

Examples from more than 100 input data required to predict the concentration of a pesticide in water, using PRZM, are shown in the following list. As can be seen in the comparison of the three parameters in the Screening Concentration In Ground Water (SCI-GROW), these higher tiered models require tremendous data input.

Snow melt factor (cm/°C)	Minimum evaporation extraction depth (cm)
Monthly daylight hours	Total hectares
Soil erodibility	Average storm duration (hr)
Topography	Number of crops
Crop interception storage (cm)	Crop rooting depth (cm)
Canopy area coverage (%)	Crop surface condition after harvest
Soil loss cover management	Crop dry weight at full canopy (kg/m^2)
Canopy height at maturation (cm)	Number of cropping periods
Day, month, year of crop emergence	Day, month, year of crop maturation
Number of pesticide applications	Number of pesticides
Target application day, month, year	Chemical application method
Pesticide application rate (kg/ha)	Application efficiency
Spray drift fraction	Core depth (cm)
Plant uptake factor	Diffusion coefficient for pesticide (cm^2/day)
Vaporization for pesticide (kcal/mol)	Henry's law constant for pesticide
Pesticide solubility	Number of soil horizons
Horizon thickness (cm)	Horizon bulk density (g/cm^3)
Soil water content in horizon	Pesticide solute dispersion coefficient
Dissolved phase decay rate (day)	Absorbed phase decay rate (day)

Vapor phase decay rate (day)
Horizon field capacity
Horizon wilting point
Percent organic carbon
Pesticide partition coefficient

Agricultural spray drift can be a significant source of pesticides in surface water. The spray drift model, AgDrift, includes generic data for screening-level assessments of drift and can be refined to consider very specific application conditions. Relevant information considered in AgDrift simulations includes pesticide and formulation, drop height, droplet size, nozzle type and orientation, wind speed, and speed of the application equipment. Both aerial and ground applications may be considered. Spray drift estimated by AgDrift can be combined with runoff/erosion estimated by PRZM, and the combined aquatic fate of these inputs can be assessed by EXAMS.

Temporal and geographic distributions of pesticide concentrations based on computer estimates (or on adequate monitoring data) are used to predict where and how frequently maximum, short-term average, or long-term average concentrations will exceed acute, chronic, or subchronic toxicity thresholds for human consumption and nontarget aquatic organisms.

The temporal and/or geographic distributions of computer-estimated or measured concentrations generally are plotted as cumulative frequency curves. Such curves are created by plotting maximum, short-term average, or long-term average pesticide concentrations against the percentage of years or sites for which equal or higher concentrations are expected. This approach allows scientists and regulators to assess whether pesticide concentration levels in surface water might pose human health or environmental concerns. Surface water that is used for human consumption and/or supports aquatic organisms is considered (Fig. 5.10).

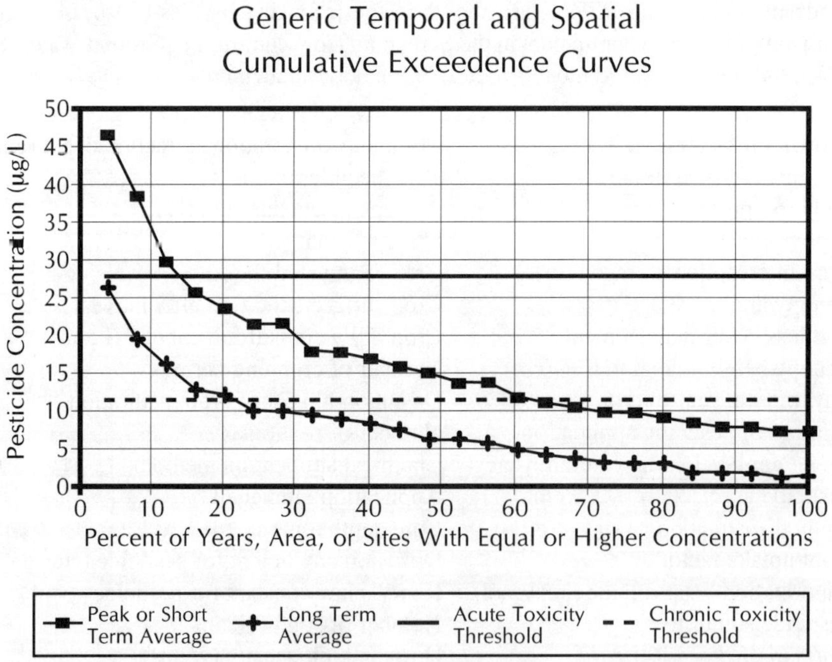

Generic Temporal and Spatial Cumulative Exceedence Curves

Figure 5.10 Chemical exceedence curves are used to show the probability of runoff and its ecological consequences.

In the past, a primary disadvantage of computer modeling was a general lack of controlled, field monitoring data to validate simulated results. Validation with appropriate data is needed to ensure accurate model estimates, and site-specific field data can be used to calibrate a model for use in other soil or climatic conditions. Due to conservative assumptions used and our knowledge of existing field and monitoring data, scientists in the Environmental Fate and Effects Division (EFED) of the EPA are reasonably confident that modeling estimates of pesticide runoff and concentrations in surface water (using PRZM) are conservative—that is, higher than actual—and therefore protective. The EPA and the plant protection industry have collaborated to validate PRZM for use in leaching and runoff modeling. Therefore, regulators are now more confident in using exposure model estimates for runoff when performing human and ecological risk assessments.

Tier 2 modeling for surface water source drinking water typically involves several standard scenarios, in this case using PRZM, AgDrift, and EXAMS. Scenarios are set up for each crop or use site, by region, and for specific crops as needed. However, the 1 ha-, 2-m-deep pond with no outlet (used for ecological modeling) is replaced by a 3.6 ha, 2.7-m-deep reservoir with inlet–outlet flow adjusted for a relatively long hydrologic residence time. The 10-ha field is replaced with a 173-ha field. Instead of assuming 100% cropped/treated (as in ecological modeling) for the watershed, a percentage of cropped acreage can be used in drinking water modeling. The differences between drinking water and ecological modeling reflect, in part, the larger watersheds required to feed drinking water reservoirs (vs. ponds fed by a smaller, localized watershed); differences in the hydrology of receiving waters are also reflected.

Field Studies The EPA does not require runoff studies routinely. But when available, the results of field and small-scale runoff studies are used to assess the effectiveness of mitigation methods. Results also are used to verify and quantify preliminary and modeling estimates of the runoff potential of pesticides and their major degradation products.

Both field and small-scale runoff studies provide data on amounts of water, soil, and pesticide transported in runoff from agricultural fields during and following rainfall.

Small-scale runoff studies are generally favored over large-scale field studies because they are large enough to consider the effects of formulation, tillage, crop cover, soil type, and slope on the transport of water, soil, and pesticides from the field, yet small enough to allow the use of weather-independent rainfall simulators. Consequently, the problem of unpredictable weather patterns in field studies is eliminated.

Small-scale runoff studies also can be much cheaper to conduct than large-scale field studies; so for the same cost, more combinations of factors affecting runoff can be studied (Fig. 5.11).

The major disadvantage of small-scale studies is that the amount of water, soil, and pesticide transported by runoff from each unit area often is substantially higher than from agricultural fields. Some of the difference is due to site-specific hydrologic factors such as landscape scale variability in sediment deposition, ponding, and infiltration that are not reflected in studies of small plot scale. Experimental conditions such as the use of high-intensity artificial rainfall in small-scale studies also may account for much of the variable output effect.

Surface Water Monitoring Surface water monitoring studies provide data on the fate and flux of pesticides in streams, rivers, lakes, and reservoirs. Pesticide concentrations in streams and rivers are highly seasonal, with peak concentrations occurring during the

Figure 5.11 (a) Rainfall is simulated to generate runoff. (b) Collecting soil and water samples from a runoff study. (c) Runoff water and sediment are sampled for pesticide residues. (d) Samples are collected and shipped to a laboratory for analysis.

first few runoff-producing storms after application, followed by rapid decline. However, pesticide concentrations are buffered in lakes and reservoirs due to longer hydrologic residence times than in rivers and streams; but pesticide concentrations may be present for long periods.

Pesticide concentrations in samples collected infrequently, or in samples collected at set sampling times not coincident with significant runoff, often do not accurately reflect peak

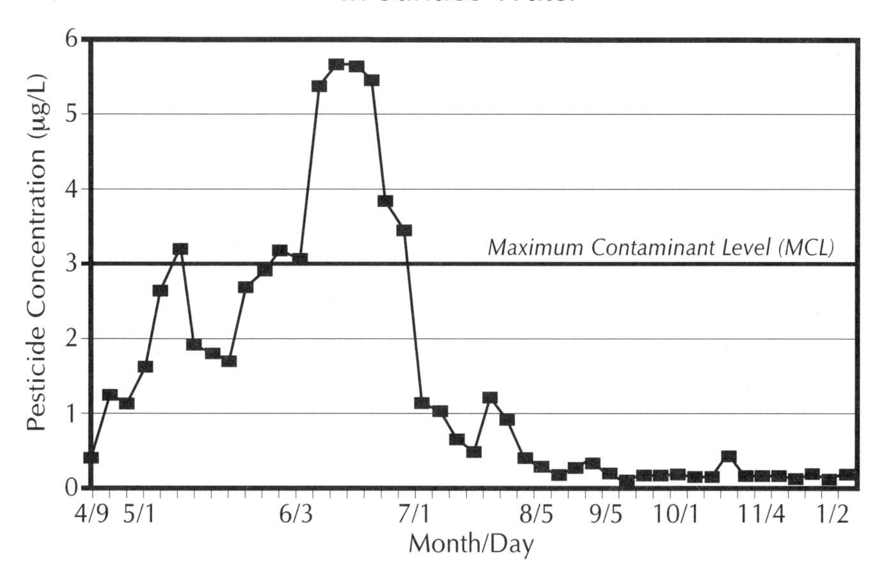

Figure 5.12 Pesticide concentrations in surface water show seasonal variations.

concentrations. Average pesticide concentrations in samples collected from a single location can vary as much as tenfold from year to year. Pesticide concentrations immediately after storms may vary by more than a thousandfold over a few days. Consequently, surface water samples taken on only a few occasions, or over a short span of time, often do not adequately represent the variation in pesticide concentration over time. The number of pesticides found in drinking water, the development of analytical methods with adequate detection limits, and the necessity of precise, timely, repetitive sampling make surface water monitoring studies quite costly. An example of changes in pesticide concentration over time in surface water is given in Figure 5.12.

Mitigation of Pesticide Exposure to Surface Water with Buffer Strips When risks are posed from pesticide contamination of surface water, mitigation is required. One solution to protecting creeks, streams, and rivers from contamination is the use of vegetative buffer strips around field borders and waterways. Buffer strips—also called grass strips, filter strips, and grass waterways—are intentionally planted to grass and often intermixed with trees and shrubs. These strips of permanent vegetation are located down slope from croplands to receive surface runoff. Their purpose is to retain pesticides, nutrients, or sediments absorbed by vegetation or soil for infiltration into the soil in the grass strip; the waterways are managed to discharge excess water.

Scientific data indicates that buffer strips do work. Their most impressive use—said to be 90–100% effective—is to trap sediments from agricultural fields. The vegetation in buffer strips slows the entrance of water; and as the water slows down it is unable to maintain sediments, which settle and become trapped in the grass. Capturing unwanted substances in this manner "filters" runoff before it can enter streams. Additionally, buffer strips form a potential habitat for a multitude of species around water bodies (Fig. 5.13).

Figure 5.13 A field demonstration study shows that leaving crop residues in the field helps to reduce erosion of soil into streams. Note the clarity of water from the lane where crop residues remain.

Some herbicides dissolve easily in water, some adsorb (bind) moderately to soil, and others adsorb tightly to soil. Thus, chemical properties of pesticides influence the effectiveness of buffer strips in reducing the amount of chemical that enters surface water. The effectiveness is greatest for pesticides that bind tightly to soil because they are deposited along with sediments in buffer strips, where they adsorb readily to vegetation. Buffer strips also can reduce the loading of moderately adsorbed and water-soluble pesticides into surface water by enhancing their infiltration into the soil beneath. Filter strips are most efficient in trapping pesticides during the early stages of a runoff event because, as more rain falls into the grass and more water flows in, less water can percolate into the saturated soil. Under these circumstances, dissolved pesticides can move in water through the grass into the receiving body of water.

Characteristics of the Treated Field

- Slope of cultivated land: Steep, short, continuous slopes *increase* both runoff and erosion to receiving water; shallow, long, and broken slopes *reduce* runoff and erosion. As the slope of the cultivated land increases, the effectiveness of filter strips in reducing pesticide loading to surface water decreases. The reason is that the greater the momentum of runoff entering the filter strip, the quicker the water flows on through. The water is "in" the buffer strip only briefly; thus, there is less time for sedimentation and infiltration to occur.
- Soil type: soil texture, organic matter, aggregation, bulk density, porosity, tillage, and erodibility of soil particles are interrelated characteristics that influence runoff versus infiltration of water. And because infiltration (along with settling of soil particles) is the primary mechanism for retaining pesticides, soils that allow infiltration (coarse-textured soils without a restricting layer) retain more.

Soils have been divided into *hydrologic soil groups* based on their runoff potential.

Hydrologic Groups

A—High infiltration rate: includes soils that are deep and well drained to excessively well drained, such as sands and gravels.

B—Moderate infiltration rate: includes soils that are deep to moderately deep, moderately well to well drained, and moderately coarse in texture.

C—Slow infiltration rate: includes soils with layers that impede the downward movement of water and soils with moderately fine or fine textures.

D—Very slow infiltration rate: includes clay soils, soils with a high water table, and soils with a shallow impervious layer.

Soils in the D group have the highest runoff potential, whereas soils in the A group have high infiltration and low runoff potential. Generally, soils with high runoff potential have less potential for ground water contamination, and soils with low runoff potential have higher potential for ground water contamination.

- Tillage type: Rainfall can cause bare soil to form a crust, increasing the likelihood of runoff. Tilling the soil after application to incorporate the pesticide greatly reduces its movement in runoff, but it also increases soil movement. Conservation tillage, which leaves residue on the soil, generally increases infiltration and decreases pesticide runoff. However, heavy rain within 24 hr of application may cause more runoff because of the pesticide's availability on the vegetative matter where it can easily be washed off (Fig. 5.14).
- Application method: Incorporation reduces runoff potential.
- Amount, intensity, and frequency of rainfall: One of the most important factors in determining the amount of pesticide runoff is the length of time between application and the first heavy rain. Intense rainfall shortly after application can lead to significant pesticide runoff, whereas a gentle rain a week or more after application will ensure infiltration. Although intense runoff events remove more pesticide from the target site, they also result in greater dilution. The pesticide concentration in runoff often is the greatest in intermediate runoff events.

Figure 5.14 There are many agronomic techniques (e.g., grass strips) that can be used to reduce soil erosion on steep slopes.

- Herbicide concentration in solution: Sorptive partition of a pesticide as measured by K_d will influence the amount of pesticide involved in water (via runoff) versus in soil particles (via erosion).

Characteristics of the Grass Strip

- Width: Most research has focused on the strip width necessary to allow soil particles to settle out. But it is safe to assume that wider buffer strips allow more opportunity for infiltration and sedimentation.
- Soil type: Because infiltration is the primary mechanism for retaining pesticides, soils that allow more infiltration (coarse-textured soils without a restricting layer) retain more water with pesticides.
- Moisture level: Saturated soils do not allow additional infiltration, even if they are very sandy.
- Vegetation density and type: The thicker the grass in the filter strips, the more water will infiltrate into the ground, provided there is not a restricted layer.
- Management of the buffer strip: The buildup of sediment at the point of entry can raise the ground level, acting as a barricade to incoming water and rendering the strip ineffective for filtering pesticides. Water then assumes a lower route, rushing through the filter strip and often forming a channel. Once water channels through a buffer strip, infiltration is limited and the water—and all that it carries—empties into the receiving surface water.

Although buffer strips offer one solution, the total solution to reducing nonpoint-source pollution may be a combination of the following:

- Conservation tillage: Can modified tillage practices reduce erosion?
- Pest scouting: Is the pest present in numbers that warrant treatment?
- Crop rotation: Can rotating crops reduce the number of pests?
- Soil testing: What nutrients and how much of them are actually needed?
- Pesticide selection: Are there pesticides that don't have water quality restrictions?

Ground Water Exposure Assessment

Techniques to Predict Ground Water Pesticide Concentrations

Tier 1 Screening Models Ground water screening models typically provide a *rank of relative leaching potential*—that is, the relative likelihood of a pesticide to leach to ground water—not an *estimate of actual concentration.* Among screening models used are the Ground Water Ubiquity Score (GUS) and the Screening Concentration In Ground Water (SCI-GROW) model, which estimates the potential concentration of a chemical in ground water.

GROUND WATER UBIQUITY SCORE GUS is useful for comparing the intrinsic leaching potential of pesticides. The GUS model is more sophisticated than simple trigger values because it uses an equation that combines both pesticide mobility and persistence parameters.

TABLE 5.1 Calculation of Tier 1 SCI-GROW and GUS Pesticide Screening Values

	Model Input			SCI-GROW Output	
Pesticide	Average $t_{1/2}$ (days)	K_{oc}	Rate (lb/Acre)	Maximum Exposure Value (ppb)	GUS (Index)
DBCP	203	88	80	321	4.7
Fluometuron	189	100	4	12	4.6
Atrazine	146	89	4	10	2.4
Bentazon	42	34	2	2	4.0
Dicamba	28	13	8	9	4.2
Simazine	98	124	10	10	3.8
Carbofuran	1	46	2	2	0.0
Metolachlor	67	200	4	2	3.1
EPTC	40	145	7	2	2.9
Molinate	21	50	8	2	3.0
Cyanazine	23	200	5	0.5	2.3
Butylate	30	304	7	0.7	2.2
Alachlor	7	124	4	0.4	1.6
2,4-D	10	49	2	0.1	2.3
Terbufos	27	617	35	2	1.7
Carbaryl	14	288	6	0.3	1.8
Diazinon	21	1,596	10	0.2	1.1
Dieldrin	2555	21,616	8	0.1	0.0
Trifluralin	115	7,000	2	0.01	0.3

Average values for only two pesticide parameters are needed to calculate GUS: the soil degradation half-life and the soil K_{oc}. (GUS $= \log_{10}(\text{half-life}) \times [4 - \log_{10}(K_{oc})]$. Pesticides with GUS values greater than 2.8 are more likely to leach to ground water than are those with GUS values between 1.8 and 2.8. Pesticides with GUS values less than 1.8 are unlikely to leach to ground water.

SCI-GROW The EPA begins Tier 1 with a SCI-GROW screening model for estimating ground water concentrations of a pesticide. SCI-GROW is a regression-based model developed by comparison of actual pesticide concentration in shallow ground water (in areas vulnerable to leaching) to pesticide half-lives in soil, K_{oc} values, and application rate. SCI-GROW provides a typical maximum pesticide concentration that may occur in highly vulnerable ground water (10- to 30-ft depth, overlaid by highly permeable soils) after a pesticide application at the maximum seasonal rate, under conditions most likely to favor movement of the pesticide to ground water; irrigation is used to ensure that, at minimum, a 30-year normal rainfall is applied at the site. The screening concentration estimated by SCI-GROW is intended as a likely upper-limit concentration in economically significant ground water in less than 1% of the pesticide use area.

Table 5.1 is an example of SCI-GROW and GUS output. It is important to recognize the simplicity of the SCI-GROW model. It uses only a few parameters to predict actual contamination of ground water.

Tier 2 Mechanistic Studies If a pesticide has a low leaching potential and passes a Tier 1 SCI-GROW or some other screening, no further assessment is required. However, if Tier 1 screening indicates significant leaching potential, a higher tier assessment is needed to further define the probable impact on ground water.

Computer modeling also has a role in the higher tier assessment of pesticide leaching and ground water contamination potential. A fate and transport model to estimate the extent of pesticide leaching in the unsaturated zone was first developed by the EPA in 1984: the Pesticide Root Zone Model (PRZM) provides site-specific leaching estimates. As with other pesticide soil fate and transport models, PRZM incorporates soil characteristics and hydrology, weather, irrigation, and crop management practices into complex mathematical formulas that estimate leaching potential. The EPA uses PRZM to make multiple-site comparisons of the leachability of a pesticide to older reference pesticides with histories of use and extensive ground water monitoring. Models such as PRZM also provide estimates of the concentration level at which a pesticide will leach, and they are useful tools for comparing different pesticides at the same site and/or for comparing different sites with the same pesticide. Model estimates interpreted with available laboratory and field data specific to the pesticide are most useful.

Pesticide leaching models such as PRZM may be compared to results from actual field studies to estimate reliability in predicting leaching potential. These models then can be used to estimate the impact of a pesticide over an entire use area with comparable soils and rainfall. Modeling may be used to develop use restrictions (e.g., soil and organic matter restrictions) to mitigate the potential for pesticides to reach ground water.

GROUND WATER FIELD STUDIES Prospective study. When analysis indicates a Tier 1 failure, the EPA may require ground water monitoring (Tier 2) studies to determine if pesticide use restrictions are necessary. They may be required for a registered pesticide when new data, for example, ground water detections at multiple sites, indicate contamination that was not anticipated initially. Ground water monitoring studies also may be required as a condition of registration when properties of a new pesticide and modeling results indicate the need for actual field data to better ascertain risks. The EPA commonly requires a prospective ground water (PGW) monitoring study to evaluate the leaching potential of new pesticides and of those without a substantial use history and ground water monitoring data.

The PGW study is typically conducted at 1–3 locations within a use area considered highly vulnerable to leaching [e.g., one with sandy soil or a shallow water table (less than 30-ft deep)]. A minimum of 8 clusters of suction lysimeters (soil water sampling devices) are installed at 3-ft, soil surface to water table depth intervals at each 1–2 ha site; and 8 clusters of monitoring wells are screened in the shallow aquifer (Figs. 5.15 and 5.16). The pesticide is applied at the maximum labeled rate, and sites are maintained using standard agronomic practices (Fig. 5.17). Typically, irrigation is applied to supplement rainfall and maintain the monthly water input at 120–130% of normal rainfall to ensure potential for water and chemical leaching. Weather stations and data loggers are used to monitor meteorological conditions and soil temperature during the study. A nonreactive, conservative tracer (e.g., bromide) is applied to track water movement and verify aquifer recharge. Typically, sites are monitored monthly for a minimum of 2 years, or until the dissipation of parent and metabolites is complete and the tracer has shown sufficient recharge of shallow ground water. Results from completed studies are used to estimate the magnitude of ground water contamination likely to occur within the proposed use area.

Retrospective study. For existing compounds with an established use history, the EPA may require carefully planned retrospective monitoring (in areas where the pesticide has

Figure 5.15 Water from wells is purged from the lysimeter for pesticide residue analysis.

been used and where soil types and hydrogeology are well defined) to provide geographically specific information on how a pesticide is impacting ground water. Data from the study can be used to determine if mitigation is necessary.

The retrospective study involves the selection of dozens to hundreds of sites representing vulnerable areas within an established use region: sites with sandy soil, shallow ground water, and/or high amounts of rainfall or irrigation. Potable water or irrigation wells can be used, as can newly installed monitoring wells. Typically, registrants install new wells to ensure proper construction and installation and to minimize the possibility of spurious detections of pesticides; they are placed in shallow, poor yielding aquifers typically not used for human consumption or irrigation.

In some cases, the EPA may require tap water monitoring studies to determine actual drinking water exposure of individuals living in the use regions. Tap water monitoring samples of "finished" drinking water are taken from private drinking wells that service

Figure 5.16 Lysimeters are set in farm feilds to sample ground water.

people's homes. Typically, the registrant takes tap water samples from hundreds or thousands of homes, and the samples are analyzed for the pesticide specific to the study as well as for any metabolites of interest.

Some tap water studies include public water supplies; drinking water obtained from surface water sources (reservoirs) may be tested in similar programs. These studies may involve a one-time sampling or periodic sampling for a year or more. Data from these studies are useful in refined evaluations of exposure and risk to the population consuming surface or ground watcr from thc rcgions sampled.

Figure 5.17 Pesticide application is made for a field scale study. Note the lysimeters in the background.

GROUND WATER MONITORING The EPA uses ground water monitoring data collected by other agencies to evaluate the impact of pesticide use on ground water quality in the public domain. One major effort in this area is thc collection of data in the Pesticides in Ground and Surface Water Database, which is maintained by the EPA's Office of Pesticide Programs. The database summarizes monitoring data (by state and county) from various state and federal agencies and other sources. The data are collected according to rigorous, prescribed processes and used by the EPA to support regulatory decisions. The U.S. Geological Survey has increased the number of pesticide analyses included in their ground water monitoring programs. USGS data are of particular value to regulators; traditionally, the data have been linked to the USGS mission to collect a full complement of hydrogeologic data to facilitate interpretation of their water resources monitoring data.

Even under optimum conditions, not all questions on the potential of a pesticide to impact ground water can be answered from field scale monitoring studies and modeling or from outside monitoring studies. Often there are situations where the likelihood of a pesticide leaching to ground water remains uncertain. Pesticide producers (registrants) cooperate more and more with state regulatory, agricultural, and environmental agencies to design monitoring programs to ensure that unexpected, undesirable environmental effects do not surface after registration. If uncertainties remain when a product is registered, the registrant may agree to conduct ongoing monitoring programs designed to provide information that can be used to head off potential ground water contamination problems.

PUBLIC POLICY ON WATER QUALITY

Government statutes address national, regional, state, and local water quality issues: chemicals leaching to ground water, drinking water standards, surface water runoff, integrity of wetlands, and protection of endangered species that inhabit aquatic environments. Some of the major laws, their purposes, and their potential impacts in relation to protecting, maintaining, and enhancing water quality are described herein.

Ground Water Protection under FIFRA

FIFRA is an extensive environmental protection law that requires the registration of pesticides for use in the United States. FIFRA requires water protection information on pesticide labels. It allows the EPA to restrict, cancel, or temporarily suspend pesticides that pose "unreasonable risk" through contamination of water. The EPA has the authority to seek national cancellation of pesticides that continually pose a threat to water quality. The EPA's efforts to prevent pesticides from reaching ground water include the following (Fig. 5.18):

- Predicting (on the basis of research data submitted by the manufacturer) a pesticide's potential to leach into ground water
- Establishing national label restrictions addressing concerns on leaching
- Requiring a restricted-use classification, triggering additional training requirements for users
- Providing each state the opportunity to develop and implement a State Management Plan for each pesticide identified as a potential leacher
- Canceling pesticides known to contaminate ground water despite aversion efforts

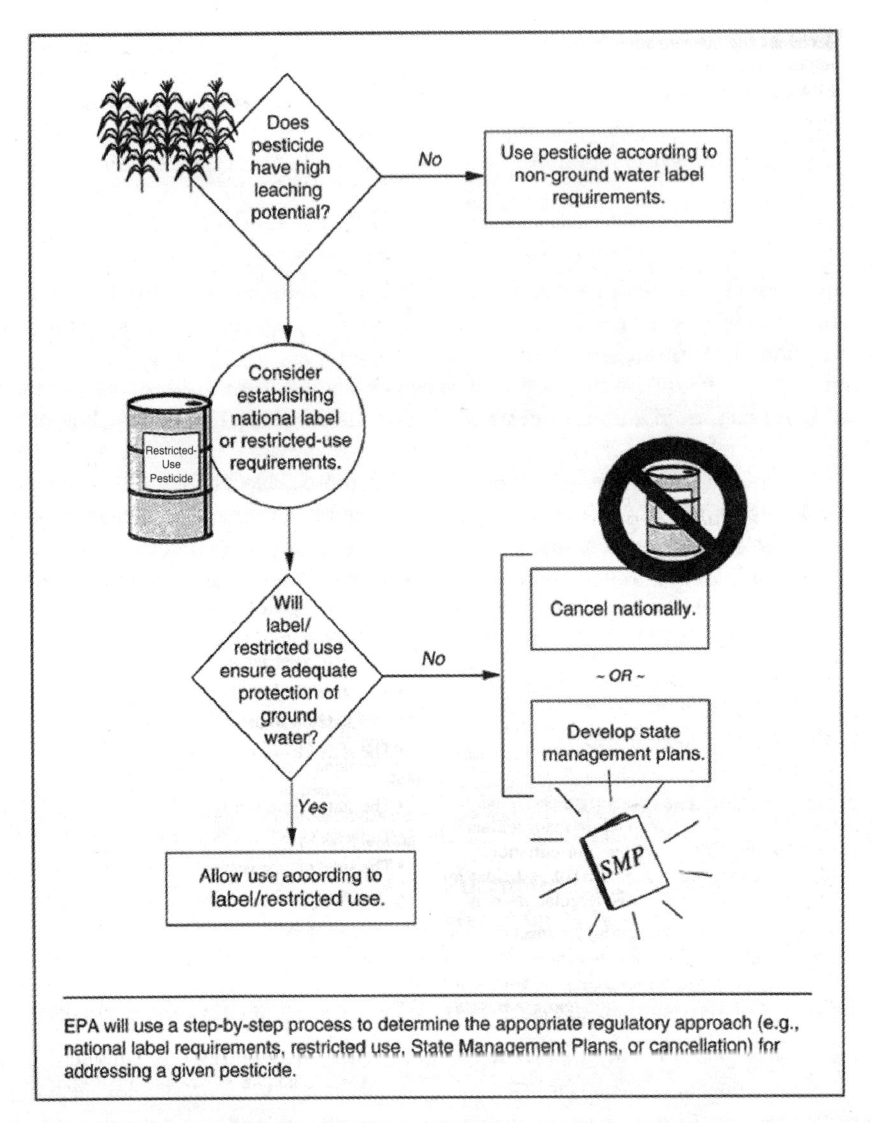

Figure 5.18 The regulatory framework that the EPA uses to judge whether or not a pesticide will require a state-specific ground water management plan.

State Management Plans In October 1991, the EPA exercised its regulatory authority under FIFRA in the release of its *Pesticides and Ground Water Strategy,* an approach to preventing pesticide contamination of the nation's ground water.

The centerpiece of the strategy was the development and implementation of State Management Plans (SMPs): a major departure from the agency's usual way of making national decisions on pesticide use, under FIFRA.

- The EPA recognizes that risks to ground water vary tremendously across the United States due to variations in hydrogeology, soil type, climate, pesticide use, crop patterns, etc.

- The premise is that, if a pesticide poses risk to ground water only in certain areas of the country, it makes sense to modify its use state-by-state rather than nationally.

The SMP approach emphasizes prevention and a cooperative federal–state partnership in which the states take the lead responsibility for protecting ground water. It gives states the flexibility to tailor pesticide management practices to local conditions and to promote the environmentally sound use of pesticides that might otherwise pose a high risk to ground water resources.

Certain pesticides determined to leach into ground water may remain registered for use *subject to an EPA-approved SMP, only.* However, SMPs are required only if label restrictions do not adequately ensure ground water protection. An SMP requirement is the most protective action the EPA can take other than cancellation of a product.

SMPs are implemented as follows:

- The EPA publishes a rule in the Federal Register declaring that the use of a certain pesticide is prohibited, as of a specific date, except in states that have an EPA-approved State Management Plan for that pesticide. States are not required to submit SMPs for every pesticide regulated through this approach, but states must have an approved SMP in place by the appropriate, preapproved date to continue the legal sale and use of the pesticide.
- The EPA asks states to develop two different kinds of SMPs—generic and pesticide-specific—and both types must contain components defined by the EPA.
- The Generic State Management Plan must contain the basic underlying framework of SMPs that is the same regardless of the pesticide addressed. The EPA is hopeful that generic SMPs will facilitate the development and approval of pesticide-specific plans.
- The Pesticide-Specific State Management Plan allows states to tailor their own strategies to prevent ground water contamination. Its basic components include a philosophy and goals toward protecting ground water:

  Roles and responsibilities of state agencies
  Legal authority
  Resources
  Basis for assessment and planning
  Monitoring
  Prevention actions
  Response to detections of pesticides
  Enforcement mechanisms
  Public awareness and participation
  Information dissemination
  Records and reporting

Surface Water Protection under the Clean Water Act

The purpose of the Clean Water Act is to restore and maintain the chemical, physical, and biological integrity of the nation's waters and to eventually eliminate pollution discharge. In 1987, the Clean Water Act turned to nonpoint-source pollution, requiring states to develop control plans. Information on specific watersheds can be found at http://epa.gov/surf/.

TABLE 5.2 Examples from a Water Quality Inventory Report

	Percentage of Total Waters Assessed
Rivers and streams	18% of 3,551,247 river miles
Lakes and reservoirs	46% of 39,920,000 acres
Estuaries	74% of 36,890 miles
Ocean coastal waters	6% of 56,121 miles
Great Lakes shoreline	99% of 5,382 miles
Wetlands	4% of 277 million acres

Levels of Overall Use Support (%)

	Rivers	Lakes	Estuaries
Fully supporting	56	43	56
Threatened	60	13	12
Partially supporting	25	35	23
Not supporting	13	9	9
Not attainable	<1	<1	0

Five Leading Sources of Water Quality Impairment

Rank	Rivers	Lakes	Estuaries
1	Agriculture	Agriculture	Municipal point sources
2	Municipal point sources	Urban runoff/storm sewers	Urban runoff/storm sewers
3	Urban runoff/ storm sewers	Hydrologic/habitat modifications	Agriculture
4	Resources	Municipal point extraction	Industrial point sources
5	Industrial point sources	On-site wastewater sources	Resources extraction disposal

Five Leading Causes of Water Quality Impairment

Rank	Rivers	Lakes	Estuaries
1	Siltation	Metals	Nutrients
2	Nutrients	Nutrients	Pathogens
3	Pathogens/low dissolved oxygen	Organic enrichment/low dissolved oxygen	Organic enrichment
4	Pesticides	Siltation	Siltation
5	Organic enrichment/ low dissolved oxygen	Priority organic chemicals	Suspended solids

National Water Quality Inventory Report: Assessing the Quality of Surface Water The Clean Water Act requires the EPA to publish a National Water Quality Inventory Report (http://www.epa.gov/ow/national) every two years (Table 5.2). The report provides a benchmark upon which water quality assessments on a number of environmental and recreational entities are based: fishing and swimming, ground water, wetlands, coastal waters, and the Great Lakes.

In general, waters are evaluated on their ability to support aquatic life, fish for consumption, shellfish harvesting, recreation, and agriculture and on their suitability as a source of drinking water. The states' assessment for overall use support for each body of water is characterized as follows:

- Fully Supports Overall Use: All designated beneficial uses are fully supported.
- Threatens Overall Us: One or more beneficial uses are threatened.
- Partially Supports Overall Use: One or more beneficial uses are partially supported.
- Does Not Support Overall Use: One or more beneficial uses are not supported.
- Not Attainable: Use support of one or more beneficial uses is not achievable.

The cumulative information contained in National Water Quality Inventory Reports is used by Congress to evaluate the effectiveness of goals outlined by the Clean Water Act. It also imparts valuable information for directing public policy decisions that, in turn, identify potential concerns and direct resources to address those concerns.

Total Maximum Daily Load The Total Maximum Daily Load (TMDL) is a calculation of the maximum amount of pollutant that a body of water can receive daily, from all contributing point and non point sources combined, and still meet water quality standards. The calculation includes a margin of safety to ensure that the body of water can be used as designated by the state and must allow for seasonable variation in water quality. The total amount is allocated among the sources of pollution emptying into the water. Water quality standards set by states, territories, and tribes identify the uses for each body of water: drinking water supply, contact recreation (swimming), and aquatic life support (fishing); the scientific criteria to support each use are also identified. This concept and implementation of TMDL is currently being developed by the EPA.

Safe Drinking Water Act

The Safe Drinking Water Act (SDWA) was established for the purpose of keeping drinking water free of contaminants. Under SDWA, each state carries out a Source Water Assessment Program. States must identify sources of public drinking water, assess the water systems' susceptibility to contamination, and inform the public of the results. The EPA's Office of Drinking Water evaluates, describes, and communicates health risks from contaminants in drinking water through Health Advisory Levels and Maximum Contaminant Levels.

Under SDWA, community water supplies must be sampled quarterly and analyzed for regulated chemicals that commonly contaminate ground and/or surface water. A running annual mean concentration of any regulated chemical exceeding its Maximum Contaminant Level (MCL) is deemed out of compliance. Actual test results for drinking water samples from community water systems can be found at http://www.epa.gov/safewater/.

Health Advisory Levels A Health Advisory Level (HAL) is the maximum level of a drinking water contaminant, in milligrams per liter (parts per million, or ppm) or micrograms per liter (parts per billion, or ppb), that would be expected not to cause health risks over a given duration of exposure. This does not necessarily mean that levels above the HAL will pose health risks but, rather, that uncertainty warrants prevention of exposure above the HAL. However, HALs are nonenforceable standards.

TABLE 5.3 HALs for a Specific Pesticide in Drinking Water

Exposure Duration	Population Segment	HAL (ppb)	Uncertainty Factor
1 day	Child	100	100
10 days	Child	100	100
7 years	Child	50	100
7 years	Adult	200	100
70 years	Adult	3	1000

The EPA standards for exposure differentiate between adults and children on the basis of body weight: 10 kg (22 lb) for children; 70 kg (155 lb) for adults. It is assumed that children consume 1 liter (about a quart) of water daily, whereas adults drink 2 liters. Multiplication of the representative body weights by the No Observed Adverse Effect Level (NOAEL) and Lowest Observed Adverse Effect Level (LOAEL) yields total daily doses on which to estimate the potential for acute and chronic adverse effects. (See Chapter 2 for more information on NOAEL and LOAEL).

Since the NOAEL and LOAEL values are derived from animal testing, there is uncertainty whether humans might be more sensitive than test animals to the contaminant. To allow for that contingency, the EPA applies "uncertainty factors" which further reduce the acceptable dose for drinking water. Typically, NOAEL values for children and adults are divided by an uncertainty factor of 100 or more. If the Health Advisory Level is calculated from the LOAEL, an uncertainty factor of 1000 or more is used. Thus, a HAL is based on toxicological evidence and conservative assumptions about the data; it is calculated as follows:

$$\text{HAL(mg/l or } \mu\text{g/l)} = \frac{\text{(NOAEL or LOAEL)} \times \text{body weight}}{\text{uncertainty factors} \times \text{water consumption}}$$

Note that the HAL represents a concentration in water that is very different from the actual dose consumed (i.e., the total mass of contaminant taken in by a person).

HALs are derived for exposure periods of 1 day, 10 days, "longer term," and lifetime. The 1-day HAL is calculated for a child exposed to the drinking water contaminant for 1 day. The 10-day HAL provides information relative to a child drinking the contaminant for 1–2 weeks. The longer term HAL is derived for both a child and an adult and assumes an exposure duration of 7 years (10% of an individual's lifetime). A lifetime Health Advisory Level is derived for an adult and assumes that the individual will be exposed for a lifetime of 70 years.

Examples of Health Advisory Levels for a specific pesticide—in this case atrazine—in drinking water are presented in Table 5.3. The NOAEL and LOAEL for this specific pesticide are 15 ppm and 150 ppm (mg/kg/day), respectively.

Maximum Contaminant Levels The SDWA directs the EPA to protect human health by establishing Maximum Contaminant Levels (MCLs) for pesticides and other potential contaminants of drinking water. MCLs are *legally enforceable standards* that may not be exceeded (see http://www.epa.gov/ost/drinkingstandards). An MCL is the highest annual average concentration of a contaminant allowed in public water supplies. Like the lifetime HAL, the MCL is a calculated value based on the assumption that the average person weighs 70 kg, lives 70 years, and drinks 2 quarts of contaminated water daily. Information on the estimated per capita water ingestion in the United States can be found at http://www.epa.gov/waterscience/drinking/percapita/. Calculation of the MCL value, however, also includes consideration of the cost, feasibility, and practicality of current technology to further reduce contaminant concentrations. MCLs may change as new technologies evolve, making the reduction of contaminant concentrations feasible.

MCLs for noncarcinogenic contaminants are calculated much like HALs. The MCL is established at a level 100–1000 times lower than the lowest dose known to affect the most sensitive test animals. For pesticides or other chemicals that are determined to be carcinogenic, MCLs must be set at the lowest feasible level.

Currently, public water utilities are required by SDWA to collect at least four samples per year from the finished water supply (tap water) for contamination analysis. If the *annual average residue* of a pesticide, as determined from the samples, exceeds the maximum contaminant level, consumer notification is required. Water utilities may have to use an alternative water supply, remove the contamination by filtration, or blend the supply with water from an uncontaminated source. Information about local drinking water systems is available at http://www.epa.gov/ogwdw/dwinfo.htm.

SUMMARY

Protection of ground and surface water quality is critical to economic viability as well as to human health and environmental quality. Although pesticides have many positive attributes, their use can impact the quality of water adversely. The detection of pesticides in water has aroused public interest in the overall impact of pesticides on environmental quality. Public concern has resulted in a wide array of federal, state, and local laws that deal with protecting the quality of water (Fig. 5.19).

In the interest of minimizing risks associated with pesticides, significant public resources have been allocated for the development and implementation of rational pesticide use policies based on solid scientific evidence and evaluation. The result is a complex matrix of state and federal policies, all of which impact the assessment and management of pesticides on the basis of risk they pose to the nation's water. Compliance involves extensive, detailed, and expensive laboratory research and field studies to determine the behavior and environmental fate of pesticides—that is, solid scientific evidence—and it follows that manufacturers must commit significant financial resources to product development en route to the marketplace.

A clear understanding of pesticide contamination of water requires knowledge of the interaction between the chemical and water. Public policy decision makers must address key management issues involving predicted environmental trigger values, modeling, field tests, and monitoring studies:

- Which watersheds and aquifers are the most vulnerable?
- What are the pesticides of concern?
- What are the sources of those pesticides?
- What can be done about them?
- Will it take legislation, best management practices, water quality testing, or other forms of management to reduce the impact of pesticides on surface and ground water quality?
- What needs to be done and who will educate pesticide users to protect the nation's water resources?

Figure 5.19 Finding solutions to water quality contamination will require those interested in recreation, agricultural, industry, and wildlife to work together.

A pesticide's route and rate of entry into the environment, as well as its degradation characteristics, are key to understanding and predicting its potential impact on surface and ground water. Preregistration research data play a significant role in determining use pattern and hazard statement language for the pesticide label. Research findings also influence the stringency of postregistration monitoring programs.

Technological breakthroughs in detecting pesticides at very low concentrations, combined with researchers' and regulators' scrutiny of water resources and the availability of validated models for pesticide environmental fate, have increased our scientific understanding of the potential of pesticides to contaminate our water resources. Effective management of pesticides to reduce contamination potential is essential.

CHAPTER 6

PRODUCT DEVELOPMENT AND REGISTRATION: BLENDING SCIENTIFIC INFORMATION INTO PUBLIC POLICY DECISIONS

Data developed for product registration are used by the EPA in its decision to register a pesticide active ingredient. The use of risk analysis is key to the data needs for registration, reregistration, and Special Review. The use of risk assessment within the registration process provides the EPA with a means to address risk at all phases of product registration. The risk analysis framework also guides manufacturers, as they develop a product, in determining what risk-related issues need to be considered to market their product. These steps include: patent research and application, testing the product using the EPA's required data requirements based on use patterns, and submission of the registration package and proposed labeling to the EPA. Product stewardship and registration must be addressed at both the state and the federal levels.

AN ANALYSIS OF RISKS

The risk assessment process is a critical component of pesticide product development and regulatory review. The principles of risk assessment applied to pesticides are fundamentally the same as those applied to bridge and highway design, pharmaceuticals, and innumerable consumer products. The process is directed toward establishing an objective basis on which to assess risk. Registration and risk assessment are based on legal use of the pesticide. State and federal inspectors also determine whether applicators are using, storing, and disposing of the products in accordance with the label. Illegal uses or misuses are dealt with by other means (e.g., enforcement, product cancellations, product restrictions, and applicator training).

In FIFRA, the U.S. Congress set the standard for making pesticide registration decisions: "... any unreasonable risk to man or the environment, taking into account the economic, social, and environmental costs and benefits of the use of any pesticide." Environment is defined to include "water, air, land, and all plants and man and other animals living therein, and the interrelationships which exist among these." Thus, both human health and

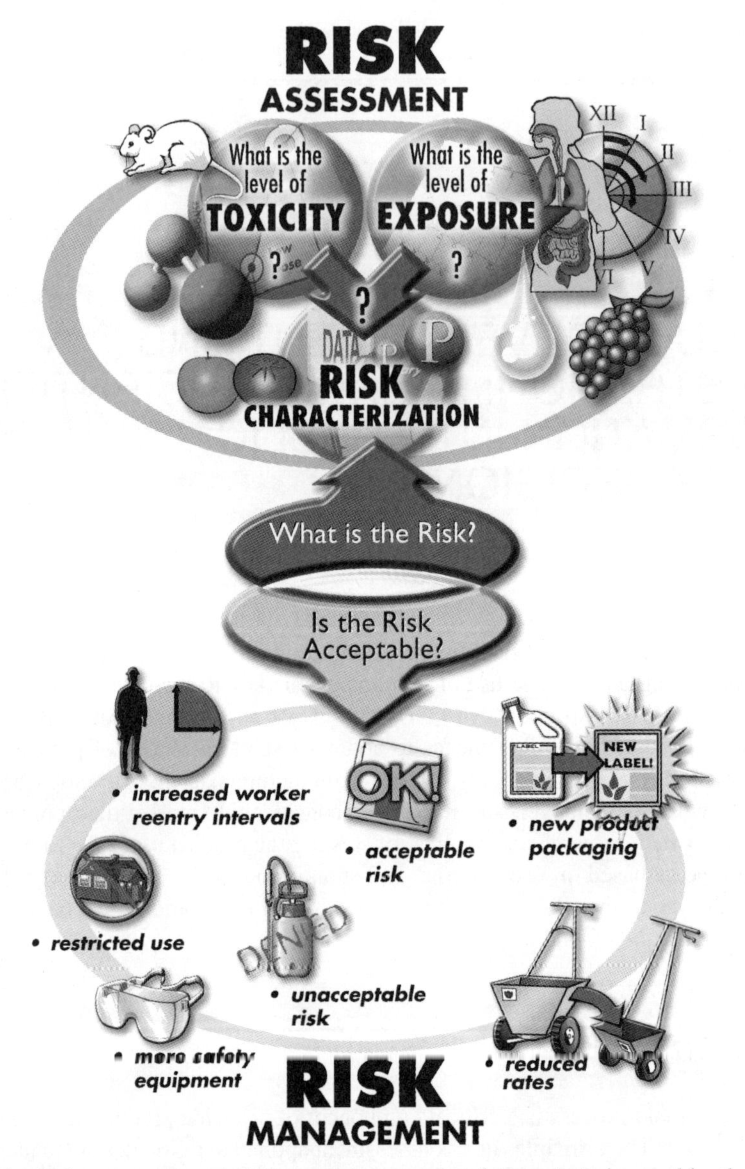

Figure 6.1 Risk assessment and risk management are regulatory strategies used by the Environmental Proctection Agency to regulate the registration and use of pesticides.

ecological risk assessments are essential to the decision-making process behind pesticide registration.

"Risk analysis" is a systematic framework for understanding and actually managing diverse risks through the processes of risk assessment, risk management, and risk communication. It allows the incorporation of scientific and public health principles into decision making and the setting of priorities (Fig. 6.1).

The EPA intends the risk assessment process to provide the pesticide industry and the public at large with methods and criteria to estimate the level of risk posed by a pesticide. It is a science-based process that aids in the overall analysis of risk. Policy defines the process, and policies are not driven exclusively by science. Risk assessment considers scientific knowledge and its inherent uncertainties. It is the process of quantifying and characterizing risk (i.e., estimating the likelihood of occurrence and the nature and magnitude of potential adverse effects).

Risk management is the process by which decisions and judgments on the acceptability of levels of risk described in the risk assessment process are made. Risk managers must weigh policy alternatives by integrating risk assessment results with social, economic, and political factors. The following are examples of risk management approaches used to reduce human and ecological risk:

- Registering the pesticide
- Restricting its use to certified applicators
- Lowering application rates
- Reducing the number of applications
- Increasing application intervals
- Using alternative application methods

These measures often take the form of label changes designed to reduce the amount of pesticide used and to lower exposure potential for farmworkers, the general public, and wildlife.

Intertwined with the process of risk assessment and risk management is effective communication of risk-based decisions. Risk communication is integral for clear conveyance of the science- and policy-based decisions concerning pesticide safety and use.

MANUFACTURER'S PRODUCT DEVELOPMENT AND COMMERCIALIZATION PROCESS

Corporate decisions on whether or not to develop potential pesticide products are based on risk assessment, marketability, and projected cost of production and return on investment. Risk assessments must be conducted periodically throughout the development and commercial life of a pesticide, frequently beginning with limited, preliminary data acquired very early in the development process. As more data become available, risk assessments are refined by virtue of an enhanced understanding of the toxicological properties and chemical fate of the pesticide as well as better exposure estimates. Scientists who develop data often serve as experts who present and interpret it for risk managers. The development team assesses data at various intervals to decide whether to cancel or continue research and plans for commercialization of the product.

Registrants often use similar or more stringent criteria than those used by the EPA and conduct a preliminary review of their own data; this assists registrants in gauging their products' prospects for registration. Preliminary reviews also may serve as indicators of the amount of time the EPA might spend evaluating the data, that is, how quickly products might be registered. If potential adverse effects are identified during any risk assessment, scientists must decide if and how the potential risk can be reduced; for example, by changes in formulation, methods of application, use rates, and marketing or by use of personal protective equipment. Strategies for reducing risk involve what is known as *risk mitigation*.

For example, suppose a pesticide applied at a rate of 1 lb of active ingredient per acre has the potential to cause unreasonable risk to foraging bobwhite quail in treated fields. A risk assessment may indicate that rates at or below 0.75 lb/acre would negate that potential. So, to mitigate risk, product development teams may be asked to reduce the proposed use rate.

Lowering the application rate of a product requires reassessment of its efficacy. For this example, an application rate of 0.75 lb/acre would control most broadleaf weeds in corn, but two resistant perennial weeds would not be controlled at that rate. Based on this information about reduced weed control, as well as data indicating risk potential at higher rates, the development team might conclude that the product would not compete successfully in the marketplace; if so, commercial development would cease.

A thorough, well-organized risk assessment process does the following:

- It defines guidelines for required tests and identifies risk standards that will be used to quantify the data.
- It stipulates full reevaluation of studies that supported initial (or former) registration of a product to verify sufficiency according to current standards and to identify any need for additional testing prior to reregistration.
- It enables manufacturers to identify and eliminate high-risk products or risks early in the development process, thus minimizing expenditures in support of a product that most likely would not be granted registration or would undergo a protracted registration process.
- It requires sound, factual documentation of the registration process as a basis on which customers, company management, and stockholders can calculate their commitment to advancement of the product.
- It is recursive, that is, it continually evaluates and reconsiders risk or new data as it evolves.

From Idea to Market

Discovery of "acceptable" pesticides requires researchers to unravel the complexities of chemical behavior in the environment and on human health. Experiments designed to do so are time consuming and costly, and they require expertise from numerous scientific disciplines (see Chapters 2, 4, and 5 for detailed information on testing requirements and risk assessment protocols).

Delivering the ideal product from the research laboratory to the marketplace requires extensive testing. New compounds showing promise as a pesticide, today, probably will not be available for use for at least five years. Thus, in considering what products to develop, a company anticipates potential regulations, future markets, public demands, and how the product will fit into tomorrow's integrated pest, crop, or systems management scheme. To justify research expenditures, a pesticide manufacturer looks for profitable products capable of entering major markets with the potential to expand into specialized markets (e.g., an insecticide for insect pests of corn expanding for use in the greenhouse as well) (Fig. 6.2).

Early pesticides were discovered largely by formulating and screening thousands of chemicals for potential pesticidal activity. Emphasis has shifted from massive screening programs to the development of "designer compounds" created through computer design and simulation. Conventional chemistry is still an option. However, the quest for future

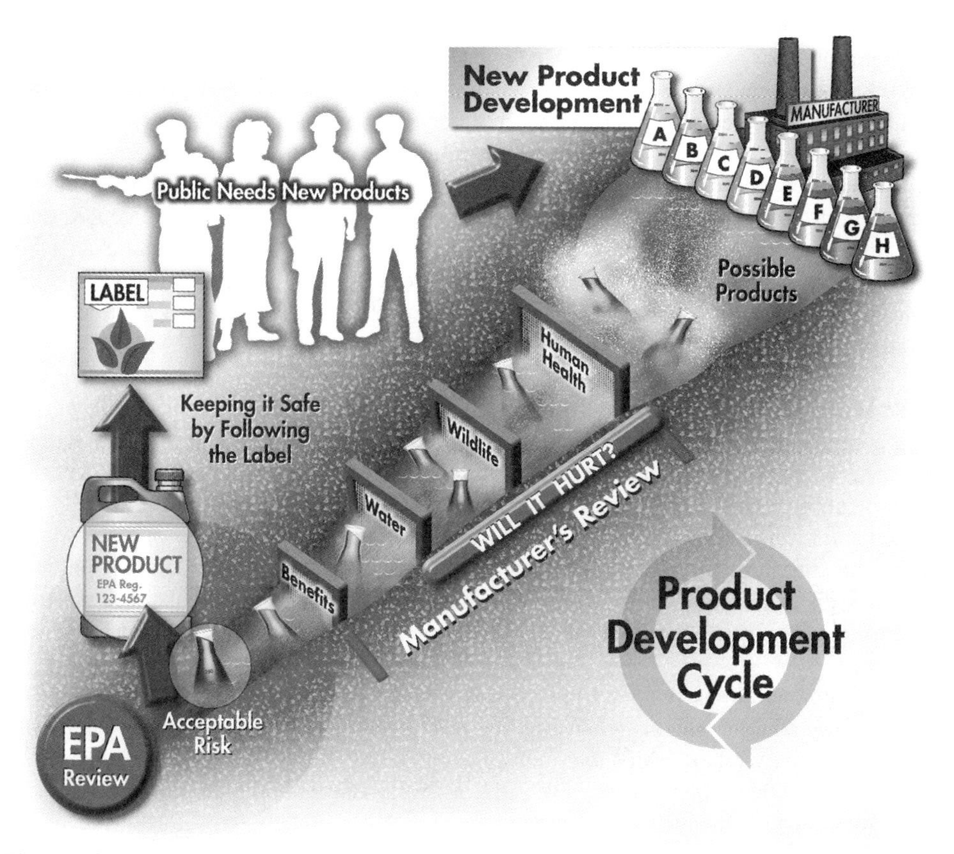

Figure 6.2 Risk assessment considerations are fully integrated into the product development cycle.

pesticide products is now channeled toward pesticides derived from bacteria, viruses, and fungi; growth or chitin inhibitors; and bioengineered genes making plants resistant to particular pests. The pesticide product for tomorrow's needs must offer solutions such as:

- better efficacy,
- compatibility with pest management practices,
- nonleaching tendencies,
- less persistence in the environment,
- reduced residues in food, and
- lower risks to humans and the environment.

Such solutions are the keys that enable the pesticide and the manufacturer to compete successfully in the marketplace. A commercially successful product allows the manufacturer to recover the costs of initial testing and registration and return a profit for reinvestment into future product development.

The development of any new pesticide begins with discovery and ends with product registration. But the steps in creating a commercial pesticide are multiple and complex.

Designer chemical development blends the skills of many specialists, beginning with synthetic chemists and biologists. Toxicologists evaluate the impact on various mammalian and other nontarget organisms. Residue, environmental, and metabolism chemists conduct additional tests. Research expands from laboratory or greenhouse evaluations to actual use sites and involves testing under expected environmental conditions.

Formulation chemists create products—emulsifiable concentrates, granulars, wettable powders, etc.—that contain the active ingredient. The type of formulation influences toxicity, use, performance, and residue activity. Only after the data demonstrate a viable product does the manufacturer seek registration from the EPA. Process chemists and engineers design systems for mass production of the active ingredient and various end use products.

Discovery Phase Most manufacturers employ a screening process to identify promising compounds for further study and unfavorable ones to be discarded. Primary screens are used to pinpoint chemicals with pesticidal properties. Their impact on growth, development, behavior, and mortality of pest insects, weeds, and diseases is carefully observed and recorded. It is important that the process uncover novel modes of action rather than mirroring those already in the marketplace. From a marketing standpoint, chemicals that lack distinct advantages in terms of efficacy and selectivity are unlikely to generate revenues sufficient to support costs associated with development, registration, and field support.

Pesticides entering screening trials may be synthesized with specific targets in mind or produced by microorganisms inhabiting the soil. Others may be discovered by examining structural activity relationships and molecular modeling. Still others may originate from industrial manufacturing waste streams and from by-products of other industries. Currently, as many as 80,000 chemicals are screened annually by a single manufacturer in search of active pesticidal compounds.

Tomorrow's marketplace will reflect increased demands for commercially successful pesticides. Both product safety and performance must be enhanced to provide superior products for the future. Because of the difficulty in discovering pesticides that meet commercial expectations, conventional synthetic chemistry is being augmented by novel approaches such as fermentation screening and combinatorial chemistry: processes which have proven successful in the pharmaceutical industry and which are now being applied effectively in the search for pesticide active ingredients.

Fermentation screening involves the search for natural toxins that can be used as pesticides. The world's greatest biodiversity exists in soil, and fermentation screening involves culturing soil organisms and testing the fermentation broth for pesticidal activity. If activity is found in initial screens, the laborious tasks of purification, isolation, and identification of the active compound are undertaken. This can lead to entirely new classes of pesticidal compounds whose safety may be enhanced because of their origin as natural products.

Combinatorial chemistry harnesses the power of computational chemistry and robotics to accelerate the processes of chemical synthesis and screening. Combinatorial chemists focus on a compound with promising activity and, with the help of powerful computer algorithms, predict a set of chemical analogs that may afford superior activity. With the help of robotics, these analogs are synthesized and sent through high-throughput screening systems. This automated process accomplishes the synthesis and screening that traditionally required teams of chemists.

Fewer than 200 of the 80,000 chemicals screened each year warrant further evaluation. Most compounds are excluded from additional testing because they are not sufficiently selective, the cost to the manufacturer is unacceptable, the potential market is too small, or they demonstrate unacceptable hazards. In order to be economically viable, a pesticide must be used commercially for one or more major crops: corn, soybeans, wheat, or cotton.

Secondary screening (on compounds that have survived primary screening) involves the use of proven, reliable predictors of biological and environmental properties to identify negative chemical attributes. Biological, environmental, and economic questions pertinent to the commercial potential of these chemicals are listed here:

- How selective is the pesticide?
- What is the water solubility of the material?
- How mobile is the compound in soil?
- How persistent is the compound in the environment?
- How toxic is the compound to nontarget organisms such as birds, aquatic organisms, insects, and plants?
- Does the material bioaccumulate, that is, does it concentrate in animal tissue?
- What formulation prototype will be tested?
- Can the material be produced in sufficient quantity to continue testing?
- What kinds of manufacturing processes may be needed?
- Is there a market, and does it meet other commercial objectives?
- Can it be produced economically?
- What are the strengths and weaknesses compared to existing products?

Research and economic development teams evaluate the relative strengths and weaknesses of each chemical and, in turn, their analyses are consolidated into product profiles. Many products are shelved because their profiles indicate undesirable health or environmental attributes, manufacturing problems, or sales potential insufficient to capture an adequate share of the market. Decisions to eliminate certain products carry an expensive price tag: two years of research costing more than $1 million for each chemical in the secondary screen.

Patent Search The value of a project depends on availability of patent protection on the proposed product, on the absence of blocking patents controlled by third parties, or at least the absence of relevant patents for which licenses are not available on reasonable terms. The legal conclusion that no third party holds a patent that would be infringed by the proposed product, or that authority to practice under any existing patent is readily available, is commonly referred to as "freedom to operate." Before investing heavily in a project, it is prudent to have a qualified patent expert conduct both a "patentability search" to evaluate the likelihood of obtaining patent protection and an "infringement search" to determine availability of freedom to operate.

Subject to payment of maintenance fees, patents granted after June 8, 1995, are assigned a term of 17 years from the date of grant or 20 years from date of application, whichever is greater. Patents granted before June 8, 1995, had a term of 17 years from the date granted.

Patent protection can significantly enhance the value of a project because it prevents competitors from simply copying a new invention, thereby benefiting from the inventors work while avoiding all of the costs the inventor incurred. An invention must be novel, useful, and nonobvious to be patentable. Novelty and nonobviousness are determined by comparing the claimed invention with what was in the public domain when the invention was made. Obtaining patent protection involves an application process in each country for which protection is desired.

In most countries, if two parties file patent applications claiming the same invention, the party who was first to file gets the patent. In the United States, the applicant who can satisfactorily prove, via a proceeding called an "interference," that he was first to invent gets the patent.

The patent application process almost always takes more than a year, depending on the work load of the relevant examining group in the Patent and Trademark Office and on the complexity of issues that arise. In the worst case, if a U.S. application becomes involved in multiparty interference, it is very possible that the process will consume a decade; in such circumstances, both the availability of patent protection and freedom to operate remain uncertain.

It is important to understand that owning a patent does not in itself confer freedom to operate. A patent empowers the patent owner to prevent unauthorized parties from making, using, selling, offering to sell, or importing the invention defined by the "claims" of the patent. The claims are numbered, one-sentence paragraphs found at the end of the printed patent.

Owning a patent does not entitle the patentee to ignore the claims in patents owned by others. It is not at all uncommon for a new invention to be patentable but also covered by the claims of an earlier patent. For example, a specific chemical compound useful as an insecticide may be patentable even if the chemical class of which it is a member has previously been patented. If both patents are issued, neither the owner of the compound patent nor the owner of the patent on the chemical class could make, use, sell, offer to sell, or import the compound in the absence of a license from the other.

Predevelopment Phase Chemicals which survive the secondary screening process are tested extensively to produce a database that supports EPA registration of the product. Manufacturers must conduct and analyze research and present their findings according to protocol outlined by the EPA. Most data submitted in support of pesticide registration results from research conducted under Good Laboratory Practices (GLPs).

Data Requirements for Pesticide Registration Regulations dealing with pesticide registration are part of the Code of Federal Regulations (CFR) Title 40 Part 158. The 40 CFR Part 158 regulations specify the types of data required by the EPA as a basis on which to make regulatory judgments on the risks associated with pesticides (see Appendix 6.1 for a list of potential tests). The following examples illustrate the kinds of 40 CFR Part 158 information which is updated periodically (http://www.epa.gov/pesticides/cfr.htm).

PESTICIDE USE SITES The pesticide registrant (in most cases the manufacturer) needs to first identify the intended use sites for the pesticide being tested; there are 14 site categories:

- Agricultural crops
- Ornamental plants and forest trees

- General soil treatment and composting
- Processed or manufactured products and food or feed containers or dispensers
- Pets and domestic animals
- Agricultural premises and equipment
- Household
- Wood or wood structure protection treatments
- Aquatic sites
- Noncrop, wide area, and general indoor–outdoor treatments
- Antifouling treatments
- Commercial and industrial uses
- Domestic and human uses
- Miscellaneous indoor uses

SPECIFIC USE PATTERNS Once the site group has been identified, the registrant must place the pesticide into one or more subgroups that more clearly identify proposed uses. The following examples represent subgroups listed for agricultural crop, aquatic, and household site groups.

- Agricultural crop uses: small fruits, tropical and subtropical fruits, vegetables, commercial greenhouses, fiber crops, forage crops, grain and edible seed crops
- Aquatic uses: food processing water systems, pulp and paper mill systems, swimming pool water, agricultural irrigation water and ditches, and estuaries
- Household uses: nonfood areas, food handling and food storage areas, household contents and space

GENERAL USE PATTERNS Specific use patterns and their subgroups are consolidated into nine general-use patterns:

- Terrestrial food crop
- Terrestrial nonfood
- Aquatic food crop
- Aquatic nonfood
- Greenhouse food crop
- Greenhouse nonfood
- Forestry
- Domestic outdoor
- Indoor

It is these general-use patterns that determine the data requirements for new product registration (Table 6.1). For example, an agricultural crop use demands an exceptionally rigorous testing program because of food safety issues and the general proximity of application sites to fish and wildlife.

TABLE 6.1 Examples of Pesticide Assignments to General Use Patterns

Pesticide Use Site	Specific Use Patterns	General Use Patterns
Agricultural crops	Vegetables	Terrestrial food crop
Agricultural crops	Greenhouses: mushrooms	Greenhouse food crop
Agricultural crops	Grain and edible seed crop: rice	Aquatic food crop
Ornamental plants and forests	Forests	Forestry
Aquatic sites	Swimming pool water	Aquatic noncrop
Household	Food handling/storage areas	Indoor

DATA REQUIREMENT TABLES The types of data that must be submitted with an application for pesticide registration are listed by general use patterns. The Code of Federal Regulations (Subpart D of 40 CFR Part 158) lists by major data subheading the types of information required:

- Environmental fate
- Insect toxicity
- Nontarget organisms
- Physical and chemical characteristics
- Plant protection
- Product performance
- Reentry protection
- Residue chemistry
- Spray drift
- Toxicology
- Wildlife and aquatic organisms

Each general use pattern is listed as a headline on each data table, and the studies required for registration appear in the column directly underneath. Under a pesticide classified for use on a terrestrial food crop, the following required studies would be listed in the toxicology table: acute oral toxicity, acute dermal toxicity, acute inhalation toxicity, primary eye irritation, primary dermal irritation, dermal sensitization, chronic feeding, oncogenicity, teratogenicity, reproduction, gene mutation, structural chromosomal aberration, other genotoxic effects, general metabolism, and domestic animal safety. The total number of studies listed constitutes the minimum data package needed to support registration.

Additional studies are listed in data tables as "conditionally required"; their necessity is dependent upon results of required studies and other product use considerations. Data requirements for agricultural food crop pesticides are rigorous, whereas the requirements for a pesticide intended for use as a household disinfectant are less rigorous.

STUDY GUIDELINES The regulations in 40 CFR Part 158 do not specify what methodologies and procedures are to be used in toxicological research. The specific study designs and measurements are provided to the registrant through an EPA publication series called Harmonized Guidelines, formerly called Pesticide Assessment Guidelines, at http://www.epa.gov/opptsfrs/home/testmeth.htm). Examples include the following:

Guideline Number	Title
810	Product Performance
830	Product Properties
835	Fate, Transport, and Transformation
840	Spray Drift
850	Ecological Effects
860	Residue Chemistry
870	Human Effects
875	Occupational and Residential Exposure
880	Biochemical
885	Microbial Pesticides

These guidelines also specify how tests are to be evaluated, analyzed, and reported. For example, step-by-step information for performing a chronic feeding study includes study design; animal selection parameters such as age, species, weight range, and numbers; dose administration route, frequency, and duration; and observations, measurements, analyses, and procedures that must be performed and recorded for each animal. Similar guidance is provided for each study required. Study guidelines are updated as scientific knowledge expands. Studies conducted outside these rigorous guidelines are generally not acceptable for regulatory purposes and must be repeated.

Decisions to Move toward Commercialization Successful development of a product requires teams of scientists working on various test components and discussing their results at every level of the testing process. Discussion on the chemical's fate and behavior, short- and long-term health effects, ecotoxicity, environmental toxicity, and production and economic information is ongoing. The benefits of the product versus its risk potential are under constant scrutiny. The project may terminate at any time during the predevelopment process if evidence suggests potential biological, environmental, or marketing problems. Only one or two chemicals out of 80,000 survive the rigors of the 7- to 9-year research and evaluation process, and those must assume the burden of recouping research and production costs for all.

Registration Review After several years of testing, the registration data package is submitted to the EPA. The data include the results of studies on acute, chronic, reproductive, and developmental toxicity (Chapter 2); of ecological studies to determine harmful effects on nontarget plants and animals (Chapter 4); and of studies on environmental fate to determine rates at which the pesticide breaks down and whether it translocates to unwanted sites (Chapter 5) (Fig. 6.3).

The final internal review and discussion are conducted by the developer to ensure validity and accuracy of all data. Each experiment must be accompanied by complete descriptions of experimental design and procedures in sufficient detail to allow full assessment by the EPA. Once the internal company review is satisfactorily completed, the scientific data can be forwarded to the EPA as part of the registration package.

The EPA may meet with company representatives to discuss research methodologies or conclusions or to request the original data. If registrants fail to follow prescribed guidelines,

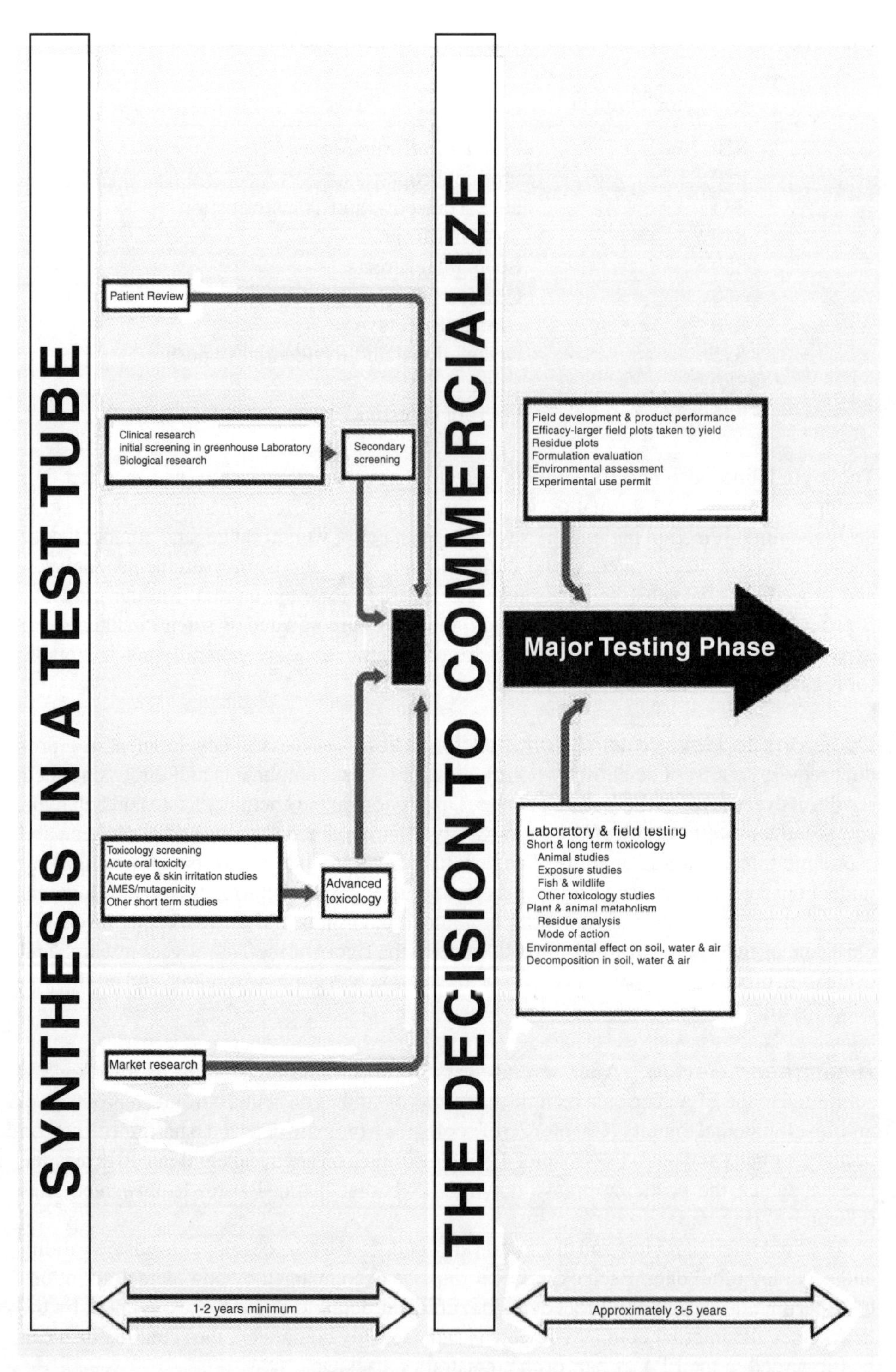

Figure 6.3 The flow of a potential pesticide product from the laboratory into the EPA's evaluation and registration process.

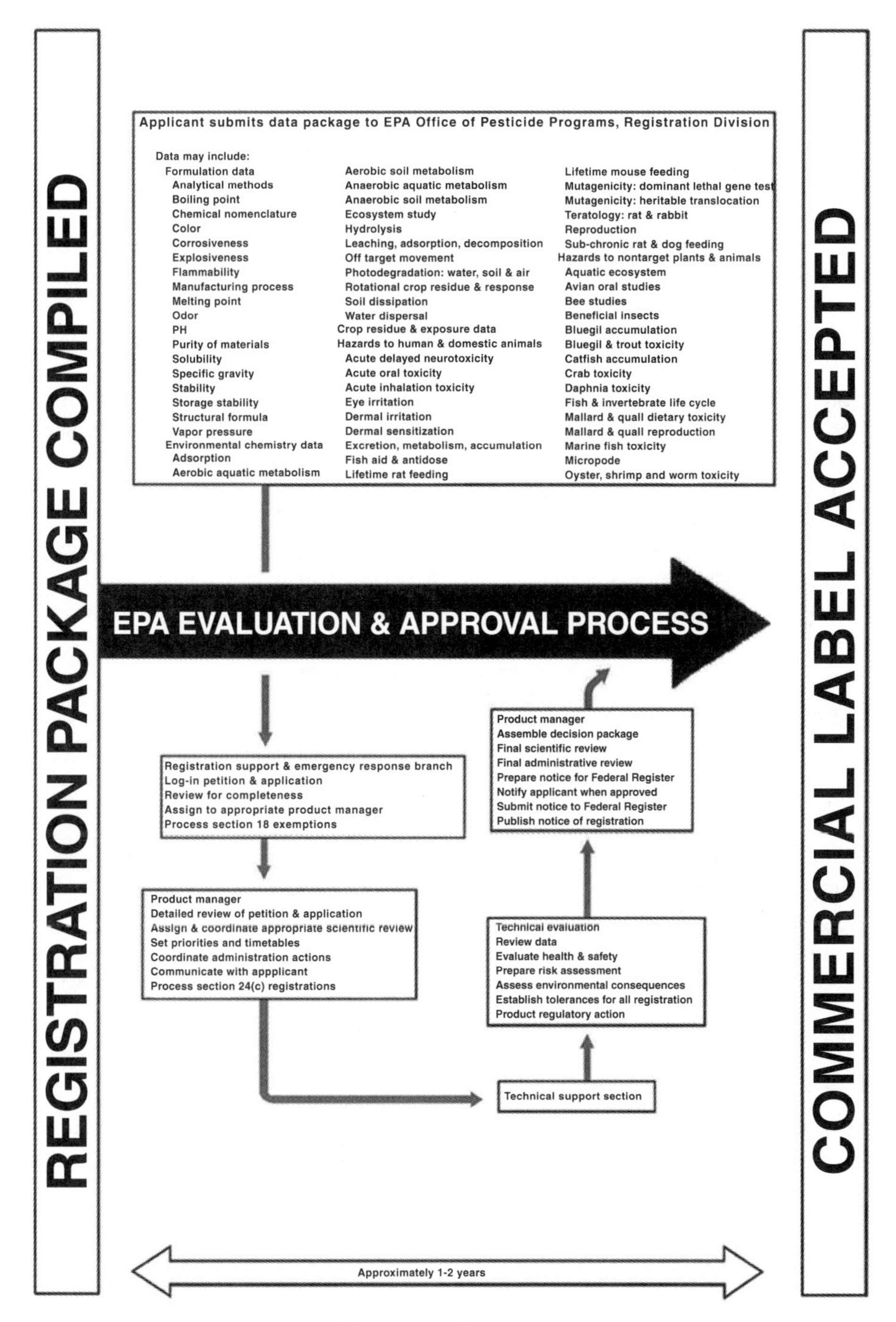

Figure 6.3 (*Continued*)

they may be asked to repeat experiments, redesign research methods, or conduct additional testing. Registration ultimately will be granted only when the EPA concludes that the potential risk to humans, wildlife, and the environment is within regulatory guidelines.

The EPA sometimes will issue a conditional registration for a product that meets the legal standard of reasonable certainty of no harm, but which still warrants additional study before a final registration decision is made. Conditional registrations are generally limited to a specific period of time, often the time needed to accomplish the additional study(ies). There may be geographic or stringent use restrictions placed on use of the product under conditional registration. Prior to expiration of a conditional registration, the EPA reevaluates use of the product based on information received from the field and on studies conducted by the registrant to determine if full registration is appropriate.

Product Labeling The label is reviewed and approved by the EPA as part of the registration process (see Chapter 7 for more details). The label provides use directions, hazard information, safety warnings, steps to take to mitigate hazard, and other information pertinent to applicators.

The Cost of Federal Registration Examples of costs in addition to those associated with conducting the required research, can include fees charged by the EPA to register the pesticide products

One-Time Registration Fee for New Products	
New chemical application	$184,000
New biological or microbial pesticides	$64,000
New use pattern	$33,000
Experimental use	$4,500
Annual Maintenance Fees	
First registration	$750
Additional registrations (50 or fewer registrations)	$1,400 (with a cap of $50,000)
Additional registrations (more than 50 registrations)	$1,400 (with a cap of $95,000)
Food Use Tolerance Fees	
First nine commodities/active	$64,025
Each additional commodity	$1,600
Request for exemption	$11,800
Repeat of existing tolerance	$8,000
Reregistration Fees	
One-time fee for each active ingredient registered before 1984	$155,000

State Registration Federal registration of a pesticide is only part of the regulatory process. Once a product is registered by the EPA, it must also be registered at the state level. Most states accept the EPA's scientific review, but several states—notably, California, Florida, and New York—have legislative mandates requiring additional data reviews. Data reviewed by these states are typically the same data required by the EPA. However, at the state level the scientific review process takes into consideration the uniqueness of that state's environment, agriculture, and public concerns. At the conclusion of the review process, a state has the same options as the EPA: the registration may be granted or denied, or more data may be required. As a condition of registration, a state may require modification of the federal label. States also have the authority to cancel their registration of a product. State lead agencies responsible for registering a product often are accountable for enforcing the proper use of pesticide products and for educating pesticide users. In support of these efforts, states charge registration fees for each product sold or used in the state. State registration fees vary considerably from state to state, but they normally are based on the number of products or a percentage of sales. Costs can range from a low of $10 per product to a high of $325 per product; some fees are based on annual sales (e.g., 1.3% of annual sales).

Product Stewardship Phase A product must bear an EPA-approved label before the registrant can introduce it for sale in the United States. The average cost of developing a single marketable product from among 80,000 chemicals screened is estimated at $100 million. Total expenditures might reach $160 million if the cost of manufacturing plants, etc., are factored into the equation. Nearly half of the original 17- to 20-year patent life is spent on research, development, and registration processes; and the pesticide that survives to earn an EPA registration number *still* is not guaranteed success in the marketplace. Manufacturers must exercise good stewardship in maintaining and supporting their pesticide product and ensuring that its use is consistent with the label. They are required under FIFRA to report any evidence of problems relative to the use of the product that are identified after registration; as a result, additional label restrictions, suspension, or cancellation of the product might be imposed—and the manufacturer's investment in the pesticide product might be lost in the process.

The Pesticide Marketplace

The EPA's Biological and Economic Analysis Division publishes an annual report, Pesticide Industry Sales and Usage, on the pesticide marketplace (http://www.epa.gov/oppbead1/pestsales/).

How Much Pesticide Is Used?

- The United States purchases 20% of the 22 billion pounds of pesticides sold worldwide.
- The United States uses 4.5 billion pounds of active ingredients.

How Many Pesticide Products Are There? Some 875 active ingredients are formulated into 21,000 registered pesticide products.

What Are the Main Products?

- Chlorine, as a disinfectant of drinking water and wastewater, makes up 50% of the total amount of pesticide used in the United States.
- The most widely used pesticide in U.S. agricultural crop production is atrazine.

- The herbicide 2,4-D is the pesticide most commonly used in nonagricultural sectors.
- The top five conventional pesticides for agricultural crop production are atrazine, metolachlor, metam sodium, methyl bromide, and dichloropropene.

What Revenues Do Pesticides Generate?

- Pesticide sales total $11 billion annually, with agricultural products accounting for 70%.
- The average farm operation spends about $4200 annually for pesticides.

Who Are the Makers and the Users? The estimated composition of the pesticide industry is as follows:

- 8 major manufacturers
- 100 additional manufacturers
- 7300 producing establishments
- 6000 to 10,000 manufacturing workers
- 2200 formulators
- 40,000 pest management firms
- 384,000 certified commercial applicators
- 70 million households use pesticides
- 960,000 certified private applicators

THE EPA DECISION-MAKING PROCESS IN PRODUCT REVIEW AND REGISTRATION

The intensive regulatory assessment process utilized by the EPA benefits regulators and the public.

- The process more readily fulfills the EPA's mission to protect public health from unreasonable adverse effects.
- The risk assessment process helps the EPA make more consistent, well-informed registration decisions by encouraging more in-depth review of technical information.
- The process provides a forum where EPA scientists and risk managers have a common basis for discussion of conclusions drawn from risk assessment.
- The process also helps guide EPA decisions on whether additional data are needed to clarify potential risk.

Use of Risk Assessments

The primary decisions that the EPA must make concerning pesticides are to

- register a new product or use,
- reregister an existing product,
- cancel a current registration,

- determine if labeling protects human health and the environment, and
- continually reevaluate safety according to new legislated standards.

Risk assessments are performed by the EPA during registration and reregistration processes. They are also conducted whenever new findings suggest that adverse effects might result from use of a previously registered product. There are four end points to which assessments are directed: Experimental Use Permit; Registration; Reregistration; and Special Review.

Experimental Use Permit Prior to the time a registration application is submitted to the EPA, a company developing a new product usually requests permission [in the form of an Experimental Use Permit (EUP)] to test the product under field conditions in numerous marketing areas. The objective is to gather data on product performance or other parameters that can be obtained only under normal use conditions as prescribed by the label. An EUP also may be required to allow testing of a registered product when adding a significant new use pattern (e.g., when adding aquatic weed control to the label of a row crop herbicide).

An EUP is required for a pesticide applied to land when the research will be conducted on a cumulative total of 10 or more acres. It is also required for a pesticide applied to water when the research will be conducted on a cumulative total of at least 1 surface acre. The upper limit for area treated under an EUP is typically 3000 acres. When applying for an EUP, the registrant submits an abbreviated but significant data package including toxicological and environmental data. The EPA reviews the data package and judges whether there exists a potential for unacceptable risk to humans or the environment under limited use conditions. If there is no such indication, the EPA issues the EUP.

In granting an EUP, the EPA imposes numerous restrictions on the use of the pesticide and on the target site. If use of the pesticide can be reasonably expected to result in a residue on food or feed, the applicant must certify that the food or feed will be disposed of in a manner which does not endanger man or the environment. An alternative to disposing of the food or feed is to submit evidence of a tolerance, obtain a temporary tolerance, or submit evidence that the use is exempt from the tolerance requirement. Additionally, the EPA may require that additional tests be conducted during the permit period.

Research conducted under an EUP is closely monitored. Unless otherwise directed, an applicant submits interim reports to the EPA every 3 months, and a final report must be submitted within 180 days of the end date of the EUP. Applicants are also required to immediately report to the EPA any adverse effects from use of or exposure to the pesticide. Permits are effective for a specified period of time, normally 1 year, but may be renewed by the applicant.

Registration During the registration process, the EPA evaluates all data in support both of active ingredients not previously registered and of new uses (for a registered product) that require label changes. When requesting product registration, the pesticide manufacturer (registrant) must submit to the EPA all data required by FIFRA. When all individual studies have been reviewed, the results are factored into human and ecological risk assessments to evaluate whether requested uses for the product present unacceptable risk to human health or the environment. Review and assessment are conducted by representatives of all disciplines within the EPA: product residue, environmental chemistry, human health, and ecological effects.

The EPA's Lower Risk Pesticide Policy The EPA has identified 5 areas for implementation of a voluntary reduced-risk pesticide initiative for pesticide manufacturers: (1) developing criteria to identify low-risk pesticides, (2) streamlining the overall registration process, (3) improving pesticide labels, (4) making more pesticide information available to support informed choices in the marketplace, and (5) encouraging the development of reduced-risk pesticides (via statutory changes which would extend the period of exclusive use of data or via patent protection).

Under this voluntary approach, manufacturers of products containing a new active ingredient thought to be worthy of reduced-risk classification must submit substantive data. Claims of reduced risk must be supported by evidence of reduced toxicity to humans or to other nontarget organisms and/or improved environmental performance. Pest resistance and suitability of the product as a component of an integrated pest management program must also be considered. The EPA encourages the registration of pesticides that pose low or reduced risks compared to alternative pesticides; consequently, applications documenting low-risk characteristics are granted priority consideration in the review process.

Reregistration More than 50,000 pesticide products have been registered in the United States since FIFRA was enacted in 1947. From the very beginning, FIFRA intended that all registered pesticides be reregistered every 5 years, but economics nullified that intent. Congress amended FIFRA in 1988 and mandated through legislative language that all pesticides registered before November 1984 were subject to reregistration by the EPA (http://www.epa.gov/pesticides/reregistration.htm).

FIFRA 1988 established a 5-phase reregistration process including the following:

- Phase 1: Identifying Active Ingredients. Pesticide active ingredients covered by FIFRA reregistration requirements are prioritized for consideration by the EPA.
- Phase 2: Manufacturer's Intentions to Reregister Products. Notice of intent is required of manufacturers to support reregistration of their products by agreeing to conduct the required studies.
- Phase 3: Availability of Current Data to Support Reregistration. Registrants certify access to raw data, reformat existing studies, summarize studies, and flag potential adverse-effects data.
- Phase 4: EPA Identifies Outstanding Data Requirements. The EPA reviews the toxicological, environmental, and chemical information submitted by registrants. Registrants are informed of additional data needed to support reregistration.
- Phase 5: EPA Makes Reregistration Decision. A pesticide is considered eligible for reregistration if its supporting data indicate that it does not cause unreasonable adverse effects to people or the environment when it is used according to product label directions and restrictions.

These amendments required that reregistration fees be paid by manufacturers submitting a product for reregistration, thus helping to provide the necessary funding. All data used to support pesticides registered before 1984 became subject to reevaluation and upgrading, if necessary, to comply with current guidelines and standards. The refined data enhance the ability of the EPA and the registrant to judge whether risk conclusions and registration decisions made during the initial product registration process meet today's standards.

The EPA produces Reregistration Eligibility Decisions (REDs) once a substantially complete set of data on a chemical case has been reviewed and no significant issues concerning the use of the pesticide remain. These REDs take the form of risk assessments and are comprised of product, toxicological, ecological, exposure, and risk characterizations leading to risk mitigation options for reregistration. The EPA must review the human health and environmental effects of these older active ingredients and determine whether they are eligible for reregistration. A pesticide must exhibit no unreasonable risk to people or the environment, when used in accordance with its approved labeling, for eligibility. REDs can be reviewed at http://www.epa.gov/pesticides/reregistration/status.htm.

All registered pesticides also must meet the safety standards of the Food Quality Protection Act (FQPA), which became effective August 3, 1996. Under this law, the EPA must consider the potential for increased susceptibility of infants and children to the toxic effects of pesticides. The EPA also must reassess existing tolerances (maximum pesticide residue limits in food), considering aggregate exposure to pesticide residues from many sources and the cumulative effects of pesticides and other compounds with common mechanisms of toxicity. For large groups of pesticides that require cumulative assessment, such as the organophosphates (OPs), the EPA is completing individual Interim REDs (IREDs) and Tolerance Reassessment Progress and Interim Risk Management Decisions (TREDs). Once the EPA completes a cumulative risk assessment and makes a risk management decision encompassing the entire group of related pesticides, the individual decision may be issued as a RED; further risk mitigation may be required at that time.

Through the reregistration process, the EPA ensures that older pesticides meet contemporary health and safety standards, that their labeling is improved, and that their risks are reduced.

Special Review The Special Review process allows the EPA the regulatory flexibility to reevaluate the registration of a pesticide. A Special Review may be initiated when new evidence suggests that the legal use of a specific active ingredient may pose unreasonable risk to human health or the environment. The Special Review process (40 CFR Part 154.7) allows the EPA legal recourse to reconsider all data, wildlife incidents, and regulatory decisions relevant to prior registration of a pesticide. A list of pesticides that are currently undergoing or have completed Special Review can be found at http://www.epa.gov/docs/SpecialReview/sr00status.pdf.

The general criteria for initiating the Special Review of a registered pesticide are

- acute toxicity to humans or domestic animals,
- chronic health effects in humans,
- hazards to nontarget organisms,
- risk to threatened or endangered species, and
- risk to critical habitat of threatened or endangered species.

The initiation of Special Review does not require meeting all of the stated criteria; it may be initiated by concerns sparked by one or more of the criteria. The process officially begins with an EPA letter of notification (called the Grasley-Allen letter) to the registrant, stating that the active ingredient is formally being placed under Special Review and giving reasons why.

Information pertinent to suspected risk associated with an active ingredient is scrutinized by EPA reviewers who, in turn, prepare a "risk review." Conclusions drawn from analysis are forwarded to the FIFRA Scientific Advisory Panel (SAP), to the U.S. Department of Agriculture, and to the Federal Register (for general public access) for comments on the scientific accuracy, data interpretation, and rationale behind proposed risk reduction measures. The SAP (http://www.epa.gov/science1/) is comprised mainly of public sector scientists, university personnel, government scientists, and consultants who are knowledgeable on the issues being considered. The SAP provides scientific advice on pesticides and pesticide-related issues regarding the impact of the following regulatory actions on health and the environment:

a. Notice of intent to cancel a pesticide registration or to change its classification under Section 6(b)(1) of FIFRA.
b. Notice of intent to hold a hearing to determine whether or not a pesticide's registration should be canceled or its classification changed under Section 6(b)(2) of FIFRA.
c. Emergency orders immediately suspending registration of a pesticide before notification of the registrants, pursuant to Section 6(c)(3) of FIFRA.
d. Regulations to be issued under Section 25(a) of FIFRA.

The role of the SAP has been expanded to that of a peer review body for current scientific issues which may influence the Office of Pesticide Programs' regulatory decisions. The panel is composed of seven members selected on the basis of their professional qualifications to assess the impact of pesticides on human health and the environment. Members are appointed by the EPA administrator and the Academy of Sciences. SAP members representing the disciplines of toxicology, pathology, environmental biology, and related sciences serve a 4-year term, with appointments made on a staggered basis. An additional 50–60 ad hoc SAP members with unique expertise also are available; 6–12 of these members usually participate at each meeting, providing input on particular issues within their areas of expertise.

If, after taking all comments under advisement, the EPA concludes that risk reduction measures are needed, there are four avenues of pursuit:

- Alteration of label language
- Classification of products containing the active ingredient for restricted-use only, thereby limiting their use to certified applicators
- Elimination of specific uses
- Suspension or cancellation of the registration

Details of the EPA's Registration Process

The Federal Insecticide, Fungicide, and Rodenticide Act (FIFRA) and the Federal Food, Drug, and Cosmetic Act (FFDCA) authorize the EPA to regulate the registration, manufacture, sale, and use of pesticides. The EPA is charged with assuring that pesticides perform their intended functions without causing unreasonable adverse effects on man or the environment. In carrying out these responsibilities, the EPA requires all pesticides to undergo a rigorous testing process and a comprehensive regulatory review to determine if adverse effects are likely to occur as a result of their use. All pesticides require an EPA registration, which is a revocable license to sell and use a pesticide product.

Congress delegates to the EPA the regulatory responsibility and decision-making authority to administer the federal pesticide registration process as specified by FIFRA. The registration process is complex and takes considerable time, resources, and expertise on the part of the EPA, the pesticide manufacturing industry, and public interest groups. This dynamic process continues to evolve as new questions are addressed and as challenges posed by the use of pesticides are met. More tests are required in response to public concern and improved technologies that yield more precise residue detection and toxicological assessment. Improved methods for hazard predictions, novel approaches to hazard reduction, and incorporation of the broadening scope of relevant scientific knowledge into industry and government policy decisions benefit the pesticide registration process.

The basic pathway for pesticide registration is: (1) research conducted by the manufacturer prior to its decision to pursue registration; (2) decisions to pursue registration and conduct required studies; (3) data submitted by the manufacturer to the EPA; (4) EPA review; and (5) a decision by the EPA either to register the pesticide based on the merits of data submitted or to deny registration (Fig. 6.3). Evaluation of registration data must provide the EPA assurance that the pesticide will perform its intended function without adverse effects on people, wildlife, and the environment.

Manufacturers seeking to market a pesticide product must apply to the EPA for registration. The EPA requires each manufacturer to support the product with specific scientific data. The review of active ingredients during the registration process is similar to the federal Food and Drug Administration's critique of a pharmaceutical drug. Product chemistry, health-related, and ecological evaluations are required. The manufacturer also must substantiate specific use patterns predicated by the pesticide label. For example, investigation of the environmental fate and characterization of any pesticide residue is required for its use on food crops, meat, or milk. At any time during the approval process, the EPA may require additional testing or clarification of existing information. The pesticide manufacturer (registrant) must pay for all costs associated with data development.

The liaison between the EPA and the registrant is the EPA Product Manager (PM) or Regulatory Action Leader (RAL) during initial registration, or the Chemical Review Manager (CRM) during reregistration. The role of the PM or RAL is to coordinate the EPA's internal review and to monitor the status of registration. The PM or RAL also helps facilitate discussion among agency scientists and resolves problems which occur during the registration process. The EPA's policy is to delegate to one PM the responsibility to document data on a specific active ingredient. The assigned PM must view the whole picture: health and safety issues, environmental and wildlife concerns, and product chemistry.

Upon receipt of a registration application, the PM/RAL directs various components of the supporting data package to the appropriate review divisions or personnel within the EPA Office of Pesticide Programs (OPP). Data analyses may include independent reviews by staff or divisions within OPP: Environmental Fate and Effects; Biological and Economic Analysis; Field and External Affairs; Health Effects; Registration; Information Resources and Services; and Special Review and Reregistration. Increasingly, reviews are conducted by teams representing each of these divisions. In addition, the reviews are conducted internally by qualified staff in the Antimicrobial Division and in the Biopesticide and Pollution Prevention Division.

The regulatory staff at the EPA ultimately is responsible for determining whether the manufacturer has adhered to protocol and scientific methods in developing the data. This determination includes an evaluation of the completeness, accuracy, and validity of data interpretation. The EPA's decisions must be scientifically based in accordance with the agency's policies and legal mandates (which can include economic, social, and environmental impacts) so that the agency can withstand scientific peer review, public scrutiny, and legal challenges. A pesticide is not registered until the EPA is satisfied that all data requirements have been met, that supporting studies are valid, and that the data allow the agency to evaluate the benefits and risks associated with use of the product.

The EPA grants a pesticide registration only after extensive deliberation and thorough scientific review. Registration constitutes approval, and it will not be granted if the product does not meet FQPA requirements.

EPA Oversight after Registration

The responsibilities of the EPA do not end at the point of registration. Product information is continually collected, assembled, and reviewed. Specifically, the EPA requires updates from the manufacturer where scientific studies and field use indicate a potential for environmental pollution, adverse impacts on human health, or toxic effects on nontarget organisms.

Reporting Adverse Information after Registration The EPA can obtain information on the adverse impacts of pesticides via the FIFRA Section 6(a)(2) reporting process. FIFRA states in Section 6(a)(2) that "if at any time after the registration of a pesticide the registrant has additional factual information regarding unreasonable adverse effects on the environment of the pesticide, he shall submit such information to the Administrator." Incident data can originate from many sources: universities, poison control centers, health departments, state and federal fish and game agencies, state departments of agriculture, EPA regional offices, or the media. Essentially, any adverse impact known by the registrant is subject to incident reporting.

The Section 6(a)(2) reporting requirement is a legal obligation and applies solely to the registrant. Reports to the EPA can originate from scientific research conducted by the manufacturer with the intention of supporting continued registration; but more often they result from data (on adverse effects) collected in field use situations. The pesticide manufacturer provides the EPA information clearly identified as a Section 6(a)(2) report; in addition, the manufacturer must identify the observed adverse effect. An example of an incident that requires a Section 6(a)(2) report is when a pesticide impacts aquatic organisms at a dose lower than previously shown.

All Section 6(a)(2) and incident reports submitted to the EPA are categorized and indexed in the Incident Data System, which is part of the Pesticide Information Network operated by the EPA. Information concerning new, potentially adverse effects is submitted to the proper division for review and analysis to determine if an immediate EPA review is warranted. For instance, the Environmental Fate and Effects Division may examine potential impacts on wildlife, whereas human health concerns might be reviewed by the Health Effects Divisions within the Office of Pesticide Programs.

An EPA work group meets regularly to discuss FIFRA Section 6(a)(2) and incident reports. Priority is given to those pesticides with the most serious problem potential to expedite review, response, and remedial action. More information on Section 6(a)(2) reporting can be found at http://www.epa.gov/pesticides/fifra6a2/.

Figure 6.4 Risk assessment is a responsibility shared among government, the public, and the manufacturer.

SUMMARY

To ensure that pesticide products available for sale in the United States do not pose unreasonable risk to human health and the environment, manufacturers and the EPA expend enormous time, energy, resources, and expertise on developing data to support registration. Data are carefully analyzed, using scientifically based criteria to address risk potential (Fig. 6.4).

Manufacturers spend millions of dollars in research, countless hours working on EPA registrations, and even more energy marketing their products. There is a lot of pressure on manufacturers to meet EPA registration requirements and to develop products needed. They must also earn enough to recoup their initial investment and make profits for the company and its shareholders.

Manufacturers must supply scientific evidence that a pesticide, when used as directed on the label, will not injure humans, crops, livestock, other nontarget organisms, or the environment and that it will not produce illegal residues on or in food or feed. The steps manufacturers use to meet these requirements are complex and lengthy.

Once the EPA receives an application for product registration, vast resources are dedicated to reviewing pertinent scientific data, to analyzing the data and proposed uses in light of various sets of risk criteria that the EPA and Congress have established over the years, and to determining any restrictions necessary to ensure that the product does not pose unreasonable risk. If new data or information becomes available, or if the agency receives evidence through the adverse effects reporting process that indicates risk potential higher than originally thought, the agency may activate an evaluation mechanism and take steps to mitigate the risk. It may take a variety of steps in the reregistration and Special Review processes to actually reduce the risk; for example, it may require additional personal protective equipment or alternate methods of application. The EPA can reduce the amount of product applied, overall, or restrict product use to ensure that those handling it are properly trained. In addition, if none of these steps is sufficient, or if the risks are judged too high, the agency also can cancel one or more uses of the product—or cancel it altogether. In some cases the registrant can voluntarily cancel their registration for the product under review.

APPENDIX 6.1. LIST OF TESTING REQUIREMENTS FOR EPA REGISTRATION

Product Chemistry Data Requirements

Product identification and disclosure of ingredients
Description of beginning materials and manufacturing process
Discussion of formulation of impurities
Certification of limits
Analytical methods to verify certified limits

Physical and Chemical Characteristics Data Requirements

Color
Physical state
Odor
Melting point
Boiling point
Density, bulk density, or specific gravity
Solubility
Vapor pressure
Dissociation constant
Octanol–water partition cocfficicnt
pH
Stability
Oxidizing or reducing action
Flammability
Explodability
Storage stability
Viscosity
Miscibility
Corrosion characteristics
Dielectric breakdown voltage
Submittal of samples

Aquatic and Wildlife Data Requirements

Acute avian oral toxicity (LC_{50}) in bobwhite quail or mallard duck
Acute avian oral toxicity (LC_{50}) in bobwhite quail or mallard duck using typical end use product

Acute avian dietary toxicity (LC_{50}) in bobwhite quail
Acute avian dietary toxicity (LC_{50}) in mallard duck
Wild mammal toxicity test
Avian reproductive toxicity in bobwhite quail
Avian reproductive toxicity in mallard duck
Simulated terrestrial field study
Actual terrestrial field study
Fish toxicity in bluegill sunfish
Actual terrestrial field study
Fish toxicity in rainbow trout
Fish toxicity in rainbow trout using typical end use product
Invertebrate toxicity fresh water LC_{50}
Invertebrate toxicity fresh water LC_{50} using typical end use product
Toxicity to estuarine and marine organisms in fish, mollusks, and shrimp
Early life stage in fish
Life cycle in aquatic invertebrates
Fish life cycle study
Aquatic organism accumulation study
Simulated field tests for aquatic organisms
Actual field tests for aquatic organisms

Mammalian Toxicology Data Requirements

Acute oral toxicity in the rat
Acute dermal toxicity
Acute inhalation toxicity in the rat
Primary cyc irritation in thc rabbit
Primary dermal irritation
Dermal sensitization
Acute delayed neurotoxicity in the hen
90-day feeding study in the rodent
90-day feeding study in the nonrodent
21-day dermal toxicity
90-day subchronic dermal toxicity
90-day inhalation in the rat
90-day neurotoxicity in the hen
90-day neurotoxicity in the mammal
Chronic feeding study in the rodent
Chronic feeding study in the nonrodent
Oncogenicity study in the rat
Oncogenicity study in the mouse
Teratogenicity in the rat
Teratogenicity in the rabbit

Two-generation reproduction study in the rat
Chronic feeding/oncogenicity in the rat
Gene mutation
Structural chromosome aberration
Other genotoxic effects
General metabolism
Dermal penetration
Domestic animal safety

Plant Protection Data Requirement

- Tier 1
  - Seed germination and seedling emergence
  - Vegetative vigor
  - Aquatic plant growth
- Tier 2
  - Seed germination and seedling emergence
- Tier 3
  - Terrestrial field
  - Aquatic field

Reentry Protection Data Requirements

Foliar residue dissipation
Soil residue dissipation
Dermal passive dosimetry exposure
Inhalation passive dosimetry exposure

Nontarget Insect Data Requirements

Honeybee acute contact (LD_{50})
Honeybee toxicity or residues on foliage
Field testing for pollinators

Environmental Fate Data Requirements

Chemical identity
Hydrolysis
Photodegradation in water
Photodegradation on soil
Photodegradation in air
Aerobic soil metabolism study
Anaerobic soil metabolism study
Anaerobic aquatic metabolism study

Leaching and adsorption/desorption
Laboratory volatility study
Field volatility study
Soil field dissipation study
Aquatic sediment field dissipation study
Forestry field dissipation study
Combinations and tank mixes
Long-term soil dissipation study
Confined rotational crop study
Field rotational crop study
Accumulation in irrigated crops
Accumulation in fish
Accumulation in aquatic nontarget organisms
Small-scale prospective ground water monitoring study
Small-scale retrospective ground water monitoring study
Large-scale retrospective ground water monitoring study

Residual Chemistry Data Requirements

Chemical identity
Directions for use
Nature of residue in plants
Nature of residue in livestock
Residue analytical method in plants
Residue analytical method in animals
Storage stability
Magnitude of the residue in potable water
Magnitude of the residue in fish
Magnitude of the residue in irrigated crops
Magnitude of the residue in food handling
Magnitude of the residue in meat, milk, poultry, and eggs
Crop field trials
Magnitude of the residue in processed food and feed
Reduction of residues
Proposed tolerance
Reasonable grounds in support of petition
Analytical reference standard

Spray Drift Data Requirements

Droplet size spectrum
Drift field evaluation

CHAPTER 7

PESTICIDE LABELS: THE CONVERGENCE OF SCIENCE, PUBLIC POLICY, AND USER RESPONSIBILITY

The pesticide label conveys general and technical information from regulatory agencies and pesticide manufacturers to the agricultural community, the commercial pest management industry, and the general public. It is the one source where scientific review, regulatory oversight, and public policy are interwoven to achieve a common objective: to clearly and precisely convey information on handling, storing, applying, and disposing of pesticides in a manner conducive to good health and environmental stewardship.

PESTICIDE PRODUCT REGISTRATION AND THE REGULATORY PROCESS

As discussed in Chapter 6, each pesticide must receive EPA registration before it can be sold, distributed, or used in the United States. The EPA Office of Pesticide Programs (OPP) grants product registration only after a thorough evaluation process wherein toxicological, environmental, and product use information is examined. In reviewing product data and issuing registration, the EPA complies with the mandate by the Federal Insecticide, Fungicide, and Rodenticide Act (FIFRA) that potential risks are minimal and that use according to label directions will not have unreasonable adverse effects on human health, wildlife, or the environment.

The review of labels and label amendments is required for most chemicals submitted for EPA registration:

- Existing chemicals, with or without data
- Products for which efficacy data must be submitted
- Chemicals under development (for the manufacturer's use, only, in testing the product)
- Products for which reregistration is requested
- Products for special local needs or experimental use

All or specific parts of the label are reviewed and the EPA either approves it or requests changes. The label review for a product that is significantly similar to an existing one generally is briefer than for a brand-new chemical. Label reviewers evaluate and verify label content to ensure that it meets all federal regulations and that it is easily understood. Only after the reviewer is satisfied with the label does it gain approval.

The pesticide label communicates critical product information to users, and the EPA mandates that users read and follow label directions. The label undergoes in-depth scrutiny during the registration process, and the EPA's decision to register a product is based in part on the premise that users will apply it according to label specifications.

Consideration is also given to misuse potential, and the EPA's reliance on good science and responsible use by the public are linked through general and specific label language. For example, a label might address potential and actual risks to wildlife by stating that drift and runoff from treated areas may be hazardous to aquatic organisms nearby.

Federal and state inspectors are also responsible for ensuring that manufacturers comply with FIFRA. Product label inspections are conducted randomly at various manufacturing and retail establishments. Inspectors review product labels to verify that they reflect the specifications approved by the EPA. If the product is considered *misbranded,* enforcement action is warranted.

TYPES OF PESTICIDE LABELING

A pesticide may be used only according to the directions on the label and labeling that accompany the product at the time of sale. Although the label is always affixed to the container itself, in some cases a detached, *supplemental* label may be provided as well. The supplemental label may contain an additional approved use site or information related to the use under Section 18 or Section 24(c) registration.

A label can be changed

- by amendment, which requires EPA review and approval. It is used any time the directions for use are modified, or when other substantive changes are made to label content.
- by notification, where the registrant certifies that only a particular item has been changed. It is often used when the change is in response to newly issued EPA guidance.
- by nonnotification, where the registrant makes a change but does not report it to the EPA. It is used to correct typographical errors and to change color, format, and other information unrelated to FIFRA.

FIFRA Section 3: Requirements of Full Registration

FIFRA Section 3 defines stringent data requirements for full federal registration of a pesticide. Most products offered for sale have FIFRA Section 3 labeling.

FIFRA Section 5: Experimental Use Permits

FIFRA Section 5 allows limited experimental use of a pesticide for generating data in support of full registration. EUP labels must reflect their experimental use status and define

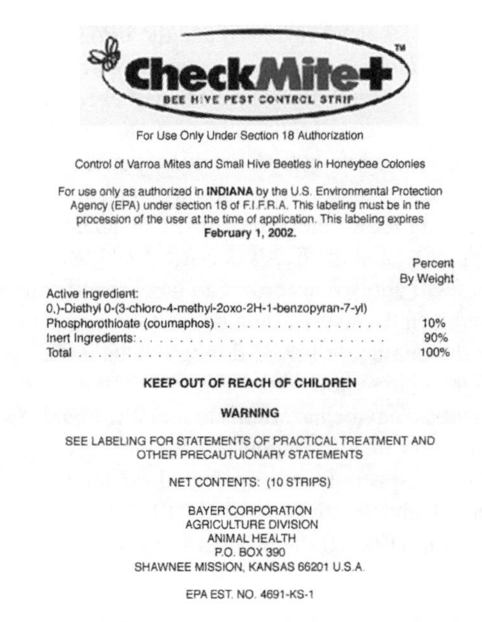

CheckMite+™
BEE HIVE PEST CONTROL STRIP

For Use Only Under Section 18 Authorization

Control of Varroa Mites and Small Hive Beetles in Honeybee Colonies

For use only as authorized in **INDIANA** by the U.S. Environmental Protection Agency (EPA) under section 18 of F.I.F.R.A. This labeling must be in the procession of the user at the time of application. This labeling expires **February 1, 2002.**

	Percent By Weight
Active Ingredient:	
0,)-Diethyl 0-(3-chloro-4-methyl-2oxo-2H-1-benzopyran-7-yl)	
Phosphorothioate (coumaphos)	10%
Inert Ingredients:	90%
Total	100%

KEEP OUT OF REACH OF CHILDREN

WARNING

SEE LABELING FOR STATEMENTS OF PRACTICAL TREATMENT AND OTHER PRECAUTUIONARY STATEMENTS

NET CONTENTS: (10 STRIPS)

BAYER CORPORATION
AGRICULTURE DIVISION
ANIMAL HEALTH
P.O. BOX 390
SHAWNEE MISSION, KANSAS 66201 U.S.A.

EPA EST. NO. 4691-KS-1

Figure 7.1 An example of a Section 18 label.

corresponding use limitations. A product with FIFRA Section 5 labeling cannot be sold or advertised, and, at any time during the experimental use period, the EPA or state regulatory agencies may monitor its permitted use.

At the conclusion of the EUP period, the applicant must submit a final report of results attained. It must include all data gathered during the testing program; a description of the method of disposal used for containers and unused product (including amounts disposed of and the method and site of disposal); and the method used for disposing of affected food or feed. (See Chapter 6 for more information on EUPs).

FIFRA Section 18: Exemptions for Federal and State Agencies

Under certain well-defined conditions of emergency, crisis, or quarantine, a state may petition the EPA to register a product for temporary use. FIFRA Section 18 gives the EPA the authority to temporarily exempt a product from full registration requirements. If granted, the emergency exemption temporarily expands the product label to include the proposed use. Reporting requirements under Section 18 include the amount of product applied and the size of the area treated. Contact http://www.epa.gov/opprd001/section18/ for more information on the Section 18 program (Fig. 7.1).

FIFRA Section 24(c): Authority of States (Additional Uses)

Under FIFRA Section 24(c), states are allowed to register additional uses of a federally registered product to meet special local needs (SLNs). Examples of special local needs are rate modifications and application timing. Section 24(c) labeling applies only to the use pattern for crops, commodities, or sites already authorized by the label issued under Section 3 (http://www.epa.gov/opprd001/24c/). An SLN registration is legal only in the state of issuance. In addition to the federal label, an SLN registration requires a special product label, and both labels must be in the possession of the user.

FIFRA Section 25(b): Exempted Pesticides

Except under very limited circumstances, pesticide products must be registered by the EPA before being offered for sale in the United States. However, one exception to this requirement is for pesticides determined under Section 25(b) of FIFRA "to be of a character which is unnecessary to the subject of this Act" *and* exempted, by regulation, from the requirements of FIFRA.

In 1996, the EPA exempted (from FIFRA requirements) certain minimum-risk active ingredients that satisfy 40 CFR Part 152.25(g)(1) exemption provisions. Currently, there are more than 30 such active ingredients listed in 40 CFR Part 152.25(g)(1). In accordance with 40 CFR Part 152.25(g)(2), exempt products with these active ingredients must contain only minimum-risk inert ingredients. These are known as "25(b) products," and they are not registered by the EPA. The labeling requirements for 25(b) products are somewhat different from those for regular pesticides, and they are not reviewed by the EPA. However, many states do require that 25(b) products be registered in their state before being offered for sale.

SOURCES OF INFORMATION FOR CRAFTING A PESTICIDE LABEL

Most pesticide label information is contained in 40 CFR Part 156.10, which interprets that found in FIFRA.

Pesticide Registration Notices

In addition to 40 CFR Part 156.10 and other sections that modify labeling in certain instances (e.g., for an EUP or for products that fall under the Worker Protection Standard), interpretive documents known as Pesticide Registration (PR) notices (http://www.epa.gov/PR_Notices/) contain information on labeling requirements. These are issued by the EPA Office of Pesticide Programs.

EPA Label Review Manual

Many documents address or explain labeling requirements: policy and criteria notices, policy memos, standard operating procedures, etc. For the convenience of those who write or review labels, the EPA OPP has compiled into a *Label Review Manual* (http://www.epa.gov/oppfead1/labeling/lrm/) the labeling requirements that they collectively address. But although the manual frequently is "the" source used in preparing or reviewing labels, it is not the original authority: it is simply a compilation of labeling requirements.

Graphics

Graphics and symbols are permitted on pesticide product labels if they are accompanied by explanatory text, if their meaning is clear, if they do not obscure or crowd required label language, and if they do not misbrand the product. Graphics or symbols may not be used in place of required text but should enhance the user's understanding of it.

The EPA generally reviews the suitability of graphics and symbols on pesticide labels on a case-by-case basis. Here are some examples of acceptable graphics:

- Arrow diagrams on how to open product containers
- Graphics that depict nozzle spray patterns and/or application patterns
- Pictograms near precautionary statements on the various routes of exposure (oral, inhalation, dermal)
- Pictures consistent with label text that indicates approved application sites
- Pictures that show appropriate protective gear
- A category of symbols typically allowed includes those that do not relate to FIFRA at all (e.g., symbols used by the Department of Transportation and firefighter associations)

Example of unacceptable graphics include the following:

- A food or flower for which the product is not labeled
- Depictions of any nonfood site for which the product is not labeled
- Pictures of a pest for which the product is not labeled
- Pictures of people using a product without the required personal protective equipment
- Drawings or photos representing the fragrance of the product

There are also graphics and symbols that are considered false or misleading:

- Pictures of children, unless the product is registered for use on children or for use in swimming pools
- Pictures of a residential use site on a product used strictly in industrial situations
- Symbols that imply that the product is "safe" or "nontoxic"

Label Claims

There are specific examples of claims considered to be false or misleading in 40 CFR Part 156.10(a)(5). Label claims are evaluated on a case-by-case basis by the EPA reviewers. Descriptions of disallowed claims are as follows:

- False or misleading statements concerning the composition of the product
- False or misleading statements concerning the effectiveness of the product
- False or misleading statements on the value of the product for other than pesticidal purposes
- False or misleading comparisons with other products
- Any statement that directly or indirectly implies that the product is recommended or endorsed by a federal agency
- True statements used to project a false or misleading impression to the purchaser
- Label disclaimers that negate or detract from labeling statements required under FIFRA

- Claims about the safety of the product or its ingredients, including terminology such as *safe, nonpoisonous, noninjurious, harmless,* or *nontoxic to humans and pets,* with or without a qualifying phrase such as *when used as directed*
- Nonnumerical and/or comparative statements on the safety of the product

In addition to claims made on the label, claims made in advertising must be consistent with those on the label; and the criteria that apply to claims also apply to product names.

ORGANIZATION OF THE PESTICIDE LABEL

Pesticides are developed by the manufacturer, registered with the EPA, and sold to the public with the assumption that users will read and follow label instructions. Specific information on use, safety, personal protective equipment, environmental precautions, storage and disposal, and other topics are found on the pesticide label. The purpose of the label is to provide clear directions to allow maximum product benefit while minimizing risk to human health and the environment. All research, testing, and regulatory processes ultimately are reflected through label language.

Every pesticide label includes the statement, "It is a violation of federal law to use this product in a manner inconsistent with its labeling." This language obligates the purchaser or user of any pesticide product to assume legal responsibility for its use. Further, courts of law and regulators recognize the pesticide label as a binding contract that requires the user to apply the product exactly as directed. Terms such as *must, shall, do not,* and *shall not* on a pesticide label place the responsibility for handling and using the product according to label directions squarely on the shoulders of the user. Any departure from label directions constitutes an illegal use of the pesticide: *The label is the law.* (See Chapter 8 for more information on legal issues).

The term *use* means more than application alone. Federal and state regulations define pesticide use to include handling, mixing, loading, storage, transportation, and disposal. It is an all-encompassing definition for every activity that involves a pesticide—from purchase to container disposal.

The pesticide label is more than just a piece of paper: it instructs the user on how to use the product safely and effectively, and it serves as a legal measuring stick. Many statements on the label result from rigorous scientific investigation and regulatory decisions. Pesticide users should read and follow pesticide label directions to ensure effective pest management, personal safety, environmental protection, and legal compliance.

Your familiarity with pesticide label content and design is crucial to selecting the most appropriate products and attaining maximum benefit from their use. Label information may seem overwhelming at first, but once you are comfortable with general label format, comprehension follows.

FIFRA and its regulations mandate that the labeling of every pesticide product include brand name, name and address of the registrant, net contents, EPA product registration number, manufacturing establishment number, ingredient statement, warning or precautionary statements, use classification (if restricted-use), signal word, use directions, and storage and disposal information. Although the EPA has standards for location and content of certain label information, layout is arbitrary; and as a general rule manufacturers design their own labels. Label content is generally divided into four major categories: safety information, environmental information, product information, and use information.

Safety Information

Child Hazard Warning The front panel of every pesticide product label must bear the statement, "Keep Out of Reach of Children." The EPA may waive this requirement only in cases where the likelihood of contact with children is extremely remote, or in cases where the product is approved for use on children (e.g., lice control products). However, the labeling of a product for use on children does not mean that children should be the *users,* that is, they should not be administering the product to or upon themselves. And in that case, "Keep Out of Reach of Children" is still important label content.

Signal Words A signal word must appear prominently on the front of every pesticide container; in essence, it serves as a one-word summary of the potential human toxicity of the product. The three signal words, in decreasing order of toxicity, are DANGER (highly toxic), WARNING (moderately toxic), and CAUTION (slightly toxic). The signal word also appears at the beginning of the precautionary statement on the product label.

The signal word is assigned on the basis of laboratory tests conducted on each product. Data are compiled from animal studies on exposure through ingestion, inhalation, and dermal (skin and eye) absorption; and the route of exposure that indicates the highest human toxicity potential determines the signal word assigned to the label. For example, if laboratory test results indicate a product to be moderately toxic if ingested, highly toxic if inhaled, and slightly toxic if absorbed through the skin or eyes, the signal word would be DANGER, based on inhalation studies.

Signal words assigned to pesticide labels reflect the single most serious *type* of toxic effect achieved during laboratory testing of the product; the signal word on a given label is based on oral *or* dermal *or* respiratory effects (Table 7.1).

Hazards to Humans and Domestic Animals Precautionary statements indicating specific hazards, routes of exposure, and precautions for avoiding human and animal injury are required on the label. Examples of precautionary warning language include the following:

- *Harmful if swallowed, inhaled, or absorbed through the skin.*
- *Do not breathe vapors or spray mist.*
- *Avoid contact with eyes, skin, or clothing.*
- *Handle concentrate in a ventilated area.*

Individual warning statements are based on the toxicity category of the product for each route of exposure.

Protective Clothing and Equipment Statements These are intended to reduce the potential for applicator exposure. Most pesticide labels contain very specific instructions on the type of clothing that must be worn during handling and mixing (Fig. 7.2).

Potential routes of exposure determine the types of protective clothing designated on the label. Generally, a long-sleeved shirt and long pants are the minimum requirements. The label specifies items such as respirators, chemical-resistant gloves, aprons, protective eyewear, and boots when they are required. (See Chapter 11 for more information on personal protective equipment).

TABLE 7.1 Signal Words Used in Labeling

	Oral Toxicity		Dermal Toxicity			
Signal Word	LD_{50}^{a}	Lethal dose for 150 lb. person	Toxicity LD_{50}^{a}	Eye Effects	Skin Effects	Inhalation LC_{50}^{b}
DANGER	Up to and including 50 mg/kg	A taste to a teaspoonful	Up to and including 200 mg/kg	Corrosive: corneal not reversible	Corrosive	Up to 2000 μg/l
WARNING	50–500 mg/kg	A teaspoonful to 1 oz	200–2000 mg/kg	Corneal opacity: reversible within 7 days; irritation present for 7 additional days	Severe irritation at 72 hr	2000 to 20,000 μg/l
CAUTION	>500 mg/kg	>1 oz	>2000 mg/kg	No corneal opacity: no irritation; or reversible within 7 days	Mild to moderate	>20,000 μg/l

[a] LD_{50} values are stated in mg of pesticide per kg of body weight. 1 mg/kg = 1 part per million (ppm).
[b] LC_{50} values are stated in micrograms of the compound per liter of air.

Figure 7.2 The use of chemical resistant gloves is important in reducing pesticide applicator exposure.

Common label language:

- *Wear full face shield, rubber gloves, apron, and waterproof footwear when pouring concentrate or when exposure to concentrate is possible.*
- *Eye protection, chemical-resistant gloves and footwear, a long-sleeved shirt, and long-legged pants or coveralls are required.*

Statement of Practical Treatment Such a statement (first aid) provides valuable information to persons at the scene of a pesticide poisoning.

Some examples:

- *If on skin or clothing: Take off contaminated clothing and immediately begin rinsing the skin; continue for 15–20 minutes, using plenty of water. Call a poison control center or doctor for treatment advice.*
- *If swallowed: Immediately call a poison control center or doctor for treatment advice. Have the victim sip a glass of water if they are able to swallow. Do not induce vomiting unless told to do so by a poison control center or doctor. If the victim is unconscious do not give anything by mouth.*
- *If inhaled: Move the victim to fresh air. If the person is not breathing, call 911; then administer CPR. Call a poison control center or doctor for further treatment advice.*
- *If in eyes: Hold eyes open and rinse slowly and gently with water for 15–20 minutes. Remove contact lenses, if present, after the first 5 minutes, then continue rinsing the eyes. Call a poison control center or doctor for treatment advice.*

Note to Physicians This informs doctors and emergency responders—not consumers—of appropriate medical procedures for poisoning victims. The statement might indicate one or more of these examples:

- *There is no specific antidote.*
- *If the product is ingested, induce emesis (vomiting) or perform stomach gavage.*
- *The use of an aqueous (watery) slurry of activated charcoal may be considered.*

Figure 7.3 The pesticide label provides very specific information on how to deal with a pesticide poisoning.

Many products also list a toll-free telephone number for emergency information. Emergency telephone numbers also are provided on the Material Safety Data Sheet (MSDS), which is available through the pesticide distributor or manufacturer (see Chapter 15 for information on the MSDS) (Fig. 7.3).

Physical and Chemical Hazard Statements These statements on a pesticide label indicate the flammability or explosive tendencies of the product. These statements state specific hazards and conditions to be avoided. Some examples follow:

- Extremely flammable
- Contents under pressure
- Keep away from fire, sparks, and heated surfaces.
- Do not puncture or incinerate container.
- Exposure to temperatures above 130°F causes bursting.

Environmental Hazard Statements

Environmental hazard statements (for example, *This product is highly toxic to bees.*) indicate the nature of potential hazards as well as precautions to avoid personal injury and damage to nontarget organisms or the environment. Potential hazards are determined by a series of tests that evaluate the toxicity of the pesticide to wildlife such as mammals, fish, birds, aquatic invertebrates, and pollinating insects (Table 7.2).

If incident data are received on fish kills, etc., the statement may read, *Extremely toxic to fish.* Or, in an effort to reduce risk, the label may prescribe measures such as, *Do not allow drift to contact nontarget plants,* or, *Do not apply directly to water or wetlands.*

TABLE 7.2 Evaluation of Toxicity of Pesticides to Wildlife

For Avian Dietary LC_{50}	
$LC_{50} < 50$ ppm	Very highly toxic pesticide
$LC_{50} = 50$–500 ppm	Highly toxic pesticide
$LC_{50} = 501$–1000 ppm	Moderately toxic pesticide
$LC_{50} = 1001$–5000 ppm	Slightly toxic pesticide
$LC_{50} > 5000$ ppm	Practically nontoxic pesticide
For Avian Acute Oral LD_{50}	
$LD_{50} < 10$ ppm	Very highly toxic pesticide
$LD_{50} = 10$–50 ppm	Highly toxic pesticide
$LD_{50} = 51$–500 ppm	Moderately toxic pesticide
$LD_{50} = 501$–2000 ppm	Slightly toxic pesticide
$LD_{50} > 2000$ ppm	Practically nontoxic pesticide
For Mammal Acute Oral LD_{50}	
$LD_{50} < 10$ ppm	Very highly toxic pesticide
$LD_{50} = 10$–50 ppm	Highly toxic pesticide
$LD_{50} = 51$–500 ppm	Moderately toxic pesticide
$LD_{50} = 501$–2000 ppm	Slightly toxic pesticide
$LD_{50} > 2000$ ppm	Practically nontoxic pesticide
For Fish or Aquatic Invertebrate LC_{50}	
$LC_{50} < 0.1$ ppm	Very highly toxic pesticide
$LC_{50} = 0.1$–1 ppm	Highly toxic pesticide
$LC_{50} = 1$–10 ppm	Moderately toxic pesticide
$LC_{50} = 11$–100 ppm	Slightly toxic pesticide
$LC_{50} > 100$ ppm	Practically nontoxic pesticide

If the pesticide has the potential to harm an endangered or threatened species or its critical habitat, environmental hazard use limitation statements may read as follows:

- *Do not apply within a 100 yards of species habitat by aerial application or within 20 yards of species habitat for ground applications.*
- *Do not apply directly to water within the shaded area as shown on the bulletin, or refer the user to an endangered species bulletin for further information.*
- *Restrictions for the protection of endangered species apply to this product.*
- *If restrictions apply to the area in which this product is to be used, you must obtain the Pesticide Use Bulletin for Protecting of Endangered Species for that county.*

Statements on environmental impact may indicate that the product . . . *may travel through soil and can enter ground water,* or . . . *has been found in ground water.* Label instructions indicate how to reduce environmental impact: *This product may not be mixed, loaded, or used within 50 feet of all wells, including abandoned wells, drainage wells, and sink holes;* or *This product has been shown to leach under certain conditions. Do not apply to sand and loamy sand soils where the water table (ground water) is close to the surface.*

Examples of Safety and Environmental Information

Child Hazard Warning	**Keep Out of Reach of Children**
Signal Word	**WARNING**
Hazards to Humans and Domestic Animals	May cause eye injury. Harmful if swallowed, inhaled, or absorbed through the skin. Do not get in eyes or on clothing. Avoid breathing vapor or spray mist.
Protective Clothing and Equipment	Wear goggles, face shield, or safety glasses when handling the undiluted material.
Statement of Practical Treatment	**If swallowed,** call a physician or poison control center. Do not induce vomiting. **If in eye,** flush eyes with plenty of water for at least 15 minutes. Get medical attention. **If on skin,** remove contaminated clothing and wash skin immediately with soap and water. **Note to physician:** Treat the patient symptomatically.
Environmental Hazard Statements	This pesticide is extremely toxic to fish and wildlife. Do not apply directly to wetlands. Do not contaminate water by cleaning equipment or disposal of wastes. Drift and runoff from treated areas may be hazardous to aquatic organisms in neighboring areas. This product is highly toxic to bees exposed directly or via residues on blooming crops or weeds.

Examples of Use and Product Information

Restricted-Use Classification	**Restricted-Use Classification** For retail sale to and use by only certified applicators or persons under their direct supervision.
Brand-Name	**Zapo Aminophos 2 EC**
Formulation	Emulsifiable Concentrate
Ingredient Statement	Active Ingredient Gratol (1.1 Dimethyl). . . . . 25% Inert Ingredients . . . . 75% Total. . . . . 100% Contains 2 lb active ingredient per gallon.
Physical and Chemical Hazard	Do not use or store near heat or flame.
Misuse Statement	It is a violation of federal law to use this product in a manner inconsistent with its labeling.
Storage and Disposal Statement	Store in original container. Do not store next to food, water, or feed. Protect from freezing. Triple rinse (or equivalent). Puncture and dispose of in a sanitary landfill or dispose of by other procedures allowed by state or local authorities.
Restricted Entry Statement	Do not reenter area for 24 hours where Zapo has been applied.
Directions for Use	Horizontal barriers may be established in areas if intended on covering, such as floors and porches. Applications shall be made by low-pressure spray (less than 50 psi). Apply the emulsion to fill dirt at the rate of 1 gallon per 10 square feet.
Agricultural Use Requirements	Use this product only in accordance with its labeling and with the Worker Protection Standard, 40 CFR Part 170.
Warranty	Notice on Condition of Sale: No guarantee, expressed or implied, is made to the effect or results to be obtained if not used in accordance with directions or established safety practices.
Name and Address	Manufactured by Acme Chemical Co. Town, State
Establishment Number	EPA Est. No. 2534-IN-1
Registration Number	EPA Registration No. 2534-65

Use Information

Directions for use often comprise the bulk of a pesticide label. They must be adequate to protect the public from fraud and personal injury and to prevent unreasonable adverse effects on the environment. They must provide guidance to the user on the pests controlled, sites of application, compatibility with other pesticides, mixing or dilution rates, application rates, equipment needed for application, timing and frequency of applications, harvest intervals, and general information for successful results.

Directions for Use These may appear on any portion of the label. Because of the detail required for specific applications, use directions for common sites, pests, and applications may be grouped together under a general heading. Information specific to individual uses may be addressed under specific headings. Considerable guidance on the content can be found in the EPA Label Review Manual.

For products falling under the scope of the Worker Protection Standard (WPS), an *Agricultural Use Requirements* box is required in the *Direction for Use* section. The content begins with the statement, *Use this product only in accordance with its labeling and with the Worker Protection Standard, 40 CFR Part 170.* The standard itself will not appear on the label, so the user will be requested to obtain the standard from the EPA, the Cooperative Extension Service, or the Department of Agriculture. In addition to this box, there are many additional requirements indicated in 40 CFR Part 170 that must appear on the label.

Agricultural Use Requirements

Use this product only in accordance with its labeling and with the Worker Protection Standard, 40 CFR part 170. This Standard contains requirements for the protection of agricultural workers on farms, forests, nurseries, and greenhouses, and handlers of agricultural pesticides. It contains requirements for training, decontamination, notification, and emergency assistance. It also contains specific instructions and exceptions pertaining to the statements on this label about Personal Protective Equipment (PPE), and restricted-entry interval. The requirements in this box only apply to uses of this product that are covered by the Worker Protection Standard.

Do not enter or allow worker entry into treated areas during the restricted-entry interval (REI) of 12 hours.

Exception: If the product is soil-injected or soil incorporated, the Worker Protection Standard, under certain circumstances, allows workers to enter the treated area if there will be no contact with anything that has been treated.

PPE required for early entry to treated areas that is permitted under the Worker Protection Standard and that involves contact with anything that has been treated, such as plants, soil, or water, is:

- Coveralls
- Chemical-resistant gloves such as Barrier Laminate or Viton
- Shoes plus socks
- Protective eyewear

Reentry or Restricted Entry Statements These indicate reentry precautions and a time interval during which reentry into a treated site is not allowed. The statement indicates the length of time that must elapse after the pesticide application before individuals may

(a)

(b)

Figure 7.4 (a) A field posted with the EPA Worker Protection sign. (b) A sign at a greenhouse informing the workers to stay out of an area that has been treated with a pesticide.

enter the treated area without personal protective clothing and equipment. Examples of reentry statements (Fig. 7.4) include:

- *Do not enter treated areas without protective clothing until sprays have dried.*
- *Do not enter or allow worker entry into treated areas during the restricted-entry interval of 24 hours.*

Storage and Transportation Statements

- Store at temperatures above 32°F.
- Do not contaminate feed, foodstuffs, or drinking water.
- Do not store next to feed or food or transport in or on vehicles containing foodstuffs or feed.
- For help with any spill, leak, fire, or exposure involving this material, call Chemtrek (800-424-9300).

Container Rinsing and Disposal Statements These include instructions for container rinsing and proper procedures for handling pesticide containers and disposing of unused products. Federal, state, and local regulations often must be consulted to determine how to dispose of unused pesticide concentrates or dilute mixtures. Container disposal statements might read as follows:

- Triple rinse (or equivalent).
- Do not reuse container.
- Offer for recycling or reconditioning.
- Puncture and dispose of in a sanitary landfill.
- Disposal by other procedures allowed by state and local authorities.
- Improper disposal of excess pesticides, spray mixture, or rinsate is a violation of federal law.

Although numerous pesticide labels still state that properly rinsed containers may be burned, almost every state has clean air laws which prohibit such disposal.

Product Information

Brand (Trade) Name The brand name under which a pesticide product is sold always appears on the front panel and often is the most conspicuous part of the label.

Name and Address of the Producer, Registrant, or Other Entity for Whom the Product Was Produced These items must be shown on the label. If the registrant's name appears on the label and the registrant is not the producer, it must be qualified by appropriate wording such as "Packed for . . ." or "Produced for . . . ."

Net Weight or Volume of the Contents The net weight or volume of the formulated pesticide product is displayed prominently on the label or stamped on the container.

Product Registration Number This appears on the label, preceded by the phrase "EPA Registration No." or "EPA Reg. No." The registration number identifies a specific pesticide product and signifies that federal registration requirements have been met. At a minimum, the registration numbers consist of two sets of digits: e.g., 491-005. The first set of digits identifies the registrant. The second set is a sequential number assigned to products submitted for registration; it represents the specific registration issued to the company by the EPA. Together, these numbers clearly identify the product.

The product's distributor number is indicated by a third set of numbers at the end of the registration number. However, not every product has this number. This number represents the number of the company that has agreed to *distribute* a registered product of another company. For example, in the registration number, 264-498-10107, 264 represents the company number of the registrant, 498 represents the specific registration for a particular product, and 10107 represents the company number of the company that has agreed to distribute company 264's product. The company number is assigned sequentially at the time of application. Product labels containing a distributor number can be changed by the registrant only, not by the distributor.

Establishment Number This number is preceded by the phrase "EPA Est." The EPA requires pesticide production sites to be registered with the EPA. A pesticide-producing establishment is assigned an EPA establishment number that clearly identifies that location.

All pesticides produced at that location must bear its EPA establishment number on the label or container. Farm service centers that repackage bulk pesticides must be registered as pesticide-producing establishments; and, as with all pesticide producers, they must keep records of their pesticide production and file annual production reports.

Ingredient Statement This usually is found on the front panel of the label. It identifies the name and percentage of the active ingredient, including microbial strains, biochemicals, pheromones, etc. It also includes the total percentage of inert ingredients. Chemical names often are complex; for example, 2-chloro-4-ethylamino-6-isopropylamino-*s*-triazine is the active ingredient in the product AAtrex. To aid communication, EPA- or ANSI-approved (American National Standards Institute) common names may be substituted for chemical names.

Inert ingredients allow active ingredients to be formulated into many different products. As part of the formulation, they determine the handling properties of a product and influence toxicity, release rates, residual activity, persistence, and method of application. There are no pest management claims for inert ingredients, even if they serve as an active ingredient in another formulation. Because precise product formulations are confidential, the total percentage (by weight) of inert ingredients usually is the only information on inerts found on the label. However, the EPA has designated that some inert ingredients be included on the label, including those on the *List of Inerts of Toxicological Concern* (http://www.epa.gov/opprd001/inerts/fr54.htm).

Formulation The formulation of the product often appears on the front panel of the label, either near the brand name or in the general information section. Information about the type of product formulation—granular, liquid flowable, dry flowable, microencapsulated, emulsifiable concentrate, etc.—provides insight about application equipment, handling properties, and performance characteristics.

Product labels often convey the pesticide formulation by a suffix to the brand or trade name. Table 7.3 lists many of these suffixes and their meanings. A suffix can also include a number that indicates the amount of active ingredient in the product. The number contained in the brand-name suffix of a solid formulation, such as a dust, granule, wettable powder, etc., describes the percentage of active ingredient on a percent by weight basis. For example, the brand name Sevin 50W tells the purchaser that the product is formulated as a wettable powder (W) and that it is 50% active ingredient by weight.

The number included in the brand-name suffix of a liquid formulation such as a liquid flowable (L) or an emulsifiable concentrate (EC) describes the amount of active ingredient on the basis of pounds per gallon. The brand name Treflan 4EC indicates an emulsifiable concentrate that contains 4 lb of active ingredient per gallon of product.

General-Use versus Restricted-Use Classification The EPA may classify a pesticide product for restricted use due to the complexity of its designated use, concerns about environmental safety, or potential human toxicity. The EPA's list of restricted-use products can be found at http://www.ianr.unl.edu/pubs/pesticides/ec2500.pdf.

A restricted-use product may be bought and used only by a certified applicator or by persons under the direct supervision of a certified applicator (Fig. 7.5). A restricted-use statement appears conspicuously at the top of the front panel of the label, as well as in the Directions for Use, to make oversight unlikely. All restricted-use pesticides are identified by language such as, *For retail sale to and use only by certified applicators or persons*

TABLE 7.3 Suffixes of Chemical Brand Names

	Describes the Formulation
Suffix	Meaning
AF	Aqueous flowable
AS	Aqueous suspension
D	Dust
DF	Dry flowable
E	Emulsifiable concentrate
EC	Emulsifiable concentrate
ES	Emulsifiable solution
F	Flowable
FL	Flowable
G	Granule
L	Liquid flowable
OL	Oil-soluble liquid
P	Pelleted
S	Soluble powder
SG	Sand granules
SL	Slurry
ULV	Ultra-low volume concentrate
W	Wettable powder
WDG	Water-dispersible granules
WP	Wettable powder
WSP	Water-soluble packet
	Describes How a Pesticide Is Used
Suffix	Meaning
GS	For treatment of grass seed
LSR	For leaf spot and rust
PM	For powdery mildew
RP	For range and pasture
RTU	Ready-to-use
SD	For use as a side dressing
TC	Termiticide concentrate
TG	Turfgrass fungicide
	Describes Characteristics of the Formulation
Suffix	Meaning
BE	The butyl ester of 2,4-D
D	An ester of 2,4-D
K	A potassium salt of the active ingredient
LO	Low odor
LV	Low volatility
MF	Modified formulation
T	A triazole
2X	Double strength
	Describes Special Locations
PNW	For use in the Pacific Northwest

Figure 7.5 Pesticide dealers must confirm that persons purchasing restricted-use pesticides have the appropriate up-to-date credentials.

under their direct supervision, and only for those uses covered by the certified applicator's certification.

Pesticides that remain unclassified are referred to as general-use pesticides and may be purchased by the general public; most pesticides used by homeowners are general-use products. Except on some chlorine products, the general-use statement is never used on labels. Identifying the restricted-use product is deemed sufficient.

Warranty Such information is provided by the manufacturer as assurance that the product conforms to its chemical description on the label and that it is fit for labeled purposes, if used according to directions, under normal conditions. The warranty does not extend to any use of the product contrary to label instructions, nor does it apply under abnormal conditions such as drought, tornadoes, hurricanes, or excessive rainfall.

PUBLIC POLICY

Consumer Labeling Initiative

Early in 1996, the EPA launched a multiyear study called the Consumer Labeling Initiative to provide information on how consumers comprehend and use labels for indoor and outdoor pesticides and household surface cleaners. *EPA's Consumer Labeling Initiative: Phase I Report* provides excellent information on consumer attitudes, an in-depth literature search, and comments from a number of professionals. A portion of the Phase I Report attempts to answer questions such as these:

- What do consumers want to know about these products?
- Do consumers understand labels?
- Do consumers follow instructions on labels?
- Do consumers find the information on labels confusing?

- How do consumers currently use label information to make purchase decisions?
- What could motivate consumers to become more likely to use label information?
- Do consumers over-use or under-use products?

Answers to these questions were obtained via personal, 45-min interviews with 135 consumers in Miami, New York, Los Angeles, Dallas, and Chicago. Some of the information generated is summarized in the following section.

Use of Product Labels by Consumers Four main factors influence the likelihood of the consumer to read a pesticide label: when the product is unfamiliar, when it is perceived as dangerous, when it is difficult to use, and/or when it is used outdoors.

- The most common part of the pesticide label read by consumers is the segment headed *Directions for Use.* Under this heading, the consumer can determine where the product is to be used, what pests it controls, and how to apply it.
- Consumers want fewer technical words and less jargon on labels. They indicate that small fonts and lack of color contrast on most labels make them difficult to read.
- Consumers have difficulty with chemical names found in ingredient statements. They tend to rely on common names (e.g., 2,4-D and carbaryl) when comparing products. Consumers in general do not understand terms such as *inert ingredients* and the statement, *It is a violation of federal law to use this product in a manner inconsistent with its labeling.*
- The label heading *Hazards to Human Health* is interpreted by consumers to mean that the product is hazardous to humans. They also suggest that the term *First Aid* replace *Statement of Practical Treatment.* Consumers indicate that they understand the meaning of the signal word DANGER but do not recognize the significance between products labeled CAUTION and those with the signal word WARNING.

The EPA has concluded from Phase I consumer labeling data that certain label changes can be made to consumer pesticide products to enhance readability and usefulness. Some examples include using common names, using the heading *First Aid* in place of *Statement of Practical Treatment,* and including general and/or emergency phone numbers. The EPA also suggests that a Product Label Consumer Education Task Force be organized to identify consumer education activities that focus on the importance of reading labels. The goal is to make every pesticide label effectively clear, concise, and informative.

Phase I of the Consumer Labeling Initiative was followed by Phase II: the implementation of Phase I findings. A variety of guidance documents were published in 1997 encouraging the use of phone numbers, common names in the ingredient statement, the term *First Aid* instead of *Statement of Practical Treatment,* and *other* ingredients in lieu of *inert.* In 1998, two statistically representative written and phone surveys were developed to further explore consumer behavior, understanding, and preferences regarding pesticide labels.

The data from these surveys and 27 minifocus groups led to label change recommendations that have been implemented by the pesticide industry; they are currently being implemented by the EPA.

The EPA began the implementation of a consumer education campaign with the slogan "Read the Label FIRST!" Additional information on this initiative can be obtained from the Consumer Labeling Initiative (CLI) Web site: http://www.epa.gov/oppt/labeling.

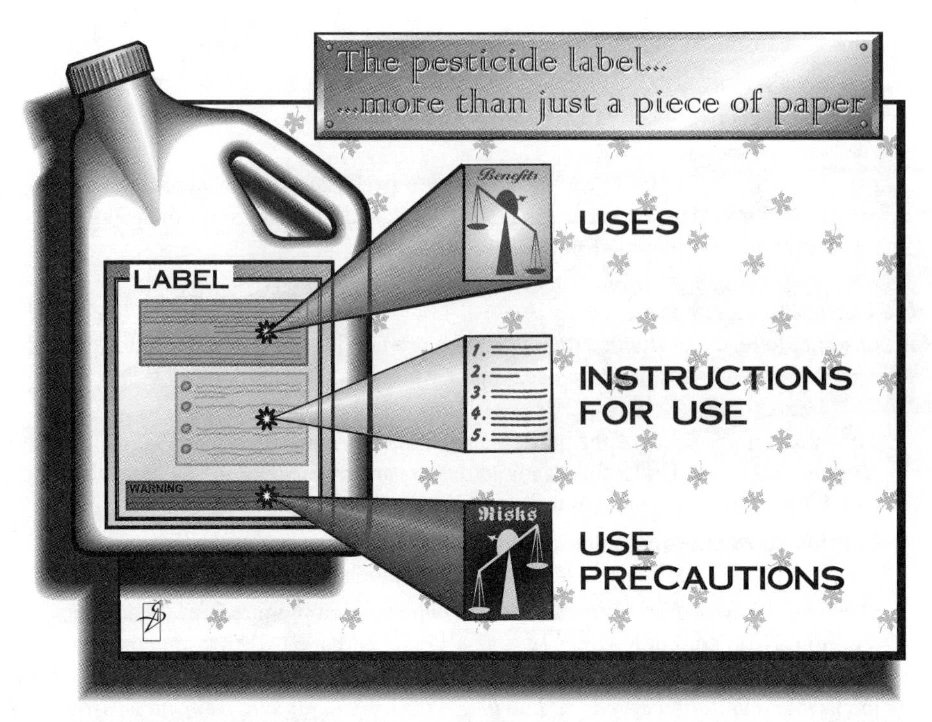

Figure 7.6 The pesticide label provides the user with specific information on how to use the product to manage the pest and on precautions the applicator must take to prevent injury to people, water, and wildlife.

SUMMARY

The information on a pesticide label neither appears randomly nor without thought, and label design and development are not left solely to the manufacturer (Fig. 7.6). There are numerous, very specific requirements, policies, and regulations governing the types of information that must appear on a pesticide label. The EPA scrutinizes each label for compliance and, if labeling and use rules are not followed once the label is on a product and the product is offered for sale, enforcement action can be levied to ensure correction. Label content is very similar for most pesticides.

CHAPTER 8

LIABILITIES AND LAWSUITS: UNDERSTANDING REGULATIONS, INSPECTIONS, AND THE COURTS

New businesses spring up daily across the country in anticipation that they will succeed where others have failed. This initial optimism can fade quickly. Businesses also close daily as many owners who once dreamed of success have those dreams shattered when they are forced out of business, often from carelessness or mismanagement. Often, only a fine line separates successful long-term businesses from the failures.

To succeed in a competitive business world, business owners must offer outstanding service and quality products at competitive prices. Successful owners continually identify their potential customers, evaluate how to best advertise their products and services to these customers, and price their products and services appropriately.

Business owners can be liable for the actions of their employees; therefore, hiring decisions are the most important decisions they will ever make. Employees must be educated and trained to deliver services, professionally, in a safe and effective manner. Progressive companies continually educate their employees to keep them abreast of changing technology (Fig. 8.1).

Staying on top in any business requires constant attention to detail. The challenge of establishing new customers and recruiting new employees is perpetual. New competition, new and unique products and services, a finicky buying public, and employee morale and retention must be examined on a continual basis.

REGULATIONS IN THE MODERN BUSINESS WORLD

Whether the pest management services are performed in a customer's field, home, business, lawn, or landscape, business owners need to understand the regulations that affect business activities. Most employers typically establish their own written policies (internal rules) for employees. These written policies are often compiled in handbooks designed to guide employees on the nature and scope of the employee–employer business relationship.

Figure 8.1 Managing a business in a competitive marketplace is complex.

Although employee policies present challenges, the real challenges are those policies and rules of government that prove most difficult for businesses to follow (Fig. 8.2). The following are some examples:

- There is an almost endless number of local, state, and federal agencies that are constantly preparing new rules or revising those already in existence.
- Most pesticide application businesses cannot afford to employ full-time, in-house legal counsel or safety specialists to help answer questions or find ways to implement necessary changes.
- New state and federal regulations are routinely published in publications that are frequently unknown or inaccessible to most business owners.
- Rules are often very difficult for the average business owner to understand because they are complex and sometimes poorly written, and often they contain confusing legal terminology.
- There is often an overlap or conflict of regulations established by different agencies; there even may be overlap within the same agency.
- Government agencies rarely provide educational assistance to the regulated community; they instead expect companies to *know* what they must do to comply.
- A person may not understand why the paperwork is necessary when they are already in compliance.

These examples demonstrate why achieving total compliance is such a challenge. But business owners also must recognize that their proper implementation of regulations can

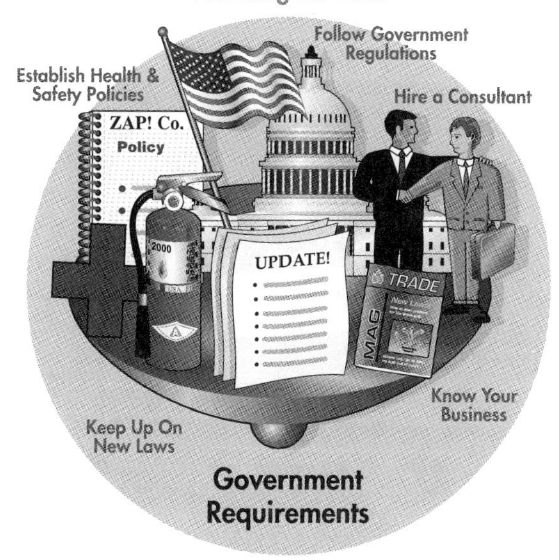

Figure 8.2 Government regulations play an important part in good business management.

improve the profitability and longevity of their companies. A well-organized regulatory program helps prevent employee injuries, property damage claims, environmental contamination, rising insurance premiums, and judicial and regulatory proceedings.

REGULATIONS: NEEDLESS PAPERWORK OR VALUED ASSET?

Many regulations require that deadlines be met, forms submitted, records maintained, inspections conducted, and plans written. Cynics often view paperwork as wasted time and money without tangible benefits. This view is particularly common to those companies forced to comply. Company owners who feel forced into compliance often waste time shuffling papers, programs, and plans. The requirements are seldom implemented, nor are they effectively communicated to the staff.

What If I Don't Comply?

Most regulations are written with the assumption that companies will comply. But, what are the odds of getting caught if one chooses not to comply? The plain truth is that most government agencies have limited staff who spend time in the field conducting compliance audits; most of their time is spent dealing with complaints. Therefore, the odds of getting caught on a *routine* inspection are slim, although disgruntled employees, prospective employees, or exemployees are common sources of complaints to regulatory officials.

Businesses should not focus on compliance from the standpoint, What if I get caught? Most regulations are enacted to prevent real problems, and it is important to realize that they result from public pressure and/or a particular difficulty that the regulated industry has faced (Fig. 8.3).

Prevention is No Accident

Every Regulation Was Written to Prevent a Tragedy

Figure 8.3 Regulations are meant to guide business owners in preventing workplace accidents.

Compliance has little to do with government edicts or mandates. Owners who appreciate the premise of a given regulation and consider its long-term benefit potential are willing to comply. The implementation of regulatory requirements often provides long-term benefits and profitability. Viewing regulations positively helps a business prosper, contributes to a safer workplace, and leads to better-trained employees. The *last* reason to comply with regulations should be to avoid enforcement actions and repercussions by regulatory agencies. Compliance makes good business sense.

LAWS AND REGULATIONS

The authority to create federal and state pesticide laws generally rests with elected officials. Usually, laws created by the elected officials provide a basic framework and intent within which to operate. More specific mandates—usually called rules and regulations—are developed and enacted by regulatory agencies. Changes in federal law often result in parallel changes in state laws and regulations (Fig. 8.4). State law must meet minimum standards assigned by federal law; and often they are even more stringent.

Regulations developed within an agency (EPA, USDA, state regulatory offices, etc.) are products of a rule-making process. This process generally involves publishing the proposed rule in either the Federal Register (if the origin of the rule is a federal agency) or its State Register (if the origin of the rule is a state agency). These publications inform the public of new regulations as well as proposed rule changes. After the comment period ends, and after review of all comments, the agency responsible for the rule makes appropriate revisions. The final rule eventually is published in the same publication. Federal pesticide regulations

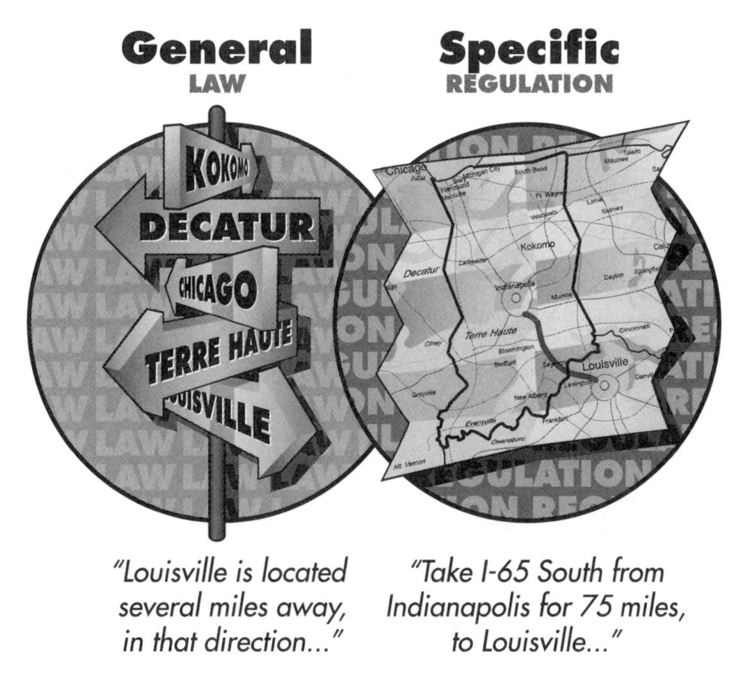

Figure 8.4 Regulations provide the specific details needed to implement laws passed by states and the U.S. Congress.

under FIFRA are compiled in the Code of Federal Regulations, Volume 40, Parts 150–190. The Code of Federal Regulations is typically abbreviated "CFR."

Reasons for Pesticide Laws and Regulations

Pesticide laws and regulations are usually founded upon a set of common goals or objectives:

- Provide for the proper and beneficial use of pesticides to protect public health and safety.
- Protect the environment by controlling the use and disposal of pesticides.
- Assure safe working conditions for farmworkers, commercial pest management personnel, and consumers.
- Assure users, including homeowners, that pesticides are labeled properly, that they are appropriate for their intended use, and that they contain all instructions and precautions necessary to ensure that benefits exceed risks.
- Assure consumers of food produced in the United States that the allowable pesticide residue on that food meets the criteria, "reasonable certainty of no harm."
- Encourage the use of integrated pest management systems that emphasize biological and cultural pest management techniques with selective use of pesticides.

Government's Role in Managing Pesticides

Congressional response to public demand for legislative reform places authority for pesticide registration, distribution, sale, application, and disposal with the EPA. Amendments to

FIFRA in 1972 assigned primary pesticide use enforcement to state lead agencies, provided they meet certain standards. Most states typically designate their department of agriculture (or equivalent) as their lead agency in pesticide matters. The EPA provides funding to and oversight of state lead agencies as they implement and enforce pesticide laws within their boundaries.

A broad array of pesticide regulatory programs has been created since the states began to assume responsibility for enforcing pesticide laws. Those programs include pesticide applicator training and certification, pesticide product registration, and pesticide use enforcement. Most states also are developing programs for direct response to environmental and public health concerns. These include pesticide disposal, organic food production, food safety standards, surface and ground water protection, endangered species protection, and occupational safety.

How a Bill Becomes a Law

The power to create new laws or modify those already in existence is among those of elected officials in the U.S. Congress and state legislatures. Voting on any new bill (law or statute) is a serious responsibility for legislators since their tenure may depend largely on their voting record. Voting yes or no on an issue can ultimately mean the difference between being reelected or not.

A bill at any level of government is a precursor to law. A majority of elected legislators must ultimately support a bill if it is to become law. The bill is enrolled and usually given a hearing within a specific legislative committee. Debate within a committee often results in modification of the original bill in hopes of gaining the majority (committee) vote required for advancement to the legislature, that is, to release the bill to the House of Representatives. The bill that is brought before the House of Representatives is subject to further discussion and amendment by the elected members. Once the bill is in final draft, each member of the House must decide whether to vote for or against it.

A similar version of the House bill is usually introduced in the Senate, where its language is scrutinized and its specifics debated; the Senate may incorporate its own changes before voting. Bills that successfully pass both houses of Congress are reconciled by a joint committee comprised of senators and representatives who confer and form a consensus on the wording of the bill (Fig. 8.5). The bill will then go back to both houses to vote on whether or not to accept the conference committee report.

A bill becomes law only when the governor (if a state bill) or the president (if a federal bill) signs it, or if a veto by one of these officials is overridden by the legislature.

Regulations Written by Career Professionals

Since elected officials rarely have the time or the expertise to micromanage laws, implementation is delegated to federal and state agencies. Agency professionals develop regulations that provide the specifics on what each person or business must do to comply.

Federal laws dealing with pesticides, worker safety, transportation, and discrimination are generally left to the EPA, the Occupational Safety and Health Administration (OSHA), the Department of Transportation, and the Equal Employment Opportunity Commission, respectively. The process is similar under state law. Laws dealing with pesticides, safety, wages, work-age limitations, fire prevention measures, and tax codes are promulgated and administered by agencies or departments within state government.

The writing and development of regulations can be a very lengthy process. Usually, government agencies first solicit input from the regulated community, their own staffs,

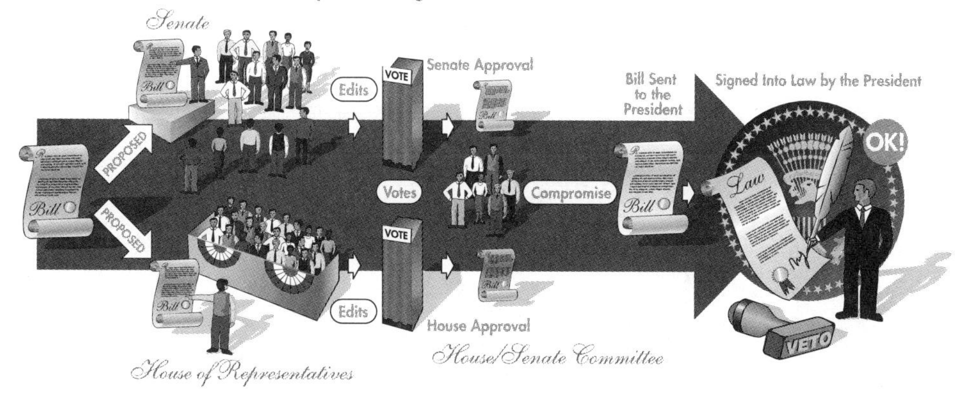

Figure 8.5 The process whereby a bill becomes law is a complex and lengthy endeavor.

environmental groups, and the public at large. Each agency devotes countless hours and resources to engage all interested and affected parties in lengthy, complex discussion on the draft rule. Questions such as the following must be addressed before the draft is written:

- Is there a need for the regulation?
- Is the regulation required by legislative enactment, or is it discretionary with the agency?
- How will the regulation impact businesses financially?
- Will records and reports be required?
- Will there be a fee?
- Will there be penalties for noncompliance?

Input of this nature is vital to assure the desired impact once regulations are enacted. The goal is to pass regulations that are fair to all parties, easily implemented and enforced, and consistent with legislative intentions.

Public Input into Regulation Development Regulatory bodies are not accountable to a constituency as are elected officials. Most agencies employ career civil servants whose jobs do not hinge on who is elected or what party is in office. The only exceptions are those at top administrative levels who are *appointed* by an elected official such as the president or the governor. These personnel serve at the "pleasure" of the person who appointed them to their administrative positions.

The deliberate process of seeking public input helps maintain the integrity of the system by assuring that the voice of the public is heard (Fig. 8.6). The process of promulgating regulations usually includes the following steps.

Legal Review Normally, a regulation must pass an internal legal review by agency staff lawyers. In addition, lawyers or staff evaluate the projected impact of the regulation on the regulated community, the related costs and benefits to the public, and whether the proposed regulation lies within the authority of the agency.

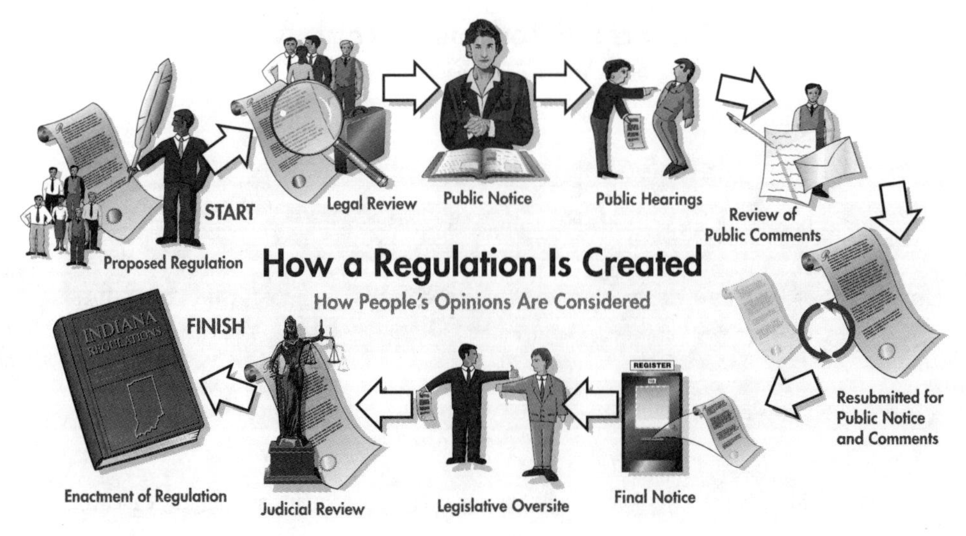

Figure 8.6 Regulations are often written with much input from the public.

Public Notice through Publication New regulations are published in either the Federal Register or the State Register. The intent is to make the public—especially any sectors not involved in early drafts—aware of what is being proposed and to provide them an opportunity to comment (Fig. 8.7).

Public Hearings Public hearings afford the key agency personnel responsible for a proposed regulation the opportunity to meet one-on-one with those who have questions and concerns about it.

Review of Written Comments Offered by the Public Agencies also review and consider all written comments on the proposed rule. Written responses are often provided.

Major Changes Require Resubmittal Regulations that require significant modifications based on written and verbal comments sometimes have to be resubmitted for public scrutiny.

Final Notice to Adopt in Register Once the suggested changes are incorporated into the regulation, the "final notice" is resubmitted to the Federal Register or State Register. It becomes an official regulation at the end of the time period designated in the final notice.

Legislative Oversight Elected officials generally have the right to require changes to regulations and to hold hearings on those regulations as a checks-and-balances process.

Judicial Review The court system is also an important voice in whether a regulation is legal, constitutional, or within the delegated authority given to a specific agency. The courts often play a major role in determining whether the regulation, as written, complies with the intent of state and federal law.

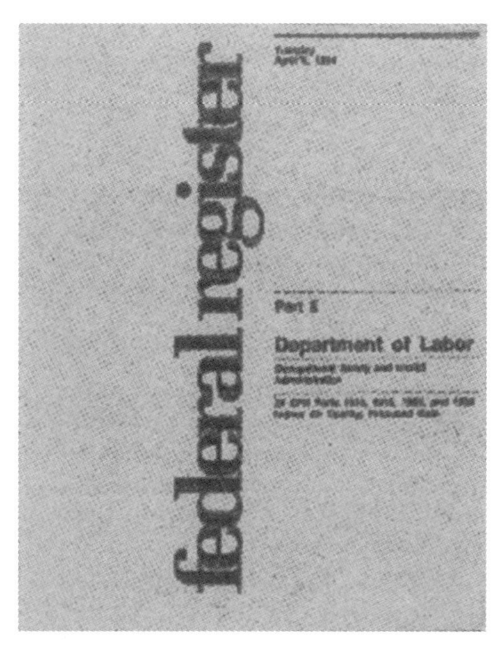

Figure 8.7 The Federal Register is the mechanism by which federal agencies inform the public on new or revised regulations.

READING REGULATIONS

All federal agencies compile their regulations into a series of publications—the Code of Federal Regulations (CFR). The books that comprise the CFR are much like those in a set of encyclopedia; but instead of identifying the books alphabetically, CFR volumes are identified by number. Each book number is specific to the federal agency responsible for enforcement of the regulations it contains. For instance, book number 29 CFR identifies regulations administered by OSHA; book 40 CFR belongs to the EPA (http://www.epa.gov/pesticides/cfr.htm); and book 49 CFR is assigned to the Department of Transportation. State regulations are compiled into the Administrative Code of each respective state.

Read the Entire Rule

As with any book, it is impossible to open the CFR to a specific paragraph and fully understand it without knowing what was addressed in preceding paragraphs, nor is it possible to know what comes after without reading further. In the case of regulations, it is easy to read a sentence out of context without reading the entire regulation.

Understand the Characters

Words have different meanings depending on how they are used in a regulation. Always read the definitions at the front of each regulation, because these are equivalent to the characters of a book. If you don't have a good understanding of the characters in the book, you won't understand who is doing what to whom. The regulation is interpreted using the definitions provided in the regulations.

Look for Help

Read the regulations to gain insight into what is being asked of you and to make sure that they do, in fact, apply to your particular business. Write down questions as you go, asking for clarification of anything that you don't fully understand. Make sure that you understand the difference between legal requirements—things that you "must" and "shall" do to comply—and mere recommendations.

Policies, often viewed as pseudo-regulations or working guidelines, are written regulatory documents that address specific points of compliance not otherwise covered by regulation. They are implemented to establish consistency in answering questions and in dealing with problems not specifically addressed in existing regulations.

Listed below are some excellent sources of regulatory information. When contacting someone regarding regulations, document

- the date of the call,
- the questions that you asked,
- the answers provided, and
- the name of the person who answered your questions.

File the information in a folder with the corresponding regulation. Use the folder for all references to that regulation. For instance, if you obtain information from a publication, be sure to place a copy of the title page and other pertinent pages in the folder.

Regulatory Agencies The best choice for answers to regulatory questions usually is the agency that wrote the rules, requires their implementation, and enforces their provisions. When consulting with agency personnel, ask them whether they have developed any documents, videos, or policies to help the regulated community understand and implement the regulations.

National and State Trade Associations Trade associations have direct contacts within regulatory agencies that govern their industries. They have representatives familiar with the various regulations in which they have a vested interest who can interpret the legalese. Be sure to ask the association for interpretive summaries, written reports, and educational materials that they have developed. Most state and national associations also have continuing education meetings and newsletters that include discussions on regulatory compliance.

Attorneys and Professional Consultants Professional consultants who know the ins and outs of your industry can be excellent sources of information on the regulations that govern it. Also, they may have contacts within various trade associations and government agencies and know other industry consultants who can answer your questions. Some consultants offer valuable hands-on training, inspection check sheets, and on-site safety and environmental audits. Attorneys also can be very helpful; but they often specialize, much like doctors. Be certain that the attorney with whom you consult has the appropriate expertise to advise you.

Product Manufacturers A good source of regulatory information on specific regulated products is the product manufacturer. Contact the company's headquarters and request to speak with one of their regulatory affairs specialists.

Training Companies Corporations sometimes offer classes dealing with a specific regulation (e.g., EPA Worker Right-to-Know), groups of regulations (e.g., personnel record management), or all regulations administered by an agency (e.g., OSHA).

Their instructors generally have both a professional and a personal understanding of the regulation; they know the steps for implementing the regulation, and they know the common pitfalls to be avoided.

These companies can provide reference materials such as books, handouts, check sheets, specific forms, model plans, and other documents that can be used as working templates.

Trade Magazines Some trade journals have to be purchased, but many are free; look for the tear-out free-subscription card inside. The value of trade magazines is that they are written by business owners, consultants, and writers familiar with your industry who can explain difficult concepts and confusing regulations in practical, easy-to-understand language.

LIABILITY FOCUSED ON THE PESTICIDE USER

Pesticides in today's pest management picture are essential components of integrated pest management. In many circumstances, pesticides may be the only effective means of controlling disease organisms, weeds, or insect pests.

Although pesticides provide many benefits, there also are inherent risks, or liabilities, associated with their production and use. It is important to balance the benefits associated with pesticides with their potential for negative impact on human and environmental safety. The risks of acute poisoning and concerns about chronic (long-term) impacts of exposure to pesticide residues in food continue to be debated. Natural resources can be degraded when pesticide residues in storm water runoff enter streams or leach into ground water. Pesticides that drift from the site of application to wildlife habitat may harm or kill nontarget plants, birds, fish, or other wildlife. The mishandling of pesticides in storage facilities and in mixing and loading areas often contributes to soil and water contamination.

Pesticide users have many legal responsibilities in managing pesticide risks. Mistakes with pesticides seldom go unnoticed. Local, state, and federal laws and regulations place the responsibility and liability for correct pesticide use squarely on the shoulders of pesticide users. Companies and individuals cannot afford to take a wait-and-see attitude: loss prevention, quality control, and sound risk management must be emphasized. Users must recognize potential pesticide problems and prevent their occurrence. One of the common mistakes made by pesticide applicators, when asked by the public about the risks inherent to the products they use, is to say that a product is *absolutely safe*. This type of statement is rarely believed, nor accurate.

Pesticide risk management must be addressed with the same level of intensity and commitment assigned to those business activities that produce income. Prudent management of pesticide risk involves recognizing the potential for problems and developing prevention strategies. It is essential to recognize that pesticide use—transportation, storage, application, disposal—carries the potential to cause harm to people and to degrade the environment (See Chapters 9 and 10 for more details). There are unique problems involved in each phase of pesticide handling. Corrective actions are required when preventive actions fail or when a change in law requires additional measures. These general concepts—recognition, correction, and prevention—do not shelter the pesticide user from liability, but often they reduce the likelihood of legal action by regulatory agencies and citizens.

INSPECTION, INVESTIGATION, AND REPORTING

Federal and state regulatory entities face increasingly complex issues on pesticide use. Environmental groups and community organizations have demanded more regulation and accountability of the pesticide industry; the result is continual reevaluation of pesticide legislation by Congress, by the EPA, and by state legislatures. With increased public concern for health and environmental issues, federal and state regulatory agencies are reassessing pesticide-monitoring policies and the methods used in risk analysis.

Federal and state governments regulate pesticides through product registration, applicator certification and licensing, and enforcement. The regulatory system controls pesticides from manufacture to disposal through laws administered by several government agencies. It incorporates such elements as product label registration, restricted-use pesticide dealer licensing, applicator certification and licensing, rules of conduct, and recordkeeping requirements. Inspections or investigations by regulatory agencies may lead to the discovery of acts or omissions that are considered improper, questionable, or illegal.

Routine Compliance Monitoring

State, federal, or local officials may conduct a routine compliance audit of pesticide activities, either announced or unannounced. They may review records and evaluate practices to ensure compliance with current laws and regulations. They may inspect the facility for good housekeeping practices and worker safety, collect pesticide samples for analysis, and review pesticide application records.

Routine inspections do not always occur at primary pesticide facilities. Analyzation of tank samples, from trucks transporting pesticides and from workers applying them, aids inspectors in determining whether label directions and pertinent regulations are being followed. Most government agencies require inspectors to explain the purpose of their activities and to inform pesticide facility personnel if a violation is suspected.

Expectations to Meet Deadline Reporting

Failure to meet deadlines imposed by law is one of the easiest ways to incur liability. Deadlines are associated with such things as community right-to-know activities, reports on state fertilizer volume sales (for fee assessment), pesticide production and repackaging, pesticide or fertilizer spill reports, pesticide and fertilizer containment facility construction, renewal of applicator certifications and licenses, and renewal of product registrations.

Complaint Leading to an Investigation

Investigations are usually initiated when a complaint is made that a pesticide or fertilizer product is being handled in a manner that violates a specific law or endangers human health or the environment. Most pesticide and fertilizer complaints are listed with a state's department of agriculture or environmental agency and/or the U.S. EPA. Under certain circumstances, other agencies such as the federal Food and Drug Administration, the Department of Agriculture, the Department of Transportation, the Occupational Safety and Health Administration, or local health departments may be called to investigate a complaint. Representatives of these agencies may appear on-site to gather information pertinent to the complaint. The agents' initial goal is to investigate the complaint and conduct personal interviews. These fact-gathering investigations may include written statements and records, photographs, and the collection of air, water, soil, or plant samples (Fig. 8.8).

It usually is beneficial to cooperate fully. If there is concern about the investigation or its scope, retain an attorney for assistance. It is important to be polite during the questioning period. If you do not know an answer, say so. Never guess or make assumptions. Consider the following points when involved in an investigation:

- Be alert and well prepared.
- Check the investigators' credentials. Write down their names, positions, agencies, and phone numbers, or obtain a business card from each individual involved. If you're still unsure whether or not an investigator has the legal authority to perform a certain function, ask him or her to show you the section of the law or regulation that specifies authority to conduct the inspection.

Figure 8.8 A state pesticide inspector draws water from a well as part of a pesticide misuse investigation.

- If the inspection is conducted under the Federal Insecticide, Fungicide, and Rodenticide Act (FIFRA), the firm or individual must be issued a Notice of Inspection identifying the purpose of the inspection or investigation and stating whether or not a violation is suspected. If the inspection is not being conducted under FIFRA, you should still determine the purpose of the inspection and whether or not a violation is suspected.
- Ask the agents these three questions: (1) Who filed the complaint? (2) What laws allegedly have been broken? and (3) What authority do you have to conduct the investigation? Some states may require agents to withhold this information if it is confidential or privileged.
- Answer specific questions, but do not answer if you believe you need to discuss the question and answer with legal counsel.
- Never be rude or show a lack of respect for the investigators.
- Do not tolerate rudeness or lack of respect for yourself or your property.
- Be cooperative to the extent possible. Remember, though, that it is the investigators' responsibility to ask questions and to uncover evidence, some of which may prove that you have violated the law.
- Request a duplicate or split sample of whatever the investigators collect. Make sure that the investigators record and handle the split sample in a manner similar to the way they handle their own. When involved with the EPA or their agents, ensure that the samples are sealed according to EPA protocol (Fig. 8.9).
- You probably will be asked to sign a receipt for any physical samples or copies of records that the investigators collect. Be sure that the information is correct before signing.
- Obtain copies of all completed forms and written information compiled by the agents.
- Ask the agents what you should expect from the agency as the investigation unfolds.
- Seek legal counsel whenever it appears advisable. The need for representation may not be obvious. Take your own notes on questions asked and answers provided. Good notes will provide your legal counsel with valuable facts pertinent to the case.
- Depending on the basis of the complaint, it may be appropriate to notify your insurer of the investigation.

Figure 8.9 Evidence collected by a state pesticide inspector as part of a pesticide-misuse investigation at a public school.

Accidental Releases into the Environment

A pesticide release can be a major source of liability. When a major release occurs, the scene is often filled with first responders (firefighters, law enforcement officers, and medical professionals) who will attempt to secure the site and prevent further contamination of the surrounding area. Officials representing state departments of agriculture and environmental agencies often will respond to gather information about the accidental release if it threatens the health of a community (e.g., fire at a pesticide distribution facility) or poses imminent danger to aquatic or human life due to surface water contamination. In addition to the cost of remediation for the accident site, fines and penalties may be assessed against the responsible parties. It is advisable to seek the services of an attorney and an environmental consultant when involved in a major pesticide accident.

Environmental Site Assessments

Farmers, commercial applicators, and even some homeowners may be required to conduct an environmental site assessment of real estate prior to transfer (sale) of the property, when borrowing funds against the value of the property or when leasing the property (see Chapter 9 for more details). Environmental assessments may be required by a lender, purchaser, or state statute. The goal of an environmental assessment is to identify possible sources of pollution and to uncover any potential problems that could impair the value of the property or make it risky as collateral in lending situations. In addition, audits also may be required at the end of a lease before the property owners or new tenants assume possession of the property. A thorough environmental site assessment also may assist in establishing a "due diligence" in forming an "innocent purchaser" defense if contamination is subsequently discovered.

ATTORNEY–CLIENT PROTECTION

The American legal system has long considered the relationship between clients and their attorneys to be strictly confidential. This edict is protected by the attorney–client privilege, which protects virtually all forms of communication between attorneys and their clients: verbal, written, taped, etc. Clients may speak freely and honestly to their legal counsel with confidence that disclosure is forbidden without their approval. Attorneys who gain their clients' trust under the attorney–client privilege are in a better position to render good legal advice than are those whose clients are afraid to tell all. Clients must realize that their attorneys need the benefit of complete, accurate information in structuring the best possible defense (or in building a case, if the client is the plaintiff). They want no "surprises" once they get to court. Therefore, make sure you fully disclose the facts to your legal counsel (Fig. 8.10).

Attorney Work Product Privilege

Attorney work product is work performed by or on behalf of an attorney. It consists of information collected by an attorney, or by someone under his or her direction, in anticipation of litigation; it includes witness interviews, factual compilations, analyses, conclusions, and other products of the attorney's efforts on behalf of the client.

The attorney work product privilege is a provision of the law that is invoked only after an attorney–client relationship has been established and only in cases where the threat of

Figure 8.10 A client can discuss anything with their attorney knowing that what they say is protected by the client–attorney privilege.

litigation exists. When considering an environmental site assessment, for example, it is wise to solicit the advice of an attorney before making even preliminary plans, because information gathered under the direction of counsel can often be legally protected by the attorney work product privilege; that is, it is protected from discovery and disclosure by the government or opposing counsel (Fig. 8.11).

Discoverable Evidence Linked to Privileges

Discovery is the litigation process where each side must disclose to the other whatever evidence or information they may have in their possession or present during formal proceedings: documents, photographs, recordings, etc. Lawyers may even require opponents to submit written responses to questions or to respond orally, under oath, to questions asked during a deposition. The purpose of this procedure—discovery—is to make attorneys on both sides fully aware of all evidence, whether favorable or unfavorable, that may be presented at the trial.

Understanding the concept of discoverable evidence is important: Only the attorney work product and communications between client and attorney are protected from discovery and disclosure in any legal proceeding. Evidence or information developed by the client prior to establishing the attorney–client relationship generally is subject to disclosure.

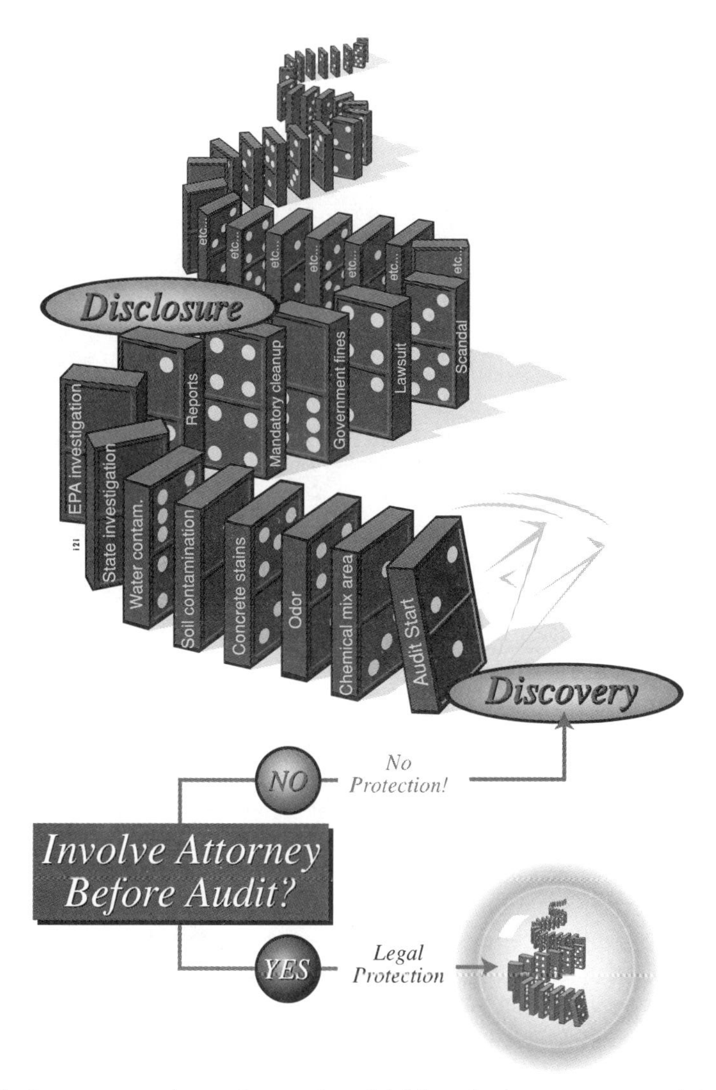

Figure 8.11 In some cases it may be very beneficial for a business owner to seek legal counsel before undergoing an environmental audit.

ADMINISTRATIVE PROCEEDINGS

Information collected by regulatory agencies is often used to determine compliance with federal, state, and local laws. Information collected during an investigation generally is not publicly available until the case is completed and an enforcement decision has been rendered.

In many instances, regulatory agencies will not continue the inquiry into alleged wrongdoing when the facts collected during the investigation do not corroborate the accusations or when a causal relationship cannot be proven. It is advisable to request a written response confirming that the investigation of the complaint was not substantiated by the evidence collected. This written response may be a valuable document to the person being investigated and often, as part of the public record, completes the investigation file.

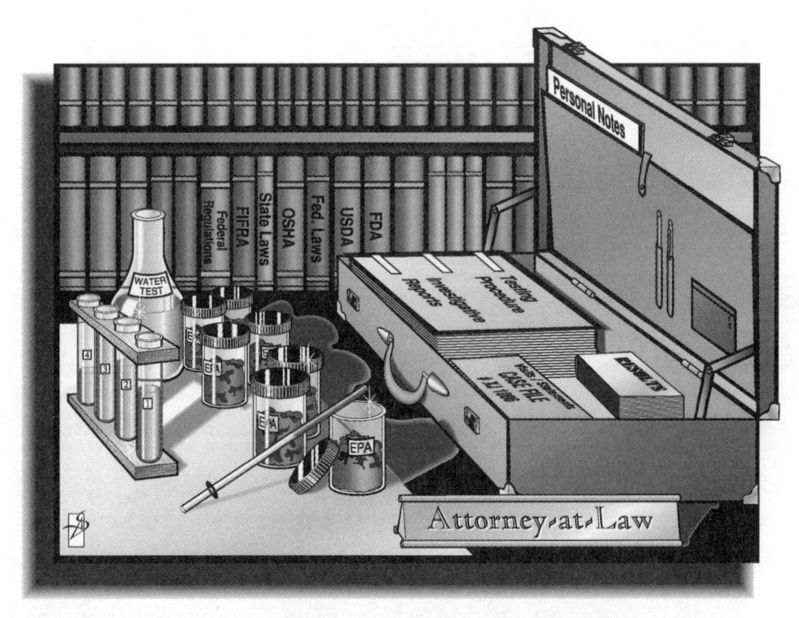

Figure 8.12 It is always best to seek advice from an attorney who is knowledgeable on environmental laws and regulations.

The information collected may indicate the pesticide complaint has validity and that a stronger civil response is required by the agency. Based on laws passed by a legislative branch and regulations promulgated through the rule-making process, the appropriate agency has authority to pursue legal remedies (enforcement actions) when a person has not complied with a particular statute or regulation.

The procedures employed by regulatory agencies to officially charge a person with breaking a law or regulation differs among states and among federal governmental units. It is important to understand the procedures used in determining criminal guilt or liability. A civil proceeding, such as an administrative hearing, is one in which defendants who are found liable may be ordered to pay for damages and/or assessed a monetary penalty. Defendants in a civil hearing can be forced to explain what they did, what they observed, and when they first observed the problem. However, attorneys cannot compel their adversaries to disclose information protected under attorney–client privilege or attorney work product privilege. The following discussion on procedures provides a general guide for someone who has been formally charged with violating rules and regulations (Fig. 8.12).

Official Charges against Defendant

The normal regulatory procedure is to send a formal enforcement letter by certified mail (or other legal service) to the person alleged to have violated the law. The enforcement letter typically contains an explanation of the agency's authority under appropriate laws and administrative codes, the facts associated with the case, and a statement of possible legal penalties. A document may be included which allows the recipient to admit or deny the allegations or which explains the methods of payment available for proposed penalties. Enforcement letters also may provide for an informal settlement conference or for a formal hearing before a judicial officer to hear the facts of the case and establish or recommend penalty. It is important to respond promptly to enforcement correspondence. In some cases, failure to respond by a certain deadline will cause an individual's rights to be waived or a default (e.g., assumption of guilt) to be automatically entered, resulting in an impairment or loss of the right to defend oneself and contest the charges. The deadline for response usually is included in the letter. An attorney should always be contacted if you are unsure whether to admit or deny a charge, how the response should be worded, or what the deadline is.

Determination of Case by Informal Meeting

When there are mitigating circumstances or other pertinent facts that might affect the outcome of a complaint, the respondent may ask for an informal hearing or settlement conference with the regulatory agency. The informal hearing procedure generally affords the person being charged an opportunity to present the facts from his own perspective and to clarify any discrepancies.

Facts should be presented in a calm and professional manner after the case file has been carefully reviewed for accuracy. Usually this should be handled by an attorney. If there is a plausible explanation for the violation, the regulatory agency will want to hear it. The position of the accused should be presented in a firm but reasonable manner. It is particularly helpful to the respondent if they can explain the steps they have taken to mitigate the problem/risk subsequent to the inspection and receipt of complaint. The outcomes of informal hearings vary but may include retraction of the enforcement letter, renegotiation of civil penalties, or no change relative to the charge letter.

Determination of Case by Formal Hearing

A formal administrative hearing is important because it allows for an independent review of the facts associated with the case. Administrative hearings generally involve legal counsel for a state or federal agency presenting the evidence and legal counsel presenting the facts on behalf of the respondent. The evidence from both sides is presented to an administrative law judge (ALJ) or to a panel of individuals comprising a commission or review board.

Specific procedures must be followed in formal hearings. If the accused decides to represent himself and is unfamiliar with the prescribed procedures, an attorney should be consulted. In some cases, trade association representatives can offer insight and, in some specific instances, may also attend the hearing with you. Certain rights may be relinquished and certain evidence may be disallowed later if the rules are not followed. The ALJ will render a decision regarding compliance or noncompliance based on oral and/or written testimony and other evidence. The ALJ typically will also determine any civil penalty to be levied.

In other federal and state agencies, an ALJ will preside over a hearing to ensure fairness to all parties, but the outcome of the case is decided on the basis of a formal public vote by commission or review board members. The vote determines if the alleged charges are substantiated by facts collected. Many commissions and review boards also are responsible for assessing penalties including: revocation or suspension of a license or a certification, levying a fine, probation, and restitution to the victim (if any) for losses or damages. Usually, the review is limited to the facts as stated in the administrative record. In some situations (e.g., violations noted in one state also occur in other states), the case is referred to an EPA regional office.

Decisions to Appeal to State or Federal Court

Generally, any respondent assessed a penalty has the right to appeal the state or federal government decision to a court of law when he or she feels that the facts or law do not support either the decision reached or the punishment administered through an informal or formal hearing. Often there are limits to what can be presented; for example, if rules of procedure are not followed in the administrative process, certain aspects of the case may be declared inadmissible.

CIVIL PROCEEDINGS (LAWSUITS)

In the United States, the role and authority of the system of federal courts is dictated by the U.S. Constitution, whereas the authority of state courts is almost purely a product of state law. Federal courts only have limited jurisdiction and are not authorized to hear most cases. Regardless of which system is utilized, state or federal, judicial or administrative, our system of justice always allows aggrieved parties at least one opportunity for an appeal to an appellate court.

A tort or civil wrong is a product of common law. Common law represents a set of rules and principles which derives its authority from the courts. Common law has its roots in judicial decisions made by the courts in England prior to the American Revolution. Whereas common law is the product of the courts, statutory law is founded upon rules of law established by a legislative body. To some extent, common law rules and principles have been incorporated into state and federal statutes.

A common law theory arises from the general legal duty that individuals in a law-abiding society owe to one another. Every adult is obligated to a certain duty of care for the personal and property rights of others when engaged in daily activities. A violation of this responsibility may give rise to a cause of action whereby the injured party may pursue a legal or equitable remedy.

Common law actions of civil wrongs or torts are known as "lawsuits." They are initiated by the person who has suffered some injury to his or her person or property—the plaintiff—as a result of the acts or omissions of another—the defendant. A lawsuit is a proceeding in court before a judge who is a member of the state or federal judicial system. An administrative hearing, by comparison, is handled by either a hearing officer or an administrative law judge who is usually employed by an administrative agency. Most appeals from administrative hearings are handled by transferring the case to the judicial system.

In certain cases, parties are also entitled to demand a trial by jury. When a jury is utilized, the function of the jury is to make all factual determinations that will dictate the outcome of the case (Fig. 8.13). Juries accomplish this by considering and weighing the evidence which was presented at trail, applying the pertinent law to the facts of the case, and determining the ultimate outcome. In a jury trial, it is the responsibility of the judge to provide the jury with instructions on what the pertinent law is that must be applied.

In a nonjury trial, the judge will assume the responsibility for evaluating and weighing the evidence. The judge will make the necessary factual determinations and render a decision on the ultimate outcome by applying the relevant rules of law and legal principles.

Common law theories of recovery, as they pertain to pesticide storage, transport, application, and disposal, typically fall into one or more of the following categories: trespass, nuisance, negligence, or strict liability.

Figure 8.13 Testimony from experts often is required in cases tried in court.

Trespass Theory

A trespass is an unauthorized entry which causes damage or injury to the property of another. Trespass to land need not involve the actual entry of one person upon the land of another. A trespass may be committed by discharging materials such as pesticides onto someone else's land. The line between this type of trespass and other types of tort liability is sometimes difficult to determine. Generally a plaintiff must demonstrate (1) an invasion affecting an interest in the exclusive possession of his property, (2) an intentional commitment of the act which results in the invasion, (3) reasonable foreseeability that the act committed could result in an invasion of the plaintiff's possessory interest, and (4) damages to the property.

Nuisance Theory

A nuisance arises whenever a person uses his property to cause injury or annoyance to a neighbor. A nuisance is an activity which arises from a person's use of their own property that causes an obstruction or injury to the right of another or of the public to the quiet enjoyment of property by producing annoyance, inconvenience, or discomfort. Pesticide use can produce potentially offensive odors and may also cause discomfort in certain individuals.

Many states have enacted "right-to-farm" statutes to protect farmers with a nuisance defense for agricultural operations which meet certain criteria, as long as they follow generally acceptable agricultural practices. Right-to-farm statutes are a statutory defense in a nuisance suit. The farmer (or industry) must meet certain conditions to qualify for this defense. Right-to-farm statutes generally will not protect farmers from acts of negligence. Many activities would fall under this exclusion. Additionally, even though those statutes may limit civil liability, they often do not limit criminal liability (e.g., it can be a criminal offense to maintain a public nuisance).

Negligence Theory

Negligence is charged to an individual who has failed to act in a reasonable and prudent manner in a situation where the individual had a duty to another person or to the public. A person who is negligent is responsible for the damages that the act or omission causes—unless some defense is available.

The standard of care imposed by law is that which would be exercised by a person exhibiting ordinary prudence under the same set of circumstances. This is often referred to as the "reasonable person" standard.

For a pesticide user to be liable, the act or omission must be legally related to the cause of injury. Generally, the user's act would have to have caused a natural and continuous sequence that produced injury which otherwise wouldn't have happened. Any person allegedly harmed by the improper application, transportation, or storage of pesticides can attempt to recover any losses under a negligence cause of action.

Strict Liability Theory

The laws also impose strict liability on some individuals making them responsible for the consequences of their activities regardless of other contributing factors or defenses they may put forth. Under strict liability, a person is liable if they performed the act, regardless of fault. A plaintiff only has to prove causation and damages. The defendant may have acted in an expert manner. Strict liability may apply for activities that have an inherently dangerous or ultrahazardous nature, such as the application of fumigants or aerial application. In some jurisdictions, even though strict liability may not apply, a heightened duty may instead be applied to pesticide applicators.

CRIMINAL PROCEEDINGS

A criminal act may be punishable by time in prison. Most environmental statutes permit judges to sentence violators to jail when the defendant has been convicted of knowingly or willfully committing an environmental crime. And, although criminal defendants cannot be forced to testify against themselves, any documents, observations, conversations, etc., not protected under the attorney–client privilege may be admitted into evidence.

LABELS AND APPLICATOR CERTIFICATION SET STANDARDS OF CONDUCT

The Label as the Primary Source of Liability

Accompanying each pesticide product is a label which provides written instructions for achieving the desired level of pest management—the benefits (See Chapter 7 for more information on the pesticide label). The label also provides detailed statements communicating the risks and standards of care associated with the use of the product—the liabilities. The instructions and precautions become a legal benchmark by which the actions of the user are compared to the expected "standards of conduct" outlined by the label.

It is the label that establishes a standard of care. The label statement, "It is a violation of federal law to use this product in a manner inconsistent with its labeling," obligates the user to follow product stewardship instructions. This label statement legally binds the user to follow label directions because FIFRA and state pesticide laws mandate that actions contrary to label directions are considered unlawful acts.

The words *use, inconsistent,* and *labeling* need further explanation in the sentence, "It is a violation of federal law to use this product in a manner inconsistent with its labeling." The pesticide label and additional written materials that accompany a product collectively comprise the pesticide labeling. Labeling also includes additional sources of information (e.g., EPA Worker Protection Standard, EPA Endangered Species Program Bulletin, state Ground Water Management Plan, company Product Use Bulletins) referenced on the label or accompanying materials.

The word *use* carries the usual connotation of pesticide application, but its legal definition is intended to include handling, mixing, loading, storage, transportation, and disposal. This all-encompassing definition covers every activity that involves a pesticide—from product purchase to container disposal.

Use of a pesticide contrary to its labeling represents inconsistency, and inconsistent use establishes potential liability. For example, label directions that say *shall not, do not,* and *must not* are mandatory and provide regulatory investigators with "enforceable language." Label terminology such as *may, can,* and *recommend* is considered nonenforceable, informational, or "advisory" language. It is very important when reading label directions to differentiate between enforceable and advisory language. Contact the manufacturer of the product, the Cooperative Extension Service personnel, or pesticide regulatory agencies for label interpretations when there is any doubt.

There are situations where pesticides may be used in variance with the precise directions on the label known as the FIFRA Section 2(ee) exemptions. These include:

- *Pests Not on the Label* Minor and occasional pests are often not listed on pesticide labels. According to FIFRA you may apply a pesticide "against any target pest not specified on the labeling if the application is to crop, animal, or site specified on the labeling, unless the Administrator has determined that the use of the pesticide against other pests would cause an unreasonable adverse effect on the environment." Some states are more stringent and do not agree with FIFRA. These states do not allow use of pesticides against an unlisted pest.
- *Dose, Rate, or Concentration Lower Than on the Label* Federal law usually allows applicators to apply a pesticide at any dosage, concentration, or frequency lower than that listed on the labeling. There are certain exceptions. For example, termiticides used for preconstruction treatment cannot be applied at a lower dosage and/or concentration than specified on the label for applications prior to installation of the finished grade. Some states prohibit a dose, rate, or concentration less than the labeled rate for termite soil treatments. However, the product's warranty may be voided when an application goes below labeled rates and the application results in less than adequate pest management.
- *Using a Method of Application Not Excluded from the Label* Employing any methodology application not prohibited by the labeling unless the labeling specifically states that the product may only be applied by the methods specified by the label.
- *Special Local Needs Registrations* Section 24(c) of FIFRA permits states to grant registrations for additional use of pesticides to meet "special local needs." Both the original label and the special 24(c) label must be in the possession of the user, and the use is only permitted in the state in which the 24(c) registration was authorized.
- *Emergency Exemptions* A state may petition the EPA for a Section 18 FIFRA "emergency exemption" from regular registration requirements. Section 18 exemptions permit states to grant temporary emergency uses for a particular product. Such emergencies involve public health risks or quarantines.

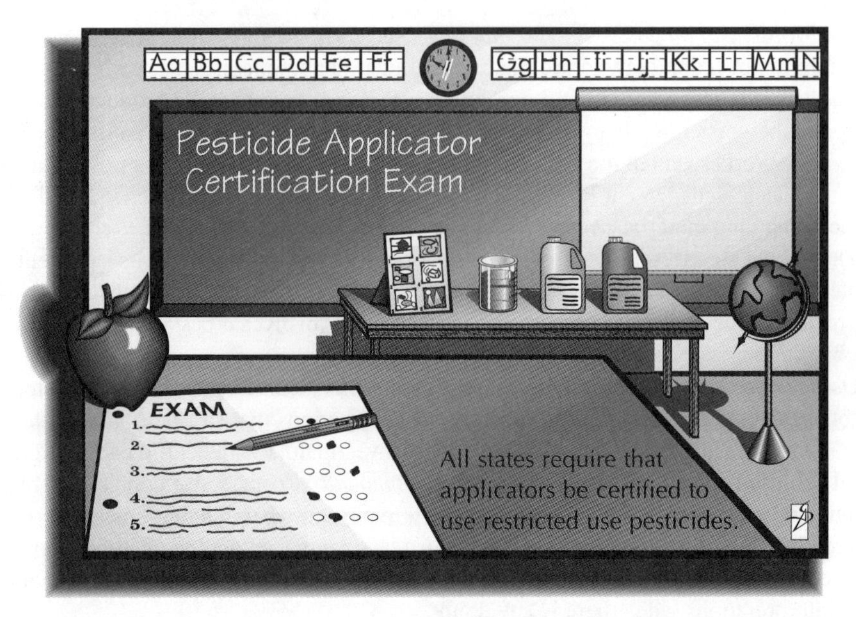

Figure 8.14 A person must pass state pesticide certification examinations before they can purchase restricted-use pesticides, or any time they commercially apply a pesticide to someone else's property.

Certified Pesticide Applicators Held to a Higher Standard of Conduct

The certification process was established to provide technical knowledge for those using restricted-use pesticides—those pesticides that pose the greatest risk of harm to people, wildlife, and the environment. Most states also require person to become certified if the business charges a fee for the application of any pesticide to the property of another.

The certification process generally involves educational training and examinations that cover pest biology, human health and safety, environmental issues (e.g., water quality and endangered species), regulatory updates, label interpretation, and other job-specific information. Nationally, there are approximately one million applicators certified to purchase and apply restricted-use pesticides. Certified applicators are required to retest periodically or to attend pesticide education programs to maintain their certification (Fig. 8.14).

Pesticide applicator certification holds the certified applicator to a higher standard of conduct and increased duty of care than a noncertified person. Private applicators—primarily farmers and ranchers—receive training from their local Cooperative Extension Service (Fig. 8.15). Commercial pesticide applicators may elect to attend training programs, or they may study on their own for state pesticide applicator certification exams. The exams are administered by the state pesticide regulatory agencies. Topics may include laws and regulations, human and environmental safety, formulations, label comprehension, pest biology, and integrated pest management. In addition to passing state certification exams, applicators in many states are encouraged to meet continuing education standards for recertification (in lieu of retesting) to ensure that they remain current on pesticide issues. The federal pesticide law (FIFRA) along with many state pesticide laws provide for higher penalties for certified applicators than for private applicators (e.g., farmers).

Figure 8.15 The Cooperative Extension Service found at state land grant schools provides many educational programs and workshops to help in the certification and recertification process.

Pesticide Drift as an Example of Liability Consider the following facts from court records. A farmer contacted his commercial agricultural pesticide applicator to spray 2,4-D and atrazine on his crop. The farmer warned the applicator that his next-door neighbors had a garden and many fruit trees.

Returning home from a doctor's visit, the neighbors discovered that their orchard and garden were damaged severely. The sister-in-law who had remained in the house stated that the field next door had been sprayed and that the smell of the spray was so strong she had to close the windows.

The neighbors exercised their legal rights and took both the farmer and the commercial applicator to court. The farmer's lawyer argued that his client should not be part of the lawsuit because he had nothing to do with the herbicide application. He argued that the commercial applicator was an independent contractor and that, as such, he alone was responsible for the drift and any damage it caused. The neighbors' lawyer argued that the farmer was partially responsible because the use of a ground boom to apply herbicides is an abnormally dangerous activity and that the act of spraying carries a high likelihood of chemical trespass. The court determined that the commercial applicator's being an independent contractor did not necessarily absolve the farmer from responsibility. The court said that parties engaged in inherently dangerous activities cannot insulate themselves from responsibility, so a determination on whether the farmer was engaged in an inherently dangerous activity was needed.

The following criteria were used by the court in determining whether the act of spraying a herbicide with a ground boom was abnormally dangerous:

- Existence of a high degree of risk of some harm to the person, land, or property of others.
- Likelihood that the harm that resulted from it will be great.
- Inability to eliminate the risk by the exercise of reasonable care.
- Extent to which the activity is not a matter of common usage.
- Inappropriateness of the activity to the place where it is carried out.
- Extent to which its value to the community is outweighed by its dangerous attributes.

The court in this instance ruled that ground-based application of herbicides in rural and agricultural environments is not abnormally dangerous. *The court also placed the responsibility on the commercial applicator because he was, in fact, licensed by the state to apply chemicals.* Other than arranging with the applicator to have his field sprayed, the court ruled that the farmer "did not otherwise act" in the spraying operation. The court ruled that the neighbors' claim against the farmer should be dismissed.

Instructions with Special Meaning

Certain actions during pesticide use may pose significant hazards to the public, the applicator, or the environment. Risks can be greatly reduced by carefully following label directions and precautions and by practicing common sense and good judgment when using pesticides. The following lists identify good practices to minimize risk during pesticide use.

Health and Safety Precautions

- KEEP OUT OF REACH OF CHILDREN is an obvious safeguard for all pesticides.
- Restricted-use pesticides may have special health or environmental implications that require prudence in mixing, application, and disposal. Read the label carefully.
- Compliance with reentry statements specifying a time span that must elapse between a pesticide application and the admittance of persons or animals into the treated area is critical to minimizing pesticide exposure.
- Terminology such as *until dusts have settled* and *until sprays have dried* represents the minimum standard for reentry. Many pesticide labels restrict entry into treated areas for 12, 24, 48, or even 72 hr after application, as mandated by the EPA Worker Protection Standard.

Application Precautions

- Maintain up-to-date certification, licensing, and insurance.
- Provide any pertinent safety training—required or not—for both noncertified and certified employees.
- Fulfill state and federal pesticide recordkeeping requirements.

- Comply with manufacturers' label instructions to maintain product warranty.
- Use pesticides only on crops and sites and at rates specified by the label. Failure to do so may result in crops with illegal pesticide residue, and such crops are subject to confiscation and destruction by state or federal authorities.
- Follow preharvest guidelines to ensure that harvested crops will meet federal and state pesticide residue tolerances.
- Comply with EPA Worker Protection Standards which assign responsibility to both employers and employees.

Environmental Precautions

- Extreme care at mixing and loading sites is critical in preventing soil and water contamination (e.g., wellhead areas).
- Have a plan and the necessary equipment in place to contain a pesticide spill, and know how to properly mitigate the damage. (See Chapter 13 for more details.)
- Understand your requirements under the Community Right-to-Know Act. Some states have posting and notification regulations for informing the general public. (See Chapter 15 for more information.)
- Dispose of pesticide containers and rinsate as specified on the label or as required by local, state, or federal requirements which may be more restrictive. Burning or burying pesticide containers on private property is illegal in most states.
- Keep pesticides on the targeted application site; avoid drift which might subject the applicator to liability claims.
- Never store, mix, load, or wash equipment near wells, bodies of water, ditches, or drains.
- Make sure that specific buffer zones are maintained according to label directions, best management practices, and watershed restrictions.
- Follow instructions from relevant endangered species bulletins.
- Maintain a controlled inventory to guard against theft and the necessity to store old or canceled products. Stay current with the industry to receive information on products due to be phased out, and be sure to use any such products during the prescribed phaseout period. If the applicator is left holding canceled pesticide products, they should be checked frequently to make sure that all containers are secure (i.e., not leaking). The storage, handling, transportation, and disposal of such waste must be conducted under the jurisdiction of the EPA and/or state and/or local operators and facilities. Failure to do so could result in criminal enforcement.
- Always triple-rinse or pressure-rinse empty containers. If you're in an area that collects rinsed empty containers for recycling, be sure to participate.
- Many states conduct "Clean Sweep" or pesticide disposal programs for unused or canceled products. This is an outstanding way to eliminate the potential liability of storing such products on your farm or business site.
- If using a pesticide subject to ground water protection regulations, comply with all provisions of the specific State Management Plan for the product (e.g., handling provisions, rate reductions, geographic restrictions).

Inconsistent Use May Violate Product Warranty

Pesticide manufacturers guarantee in the warranty section of the label that the product conforms to the chemical description noted in the ingredient statement. The warranty also will specify that the product will perform as represented on the label when used according to directions. However, the label language also will indicate that the buyer or user of the product assumes all liability when the product is used in a manner inconsistent with label directions and precautions; therefore, misuse voids the manufacturer's warranty. However, even though it is not illegal under federal law, warranties can be revoked when applicators choose to use rates below those specified on the pesticide label.

Following is an example of a warranty listed on a pesticide label under the title LIMIT OF WARRANTY AND LIABILITY.

> Monsanto Company warrants that this product conforms to the chemical description on the label and is reasonably fit for the purposes set forth in the Complete Directions for Use Label booklet ("Directions") when used in accordance with those Directions under the conditions described therein. NO OTHER EXPRESS WARRANTY OR IMPLIED WARRANTY OF FITNESS FOR PARTICULAR PURPOSE OR MERCHANTABILITY IS MADE. This warranty is also subject to the conditions and limitations stated herein.
>
> Buyer and all users shall promptly notify this Company of any claims whether based in contract, negligence, strict liability, other tort or otherwise.
>
> Buyer and all users are responsible for all loss or damage from use or handling which results from conditions beyond the control of this Company, including, but not limited to, incompatibility with products other than those set forth in the Directions, application to or contact with desirable vegetation, unusual weather, weather conditions which are outside the range considered normal at the application site and for the time period when the product is applied, as well as weather conditions which are outside the application ranges set forth in the Directions, application in any manner not explicity set forth in the Directions, moisture conditions outside the moisture range specified in the Directions, or the presence of products other than those set forth in the Directions in or on the soil, crop or treated vegetation.
>
> This Company does not warrant any product reformulated or repackaged from this product except in accordance with the Company's stewardship requirements and with express written permission from this Company.
>
> The exclusive remedy of the user or buyer, and the limit of the liability of this company or any other seller for any and all losses, injuries or damages resulting from the use or handling of this product (including claims based in contract, negligence, strict liability, other tort or otherwise) shall be the purchase price paid by the user or buyer for the quantity of this product involved, or, at the election of this company or any other seller, the replacement of such quantity, or, if not acquired by purchase, replacement of such quantity. In no event shall this company or any other seller be liable for any incidental, consequential or special damages.
>
> Upon opening and using this product, buyer and all users are deemed to have accepted the terms of this LIMIT OF WARRANTY AND LIABILITY which may not be varied by any verbal or written agreement. If terms are not acceptable, return at once unopened.

SEEKING PROFESSIONAL ADVICE

Environmental and Safety Consultant

Consultants who deal with regulatory issues such as regulatory compliance, safety analysis, and employee training are specialized professionals who are available to help you run your business. They should be hired to meet your specific needs. You should be very clear about

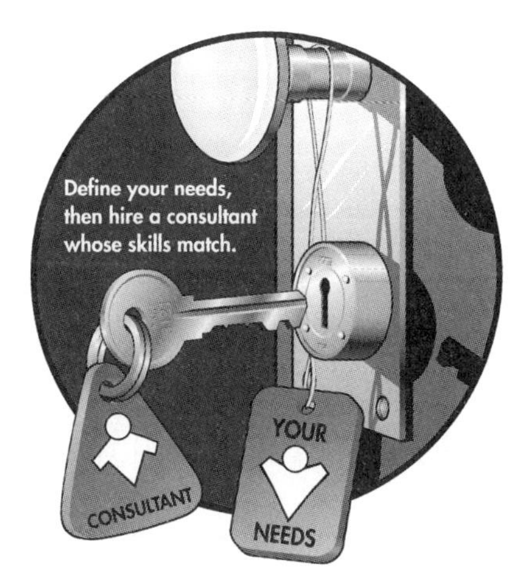

Figure 8.16 Professional consultants can provide business owners with valuable advice on business management.

what you expect the consultant to do, what products and services you desire or need, and what you are willing to spend (Fig. 8.16). Hire a consultant only if you are truly serious about implementing the changes that they recommend; otherwise, the money spent for their services is wasted!

You should consider the following when interviewing safety and environmental consultants.

- Ask trade associations for their recommendations. By checking with associations first, the field of choices can often be narrowed.
- Hire consultants who have experience with your industry. Consultants who have worked for businesses similar to yours will know where to look for problems and what questions to ask. Familiarity and experience with your type of business usually leads to better advice.
- References are important. Current and former clients of a consultant can give you valuable insights into that consultant.
- Communication is essential. Consultants must be able to communicate well with the business owner and company employees. Communicating effectively is just as important as knowing what to do. Avoid consultants who are poor communicators.
- Look for consultants who are versatile. Consider hiring consultants who provide additional services such as employee training, regular updates, newsletters, and written plans. Continuity can be important, since the consultant will already be familiar with your operations. Bringing in a new consultant for different services means that you will have to "retrain" the new person on your specific operations. This costs money.
- Find out what level of experience the consultants have with governmental agencies. Consultants should be willing to help intervene on your behalf with agencies. The consultant needs to be available if and when government regulators begin asking questions that you are having difficulty answering.

- Avoid consultants who simply offer "cookie cutter" programs. Consultants should visit the site and customize programs. In the field of regulatory compliance, "one-size fits all" programs seldom succeed.
- The consultant should offer you a fixed price or an estimate with a capped maximum for their products or services. The price should include providing you with copies of laws and regulations, written plans and programs, inspection check sheets, annual on-site reviews, over-the-phone advice, and employee training.
- Make sure the consultant carries professional liability insurance, sometimes called "errors and omissions." This is the insurance that protects you, as well as the consultant, if the consultant is negligent in his or her advice or activities.
- Get a written agreement with the consultant that clearly spells out the obligations of both parties.
- Confer with your attorney about the hiring of the consultant. The attorney can draft or revise the agreement and can also discuss the possible benefits of his or her involvement in the work of the consultant. Your attorney may be able to help you keep any damaging information confidential.

Attorney

If you need the services of an attorney—whether to provide legal advice regarding your potential liabilities, to review your insurance policies, or to defend you in a lawsuit or administrative hearing—keep the following in mind:

- Today it is common for lawyers, like physicians, to develop specialties. Therefore, not all lawyers will be equally competent to handle your problem. Seek out those with specialized experience or training in handling situations and businesses similar to your own.
- Be certain to discuss rates to be charged, the nature of the services which the attorney expects to perform, and the estimated total cost for the services. Also, it is important to understand what costs or expenses are your responsibility. Written fee agreements are recommended.
- If the services are to be billed on a hourly basis, the lowest rate quoted will not always result in the lowest total bill. Attorneys with more experience and training may charge higher rates (on an hourly basis) but handle the matter more efficiently and effectively.
- Make certain that you are comfortable with your attorney's good judgment; you'll often have to rely on it when making legal, tactical, and strategic decisions.
- It is critical that you candidly discuss your legal problems with your attorney. If you do not disclose everything you know, your attorney may be hindered in representing you; and you'll run the risk that you and/or your attorney will be surprised at an inopportune time. Remember, the attorney–client privilege may protect you from damaging disclosures.
- Hiring an attorney at the first sign of a problem will often save you money, since prevention may still be possible.
- Don't fall into the trap of thinking that you can defend yourself. Rules of procedure and evidence have been established at most levels of administrative hearings and in

all courts, and they must be followed. If you do not know these, the facts of your case most favorable to you may never be heard.

- You can fire your attorney, but you may still be responsible for reasonable fees.
- Contact your local bar association if you feel you have been improperly treated by your attorney.

SUMMARY

Regulations constantly change. Yesterday's "hard and fast rule" quickly gives way to new interpretations today and new regulations tomorrow. Governmental agencies must update their regulations to keep the regulations on the books current with the businesses they regulate, if they are to protect the public and the environment and to treat businesses on a "level playing field." It is difficult for most businesses to stay "up-to-the-minute" on all laws, regulations, and policies. It is also difficult for regulatory agencies to keep publications current because as soon as they are printed, they start to become dated. Regulatory agencies are generally much more interested in gaining industry compliance with their rules than they are in enforcing violations, they are therefore usually anxious to help answer questions and to provide compliance assistance.

Pesticide users are obligated to store, handle, and dispose of pesticides in a responsible manner to protect nontarget areas, public safety, and the environment. Improper pesticide use leaves the user vulnerable to a myriad of liabilities (Fig. 8.17). Pesticide labels provide

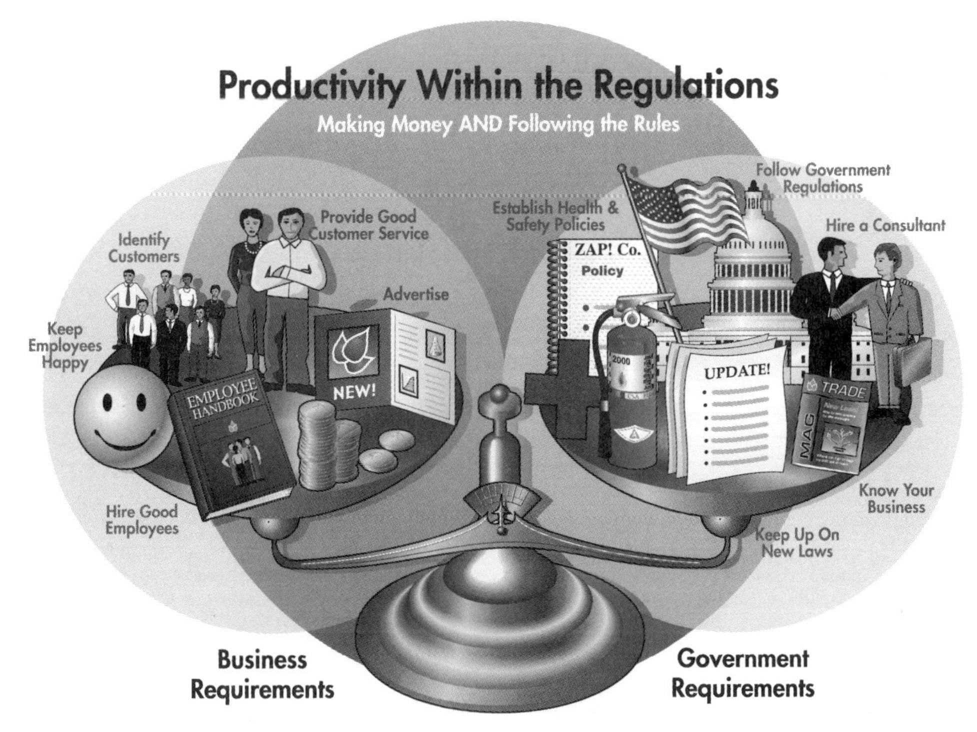

Figure 8.17 Today's business owners must offer competitive services and follow numerous regulations to remain successful over the long term.

the user not only with specific instructions for using the product to its greatest advantage, but also with a mechanism to minimize liabilities by following label information. It is a violation of both federal and state law to use a pesticide in a manner inconsistent with its labeling. The label is the key element in determining whether or not the user is in compliance with the law. The pesticide user should be aware of state laws and regulations pertaining not only to pesticide use, recordkeeping, and containment but also to laws governing workplaces, transportation, etc. With this knowledge, the user can evaluate and minimize the potential for liability.

CHAPTER 9

ENVIRONMENTAL SITE ASSESSMENTS: MANAGING THE FACILITY AGAINST CONTAMINATION

What is a business worth? Traditional benchmarks such as assets, liabilities, operating income, market share, goodwill, and customer satisfaction are used to gauge the financial success of a business. A new factor among these traditional measures of business success is environmental liability.

Environmental liability often is determined by conducting an environmental assessment which can disclose a history of real property use and abuse. Environmental assessments are now performed routinely at commercial pesticide and fertilizer application businesses, farms, golf courses, lawn care companies, pest management firms, and other businesses. Such assessments supplement the findings of financial institutions, independent consultants, and internal auditors who evaluate and report on financial records. The availability of liability insurance coverage and its cost often are based on the documentation of good environmental stewardship (Figs. 9.1 and 9.2). Assessments also are used in determining the value of a potential investment.

TRANSFER OF PROPERTY AND ENVIRONMENTAL DISCLOSURE DOCUMENTS

Many states have a law that requires a seller or transferor of real estate to prepare and deliver an Environmental Disclosure Document (EDD) prior to the completion of a transfer. Failure to disclose an environmental defect, such as a leaking storage tank, in an environmental disclosure document may excuse a purchaser from an agreement and/or expose the seller to damages or penalties. Increasingly, buyers and lenders are requiring disclosure documents even when there is no state law compelling the seller or transferor to supply them. Several states require a seller/transferor of certain categories of property to deliver an Environmental Disclosure Document to other parties to the transfer—the transferee/buyer and the buyer's lender (if identified)—30 days before the transfer occurs. If an environmental defect previously unknown to the buyer or lender is disclosed in the EDD, the buyer and lender

Figure 9.1 Spilled pesticide contained on a concrete slab.

may be excused from the agreement to purchase or finance, respectively. Environmental defects are conditions that (1) constitute a material violation of an environmental statute, regulation, or ordinance; (2) require remedial activity under an environmental statute, regulation, or ordinance; (3) present a substantial endangerment to public health, public welfare, or the environment; (4) have a material or adverse effect on the market value of the property or of an abutting property; or (5) prevent or materially interfere with another party's ability to obtain a permit or license required under an environmental statute, regulation, or ordinance.

Real estate occupied by a company that operates on an otherwise profitable basis may actually be a high risk for acquisition, or poor collateral for a loan, when a problematic environmental assessment is factored into the bottom-line equation. Mismanagement and poor housekeeping practices with pesticides, antifreeze, oils, batteries, fuels, chemical waste,

Figure 9.2 The improper storage of used oil along a creek.

and rinsates, or a lack of credible documentation on waste handling, storage, and disposal may offset positive factors such as market share, assets, and future profits.

In addition to placing company personnel and perhaps the public at risk, improper handling of pesticides and fertilizer products can result in significant environmental contamination. Many pesticide and fertilizer applicators are shocked by the magnitude of their environmental liabilities. The costs to clean up a property, especially ground water, where improper chemical storage and handling has caused contamination can range from thousands to millions of dollars. These costs take away from the bottom line. Such liabilities can be devastating to a company's profitability and to its public image. For instance, how confident is the public with a company that spends four to five years' profit to clean up environmental contamination that could have been prevented? Even more critical to the financial future of the company is the negative view that investors will have with such gross mismanagement (Fig. 9.3).

Real estate normally is regarded as an asset, but a property can become a liability if the costs associated with an environmental cleanup exceed its fair market value. It is unlikely

Figure 9.3 Contaminated soil from spilled diesel.

Figure 9.4 A ruptured fertilizer tank releasing its contents.

that the owner of such a property will be able to sell it unless and until the contamination is remediated to the satisfaction of governmental officials, the prospective buyer, and the insurance carriers involved. Alternatively, a buyer might commit to paying for the cleanup and, in doing so, use it as leverage to negotiate the purchase price. In some situations, money may be held in escrow until the site is cleaned; with any money not spent on the cleanup returned to the seller. In any of these scenarios, the seller ultimately pays for the environmental cleanup of the property (Fig. 9.4).

Contaminated soil or water can lead to short- and long-term consequences. For example:

- The likelihood that governmental agencies will take action against past, present, or future owners of the property will increase.
- Persons found guilty of willfully, unintentionally, or ignorantly committing an environmental offense may be subject to criminal actions and sentenced to time in jail.
- Fines can be levied against those responsible.
- Retirement income for owners of small- to medium-sized companies may have to be forfeited to pay for remediation.
- The cost of remediation may exceed the real estate value.
- Substantial environmental cleanup costs may force a company into bankruptcy.
- Owners may face higher premium rates for (or be denied) liability insurance.
- Prospective buyers may encounter difficulty acquiring financial assistance from lending institutions.
- Owners may be unable to sell the property.
- Individuals exposed on the property, or as a result of contamination resulting from activities on the property, may file personal injury actions or worker's compensation claims for conditions derived or alleged from exposure.
- Persons exposed to soil and water contamination may file legal claims against past and present owners.

Is there an environmental time bomb at your facility? Can you afford to wait for the explosion, or should you take action, now, to defuse it? Pest management businesses are scrutinized by the public and the media. Significant mistakes seldom go unnoticed. The purpose of this chapter is to explain why people and organizations involved in all aspects of pest management—both urban and rural—should place a premium on good environmental site management.

THE "PHASES" OF ENVIRONMENTAL SITE ASSESSMENTS

Presale or preloan environmental assessments—whether administered by a loan officer, an environmental consultant, or an environmental attorney—usually involve a line of questioning aimed to (1) disclose historical uses of and pest management practices employed on the property and (2) determine whether or not the property is in compliance with current environmental regulations (Fig. 9.5).

Phase 1 Environmental Site Assessment—Investigation

A Phase 1 assessment includes a search for and a review of private and public documents related to the history of the property, a visual inspection of the site, and mapping of the physical characteristics of the property. This in-depth research provides crucial details about the history of pesticide use and that of other chemicals on the property. Documents that piece together the history of the property may include title transfers for the preceding 50 years, site plans, old and new photographs of the site, and topographic maps. Public records from municipal fire departments, utility companies, county health departments, and environmental regulatory agencies may provide details of compliance with environmental

Figure 9.5 Environmental audits can include investigation, sampling, and remediation depending on whether or not pesticide contamination is found.

Figure 9.6 A truck unloading fertilizer onto a moving belt. Note the spilled material on the ground that will need to be cleaned prior to a rain.

laws and regulations. Interviews with past and present landowners, tenants, employees (both current and former), and neighbors may render additional information.

A visual inspection of the property is a crucial component of Phase 1 Environmental Site Assessments. Evidence suggestive of potential pollution problems might include signs of improper disposal such as bare earth, ditches barren of vegetation, distressed vegetation in field margins, burn piles containing pesticide containers, or discarded pesticide containers; soil or concrete staining from spills or leaks at the chemical production or storage facility; records which disclose costly disposal of stored pesticides that have been canceled from further use; water contamination from uncapped wells or storage of chemicals near wells; strong chemical odors at the mixing and loading site; abandoned pesticide-container dumps; and old, underground tanks (either active or abandoned) (Fig. 9.6).

A favorable Phase 1 assessment does not guarantee contamination-free property. Instead, it means that the assessment did not reveal evidence of environmental contamination. The assessment demonstrates that the buyer has exercised due diligence to make discovery and, if the audit findings are noted with the transfer, serves to distance the buyer from liability for past actions on the property. Typically, a lender or buyer will not request additional testing or investigation when the Phase 1 environmental assessment reveals no credible evidence of environmental abuse of the property. If contamination is suspected, a more detailed, Phase 2 assessment (e.g., soil and water sampling) may be necessary to confirm the presence of contamination and to determine its extent. It is also conceivable that the lender will simply deny the loan or that the buyer will lose interest in purchasing the property when an environmental audit indicates contamination problems—or even a potential for them.

Phase 2 Environmental Site Assessment—Sampling

In a Phase 2 environmental assessment, any visual characterization of and background information on the site are supplemented by the collection and analyses of air, soil, and/or water samples. Sampling methods used to determine the presence and level of contaminants often are federally or state prescribed. A Phase 2 environmental assessment may be very expensive and requires a greater level of technical expertise than does the Phase 1 assessment.

Figure 9.7 Taking a vertical core soil sample around a pesticide mixing and loading site.

Soil samples at the surface or subsurface may be collected across the property or at a specific site of concern. Samples are collected at various depths and sent to laboratories equipped for analyses of environmental contamination. The analytical results often are presented on a map indicating where each sample was taken and what levels of contamination exist across the surface and at certain soil depth intervals (e.g., 0–1 ft, 1–2 ft). When the deepest samples indicate contamination, a second round of soil sampling may take place to determine the level at which contamination is no longer detected (Figs. 9.7–9.9).

Surface water samples may be collected from ditches, ponds, creeks, and even reservoirs. Environmental auditors often will focus on wells. Ground water may be sampled from wells already on-site or from shallow monitoring wells drilled into the water table.

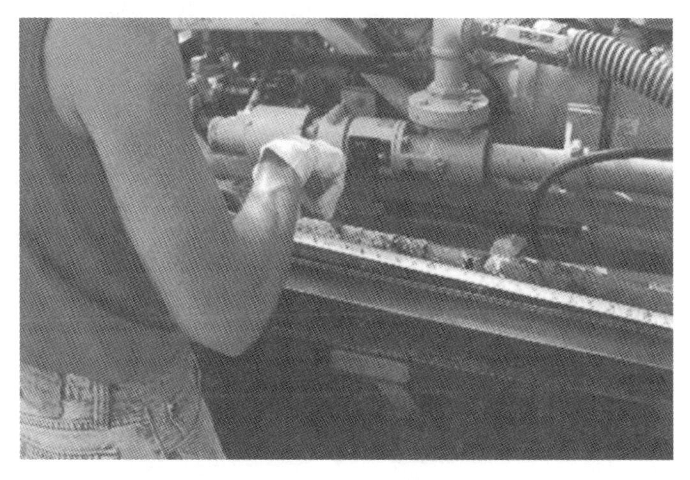

Figure 9.8 Each section of a core soil sample is identified according to depth, which will allow the investigator to determine if the suspected problem is at the surface or lower in the soil profile.

Figure 9.9 Each sample is clearly marked so that the analysis can be identified at a specific spot and depth.

A thorough review of environmental sampling data will reveal the identity of any contamination and "hot spots" of pesticide concentrations; it will facilitate a general assessment of the property. Care must be exercised to request laboratory analyses (in addition to scheduled regulatory analyses) for all pesticides and other contaminants known to have been present on the property at any time. Combining the Phase 1 and Phase 2 information into a Site Assessment Document provides a benchmark that enables consultants, environmental engineers, and regulatory authorities to determine what actions, if any, might be necessary under Phase 3.

Phase 3 Environmental Site Assessment—Cleanup

Phase 3 requires the property owner and any other potentially responsible parties, along with the consultant firms, to meet and discuss with the regulatory agencies a plan outlining specific objectives for site remediation. The details are very much driven by site-specific information, the contaminant, legal statutes, cost, and technologies available to aid the procedure (Fig. 9.10).

Federal and state guidelines generally dictate to environmental firms how they must handle, remove, transport, and dispose of soil, water, concrete, and other contaminated items. These standards, along with site expertise developed by environmental engineers, are used to draft a Site Remediation Document which has as its core one or more methods that will achieve remediation of the site: soil excavation, placement of ground water monitoring systems or extraction wells, on-site treatment of contaminated soil and water, construction of subsurface barriers or retaining walls, bioremediation, incineration, and/or clay caps. In addition, a site-specific cleanup standard is listed for approval by regulatory agencies. The Site Remediation Document also describes in some detail what health and safety practices must be followed: the monitoring of surrounding areas for potential impact during remediation, transportation routes to be followed when contaminated materials are removed from the site, the name of the disposal company that will

Figure 9.10 Pesticide-contaminated soil being removed from a site.

be accepting the pesticide waste, and the time frame for filing reports with regulatory agencies.

IMPORTANT COMPONENTS OF RISK MANAGEMENT

Business owners and stockholders benefit when their companies profit and grow. And maximum growth is achieved by retaining existing customers while securing new ones, by offering consistently better products or services, and by decreasing operation expenses. Taking shortcuts may yield short-term savings; but to guard against long-term liabilities, companies need to emphasize safe workplaces, good management practices, and environmental compliance. They should be proactive, not reactive!

Selling business property—either urban or rural—might be the most significant, long-term financial move that an owner will ever make. Poor environmental management practices put long-term—even lifelong—investments of time, energy, and funds at risk. Failing to recognize and address environmental site problems in a timely fashion and failing to correct unacceptable environmental practices can lead a profitable business into bankruptcy.

The major benefit of an environmental site assessment—whether performed by an outside consultant or by in-house personnel—is the identification of existing liabilities and any potential for future liabilities. Ideally, following an assessment, steps are taken to reduce or eliminate practices determined to be conducive to liability. They (site assessments) are a management tool.

Environmental site assessments provide insight into past and present practices that contribute to inadequate environmental protection. Likewise, they direct prudent business people toward total compliance and a contamination-free, environmentally sound facility. Following are questions that owners and managers should ask themselves:

- What was the condition of the property prior to my gaining control of it?
- Did my predecessor or a prior business leave the site contaminated?

- Have my neighbors' practices adversely impacted ground water beneath or surface water runoff onto my property?
- Have my own practices adversely impacted the environment on or near the property?
- What prior actions and practices of my own may have increased my liability potential?
- Do any of my current practices need to be changed to reduce the potential for future liability?
- What steps can I take to reduce future contamination and to reduce existing contamination?
- Do I need legal advice?
- Does my state offer environmental assessment assistance programs?

Reviewing and Assessing Past, Present, and Future Liabilities

Environmental assessments are essential to anyone planning to purchase, sell, or lend money on a business involving pesticide storage, handling, transport, application, or disposal. Taking title to a property sight unseen and without benefit of an environmental assessment is viewed as a willingness to assume responsibility for all existing liabilities and the irresponsibility of others.

A buyer who purchases property without the benefit of an environmental site assessment could inherit a wide range of liabilities resulting from previous owners' practices. Lack of knowledge does not protect the buyer from environmental liabilities caused by the seller or predecessors of the seller. Buyers should be wary of a low price tag on a property because environmental liabilities—which may well translate to hidden costs—may be part of the "great deal."

Banks and other lenders have a vested interest in the environmental status of property that they mortgage because any problems become theirs if the mortgagee defaults; that is, if they repossess a property, they likewise assume any existing environmental liability because most environmental statutes do not require evidence of improper conduct by the owner. Therefore, lenders are wary of approving loans without benefit of an environmental assessment, and most lending arrangements now specifically require periodic compliance audits.

The possibility of foreclosure on contaminated property affects loan acquisition for all industries that store and mix pesticides, fertilizers, and fuels on-site. Lenders are wary of soil and water contamination (especially of unconfined aquifers) and generally take precautions to preclude it from their real estate transactions. In doing so, they complicate initial loan acquisition for purchasers, and companies whose practices are found to cause adverse environmental impacts have difficulty obtaining subsequent loans for improvements or expansion.

History—What Has Already Happened? Many practices that were considered the norm years ago are now considered unacceptable. Certain seemingly acceptable practices, repeated over time, can trigger serious environmental problems that necessitate expensive solutions. Minor contamination is still contamination; and it can build up over a long period of time and eventually reach surface or ground water. Less than environmentally sound practices, coupled with the repetitive use of pesticides that may linger in the environment, can dramatically increase the seller's liability and seriously diminish the property value. The buyer may gain the upper hand, using the decline in property value to negotiate a lesser price, thereby cutting into the equity of the property (Fig. 9.11).

Figure 9.11 Investigations often focus on past uses of the property to help determine the likelihood of contamination.

One purpose of an assessment is to gauge the probability of a contamination problem. The likelihood of a problem may be directly related to the actions of previous owners or occupants of the site. For instance, farm acreage used solely for crop production generally is of less concern than an area on the same property that contains an underground fuel tank or an area where pesticides are or were previously mixed. Of even greater concern are commercial application business locations, both past and present. Older businesses generate a higher level of concern than those established in our modern climate of increased environmental awareness and responsibility.

The greater the concern about a particular site, the more important the environmental assessment. Properties with a high probability of contamination should be researched thoroughly. Professional site assessors can access a variety of land records and governmental databases (e.g., superfund sites at http://www.epa.gov/superfund/sites/) to develop the history of a site. This allows them to determine who previously owned the property, when they owned it, what type of business they conducted on it, and which chemicals and other products they used. Aerial photographs are available for most sites and often provide a historical view of land use. The U.S. Geological Survey has a clearinghouse for aerial and satellite photographs.

Landownership records may provide only partial information about a site. For example, the site assessor may find that XYZ, Inc., was the prior owner; but that information, alone, does not disclose prior use of the property. The assessor may need to interview former XYZ managers, employees, neighbors, etc., who have personal knowledge of how XYZ conducted its operations.

Once a preliminary history of the site is determined, the assessor should inspect the property, paying specific attention to areas of concern (e.g., former pesticide-mixing areas especially near wellheads). An experienced professional assessor should be able to distinguish normal from unusual by visual inspection. If unacceptable conditions are present, the site assessor may elect to take soil and water samples. Personal interviews, government and business record searches, visual inspections, and soil and water samples yield information that the assessor can use in reaching a conclusion about the risk—or risk potential—associated with the property.

Present Practices—What Is Happening Currently? Obviously, what's done is done: Property owners have no control over what took place at the site under previous ownership. They can, however, identify and clean up problems and thereby minimize the potential for future risk—and they certainly can govern activities under their ownership. It is important that owners and managers control their operations, minimizing exposure potential as a priority and maintaining property value.

Assessment of current practices and regulatory compliance at a facility determines how well the site is being managed from an environmental standpoint. Business owners should

- know their role in environmental stewardship,
- devise a plan to fulfill that responsibility, and
- execute the plan.

Based on information gathered during an assessment, the assessor may be able to predict whether or not the storage and use of pesticides, fertilizers, solvents, petroleum products, etc., will eventually have a negative impact on the property—tomorrow, next year, or 10 years down the road (Fig. 9.12).

Property owners and corporate decision makers use site assessments routinely to identify existing liabilities and liability potential. Many businesses assign specific employees to perform internal assessments on a continual basis, usually annually or semiannually; others hire

Figure 9.12 A review of present activities can identify pesticide storage, mixing, and loading activities that may be contaminating soil and water.

outside firms or private consultants such as environmental or health and safety specialists. Comprehensive assessments include inspection of storage and handling practices as well as pesticide and pesticide-container disposal methods.

Prudent business people combine good capital investments with sound operation procedures. For instance, investing in the construction of a quality containment facility and establishing a standard of impeccable handling practices for employees go hand in hand with environmental stewardship. For example, a concrete containment area where well-informed, responsible employees can mix and load without jeopardizing the surrounding area or ground water goes a long way toward environmental safety.

The Future: Will Rectifying Prior Problems Reduce Liabilities? Environmental site assessments are conducted to identify and deal with past and present contamination issues. These issues are raised at the time a business is offered for sale and during inheritance proceedings, mergers, and regulatory activities. Whether you are "buying" liabilities from the past or incurring new ones, the future sale or transfer of the property is impacted.

A prudent buyer should never purchase property on which chemicals have been used or stored without the benefit of an environmental site assessment. The expenditure for a thorough, professional assessment up front may save thousands of dollars in the end; real estate titles should never be transferred without one. Failure to rule out liability potential could have disastrous financial consequences: possible property and business devaluation due to media attention, and significant cash outlay for remediation (Fig. 9.13).

Occasionally, owners conduct internal environmental assessments and report their findings to potential buyers as assurance that the property is not contaminated. Although this may be legitimate from the seller's point of view, the buyer's best interest is served by

Figure 9.13 Companies must work at reducing all sources of pesticide contamination so that in the future the business can be sold without having to worry about an environmental pollution issue.

requesting a professional third-party assessment. Ultimately, either the buyer or the seller may use the results as leverage for an adjustment of the purchase or sale price, respectively. For instance, if a seller conducts an internal assessment and a follow-up one conducted by a professional consultant yields identical or similar positive findings, the seller would be in a good position to ask premium dollars for his property.

CONDUCTING ENVIRONMENTAL SITE ASSESSMENTS

Who Conducts Environmental Site Assessments?

Anyone with a financial stake in a property may request an environmental assessment. The person or firm designated to perform it depends largely on its purpose. Buyers usually hire outside consultants to conduct environmental assessments to verify the absence of liability before signing a purchase agreement, whereas the seller's objective might be to confirm the worth of his property in support of the sales price. Hiring a third party is perceived as important due to the extensive responsibility assumed in drawing conclusions about the liability or liability potential of a property.

The professional consultant's report is intended to protect the interests of the buyer. The amount of work conducted by an environmental consultant depends on the scope of the project. The consultant's job is to make visual observations, conduct a government records check, interview employees, identify past uses of the property, and develop a report summarizing the findings. Based on the findings, the consultant may recommend soil and water samples to confirm or rule out the presence of environmental problems. The buyer is made aware of environmental problems through the consultant's report, but no consultant will guarantee that the land is clean.

The in-house environmental specialist's report is intended to protect the interests of his or her company. Many businesses employ individuals to conduct their own environmental assessments of corporate facilities and potential acquisitions. Findings are reported to facility and corporate management personnel, and problems noted become the focus of future inspections. Corporate auditors keep operations on cue with environmental policies and procedures and assure compliance with local, state, and federal laws.

Insurance companies often conduct site inspections of pest management businesses that they insure. The focus is to rule out environmental contamination before renewing the policy: The insurance assessor's report is intended to protect the insurance company. But the insured can benefit as well.

When insurance assessments detect areas of environmental concern, owners are alerted to the need for remediation. Many insurers can aid property owners in addressing problems before they create a liability. By helping companies identify and address their environmental responsibilities, the insurance company reduces its own risk in underwriting the coverage.

Lenders have become experienced in the practice of conducting environmental site assessments, using consultants and occasionally their own staff. They often begin by requiring business owners to complete a questionnaire. If red flags go up, an on-site inspection may be conducted, although the decision to inspect is highly dependent on the type of business borrowing the money, the amount being borrowed, the collateral, and the lender's procedural policies.

Site visits are customary when a business has become insolvent and the lender is considering foreclosure. At such a point in time, it is wise for the lender to know whether or not the site has environmental problems; if it does, foreclosure may not be in the lender's best interest. In other words, in taking possession of a contaminated property, responsibility for remediation—and the cost thereof—would default to the lender. The lender's site assessment report is intended to protect the financial institution.

Government agencies such as the U.S. Environmental Protection Agency and state departments of agriculture and natural resources conduct site assessments, as do some fire marshals, local fire departments, etc. The scope of their investigations typically is limited to enforcement of a select set of environmental and safety regulations. Regulatory and environmental assessment reports are intended to protect the public and the environment.

Environmental site assessments have become routine to protect property interests and reduce liability for all parties concerned. It is important for owners and managers of pest management businesses, small or large, to conduct periodic internal assessments. State disclosure laws regarding such assessments should be thoroughly understood. Self-assessments and those ordered and paid for by owners and managers are the only ones intended to protect the interest of the business.

Self-Assessments and Legal Advice

Although legal opinions are not always needed, it is important to be sensitive to the legal implications of an environmental site assessment.

Environmental Problems Existing Prior to the Self-Assessment Managers often know up-front whether serious problems exist. If a problem serious enough to bear legal ramifications—possible lawsuit, criminal action, or administrative hearing—is known to exist, it is essential to seek counsel prior to organizing a self-assessment. The determination of whether or not there is a problem should be made with the assistance of an attorney. Nothing about the problem should be committed to writing (or taped) until an attorney has been retained. Information generated during a pesticide investigation may need to be protected from disclosure under the attorney–client privilege or as attorney work product.

Housekeeping Self-Assessment The housekeeping self-assessment focuses primarily on transfer areas and deals with routine, everyday problems relating to chemical storage, handling, mixing, and disposal. Housekeeping assessments are typically documented with detailed notes. Those that are routine and well-documented demonstrate the company's intention to identify and address problems, both real and perceived, before they become environmental issues. Housekeeping assessments also may be useful in dealing with regulatory agencies, potential buyers, insurance representatives, and attorneys.

Unanticipated, even serious, problems may be discovered during any environmental site assessment; and especially when it happens during a self-assessment, such findings should not be discussed with employees. The assessment activity should be halted and an environmental attorney contacted, immediately. The attorney should provide guidance on the magnitude of the problem. If it is serious, available options will be explained and the client advised on how to proceed. The attorney may want to intercede on behalf of the client to protect certain evidence from later disclosure under the attorney–client privilege or attorney work product privilege.

Figure 9.14 Employees can identify current problems and offer practical solutions to solve them.

Self-Assessment: Preparation for an Internal Audit

Preparation is critical for managers who conduct their own assessments. A well-organized and thought out plan makes the visual assessment more meaningful. Such plans often include the information contained in the following sections. A check sheet or spreadsheet that details what information will be obtained through the site assessment is beneficial.

Interviews with Past and Present Employees Employees (and sometimes neighbors) can provide managers valuable information regarding site history and past and present practices. In such interviews, quiz employees about both company policy and actual activities, and ask how they might change practices for the better (Fig. 9.14).

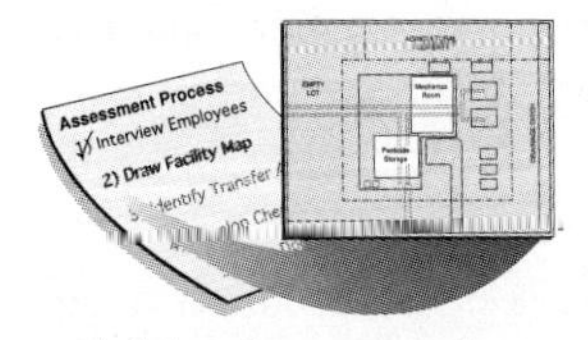

A Map of the Facility Develop a detailed map of the property, citing areas of significance, and obtain a blueprint (or draw a graph) of each structure on-site. All permanent buildings, other structures, and activity-specific areas should be identified and each location assigned a name which best describes the activities that occur there:

- Mechanic's shop
- Pesticide storage room (or building)
- Fertilizer blending operation
- Dry fertilizer area
- Pesticide containment area
- Trash dumpster
- Aboveground and belowground fuel tanks
- Anhydrous ammonia tanks
- Holding tanks (for chemicals, fertilizer, waste oil, pesticide rinsate, etc.)
- Parking areas

Figure 9.15 A drainage pipe flowing from a pesticide application business into a creek can greatly increase the impact to stream quality when chemicals are mismanaged on-site.

- Railroad lines (and areas where railroad cars carrying bulk fertilizers are loaded or unloaded)
- Former and current dump areas
- Chemical transfer areas

Clearly identify on the map any active and abandoned wells, soil type, and depth to ground water. Every spigot or location where water is accessed should be marked with its use. The map should provide information regarding the direction(s) in which surface water flows following a heavy rain (Figs. 9.15–9.17). Be sure to indicate the presence of drainage tiles and the points on the property where surface water runs off. Identify the nearest creek or ditch downgrade from the property boundaries. Identify all drains

Figure 9.16 A pesticide storage building and used oil containers being stored are next to an outside drain.

Figure 9.17 It is important that all pesticide spills be cleaned to prevent the chemicals from dissolving in rainwater and leaving the site through various drains.

inside buildings. If known, mark the location of the septic system and the leachate field. All of these items are points of access for materials to reach soil and water. Identify utilities such as telephone poles, electrical transformers, pipelines, and sewers on the property.

After a map is developed, but before it is used for inspection and marked on, make photocopies of the clean original for subsequent use. When each site assessment is conducted, make sure the map is marked with

- the date of the inspection,
- the name of the person(s) conducting the assessment, and
- a clear, concise scale (and legend, if necessary).

Update the original map as new buildings are added or demolished or when other modifications are made on-site.

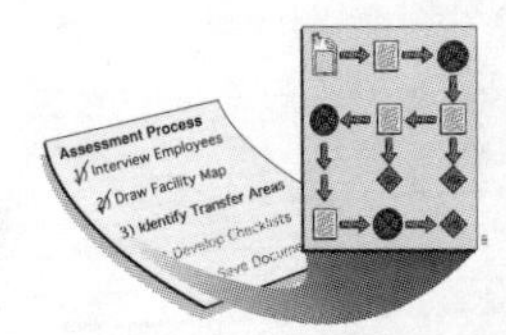

Transfer Areas Identify (and mark on the map) all transfer areas, both current and past. This takes deliberate thought and interrogation. It is easy to overlook locations that might fit into the definition of "transfer area"; For example, the spot where unrinsed pesticide containers are stored until they are properly rinsed and the equipment parking area. Develop a diagram tracing the route of each pesticide, fertilizer, fuel, solvent, etc., from its arrival at the facility (or property) to its point of final departure. The diagram should include routes of the concentrated product, any altered (diluted) portion of the product, and any waste from the product (unused dilutions; rinsates; excess product; clothing, equipment, and containers contaminated by the product, etc.). Each point along the flow chart—wherever the chemical is stored, mixed, handled, applied, or disposed—becomes a transfer area and should be so designated in the site assessment. Label each point according to the activity that occurs there (Figs. 9.18–9.21).

Figure 9.18 Transferring herbicides from a tanker truck into storage tanks requires attention to details.

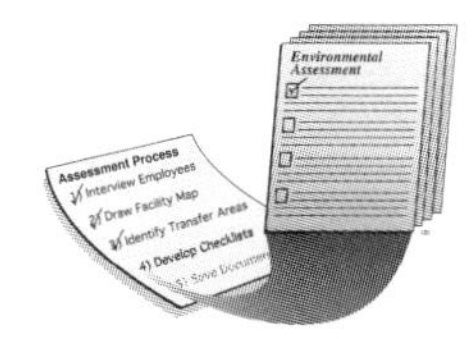

Check Sheets/Checklists Check sheets are critical in assuring that the site assessor doesn't miss anything. Devise a check sheet—or select an existing one—that allows ample space to record observations and jot down notes and corrective actions required. Self-assessors will find that their check sheets will evolve and expand over time as they become more comfortable with the procedure.

Check sheets should cover all questions and comments that arise during an environmental assessment; there should be itemized space, as well as blank space for notes, so that the assessor may record all information gathered during the inspection and evaluation process. Transfer areas may become checklist headings, with additional items listed under each. In addition to the check sheets provided in Appendix 9.1, sources such as the Fertilizer

Figure 9.19 A transfer area can include pumping rainwater that has mixed with spilled pesticide over the side of a pesticide containment facility.

Figure 9.20 Burning plastic and paper pesticide containers is not allowed under most state laws.

Institute, trade magazines, newsletters, newspapers, the Tennessee Valley Authority, and the American National Institute of Standards often offer checklists that yield helpful ideas in customizing one just right for a given facility or property. Also, insurance companies and regulatory officials sometimes offer written guidelines that can be used as the basis for (or a supplement to) check sheet composition.

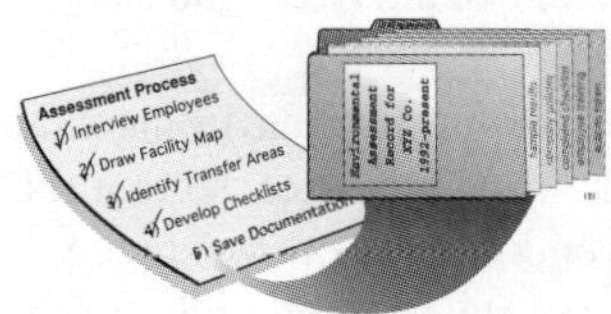

Maps, Check Sheets, and Documentation Files A complete record of assessment should be maintained by the facility or company manager (or the property owner) for the purpose of answering regulatory inquiries and questions from prospective purchasers, investors, and insurance carriers.

Save all maps, check sheets, and documentation files as evidence of prudent operational procedures. These documents should be saved for as long as the manager and/or owner maintain their relationship with the property.

Figure 9.21 Water supplies must be protected against backsiphoning.

However, if serious problems are detected, an environmental attorney should be consulted on whether it is still advisable to retain documentation.

It is advisable to confer with an attorney before you discard old business documentation, including but not limited to documentation on environmental problems. In some cases, the loss or destruction of documents may permit a judge or jury to assume that the documents were damaging to the business that lost or destroyed the documents. It is advisable that every business adopt consistent document retention policies and strictly adhere to those policies so that they are not accused of willfully and deliberately destroying harmful evidence.

When to Conduct an Environmental Site Assessment

Ideally, environmental assessments should be conducted during the busiest months of the year—when everyone's time is limited. This allows the assessor to observe how chemical handling practices (and perhaps the environment) are affected when the staff is juggling multiple priorities. Weaknesses are most likely to surface when time is short, sending up red flags.

Some managers prefer to review their operations during the off-season, when there is more time to ask questions, identify and correct problems, and train employees. But it may be hard to address a contamination problem months after its inception; the source of the problem may be elusive, making it difficult to establish cause and effect and to prescribe remediation.

Another alternative is to conduct both a targeted mini-assessment and a complete assessment on an annual basis. Consider scheduling a complete assessment during the slower months, when time is ample, and an internal mini-assessment during the busy season.

Mini-assessments should target high-risk operations (e.g., mixing and loading, chemical delivery, fertilizer outload, and pesticide storage areas) and take only a few hours. It is suggested that they be conducted without notice, as high-risk areas often become environmental hot spots during the busy season as staff members scramble to meet production goals. Employees must be made aware of their ongoing role in protecting the business, the property, and the environment. Focused, unannounced mini-assessments tend to reveal the everyday, busy-season truth about how chemicals are handled (Figs. 9.22–9.24).

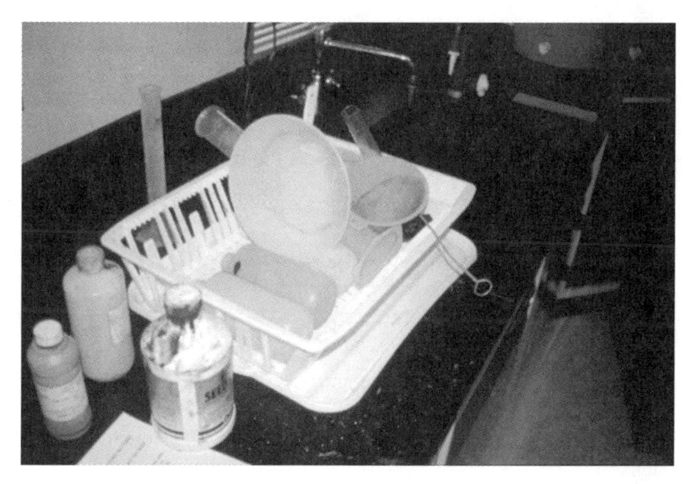

Figure 9.22 Mixing pesticides around sinks is a poor practice.

Figure 9.23 Hoses should never be allowed to contact the pesticide mixture in a tank.

How to Conduct an Environmental Site Assessment

Clearly, the environmental assessor—whether a consultant or a company employee—needs a sharp eye for intricate details and potential problems. It is important to recognize subtle indications and contamination "fingerprints" indicating that a chemical may be or has been released into the environment.

Ask yourself if you would fix a particular problem or modify a practice if you knew the state agricultural or environmental inspector—or a potential buyer—were visiting tomorrow. If the answer is yes, then it is worth noting and changing. Ask yourself, What are the potential negative consequences of this situation?

Armed with a detailed map and check sheet, pay specific attention to anywhere chemicals can come into contact with soil or water, or where they may have in the past (Figs. 9.25 and 9.26). Don't overlook the small spill. Repeated small leaks and spills often overlooked—

Figure 9.24 Prevention is the key to avoiding most chemical spills. Note the close proximity of the truck to the fuel tank.

Figure 9.25 Companies need to find solutions when their employees place water hoses in tanks.

tend to create big environmental problems, eventually. In addition to the environmental liability, lost product translates to decreased profit: It gets you twice! When confronted with a problem, record observations in writing, directly on the map, or on audiotape or videotape.

Treat existing and potential problems with the same concern, because today's problems were only potential problems yesterday. The only way to break the contamination cycle is to address problem potential as soon as it is recognized.

Where to Look

Most self-assessments, as well as those conducted by consultants, buyers, and regulatory personnel, focus attention on operations in which a chemical product is transferred, moved,

Figure 9.26 Handling pesticides around water requires additional protective measures to guard against spills reaching the water.

or stored. Transfer areas—hot spots—are earmarked for inspection because activities that occur there are more likely to result in soil and water contamination than activities elsewhere on-site. Examples of hot spot activities include the following: unloading pesticides from delivery trucks, loading spray rigs, pouring pesticides into backpack sprayers, cleaning tanks, storing used motor oil, and mixing fertilizer.

How to Evaluate Contamination

Visual inspection and research into past practices on-site comprise the first step in evaluating the environmental quality of real estate. Evidence that chemicals have interacted with soil often can be detected without the aid of laboratory analysis. However, the extent and magnitude of a problem can be established only through soil analysis of both contaminated and uncontaminated areas.

Walk the property at a slow pace, looking for obvious signs of contamination: soil staining, dead vegetation, accumulation of batteries, and poor housekeeping. Be especially alert to chemical odors. Make a note to walk the facility from a different direction when conducting the next audit.

Accurate visual assessment hinges on inspectors' insight into operations conducted at the site; assessors must be aware of all activities performed on-site to earmark all transfer areas for inspection. They must be able to differentiate between normal practices done properly and normal practices done improperly. They must recognize evidence of chemical release—even trace amounts—into the environment: for instance, a slow-leaking valve that drips ever so slightly, yet continuously, onto the soil over an extended period of time. It can also build up to reportable quantities which is why it is a good idea to constantly change the mixing and loading sites.

The following criteria are typically used to evaluate the environmental condition of real estate.

Discoloration and Staining Discoloration and staining may occur: for example, black diesel stains on soil, yellow pesticide stains on concrete, and black oil sheen around drains. Such evidence may indicate previous spills (large or small) or ongoing releases of very small quantities (Fig. 9.27).

Stressed or Dead Vegetation High levels of most chemicals, including fertilizers, can kill vegetation. Even releases at very low concentrations can stress plants. They may appear different (smaller, thinner, distorted) than the same species growing just a few feet away, for example, larger plants may be indicative of fertilizer runoff. A path of dead or stressed plants may indicate damage caused by a contaminant in surface water runoff (Figs. 9.28 and 9.29).

Unpleasant Odors Odors are most evident when contaminated soils are disturbed, releasing trapped gases. Common occurrences include the release of ammonia or diesel fumes when contaminated soil is exposed during dumpsite excavation. Anaerobic (without oxygen) conditions prevent or slow the breakdown of chemicals in soil beneath the surface; but when the contaminated soil is turned, the release of odorous gases occurs. Spills inside storage rooms can also produce very strong odors (Fig. 9.30).

Figure 9.27 Used oil stains on the ground.

Figure 9.28 Herbicide residues accumulating in low-lying areas in sufficient quantity to kill all vegetation.

Figure 9.29 Application equipment parked in the same area can kill plants when rain washes off residues from the surface of the equipment.

Figure 9.30 A forgotten pesticide container leaking in a storeroom.

Unsightly Housekeeping Practices A messy site does not prove that there are environmental problems. But it does provide the assessor with a feel for how chemicals are currently (and were in the past) handled and stored. A messy, dirty site always receives more negative attention from an environmental assessor than one that is neat and clean (Fig. 9.31).

Sampling and Employee Questioning Many environmental problems are obvious, but some are not. Problems that are difficult or impossible to detect during a visual inspection include (Fig. 9.32):

- Leaking underground fuel tanks, piping, and plumbing
- Indoor drains that connect to leach beds or field tiles
- Soil contamination beneath concrete containment, mixing, and loading pads
- Old facilities that have been buried or had buildings placed over them
- Ground water contamination

Figure 9.31 Used oil must be managed to prevent spills and leaks.

Figure 9.32 Open drains inside pesticide storage rooms and where application equipment is parked means that any leaks could find their way into the drain and out of the facility.

Environmental assessors have to question past and present employees about previous activities at a site to get a feel for what problems or potential problem might exist. Accurate assessment of "hidden" elements requires soil and water analysis to determine whether or not there is contamination; what areas, if any, are contaminated; and the levels of detectable contamination.

How Long Should the Assessment Take?

The length of time required for the visual portion of most environmental assessments varies with the qualifications and experience of the assessor. Generally, the visual inspection can be completed in 2–8 hr, depending on the size of the facility. More time is needed to check records, conduct interviews, and take soil samples; these can take weeks or months to complete. Environmental site assessments of commercial, agricultural pesticide and fertilizer application businesses generally take more time, due to the diversity of their operations (pesticides, fertilizers, mechanic's shop, fuel tanks) and because they occupy more acreage.

Visual inspections of most other pest management businesses (e.g., lawn care, right-of-way weed control, golf course maintenance, aquatic weed control) commonly require only 2–4 hr. Their facilities typically support fewer operations and occupy minimal acreage.

INCORPORATING SITE ASSESSMENT INFORMATION INTO BUSINESS PRACTICES

Transforming Liabilities into Assets

Environmental site assessments that focus on daily practices need not be time consuming or expensive. Continual visual observation by managers usually allows quick remediation of any problems that arise; problems caught early generally are not expensive to correct. Sometimes, a simple procedural change may be all that is necessary. But it should be noted that the gathering of information through routine, in-house assessment—or through more in-depth, annual review—is only the first step in finding solutions to problems.

Figure 9.33 An accident waiting to happen.

It is the responsibility of property and facility managers to practice sound environmental stewardship; for example, they are obligated to prevent the release of pesticides and fertilizers into soil and water. Those who fail to correct problem practices are, in essence, condoning them; therefore, they must be willing to accept any resulting liability. Unfortunately, management often elects to ignore problems without seeking legal advice from knowledgeable environmental counsel. Failure to take action may result in property that cannot be sold, environmental cleanup costs, charges of willful violation of the law, fines, or even criminal sanctions.

The optimal approach is to take remedial action to eliminate or significantly diminish problems or problem potential identified through site assessment. For example, everyone would agree that an aboveground diesel fuel tank located next to a creek represents an accident waiting to happen. The problem is the location of the tank, so the solution is to move the tank away from the creek: a simple, direct, and manageable approach.

Here is another example: On the edge of a 6-ft loading dock is a minibulk pesticide container on a wooden pallet. A proactive precautionary action to lessen the probability of a spill would be to relocate the pallet and minibulk to the ground or into containment. A reactive approach would be to leave the minibulk on the loading dock, taking no action unless an accident actually occurs—e.g., a forklift operator could accidentally knock the minibulk off of the loading dock, creating a spill. Proactive initiatives are the key to preventing existing operational practices from creating liability (Fig. 9.33).

Developing Practical Solutions

Environmental site assessments not only identify existing problems and problem potential, they also influence business decisions such as allocating money for remedial action. Solutions to problems—such as deciding where to locate mixing and rinsing pads—may require significant thought, planning, and expenditure. Others have quick fixes, such as educating workers to cap pesticide containers left outside to keep them from collecting rainwater.

Environmental site assessments—simple or complex, in-house or otherwise—provide a snapshot of facility operations. The more thorough the assessment, the better the focus on prevention and solution.

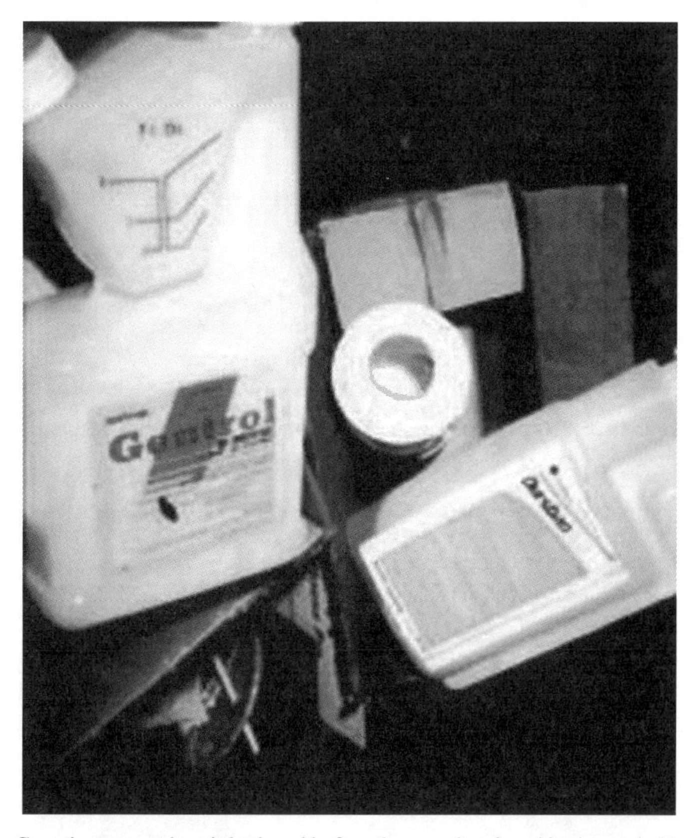

Figure 9.34 Containers must be triple rinsed before they can be placed in the trash. Note the powder in one container and the liquid in another.

For example, one morning a manager decided to visually inspect her facility. She discovered many unrinsed, empty pesticide containers and others that were partially full of product, in the dumpster. She also discovered undrained, used oil filters. In the mixing and loading area she found obvious pesticide stains on the soil and detected a strong, pungent odor. The manager called her staff together that very day to discuss the findings and to inform them that practices would have to change (Figs. 9.34–9.36).

Simply and clearly, problems need to be corrected. Managers should personally address serious problems such as soil contamination. Solutions to simpler, less urgent problems, such as triple-rinsing containers and draining oil filters, can be assigned to other staff members (Fig. 9.37). Never overlook or underestimate the value of a well-educated staff for preventing and solving problems.

Practices which cause major problems such as soil contamination around a mixing area must be stopped immediately. Other examples of risky practices that a walk-through assessment might reveal are washing equipment and rinsing containers near wells, ditches, or creeks. Others include aboveground fuel tanks placed near roads or wells, and used antifreeze and solvents disposed of in floor drains leading to outside septic systems or tiles. Sometimes interim solutions are necessary to allow time for incorporation of permanent practices.

Figure 9.35 An unrinsed pesticide container retrieved from the garbage.

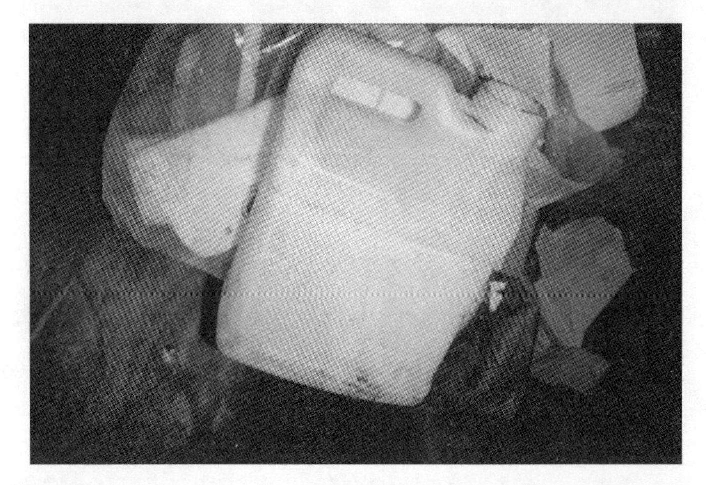

Figure 9.36 A container that was partially rinsed and thrown in the trash.

Figure 9.37 Oil filters being properly drained.

The entire assessment process must be documented. Managers and supervisors should determine a time frame for follow-up inspections to verify that remedial procedures are incorporated. Documentation of inspections and corrective procedures can be useful in communicating with company assessors, regulatory investigators, insurance companies, community leaders, potential buyers, and employees.

Company Policies Should Address Environmental Problems

Every aspect of every business is unique: facilities; operations; employee knowledge and education; and pesticide, fertilizer, fuel, and other chemical management policies. Therefore, written and practical policies should be company driven, site specific, and tailored to the company's environmental goals.

Both written and practical company policies should lend guidance to managers and employees on handling potentially hazardous materials in an environmentally sound manner. They should convey to employees specifically how to perform tasks to prevent accidents and contamination. Written policies and procedural practices must be clear and concise, yet detailed enough to cover the nuts and bolts of specific issues. Policies provide other benefits as well:

- Improvement of long-term profitability by avoiding situations that require expensive cleanups, penalties, and fines
- Reduction of environmental, legal, and regulatory liability
- Elimination or proper management of waste
- Safety of employees
- Clarification of expectations of employees
- Identification of employee education topics
- Communication of environmental stewardship to the community

Written Policies Can Be Sources of Liability A written policy that is not followed can be worse than no written policy at all! During legal proceedings, attorneys may cite a company's written policy as evidence of responsibility awareness. Armed with the written policy stating what should have been done, they may argue that the company's failure to follow its own procedural guidelines demonstrates a conscious decision to disregard that responsibility. They may charge that the company's failure to implement the policy led to the problem of concern and that the company acted recklessly, if not intentionally, by not following its own operational guidelines. The result? The company is held liable for environmental contamination.

A company's written environmental policy can even be used as the foundation for criminal charges. Given the fact that written policies demonstrate awareness, ignoring them may lead a judge or jury to conclude that the company knowingly and willfully violated environmental laws. Willful actions that lead to an environmental problem (e.g., disposing of contaminated soil along a creek bank) can be deemed criminal offenses. And the severity of civil penalties and fines increases dramatically in cases of willful violation. Owners and managers must recognize that they will be held responsible for environmental stewardship even when subordinates commit an offense.

Figure 9.38 A commercial agricultural retail facility mixing/loading site that is being managed according to industry standards.

Tailoring Policies for Your Business Because written policies are created for various purposes—educational, regulatory, legal—managers must devote serious thought to defining overall goals before putting a policy in place. Employees need to be brought into the process; that is, they should be allowed input into policies that they will be required to implement. Oftentimes, they can offer workable methods for achieving and maintaining environmental stewardship—and their input can be very instrumental in their embracing company policy.

There is no easy way, best method, or simple strategy for writing environmental policies. Each policy is dependent on the environmental issue being addressed, the company's philosophy, the allocation and commitment of resources, and legal and business considerations. Following are examples of statements that might appear in a written policy.

Environmental Mission Statement An environmental mission statement is a brief but profound declaration of company attitude toward, responsibility for, and commitment to environmental stewardship (Figs. 9.38 and 9.39).

Problem Statement The problem statement specifies a given practice, such as disposal of used oil in a dumpster, as the primary concern. It clearly identifies the problem and explains the significant environmental and legal ramifications.

Solution Statement This is the most important section of an environmental policy. The solution statement details how the identified problem will be addressed and what specific steps are necessary to eliminate or reduce the likelihood of recurrence. The solution statement may include procedural steps, educational activities, and written documentation.

Benefits Statement The benefits statement details the anticipated advantages of the policy—to the business itself, to its employees, and to the community—as a result of implementation of the defined procedures (Fig. 9.40).

Figure 9.39 An example of a pesticide storeroom and its attached mixing area at a golf course.

Compliance Statement It is important that each employee understand the consequences of failing to follow written procedures. For example, the first incident of failure to comply might draw an oral warning; the second, a written warning; the third, time off without pay; and the fourth, employment termination. A flagrant infraction of any environmental or safety policy might invoke termination, even in the absence of previous infractions (Fig. 9.41).

Date, Person Responsible, and Accessibility Statements The date the policy was written (or updated) should appear within the document. The policy should list the names of personnel responsible for answering questions relative to company policy. All policies should be contained in a well-indexed, well-organized format (e.g., in a binder or on a computer); they should not be scattered among multiple departments (see Chapter 12 on bulletin boards). Managers should update their policy file, as necessary; but, in all cases, complete but separate copies of all policies—including those that have been updated and

Figure 9.40 A fertilizer blending operation where loading is done over an impervious surface.

Figure 9.41 Giving notice to employees not to wash vehicles near the facility's well.

those that are no longer in force—should be kept in a secure location, accessible only to company officials and legal counsel. Such information can become valuable reference material if a company is asked to identify the policy (or version thereof) that was in place at a specified time.

Policy Implementation through Employee Training

Procedural policies are difficult to implement. But with explanation and training, encouragement, enforcement, review, and teaching by example implementation becomes routine. Training and education should equip employees with the knowledge necessary to make informed decisions on the products they use, store, and dispose of as a function of their employment. Continual education improves the probability that pesticides will be handled judiciously, protecting both the environment and the interests of the company. Annual employee evaluations can be used to reinforce the importance of company policy, especially if employees realize that pay raises, promotions, and continued employment are contingent upon their implementation of prescribed procedures.

Employee education processes must go beyond mere communication of company policy. They must cultivate and enhance employee understanding, skill, and motivation.

For instance, obligating employees to triple rinse pesticide containers is one thing, but helping them comprehend why it is important encourages their cooperation. For instance, they might not appreciate why triple rinsing is necessary. But employees are likely to give their all if employers do the following:

- Make them understand: Convey that by carrying out this simple task they, personally, are having an economic, technical, regulatory, and environmental impact.
- Equip them with the skills required: Teach them techniques that facilitate the procedure.
- Motivate them: Offer incentives, or stress that their annual pay raises are contingent on compliance.

Employees must realize that their personal actions play an important role in the success and profitability of the business—that they are stakeholders in the company. Pesticide applicators frequently use personal judgment and apply skills independently, both at their business headquarters and on job sites, so it makes sense that employees be trained to reflect company policy. And it may be beneficial for management to canvass employees, collectively and independently, for suggestions on policies and revisions.

All employees, including office staff, should receive training at least annually on all written company policies as well as OSHA and other applicable regulations. Written policies can be handed out with paychecks, with instructions for each employee to read them prior to the next scheduled safety and environmental meeting. This too is an excellent source to put on the employees bulletin board. Scheduling monthly, or as often as necessary, meetings where issues, policies, and problems are addressed is a simple way to instill safety awareness.

Safety meetings that focus on environmental policies and preventive practices often generate the best results when they are held outdoors with an instructor demonstrating the "new" approach. Hands-on training is by far the best avenue for teaching employees how to handle chemicals properly, and employees retain the information longer. In either case—classroom education or hands-on training—learning activities underscore the importance of company policies and demonstrate that employee participation is essential to the company's success.

It is important to document attendance and participation in training sessions, using the employee's signature. Training documentation—whether an internal safety meeting or outside training session—should also include the agenda, title of each program, speakers, and the date. Such training documentation will assist the manager when questions are raised about the company's training programs. Tests help give documentation credibility.

Routine Inspections Support Policy, Education, and Prevention

Even if a company prepares environmental policies, educates and trains its employees, and corrects problems identified during assessment, the benefits will be greatly diminished unless follow-up evaluations are conducted. Continual environmental assessment and monitoring of work activities are just as important as the ongoing activities that generate income, such as sales training, product performance, and quality service. Failure to conduct routine environmental assessments could mean that past investments of time, energy, and financial resources were wasted. Companies must be vigilant to assure that previous environmental mistakes are not repeated. For instance, cleaned soil can become a new liability if contamination recurs; whatever practices were responsible for the initial problem (e.g., continual small spills at a mixing site) must be corrected.

Managers should never dismiss an opportunity to assess their facility or their employees. Early intervention is important in protecting the environment, so daily observance is paramount. It is a good idea to carry a notebook and jot down things that need attention when they are first observed.

Things such as failure to triple-rinse containers in a timely manner or worn out sight gauges on equipment can be detected—and corrected—before they become an issue.

There are numerous reasons why employees may fail to follow prescribed company policies. Although some may offer flimsy excuses, others' reasons may be more concrete—or at least more understandable. For example, some may cut corners under pressure to meet daily production schedules. Finding themselves between the proverbial rock and a hard place and recognizing that there is no way to accomplish all that is expected of them and stay on schedule, they skip procedural steps that aren't too obvious (maybe they won't get caught!). Managers must be cognizant of the pressure that prescribed procedures and production schedules impose; they must be alert to signs of compromised performance; they must step back and ask themselves if this is the result they want; they must consider the impact of resulting environmental liability on the bottom line; and they must ask themselves, Do I cut corners, too? Am I condoning inferior performance, by example?

Since changes in work practices may occur gradually, over extended periods of time, it is critical to monitor procedural policies and the impact of training on employee performance. Company rules and practices, as well as personnel training, need to be reviewed and supplemented continually. Those companies that do so, and those that stay abreast of their operations in relation to industry progression, will benefit. In other words, as practices change to meet an ever-increasing demand to perform more efficiently, effectively, and safely, management must adjust accordingly. The result will be managers with fewer worries, employees who are better educated and trained, reduced environmental liabilities, and happier neighbors who are less likely to contest facility operations.

LEGAL IMPLICATIONS ASSOCIATED WITH ENVIRONMENTAL SITE ASSESSMENTS

It is important to understand that site assessments may compromise or threaten your legal rights, that is, to be aware of associated legal issues: what to do and what not to do prior to conducting (or hiring someone to conduct) an environmental site assessment.

Environmental attorneys can offer advice on how to conduct assessments and how to record the findings in a manner that will safeguard their clients' rights. Those who practice environmental law will know

- the applicable laws,
- what liabilities are triggered by applicable laws,
- the consequences of action or inaction on the part of their clients, and
- the options potentially available for solving environmental problems.

Seek the advice of an attorney who practices environmental law, because the gathering, recording, and reporting of information derived from a site assessment may carry additional

Figure 9.42 Take care not to document information without first consulting an attorney.

reporting responsibilities and legal ramifications. Never put in writing or record electronically (by computer, video-, or audiotape), and never discuss with employees or anyone else, any information relating to environmental problems discovered during an assessment; that is, don't document anything that you don't want a judge, jury, or regulatory agency to see or hear: Always discuss the situation with an attorney first (see Chapter 8 for details on hiring an attorney). In choosing an attorney, contact the state bar association to see if it has an environmental designation for attorneys specializing in that field (Fig. 9.42).

Identifying an environmental problem and taking no remedial action can be worse than not identifying the problem at all. Regulators may view failure to address an identified environmental problem as a willful violation. It also subjects the violator to criminal sanctions. You may need to ask your attorney certain questions, such as the following:

- How can we protect our legal rights and still conduct (or hire someone to conduct) an environmental site assessment?
- Should I write down or otherwise document whatever is discovered during a site assessment?
- Can information from an environmental site assessment be used against me?
- If there is a perceived problem, is it under local, state, or federal jurisdiction?
- Which agencies will be involved if we report the problem?
- Is the problem serious or minor?
- Is it better to initiate corrective action prior to meeting with the authorities?
- What are the penalties associated with not reporting or correcting the problem?

Be Aware of Legal Rights under Assessments Conducted by Outside Parties

Owners/operators should assess their own property first and correct any existing problems before allowing outsiders to perform an environmental assessment of the property.

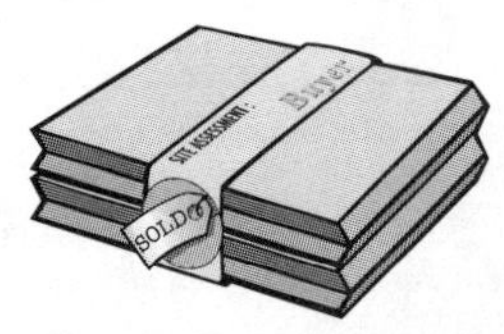

Buyer Assessments When an environmental assessment is performed for a prospective buyer of property, it is very difficult to later argue that the results of such an assessment are privileged and cannot be disclosed to regulators or other parties. This usually will not be a concern if the assessment reveals minor problems. However, if major problems are found and the prospective buyer decides not to purchase the property, the assessment prepared on behalf of the buyer could later cause problems for the seller.

For this reason, it is recommended that an attempt be made to identify and resolve major or serious problems through self-assessment before attempting to sell. The results of a self-assessment can be kept confidential more easily. If an attorney is consulted, it may be possible to have the assessment performed in a manner in which the findings can be protected from disclosure, thereby protecting the owner, through statutory or other types of legal privilege or protection, from later prosecution.

Lender Site Assessments When lenders are concerned about environmental liabilities associated with a property, borrowers may have to answer certain environmental questions. The questions posed to the borrower often depend on the type of loan in question, the amount of money being borrowed, security interests and collateral applied toward the loan, etc. Answers to the environmental questions may influence the lender to require an environmental site assessment. Following are examples taken directly from a lender's questionnaire.

- Have there been any federal, state, or local environmental enforcement actions against the borrower?
- Is the borrower presently subject to any court or administrative orders (regarding environmental matters) that require continuing compliance?
- Has the borrower or the site ever had a reportable spill of petroleum or hazardous substances?
- Does the real estate contain asbestos or PCBs?
- Has the lender been provided a report showing the ownership of the real estate for the previous 40 years?
- Are there any underground storage tanks on the property?
- Have any underground tanks ever been removed from the property? If so, who removed them (provide documentation)?
- Has the borrower ever conducted an environmental assessment?
- Have any claims for environmental problems been made under the borrower's insurance policies?
- Is the company or the site out of compliance with any of the environmental permits that it holds?

- Are storage containers of fuels, chemicals, pesticides, solvents, cleaning fluids, animal waste, or other waste material stored on the property? If yes, is there any evidence of spills, leaks, or discharges into surface or ground water?
- Are their any areas at the property where the ground is stained or where there is dead or stressed vegetation?
- Does the operation generate chemical or animal waste as a part of its operation?
- Has trash and/or waste ever been burned on the property?
- Are there any open or buried dumps on the property?
- Have there been any health-related complaints or claims filed by workers at the property?
- Are there currently any canceled pesticides on the property?
- Are there any hazardous waste problems or other environmental impairments existing on the property, from either current or previous operations?

It is a federal offense for anyone to be less than truthful when answering questions related to financial transactions. But providing a "yes" answer often compels the applicant to provide documentation. For instance, one lender has the following statement on its environmental questionnaire: "This checklist is designed to alert the user to environmental issues and conditions which may adversely affect the borrower or the bank. Any 'yes' answer to the questions above needs satisfactory explanation and documentation in the file."

Borrowers should be aware that lenders generally have questionnaires, and they should ask for a copy, early on. The entire questionnaire and/or difficult questions and any that make the borrower uncomfortable should be discussed in detail with an attorney. Seeking legal advice is important because the lender's questionnaire may also be used as evidence.

Insurance Assessments Insurance companies may conduct their own assessments to confirm that the policyholder remains a good risk. The insurance underwriting process allows insurers to go on-site to evaluate a property (see Chapter 14 for more details on insurance). Generally, the insurance assessment includes a visual inspection and the issuance of a questionnaire. Again, any environmental problems detected may be used, later, as evidence in a court of law; therefore, the best advice is to clean up the site and seek legal counsel prior to an assessment. Further, it is important to fully understand all insurance contracts, that is, to know explicitly what coverage is purchased. In particular, the insured must confirm coverage that would pay in the case of regulatory action or lawsuit challenges. Keep previous, obsolete policies as part of the permanent company file; in certain instances, old insurance policies entitle the insured to compensation, retroactively, for problems discovered after the fact.

Regulatory Assessments Environmental assessments performed by regulatory personnel can be stressful because they usually are unannounced. Regulatory agencies have the legal right to enter any property during normal business hours to conduct an inspection. The inspection may be routine or it may have been triggered by a complaint. In any case, the inspector representing the regulatory agency should provide the property owner with the following information. If the inspector does not volunteer this information, you have the right to it and should ask.

- The inspector's name and some form of identification (a card, badge, or other credential)
- The agency represented (name, address, phone number)
- The name and phone number of the inspector's supervisor
- Legal authority for the inspection
- Information on whether the inspection is routine or the result of a complaint
- The name of the person who filed the complaint
- Information about what was alleged by the complainant
- Disclosure of what is or is not found during the inspection
- Any violations detected
- Any enforcement action that is likely to result from the inspection
- A date by which a written copy of the findings is to be expected

It is the regulatory official's job to ask questions through legal means, but that does not mean that the owner/operator must volunteer information to support alleged complaints. Answering some questions could be construed as admission of guilt; these and certain other types of questions should be addressed through an attorney. Any information disclosed during a regulatory inspection is considered discoverable evidence and could be used against the company in a civil or criminal case.

Companies should have a plan in place for handling impromptu inspections. It is particularly important to plan who will represent the company by accompanying the inspector. All employees should be well aware of their role (i.e., their responsibilities) during an inspection. All samples drawn by the inspector should be split with the company for separate testing.

SUMMARY

We live in a rapidly changing world, and we must change with it or be left behind. Government regulations, public scrutiny, and industry standards have mandated that pesticide and fertilizer industries change. Many environmental practices considered routine in the past are unacceptable today. The days when small fertilizer and chemical spills, fuel releases, and/or minor misuse were overlooked are over: The risks are too great.

An environmental site assessment can be a valuable tool (Fig. 9.43). It can increase the bottom line and the value of a business by reducing liability. Lenders require verification of a property's environmental condition before lending money toward its purchase. If a regulator takes soil and water samples in which fertilizer or pesticides are detected above reportable quantity levels, someone is going to have to pay for the cleanup. The law provides that past property owners, years back, can be held liable for environmental cleanup, and it can cost thousands—sometimes millions—of dollars. A timely environmental site assessment can uncover problems—major, minor, potential—for remediation to reduce, prevent, or eliminate contamination. Identifying problems in a timely manner and taking appropriate remedial action can add to the bottom line. That is, early remedial intervention by way of assessment can save a company hundreds of thousands of dollars down the road. Environmental cleanup isn't cheap, so it makes sense, monetarily and otherwise, to avoid the necessity.

Prudent managers maintain an awareness of the tools of compliance and environmental stewardship and, more important, they implement them. They assess their facilities

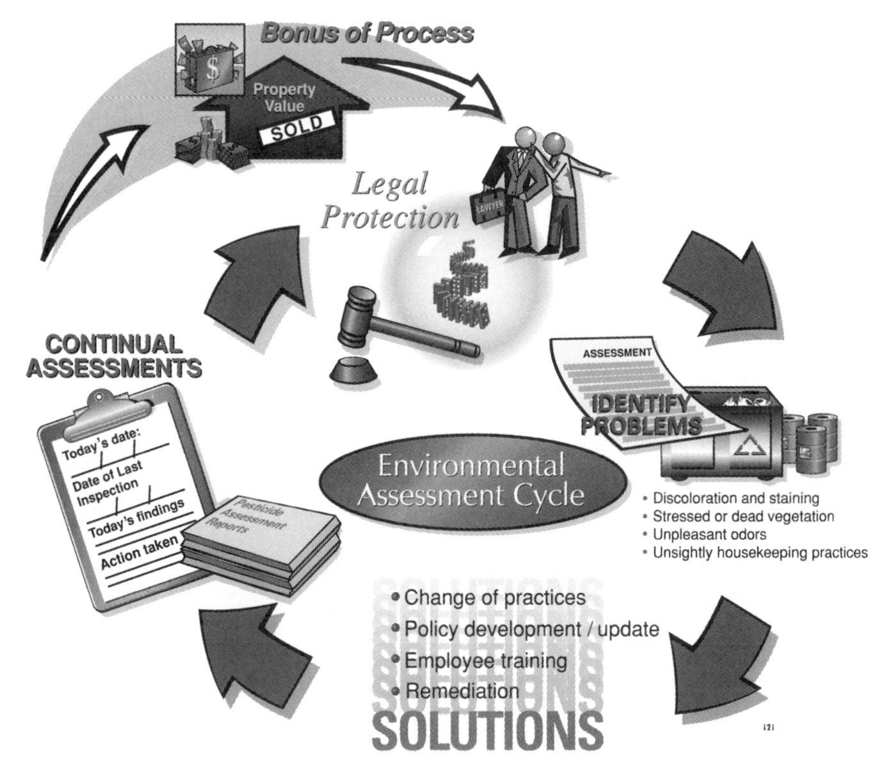

Figure 9.43 Environmental audits are an important tool in managing a profitable business.

and properties continually to ensure accordance. The effective management of a business requires sensitivity to information that site assessments present, and personnel must be cognizant that negative data might be admissible as evidence in criminal or civil proceedings. Seeking legal counsel early on ahead of the assessment process makes good business sense; it affords the company legal protection from disclosure under the attorney–client privilege and the work product privilege.

An in-house environmental assessor's work is never done. It requires constant awareness of the daily, monthly, and yearly operations that affect the economic and environmental success of the business. Conducting the assessment and documenting problems constitute only the first step. Formulating a solution, writing company policy, and training and monitoring employees round out the equation.

Environmental risks assumed in operating a chemical application business in today's world are greater than ever before. Smart business owners recognize that they must maintain control, and site assessment is a mechanism for evaluating and managing those risks.

APPENDIX 9.1 ENVIRONMENTAL AUDIT CHECK SHEETS

It is suggested that owners and managers develop their own, company-specific checklists. The following may be used as is or as a starting point in tailoring others to suit specific situations; each of the samples can be completed in 45 min or less.

The samples are divided into six categories:

- Facility Overview
- Waste Management
- Fuel Storage Tanks
- Fertilizers
- Pesticides
- General Awareness

FACILITY OVERVIEW—CHECK SHEET #1

Self-Assessment Report

(Name of Company) (Branch)
(Street, P.O. Box, City, State, Zip) (Date)
(Person Conducting the Assessment)

Drainage Ditches

1. Is grass along drainage ditches or on-site gullies dead or stressed? .... Yes No N/A
2. Does the path of surface water runoff contain dead/stressed vegetation? .... Yes No N/A
3. Do ditches show soil discoloration or have noticeable odors? ..... Yes No N/A

Private and Community Wells

4. Does the well water have an unusual color, taste, odor, or sheen? .... Yes No N/A
5. When equipment is filled with water directly from the well, is the well protected against backflow? .... Yes No N/A
6. Is a fixed air gap used between the water source and the equipment? .... Yes No N/A
7. Is a fixed air gap used between the water source and a water holding tank? .... Yes No N/A
8. Is an antibackflow device installed in the water system plumbing? .... Yes No N/A
9. Is the soil surrounding the well casing elevated? .... Yes No N/A
10. Is the well casing sealed from surface water? .... Yes No N/A
11. Are all abandoned or out-of-service wells sealed? .... Yes No N/A
12. Are chemicals stored away from the well? .... Yes No N/A
13. Are well construction logs on file? .... Yes No N/A
14. Has the well been tested for pesticides and nitrates? .... Yes No N/A
15. If #14 is answered yes, are the results on file? .... Yes No N/A
16. If tests (see #14 and #15) are positive, is the facility using bottled water? .... Yes No N/A

Burn Piles

17. Do burn piles exist on the property? .... Yes No N/A
18. If #17 is answered yes, are pesticide containers or used oil filters found in burn piles? .... Yes No N/A
19. Are burn piles near creeks, ditches, or wells? .... Yes No N/A

Junk and Refuse Piles

20. Does the property have (or has it ever had) open or buried dumps? . . . . Yes No N/A
21. Are any chemicals, solvents, paints, tires, or batteries buried on-site? . . . . Yes No N/A
22. Are any used 55-gallon drums stored on-site? . . . . Yes No N/A
23. Are 55-gallon drums reused to store rinsate, used oil, and so on? . . . . Yes No N/A
24. Is there any type of containment for 55-gallon drums? . . . . Yes No N/A
25. Do discarded tanks contain unidentified rinsate or sludge? . . . . Yes No N/A

Commercial Dumpsters

26. Does the trash contain used oil or oil filters that have not been drained? . . . . Yes No N/A
27. Does the trash contain unrinsed or partially full pesticide containers? . . . . Yes No N/A
28. Does the dumpster area have dead vegetation, stains, or odors? . . . Yes No N/A
29. Is the dumpster covered to keep rainwater out? . . . . Yes No N/A
30. Is the dumpster located in an area where runoff is directed into a creek? . . . . Yes No N/A
31. Is the dumpster left unlocked when not in use? . . . . Yes No N/A
32. Is the dumpster equipped with a drain containing a plug? . . . . Yes No N/A

WASTE MANAGEMENT—CHECK SHEET #2

Self-Assessment Report

(Name of Company) (Branch)

(Street, P.O. Box, City, State, Zip)

(Person Conducting the Assessment) (Date)

Is documentation of the disposal or recycling of the following wastes (#1–16) available?

1. Used oil . . . . Yes No N/A
2. Lubricants . . . . Yes No N/A
3. Used oil filters . . . . Yes No N/A
4. Hydraulic fluids . . . . Yes No N/A
5. Cleaning solvents . . . . Yes No N/A
6. Antifreeze . . . . Yes No N/A
7. Paints . . . . Yes No N/A
8. Fuel filters . . . . Yes No N/A
9. Freon/refrigerant . . . . Yes No N/A
10. Vehicle batteries . . . . Yes No N/A
11. Pesticide containers . . . . Yes No N/A
12. Spent acids . . . . Yes No N/A
13. Pesticide rinsate . . . . Yes No N/A
14. Petroleum . . . . Yes No N/A
15. Sludge from tank bottoms . . . . Yes No N/A
16. Secondary containment rainwater . . . . Yes No N/A
17. Is used oil stored in labeled drums or other containers? . . . . Yes No N/A
18. Do all 55-gallon drums stored outdoors have caps in place? . . . . . . Yes No N/A
19. Are waste materials stored in labeled drums? . . . . Yes No N/A
20. Does the mechanic's shop have a floor drain? . . . . Yes No N/A
21. Are odors, spills, or poor housekeeping practices detected in the mechanic's room? . . . . Yes No N/A

FUEL STORAGE TANKS—CHECK SHEET #3

Self-Assessment Report

(Name of Company) (Branch)
(Street, P.O. Box, City, State, Zip)
(Person Conducting the Assessment) (Date)

Aboveground Fuel Tanks

1. Are fuel tanks located near a well or ditch? .... Yes No N/A
2. Are fuel tanks located uphill from the well? .... Yes No N/A
3. Are fuel tanks protected (e.g., by a barrier) from vehicular damage? .... Yes No N/A
4. Is the dike covered to prevent storm water accumulation? .... Yes No N/A
5. If #4 is answered no, is the pad kept clean so that water can be discharged at the site? .... Yes No N/A
6. Are fuel tanks located in containment? .... Yes No N/A
7. Is the containment impermeable? .... Yes No N/A
8. Have previous cracks in the containment been repaired? .... Yes No N/A
9. Are there presently any cracks in the containment? .... Yes No N/A
10. Are there visible signs of spills or leaks around fuel tanks? .... Yes No N/A
11. Are petroleum odors apparent when the soil is disturbed around fuel tanks? .... Yes No N/A
12. Is electrical wiring in conduit? .... Yes No N/A
13. Are No Smoking and Health Hazard (NFPA diamond) signs present and enforced? .... Yes No N/A
14. Are fuel tanks stationed at least 25 ft from any structure? .... Yes No N/A
15. Have all reportable incidents of leaks and spills been properly cleaned? .... Yes No N/A
16. If #15 is answered yes, is documentation of the cleanup on file? .... Yes No N/A

Belowground Fuel Tanks

17. Are there belowground storage tanks on-site? .... Yes No N/A
18. Have storage tanks been removed from the site within the last 3 years? .... Yes No N/A
19. If #18 is answered yes, are the records on tank removal on file? .... Yes No N/A
20. Has there been a reportable leak incident on-site? .... Yes No N/A
21. If #20 is answered yes, are leak and spill cleanup records available? .... Yes No N/A
22. Are leak detection equipment and secondary containment systems installed? .... Yes No N/A

23. Have tanks been hydrostatically tested for leaks? . . . . Yes No N/A
24. If #23 is answered yes, are the results of the hydrostatic test and the closure report on file? . . . . Yes No N/A
25. Is an inventory of fuel tank volume taken between tank filling and use? . . . . Yes No N/A
26. Are tanks properly registered and permitted? . . . . Yes No N/A

FERTILIZERS—CHECK SHEET #4

Self-Assessment Report

(Name of Company) (Branch)
(Street, P.O. Box, City, State, Zip)
(Person Conducting the Assessment) (Date)

Dry Fertilizer

1. Are products stored under roof? . . . . Yes No N/A
2. Is rainwater diverted away from contact with product? . . . . Yes No N/A
3. Is fertilizer loaded over a concrete surface? . . . . Yes No N/A
4. Is the loading area swept on a regular basis? . . . . Yes No N/A
5. Are areas under conveyor systems cemented and contained? . . . . . . Yes No N/A
6. Are conveyors covered to keep out water and reduce fugitive dusts? . . . . Yes No N/A
7. Is there spilled fertilizer near rail unloading areas? . . . . Yes No N/A

Anhydrous Ammonia Storage Tanks

8. Are tanks equipped with remotely operated emergency shutoff valves? . . . . Yes No N/A
9. Are there visible signs of leaks? . . . . Yes No N/A
10. Are hose expiration dates current? . . . . Yes No N/A
11. Are emergency respirators in good condition? (There should be two canisters and two respirators readily accessible.) . . . . Yes No N/A
12. Is the jump tank/emergency shower located within 10 sec of tanks? . . . . Yes No N/A
13. Are tanks properly labeled on two sides with "Anhydrous Ammonia" or "Caution—Ammonia," first aid information, and product hazards? . . . . Yes No N/A
14. Are tanks and piping protected from vehicle impact? . . . . Yes No N/A

Liquid Fertilizer

15. Is the fertilizer storage area near a well? . . . . Yes No N/A
16. Is bulk fertilizer in containment? . . . . Yes No N/A
17. Will the containment facility hold the contents of the largest tank, plus 6 inches of rain (if containment is not under roof)? . . . . . Yes No N/A
18. Is the concrete in the containment area sealed (no cracks)? . . . . Yes No N/A
19. Have previous cracks in concrete been repaired? . . . . Yes No N/A
20. Is the containment facility free of fertilizer residues? . . . . Yes No N/A
21. Is there trash in the containment area? . . . . Yes No N/A
22. Is contaminated water standing in the containment area reused? . . . Yes No N/A

23. Are there stains, dead vegetation, or odors in proximity to the containment area? . . . . . Yes No N/A
24. Arc bulk tanks equipped with locks? . . . . . Yes No N/A
25. Are sight gauges equipped with locking valves? . . . . . Yes No N/A
26. Are sight gauges secured to tanks every 10 ft? . . . . . Yes No N/A
27. Are spill buckets used to catch drips when coupling/uncoupling hoses? . . . . . Yes No N/A
28. Are tanks labeled with guaranteed (content) analysis (e.g., 10-34-0 or 28-0-0) and health/physical hazards? . . . . . Yes No N/A
29. Is the mixing area well lighted? . . . . . Yes No N/A
30. Is the storage area well ventilated? . . . . . Yes No N/A
31. Can and/or do fertilizers enter sewer systems? . . . . . Yes No N/A
32. Are brooms, shovels, personal protective equipment (PPE), and absorbent materials available for spills? . . . . . Yes No N/A

PESTICIDES—CHECK SHEET #5

Self-Assessment Report

(Name of Company) (Branch)
(Street, P.O. Box, City, State, Zip)
(Person Conducting the Assessment) (Date)

Chemical Mixing Areas

1. Is the mixing area near a well? . . . . Yes No N/A
2. Is there dead or stressed vegetation or stained soil near the mixing area? . . . . Yes No N/A
3. Does disturbance of the soil surface produce an odor? . . . . Yes No N/A
4. Is the pesticide storage area properly marked on the outside of the building? . . . . Yes No N/A
5. Is the mixing area well lighted? . . . . Yes No N/A
6. Is the mixing area well ventilated? . . . . Yes No N/A
7. Is the mixing area under roof? . . . . Yes No N/A
8. If #7 is answered no, is rainwater or wash water runoff contained within the mixing area? . . . . Yes No N/A
9. Is contaminated water standing in the containment area reused? . . . Yes No N/A
10. Is water released as storm water clean? . . . . Yes No N/A
11. Are brooms, shovels, personal protective equipment (PPE), and absorbent materials available for spills? . . . . Yes No N/A
12. Does the chemical storage area have a concrete floor? . . . . Yes No N/A
13. Is the concrete sealed? . . . . Yes No N/A
14. Have previous cracks in the concrete been appropriately repaired? . . . . Yes No N/A
15. Are pesticides mixed, transferred, or handled in a contained area? . . . . Yes No N/A
16. Are there underground pipes leading into and out of the pesticide storage area? . . . . Yes No N/A
17. Are there noticeable spills, stains, and odors around internal drains? . . . . Yes No N/A
18. Is the concrete curbed to prevent off-site movement of water used during a fire or spill? . . . . Yes No N/A
19. Is an air gap used at all times when filling water tanks? . . . . Yes No N/A
20. Are canceled pesticides stored on-site? . . . . Yes No N/A
21. If #20 is answered yes, are they contained? . . . . Yes No N/A
22. Are pesticides stored in their original, labeled containers? . . . . Yes No N/A
23. If #22 is answered no, are labels attached to the nonoriginal containers? . . . . Yes No N/A

24. Are storage areas free of leaking or broken pesticide containers? . . . . Yes No N/A
25. Can pesticides enter sewer systems? . . . . Yes No N/A
26. Are pesticide containers and/or equipment mixed or cleaned over sinks? . . . . Yes No N/A
27. Is the pesticide storage room locked when unattended? . . . . Yes No N/A
28. Are brooms, shovels, personal protective equipment (PPE), and absorbent materials available for spills? . . . . Yes No N/A
29. Is the pesticide storage area heated? . . . . Yes No N/A
30. If #29 is answered yes, is the electrical wiring and furnace/heat source in good condition? . . . . Yes No N/A

Pesticide Containers

31. Are pesticide containers triple rinsed immediately after emptying? . . . . Yes No N/A
32. Are rinsed and unrinsed containers kept dry (out of rain) pending disposal? . . . . Yes No N/A
33. Are pesticides stored in an inside storage area with a concrete floor? . . . . Yes No N/A
34. Are all unrinsed containers stored with lids? . . . . Yes No N/A
35. Are pesticide containers burned on-site? . . . . Yes No N/A
36. Are pesticide containers reused for any purpose? . . . . Yes No N/A
37. Are pesticide containers punctured after being rinsed? . . . . Yes No N/A

Application Equipment and Parking Areas

38. Is the parking area for application vehicles secured against unauthorized access? . . . . Yes No N/A
39. Does the parking area for application equipment exhibit dead vegetation, stains, or odors? . . . . Yes No N/A
40. Are application vehicles and equipment washed on-site, over containment? . . . . Yes No N/A
41. Is water used to wash equipment captured and reused? . . . . Yes No N/A
42. Is the wash area free of dead vegetation, odors, and staining? . . . . . Yes No N/A
43. Have hoses been replaced according to manufacturers' specifications? . . . . Yes No N/A
44. Do hose fittings, valves, or connections leak? . . . . Yes No N/A
45. Are application vehicles equipped with shovels and spill bags? . . . Yes No N/A
46. Does all application equipment have air gaps? . . . . Yes No N/A
47. Is all application equipment parked over containment areas? . . . . . . Yes No N/A
48. Are service vehicles parked near or over open floor drains? . . . . Yes No N/A

49. If #48 is answered yes, are floor drains equipped with turnoff valves?. .... Yes No N/A
50. Is routine vehicle maintenance (oil and antifreeze changes) conducted on site? .... Yes No N/A
51. Are all spent fluids handled and documented by a disposal firm? .... Yes No N/A

Bulk Pesticide Tanks and Minibulk Containers

52. Are bulk chemicals (>55 gallons) stored under roof? .... Yes No N/A
53. Do bulk pesticide tanks have secondary containment? .... Yes No N/A
54. Can the containment area hold the contents of the largest tank, plus 6 inches of rain (if not under roof)? .... Yes No N/A
55. Have previous cracks in concrete been appropriately repaired? .... Yes No N/A
56. Is the containment area free of pesticide stains? .... Yes No N/A
57. Is the containment area free of trash? .... Yes No N/A
58. Is contaminated water in containment reused? .... Yes No N/A
59. Is there dead or stressed vegetation, stains, or odors in proximity to the containment area? .... Yes No N/A
60. Are bulk tanks equipped with locks? .... Yes No N/A
61. Are sight gauges equipped with locking valves? .... Yes No N/A
62. Are sight gauges clamped/secured to tanks? .... Yes No N/A
63. Are spill buckets used to catch drips from valves whenever they occur, particularly when coupling/uncoupling hoses? .... Yes No N/A
64. Are bulk tanks and minibulks properly labeled (e.g., product label, correct EPA establishment number, and net contents attached)? .... Yes No N/A
65. Are there signs of leakage in or near minibulk storage areas? .... Yes No N/A
66. Are empty minibulks stored under cover? .... Yes No N/A
67. Are minibulks containing product stored in secondary containment? .... Yes No N/A

GENERAL AWARENESS—CHECK SHEET #6

Self-Assessment Report

(Name of Company) (Branch)
(Street, P.O. Box, City, State, Zip)
(Person Conducting the Assessment) (Date)

1. Has there ever been a reportable spill of petroleum or other hazardous substance on-site? . . . . Yes No N/A
2. Has a wildlife or fish kill ever been recorded nearby? . . . . Yes No N/A
3. Does the site have unusual or noxious odors? . . . . Yes No N/A
4. Are there on-site pits, ponds, or lagoons (new or old) used to hold chemical or animal waste or storm water runoff? . . . . Yes No N/A
5. Has the firm ever been the focus of federal or state environmental enforcement actions? . . . . Yes No N/A
6. Is the business subject to any court or administrative order (regarding environmental matters) that requires continual compliance? . . . . Yes No N/A
7. Has the business ever entered into consent decrees or administrative orders relating to an environmental law? . . . . Yes No N/A
8. Has any previous environmental assessment been conducted? . . . . . Yes No N/A
9. Is anything on the property painted with lead-based paint? . . . . Yes No N/A
10. Has sandblasting of lead paint been conducted on site? . . . . Yes No N/A
11. Have claims for environmental problems been filed with insurance carriers? . . . . Yes No N/A

CHAPTER 10

OCCUPATIONAL USE OF PESTICIDES: HANDLING PRODUCTS IN THE WORKPLACE

Today's pest management industry is highly competitive. Employers need highly skilled individuals to perform a multitude of tasks associated with the manufacture, distribution, handling, application, and storage of pesticides. Good managers recognize that employees do not become skilled overnight: experience is essential to proficiency. They know that educational training programs help develop skills, improve worker competency, and promote job awareness and productivity.

Delivering competitive, quality pest management services requires more than simply knowing how to manage pests. Employee safety training programs are critical in reducing transportation accidents, demonstrating and emphasizing the proper use of personal protective equipment, and averting worker injuries from pesticide spills or releases. There are other benefits, too. Employees who fully understand their jobs effectively reduce losses from regulatory fines, hazardous waste cleanups, and property damage lawsuits. They are less likely to be injured on the job, thereby limiting worker compensation claims and employee litigation. Time and money spent on providing safety and job performance training pay big dividends by reducing production losses, boosting company profitability, and improving employee morale and retention.

EMPLOYER'S COMMITMENT TO PESTICIDE SAFETY

Pesticide-related business owners need to conduct regular, in-depth audits to identify all potential chemical hazards on the premises and those likely to present themselves at job sites. Pesticide storage, container rinsing and disposal, excess product disposal, and the transportation of hazardous chemicals all carry risk potential. Audits alert employers to the presence of particularly hazardous chemicals and practices that should be corrected through alternative chemical selection and the implementation of safety programs to reduce hazards.

Putting safety programs into place requires an aggressive and sustained commitment from management, and educating the work force facilitates the establishment and implementation

Figure 10.1 There are many ways that mismanaged pesticides can pose risk to employees and the community.

of safety strategies. Workers must understand that following safety protocols developed by their employer is mandatory for long-term employment. Openly sharing information with employees and mandating safety compliance communicates concern for their safety and well-being (Fig. 10.1).

EMPLOYEES MUST TAKE SAFETY SERIOUSLY

Employees play a major role in fostering safety in the workplace. Employers cannot maintain a safe working environment without the full cooperation and active involvement of their workforce. Employees must be trained to recognize potential safety problems, to avert those problems, and to exercise appropriate steps to prevent exposure to pesticides and other hazardous materials. Written company policies should state that much of the responsibility for maintaining safety in the workplace rests on the employees themselves, and it should be stressed that a pesticide safety program is only as good as their commitment to it.

Employees must be constantly alert to safety concerns and understand that the greatest threat to their well-being is their own failure to follow prescribed practices. Not only should workers recognize potential problems and how to prevent them, they also must be prepared to respond to chemical emergencies: warehouse fires, accidental poisonings, tank leaks, broken hoses, improper operation of valves on service vehicles, etc. Preventive strategies and a thorough knowledge of proper reactionary steps are essential components of any pesticide safety policy.

Failure to recognize the potentially serious consequences of a pesticide emergency can prove catastrophic. Employees who ignore safety practices increase the likelihood of injury

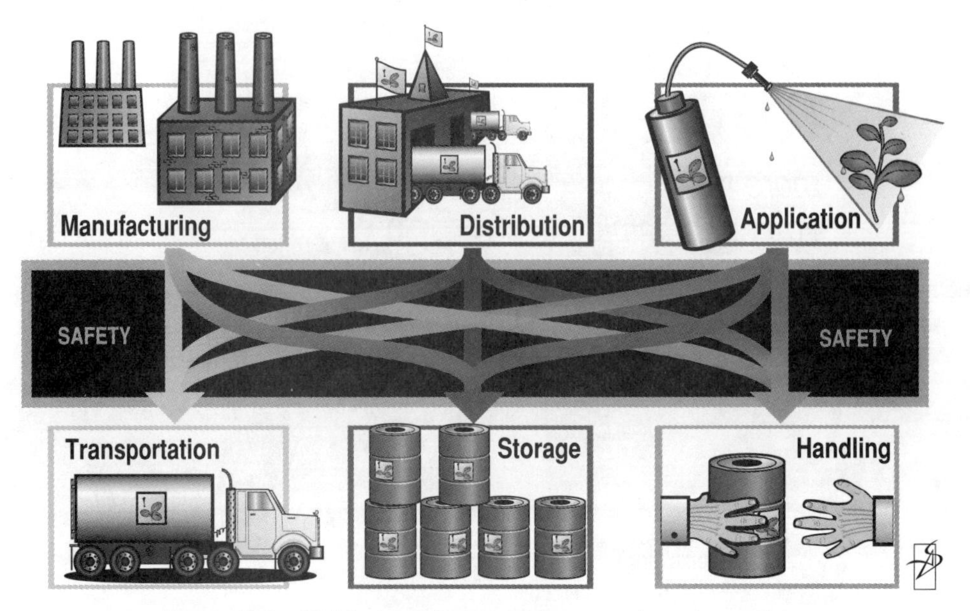

Figure 10.2 All jobs involving pesticides carry a hazard potential.

to themselves and others. Loss of income, the cost of medical treatment, and job loss resulting in financial strain are potential employee consequences of even a minor safety problem. Also, negative publicity from a safety-related incident can damage the company's reputation and the future success of the business.

PESTICIDE SAFETY IS A SHARED RESPONSIBILITY

All jobs involving pesticides—indoors or outdoors, in manufacturing or in pest management services—carry a hazard potential. Responsibility for the safe transportation, storage, and handling of pesticides and other hazardous materials must be assumed at every level: manufacture, distribution, and application. Each segment of the pesticide industry bears the responsibility to conduct its business in a manner whereby employees are thoroughly trained and safe work practices are in place (Fig. 10.2).

Both labor and management bear safety responsibilities. Both must commit to job safety each and every day through practice and communication. Today's increased focus on and attention to safety considerations requires everyone associated with pesticides (custodian, office manager, supervisor, applicator, etc.) to have some knowledge of the products being stored and handled within the business. The responsibility for using pesticides safely—that is, taking the steps necessary to protect human health and the environment—cannot be taken lightly by either labor or management (Fig. 10.3).

PESTICIDE PRODUCT SELECTION CONSIDERATIONS

There is a seemingly endless variety of pesticide products sold in the urban and agricultural marketplace. Casual observation in any distribution outlet, hardware store, or lawn and garden center will reveal numerous variations—even in products manufactured by the same

Figure 10.3 Training programs are important in teaching employees how to prevent problems associated with the use of pesticides.

chemical company and even in products that contain the same ingredients. Manufacturers often produce various pesticide products to meet different pest management needs.

The combination of an active ingredient with a compatible inert ingredient is referred to as a formulation. Pesticides are formulated for a number of different reasons. A pesticide active ingredient in relatively pure form, ready for the manufacturer's use, rarely is suitable for field application. An active ingredient usually must be formulated in a manner that

- increases pesticide effectiveness in the field,
- improves safety features, and
- enhances handling qualities.

The formulation gives the product its unique physical form and specific characteristics that enable it to fill a market niche. For most practical purposes, the terms *formulation* and *product* can be used interchangeably.

Selection of the proper formulation is critical in any pest management process involving pesticides. It is an important management decision that impacts profitability, human safety, and environmental quality (See appendix 10.1 for more information or evaluating pesticides). An understanding of the properties of various formulations is significant to the applicator who mixes, loads, and applies the product—as important, in fact, as it is to the supervisor. Applicators come into close contact with both the concentrated and the diluted product, and knowledge of the safety properties (e.g. skin hazards, inhalation hazards, etc.) of the formulation is essential to their own protection. Familiarity with the attributes of a formulation also helps applicators protect environmental quality by raising their awareness of potential adverse effects on areas surrounding the application site.

Formulation Selection Considerations

The importance of formulation type is generally overlooked. A well-considered decision to use the most appropriate formulation for a given application includes analysis of the following factors.

Applicator Safety Different formulations present various degrees of hazard to the applicator. Some products are easily inhaled, whereas others readily penetrate the skin or cause injury when splashed in the eyes.

Environmental Concerns Special precautions need to be taken with formulations that are prone to drift, to volatize, or to move off-target into water. Wildlife also can be affected to varying degrees, depending on formulation. Birds may be attracted by granules, whereas fish or aquatic invertebrates can prove especially sensitive to other formulations.

Pest Biology The growth habits and survival strategies of a pest often determine what formulation provides optimum contact between the active ingredient and the pest.

Available Application Equipment Pesticide formulations (e.g., liquid vs. dry) require specific types of application equipment for personal safety, spill control, and, in special cases, containment.

Surfaces to Be Protected Applicators must be aware that certain formulations can stain fabrics, discolor floor coverings, dissolve plastic, or burn foliage.

Cost Product prices may vary substantially, based on both the ingredients and the complexity of delivering active ingredients in specific formulations.

An Overview of the Formulation Process

The active ingredients in pesticide products come from many sources. Some, such as nicotine, pyrethrum, and rotenone, are extracted from plants. Others have a mineral origin (sulfur, borates, boric acid), and a few are derived from microbes (e.g., *Bacillus thuringiensis).* However, the vast majority of active ingredients are synthesized (man-made) in the laboratory. Synthetic active ingredients may have been designed by an organic chemist or discovered through a screening process of chemicals generated by various industries.

Regardless of their source, pesticide active ingredients have different solubilities. Some dissolve readily in water, others only in oils. Some active ingredients may be relatively insoluble in either water or oils. These different solubility characteristics, coupled with the intended use of the pesticide, in large measure define the types of formulations in which an active ingredient may be delivered.

It is preferable from the manufacturer's perspective to use the active ingredient in its original form, whenever possible (e.g., a water-soluble active ingredient formulated as a water-soluble concentrate). When it is not possible, the manufacturer can alter the active ingredient to change its solubility characteristics; this can be accomplished in a manner that does not detract from the active ingredient's pesticidal properties.

Usually, an active ingredient is combined with appropriate inert materials prior to packaging. Familiarity with the different terms and processes described herein will lead to a greater understanding and appreciation of the advantages and disadvantages of many commonly used pesticide formulations.

Sorption In some cases it may be necessary or desirable to adhere a liquid active ingredient to a solid surface such as a powder, dust, or granule. This process, called *sorption,* can be accomplished by two possible mechanisms:

- *Adsorption* is a chemical–physical attraction between the active ingredient and the surface of the solid.
- *Absorption* represents the entry of the active ingredient into the pores of the solid.

Solution A solution results when the solute (a solid, a liquid, or a gas) is dissolved in the solvent (a liquid). The components of a true solution cannot be mechanically separated; once mixed, a true solution does not require agitation to keep its various parts from settling. Solutions are frequently transparent.

Suspension A suspension is a mixture of finely divided, solid particles dispersed in a liquid. The solid particles do not dissolve in the liquid, and the mixture must be agitated to maintain thorough distribution. Most suspensions have a cloudy appearance. Labels on suspensions direct the user to shake the product well before using. These products also form a suspension when mixed with water for application as a spray. Explicit label information describes the need for sufficient agitation to keep the product dispersed in the spray tank; some solutions need only slight agitation, whereas others require more aggressive agitation.

Emulsion An emulsion is a mixture attained by dispersing one liquid into another. The pesticide active ingredient is dissolved in an oil-based solvent, creating a product that is then mixed with water, forming an emulsion. An emulsifying agent is added to the formulation to discourage separation; but each liquid retains its original identity, and some degree of agitation generally is required to keep the emulsion from separating.

Solid Formulations

Solid formulations can be divided into two types: ready to use, and concentrates which must be mixed with water to be applied as a spray. Three of the solid formulations are ready to use: dusts, granules, and pellets. Three others are mixed with water: wettable powders, dry flowables, and soluble powders.

Dusts Dusts are manufactured by the sorption of an active ingredient onto a finely ground, solid inert material such as talc, clay, or chalk. They are relatively easy to use because no mixing is required and the application equipment (e.g., hand bellows and bulb dusters) is lightweight and simple to operate. Dusts can provide excellent coverage, but their small particle size creates an inhalation and drift hazard; therefore, they are seldom used in large-scale outdoor situations. Dusts are more commonly used as spot treatments for outdoor insect and disease management. Indoors, dusts facilitate delivery insecticides into cracks and crevices, behind baseboards and cabinets, and in wall voids—away from people and pets. Commercial pesticide applicators use dusts effectively for managing insect pests in residential and institutional settings (Fig. 10.4).

Granules The manufacture of granular formulations is similar to that of dusts, except that the active ingredient is sorbed onto larger particles, usually clay, sand, or ground plant materials. Granules are defined by size; they pass through a 4-mesh (number of wires per inch) sieve but are retained by an 80-mesh sieve.

Granules are applied dry and usually are intended for soil applications where they have the advantage of weight to carry them through foliage to the ground below. The larger

Figure 10.4 Dust formulation.

particle size of granules, relative to dusts, minimizes the potential for drift. There is also a reduced inhalation hazard, although small particles can be associated with the formulation—especially when a bag is being emptied. Granules also have a low dermal hazard. Their primary drawbacks are their bulk and the problems it presents in handling, occasional difficulties in achieving uniform application, and their attractiveness to nontarget organisms such as birds. Some granules have to be incorporated into the soil to be effective (Fig. 10.5).

Pellets Pellets are very similar to granules, but their manufacture is different. The active ingredient is combined with inert materials to form a slurry (a thick liquid mixture). The slurry is then extruded under pressure through a die and cut at desired lengths to produce particles that are relatively uniform in size and shape. Pellets are typically used in spot

Figure 10.5 Granular formulation.

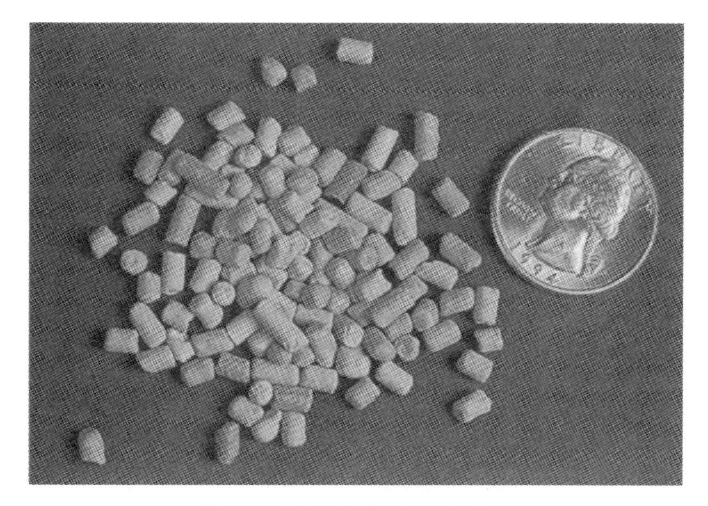

Figure 10.6 Pellet formulation.

applications and provide a high degree of safety to the applicator. They do carry the potential to roll on steep slopes and thereby harm nontarget vegetation or contaminate surface water (Fig. 10.6).

Wettable Powders Wettable powder is a common formulation composed of finely divided solids, typically mineral clays, to which an active ingredient is sorbed. They are diluted with water to form a suspension for application as a liquid spray. Wettable powder formulations generally contain wetting and dispersing agents to help distribute the powder throughout the tank. They are ideal for applying active ingredients that are not readily soluble in water.

Wettable powders tend to pose a lower dermal hazard than liquids, and they do not burn vegetation as readily as oil-based formulations. However, they do present an inhalation hazard during mixing and loading, due to their powdery nature.

There are disadvantages associated with all formulations that form a suspension in the spray tank: They require agitation to prevent settling; they can be abrasive to equipment; and they may plug strainers and screens (Fig. 10.7).

Dry Flowables Dry flowables—sometimes called water-dispersible granules—are manufactured the same as wettable powders, except that the powder is aggregated into granules. During the mixing and loading process, dry flowables pour easier from the container than wettable powders; and because of their larger particle size they present less of an inhalation hazard. Dry flowables form a suspension when diluted with water in the spray tank for application as a liquid (Fig. 10.8).

Soluble Powders Soluble powders, although not particularly common, are worth mentioning for purposes of contrast with wettable powders and dry flowables. They are not readily available because very few solid active ingredients are soluble in water; those that are water soluble dissolve in the spray tank, forming a true solution. Soluble powders provide most of the same benefits as wettable powders, without the need for agitation once

(a)

(b)

Figure 10.7 Wettable powder (a) before mixing and (b) after mixing.

(a)

(b)

Figure 10.8 Dry flowable (a) before mixing and (b) after mixing.

they are dissolved in the tank; and they are nonabrasive to application equipment. Soluble powders, as with any finely divided particle, can present an inhalation hazard during mixing and loading (Fig. 10.9).

Liquid Formulations

Following are descriptions of four common liquid formulations that are to be mixed with a carrier. The carrier generally is water, but in some instances labels may permit the use of crop oil, diesel fuel, kerosene, or some other light fuel oil.

Liquid Flowables The manufacture of liquid flowables mirrors that of wettable powders, with the additional step of mixing the powder, dispersing agents, wetting agents, etc., with water before packaging. The result is a suspension that must be further diluted with water for application as a liquid spray; it has all the advantages of a wettable powder. An important benefit of liquid flowable formulations over wettable powders is that, since the powder is already suspended in water, they pose no inhalation hazard during mixing and loading.

Liquid flowables form a suspension in the spray tank but usually do not require constant agitation during application, due to the extremely small size of the suspended particles. Occasional agitation usually is sufficient to prevent settling in the tank. One problem with liquid flowables is the difficulty in removing all of the product from the container during mixing, loading, and container rinsing (Fig. 10.10).

Microencapsulates Microencapsulates consist of a solid or liquid inert combined with an active ingredient and surrounded by a soluble plastic or starch coating. The resulting capsules can be aggregated to form dispersible granules (see section Dry Flowables), or they can be dissolved in water to form a liquid formulation. Encapsulation enhances applicator safety while providing timed release of the active ingredient. Liquid microencapsulates are further diluted with water and applied as sprays. They form suspensions in the spray tank and have many of the same properties as liquid flowables (Fig. 10.11).

Emulsifiable Concentrates Emulsifiable concentrates consist of an oil-soluble active ingredient dissolved in an oil-based solvent to which an emulsifying agent has been added. The concentrate is then mixed with water, forming an emulsion for application as a spray.

Emulsifying agents are long-chain chemicals that orient themselves around oil droplets and bind the oil and water surfaces together to prevent separation. This facilitates the carriage of oil-soluble active ingredients in water to form an emulsion. Some agitation is typically required to maintain dispersion of the oil droplets.

Emulsifiable concentrates are not abrasive to application equipment, nor do they plug screens and strainers. But they do have disadvantages: They readily penetrate oily barriers such as human skin, thereby posing a dermal hazard; some have an offensive odor; they can burn foliage; and they cause the deterioration of rubber and plastic equipment parts (Fig. 10.12).

Solutions Solutions consist of water-soluble active ingredients and water, in concentrated form. The concentrates must be further diluted with water for application. They form a true solution in the spray tank and require no agitation once they are thoroughly dissolved. They are not abrasive to equipment and will not plug strainers and screens.

(a)

(b)

Figure 10.9 Soluble powder (a) before mixing and (b) after mixing.

(a)

(b)

Figure 10.10 Liquid flowable (a) before mixing and (b) after mixing.

(a)

(b)

Figure 10.11 Microencapsulate (a) before mixing and (b) after mixing.

(a)

(b)

Figure 10.12 Emulsifiable concentration (a) before mixing and (b) after mixing.

Although solutions are not particularly common, several major herbicides with wide-scale use are formulated as concentrated solutions, including products containing paraquat, glyphosate, and 2,4-D. Aside from lack of availability, solutions have few disadvantages; however, some that are produced as dissolved salts can be caustic to human skin and the eyes (Fig. 10.13).

Miscellaneous Liquid Formulations Most liquid formulations are designed for mixing with a carrier before application. However, some liquid products are sold ready to use. Liquids generally have a low concentration of active ingredient.

Low- and ultra-low-volume concentrates used in specialty situations (e.g., space spraying and fogging) often are applied undiluted. Dermal hazards are a problem during mixing and loading because of the high concentration of active ingredient. Concentrated low- and ultra-low-volume formulations require special equipment for delivery of the product in the form of very tiny droplets. Consequently, although they provide excellent coverage, drift potential and inhalation problems during application can be quite high.

Aerosols and Fumigants

These two formulations have very different properties and uses, yet they are frequently confused. Aerosols move the active ingredient to the target site in the form of a mist composed of very small particles (solids or liquid drops). The particles can be released under pressure or produced by fog or smoke generators. Aerosols provide thorough coverage and are especially effective for indoor insect management; however, they can be difficult to confine to the target area. Two additional problems posed by aerosols are their inhalation potential and their tendency to move off-target.

Fumigants deliver the active ingredient to the target site in the form of a gas. Some fumigants arc solids that turn into gases in the presence of atmospheric moisture. Others are liquids under pressure that vaporize when the pressure is released. Fumigants can completely fill a space, and many have tremendous penetrating power. They can be used to treat objects (e.g., furniturc), structurcs, commodities, and even soil for pest insects and other vermin. Fumigants are among the most hazardous pesticide products to use due to their extreme inhalation danger.

Additional Formulation Components

Synergists Synergists are chemicals that boost the pesticidal activity of an active ingredient. The combination provides a greater degree of pest management than would be expected from the simple additive effects of each.

Synergists are used with a variety of pesticides, including insecticides, nematicides, and fungicides. They have little, if any, pesticidal activity when used alone; but, nonetheless, it is EPA policy to list synergists as active ingredients on product labels.

Piperonyl butoxide is a synergist commonly used with pyrethrin insecticides. It is believed to function by slowing the insect pest's ability to metabolize (detoxify) pyrethrin, resulting in a death rate higher than would be expected without the additive; that is, fewer insects recover from exposure to the insecticide.

Adjuvants An adjuvant is any compound that facilitates the action of a pesticide or that modifies characteristics of pesticide formulations or spray solutions. These *pesticide*

(a)

(b)

Figure 10.13 Solution (a) before mixing and (b) after mixing.

additives include materials that lower the surface tension of water (i.e., surfactants) in the spray mixture and substances that increase the wettability of the spray solution on contact surfaces.

Adjuvants are used in pesticide spray solutions as

- wetting agents,
- penetrants,
- spreaders,
- cosolvents,
- drift retardants,
- stickers, or
- stabilizing agents.

It is obvious that the term *adjuvant* has a wider meaning than *wetting agent* or *surfactant.* There are many adjuvants that have little, if any, effect on pesticidal activity. These types of adjuvants include

- antifoam agents,
- buffering agents,
- drift retardants, and
- compatibility agents.

Adjuvants are included in pesticide formulations as part of the total product sold by the manufacturer or as an additive to be mixed with pesticide products in the spray tank. Adjuvants are classified according to their type of action.

- *Activator agents* include surfactants, wetting agents, penetrants, and oils. Activators are the best-known class of adjuvants because they are normally purchased separately by the user and added to the pesticide solution in the spray tank.
- *Spray modifier agents* include spreaders, stickers, spreader/stickers, film formers, deposit builders, thickening agents, and foams. Spray modifiers generally are formulated into the pesticide product by the manufacturer.
- *Utility modifier agents* include emulsifiers, dispersants, stabilizing agents, coupling agents, cosolvents, compatibility agents, and antifoam agents. Utility modifiers, like spray modifiers, generally are formulated into the pesticide product by the manufacturer.

PURCHASING PESTICIDES BY PHONE

Many state departments of agriculture have issued formal warnings about purchasing pesticide products by phone. One particular product was being telemarketed aggressively to farmers as a selective herbicide for use on agricultural crops or as a nonselective product with up to seven years' residual soil activity. In this case, an applicator following label directions on this unregistered product would purchase 196 gallons to treat 1 acre of Johnsongrass. At \$179.80 per 2.5-gallon jug (as quoted to one farmer), it would cost \$14,096 to treat one acre!

Pesticide applicators should consider the following points:

- Purchase products from local businesses whenever possible. Avoid dealing with vendors whose only address is a post office box.
- Demand that phone solicitors provide the EPA registration number—and an advance copy of the label—before agreeing to accept shipment of any pesticide product.
- Herbicides sold through the mail or over the phone often are not registered for sale or use in the state—that is, it is illegal to buy or use them.
- Contact the pesticide regulatory agency in your state and ask whether the product being sold over the phone is registered with the EPA and the state.
- Telephone solicitors often misrepresent the products they sell, either recommending them for inappropriate uses or exaggerating the length of their residual activity.
- Avoid vendors who claim they can provide a product "just like" one of the best selling products—for half the price.

FOLLOWING LABEL DIRECTIONS

The pesticide label is a binding contract between the user and the public. The EPA is responsible for approving label wording to specify use of the product in a manner that will not adversely affect human health or the environment. Since federal law prohibits using a pesticide "in a manner inconsistent with its labeling," any applicator who misuses a pesticide—and in most cases the applicator's employer—is breaking both federal and state law. (See Chapters 7 and 8 for more information.)

Although it is a violation of federal and state law to use a pesticide "in a manner inconsistent with its labeling," it is not always easy to determine whether a particular use is "inconsistent." A statement such as "Do not treat vegetable gardens" is clear-cut: any applicator who treats a vegetable garden with the product is breaking the law.

But label wording is not always that precise. For example, all termiticides have a section on the label regarding "re-treatment restrictions." When there is evidence of reinfestation subsequent to the initial treatment, some termiticide labels state that "the application should be made as a spot treatment to these areas." Based on this statement, would it be a label violation to re-treat the entire structure? Not under federal law, because the EPA considers the word "should" in a label statement to be *advisory* rather than mandatory; and advisory statements *suggest but do not require* that a direction or precaution be followed.

Advisory statements are not enforceable, and statements containing words or phrases such as *should, may, it is recommended that,* or *it is advisable to* are advisory. Mandatory statements are enforceable; they require that certain directions or precautions be followed. They either contain key words such as *must, shall,* or *will,* or imperative expressions such as *do not, use only,* or *for use only by.*

Label Directions by Reference

Sometimes information considered necessary by the EPA for safe application of pesticides is difficult to convey in label language. In such cases, the directions may be incorporated by "reference" on the label. Instructions for protecting endangered species, for example, may be provided through bulletins or pamphlets for specific counties. The EPA also has adopted use by reference for ground water protection and the EPA Worker Protection Standard. Directions by reference on a pesticide label are subject to the same mandates as directions stated directly on the pesticide label.

Reading the Pesticide Label

Pesticide applicators such as commercial pest management technicians and growers should be familiar with the formulations they use, even if they are not involved in the product selection process. As stated previously, formulation directly influences human and environmental hazard potential. Inattention to the formulation being used can mean the difference between a routine application and one that becomes a source of environmental contamination or serious human exposure.

The off-season is a good time to read pesticide labels. Take the time to become familiar with label language and to gain an understanding of label directions and precautions on products that you will be using. But be certain to read the label of each and every product you use—again—at the time of use. Labels do change, even on the same product.

Make reading the label your first priority each and every time you reach for a pesticide product, even if you think you know it thoroughly. Reading, understanding, and complying with label instructions assures effective pest management with minimum risk to human health and the environment.

Read the Label before Purchasing the Pesticide

- Make sure the product is registered for your intended application site.
- Confirm that no prohibitions exist against the use of the pesticide as you intend to use it.
- Review the environmental precautions to ensure that your application site is not excluded due to extenuating circumstances such as the presence of endangered species or sensitive areas such as wellheads, playgrounds, hospitals, or nursing homes.
- Determine that you have the necessary application equipment, and make sure that you know how to operate it properly.
- Review the requirements for personal protective equipment. Do you have it? If not, get it. And use it. If a respirator is necessary, read the instruction manual and learn how to fit it properly. The right fit is essential; anything less could compromise your safety.

Read the Label before Mixing and Applying Pesticides

- Understand how to mix and apply the chemical properly.
- Determine what first aid and medical treatment is appropriate, should an accident occur.
- Follow application directions.

Read the Label When Storing Pesticides

- Know how to store the pesticide properly.
- Understand and take precautions to prevent fire hazards.
- Be sure storage areas are posted properly and locked.

Read the Label before Disposing of Pesticides

- Understand how to rinse pesticide containers properly and dispose of the rinsate correctly.
- Gather all information about how to dispose of surplus pesticides and containers.

Read the Label to Educate Your Employees Provide the following information to employees:

- Where and in what form pesticides may be encountered during work activities
- Potential hazards from toxicity and exposure
- Routes through which pesticides can enter the body
- How long to stay out of treated areas
- What personal protective equipment to wear

MOVING PRODUCTS BETWEEN THE FACILITY AND THE JOB SITE

Fleet vehicles serve as rolling "offices" for pesticide applicators. They transport equipment and materials to the worksite. They take applicators from job to job, facilitating timely, dependable service to residential and industrial clientele. They advertise for the company: good or bad!

Keeping vehicles clean and neat does make a difference—not only in how the public views your *company,* but in how the pest management *profession* is regarded as well. It is important to remember that your vehicles' appearance can generate new customers or turn away old ones: Your vehicles project an image. The care that you devote to maintaining your fleet goes a long way in conveying your commitment to quality service. Conversely, rigs that are stained with pesticides, that have granules and debris in the bed, that carry old and poorly serviced application equipment, or that simply present an unkempt appearance raise doubts about your business and the quality of your work. There are no positive effects from a shoddy fleet of service vehicles, so don't compromise your business potential. Demonstrate your commitment to personal safety and environmental quality, and advertise your company proudly (Fig. 10.14).

Maintain your vehicles on a regular basis, and train your drivers to conduct daily inspections to head off major repairs. Most breakdowns can be prevented through scheduled maintenance and prompt attention to minor problems. It is very simple: Trucks on the move make money, and trucks broken down, don't. A breakdown can cause costly disruption of business and, in some cases, loss of clientele who are unable or unwilling to wait till you are back up and running.

Vehicular liabilities include traffic accidents and/or spills. Even minor fender benders can take a vehicle out of service for weeks. In addition to the inconvenience, accidents can cause bodily injury and trigger numerous other situations:

- Serious financial loss from lost production
- Contamination from spilled fuel, pesticides, and fertilizers—and cost of cleanup
- Undesirable media attention
- Lawsuits

Figure 10.14 The neat appearance of a service truck can generate new customers.

- Damage to your business reputation
- The need to purchase new vehicles
- Destruction of the business if uninsured (or underinsured)
- Increased insurance rates

Vehicular accidents on the road and at job sites do happen; and the more vehicles a company puts on the road, the greater the likelihood of an accident. One way to reduce the frequency of accidents is to train employees to drive defensively. Another is to hire individuals with good driving records: past performance is an indicator of what to expect. Company owners and managers, office staff, and applicators who are trained to work as a team when a chemical spill occurs (whether or not a vehicle is involved) can effectively reduce the severity of damage.

Key elements in managing a safe and productive service fleet include

- routine inspection and maintenance,
- defensive driving,
- reacting to emergencies as a team, and
- the review and critique of past incidents to determine where improvements are needed.

Vehicle Maintenance and Safety Inspections

The purchase of vehicles and pesticide application equipment constitutes a sizable investment: new trucks start at about $30,000; and, in some industries, it is common to pay $150,000 for the latest spray rig. So it is important to recoup the cost, and to do that a company must keep its fleet in good working order and on the road. Vehicles and equipment that are not inspected and serviced routinely are more likely to require costly, time-consuming repairs—and much more likely to require early replacement than those that receive proper care.

You never appreciate the dependability of a truck until it breaks down on the job. It is then that you ask the mechanic, How bad is it? What's it going to cost? and How long will it take to fix it? Depending on what the mechanic says, you're left with two options: trade it in, or scramble to get a loaner so that business won't fall behind.

There are many reasons why fleet vehicles should be pampered:

- New trucks cost $20,000–35,000 so you need to get your money's worth out of them.
- Preventive maintenance can be budgeted; breakdowns can't.
- Companies lose up to $1000 a day when a vehicle is in for repairs.
- Technicians continue to earn their wages and benefits, even when their company vehicles are out of commission.
- Repair costs can be astronomical: $3500 for a new motor; $500 for a new set of tires; and $1000 for repairs to the cooling system. Preventive maintenance costs pennies on the repair dollar.
- Customers become unhappy when you don't show up as scheduled.

The Every Day Visual Inspection It should be the responsibility of every technician to perform a 360-degree, 5-min inspection of their vehicle each morning. These quick inspections can keep the fleet in top shape by catching small problems before they become large and expensive ones. Consider inspecting the following items each morning before the technicians leave on their routes:

- **Oil level** Check the oil when the motor is cold.
- **Coolant level** Make sure the coolant reaches the "cold" level when the motor has been sitting overnight.
- **Brake fluid level** Check the brake fluid reservoir to see if the brake fluid is at the full mark.
- **Transmission fluid level** Look for reddish fluid under a vehicle that has been parked overnight, as an early warning sign of a transmission problem.
- **Tire pressure** Look for the numbers on the tires that represent maximum tire pressures; or check the identification tag on the driver's door, which lists a range of pressures based on the weight that is to be carried by the vehicle.
- **Lights** Turn on all lights and turn signals to verify that they function properly.

Technicians should record their inspections on a check sheet that is turned in daily. Place each inspection in a folder marked with the vehicle's identification number. This will allow you to track how the vehicle is performing over time. Take the vehicle to a mechanic when a technician has to continually add fluid or when leaks develop. Finally, show the inspection reports to your insurance carrier to request a discount for preventive maintenance.

Tip of the Day Do your technicians actually perform the inspections, or do they just fill out the paperwork? One way to check is to tape a $5 bill to the oil stick, then ask the employees if they checked their oil. If everyone says yes, but no one mentions the $5, guess what! Use this example as a training tool.

TABLE 10.1 Comparison of Maintenance Recommendations: Traditional vs. Owner's Manual

	Traditional	Manual
Change oil	3000–4000 miles	7500 miles
Replace coolant	2 years	5 years/150,000 miles
Change transmission fluid	2 years/24,000 miles	50,000 miles
Replace fuel filter	12 months	2 years/24,000 miles
Change pcv valve	30 months/30,000 miles	—
Replace air filters	24 months/24,000 miles	—

Routine Servicing by Professional Mechanics The traditional maintenance schedule should be followed by those who believe that taking preventive measures at closer intervals (e.g., fewer miles or months than indicated by the owner's manual) extends the life of their vehicles. This is a good strategy if your fleet carries heavy loads, tows trailers, or pushes snow or when they are driven consistently under stop-and-go conditions. When the truck is driven under normal conditions, consider using the recommendations listed in the owner's manual (Table 10.1).

There's no reason why the odometer should not reach 200,000 miles. Although many companies have a policy of replacing vehicles with much lower mileage (for better trade-in values), it should be the goal of every business to manage their fleet as if they were going to keep every vehicle for an extended period of time. There's no doubt about it: Vehicles run better, longer if they are inspected daily. This, of course, translates into increased revenue.

A few moments spent checking equipment and supplies can also make a difference. Make sure that you have everything you need for the day's job, including company paperwork, to avoid time-consuming trips back to the office. Examine pesticide containers and application equipment before leaving for the job site, making sure that all lids are on tight and that the equipment is stored securely. This could mean the difference between a productive day's work and an unproductive day spent cleaning up a spill. A sample vehicle and trailer inspection check sheet is provided in Appendix 10.2.

Unannounced Company Vehicle Inspections Many supervisors make impromptu inspections of company vehicles to ensure that employees are maintaining them appropriately. They take into account the overall appearance of a vehicle, checking to see

- if the outside is clean and free of rust;
- if the cab and dash are clean and free of debris such as food wrappers, beverage containers, or disheveled paperwork, clipboards, or pens; and
- if there are unsecured items such as boxes and clipboards (inside the cab) and equipment or pesticides (inside or out) that could pose danger in the event of an accident.

These inspections demonstrate that the company is serious about vehicle maintenance and appearance, and they can signal the need to schedule repairs or vehicle replacement.

Another approach to impromptu inspections is to choose a vehicle at random and have its assigned driver conduct a bumper-to-bumper inspection—then and there, and in the presence of the other drivers—using a form provided by the company. As deficiencies are noted, make sure that all drivers understand the significance of addressing them immediately. Conduct these inspections on a different vehicle each time.

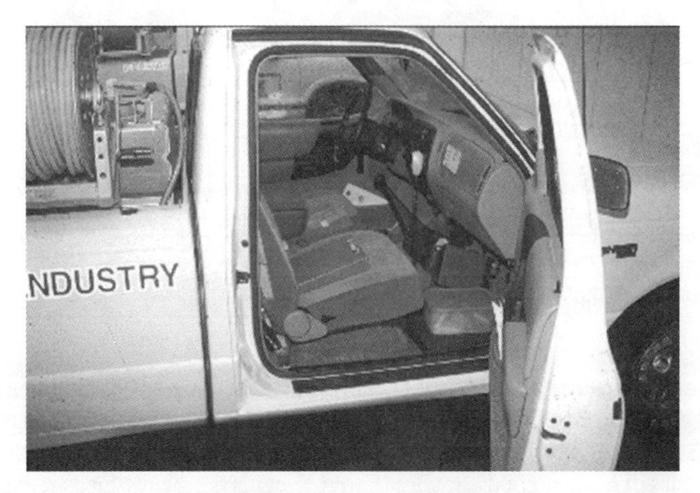

Figure 10.15 An employee who maintains their truck should be rewarded and recognized by the company.

Some companies offer driver bonuses or other rewards for maintaining the best kempt vehicle in the fleet. Incentives reinforce company standards on vehicle maintenance and appearance, and they reward employees who maintain their vehicles in accordance with company expectations.

Recognition for a job well done can be as simple as a gift or plaque awarded during a company meeting or as significant as points toward a monetary bonus at year's end. But whatever the incentive, it must be implemented objectively so that all employees have equal opportunity for recognition (Fig. 10.15).

Loading and Unloading Pesticides and Equipment

Accidents can occur when transporting pesticides even a short distance. Careless transportation can result in damaged containers and spills that might result in personal injury or soil or water contamination. Equipment has been known to fall off moving vehicles or to break as it slides back and forth in the bed of a truck. Consider the following easy-to-implement ideas for loading and unloading pesticides and equipment.

- Install self adhesive, nonslip strips on bumpers.
- Implement and follow rules for heavy lifting.
- Carry only enough chemicals for the day's work (Fig. 10.16).
- Inspect each container before loading, and confirm that
  - — none are damaged;
  - — labels are attached and legible;
  - — all caps are tightly closed and properly sealed;
  - — the outside is not contaminated with pesticides.
- Make sure that a current Material Safety Data Sheet for each pesticide being transported is on-board.

Figure 10.16 Vehicles transporting pesticides should carry only the amount necessary for the day's work.

- Transport pesticides in their original labeled containers.
- Protect pesticide bags from punctures, tears, and moisture during transport.
- Secure all pesticide containers to prevent rolling and sliding (Fig. 10.17).
- Make sure that booms are locked into place and that all equipment is secured; bungee cords make excellent tie-downs.
- Securely attach permanent tanks to the vehicle.
- Depressurize hoses that are not in use.
- Position tanks to protect gauges from damage.
- Do not carry pesticides in the passenger compartment of the vehicle.
- Lock removable or hazardous items into an appropriate container or compartment.

Figure 10.17 Pesticides must be properly secured when they are being transported.

Safe Driving Practices

Safe driving is as important as technical training in the pesticide application business. Employers should obtain employees' (and prospective employees') accident and traffic violation records from the Bureau of Motor Vehicles as a measurement of safe driving skills. These files can be valuable predictors of how safely employees might drive on behalf of the company. But, even if the records are clean, employers should offer driving skills training to each and every employee who drives a company vehicle.

Driver training should be mandatory for each new employee prior to being assigned a company vehicle, and refresher courses should be mandatory for experienced drivers. The training can be as informal as a company meeting on safe driving issues or as elaborate as formal training delivered by a loss control specialist, perhaps someone from an insurance company. There are good videotapes on the subject as well.

The pest management business is hectic. Even the best-laid plans go awry sometimes, and inevitably there are delays caused by road construction, congested traffic, or trouble locating the job address. But it should be a well-enforced company policy that lost time is not to be recouped by driving carelessly; the risk of an accident is too great. Let your drivers know in no uncertain terms that it is preferable to be late for an appointment than to compromise safety in getting there.

To ensure that employees remain conscientious about their driving, companies sometimes have supervisors ride along on "commentary" drives. That is, drivers are asked to talk through whatever they see as they are driving and to vocalize how they are adjusting their driving accordingly. A driver might say that he sees a stop sign up ahead and that he is slowing down to facilitate a safe stop. Or, he might comment that he sees a child playing near the street and is decreasing his speed to ensure the child's safety. He might acknowledge a school zone and his obligation to drive especially slow and cautiously. Such vocalization demonstrates the drivers' thought processes and allows the supervisor to identify areas that need improvement.

Employers also can encourage comments from the public on their employees' prowess as drivers, or lack thereof, by posting the company name, phone number, and vehicle number on every vehicle in the fleet. There also are decals that say, How's my driving? followed by the company name and phone number. Placing this information on all company vehicles for public feedback on how your employees are driving also serves as advertisement for the company. People tend to take note not only when vehicles are moving, but also when they are parked at the job site or at a restaurant at lunchtime.

Tips for Safe Driving There are myriad elements that contribute to safe driving: some we perform routinely, almost unconsciously; and some we don't. Note any among the following that might enhance your own good driving habits and those of your employees.

- Schedule jobs for spans of time instead of on-the-hour; for example, tell the customer to expect you between 9 and 10 a.m. instead of at 9 a.m. You can strive for 9 o'clock, but you have actually allowed yourself 60 min to arrive "on time," thereby eliminating any sense of urgency conducive to unsafe driving en route. But if you see that you cannot arrive within the time frame quoted, call the customer and let him or her know.
- Know how to get where you're going. Use computer software that maps efficient routes and estimates the time it will take to reach each destination. Never try to read a map while driving.

- Train employees on the safety functions of all vehicles they may drive.
- Do not use or allow the use of cellular phones while driving.
- Drive courteously and cautiously at all times.
- Activate turn signals well in advance of your turn—*before braking*. Don't you hate it when the guy in front of you brakes, slows, and *then* signals! Oftentimes, these drivers are already in the act of turning when they signal; a lot of good that does, right? The thing to remember is that signaling tells drivers behind you that you'll be turning soon; therefore, they are prepared for you to brake. Vice versa just doesn't make good (safe) driving sense.
- Be cognizant of the fact that application rigs require more gradual braking than other vehicles, simply to keep chemicals and equipment from shifting. You'll need to signal sooner.
- Accelerate gradually to avoid shifting the load.
- Discuss passing and defensive driving techniques, blind spots, and tailgating. If you hit another vehicle from behind, you're at fault. Always.
- Know how to respond if your tires go off the edge of the road: Don't overcorrect. Pull back onto the highway gradually.
- Know how to drive in bad weather.
- Know what steps to take in the event of an accident and/or a chemical spill.
- Never transport pesticides in the passenger space of a closed vehicle, and never allow passengers or pets to ride with pesticides (e.g., in the bed of the truck).
- Never allow unauthorized persons to ride in company vehicles.
- Establish a written policy requiring drivers and passengers to wear seatbelts.
- Do not allow smoking in company vehicles.
- Drive with headlights on at all times.
- Drive at the posted speed limit.
- Keep the radio at a low volume to facilitate hearing emergency sirens and other drivers' horns.
- Check side and rearview mirrors often to remain aware of conditions around you.
- Use the 4-sec rule to stay a safe distance behind the vehicle in front of you. Fix your eye on a stationary object up ahead (e.g., a building, a stop sign, or a billboard). As the vehicle in front of you reaches it, start counting: one thousand one, one thousand two, one thousand three, one thousand four. If you reach the object before counting to one thousand four, you're following too closely.
- Devote an employee training program to the review of the state driver's manual; a refresher on the rules of the road and traffic signs never hurts.
- Avoid eye contact with other drivers who try to confront you.
- Slow down for rude drivers who cut you off and those driving erratically.
- Stop a safe distance from the vehicle in front of you in traffic; you should be able to see its rear tires.
- If there is room in the parking lot, pull through a parking space into the one facing it so that you can drive forward when leaving, instead of backing.
- Place emergency cones or triangles at both the front and the rear of the vehicle if parked along the street.
- Have rearview mirrors installed on the back of the spray rig.

Figure 10.18 Pesticide spills must be cleaned immediately.

Emergency Response to Pesticide Spills

A pesticide spill can happen to anyone, so preplan how to respond on the contingency that it *will* happen to you. Your state of emergency preparedness will have a direct impact on the severity of the spill, that is, the degree of injury and/or contamination it causes. Develop a written spill management plan. Fill in pertinent phone numbers and keep a copy in your vehicle, at the office, and at home. Be prepared to take proper action in the event of a pesticide spill. See Chapter 13 for more information on emergency planning.

Small Spills Small liquid pesticide spills from leaky application equipment should be cleaned up immediately, no matter where they occur. Confine the spill and absorb the chemical with kitty litter or another absorbent material. Once all of the liquid is absorbed, bag the absorbent and apply it to a labeled application site or dispose of it according to state and federal regulations (Fig. 10.18). Even minor spills should *never* be washed down the driveway into storm drains: Doing so is illegal.

The Team Approach to Dealing with Major Spills Don't panic! Exercising common sense and keeping calm will help you through the worst of spills. To be effective, a company's spill response efforts must be built around all employees—supervisors, office staff, and applicators—working as a team to solve the crisis. But there should be one person on each crew assigned to take charge in the event of an emergency. All employees should be professionally trained and certified to administer CPR.

When There Is a Spill at the Job Site: What the Employee in Charge Should Do

IF A VEHICLE IS INVOLVED, TURN OFF THE ENGINE. This lessens the likelihood of the vehicle moving or catching fire.

ASSESS THE SITUATION. Your actions within the first 15 min will set the tone for what happens over the next few hours.

FIND OUT IF ANYONE IS INJURED. If so, or if fire is a threat, have someone call 911 immediately. Speak to the person loudly and directly, making sure he understands that he is responsible for making the call. Then return to handling the spill itself. Have someone else call company headquarters, advising them of the spill and asking that a member of management be sent to the scene.

DO NOT MOVE INJURED PARTIES UNLESS THEIR LIVES ARE IN DANGER. Moving them could worsen their injuries. Ask the victims their names and see if they know where they are. If they can answer these simple questions, let them know that help is on the way. If the victims are awake but seem dazed, they may be in shock. Administer first aid only if you have been trained to do so; otherwise, it is best to wait for medical help. This is a judgment call that you will have to make on the spot.

IF SOMEONE IS UNCONSCIOUS, CHECK FOR A PULSE. Place your fingers on the side of the victim's neck, or place your ear over the victim's nose and mouth to see if he is breathing. If you cannot feel a pulse or detect that he is breathing, start CPR immediately. If you can detect a pulse, have someone stay with the victim while you continue to deal with the spill.

STOP THE LEAK AT ITS SOURCE IF YOU ARE QUALIFIED TO DO SO. It may be as simple as turning off the pump or shutting down the motor. Other times, rags can be stuffed into ruptured hoses or punctured tanks.

KEEP PEOPLE AWAY FROM THE SPILL. Tell everyone to stand clear, and assign a few people to keep bystanders out of the area while you continue with emergency procedures. If barrier tape is available, stretch it around the perimeter of the spill site to keep people at a distance.

WEAR APPROPRIATE SAFETY EQUIPMENT. Protect yourself, first. Put on gloves, safety glasses, rubber boots, and whatever else is needed to deal with the chemical at the spill site.

CONTAIN THE SPILL. If possible, build a soil or pillow berm to keep the chemical from reaching drains or waterways. It is okay to let it pool on the roadway, turf, or soil because it can be removed easily once the emergency is under control. But if the pesticide reaches surface water, recovery can be quite difficult.

It is important to understand that there are three types of drains: septic, storm, and combined. Septic drains have solid covers, whereas storm and combined sewers have covers with openings to allow water in. If an open drain smells like sewage, it is a combined system; if not, it is a storm drain—and it is extremely important to know the difference.

In a storm drain, the outflow pipe protrudes from the middle to allow sediment, trash, and other objects to settle to the bottom of the catch basin; as the basin fills, the water rises and flows into the pipe. So if the quantity of chemical that spills into a storm drain is smaller than the capacity of the catch basin, the spill is totally contained therein. The chemical can then be pumped out of the basin and applied to a site listed on the pesticide label.

If the catch basin is partially full of water prior to the spill, or if the quantity of chemical spilled is greater than the basin capacity, the spill overflows into the next catch basin and is possibly contained there. The likelihood of containment is great when catch basins receive a spill.

In septic or combined drains, the pipe runs along the bottom; thus, any pesticide that enters it comes into contact with moving water. The only way to stop its advance is to determine the direction of the flow and block its entrance to the treatment facility.

Do not enter the sewer vault to block the line. Sewers are dangerous. No one should enter a sewer unless they have completed a formal confined-space training program and unless they are wearing safety equipment. You might be able to block the line by dumping sand or other absorbents by the outflow pipe, but success is doubtful. Notify the city immediately if a pesticide spill enters the sewer system. Prompt notification may allow waste managers enough time to block the flow and to prevent the pesticide from entering their biological filtering system.

BE RESPECTFUL TO EMERGENCY RESPONDERS. As the employee in charge at the spill site, you likely will be the first to deal with police officers, firefighters, and paramedics. You should cooperate fully, answering any questions they may have. Make sure to have Material Safety Data Sheets on the spilled chemical available for emergency personnel to review. It is important to know the diluted concentration of the spilled chemical as well, since emergency actions will depend on it. Emergency responders may insist that you move out of the cordoned area if you have not received HAZWOPER training.

The incident commander (usually one of the emergency responders) is in charge at the scene, upon arrival. His or her job is to deal with personal injuries and to prevent environmental contamination while ensuring the safety of everyone involved. It is up to him or her to decide what actions to take to protect the public, and he or she may or may not solicit your input. You should know, however, that the incident commander's first instinct may be to wash the spilled pesticide into drains. If that is the case, you should politely ask the incident commander to reconsider, pointing out that if the pesticide is washed down a drain it is likely to contaminate surface water or the sewage treatment plant. Ask the commander to wait for a company management representative to arrive on-site (or to speak with him or her on the phone) before implementing a decision to wash the pesticide away. Note that fleet vehicles should be marked with a decal that reads, *Please do not spray or apply water to this vehicle. It contains pesticides* (Fig. 10.19).

Figure 10.19 An example of a sign on a service vehicle asking emergency responders not to spray the truck with water.

DO NOT ADDRESS THE MEDIA. You do not have the time—nor, perhaps, the authority—to speak with reporters. It is of utmost importance that you devote your undivided attention to managing the spill.

Your focus on the problem at hand is essential, but nevertheless you should not respond to media requests by saying, *No comment.* It is advisable to state that a member of management is en route to the scene and will be glad to answer any questions. Suggest that the reporter speak to the incident commander; then excuse yourself to attend to details, and walk away.

DO NOT ADMIT GUILT, DO NOT STATE THAT YOU HAVE AMPLE INSURANCE, AND DO NOT INDICATE THAT YOUR INSURANCE WILL PAY FOR REMEDIATION OF ANY PROBLEMS THAT RESULT FROM THE SPILL. These topics should be addressed by management personnel—and even *they* cannot speak for the insurance company. Only your insurer can determine how the situation will be handled, according to policy provisions. (See Chapter 14 for more information on insurance.)

When There Is a Spill at the Job Site: What the Office Staff Should Do If you work in the office of a pest management firm, you need to know what to do if you receive an urgent call from an employee in the field. Your company should provide you with an emergency checklist (Appendix 10.3) to guide the conversation.

OFFER ASSISTANCE. Keep the caller calm and let him know that you are sending management personnel to the site.

COLLECT AS MUCH INFORMATION AS POSSIBLE. Ask questions, such as:

- What is the location?
- Is anyone injured?
- Has someone called 911?
- Are there any emergency responders already on the scene?
- Has anyone been taken to the hospital? (Fig. 10.20.)
- What was spilled and how much?
- In what direction is the spill flowing?
- What are the possible receptors of the spill?

LOCATE PERTINENT MSDS IN THE COMPANY FILES. While talking to the employee on the phone, write down the name of the spilled chemical and ask a co-worker to pull the MSDS. Do this even if the employee at the scene already has a copy, just in case it is unreadable for some reason, or in case an extra copy is needed at the spill site (Fig. 10.21).

CALL 911 TO CONFIRM THAT THEY HAVE BEEN CONTACTED. Once you hang up, call 911 to verify that they have already been contacted. If they have not been notified previously, you can relay the information you received from the employee on-site.

CALL THE MANUFACTURER. A representative of the manufacturer will know the most about dealing with human or environmental exposure to the spilled chemical. Have someone else in the office look up the chemical company's 800 number on the MSDS and call it while you

Figure 10.20 It is important to know whether the accident caused personal injury.

keep the employee on the line to answer questions posed by the company representative. Accurate answers may be critical to appropriate emergency assistance on behalf of the manufacturer. Make sure that the caller records the name of the representative contacted.

When There Is a Spill at the Job Site: What Management Should Do

GO IMMEDIATELY TO THE SPILL SITE. A member of company management must go immediately to relieve the employee in charge at the spill site. Your presence at the site is essential if consequences of the spill require calling 911.

TAKE SUPPLIES FROM THE OFFICE. Take MSDSs and labels as well as whatever spill control items you have, for example, shovels, pillows, drums. You should have a spill kit and a spill response drum ready at all times to minimize reaction time during an emergency.

Figure 10.21 Material Safety Data Sheets should be stored alphabetically by name.

SUMMON OTHER EMPLOYEES TO THE SCENE. Instruct the office staff to direct other company employees to the spill site. Their rigs should be equipped with spill recovery materials, and their on-board tanks can be used to receive waste liquid pumped from the spill. Their presence could facilitate the all-important first step in the cleanup activity.

BE PREPARED TO DEAL WITH THE AUTHORITIES. You will have to interact with multiple agencies, the media, and bystanders. Accident scenes are crowded with authorities, each trying to satisfy their own regulations. Identify yourself to the incident commander and let him know that you will take over for the employee in charge and field any questions he might have; then ask to speak to the employee for a briefing on what has occurred. Ask the employee to describe what took place and what has already been done. This firsthand account will allow you to deal with the problem more efficiently and more effectively.

CALL THE MANUFACTURER'S EMERGENCY 800 NUMBER FOUND ON THE MSDS. Even if your office staff has already called the 800 number, call again to provide details that may have been left out during the first call. The manufacturer's representative will ask if someone has been taken to the hospital and, if so, they will have a doctor who is familiar with the product and its effects call the hospital to expedite care of the injured. The representative that you speak with may also be an excellent contact person for the incident commander.

ADDRESS THE MEDIA. It is important not to ignore reporters. Explain early on that a statement will be forthcoming once you have all of the facts. And follow through. Give them the information they need to do their job: to inform the public of what is taking place in the community. Be prepared to answer all questions in an honest, forthright manner. Stress that you are working with local emergency responders to resolve the problem as quickly as possible.

CALL YOUR INSURANCE CARRIER. Have someone from your office or an employee on-site call your insurance company. The company will want to send personnel to the site to assess the situation. But any conversation between you and the insurance representative concerning what is covered and the dollar amount of your coverage should be confidential. Conduct the discussion where you are certain that it cannot be overheard. Do not offer such information to the media.

CALL OTHERS WHO NEED TO BE APPRISED OF THE SITUATION. Refer to the emergency call list in Appendix 10.3 that was compiled during your preplanning phase. The phone numbers should be checked periodically to make sure everything is current.

NOTIFY THE APPROPRIATE AUTHORITIES. Reporting requirements vary with state and federal regulations and with the product and amount spilled. But failure to report a spill when you are required to do so is a serious offense that carries significant penalties. Some companies and regulatory agencies have a 100% reporting policy: If it is spilled, it is reported.

Know who to call in your state when pesticide spills occur. The person who places the call should be someone who has the authority to make decisions, assign personnel, and allocate money to deal with the spill.

CLEAN UP THE SPILL. The local, state, and federal agencies that respond to the spill will not help with site remediation. If cleanup of the spill is beyond your capabilities or training level, call a professional contractor. Have an arrangement in place with a local contractor who will respond to the spill with dump trucks, backhoes, and other heavy equipment, as needed.

PROVIDE INFORMATION TO REGULATORY AUTHORITIES AS ACCURATELY AS POSSIBLE. Include the following:

- Name and telephone number of your contact person
- Exact location and time of spill
- Identity of substance spilled
- Estimated amount of substance released
- Where the substance was released: air, land, or water
- Potential for off-site movement
- Anticipated human and/or animal health risks
- Injuries that required medical attention
- Response of the company up to the time 911 was called
- Remedial action planned

TAKE GOOD NOTES AND KEEP THEM. Under most circumstances, if your company is involved in a spill you are required to submit a report to the agency responsible for ensuring that the problem gets resolved. And your insurance company will need to collect information for use in settling claims or for litigation purposes. Always prepare a written incident report. Record the names of everyone you speak with concerning the incident, along with the date, location, and approximate time that each conversation took place. Also, keep good records of all steps involved, from the cause of the spill to final resolution of the problem.

PESTICIDE STORAGE FACILITIES

Commercial companies may store pesticides in the same building that houses the main business, in an equipment storage building, or in a building used solely for pesticide storage. But, no matter what type, all storage facilities command certain precautions.

The Storage Facility: The Outside View

In many localities, pesticide storage facilities are constructed according to standard building codes. But there are many additional considerations when building a storage facility.

In general, storage buildings should

- be located as far as possible from structures outside your property (e.g., neighbors' houses and outbuildings);
- be separate structures designated for pesticide storage;
- be diked to prevent material from moving into areas of the facility used for offices, etc., if a spill occurs;
- be situated so that runoff from spills and leaks will not reach drains or contaminate surface water, wells, etc., and where spills are easily contained;

- be located in reasonable proximity to emergency response services;
- have their own water source, with protection against back siphoning;
- be located where there is ample space for parking company vehicles;
- be accessible only to authorized individuals;
- not be close to any roads;
- not be constructed in areas known to flood.

Security against Theft Secure the storage facility (separate building, room, cabinet) against theft, vandalism, and unauthorized access. Nearly one in ten agricultural retailers has experienced a pesticide theft—at an average loss of $29,000! Stealing pesticides has become a lucrative business since many of the newer, more popular products come in small, expensive packages. Thieves seem to target pesticides in small units that retail for perhaps $1,200 each; so a good heist may mean simply filling the front seat of a truck!

It must be assumed that pesticide thieves have plenty of customers willing to purchase stolen products, either knowingly or unknowingly. Once stolen, pesticides are nearly impossible to trace, so don't make it easy for them. Consider the following:

- Have all employees turn in keys when they leave the company, or change the locks.
- Install perimeter fencing to deter entry.
- Make sure that the storage facility, itself, and its doorways, loading dock, and parking lot are well lit, as well as entrances to and exits from the property.
- Keep the grass mowed and weeds controlled around the facility to minimize hiding places for prowlers.
- Keep high-dollar chemicals out of sight of visitors to your facility.
- Store expensive pesticides in a separate, locked room, away from exterior walls.
- Use permanent markers to write your name and the date received on all pesticide containers.
- Request law enforcement officials to patrol your area.
- Ask customers to report suspicious activities to the police.
- Install a security camera.

Post Signs on Buildings Post the storage area with warning signs labeled "DANGER—PASTICIDES—KEEP OUT" on walls, doors, and windows. Some companies put the names and phone numbers of employees on signs to indicate to emergency responders who needs to be called during an emergency. Signs should be legible at least 50 ft from the building (Fig. 10.22).

Exterior Storage of Drums Avoid exterior storage if possible. But if you must store drums outside, place them on their sides to avoid accumulation of rainwater in the top or bottom recessed areas. Be sure they are empty.

Labeling of Outside Storage Tanks All tanks should be labeled with contents: water, pesticides, anhydrous ammonia, propane, diesel fuel, etc.

Figure 10.22 Warning signs are an important risk management tool.

Managing Pesticide Water in Containment Storage Areas Rainwater is clean when it first hits your concrete containment area. It becomes contaminated when it contacts or accumulates in areas where chemical spills have not been cleaned. Clean water does not present a disposal problem, but water contaminated with any type of chemical (pesticides, fertilizers, fuels) can. Cleaning spills immediately and preventing water from accumulating prevents contamination. But if contamination does occur, proper removal and disposal are critical.

Containment capability for pesticide spills is becoming an industry standard and, in some cases, a regulatory requirement. Outdoor containment areas are often roofed to allow more flexibility in handling minor spills and leaks, and roofing nearly eliminates having to manage contaminated rainwater. The following are guidelines for pesticide containment areas:

- The areas immediately surrounding pesticide storage facilities should be diked or otherwise contained to prevent chemicals from escaping into the environment.
- When pouring concrete, consider the following parameters and ask the commercial contractors if the following criteria will meet your specific needs.
  a. 4000–4500 psi concrete (strength necessary to withstand a minimum of 4000 pounds of compression): 6 bags per yard cubed
  b. 3–7% air entrapment (need air pockets inside for contraction, which helps prevent cracks: with too much air, concrete cracks and leaks easier; without enough air, it also cracks)
  c. 6″ minimum to 12″ maximum thickness
  d. Low water content (if it is too runny it affects strength and quality, i.e., it deteriorates faster)
  e. 4- to 5-inch slump (how far it drops: the more it drops, the runnier the material and the weaker the concrete)
  f. Control joints (to control cracks)
  g. Rebar reinforcements, not mesh

h. Concrete inspection (must meet specifications)
i. 95% compaction on subgrade

- It is always best to apply a coat of sealant to concrete on which pesticides will be stored.
- Once a year, fill the sump nearly to the top. Mark the water level with chalk and watch to see if it drops. If it does, the sump area needs to be sealed.
- Keep containment and mixing areas clean. Even one drop of chemical in a containment area is too much, so always take the time to *clean up spills when they happen—not later.*
- Always place drip pans under loading and unloading valves.
- Make sure that individuals who deliver products to your facility do not leave chemical tracks. If the driver is spilling chemical during the process of unloading, insist that he quit immediately; call his company headquarters and ask that they send another driver to unload the chemical. Make it clear that you will take your business elsewhere if they cannot do better.
- If water accumulates in your chemical containment area, the simplest solution is to reuse it, that is, use the water as a make-up solution to fill spray tanks. Clearly label the container and make sure that the contents are used as soon as possible, according to label directions.

Winterization of Bulk Tanks October and November are busy months for commercial agricultural dealers as they receive pesticides and fertilizers for application the following spring. They take advantage of lower prices by accepting delivery during the off-season. Such early deliveries also allow dealers to stock volumes of material in anticipation of customers' early requests.

The downside of early delivery is that chemicals often are held outside during the winter. But a thorough inspection, in advance, can take the worry out of outdoor winter storage. Consider the following:

- Drain all hoses. If they absolutely cannot be drained, a good alternative is to put a small amount of antifreeze in them—but do not forget to drain them in the spring before pulling chemical from the tank.
- Drain pumps to prevent the buildup of sticky residues. Antifreeze keeps seals moist.
- Bring pumps and other equipment indoors to extend their working life.
- Fit locks on valves with rubber caps and/or spray with rust inhibitors.
- Lock and secure external site gauges on tanks because winter winds can cause them to break. Replace exterior site gauges with internal tank floats.
- Inspect and repair tanks and plumbing. Inspect tank hatch gaskets, main valves, connecting nipples, plugs, hoses, and tank exteriors.
- Replace mild steel valves with stainless steel because some fertilizers can weaken mild steel.
- Make sure each large tank containing a pesticide is labeled with the product name and the manufacturer's EPA establishment number; and attach the product label. The contents must be stated in gallons on the tank when it is filled.
- Tanks holding water should be so marked: WATER.

- Visually inspect the inside of each tank. Be certain that anyone who enters tanks has been trained in accordance with OSHA's confined-space regulations.
- Calculate the amount of pesticide or fertilizer in each tank, and recalculate the amounts on a periodic basis. Periodic measurements can be used as indicators of leaky tanks when snow cover prevents visual detection.
- Remove all debris from within the dike. Thoroughly wash the containment area (e.g., dikes and loading pads) so that the accumulation of clean water from melting snow can be pumped onto the ground.
- Repair and seal any cracks in the containment base or wall.
- Remove standing water to minimize freezing and thawing damage.
- Document tank and pad inspections, repairs, and maintenance on a weekly basis, and keep them as part of the permanent file.
- Educate employees, during the winter, on operational changes, spill control, spill management contingency plans, and company policies.
- Properly agitate all chemicals before selling them in the spring because of potential product separation.
- Make sure that each bulk fertilizer and pesticide storage location maintains, on-site, a bulk repackaging agreement with each product registrant.

The Storage Facility: The Inside View

A major component of good management practices is the safe indoor storage of pesticides. Whether maintaining small amounts of pesticides in a locked cabinet or large inventories at highly sophisticated sites dedicated solely to chemical storage, it is important to limit the probability of accidental human or environmental exposure. Careful attention to your pesticide storage area will decrease the potential for accidental spills, environmental contamination, and disposal of products past their shelf life (Fig. 10.23).

The cabinet area or building designated for pesticide storage should be well planned and maintained. It should remain locked to all unauthorized persons, especially children. Only those individuals who perform an active role in the pesticide application process should

Figure 10.23 Pesticide storage areas should be locked to prevent theft.

have access to the storage site, and they should be required to relock the facility upon departure.

A well-planned storage unit should contain barriers (such as dikes or other methods of containment) to prevent accidental leaks or spills from becoming a source of contamination to ground and surface water, soil, or wildlife habitats. From a monetary perspective, your pesticide storage facility represents protection against economic loss: a safeguard for your chemical and property investment. Good inventory control prevents overpurchase, prolonged storage of outdated pesticides, and the need to dispose of canceled materials. Finally, a properly maintained storage site can be instrumental in assuring the smooth transfer of property titles.

It is best to build and/or designate a separate building specifically for storing large quantities of pesticides. If a separate facility is not an option, a precise area within an existing building should be specified for pesticide storage. The following guidelines and those in Appendix 10.4 will help ensure safe and environmentally friendly storage of pesticide products.

- The interior of the storage area
  - — should be well lighted and dry;
  - — should have an impervious cement floor to facilitate cleanup of spills (floor must be maintained to prevent cracking);
  - — must have sealed floors if they are concrete (because concrete is porous);
  - — should not contain floor drains or sump pumps (any existing floor drains should be capped);
  - — should be equipped with exhaust fans and timers to prevent vapor accumulation and heat buildup (fans should be vented so that no people, animals, or plants are exposed to the fumes);
  - — should be insulated to help maintain an even room temperature. Pesticides should never freeze or become excessively hot. Specific temperature information generally is provided on the pesticide label; the range normally recommended for liquid pesticides is 40–100°F;
  - — should contain metal shelves (with lips) for storing chemicals off the floor. Wooden shelves without trays or liners are unacceptable because they can absorb spilled pesticides; large metal drums and nonmetallic containers should be kept on pallets);
  - — should include an area for storing empty, properly rinsed containers awaiting disposal;
  - — should accommodate the shelving of solids above liquids.
- Pesticides should always be stored on the ground floor. Buildings used for pesticide storage should not contain office space unless the pesticides can be completely isolated and good ventilation maintained.
- Store pesticides in their original container, with the original label attached. However, if a pesticide container is leaking, transfer the chemical to a sturdy new container that can be sealed. Attach the original label to the new container, if possible; if not, label the new container with specific information immediately.
- Store liquid pesticides and highly toxic pesticides (those that carry the signal word DANGER on the label) on low shelves to minimize the potential for exposure if the containers are broken or begin to leak.

- Containers should not extend beyond the edge of the shelving.
- Separate pesticides by classification (herbicides, insecticides, fungicides, etc.) within the storage facility to prevent cross-contamination and to decrease the likelihood of accidental misuse.
- Separate flammable pesticides from other pesticides.
- Never keep seed, fertilizer, feed, drinking water, veterinary supplies, protective equipment, or foodstuff in a pesticide storage or display area.
- Purchase only the quantities of pesticides required for a single season to minimize the need for off-season storage.
- Keep the storage site neat and tidy. Pesticide handlers must be able to
  - — see pesticide labels,
  - — detect leakage or corrosion, and
  - — get to leaks or spills to clean them up.
- Store dry products above liquids to prevent contamination.
- Store protective equipment and clothing in a nearby location that provides immediate access but is away from pesticides and their fumes, dusts, or possible spills.
- Provide an immediate supply of clean water, and have an eyewash dispenser immediately available for emergencies. Soap and a first aid kit are also necessary.
- Establish procedures to control, contain, and clean up spills. Familiarize and train everyone with the procedures.
- Provide tools (shovel, broom, dustpan) and absorbent materials (clay, sawdust, shredded paper, kitty litter) to clean up spills (Fig. 10.24).
- Mark pesticide containers with the date of purchase and rotate inventory to ensure that the oldest material is used first.
- Keep
  - — an accurate and up-to-date stored pesticide inventory (Fig. 10.25),
  - — a file of product labels available for reference,
  - — a file of Material Safety Data Sheets,
  - — a floor plan of the storage facility (showing the exact location of pesticides), and
  - — emergency phone numbers (police, fire, ambulance, poison control center) in the storage area and in the office.
- Keep pathways and aisles clear.
- Clean up wet areas immediately.
- Develop a fire emergency plan in consultation with the local emergency planning committee, fire and police departments, and your local sheriff. Notify the appropriate officials of the types of pesticides and quantities stored (See Chapter 13).

Indoor Storage of Pesticides: Protection from Extreme Temperatures Assess your storage facility at the end of the application season and take an inventory of the pesticides you have on hand. Make sure that all products are protected from freezing— which could easily translate into wasted inventory. Exposure to extreme temperatures, hot or cold, can adversely affect pesticide performance. Product containers that have been in storage for a long time are especially vulnerable to bursting or tearing when exposed to extreme temperatures, thereby increasing the likelihood of spills.

Figure 10.24 All pesticide application businesses must have materials on-site to clean up pesticide spills.

The temperature at which a pesticide freezes depends on its formulation, its solvents, and its inert ingredients: some products freeze at 32°F; others freeze at higher or lower temperatures. Read the label of each pesticide that you have in storage and identify a temperature range that would assure safe storage overall. Keep in mind the following points to minimize pesticide storage concerns:

- Estimate needs as closely as possible and purchase only the amount of pesticide needed for one application season.
- Return unopened containers to the dealer or distributor for a refund.
- Ask your dealer or distributor if they will store your pesticides over the winter.

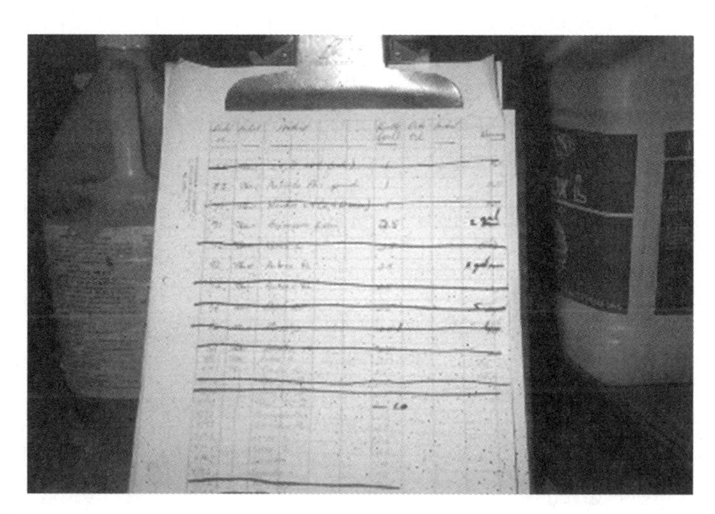

Figure 10.25 An up-to-date inventory of pesticides serves many useful functions for a company.

- Rotate your stock on a regular basis, always using leftovers and the oldest containers first.
- Store pesticides away from extreme temperatures: $>100^\circ$F and $<30^\circ$F.
- Keep pesticides out of direct sunlight to avoid overheating and ultraviolet degradation.

Fire Prevention Strategies Pesticide-related fires are dangerous and pose environmental risks. When a pesticide fire is extinguished with water, the area becomes flooded with pesticide-contaminated water that can enter drains, ditches, streams, creeks, or rivers, making cleanup procedures complicated, lengthy, and costly.

Being prepared to deal with a fire is critical (See Chapter 13 for more information on dealing with emergencies). Once a fire occurs, lack of preparedness places you at the mercy of those responding—a position that most owners find undesirable. Examples of proactive steps that can minimize damage and adverse repercussions are listed next.

- Learn from those who have experienced a fire; ask what advice they have to offer. Experience is the best teacher.
- Store pesticides away from your main building.
- Seal floor drains and add curbing to prevent water and chemicals from leaving the building.
- Remove combustible debris from your pesticide storage and shop areas.
- Ban smoking in and near buildings that house pesticides.
- Install an adequate number of smoke alarms, based on the square footage of your facility, and space them appropriately.
- Keep fire extinguishers fully charged and perform routine checks to assure their readiness; make them available throughout the facility.
- Train employees in the proper use of fire extinguishers.
- Instruct employees about when to put out a fire and under what conditions they need to evacuate the premises.
- Implement a fire emergency plan and train employees on how to react to a fire. Retrain all personnel, at least annually, on safety procedures and the steps involved in evacuation.
- If you have employees who are trained and authorized to fight a fire in your facility, caution them against toxic fumes and smoke that may be released into the air; likewise, caution emergency responders upon their arrival at the scene of a fire. For specifics, periodically review the MSDS for each product stored.
- Install fire alarms that are linked directly to the closest fire department.
- Provide local fire departments with an inventory list of products stored on the premises.
- Invite your local emergency planning committee and fire department to visit your site, annually, to share their expertise on fire prevention strategies. This also presents a prime opportunity for you to familiarize emergency responders with the layout of your facility, which could pay big dividends in the event of an emergency on the premises.
- Videotape your inventory and personal property, periodically, for documentation purposes.

HANDLING AND APPLYING PESTICIDES IN THE WORKPLACE

Avoiding the misapplication of pesticides is crucial. All pesticides can be, to some degree, harmful to humans, wildlife, plants, and other nontarget organisms. But application strictly according to the product label prevents most problems. Deviation from the label, that is, *misapplication,* is more likely to yield adverse health and environmental effects.

Pesticide Drift

Drift is the movement of a pesticide away from the application site to a nontarget area. Preventing pesticide drift is the responsibility of the applicator, and when it occurs it is considered a misapplication under state and federal pesticide laws.

Drift is most common with the use of power sprayers for outdoor applications, where wind can blow the spray or dust into nontarget areas. Other factors such as temperature, humidity, pressure, volatility of the product, droplet size, and additives also contribute to pesticide drift. Drift can occur during indoor applications as well, where fans, air conditioners, and breezes from open windows create an air current strong enough to move the pesticide off-target.

Outdoor pesticide drift can result in plant, animal, and human exposure in areas adjacent to the application site. It can deliver damaging amounts to nearby plants such as flowers, shrubs, and trees. Drift can deposit pesticide quantities sufficient to result in illegal residues on garden vegetables. It can contaminate things such as toys, laundry on the line, pet bowls, beehives, and fish ponds; in addition, outdoor applications can drift through screened windows and expose people indoors.

The applicator must know and understand the product label and make the decision whether or not to apply a pesticide; factors such as wind speed and direction, type of application equipment, and proximity of the target site to sensitive areas must be considered. If the applicator decides to make an application, it is his responsibility to select the best method, based on prevailing conditions.

Guidelines for Reducing Spray Drift

- The top priority is to avoid spraying when the wind is blowing toward sensitive areas.
- Do not spray outdoors when it is windy. Although some labels state a maximum allowable wind speed, most do not. It is up to the applicator to decide whether or not to spray; if drift occurs, the applicator is liable no matter what speed the wind was blowing at the time of application.
- Consider other weather conditions: the higher the temperature, the greater the risk of drift; the lower the humidity, the greater the risk of drift.
- Use the lowest application pressure possible.
- Increase the spray volume.
- Lower the spray boom.
- Know the volatility of the product. Volatility increases as temperatures rise.
- Choose a nozzle that produces a coarse spray. The larger the droplet, the quicker it falls and the less likely is drift to occur.
- Consider adding a drift-reduction agent (thickener) to your spray tank.

- If possible, choose a formulation and an application method that minimize drift potential.
- Look out for pesticide-sensitive sites such as beehives; lakes, streams, and ponds; schools and playgrounds; houses; hospitals; crops; pets; landscape plants; and goldfish ponds. Think about the consequences of misapplication, and act accordingly.
- When power-spraying in attics, crawlspaces, and other indoor areas, check for air intakes systems and turn off all ventilation during treatment.

Pesticide Runoff

Runoff occurs when a pesticide travels (usually downhill) to a nontarget area. It can occur if a pesticide is overapplied or if it rains shortly after treatment. If pesticide runoff reaches lakes, ponds, or streams, the results can be devastating to certain forms of aquatic life.

Some pesticides are more toxic to fish than others. Whether or not a pesticide will harm fish or other aquatic life depends on

- its toxicity,
- the duration of exposure,
- the amount (and concentration) of pesticide that enters the water,
- how quickly the pesticide breaks down in water, and
- the specific organisms involved.

Applicators must always check the pesticide label for statements under *Environmental Hazards.* Some pesticides have statements such as *This product is extremely toxic to fish; Do not contaminate water when disposing of equipment wash waters; Do not apply directly to any body of water;* or *Use with care when applying in areas adjacent to any body of water.*

Even if pesticide runoff does not kill fish directly, it can affect them indirectly by killing aquatic plants and insects that they depend on. When aquatic plants are killed, fish lose habitat or food supply, or they may suffocate from the loss of dissolved oxygen that plants provide. Fish can fall victim to secondary poisoning when they feed on insects that have been in contact with pesticides.

Pesticide Precautions near Ponds, Lakes, or Streams

- Consider nonchemical methods of control before deciding to apply pesticides.
- Choose a product that is less toxic than other pesticides to fish and other aquatic organisms. Check the product label for warning statements on environmental hazards.
- Apply the product responsibly, according to the label. Do not exceed the labeled application rate. Mix only the amount needed for the job. Follow application precautions regarding weather conditions.
- When applying a pesticide near water, check the label to determine if an untreated buffer strip must be left between the application site and the stream.

- Avoid pesticide drift and runoff toward surface water. Do not spray in windy or wet conditions conducive to movement of the pesticide off-target. Check the weather. Don't spray outside if rain is forecast.
- Do not clean or empty spray equipment where pesticide rinsate could enter a pond or stream.

Handling a Fish Kill Complaint: Consider the Following Scenario The Browns' home was built on a hill overlooking their very own pond—their getaway spot. Fun and relaxation was just a cast away. Needless to say, the Browns were astounded when they took a walk down to the pond and discovered dead fish floating on the surface and lying along the bank. Their whodunit questions brought them face-to-face with the grower whose farm field bordered the pond. They told him that they saw a pesticide rig in his field just a few days back and that they suspected his farm chemicals caused the fish kill. They asked the grower if he intended to reimburse them for the dead fish.

One option for the grower would have been to cry uncle and open his checkbook. But he realized that the recent pesticide application in his field might *not* have caused the fish kill. Some pesticides do kill fish, but most don't. And even if the pesticide he used was toxic to fish, it still would have had to move from the field to the pond somehow. The most likely avenue would have been runoff during a heavy rain; and, even then, the terrain would have to have favored drainage into the pond.

In a situation like this, a better option than immediately pleading guilty is to go to the pond to see for yourself if there are any clues on what might have caused the fish kill. Keep an open mind: your pesticides may or may not have killed the fish. The fish could have died from inadequate dissolved oxygen in the water; from a petroleum spill or algae bloom toxicity; from fertilizer runoff, fish diseases, or parasites; from septic system discharge; or from an aquatic weed control application.

Take detailed notes, ask lots of questions, and videotape the site: a lawsuit is always a distinct possibility. The following should be noted:

- Name, address, and telephone number of the pond owner
- Location of the pond
- Name, address, and telephone number of the person who discovered the fish kill, if different from the owner
- Date and time the fish kill was discovered
- Preceding weather conditions: recent temperatures, cloud cover, precipitation, and wind speed and direction
- Date and time that you visited the site
- Approximate number of dead fish
- Average size of the dead fish: small or large
- Species of the dead fish and those that are stressed
- Unusual appearance of the dead fish (e.g., flared gills, open mouths, curved spines)
- Unusual behavior such as fish swimming at the surface or jumping onto the bank, snails climbing onto vegetation to get out of the water, or tadpoles gulping near the surface
- Color and odor of the water

TABLE 10.2 Determining Probable Causes of Fish Kills

Criterion	Symptoms of Oxygen Depletion	Symptoms of Algal Bloom	Symptoms of Pesticide Toxicity
Fish behavior	Gasping and swimming near the surface	Erratic swimming	Erratic swimming
Size of fish	Large fish killed first	Small fish killed first	Small fish killed first
Species selectivity	None if oxygen low; carp and bullheads may survive partial depletion	None; all species affected	Usually; one species killed before others
Time of fish kill	Night and early morning hours	Bright sun (9 A.M.–5 P.M.)	Any hour, day or night
Plankton abundance	Algae dying	Abundance of one alga species	Herbicide may kill algae
Dissolved oxygen	< 2 ppm, usually less than 1 ppm	12–14 ppm	8–10 ppm
Water pH	6.0 –7.5	9.5 and above	7.5–9
Water color	Brown or gray or black	Dark green brown or golden	Normal

Source: Adapted from Field Manual for the Investigation of Fish Kills, 1990. Fish and Wildlife Service, NTIS.

- Water pH and amount of dissolved oxygen in the pond
- Direction of drainage from the treated field: Does the field that was sprayed drain into the pond?
- Other fields that may drain into the pond

Use Table 10.2 as an aid in determining probable cause(s) for fish kills.

Fish and Water Samples Take fish and water samples to determine whether the pesticide applied to your field did actually enter the pond. Document when, how, and where the samples are taken.

FISH SAMPLES Collect as many dead and dying fish species as possible. Wrap each fish in aluminum foil with the dull side toward the fish; put the samples on ice immediately.

WATER SAMPLES Collect 1 gallon of water in glass bottles, preferably amber-colored bottles. Rinse the bottles with pond water before taking the samples. Place samples on ice, out of sunlight. Call the land grant school in your state, the state department of agriculture, the manufacturer of the product, or a private firm to have the samples analyzed.

Making a Commitment to Pay It's an absolute requirement to inform your insurance carrier of the situation; failure to do so might nullify your coverage. Your insurance policy states that you cannot tell anyone that the insurance company will pay to settle the claim. Let the insurance adjuster do his job; if you promise payment of a certain amount, consider it out of your own wallet.

Use common sense when making pesticide applications near water. Follow label directions and watch the weather, but do not jump to conclusions just because fish are floating in a nearby pond after your application.

Protection of Ground Water

Ground water is the source of drinking water for approximately half the households in the United States. It can become contaminated with pesticides when rain carries dissolved chemicals down through the soil in a process known as *leaching*. Pesticide contamination also can occur when ground water is backsiphoned from pesticide tanks; when chemicals enter wells during termite treatments; or when pesticides, particularly concentrates, are not disposed of properly. Decontamination can be difficult or impossible.

Pesticide applicators can protect ground water from potential pesticide contamination by following these guidelines:

- Identify soil type. Coarse or sandy soils low in organic matter pose the greatest risk for leaching.
- Identify nearby water sources. Treat carefully, if at all, around wells, cisterns, springs, canals, streams, and other routes to ground water.
- Know the depth of ground water in the general area. It can be a few inches or hundreds of feet below ground. Check with your Cooperative Extension Service or the local health department.
- Choose pesticides least likely to leach. Check the label for special precautions and warnings about ground water contamination.
- Measure and calibrate carefully. The more pesticide applied, the greater the risk. Mix only the amount needed.
- Avoid spills and backsiphoning. Use caution and a backflow preventer when filling tanks.
- Never drop the hose into the tank when filling because there is always the possibility that the pesticide could backsiphon into the well or city water supply. Air gaps on spray equipment are essential.
- Apply at the right time. Do not apply before rainfall or irrigation or when soil is saturated or frozen.
- Dispose of pesticides and containers properly. Triple-rinse containers and deposit the rinse water in the spray tank.

Avoiding Secondary Poisoning

Secondary poisoning occurs when predators or scavengers feed on poisoned plants or animals and become poisoned themselves, by the toxic residues in their food source; the greatest risk of secondary poisoning comes from rodenticides and avicides. Potentially, the legal and financial impact on an applicator could be substantial. Killing an endangered species could result in penalties under the Endangered Species Act; killing a protected migratory bird could warrant penalties under the Migratory Bird Treaty Act; and poisoning a family pet could draw penalties under FIFRA, and possibly a lawsuit.

Applicators need to recognize situations where the potential for secondary poisoning is significant and take precautions to reduce risk. Choose pesticides with minimal secondary poisoning potential. Monitor for predators and scavengers at risk, and pick up poisoned rodents, pigeons, starlings, etc. Select nonchemical control methods in high-risk situations.

General Instructions for Dealing with Injured or Poisoned Wildlife The handling of listed endangered species requires a federal permit except for employees or agents of a state or federal conservation agency who are acting in an official capacity. If you discover injured or dead wildlife, do not handle it. Call your state conservation agency or the U.S. Fish and Wildlife Service for information and instructions. Both offices should be listed in your telephone directory under government agencies.

The primary objective for sick or injured wildlife is effective treatment and care; for carcasses, the objective is preservation for proper diagnosis of the cause of death. If pesticides are suspected or known to be the cause—whether the species is endangered or not—information on pesticides known to have been used in the area will be useful: product name, EPA registration number, date of application, conditions before and after application, etc.

Mixing Pesticides and Loading Application Equipment

Some pesticide products come ready to use: baits, granules, pressurized aerosols, dusts, and some home and garden liquids. Commercial applicators, however, often use products that must be diluted with water or oil for application, and the mixing and loading process is one of the most hazardous aspects of an applicator's job, due to the risk of exposure to concentrated materials. Pesticide poisoning can be acute, such as a single exposure where concentrate is spilled on the body, or it can be chronic, where poisoning occurs gradually from continual, low-level exposure. Sloppy safety practices and unavailability of personal protective equipment are typical contributors to chronic poisoning. Applicators must be trained to avoid exposure during mixing and loading.

Safety Guidelines for Mixing and Loading Concentrated Pesticides Labels often provide information on how to mix the pesticide with other products. When in doubt, consider the following process to determine the compatibility of the materials you intend to combine.

Step 1: Mixing Order of Products When mixing pesticide formulations, add products in the following order: wettable powders, water-dispersible granules, flowable liquids, suspension concentrates, emulsifiable concentrates, and surfactants.

Step 2: Compatibility Test Mix the products, in the order listed in Step 1, to a 1-quart glass jar. Add each dry product at about 1.5 level teaspoons and the liquid materials at 0.5 liquid teaspoon. If the components do not form a sludge nor a gel nor layers after 5–10 min, the materials are compatible; if they are not, you will need to add compatibility agents to achieve suspension of the products.

- Application equipment often is filled the night before an application to get a head start the next morning. But never add concentrates to the tank or granules to the hopper until right before application.
- Wear personal protective equipment (PPE) as prescribed by the label to avoid exposure during mixing and loading. Consider wearing eye protection and protective gloves, even if they are not required.
- Mix and load outdoors or in a well-ventilated area away from people, animals, food, and other items that might be contaminated. Choose a site that can be easily cleaned and decontaminated in the event of a spill.
- Open containers carefully to avoid splashes, spills, or drift (dusts).
- If a concentrate is spilled on the ground or splashed on your clothes, take action immediately to minimize contamination and exposure.

- Stand upwind from the opening of the equipment being filled, with your head well above or to the side of the tank opening.
- Use a backflow preventer or an air gap when adding water to the tank.

Inspection, Calibration, and Maintenance of Application Equipment

Application equipment failure and poor calibration are major causes of pesticide accidents, misuse, and poor performance. Every piece of application equipment must be inspected and maintained regularly to assure safe operation, and most require periodic calibration.

Problems with powered application equipment, in particular, can result in major environmental contamination or human exposure. High application pressures mean increased risk of leaks, splash-back, airborne residues, and drift, including drift back onto the applicator. Powered equipment has multiple components that can fail without proper maintenance. An equipment failure such as a burst fitting or a split hose can drench an applicator or contaminate a large area in a matter of seconds.

All application equipment should be inspected before leaving headquarters each morning. Check for cracked, split, or damaged hoses; cracked fittings; broken regulators and gauges; damaged tanks; and any other defects or signs of wear. Check oil and water levels in gasoline-powered engines. This is also a good time to lubricate fittings and perform other simple nonshop maintenance.

Calibration is the only way to know that a particular piece of equipment is applying the correct amount of pesticide. Improperly calibrated equipment can result in misuse because of overapplication or ineffective treatment because of underapplication. Equipment must be recalibrated regularly to compensate for wear in pumps, nozzles, and metering systems and to adjust settings thrown off by vibration and use.

Methods of calibration vary, depending on the equipment, pressure, rate of travel, nozzle, etc. Applicators need to closely follow the equipment manufacturer's recommended calibration method—and frequency—pertaining to their particular use. Even flow meters require periodic calibration. Electronic flow meters with a built-in automatic calibration capability still require periodic checks of actual volume applied to verify accuracy.

All equipment should get periodic shop maintenance (oil, filter, lube, belt replacements, etc.) as recommended by the manufacturer. Use only replacement hoses, pipes, and fittings that are rated for the maximum pump pressure of the equipment. Before servicing powered equipment, disconnect electric power, release all pressure, and drain all pesticide liquids from sprayers. In some cases, you may also need to remove pesticide dust from a duster before servicing.

After use, application equipment must be cleaned inside and out: inside, because pesticides can be corrosive and can clog nozzles and lines; outside, to remove pesticide residues that can contaminate anyone who touches the equipment.

MANAGEMENT PRACTICES FOR PESTICIDE DISPOSAL

Removing Pesticide Residues from Containers

Product labels direct applicators to triple rinse each empty pesticide container, pour the rinsate into the spray tank, and apply the product according to the label. This method reduces the potential for environmental damage by converting pesticide containers from

Figure 10.26 Notice that the black stripes of the triple-rinsed solution can be seen through the bottle indicating that most of the pesticide has been removed.

hazardous waste to solid waste. Also, triple rinsing ensures that all of the pesticide product is incorporated into the tank mixture so that the customer gets his money's worth.

At a time when pesticide applicators are overwhelmed by scientific information, the benefits from simple techniques such as rinsing pesticide containers often are overlooked. Examine the consequences of improper management of pesticide containers:

- Drinking water can be contaminated if improperly rinsed containers are deposited in landfills.
- Local, state, and federal laws may be violated, causing legal problems for the applicator.
- Expensive material can be left in each unrinsed container.
- Pesticides from unrinsed containers stored in a dumpster can be washed into sewers, creeks, and ditches by storm water, leading to expensive cleanup and fines.
- Local landfills can refuse to take waste from your company.
- Companies that dispose of unrinsed containers in landfills can be held partially liable for cleanup expenses under federal law.

Triple rinsing is defined as the "flushing of containers three times, each time using a volume of the normal diluent equal to approximately ten percent of the container's capacity, and adding the rinse liquid to the spray mixture." Pesticide labels on metal, plastic, and glass containers reflect this federal definition when directing applicators to triple rinse or the equivalent (Fig. 10.26). The following instructions explain two commonly accepted residue removal techniques: triple rinsing and pressure rinsing.

Triple-Rinsing Containers

1. The same personal protective equipment worn while handling the pesticide concentrate during the mixing process should be worn while rinsing containers.

2. The procedure for rinsing containers should begin immediately after emptying the contents into the application equipment. Allowing the residue to dry in empty containers for even a few hours will reduce the effectiveness of the procedure. If you cannot rinse them immediately, leave the caps on the containers until you are prepared to do so; this will help prevent the pesticide residue inside from drying.
3. Pour the pesticide into your spray solution and let the container drain for an additional 30–60 sec. This step greatly enhances the removal of residue during the triple-rinsing process (Fig. 10.27).
4. Add clean water (or other diluent as specified on the label) equal to 10–25% of the container's volume, and secure the cap.
5. Shake or roll (e.g., 55-gallon drum) the container so that the interior surfaces are rinsed.
6. Pour the rinsate into the spray mix and allow the container to drain for an additional 30 sec. This completes the first cycle.
7. Repeat the procedures outlined in Steps 4–6. This completes the second cycle.
8. Again, repeat Steps 4–6. If the rinsate still appears cloudy or milky, keep repeating until the water looks clear, indicating a thorough rinse. If the pesticide is an emulsifiable concentrate (EC) or a liquid flowable (LF), then multiple rinses are always advisable.
9. Render all plastic and metal containers unusable by puncturing or crushing them (Fig. 10.28).
10. The final step is to either dispose of the containers in a sanitary landfill or offer them for recycling.

Pressure Rinsing

1. The same personal protective equipment worn while handling the pesticide concentrate during the mixing process should be worn while rinsing containers.
2. The pressure-rinsing procedure should begin immediately after emptying the contents into the application equipment. If you cannot rinse the containers immediately, leave the caps on until you are prepared to do so. This will help prevent the pesticide from drying inside the containers.
3. Pour the pesticide into the spray solution and drain the container for an additional 30–60 sec.
4. Keep the container positioned over the spray tank as if pouring the concentrate. Puncture the bottom of the metal container or the side of the plastic container with a probe device (see manufacturer's suggestions for specific instructions and guidelines). This renders the containers unusable (Fig. 10.29).
5. Allow water to flow into and through the empty pesticide container, under pressure, until the water is clear. Slowly rotate the probe back and forth. This procedure takes 30–60 sec.
6. Dispose of the containers in a sanitary landfill, or offer them for recycling according to state regulations.

Figure 10.27 (a) A pesticide container must be completely emptied prior to triple rinsing. (b) Fill the container partially with water. (c) Shake the container from side to side to allow the water to remove the residue from the sides of the container. (d) Pour the water containing the residues back into the application equipment.

Figure 10.28 (a) The plastic container is made unusable by puncturing the bottom or the side. (b) Plastic containers that are tripled rinsed can be deposited in the trash.

Figure 10.29 (a) Triple-rinsing containers with a pressure nozzle. (b) The water passes over the entire container and drains into the application tank.

Figure 10.30 Plastic containers which are properly rinsed can be disposed of in a sanitary landfill.

Disposal of Triple-Rinsed Pesticide Containers

Disposal options depend greatly on container construction (metal, plastic, paper, glass) and the availability of facilities for disposing of or recycling the pesticide containers.

Disposal in a Landfill Properly rinsed pesticide containers can be deposited in most landfills that accept common household trash. However, sites that accept household refuse generally are prohibited from accepting waste classified as hazardous by the federal Resource Conservation and Recovery Act. Since pesticide containers that are not properly rinsed fall into this category, pesticide applicators frequently have difficulty disposing of unrinsed or improperly rinsed containers (Fig. 10.30).

Disposal on Private Land Federal laws generally do not prevent the burning or burying of containers on private property, but most state laws do prohibit these disposal methods.

Metal Container Reconditioning Dispose of triple- or pressure-rinsed metal containers as previously mentioned. In addition, properly rinsed, empty metal containers have economic value as scrap metal. They may be taken to a scrap metal facility.

Pesticide Container Collection Programs Recycling programs aimed at reducing the number of plastic pesticide containers thrown away in landfills have been conducted by many state agencies and industries associated with agriculture. Pesticide applicators bring properly rinsed pesticide containers to a collection site for inspection. Containers that meet inspection standards are passed through a chipping machine and reduced to recyclable plastic pellets.

Pesticide container collection programs have been responsible for eliminating millions of pounds of plastic that otherwise would have been deposited in landfills. The plastic recovered from pesticide containers also might be used as fuel for cement kilns or to make flowerpots, plastic fence posts, drainage tiles, guardrails, pallets, roadside sign posts, or sewage lines (Fig. 10.31).

Figure 10.31 Many states have special recycling programs for the collection of triple-rinsed containers.

Bags and Aerosol Cans

Triple and pressure rinsing pesticide containers are not viable options in certain situations. Thorough removal of pesticide products packaged in bags and aerosol spray cans may be accomplished by taking the following steps.

Multilayered Bags

1. Empty the contents of the bag into the tank or hopper.
2. Shake the bag to remove as much product as possible.
3. Cut the sides and folds to fully open the bag, and add any remaining pesticide to the tank or hopper.
4. Dispose of the cut and flattened bag in a sanitary landfill.

Aerosol Spray Cans

1. Spray remaining contents on the proper site as directed by the label.
2. Deposit the empty container in a sanitary landfill.

Containers That Do Not Require Disposal

Pesticide packaging is receiving tremendous attention in today's market. The traditional manufacturing approach is to package liquid formulations in nonrefillable plastic containers and to package granular materials in multilayered paper bags. States' prohibition against the burning of pesticide containers and pressures on the applicator to stop disposing of plastic and paper containers in landfills have placed a premium on alternative packaging. The interest in and research on container management to stem the flow of plastic containers entering solid-waste landfills has spurred the development and implementation of a new generation of pesticide packaging and recycling programs.

Returnable and Refillable Containers Millions of 2.5-gallon plastic containers have been replaced with stainless steel tanks or plastic containers that hold larger volumes (5–250 gallons). In some states, recycling programs have been implemented for minibulk pesticide containers. These minibulk containers are transported to the site of application and returned empty to the dealer or manufacturer for reprocessing and refilling from a larger storage tank. They are normally tamperproof, dedicated to a specific formulation, easily transported, and recyclable.

Water-Soluble Packaging Pesticide manufacturers are converting many products from liquids to water-dispersible dry formulations or incorporating them into a gel matrix. Both are packaged in water-soluble pouches, with the product pouch enclosed in a moistureproof bag or carton. The applicator tears the outside protective cover and places the water-soluble bag into the spray tank. The bag dissolves and releases the dry or gel formulation into the water. The benefits of water-soluble packaging include limited exposure to the concentrated pesticides, elimination of the container-rinsing process, enhanced emergency spill response, and reduction of the amount of waste placed in landfills.

Closed Granular Chemical Handling System Granular formulations are packaged in multilayered paper bags. Recycling the bags is difficult because the paper, foil, and, plastic layers prove difficult in separating. One innovative approach has been to place the granular material into a closed pesticide handling system that mounts directly to the lid of the farmer's planter box; when empty, the container is returned to the supplier for refill. The solutions provided by the closed granular container systems are threefold: reduced applicator exposure; elimination of the multilayered paper bag; and less waste going to the landfill.

What to Do with Unwanted Pesticides

It is a frequently asked question: Who will take my old pesticides? But of primary importance is knowing what *not* to do. Never bury outdated or unwanted pesticides on your property. You may be required to answer questions about chemical disposal methods if you ever want to use the property as loan collateral or if you decide to sell the property; and, in this case, forthright disclosure will reduce the value of the property dramatically. Burying pesticides is a no win situation. If you confess, you would likely be required to pay for and pass an environmental audit before you could borrow against the value of the property, purchase insurance on it, or sell it. With so much at stake, you would be wise to never bury pesticides!

Ideally, disposal of excess pesticides can be eliminated by planning the job and buying only the amount of product that you will need. If that is not feasible and the excess chemical cannot be stored safely, follow these disposal guidelines.

Is It a Registered Product? The easiest way to determine whether a pesticide is still registered is to get the EPA registration number from the product label and call your state regulatory agency which is normally the state department of agriculture (http://aapco.ceris.purdue.edu). An unopened product that is still registered is easiest to deal with. The best option is to ask the chemical dealer from whom you purchased it to take it back for a refund or credit to your account. However, if the product is more than two years old, or if the seal on the container is broken, the likelihood that the dealer will take it back is remote.

Any old, *currently registered* pesticide that your dealer will not accept should be mixed and applied according to the label. Even if the product is no longer effective, it is still legal to apply it to a labeled site, following label instructions. The least favorable option is to give it to someone else who has a use for it. *BUT*: If it is a restricted-use product, make sure that the recipient is certified to use it. Get a written release stating you are not guaranteeing the product's effectiveness.

What Disposal Options Are Available if the Product Is No Longer Registered?

Disposing of an unregistered pesticide can be complicated and expensive; application is not an option, but there are perhaps two possibilities (Fig. 10.32):

- Contact your local solid-waste district and/or your local hazardous-waste hauler and ask what they would charge to dispose of the chemical.
- Inquire about "tox-away," hazardous waste, and amnesty days in your community. These activities allow those holding unregistered, unusable pesticides an opportunity to dispose of them legally, and at a reasonable price.

Make Sure That Old Pesticides Are Securely Contained While You Contemplate Your Options. Many products left in long-term storage date back nearly a quarter of a century, and sometimes old cans leak and bags break. If that is your situation, transfer all liquid chemicals from leaking containers into containers that will hold the entire volume. Wrap old bags of pesticides in heavy-duty plastic bags to ensure containment, and properly label all containers (Fig. 10.33). As an extra precaution, place old, packaged chemicals in another container (e.g., 5-gallon plastic bucket or 55-gallon drum) to guarantee containment if a leak occurs.

Think Smart About What You Purchase. An important priority in purchasing chemicals is to estimate the amount you will need for a single season. Once you have determined the amount to purchase, check into the various options available.

- Delivery as needed. Shop around, because some pesticide dealers offer to deliver products, as needed, to avoid having to buy back unopened chemicals at the end of the spray season.
- Minibulk containers. Using minibulks is a good choice as well because you can be credited immediately for the amount of product left in the tank when you have finished spraying.
- Prepackaged pesticides. If you elect to purchase packaged chemicals, make arrangements with the dealer *at the time of purchase* to take back any unopened containers for refund or credit at the end of the spray season. A less desirable option is to use leftover products the following year, while they are still registered and effective.

SYMPTOMS OF PESTICIDE POISONING

Pesticides are classified according to their biological activity and chemical structure, and biological effects differ greatly among pesticide classes. Detailed information on other pesticides is obtainable from the EPA publication, *Recognition and Management of Pesticide*

Figure 10.32 These products were turned in during a state-sponsored "pesticide disposal day."

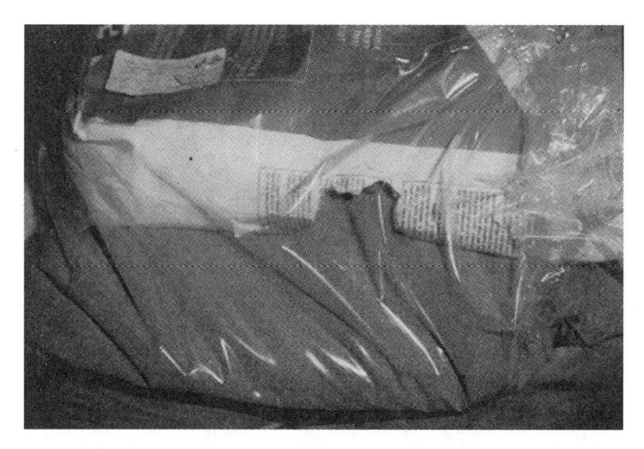

Figure 10.33 Granular and other dry materials which are no longer registered for use should be placed inside another bag to help prevent tears or spills.

Poisonings (EPA-735-R-98-003) at http://www.epa.gov/pesticides/safety/healthcare/handbook/handbook.htm.

Pesticide Toxicity

All pesticides are designed to disrupt essential metabolic processes of the target pest; these processes may be neural, hormonal, cellular, or structural. To relate the specificity of a pesticide to possible human effects, the similarity of mode of action in the pest to that in humans must be considered. Pesticides such as growth regulators that affect a pest in a unique manner have little effect on humans. Conversely, pesticides that are toxic to pest systems similar to those in man—the nervous system is a good example—may pose a greater potential hazard to humans.

The symptoms of pesticide poisoning are specific to the pesticide or pesticide *class;* the applicator should be aware that poisoning symptoms described on the pesticide label are associated with that particular *class* of pesticides. General symptoms of acute chemical poisoning include headache, nausea, dizziness, eye irritation, and skin rashes. If you experience any of these symptoms—or any of those listed on the label—while handling or applying a pesticide, stop what you are doing, leave the area, and seek help (Fig. 10.34).

Chronic exposure to pesticides and other hazardous chemicals can result in delayed or long-term health effects. Chronic effects may include deterioration of the nervous system and body organs, especially the liver; cancer; and changes or alterations in the reproductive system. As with acute toxicity, chronic toxicity is dose related. Health effects appear first in those populations with the most exposure: pesticide production workers and applicators. Both should take appropriate protective measures and wear prescribed protective clothing and gear to minimize long-term exposure (See Chapter 11 for more information on protective equipment).

Pesticide Exposure

Pesticide *toxicity* is the first component of the hazard equation; the second is pesticide *exposure*. There are three routes of exposure: *dermal* (absorption through the skin or eyes); *respiratory* (inhalation into the lungs); and *oral* (ingestion by mouth).

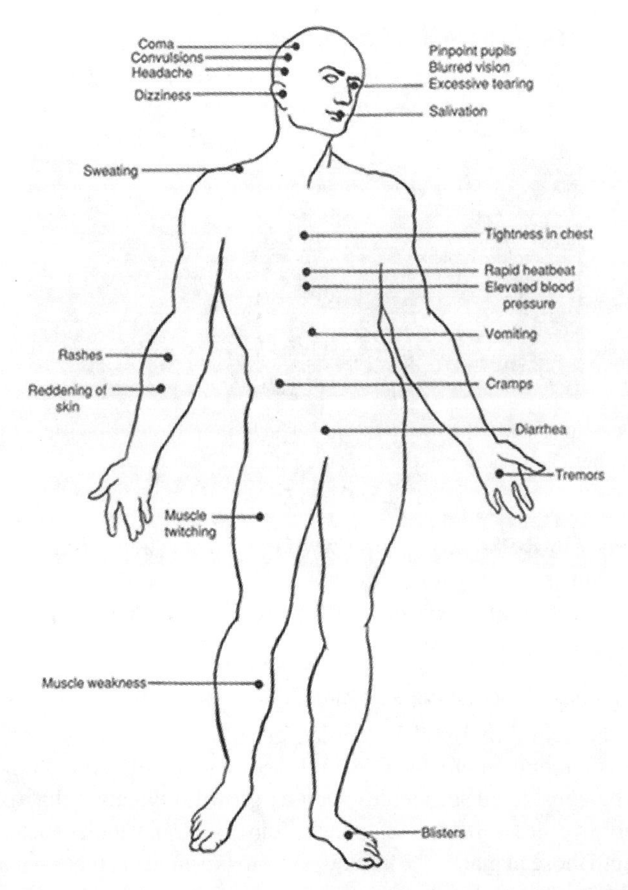

Figure 10.34 The symptoms of pesticide poisoning vary depending on the amount of exposure and the toxicity of the pesticide. Source: Pesticide Education Manual: A Guide to Safe Use and Handling. Pennsylvania State University.

Dermal exposure is the most common among pesticide applicators, where contact with concentrates during mixing and loading represents the greatest risk. The degree of dermal absorption depends on the properties of the pesticide, its formulation, and the parts of the body exposed.

The forearms and hands are the most likely sites of pesticide accumulation during routine pesticide applications. Hands left unwashed after pesticide use can contaminate other parts of the body with which they come into contact. Eyes are highly absorptive and therefore extremely sensitive to pesticides. Direct eye injury can occur when pesticides are accidentally splashed in the face.

Respiratory exposure by inhalation occurs during the handling of powders, dusts, aerosols, fine sprays, and gases (fumigants). The lungs provide a point of rapid entry into the bloodstream.

Oral exposure generally results from improper storage or handling, that is, when someone ingests a product that has been transferred to a food or drink container. Keep pesticides in their original, labeled packaging; never transfer them into bottles or food cartons of any kind. This is extremely important in case of poisoning because unmarked containers provide no instructions to medical personnel regarding the name of the pesticide, treatment recommendations, and emergency phone numbers. Always keep pesticide containers tightly closed and out of the reach of children and animals.

Indirect oral exposure occurs when pesticide residues are passed from contaminated hands to food or mouth while eating or from hands to mouth while smoking.

Carbamate and Organophosphorous Insecticides

Carbamate and organophosphorous insecticides act as neurotoxins. The mammalian toxicity of pesticides in these classes ranges from 1 mg/kg (highly toxic) to 4000 mg/kg (slightly toxic). A large percentage of carbamate and organophosphorous insecticides fall into the high- to moderate-toxicity categories because they target the insect's nervous system, which is similar to that of mammals. Carbamate and organophosphorous insecticides interfere with the signaling between nerve cells and between nerves and the muscles they activate.

Because muscles are responsible for the movement of the diaphragm during breathing, severe poisoning by organophosphates or carbamates can cause the victim to stop breathing and die. In normal muscle movement, nerves signal muscles to contract. At the point of contact between nerve and muscle (neuromuscular junction), the nerve (upon receiving a signal from the central nervous system) releases acetylcholine, which signals the muscle to contract. The acetylcholine is then counteracted by the enzyme acetylcholinesterase and the muscle relaxes. Conversely, if the acetylcholine is not removed, the muscle remains contracted. Carbamate and organophosphorous insecticides are acetylcholinesterase inhibitors; that is, they prevent the acetylcholinesterase from counteracting the acetylcholine from the neuromuscular junction. If the concentration of a neurotoxic pesticide is high enough, the muscle is permanently contracted; and if the muscles that move the diaphragm are permanently contracted, breathing stops.

Poisoning Symptoms of Carbamate and Organophosphorous Insecticides

Acetylcholine also is used as a signal from one nerve to another within the central nervous system; this action, too, can be disrupted by acetylcholinesterase inhibitors. The disruption can cause headache, dizziness, nausea, and sometimes vomiting in mild poisonings; therefore they often are misdiagnosed as the flu.

In severe poisoning, the flulike symptoms may be accompanied by restlessness, anxiety, or convulsions. The symptoms often progress to include muscle twitching, weakness, tremor, loss of coordination, diarrhea, and incontinence; hypersecretion (sweating, tearing, salivating) also may occur. If the victim suffers convulsions, incontinence, depressed respiration, or unconsciousness, the poisoning is considered life threatening.

If the insecticide is inhaled, effects on the lungs may present as coughing, wheezing, and tightness in the chest.

Treatment of Carbamate and Organophosphorous Insecticide Poisonings

Inhibition of acetylcholinesterase allows excess acetylcholine to accumulate, leading to overstimulation of acetylcholine receptors. To treat the symptoms caused by overstimulation of the acetylcholine receptors, the drug atropine, which blocks stimulation of the acetylcholine receptors, is administered by the attending physician. Atropine is given until the symptoms are alleviated, and it is maintained until the symptoms no longer return.

Carbamates are reversible inhibitors of acetylcholinesterase, meaning that, over time (several hours), the carbamates will be metabolized and the acetylcholinesterase will no longer be inhibited. If the vital signs can be maintained, the victim eventually will recover.

It is especially critical to maintain an oxygen supply to the body. The primary treatment for carbamate poisoning is to maintain levels of atropine adequate to block the acetylcholine receptors until acetylcholinesterase is no longer inhibited and the acetylcholine levels return to normal.

Organophosphorous insecticides can irreversibly inhibit acetylcholinesterase. As in treatment of carbamate poisoning, atropine is used to effectively block overstimulation of acetylcholinesterase receptors caused by organophosphorous insecticide poisoning. However, unlike carbamate poisoning, organophosphorous insecticide–inhibited acetylcholinesterase will not fully return to its uninhibited form without additional treatment. Pralidoxime compounds can facilitate the regeneration of uninhibited acetylcholinesterase from the organophosphorous insecticide–inhibited form. However, pralidoxime compounds by themselves inhibit acetylcholinesterase and exacerbate the poisoning symptoms. Thus, the treatment for organophosphorous insecticide poisoning requires the use of atropine to block the overstimulation of the acetylcholine receptors until the symptoms are alleviated. Pralidoxime is then administered to regenerate acetylcholinesterase.

Pralidoxime compounds should never be used in treating carbamate poisoning because they inhibit acetylcholinesterase. In carbamate poisoning, pralidoxime compounds are not useful for regenerating uninhibited acetylcholinesterase. Additional cholinesterase inhibition by pralidoxime compounds would exacerbate the carbamate poisoning while providing no therapeutic affect.

Medical Tests for the Effects of Carbamate and Organophosphorous Insecticide Exposure Because of irreversible inhibition, chronic exposure to organophosphorous insecticides can depress an individual's level of acetylcholinesterase. In the course of a spray season, continual depletion of acetylcholinesterase can occur, making the applicator more vulnerable to damaging effects from both organophosphorous and carbamate insecticides. Medical tests are available to determine acetylcholinesterase levels; but because these levels vary among individuals, a baseline acetylcholinesterase level must be established for each applicator, prior to exposure. Once a person's base level of acetylcholinesterase has been determined, a simple blood test can reveal any decline resulting from exposure to organophosphorous insecticides. Individuals who demonstrate reduced acetylcholinesterase levels should not apply organophosphates until their levels return to normal. The body produces acetylcholinesterase on a continuous basis, and levels can be restored within weeks. Persons who handle organophosphate insecticides for an extended period of time should consider (in consultation with their physician) an ongoing regimen to monitor their levels of acetylcholinesterase.

Pyrethrin and Pyrethroid Insecticides

Naturally occurring pyrethrins and synthetic compounds (pyrethroids) which act similarly are neurotoxins, but they are *not* acetylcholinesterase inhibitors: They affect the electrical signal that travels within a nerve. Insects and fish are much more susceptible to pyrethroids than are mammals, in which severe pyrethroid poisoning is rarely seen. Pyrethroids can irritate skin and eyes and produce allergic reactions, so proper clothing and eye protection are important when handling liquid formulations. Some individuals report tingling, stinging, burning, itching, or numbness after dermal contact with pyrethroids, either immediately after or within 2–4 hr of exposure. They do not cause sensitization and symptoms generally disappear after 24 hr.

Plan of Action for Acute Pesticide Poisonings

A pesticide user should establish a plan of action to follow in case of a pesticide-related accident. Advanced planning and preparation should be routine. Make sure all employees are familiar with appropriate emergency procedures.

Contact Medical Personnel The goals with any poisoning emergency are to prevent further exposure and to make sure the victim is breathing; then call emergency medical personnel.

Maintain Vital Signs Administer first aid while help is on the way. The cause of death of most pesticide-poisoning victims is respiratory failure, but victims often recover if their oxygen supply is maintained. Therefore, the maintenance of vital signs is imperative, and cardiopulmonary resuscitation techniques may be required. Always provide attending medical personnel with a copy of the pesticide label.

Eliminate Further Contamination

Ingested Pesticides If an individual swallows a pesticide, act immediately: do not wait for symptoms to appear. Read the pesticide label or the Material Safety Data Sheet to find out whether or not vomiting should be induced; then read it again to be sure before purposely causing the victim to vomit. Never induce vomiting if the victim is unconscious or convulsive.

In cases where vomiting *is* desirable and *can* be induced safely, fast action can mean the difference between life and death for the poisoning victim. Always keep syrup of ipecac on hand to induce vomiting if necessary. Have the victim assume a forward kneeling position, or place the victim on his or her right side if lying down; the victim is least likely to aspirate vomit into the lungs in one of these positions.

Gastric gavage—performed by a physician—is another method for removing stomach contents, but it must be performed as soon as possible and no longer than 2 hr after ingestion of the pesticide. After 2 hr, the pesticide will have passed into the intestine, thus requiring a different approach to effect removal of the poison. Physicians can administer absorptive charcoals to intercept absorption of the pesticide by the intestine and promote its elimination in feces.

It is important to remember to consult the pesticide label before administering first aid because certain situations can be made worse by causing the victim to vomit. Vomiting should not be induced if the pesticide formulation contains organic solvents or corrosives. These materials can cause serious, permanent damage to sensitive tissues of the esophagus—or the lungs, if aspiration occurs.

Pesticides on the Skin: No Visible Burns Wash the pesticide off the victim as soon as possible to prevent continued exposure and injury:

- Remove clothing and drench the skin with water (shower, hose, faucet, pond, etc.).
- Cleanse the skin and hair thoroughly with soap and water. Be careful not to injure the skin while washing.
- Dry the victim and wrap him or her in a blanket.

Chemical Burns of the Skin Taking immediate action is extremely important:

- Remove contaminated clothing.
- Wash the victim with large quantities of cold running water, being careful not to abrade the skin.
- Immediately cover the affected area loosely with a clean, soft cloth.
- Do not use ointments, greases, powders, or other drugs, even if they are recommended as first aid treatments for chemical burns.

Pesticides in the Eye It is very important to wash out the affected eye as quickly but as gently as possible.

- *Do not* use chemicals or drugs when flushing the eye.
- Hold eyelids open and wash the eyes with a gentle stream of clean water (at body temperature, if possible).
- Continue washing for 15 min.

Inhaled Pesticides If the victim is in an enclosed area, wear an appropriate respirator when removing the person from the contaminated area.

- Immediately carry the victim to fresh air.
- Loosen all tight clothing.
- Administer CPR if breathing is irregular or has stopped.
- If the victim is convulsing, watch breathing and protect the person from falling and striking his head. Pull the chin forward so that the tongue does not block the air passage.
- Keep the victim as quiet as possible.
- Prevent chilling. Wrap patient in a blanket but do not overheat.

ANSWERING CUSTOMER QUESTIONS

Todays well-informed public expects the person hired to manage their property to answer their questions truthfully and professionally. It's important that you never underestimate the importance of a question; every question asked is important to the person who is asking, and that person deserves the best answer that you can possibly provide. As a guest on your clients' property, your ability to respond knowledgeably to questions is reassuring; it gives them a sense of confidence in your technical abilities.

Answering questions is at first a difficult proposition: Expect to be stumped! Even the most seasoned professional cannot spontaneously answer every question asked; but as you become more knowledgeable in your profession, handling questions will become easier.

Key Steps in Dealing with the Public's Questions

Let the Customer Finish the Question. Give customers the opportunity to ask their entire question *without interruption.* This is important because the *real* question often does not emerge until the very end. Interrupting conveys defensiveness and can jeopardize customers' trust in you.

Keep It Simple, Stupid! Simplicity is best when answering questions. Always be aware that your customers don't really care how much you know. If they really cared to know all there is to know about the job you're doing, they'd study it themselves and manage their own property. All they want are answers to their specific questions; keep them short and to the point.

Use Everyday Language—Words That Come Naturally to You. Know how to communicate with customers. Don't use unfamiliar jargon and scientific words that they may not grasp. There are no perfect scripts or how-to steps in answering clients' questions, but it is important that you feel comfortable with what you tell the *customer.* Using your own words and examples goes a long way in gaining—and maintaining—their confidence in you as a professional.

Never Admit Guilt. Never admit that you did something wrong because you are not always in the best position to make that determination. Collect information from the customer and assure them that you or your supervisor will get back to them. If you admit guilt, then you're guilty no matter what other facts are collected. There's no reversing it.

Always Be Honest and Open. There's no doubt that being honest and open is a great attribute. Customers appreciate sincerity, and there's no better tool for establishing trust.

Be Polite. Courtesy helps your customers feel more relaxed. Make eye contact when answering their questions, and always treat them with respect.

Never Answer Questions about Your Customers. You should always respect the confidentiality of your customers. If the person asking the question wants that information, they should contact the person directly.

Know Your Limits. When you're really out on a limb and get that sinking feeling in the pit of your stomach, don't hesitate to tell customers that you'll need to do some checking to answer their question. Sometimes a quick call to the office will do the trick, and sometimes you may have to get back to the customer at a later time. Either way, it is more important to provide a correct answer than a quick one.

Achieve a 100% Follow-Up. Always write down in a dedicated notebook the customer's name and phone number, the date, and the question they asked that still needs answering. Call them back that day or the next day, at the latest.

HANDLING COMPLAINTS

Suppose you did everything correctly during the application and you still receive a complaint. Will ignoring the complaint make it go away? Follow these steps to resolve the issue successfully.

Immediately Respond in Person

Act quickly on complaints to minimize damage and to maintain a good relationship with your customer. Nothing escalates a problem faster than the customer's opinion that you are discounting or ignoring the problem. If you do not deal with the problem, the customer will find someone who will: perhaps a regulator, the health department, or his or her attorney! And if an incident results in disciplinary or legal action, the fact that you did not respond quickly will be held against you.

Act immediately if someone complains that a treatment is making them ill or that there is too much odor from a treatment, or if any other effect on human health is suggested. You must resolve these kinds of problems to prevent illness or injury. The longer the customer remains dissatisfied or worried, the more likely he or she is to contact a physician or a lawyer specializing in personal injury or chemical sensitivity.

Listen but Do Not Argue

Respond gracefully to complaints. Handle each complaint as you would want it handled if your role and the customer's were reversed. Do not brush off a complainant as a trouble-maker. Respect the customer's opinion and point of view, and listen carefully. Make eye contact and focus on what he or she has to say. Ask questions, but do not let the customer draw you into an argument: be pleasant. Once you know the customer's concerns, make it clear to him or her that you want to resolve the problem; but do not overcommit.

Inspect the Damage

A supervisor or manager should personally inspect the alleged damage. Take photographs and make sketches. Do not limit your inspection to the damage or site that your customer identifies. Be thorough. Look for other damage that might have occurred and for conditions on-site that might have contributed to the problem; your liability could be limited due to extenuating circumstances. Inform your insurance carrier that a claim may be filed; a representative of the insurance company will want to conduct an inspection as soon as possible.

If you can correct the problem on the spot, do so. If not, provide the customer with the name and telephone number of the person who will contact them for remediation.

Document the Situation

A written record should be prepared, describing the complaint or incident in detail. Often called a *record of complaint* or an *incident report,* this written record can help you identify the cause of the problem so that changes can be made to prevent its recurrence. It can be the basis for insurance and other reports mandated by law, and it also can serve to protect the company from lawsuits and regulatory actions.

The report should be long on descriptions, detail, and facts and short on speculation and personal bias. Report what happened descriptively and factually. Do it immediately upon notification of the problem so that details are not forgotten. Include the following information.

Background Information Detail the circumstances that led up to the accident: the type of application that was made; any peculiarities germane to the application; and unusual situations encountered during treatment. State the names and affiliations of people involved; the address of the customer or the accident location; the purpose of the job; and the exact date and time of the incident.

Description of Accident Provide details about what happened, in chronological order. Avoid personal opinion and speculation: stick to the facts. Describe any damage, illness, or injury that occurred.

Outcome Describe any actions taken after the accident: what was done to repair damage, clean up, tend to injuries, etc. Give names and affiliations of eyewitnesses, medical personnel, police officers, and others involved. List what still needs to be done to resolve the matter.

Companies often have preprinted incident report forms with labeled boxes and checklists that simplify the reporting process. However, a detailed memo or letter is just as effective.

Follow Up on Your Promises

If you promise to take some form of action, to return to the site, or to call at a certain time, be sure to do so. If you cannot follow through as promised, notify the customer in advance and make other arrangements. If you assign another employee to act on something related to the complaint, check to make sure it gets done. A broken promise can damage the customer relationship beyond repair. As mentioned earlier, nothing escalates a problem faster than the customer's feeling that you are unresponsive.

Resolve Difficult Complaints

Most complaints, handled properly, can be resolved as we have outlined. often, the complainant merely wants the chance to be heard and for you to acknowledge the inconvenience or damage he has suffered as a consequence of your actions.

On the other hand, a poor response to a complaint—or no response—can turn even the most minor incident into your worst nightmare. Always show concern for the complainant's point of view and express your willingness to make things right: Act responsibly.

Your immediate response, your attentiveness, your understanding attitude, and your prowess as a professional will serve you well in settling complaints, big or small. Never argue with a complainant, and always gather all the facts before drawing any conclusions or committing to financial restitution.

Despite your concerted effort to make things right, some complaints are not easily resolved. The complainant may demand corrective action over and above what you consider reasonable. You may have to bring your insurer into the picture, and you might need an attorney. Generally, complainants do not view you or your insurance company as a neutral party; therefore it is sometimes helpful to enlist the services of a disinterested third party to assess the damage (e.g., a local landscaper or a university extension specialist).

Figure 10.35 Responsible use means following the label directions.

Do Not Make the Same Mistake Again

Learn from your mistakes. when there has been an accident or complaint, incorporate future avoidance into your training program. One of the advantages of detailed, written incident reports is that they can be used to develop focused training. Conduct a company training session and discuss the specific causes of the incident. Technicians and supervisors can recount the incident and discuss proactive measures to prevent recurrence.

SUMMARY

Pesticide-related businesses must commit to safety in the workplace. There is countless potential for chemical hazards on company premises and at customers' facilities. In addition to those associated with pesticide application, hazards exist in mixing, loading, storing, and transporting pesticides (Fig. 10.35).

Proper storage of pesticides and disposal of pesticide containers are as important at the job site as they are at headquarters—perhaps even more important. Haphazard practices at the application site are more likely to suffer public scrutiny and present safety issues than are similar procedures back at the office.

Security of pesticides at the job site is probably the main failing of most pesticide businesses. It is imperative that pesticides transported to the job site be locked securely into their designated compartment in the vehicle, preferably out of sight. Pesticides must be stored for transit in such a way that they do not present an attractive nuisance at the job site. Risks from careless placement of pesticides in or on fleet vehicles include accidental exposure of anyone who might come into contact with the containers; for instance, by reaching into the bed of the pickup, and theft.

Leaking equipment and service vehicles that are messy from leakage—past or present—create a negative image of your business and may constitute liability concerns. Inspect pesticide application equipment daily and clean up spills as they occur. If a leak occurs on the job, stop the application immediately: contain the spill, fix the leak, and clean the spill site appropriately. Never continue an application with faulty application equipment.

Pesticide safety requires an aggressive and sustained commitment from management. Companies need to conduct regular, in-depth audits to identify all the hazards associated with the products they use. They also need to institute employee safety programs to reduce health and environmental risks. But both labor *and* management have safety responsibilities, and both must commit to safety through practice and communication.

The responsibility for using pesticides safely, and in that way protecting human health and the environment, cannot be taken lightly. Your commitment to your customers must always be to deliver the full benefit of the products applied, in the safest manner possible.

APPENDIX 10.1 PESTICIDE SELECTION BASED ON FORMULATION CHARACTERISTICS

Evaluation of Pesticides

The selection of pesticide products should include a cost–benefit analysis. You must consider the pest problem, application equipment requirements, efficacy of products available to manage the pest, price, environmental impact, and personal safety. Quality pesticide dealers can advise you on product selection and offer use recommendations.

The following demonstrates a sound method for evaluating and comparing pesticide products. It encourages a comprehensive approach to product selection.

Pest Identification

Pest identification and an understanding of pest biology are crucial to the success of your pest management program. Before deciding on treatment, you must determine what pest is causing the problem. Once the pest is identified, consider its specific life cycle and find out its most vulnerable stage of development (i.e., the point in the pest's life cycle at which the pesticide will be most effective).

This expertise is available from county extension educators, university research and extension specialists, university plant and pest diagnostic laboratory personnel, master gardeners, and representatives from pesticide suppliers, dealerships, and retail outlets.

Pesticide List Development

Integrated pest management (IPM) is the control strategy of choice for homeowners, growers, and commercial applicators. IPM is an approach that blends all available management techniques—chemical and nonchemical—into one strategy: Monitor pest problems, use nonchemical pest management where feasible, and resort to pesticides only when pest damage exceeds an economic or aesthetic threshold.

Labels and regulations change and new products are introduced routinely. Therefore, the pesticide selection process should be repeated just prior to each growing season.

Pesticide selection requires planning—and knowledge of the choices available. Begin by developing a comprehensive pesticide list for specific crop, turf, or home garden pests. Solicit management recommendations from various helpful sources: the Cooperative Extension Service, consultants, agrichemical and urban pesticide dealers, product manufacturers, garden and nursery center personnel, association newsletters, trade journals, and expert applicators. After developing the list, obtain labels for use in analyzing each product's strengths and weaknesses. Labels generally are available from retail outlets and their suppliers or on the Internet; also, dealers and magazines often have charts and information on product comparison.

Product Profile Worksheet

Prepare a product profile worksheet for each product under consideration, and attach each label to the corresponding worksheet. The completed worksheet becomes a ready reference for information from the product label and various experts that you have contacted.

PRODUCT PROFILE WORKSHEET

Product name____________________ Crop to be planted____________________

EPA Registration number__________ Pest to be managed____________________

Signal word______________________ Date of label review____________________

Product Cost and Restrictions

Costs and Efficacy	Certification Requirements
Registered for crop/pest complex (Y/N)	Federal restricted-use pesticide (Y/N)
Cost per acre $____, per 1000 sq ft $____	State restricted-use pesticide (Y/N)

Application costs (per acre, per field, per yard) $______________________________

Relative efficacy ________________%

Specific Recommendations for Use (If yes, list specific information requested.)

State restrictions on use (Y/N)______________________________

Buffer zones from sensitive crops (Y/N)______________________________

Application equipment requirements (Y/N)______________________________

Application timing requirements (Y/N)______________________________

Potential crop injury (Y/N)______________________________

Preharvest interval (Y/N)______________________________

Frequency of applications ______________________________

Rate(s) per acre or per 1000 sq ft______________________________

Compatibility with other pesticides/fertilizers (Y/N)______________________________

Limitations on use with surfactants (Y/N)______________________________

Wind speed (Y/N)______________________________

Temperature (Y/N)______________________________

Height above crop canopy (Y/N)______________________________

Rotational crop restrictions (Y/N)______________________________

Grazing restrictions (Y/N)______________________________

Special comments concerning production

____

____

____

____

____

Applicator/Worker Safety

*Signal word*______________________________

Personal protective equipment (List specific clothing and equipment.)

- Mixer/loader ______________________________
- Applicator______________________________
- Worker______________________________

Reentry or Restricted Entry Internal Requirements

- Mixer/loader ____________
- Applicator ____________
- Worker ____________

First aid advice ____________

____

Posting/Notification

- Oral (Y/N)
- Written (Y/N)
- Posted (Y/N)

Water Quality

- Ground water advisory statements (Y/N) ____________
- Surface water advisory statements (Y/N) ____________

Pesticide physical properties—(Consult university personnel, manufacturer representatives, and the USDA Soil Conservation Service for specific values.)

- Soil adsorption value ____________
- Hydrolysis half-life value ____________
- Water solubility value ____________
- Soil half-life ____________

Site characterization

- Classification of soil ____________
- Percentage soil organic matter ____________
- Depth to ground water ____________
- Number of abandoned wells ____________
- Sinkholes ____________
- Rivers, streams, lakes, ponds ____________

Movement Off-Target

- Buffer zones (Y/N) ____________
- Specific adjuvant information (Y/N) ____________
- Wind speed restrictions (Y/N) ____________
- Sensitive areas identified (Y/N) ____________
- Nozzle type, size, pressure (Y/N) ____________

Wildlife Species and Habitat

Endangered species named (Y/N) ____________

Toxicity statements

- Fish (Y/N) ____________
- Birds (Y/N) ____________
- Pollinators (Y/N) ____________
- Other wildlife (Y/N) ____________

Wetlands restrictions (Y/N) ____________

Product Cost and Restrictions

The product comparison process is designed for a broad evaluation of products. First, a general assessment is developed through the product profile worksheet. Then each statement in the following five worksheets is addressed by reviewing the product profiles and assigning a relative rank of 1 for highly acceptable, 2 for acceptable, or 3 for unacceptable.

WORKSHEET #1: PRODUCT COST AND RESTRICTIONS RANKING

(Consult your product profile worksheets.)

	Product Name			
Write in brand name of each product being considered:	______	______	______	______
1. *Certification requirements*	______	______	______	______
2. *Economic factors*	______	______	______	______
3. *Specific recommendations for use*	______	______	______	______
4. *Past experiences with product*	______	______	______	______
5. *Past experiences with company*	______	______	______	______
6. *Product availability*	______	______	______	______
Add the columns for each brand and place the sum in the *Total* space.				
Total *(Add lines 1–6):*	______	______	______	______

If the *Total* is 6–9 the suggested *Overall Ranking* is 1.
If the *Total* is 10–14 the suggested *Overall Ranking* is 2.
If the *Total* is 15–18 the suggested *Overall Ranking* is 3.

Overall Ranking:	______	______	______	______

1 = highly acceptable; 2 = acceptable; 3 = unacceptable

Transfer the overall ranking to the pesticide evaluation criteria chart near the end of Appendix 10.1.

Applicator and Worker Safety

Label directions that call for use of personal protective equipment (PPE) or closed handling systems have great bearing on pesticide selection.

WORKSHEET #2: APPLICATOR/WORKER SAFETY RANKING

(Consult your product profile worksheets.)

	Product Name			
Write in brand name of each product being considered:	______	______	______	______
1. *How acceptable is the signal word?*	______	______	______	______
2. *How acceptable are the product's PPE requirements?*	______	______	______	______

3. *How acceptable are the product's reentry requirements?* ______ ______ ______ ______
4. *Do you have the ability to administer first aid in case of an accident?* ______ ______ ______ ______
5. *Level of safety training that dealer or manufacturer can provide* ______ ______ ______ ______
6. *How acceptable are posting and notification requirements?* ______ ______ ______ ______
7. *Your past performance in following "Hazards to Humans and Domestic Animals" precautions* ______ ______ ______ ______

Add the columns for each brand and place the sum in the *Total* space.

Total *(Add lines 1–7):* ______ ______ ______ ______

If the *Total* is 7–11 the suggested *Overall Ranking* is 1.
If the *Total* is 12–16 the suggested *Overall Ranking* is 2.
If the *Total* is 17–21 the suggested *Overall Ranking* is 3.

Overall Ranking: ______ ______ ______ ______

1 = highly acceptable; 2 = acceptable; 3 = unacceptable

Transfer the overall ranking to the pesticide evaluation criteria chart near the end of Appendix 10.1.

Water Quality

Four factors influence ground water vulnerability to pesticide contamination:

- Chemical properties of the pesticide (low soil adsorption and persistence)
- Soil type (sandy or gravel texture, low organic matter content)
- Site characteristics (shallow water table, sinkholes, abandoned wells)
- Management practices (improper chemical storage, handling, use)

Pesticides can contaminate surface water indefinitely as runoff from treated urban and rural landscapes reaches streams, rivers, lakes, etc. Options for protecting surface water near application sites include no-spray strips around surface water supplies, wells, and irrigation ditches; grass waterways and buffer zones to filter runoff; and use of conservation practices on erodible lands, plow berms, and sinkholes. Always follow chemigation laws that require vacuum breakers, gate valves, etc., for surface water protection.

Water quality evaluations of ground and surface water may be difficult to complete, but not impossible. The best method for selecting pesticide products relative to water quality is to find specific soil–water adsorption coefficients and water hydrolysis, and soil half-life values. These can be obtained from manufacturers, university personnel, and the U.S. Department of Agriculture Soil Conservation Service.

Pesticide leaching and runoff are least likely when the soil–water adsorption coefficient (K_d) is greater than 5. The K_d value is simply a measure of how tightly the pesticide binds to soil particles: the greater the K_d value, the less likely a chemical is to leach or contribute to runoff. A very high value means that it adsorbs tightly to soil and organic matter and does not move throughout the soil: Higher is better.

Pesticides are least likely to leach when their water solubility is less than 30 ppm. In other words, the less a chemical dissolves in water, the less likely it is to move with water through the soil: Lower is better.

Pesticides are less likely to leach when their hydrolysis (breakdown in water) half-life is less than 6 months and their soil half-life is less than 3 weeks. The longer a chemical can remain in water or soil without breaking down, the more likely it is to leach through the soil: Shorter is better.

WORKSHEET #3: WATER QUALITY RANKING

(Consult your product profile worksheets.)

	Product Name			
Write in brand name of each product being considered:	_______	_______	_______	_______
1. *Product's characteristics for ground water contamination*	_______	_______	_______	_______
2. *Product's characteristics for surface water contamination*	_______	_______	_______	_______
3. *Application site characteristics for ground water contamination*	_______	_______	_______	_______
4. *Application site characteristics for surface water contamination*	_______	_______	_______	_______
5. *Your ability to meet ground water advisories on labels*	_______	_______	_______	_______
6. *Your ability to meet surface water advisories on labels*	_______	_______	_______	_______
Add the columns for each brand and place the sum in the *Total* space.				
***Total** (Add lines 1–6):*	_______	_______	_______	_______

If the *Total* is 6–9 the suggested *Overall Ranking* is 1.
If the *Total* is 10–14 the suggested *Overall Ranking* is 2.
If the *Total* is 15–18 the suggested *Overall Ranking* is 3.

Overall Ranking:	_______	_______	_______	_______

1 = highly acceptable; 2 = acceptable; 3 = unacceptable

Transfer the overall ranking to the pesticide evaluation criteria chart near the end of Appendix 10.1.

Movement Off-Target

Pesticide particle drift and volatilization pose risks to neighbors, field-workers, and the environment. Keeping products on-target increases the effectiveness of pest management while minimizing personal injury and adverse reactions of susceptible nontarget plants, organic farms, domestic animals, and wildlife. The proximity of an application site to sensitive areas such as nursing homes, subdivisions, schools, day-care centers, parks, playgrounds, hospitals, etc., is critical in developing your pest management strategy. Take every safety precaution available.

There are two major considerations for the applicator who is concerned about drift: Select application equipment and accessories that minimize drift potential, and choose products that are easily managed to prevent drift.

Management decisions that help prevent off-target movement include the following:

- Allow for buffer zones and planting setbacks.
- Incorporate pesticides into the soil.
- Operate application equipment at slow speeds.
- Choose the application method least likely to facilitate drift.
- Apply sprays as low as possible to the ground or target pest.
- Use low equipment pressure.
- Select nozzles that reduce drift potential (e.g., nozzles that can deliver large droplets).
- Alter the time of application to avoid heat, humidity, and wind.

Products formulations vary in their tendency to move off-target. Evaluate the attributes of each product to determine the best choice for your application site. If a product is selected for its drift control characteristics, it is especially important for the applicator to understand the label directions on wind speed, adjuvants, nozzle selection, etc.

WORKSHEET #4: MOVEMENT OFF-TARGET RANKING

(Consult your product profile worksheets.)

	Product Name			
Write in brand name of each product being considered:				
1. How clear are label directions on drift prevention?				
2. How acceptable are buffer zone restrictions?				
3. How acceptable are wind speed and temperature restrictions?				
4. Your ability to prevent drift onto sensitive areas				
5. Prior experience with product				
Add the columns for each brand and place the sum in the *Total* space.				
Total *(Add lines 1–5):*				

If the *Total* is 5–7 the suggested *Overall Ranking* is 1.
If the *Total* is 8–12 the suggested *Overall Ranking* is 2.
If the *Total* is 13–15 the suggested *Overall Ranking* is 3.

Overall Ranking: ________ ________ ________ ________

1 = highly acceptable; 2 = acceptable; 3 = unacceptable

Transfer the overall ranking to the pesticide evaluation criteria chart near the end of Appendix 10.1.

Wildlife Species and Habitat

Pesticide labels may stipulate special precautions to protect endangered species and wildlife habitat.

WORKSHEET #5: WILDLIFE SPECIES AND HABITAT RANKING

(Consult your product profile worksheets.)

	Product Name			
Write in brand name of each product being considered:	________	________	________	________
1. *How acceptable are endangered species restrictions?*	________	________	________	________
2. *Your ability to avoid exposure to wildlife*	________	________	________	________
3. *Your ability to reduce mortality to pollinators*	________	________	________	________
4. *Wetlands restrictions can be managed*	________	________	________	________
Add the columns for each brand and place the sum in the *Total* space.				
***Total** (Add lines 1–4):*	________	________	________	________

If the *Total* is 4–6 the suggested *Overall Ranking* is 1.
If the *Total* is 7–9 the suggested *Overall Ranking* is 2.
If the *Total* is 10–12 the suggested *Overall Ranking* is 3.

Overall Ranking: ________ ________ ________ ________

1 = highly acceptable; 2 = acceptable; 3 = unacceptable

Transfer the overall ranking to the pesticide evaluation criteria chart near the end of Appendix 10.1.

Pesticide Comparison and Selection

Selecting a pesticide is not a simple process. There may be numerous products registered for use on the same crop and the same pest that you need to manage, but you also must consider relative toxicity, approved application methods, environmental safety, and cost. And when you are assessing products' relative strengths and weaknesses, do not forget to factor in your own ability to handle and apply pesticides safely.

Information necessary to make an informed product selection can be compiled through

- identification of the pest species,
- development of a list of pesticides registered for the use you intend, and
- completion of the product profile worksheet and the pesticide evaluation criteria chart provided herein.

Pesticide Evaluation Criteria Chart

	Product Name			
Write in brand name of each product being considered:	A	B	C	D
1. *Product Cost/Restrictions*	2	3	2	1
2. *Applicator/Worker Safety*	1	3	3	2
3. *Water Quality*	2	1	1	3
4. *Movement Off-Target*	2	1	1	2
5. *Wildlife Species and Habitat*	2	1	1	3

1 = highly acceptable; 2 = acceptable; 3 = unacceptable

To develop the example, we first completed a product profile worksheet on each of four products. Then, five worksheets—product cost and restrictions, applicator/worker safety, water quality, movement off-target, and wildlife species and habitat—were used to summarize each product profile worksheet. Ranking values from the five worksheets then were transferred to the pesticide evaluation criteria chart.

Comparison of data compiled in this manner ultimately results in a single product standing out as best-suited for the situation. The same approach works for any criteria and combination of factors that you deem important. For purposes of this discussion, our pesticide choice is based on the product cost/restrictions, water quality, and applicator/worker safety criteria.

First, consider product cost/restrictions. Reading across the chart, you will see that Brand B has the worst ranking (3, unacceptable), Brand D has the best ranking (1, highly acceptable), and Brands A and C are ranked in the middle (2, acceptable). So, if price and use restrictions are the only criteria considered, Brand D is the best choice.

The second criterion is environmental impact on water quality. The product profile worksheets indicate that application site characteristics are conducive to leaching into ground water. Thus, it is important to select a pesticide product that minimizes the potential for ground water contamination. The chart shows that from a water quality perspective Brands B and C are the best choices (1, highly acceptable), Brand A's characteristics are acceptable (2), and Brand D's are unacceptable (3).

Now we have two criteria for comparison: product cost/restrictions and water quality. If we place more importance on environmental safety than price and restrictions, we can eliminate Brand D immediately because its water quality factor is unacceptable. That leaves Brands A, B, and C from which to choose.

The water quality factor of Brand A is acceptable (2), as is its cost factor, so Brand A is a potential choice. The water quality factor of Brand B is highly acceptable (1), but its cost factor is unacceptable (3); therefore it is not a candidate. That leaves Brands A and C.

Since we have chosen to place more importance on water quality than on product cost/restrictions, Brand C is our choice at this point in the evaluation process due to its superior water quality rating of 1 (compared to Brand A's rating of 2). But still we must consider our third criterion, applicator/worker safety—and this is where personal ability or latitude to change pesticide management practices might allow the selection of a superior product over an average product.

The product profile worksheet for Brand C reveals an unacceptable ranking (3) on applicator/worker safety. That ranking is based on the fact that respirators and chemical-resistant gloves and boots are stipulated for mixers and applicators, and on the fact that the applicator is required to prevent entry of workers into treated areas for 24 hr after application. Label requirements for Brand A, on the other hand, stipulate only that *mixers* wear chemical-resistant gloves.

If pesticide handling procedures can be adapted to meet label requirements, Brand C is an ideal choice because the pesticide evaluation criteria in general are rated highly acceptable (1). However, if the applicator is unwilling to implement Brand C safety procedures, Brand A is a better choice.

The best decision is an informed decision, and that is what professionalism and product stewardship are all about. Remember, you have control over pesticide selection and your ability to use and apply the product safely.

APPENDIX 10.2 VEHICLE AND TRAILER INSPECTION CHECKSHEETS

It is called the 3-min, 360-degree, walk-around inspection. Drivers should conduct this inspection each morning before leaving company headquarters. Develop an inspection form and have the employees use it every day. An example follows.

Vehicle Maintenance Checklist

Company:____

Address:____

Date:____

Inspected by:____________________________________

Notes:____

____

Time:____

Truck Number:__________________________________

Mechanical lights

Front

High-beam headlights

Low-beam headlights

Turn signals

Running lights

Emergency flashers

Rear

Tail-lights

Brake lights

Turn signals

Running lights

Emergency flashers

Backup lights

License plate lights

Other

Horn in good working order

Seat belts in good working order

Brakes in good working order

Windshield free of obstructions

Mirrors

Parking brakes

Reverse-gear warning beepers

State inspection stickers current

Title, tax, and insurance papers in clearly marked envelope

Wipers

Wiper blades in good condition

Washer fluid dispenser filled

Washer fluid pump in working order

Neatness and supplies

Clean cab—no food wrappers or trash

Paper work for day

Maps

Extra change of clothes or coveralls

Reference materials

DOT Emergency Response Guidebook

Emergency phone numbers

Record of on-board pesticides

Label and MSDS for each product

First aid kit on board

Check gauge on fire extinguisher

Proof of insurance coverage, vehicle and trailer registrations, and accident report forms in glove compartment

Pesticides and application equipment stored safely outside of cab

Under the Hood

Check engine oil level

Check coolant level (cold)

Check automatic transmission fluid

Check battery and connections

Check belts and hoses

Check power steering fluid

Sides and Back

Reflective tape

Company name and phone number

Tires

Pressure

Tread wear

Cuts and cracks

Spare tire properly inflated

Jack and tools

On-Board Pesticide Containers

Containers properly sealed and secured

Legible labels on all containers

Granules in proper containers

Bait formulations in secured containers

No oversupply of pesticides

Empty containers properly rinsed and positioned for removal at end of day

Spill Control

- Absorbent materials and rags on board
- Shovel, broom, plastic bags on board

Equipment Check

- Sprayers (not pressurized)
- Supplies in moistureproof containers
- Lids fit securely on pesticide tanks
- Spray hoses in good condition
- Pressure gauges operable
- All sprayers cleaned and secured
- Water containers labeled
- Respiratory gear on board
- Hard hat on board
- Chemical-resistant clothing on board
- Goggles, gloves, boots
- Apron and coveralls
- Equipment secured
- Emergency equipment
  - Blanket
  - Flashlight
  - Flares/triangles
  - First aid kit
  - Jumper cables
  - Hammer, pliers, screwdrivers
  - Duct tape
  - Extra fuses

Leaks

- Check for leaks on driveway or ground
- No spills in truck bed

Appearance

- Clean appearance
- Interior cab neat

Trailer Inspection Checksheet

Trailer Number:____________________

Date:____________________

Inspected by:____________________

Notes:____________________

Uncouple and check hitch/coupling device
Lubricate ball and socket daily
Check electrical connections
Check lights
Check tires
Check suspension
Check brakes
Supplies secured

APPENDIX 10.3 EMERGENCY CALL LIST

Fill in the numbers of those listed here *Now* to have them accessible during an emergency.

- Prearranged contractors
- State emergency response commission
- Local emergency planning committee (federal reportable quantities and local ordinances)
- State department of natural resources (potential for fish and wildlife exposure)
- State department of environmental management
- National Response Center (federal reportable quantities)
- National Pesticide Telecommunication Network
- Local board of health
- State police
- Sheriff
- Fire department
- Ambulance
- Physician
- Veterinarian
- Insurance
- Lawyer
- Chemical dealer
- Local electric company
- Natural resource conservation officer
- Pesticide manufacturer (consult MSDS)
- Chemtrec: 800-424-9300
- INFOTREC
- Poison control center
- University animal hospitals
  - — Large animals
  - — Small animals
- Department of Agriculture
- Department of Environmental Management
- Other

APPENDIX 10.4 PESTICIDE STORAGE CHECKLIST

Safety is the key element in proper pesticide storage. If you answer no to any of the statements listed here, you should correct your storage facility immediately. Routine inspections help to identify problems before they become serious issues.

	Date		Date		Date	
	Yes	No	Yes	No	Yes	No
General Information						
Clean, neat pesticide storage site						
Current, on-site pesticide inventory						
Posted emergency phone numbers						
Labels and MSDS on file						
Accurate storage inspection log maintained						
Pesticide Containers						
Containers marked with purchase year						
(old pesticide inventory to be used first)						
Insecticides, herbicides, and fungicides segregated						
Flammables segregated						
Pesticides stored in original containers						
Labels legible and attached to containers						
Container caps tightly closed						
No reused pesticide containers present						
Pesticides stored off floor and low to ground						
Dry formulations stored on pallets						
Feeds stored separately from pesticides						
Used containers rinsed and punctured						
Rinsed and unrinsed containers separated						
Spills and Disposal						
Storage area free of spills or leaks						
Shovel and absorbent materials						
Floor drains capped (if present)						
Safety Information						
"No smoking" signs posted						
Safety equipment separated from pesticides						
Fire extinguisher in good working order						
Storage room locked						
Storage room posted: "Pesticides-Keep Out."						
Storage site well lighted and ventilated						

CHAPTER 11

PERSONAL PROTECTIVE EQUIPMENT: SELECTION, CARE, AND USE

Decisions on personal protective equipment (PPE) present a challenge for pesticide users on farms and in gardens, pest management businesses, and greenhouses. A broad range of pesticides with varied toxicities, formulations, and PPE requirements are available. This is compounded by the equally broad range of PPE available. The person responsible for selecting equipment and defining procedures that best fit a particular situation has to make many choices whether employer or employee. This chapter discusses links among attitudes, organizational philosophies, and personal safety. It explains why PPE is necessary, offers general guidance for selection, and provides suggestions for care of PPE (Fig. 11.1).

SAFETY LINKED TO TOXICITY AND EXPOSURE

Pesticide selection hinges on expected performance and costs, but the risk associated with use also is very important. Risk depends on the pesticide toxicity and the amount of exposure the applicator gets. Two questions must be answered for proper pesticide selection. First, how safe (toxic) is the chemical? And, second, how much exposure will the applicator have?

Many complicated and competing factors affect the answers to these two questions. But the basic elements in minimizing risk are choosing a pesticide with low toxicity and taking measures to reduce the potential for exposure. To reduce pesticide exposure, it is important to know how to use, select, and maintain PPE.

Understanding Toxicity

Pesticide toxicity is something the applicator cannot change. Laboratory tests conducted by manufacturers and reviewed by the U.S. Environmental Protection Agency (EPA) determine a product's toxicity and identify its strengths and weaknesses. Pesticide product testing also yields a toxicological profile of the active ingredient(s).

Figure 11.1 Companies must continually stress the need to wear personal protective equipment.

Evaluating Product Toxicity Answering the following questions may help the user in evaluating the toxicity of a pesticide.

What Hazards Are Associated with the Formulation? Certain risks are more common in some formulations than in others. For instance, dusts are easily inhaled; emulsifiable concentrates are readily absorbed through the skin; and aerosols may irritate the eyes and pose inhalation problems. The formulation of a product may dictate the safety precautions required, including use of PPE.

Is the Product Concentrated? Pesticide users are at greater risk from concentrated pesticides than from diluted, ready-to-use products. Concentrated products are more toxic than dilutions of the same product because they contain a much higher percentage of the toxic chemical. Also, concentrated products require mixing with water or other carriers and therefore more handling before application. This extra step increases the handler's potential for exposure from splashes and spills. The mixing and loading of a concentrated pesticide generally requires more PPE than does application of a diluted product.

What Is the Signal Word on a Pesticide Product? The relative toxicity of a pesticide is reflected on the label by signal words:

- DANGER (most toxic);
- WARNING (moderately toxic);
- CAUTION (least toxic).

Signal words also can reflect the toxicity of the product's nonlethal effects such as skin and eye irritation.

The most dangerous route of entry determines the signal word. For example, if product XYZ is moderately toxic when absorbed through the skin, highly toxic if inhaled, and only slightly toxic if swallowed, its signal word would be DANGER based on inhalation studies. Two pesticides labeled with the same signal word may or may not require the same PPE because that distinction is based on the most vulnerable route of entry as determined during the testing process. A pesticide with the signal word CAUTION usually will require less PPE than one carrying the signal word WARNING or DANGER. Labels of very toxic pesticides carry the signal word DANGER, accompanied by precautionary statements mandating the use of additional PPE.

Is the Product Classified for Restricted Use or General Use? A pesticide product may be classified for restricted use based on its potential effects on human health, wildlife, or the environment. Labels of restricted-use products carry specific precautions to safeguard those entities. Generally, restricted-use pesticides are to be used only by individuals licensed through state pesticide applicator certification exams and licensing requirements, although in some states unlicensed applicators are authorized to use them under the direct supervision of a licensed applicator. A pesticide that is classified for restricted use due to human health concerns requires more PPE than an unclassified (general-use) product. However, even general-use pesticides can be harmful if misused.

What Are the Acute Effects of the Pesticide? It is important to be familiar with the symptoms of acute (sudden, short-term) pesticide poisoning in humans. Symptoms of acute poisoning can range from fatigue and headache to dizziness, nausea, eye irritation, diarrhea, general weakness, and/or chest discomfort. If you (anyone) experience these symptoms and suspect they are pesticide related, seek medical attention immediately. Signal words on pesticide labels, as previously explained, show how toxic a pesticide is. With a DANGER label, if a 150-lb malc cats a taste to a teaspoonful of the product, he may not live. For WARNING label chemicals, it may take only a teaspoon to a tablespoonful. For CAUTION labels, the lethal dose may be an ounce to a pint. It takes less than this to kill a child.

Where Is More Detailed Information About Health Effects Available? In addition to the product labels, manufacturers are required to issue Material Safety Data Sheets (MSDS) for every pesticide product. The MSDS documents the effects of both acute and long-term or chronic exposure to the particular pesticide product. Statements of indications for both acute and chronic exposure must be presented; more detail may be given than is possible on a label. The MSDS also gives information on routes of exposure—dermal, inhalation, oral, or eye. The toxicological information that is summarized in an MSDS forms the basis for label requirements specifying certain PPE. MSDSs are prepared for all product formulations, but few manufacturers write them for end use dilutions. Chapter 15 provides more detailed information on MSDS.

Applicator Safety Based on Exposure

Applicator attitudes about personal safety and PPE for pesticide use often reflect the views of co-workers and supervisors. When management follows safe practices and wears required PPE, other workers are likely to follow their examples. Companies that emphasize pesticide safety and consistent use of label-required PPE are less likely to jeopardize their employees' health and their own corporate profits. Also, they are less likely to face pesticide-related lawsuits and penalties for noncompliance. Penalties and lawsuits could run into the millions. On-the-job training, hands-on demonstrations, work evaluations, and compliance checks by supervisors can stimulate consistent use of PPE.

Evaluating Risk Potential Based on Applicator Exposure The commitment to pesticide safety of an individual or a company is the key factor determining whether or not the use of a particular product constitutes risk. Applicators who have been trained to wear PPE properly and who have it available as needed can protect their health. The following questions will assist in evaluating the potential for applicator exposure.

How Many People Work for the Firm? The more applicators/people involved, the greater the chance that someone will neglect to follow safety precautions. In this case, accidents may happen.

How Experienced Are the Applicators? Inexperienced applicators may not understand the importance of PPE no matter if they are young, old, full- or part-time. All workers must be educated about PPE and trained to use it properly. Complacency must be addressed. Random inspection should be an ongoing activity to emphasize the importance of PPE and other safety considerations.

Are All Applicators/Workers Trained Adequately? Realistically, part-time employees often do not have the same quality of safety training that is available to full-time workers. This is not acceptable. The owner/manager of an operation has the responsibility to train all workers, regardless of the number of hours they spend on the job.

For What Length of Time Does Each Applicator Use Pesticides? Applicators whose primary duty year-round is applying pesticides obviously are more likely to be involved in a pesticide-related accident than are those who apply pesticides less often. In addition, those who apply pesticides over a long period of time become the subjects of chronic exposure, whereas those who apply chemicals infrequently are less likely to experience cumulative effects.

Answers to the preceding questions should offer valuable insights into risks posed to workers who use pesticides. Employers who admit that safety has not been a high priority should consider immediate implementation of a program to protect the health and safety of their employees.

Just as important, high-risk pesticides—those that are restricted for health reasons, those that are acutely toxic, and those linked to chronic concerns—should not be used until a safety education program is under way. One option is to select the least toxic pesticide that is effective against the target pest, thereby reducing risk potential.

PERSONAL PROTECTIVE EQUIPMENT: PROTECTING YOUR HANDS

Investing in gloves can pay dividends in safety. Protecting hands and arms with chemical-resistant gloves and a long-sleeved shirt can greatly reduce pesticide exposure during mixing, loading, and application.

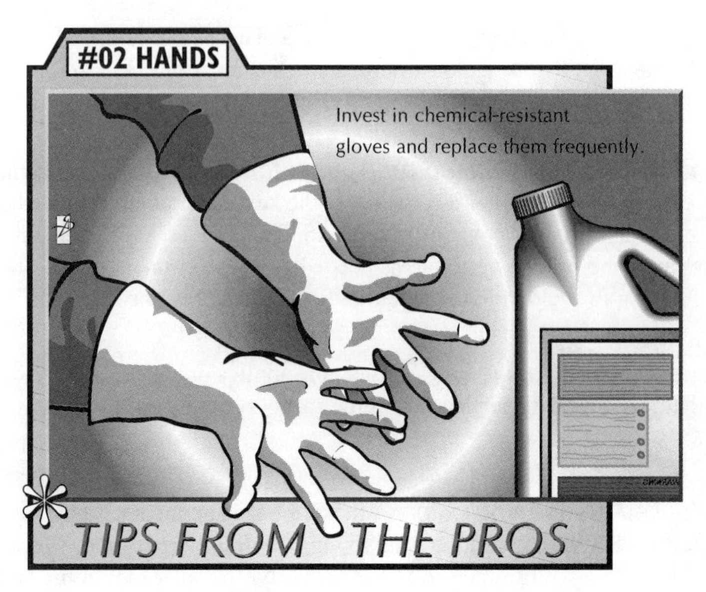

Figure 11.2 Use of chemical-resistant gloves is an important safety measure.

Hand Protection and Precautionary Statements on Pesticide Labels

PPE requirements for hand protection are listed on most labels, generally specifying chemical-resistant gloves to guard against pesticide contact. The labels of some pesticide formulations (e.g., wettable powders and granules) may suggest the use of waterproof, not chemical-resistant, gloves; others, such as those found on many ready-to-use products, may not require any gloves. Safety specialists believe, however, that applicators should wear chemical-resistant gloves routinely; it's a simple risk reduction practice that goes a long way in eliminating exposure—and it's a habit that's easy to form.

Selecting Gloves for Hand Protection

Safety manufacturers design gloves to meet the specific needs of applicators handling hazardous chemicals; consideration is given to thickness, cuff length, fabrics and coatings, etc. For example, gloves advertised as chemical-resistant, waterproof, and effective against dry formulations such as wettable powders, water-dispersible granules, microencapsulated granules, and dusts may be easily damaged in contact with solvents in some liquid formulations (e.g., emulsifiable concentrates) (Fig. 11.2).

Glove Selection Tips Taking the time to select the right kind of gloves is important: Make an informed decision.

Glove Materials Must Resist the Product's Active Ingredient and Its Solvents Glove descriptions may indicate the amount of time it takes for a chemical to move from the outside surface of the glove to its interior—generally referred to as breakthrough time. Gloves may provide hand protection for a few minutes as in the case of disposable, single-use gloves, or for hours in the case of gloves designed for repeated use. Reusable gloves should have a breakthrough time of at least 240 min (4 hr).

Gloves Should Be Thick Enough to Afford Protection during the Total Time Required for Completion of the Task at Hand Glove thickness is described in millimeters (1 ml = 0.001 inch) and the thicker the chemical-resistant glove, the more protection it will provide and the longer it will last under normal use conditions. Thick gloves generally offer more protection against chemical breakthrough but may restrict hand movement; select gloves that are at least 14-ml thick as a trade-off between thickness and manual dexterity.

Gloves Should Allow Adequate Grip So That Applicators Can Safely Carry Out Their Jobs (e.g., change nozzles and screens) Various glove surface textures—etched, raised, roughened—influence the wearer's ability to grip.

Gloves Should Be Comfortable: The Right Fit Makes the Difference The appropriate glove size can be determined by measuring the circumference of the hand (palm and back); for example, a hand circumference of 8 inches indicates a size 8 glove. But many gloves are sold as small, medium, large, etc. A hand circumference of 5–7 inches may need an extra small glove. However, gloves should not be so tight that the material is stretched thin. They will be uncomfortable and more likely to tear or break.

Hand Circumference—Glove Size

7–8 inches—small

8–9 inches—medium

9–10 inches—large

10–12 inches—extra large

Gloves Must Be Long Enough to Adequately Protect the Hands and Arms from Chemical Splashes Glove length is measured from the tip of the middle finger to the farthest edge of the glove. Lengths of 10–12 inches provide hand and lower forearm protection; 13–14 inches, middle forearm protection; 18 inches, elbow protection; and 32 inches, protection to the shoulder.

Gloves Must Be Unlined Cotton, leather, and canvas gloves, as well as gloves lined with these materials, should not be used with pesticides; nor should baby powder or talc be used to make gloves easier to put on and remove. The reasoning is that if the pesticide accidentally leaks inside the gloves, linings and absorbents would actually soak up the chemical and hold it against the skin, thereby increasing the degree of exposure.

Glove Use and Maintenance Tips

- Get new gloves regularly as needed, and provide them as requested by the applicators you supervise.
- Discard disposable gloves after use.
- Adjust gloves according to the task. When spraying overhead, gloves should be folded with a cuff or extended over long sleeves so the pesticide cannot be funneled into the gloves. Conversely, spraying downward requires long sleeves to extend over the gloves to prevent the pesticide from being channeled into the gloves.
- Keep an extra, clean pair of gloves in a zip-close plastic bag in a location free of pesticides but easily accessible in case the pair you're wearing gets torn or contaminated.

- Wash reusable gloves with soap and water before removing them.
- Store contaminated reusable gloves in a zip-close plastic bag until final cleanup or disposal.
- Wash chemical-resistant gloves with soap and water and hang by the fingertips to air dry.
- Keep pesticide-contaminated gloves separate from other safety equipment.
- Inspect gloves for visual signs of wear before each use. Watch for clues that new gloves are needed: color changes, thin spots, cracks, softening, swelling, bubbling, and stiffening. Discard all gloves with cuts, holes, abrasions, or staining.
- Never try to patch gloves with duct or electrical tape.
- Triple rinse gloves before disposal, then cut off the fingers to prevent reuse.
- Do not put pesticide-contaminated gloves in your pockets because chemicals can transfer from gloves to clothing.
- Never leave pesticide-contaminated gloves behind the seat, on the floorboard, above the visor, on the stick shift, or in the bed of your truck because pesticide residues may transfer to such surfaces.
- Never wear pesticide-contaminated gloves when feeding or watering livestock, harvesting fruits and vegetables, or cleaning animal pens.
- Wash your hands after wearing gloves.

PERSONAL PROTECTIVE EQUIPMENT: PROTECTING YOUR EYES

Eye protection is advisable when handling and applying all pesticides, and some product labels specify its necessity. Pesticides in contact with the eye can cause a wide range of symptoms, from slight irritation to blindness; and although no pesticide-to-eye contact should be considered minor, concentrates generally are more injurious than diluted products.

Eye Protection and Precautionary Statements on Pesticide Labels

A pesticide label may require the use of specific types of eye protection, such as chemical-resistant goggles, or it may only suggest general protective eyewear. The EPA defines protective eyewear as safety spectacles with side shields and brow guards; goggles; face shields; or full-facepiece respirators.

Example of pesticide label information that addresses eye protection include:

Avoid contact with skin, eyes, or clothing. Causes moderate eye irritation. Wear eye protection. If pesticide gets into eyes, flush with plenty of water. Get medical attention if irritation persists.

Do not get in eyes, on skin, or on clothing. Corrosive and causes irreversible eye damage. Applicators must wear protective eyewear. If in eyes, hold eyelids open and flush with a steady, gentle stream of water for 15 min. Get medical attention.

Selecting Eye Protection Equipment

Selecting safety equipment for preventing eye exposure to pesticides requires an understanding of the potential sources of exposure and of the kinds of eye protection available.

Figure 11.3 Eye protection equipment should be fitted properly.

Product formulation determines the probability of exposure: dusts escaping from bagged products; mists from air blast or fogging sprays; accidental liquid splashes. There are three types of eye protection equipment (excluding the full-facepiece respirator) designed to provide eye protection from airborne particles or chemical splashes: safety spectacles (glasses); goggles; and face shields (Fig. 11.3).

Safety Spectacles Regular corrective (prescription) eye glasses are not enough protection and contact lenses should not be worn to work with pesticides.

Spectacles are the most basic form of eye protection equipment. Safety spectacles are unlikely to fog and can be fitted for comfort. Safety spectacles generally are recommended for protection from only the least toxic products, under minimal exposure conditions, and they should have scratch-resistant lenses for maximum visibility and durability. Versions with brow and side shields offer some protection from overhead and side impact, but most safety specialists do not recommend them for protection against chemical splashes because they do not provide full eye coverage. Selecting the correct size is essential to comfort and satisfactory use.

Goggles Goggles with polycarbonate lenses protect eyes from unforeseen hazards, but some fail to provide adequate protection against liquid pesticide splashes. For instance, those manufactured with perforations or air holes for increased air circulation are not considered splash resistant because liquids may leak through the small air holes. Goggles with indirect vents or no vents are considered splash resistant. Most goggles have adjustable straps to fasten them securely to the face, and some have fog-free lenses.

Face Shields Face shields are intended as secondary means of eye protection and are designed to be worn over safety spectacles or goggles for full-face protection. They should not be worn without safety spectacles or goggles.

Selection Tips

- Personal spectacles, goggles, and face shields should be assigned and fitted to each applicator. This equipment should not be shared.
- Eye protection equipment should permit peripheral (side) vision.
- When purchasing eye protection, make sure it complies with ANSI Z87 for occupational eye and face protection.
- Most safety spectacles can be purchased with prescription lenses.

Maintenance Tips

- Clean the contaminated protective eyewear immediately after use. Rinse under running water to remove dust from the lenses, and air dry, rather than wiping, to prevent scratches.
- Regularly soak equipment in a solution of warm water and detergent for 10–15 min. after rinsing under running water. Goggles should be disassembled so that the headband, lenses, and frame get thorough cleaning. Avoid using scratchy materials or still brushes for cleaning.
- Do not leave eye protection equipment on the seat or floor of a truck, tractor, or other application equipment between uses.
- Do not put eyewear above the visor or hang it from the mirror of a vehicle; instead keep a zip-close bag or plastic box handy for storage between wearings.
- Store clean goggles and safety spectacles for longer periods in clean plastic zip-close bags or carrying cases, then store them in a sturdy box or compartment to prevent scratching, breaking, or crushing.
- Avoid leaving or storing eyewear in the sun since heat and ultraviolet radiation can affect the performance of plastics.
- Store all eye protection equipment away from possible contamination sites such as chemical storage areas.
- Make sure that safety glasses do not slip when a person is sweating.
- Replace spectacles, goggles, or face shields when they are pitted, scratched, stained, or cracked.
- Replace headstraps when they do not hold the eyewear snuggly to the face or when they show signs of deterioration and/or staining.
- Discard damaged protective eyewear according to guidelines for disposal of PPE.

PERSONAL PROTECTIVE EQUIPMENT: PROTECTING YOUR LUNGS

Respirators are the most complicated of all personal protective equipment, and mastering their proper selection and use requires time and effort. For example, are you confident that your respirator is the right one for the job? How would you deal with an employee who has a beard? Do you perform a seal check each time you put on a respirator? Can dust masks be used for protection against organic vapors? Have you ever wondered why you could smell or taste the chemical even when wearing a respirator? The knowledge required to answer questions such as these can mean the difference between health and sickness . . . between life and death.

Respiratory Protection and Precautionary Statements on Pesticide Labels

The statement *Avoid breathing dusts or spray mists* is found on most pesticide labels. Some provide more information, such as outcomes and corrective actions:

> *. . . may cause respiratory tract irritation. If inhaled, remove to fresh air. Get medical attention if breathing difficulty develops.*

Pesticide manufacturers label more toxic products with specific respiratory protection statements when studies show that use of the product without a respirator may pose a risk.

Respiratory protection is intended to remove contaminants from the air you breathe when handling pesticides. Labels are very clear in directing the user to respirators approved by the National Institute of Occupational Safety and Health (NIOSH) and the Mine Safety and Health Administration (MSHA). Following are examples of label directions:

> *. . . a dust–mist filtering respirator (MSHA/NIOSH approval number prefix TC-21C).* (TC = tested and certified.)
>
> *. . . a respirator, approved for pesticides, having an organic vapor cartridge and pesticide prefilter (MSHA/NIOSH approval number prefix TC-23C), or pesticide canister (MSHA/NIOSH approval number prefix TC-14G).*

NIOSH no longer certifies respirators for protection against pesticides. Newer respirators suitable for use for protection against pesticides will carry a NIOSH approval for Organic Vapor and Particulate. The approval prefix on these respirators is TC-84A, but that prefix is not specific to respirators suitable for pesticides and therefore should not be used as sole criteria for respirator selection. Some newer particulate filters are not suitable for oil and the pesticide emulsifier may prohibit their use. The EPA has not yet mandated the newer respirator descriptions on pesticide labels, so caution must be used in selection. Contact the respirator manufacturer for guidance. Older respirators approved for pesticides may continue to be used as long as they still work.

Selecting Respiratory Protection Equipment

Respirators are required when using some pesticide products, but most pesticide labels do not specify their use. Wise applicators, however, will elect to wear one when applying pesticides in enclosed areas such as greenhouses, crawl spaces, and grain bins—even when the label does not require it. Understanding each type of respirator's uses and limitations is the first step in selecting the right respirator for the job. Following are brief descriptions of the different types available.

Air Purifying Respirators Air purifying respirators do not supply fresh air. They filter the surrounding air as the applicator inhales and breathes through the mask. The air purifying respirators are designed to absorb or trap and block contaminants so that the air reaching the respiratory tract is less contaminated. Air purifying respirators may be disposable and look similar to the common dust masks, but are capable of filtering out smaller particles. They many contain activated charcoal in the filter material. Air purifying respirators are available in half- or full-face styles. Reusable styles of air purifying respirators have disposable cartridges that are labeled according to their intended purpose—whether for particulates, ammonia, or pesticide vapors. Warning! Read and follow the respirator manufacturer's instruction manual before using the respirator.

Air purifying respirators cannot be used where the oxygen level is below 19.5%, nor in atmospheres that are immediately dangerous to life or health (e.g., fumigation in an enclosed area).

Note: Breathing becomes more difficult when the filters of air purifying respirators are clogged. Move to a safer environment and change the respirator (disposable), filter, or cartridge immediately when you sense a different taste or odor. For work in a dangerous atmosphere, better protection will be provided by a supplied-air respirator.

Full-Face Respirators Full-face respirators work in the same manner and use the same cartridges and/or prefilters as half-mask respirators, and they offer two advantages: They protect the face and eyes against airborne particles and splashes; and they offer a higher level of respiratory protection.

Half-Mask Respirators There are two common types of half-mask respirators. One relies on filters (paper/fiber) to trap dusts, mists, and other particles. The second type relies on cartridges that are designed for specific uses. Often the cartridges have prefilters or more than one layer of filter material so that they can remove both chemical vapors and dusts or particulates.

The filtering-facepiece respirator, more commonly referred to as the dust–mist respirator, removes dust, pollen, particles, spray mists, and some nuisance odors as air passes through the fabric of the mask.

The pesticide applicator's comfort, utility, and performance can be improved by using filtering respirators made with some of the following features:

- Nonallergenic materials
- A metal nosepiece to facilitate fit
- An exhalation valve that allows moist air to escape the mask
- Activated charcoal to remove some odors and vapors
- Two straps to ensure a tight fit

Workers who elect to wear a respirator in situations where it is not required by the pesticide label should choose an appropriate NIOSH-approved respirator over a general purpose dust mask. Some filtering respirators do not meet NIOSH standards and therefore should not be used when pesticide labels require NIOSH-approved respirators.

Powered Air Purifying Respirator Most half-mask and full-face respirators require the user to physically draw in air. Powered air purifying respirators are equipped with a battery-operated blower that forces air into the mask to assist breathing. The forced air cools the atmosphere in the mask and reduces stress on the heart and lungs. Powered air purifying respirators with hoods or helmets may be worn by people with beards.

Supplied Air Respirators Supplied air respirators are highly specialized for use in atmospheres that are more toxic than those where air purifying respirators are permitted. When equipped with a self-contained emergency air supply, they can be used in atmospheres that are hazardous or where the oxygen level is below 19.5%. The air supply is provided through a hose hooked to a stationary cylinder of compressed air or from a portable air pump. Supplied air respirators are more expensive than air purifying respirators and require continual maintenance; applicators must be specially trained to use them.

Respiratory Protection Equipment Selection Tips

Determine the Form of the Contaminant That Constitutes the Risk There are two forms of chemical contaminants: particulates (dust, mist, and smoke) and vapors (including gases). Therefore, before selecting a respirator for protection against a specific product, consult the product label to determine what physical form "poses the risk."

Verify the Model of Respirator That Will Remove the Contaminant in Question Respirators approved by NIOSH have an attached label bearing an approval number with the prefix "TC." The product label prescribes a specific respirator by NIOSH approval number. For example, TC-23C respirators are approved for use when contaminants such as organic vapors, acid gas, or ammonia might pose a risk, but all particulate filters now being made are approved with the new designation TC-84A. Although it will be some time before manufacturers begin specifying this new class on pesticide labels, prudent applicators who purchase respirators in the meantime should proceed in selecting a TC-84A model.

Safety equipment catalogs usually list the NIOSH approval number as part of the respirator description. If ordering from a catalog that does not list the number, look for "organic vapor" cartridges and filters. If it is still unclear which to choose, call the toll-free number listed in the catalog and ask to speak to a safety specialist; or call the respirator manufacturer directly. All respirator manufacturers use the same color-coding system in designating cartridges for use in specific applications.

The various uses are often designated by color in labeling:

- Black = organic vapor, suitable for pesticides if equipped with an additional particulate filter
- Green = ammonia
- Olive = any gas or vapor that does not have a specific color associated with it. This would include multicontaminant cartridges.
- Magenta = any particulates

Fit Test Every Wearer, Every Time A qualitative or quantitative fit test must be performed as required by OSHA to determine correct facepiece size (small, medium, or large). There are several fit test protocols and fit testing can be performed by the employer, a trained worker, or a safety professional. OSHA lists specific protocols for conducting fit tests in their regulation. Manufacturers of respirators also provide fit test kits. Once a respirator giving an adequate fit has been chosen using an approved protocol, OSHA also mandates that a seal check be conducted each time a person puts on their respirator. Instructions on conducting fit checks generally accompany half-mask and full-face respirators. The procedure takes only a few seconds and must be exercised by the trained worker according to the following steps:

Put on the respirator and tighten all straps.

For a positive-pressure seal check, cover the exhalation valve (the single opening near the chin area of the respirator) with the palm of the hand and exhale (blow) into the mask for 5–10 sec. This causes pressure within the mask to increase, causing a slight bulging of the facepiece if fitted properly; if there is a leak, you will feel it.

For a negative-pressure seal check, block the filter inlets (some models have one inlet; others have two) with the palms of the hand and inhale for 5–10 sec, causing a slight collapse of the facepiece.

A good seal is indicated when the facepiece remains slightly bulging (positive pressure) or slightly collapsed (negative pressure) during the seal check. An unsatisfactory seal indicates that air is moving through openings between the respirator and the face, requiring that a second positive- and negative-pressure seal check be performed.

Remember: There is no such thing as a respirator working "well enough." A respirator with a leaking facepiece—even a small leak—is not protecting the wearer.

- Make sure half-mask respirators allow space for safety eyewear, especially when working with eye or skin irritants.
- Require everyone who wears a respirator to be clean-shaven. Plainly and simply, bearded workers cannot be protected against inhalation exposure by wearing a respirator. Even minimal beard growth prohibits a tight seal.
- Require everyone who wears a respirator to be preexamined by a physician (e.g., OSHA questionnaire and/or visitation to physician). Respirators make breathing more difficult, and the resulting stress on the heart and lungs may worsen preexisting conditions and increase blood pressure. No one should wear a respirator without medical clearance, since it is an OSHA requirement. A thorough physical examination and a review of medical history by a qualified physician are a must for using a self-contained breathing apparatus.

Maintenance Tips

- The manufacturer's instructions that accompany each respirator should be reviewed and retained as a reference.
- Conduct a positive- and negative-pressure seal check each time the respirator is put on.
- Change cartridges after 8 hr of use or more frequently if the wearer can smell or taste the pesticide or other chemicals; change particulate respirators and filters when breathing becomes difficult.
- Discard cartridges or filters which get wet.
- Look for tears in straps, face-shield scratches, missing parts, and general cleanliness of the respirator each time you put it on. Daily inspections are critical.
- Disassemble and sanitize reusable respirators at the end of each work shift. Use a soft brush or cloth and warm water with a germicidal detergent. Rinse the respirator in clean water and air dry.
- Never wash cartridges and filters; they must be discarded.
- Place reusable respirators in a zip-close bag or airtight container for storage; and just to be on the safe side, change the storage bag or container frequently to prevent buildup of any pesticide residue which might have been missed.
- Store respirators properly to prevent distortion (e.g., crushing) of the facepiece.
- Store facepieces, cartridges, prefilters, and replacement parts away from pesticide storage areas.
- Protect respirators in storage from dust, direct sunlight, extreme cold and heat, excessive moisture, and damaging chemicals.

- Purchase replacement parts such as filters and cartridges that are specifically designed for the respirator model used. Never interchange cartridges, filters, or other replacement parts from different respirator manufacturers. Users should never try to repair respirators. Consult the manufacturer or distributor from whom the respirator was purchased and make arrangements for a professional to handle repairs.

OSHA Respirator Regulations

Employers who require employees to wear respirators are required to comply with the updated regulations found in 29 CFR Part 1910.134. The following are some of the respiratory requirements that employers must provide to their employees.

- Respirators shall be provided by the employer when such equipment is necessary to protect the health of the employee.
- The employer shall provide respirators suitable for the purpose intended.
- The employer shall be responsible for the establishment and maintenance of a respiratory protection program. The written plan shall include each of the following:
  a. Procedures for selecting respirators for use in the workplace.
  b. Medical evaluations of employees required to use respirators.
  c. Fit-testing procedures for tight-fitting respirators.
  d. Procedures for proper use during routine and emergency situations.
  e. Procedures for cleaning, disinfecting, storing, inspection, repairing, discarding, or otherwise maintaining respirators.
  f. Training of employees to deal with respiratory hazards in routine and emergency situations.
  g. Training on proper use, adjustment (to fit), and removal of respirators.
  h. Procedures for regular evaluation of the effectiveness of the respiratory protection program.
  i. An example of a written respirator program is provided in Appendix 11.1.
- The employer shall provide respirators, training, and medical evaluations at no cost to the employee.
- The employer shall provide a medical evaluation to determine the employee's ability to use a respirator before the employee is fit tested or required to use the respirator in the workplace.
  a. An employee may refuse to participate in medical evaluations. If the employer allows this, the employee cannot wear a respirator nor work in an area that requires its use.
  b. The employee shall identify a physician or other licensed health care professional who will perform medical evaluations using a medical questionnaire or a medical examination.
  c. Employees giving a positive response to any question among questions 1 and 8 in section 2 of OSHA's medical evaluation form are required to have a follow-up medical examination.
  d. Employers shall provide employees an opportunity to discuss their completed questionnaires and/or medical examination results with a primary health care physician.

- Employers shall establish and retain written information regarding medical evaluations, fit testing, and the respiratory program.
- The employer shall establish a record of the qualitative or quantitative fit test administered to an employee, including:
  a. Name of the employee
  b. Type of fit test performed
  c. Specific make, model, style, and size of respirator tested
  d. Date of the test
  e. Pass/fail results
- Fit testing is required annually, or whenever a different respirator is used, or whenever there is a change in an employee's physical condition that could affect respirator fit.
- The entire written respiratory protection program shall be made available to all employees.
- The training must include information on
  a. why the respirator is necessary and how improper fit, use, or maintenance can compromise its protective effect;
  b. the limitations and capabilities of the respirator;
  c. how to use the respirator effectively in emergency situations;
  d. how to inspect, put on, use, and remove the respirator, and how to check the seals in the respirator;
  e. the procedures for maintenance and storage of the respirator; and
  f. how to recognize medical signs and symptoms that may limit or prevent the effective use of respirators.
- Training must occur at least annually, and more often as the need arises.

PERSONAL PROTECTIVE EQUIPMENT: PROTECTING YOUR BODY

Wearing the clothing and personal protective equipment specified on a pesticide label is important in preventing or minimizing exposure. PPE protects the body against splashes that may occur while pouring concentrates from a container, as well as spray mists that may drift onto the applicator under certain conditions.

Even under minimum exposure conditions, the selection of clothing should be given careful thought. A long-sleeved shirt, long pants, shoes, and socks are essential. Layers of clothing help reduce exposure by impeding the penetration of the pesticide.

Additional protective clothing sometimes is required to supplement the protection offered by ordinary work clothes. For example, a product label may require that cotton coveralls be worn over work pants. Chemical-resistant boots and an apron may be required clothing for mixing some pesticides. Proper clothing, when used with other safety measures, will reduce the likelihood that an applicator will be accidentally exposed.

Forearm and Leg Protection

Protecting the skin on the arms and legs is an important way to reduce pesticide exposure. Arms may become contaminated during mixing and loading operations if gloves are too short. Lower pant legs may become contaminated during lawn care applications.

Sleeve guards are designed for protection of the arms when wearing a long-sleeved shirt. Guards cover the arm from the wrist to above the elbow and have elastic in both ends to hold them close and block the pesticide from the openings. Chaps are designed to cover the legs; they fasten to the belt at the waist and are held to the ankle with elastic. Chaps provide additional protection during turf and nursery applications where the lower leg is most vulnerable to exposure. Long, protective pants are also available.

Sleeve guards, chaps, and protective pants are available in several materials. Some are disposable; others are reusable. If reusable, they should be cleaned thoroughly after each use.

Footwear

Some labels may only direct applicators to wear shoes plus socks. Others may specify chemical-resistant footwear, plus socks. There are many types of chemical-resistant footwear designed to be worn over socks; others are to be worn over ordinary shoes. Rubber and neoprene are commonly used in making chemical-resistant footwear. Leather and canvas are not recommended for footwear because they can absorb pesticides and trap them next to the skin, and they are not considered chemical-resistant.

The structure and design of footwear also are important. For example, steel toes protect the feet, and antiskid soles and tread help prevent falls. Tall boots provide leg protection; disposable booties can be worn over shoes for short-term, minimal protection from exposure, primarily in dry conditions.

Pant legs should hang over the outside of footwear so that pesticides cannot be funneled down the pant leg into the shoes or boots. Footwear should be washed in soapy water after each day's use. As with all PPE, footwear worn during a pesticide application should not be worn indoors; if it becomes damaged or is leaking, it must be discarded.

Headgear

Spraying upward into trees, using an air blaster in an orchard, or using a handheld sprayer in greenhouses may expose the head, shoulders, and back to pesticides. Labels direct applicators spraying overhead in any situation to wear chemical-resistant headgear.

Well-designed headgear provides protection for the scalp, neck, and ears and can help prevent pesticides from reaching the body through neckline openings in clothing. Headgear can take many forms. It may be stiff and impact resistant or soft enough to tie around the face as a hood. Many coveralls and jackets have attached hoods. Separate hoods that drape over the shoulders also are available made of various water-repellent, waterproof, or chemical-resistant materials. If a hat with a brim is chosen, the brim should be stiff enough to hold its shape without collapsing and interfering with vision, and the hat should stay on the head despite weather conditions—wind or rain. Although there are no requirements for sun protection under the Worker Protection Standard or with regard to pesticide use, the incidence of skin cancer among farm operators and workers suggests that headgear should offer as much sun protection as possible because skin cancer is associated with repeated ultraviolet exposure from the sun.

Aprons

Pesticide users can decrease the likelihood of frontal exposure by wearing chemical-resistant aprons. The product label may state that a chemical-resistant apron is to be worn when mixing, loading, or cleaning equipment. Pesticides bearing the signal word DANGER and

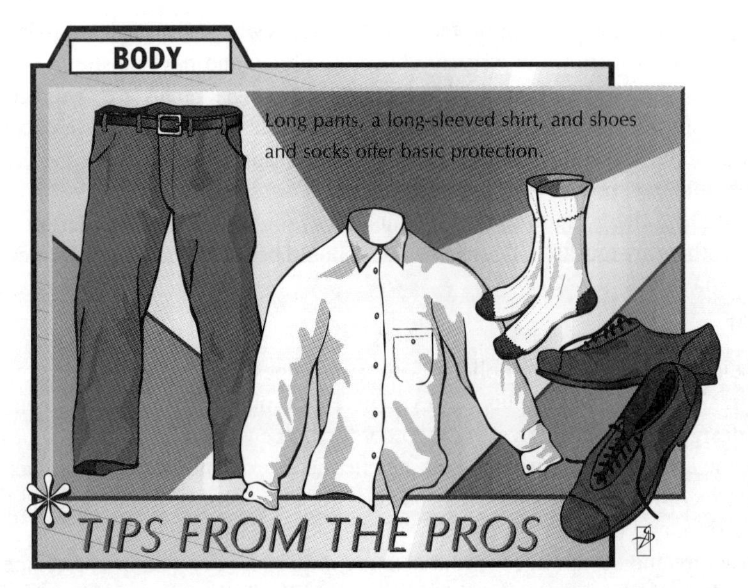

Figure 11.4 Protective clothing can prevent dangerous exposure to pesticides.

products with chronic toxicity potential may require an apron in addition to other protective clothing to protect against spills. Chemical-resistant aprons are available in materials such as butyl, neoprene, and nitrile. Aprons backed with cotton or other materials are resistant to tearing; but backing materials can absorb pesticides, making cleanup difficult or impossible.

Although aprons offer protection against pesticide exposure, many workers do not wear them. In some cases, aprons can lead to accidents when the user must climb ladders or work around machinery. Seek the advice of workers who will be expected to wear aprons, and use their input in selecting the best apron for the job, or perhaps an alternative such as coveralls (Fig. 11.4).

Coveralls

Labels may specify the use of coveralls to provide an extra layer of protection against pesticides penetrating to the skin. They are available in reusable, limited-use, and disposable styles.

Washable, reusable cotton or cotton blend coveralls are acceptable when using low-toxicity pesticides (those labeled CAUTION). However, because trace amounts of pesticides cannot be removed from cotton fabrics, annual or more frequent replacement is suggested. Cotton coveralls should never be used when handling liquid pesticide concentrate; whenever a concentrate comes into contact with clothing that is not chemical-resistant, the garments must be properly discarded.

Reusable coveralls or two-piece rain or splash suits designed to be worn over regular work clothes can be purchased with coatings of PVC, rubber, neoprene, nitrile, or polyurethane over a durable cotton, nylon-scrim, or polyester backing. They differ in chemical resistance and may be washable. Manufacturers' labels can guide the selection process and should explain care recommendations, although insufficient research has been conducted on the effectiveness of specific care methods in removing pesticide contamination.

Limited-use or disposable coveralls are designed to be worn for one workday, then discarded. Marketers use the term "limited use" to suggest that these coveralls can be worn more than once, over a period of time that might equal one working day. Limited use definitely does not mean repeated wear over a period of weeks or until the coveralls rip or tear. The difficulty with the idea of "limited use" is that the items cannot be cleaned between uses, because they are not washable. Some disintegrate upon washing; others may not but still may be contaminated. Never wash disposable nonwoven coveralls because laundering tends to reduce their liquid repellency and could move contaminants to the inside of the garment. After wearing, it is very difficult to manage taking them off and putting them on again without contaminating your hands. Where to store them between uses without contaminating other items is also a concern. Therefore, limited-use nonwoven coveralls must be discarded when they are contaminated or damaged.

Limited-use or disposable coveralls are promoted by their brand names which do change from time to time. They are available in several styles with different levels of protection, depending on their construction, fiber content, fabric structure (thickness, lamination, etc.), and finishes. For maximum protection, seams should be sealed, rather than just stitched, and closures should have overlapping flaps. Some examples are listed here:

- Standard Tyvek spunbonded polyolefin is breathable and offers about the same protection from granules and dust as do regular cotton coveralls. It is better at resisting light spray or mist, but it is not suitable for use with concentrated materials. Tyvek is available in a blue-denim color as well as white.
- Tyvek QC is polyethylene coated to repel water and offers better chemical resistance than do regular Tyvek or cotton. It is recommended for use with CAUTION- and WARNING-label pesticides and comes in colors.
- Tyvek laminated with Tychem SL or Saranex-23P can be used with DANGER-label pesticides. Saranex-coated Tyvek is bright yellow; Tychem-coated Tyvek is white. These materials lack breathability and, regardless of coating, Tyvek is not recommended for chlorinated hydrocarbons.
- Pro-Shield 1, 2, 3 spunbonded polypropylene coveralls are advertised for comfort, breathability, and barrier resistance to liquids. Pro-Shield 2 and 3 are laminated to a microporous polypropylene film. Worn over regular work clothing, these coveralls are useful in low-exposure situations with low-toxicity pesticides, but the manufacturers do not claim to offer true chemical protection.
- Pro-Shield Nex Gen is advertised as a liquid-proof barrier material for nontoxic liquid and spray applications. Independent testing showed it was a barrier to diazinon and malathion.
- Kleenguard Ultra also is a spunbonded polypropylene laminate that offers splash and barrier resistance for CAUTION-label pesticides. It has a microporous film in the middle layer for greater comfort as heat and perspiration can escape.
- Kleenguard Heavy Duty offers protection from fine dusts and liquid splashes. Its middle layer is a matrix of microfibers.
- Kleenguard Extra Protection is promoted for agricultural spraying. It has an antistatic coating and provides protection from oils and alcohols.
- Tempro is a disposable coverall material that is breathable and water and grease repellent and that has been treated for flame retardance.

Trade names such as Responder and Chem-Tech (neoprene-coated polyester fabric) are used for hazardous chemical exposure and cleanup.

Coveralls for pesticide application should be selected to provide appropriate chemical resistance against the pesticide and its formulation. Applicators should also be aware of other safety considerations.

Coveralls should fit well. A snug fit at the neck will prevent pesticides from filtering down the back and chest. Coveralls also should have adequate torso length to allow ease of movement. However, they should not be so loose or baggy that they are susceptible to entanglement in the moving parts of machinery. Coverall materials are very strong and do not tear easily if caught in machinery. Entanglement of clothing can lead to traumatic injury, unrelated to pesticides. Coveralls may have attached hoods or separate head protection.

If disposable coveralls are too long in the sleeves or pant legs, they can be cut off because the fabrics do not ravel. If woven fabric coveralls are shortened, they will need hemming to prevent raveling of strong yarns that can be an entanglement hazard.

PESTICIDE LABELS PROVIDE VALUABLE SAFETY INSTRUCTIONS

The EPA's pesticide registration process takes into consideration a product's label information on the use of personal protective equipment. Pesticide applicators must rely on label information to know what PPE to use when handling a given product. Wearing inappropriate PPE, or failure to wear any at all, can lead to overexposure. Physical illness from overexposure can mean absence from work, sometimes triggering loss of income. It can mean mounting medical expenses and stress. Physical illness from overexposure can result in pain and suffering—sometimes even death.

If PPE label instructions are ignored and routine overexposure does occur, lawsuits and penalties may follow. If failure to comply with label directions is proven, regulatory action against the applicator and/or the affiliated company may result.

Protective clothing and equipment statements are located on the label (or labeling) of a pesticide product under Hazards to Humans and Domestic Animals or as "Agricultural Use Requirements" under Directions for Use. The Worker Protection Standard (WPS) requires that labels of pesticides used on farms or in forests, nurseries, or greenhouses list the types of PPE the user must employ. It is important to realize that PPE required by a pesticide label represents minimum protection; the applicator may elect to use more protective clothing or equipment than is stipulated on the label.

The EPA requires that all pesticide labels bear the following statement under the heading Directions for Use: *"It is a violation of federal law to use this pesticide in a manner inconsistent with its labeling."* Legally, the statement places the responsibility for using the product according to label directions directly on the buyer and user. The intent is to communicate to users—homeowners, farmers, commercial applicators—that they are personally responsible for judicious use of the pesticide (See Chapter 7 for more information on pesticide labels).

EPA Chemical Resistance Category Chart

Pesticide labels refer to chemical resistance categories (A through H) for personal protective equipment (Table 11.1). Items in these categories are made of materials that the pesticide cannot pass through during the times indicated by the chart. The categories are based on the solvents used in the pesticides, not the pesticides themselves. Therefore, there will be

TABLE 11.1 EPA Chemical Resistance Category Chart

Category Listed on Pesticide Label	Type of Personal Protective Material							
	Barrier Laminate	Rubber [a]				Polyethylene	Polyvinyl Chloride[a]	Viton
		Neoprene	Butyl	Nitrile	Natural			
A	High[b]	High	High	High	High	High	High	High
B	High	Slight	High	Slight	None[e]	Slight	Slight	Slight
C	High	High	High	High	Moderate	Moderate	High	High
D	High	Moderate[c]	High	Moderate	None	None	None	Slight
E	High	High	Slight	High	Slight	None	Moderate	High
F	High	Moderate	High	High	Slight	None	Slight	High
G	High	Slight[d]	Slight	Slight	None	None	None	High
H	High	Slight	Slight	Slight	None	None	None	High

[a] Recommendation based on PPE at least 14 ml or greater in thickness.
[b] High: Highly chemical resistant. Clean or replace PPE at end of each day's work period. Rinse off pesticides at rest breaks.
[c] Moderate: Moderately chemical resistant. Clean or replace PPE within an hour or two of contact.
[d] Slight: Slightly chemical resistant. Clean or replace PPE within 10 min of contact.
[e] None: No chemical resistance. Do not wear this type of material as PPE when contact is possible.

instances where the same pesticide in two different formulations (e.g., wettable powder and emulsifiable concentrate) will require PPE from two different chemical resistance categories.

The following is an example of label information referring a user to the EPA Chemical Resistance Category Chart: *The chemical resistance selection category for this product is H. For more information about PPE materials that are resistant to this product for various lengths of time, consult an EPA chemical resistance category chart.*

The EPA chemical resistance category chart (Table 11.1) shows that pesticide products in the H category require barrier laminate or Viton materials for a full day's exposure, whereas butyl, nitrile, and rubber offer protection for 10 min of contact with the chemical. No other materials offer chemical resistance to, nor protect against, category H chemicals.

Interpreting PPE Statements on Pesticide Labels

Listed next are examples of wording found on pesticide labels relative to personal protective equipment required when handling the product; following each example are EPA interpretations of those label statements. When confused about what a pesticide label means, consult these guidelines. In most cases, the first line of a label statement addressing personal protective equipment indicates what is required for minimum protection; use of additional or more sophisticated equipment, for added protection, is left to the discretion of the user.

Long-sleeved shirt and long pants

Long-sleeved shirt and long pants, or
Woven or nonwoven coverall, or
Plastic- or other barrier-coated coverall, or
Rubber or plastic suit

Coverall over short-sleeved shirt and short pants

Coverall over short-sleeved shirt and short pants, or
Coverall over long-sleeved shirt and long pants, or
Coverall over another coverall, or
Plastic- or other barrier-coated coverall, or
Rubber or plastic suit

Coverall over long-sleeved shirt and long pants

Coverall over long-sleeved shirt and long pants, or
Coverall over another coverall, or
Plastic- or other barrier-coated coverall, or
Rubber or plastic suit

Chemical-resistant apron over coverall or over long-sleeved shirt and long pants

Chemical-resistant apron over coverall or over long-sleeved shirt and long pants, or
Plastic- or other barrier-coated coverall, or
Rubber or plastic suit

Chemical-resistant protective suit

Plastic- or other barrier-coated coverall, or
Rubber or plastic suit

Waterproof suit or liquid-proof suit

Plastic- or other barrier-coated coverall, or
Rubber or plastic suit

Protective eyewear

Shielded safety spectacles, or
Face shield, or
Goggles, or
Full-face respirator

Goggles

Goggles, or
Full-face respirator

Waterproof gloves

Any rubber or plastic gloves sturdy enough to remain intact throughout the task being performed

Chemical-resistant gloves

Barrier-laminate gloves, or

Other gloves that glove selection charts or guidance documents indicate are chemical resistant to the pesticide for the period of time required to perform the task

Chemical-resistant gloves such as butyl or nitrile

Butyl gloves, or

Nitrile gloves, or

Other gloves that glove selection charts or guidance documents indicate are chemical resistant to the pesticide for the period of time required to perform the task

Shoes

Leather, canvas, or fabric shoes, or

Chemical-resistant shoes, or

Chemical-resistant boots, or

Chemical-resistant shoe coverings (booties)

Chemical-resistant footwear

Chemical-resistant shoes, or

Chemical-resistant boots, or

Chemical-resistant shoe coverings (booties)

Chemical-resistant boots

Chemical-resistant boots

Chemical-resistant hood or wide-brimmed hat

Rubber- or plastic-coated safari-style hat, or

Rubber- or plastic-coated firefighter-style hat, or

Plastic- or other barrier-coated hood, or

Rubber or plastic hood, or

Full hood or helmet that is part of some respirators

WASHING REGULAR WORK CLOTHING

All clothing worn while handling pesticides should be considered contaminated, whether or not it is obvious. Work clothing worn while mixing, loading, or applying pesticides (even that worn during granular applications) must be stored and washed separately from the rest of the laundry.

Labels seldom explain how to launder pesticide-contaminated clothing. At best, a label might instruct the user as follows—and all of it is good advice: *Discard clothing that has been drenched or heavily contaminated with product concentrate. Do not reuse it. Follow manufacturer's instructions for cleaning PPE. If there are no such instructions for washables, use detergent and hot water. Keep and wash PPE separate from other laundry.*

Success in washing pesticide residues from common work fabrics depends on the pesticide and its formulation, the fabric thickness and fiber content, the laundering method, and the length of time between contamination and laundering. The longer the wait, the less is removed. Follow these safety tips when washing work clothes worn while using pesticides.

Laundering Pesticide-Contaminated Clothing

- Empty pockets and cuffs of clothing worn for granular applications, outdoors, to remove trapped granules before the clothing is stored to be washed. But don't sit on the back step to do this! Pesticide debris of this nature must be emptied onto an appropriate (application) site; that is, if a pesticide is not labeled for use on turf, it is likewise inappropriate to empty even small quantities of the product from pockets and cuffs onto the lawn. Also, stay clear of children's play yards, pet and livestock facilities, and any other sensitive areas.
- Discard clothing that has been drenched by a concentrate because the pesticide probably could not be removed to a safe level. Some studies indicate that heavily contaminated clothing still has detectable residues after 10 washings, so be on the safe side—*discard drenched clothing*.
- Wash contaminated clothing the day contamination occurs.
- Store contaminated clothing in a trash bag or hang it away from the family living space, in a workshed or on a clothes line outdoors.
- Never put contaminated work clothes in cloth bags or laundry baskets with other family laundry.
- Remember that socks and undergarments also may be contaminated and should be stored and washed separately from the family laundry.
- Open the washer door before handling pesticide-contaminated clothing to avoid contaminating the outside of the washer.
- Wear rubber or chemical-resistant gloves to handle contaminated clothing. Pesticides can transfer from fabric to the skin and from one fabric surface to another. Do not use these gloves for other purposes; residues on gloves can transfer to water, fabrics, and other surfaces.
- Prerinse or presoak contaminated clothing in a separate tub or use the rinse cycle of the washer. When using a washer, drain the prerinse or presoak water and refill the washer for washing.
- Pretreat contaminated clothing with a solvent-based prewash spray, especially if the contaminant is an emulsifiable concentrate.
- Wash pesticide-contaminated clothing separately from the family laundry. Research has shown that pesticide residues can be transferred from contaminated clothing to uncontaminated fabric in the wash water.
- Wash contaminated clothing and other PPE daily, as soon as possible after wearing. Delay in laundering will reduce the likelihood of total residue removal. Pesticides may bond to oily soil (if present) in clothing, making them more difficult to remove, therefore increasing the likelihood of pesticide residues remaining in laundered clothing.
- Use a heavy-duty liquid laundry detergent for best performance, especially in hard water and when the contaminant is an emulsifiable concentrate.
- Wash only a few items at a time to allow plenty of room for agitation and ample water for dilution of residues. Use the highest water-level setting, even with small loads.

- Use only hot water, not warm. Set the water heater at 140°F if there are no children or elderly family members who might be in jeopardy of scalding.
- Use at least a 10-min wash cycle.
- Use a cold-water rinse to conserve energy; no benefit from a warm rinse has been demonstrated.
- Check clothes for signs of stains or odors after laundering and rewash, if necessary, before drying.
- Line dry outdoors, if possible, because some pesticides are broken down by sunlight.
- After washing contaminated clothing, run the washer through a complete cycle with hot water and detergent, without clothes, before washing family laundry. This helps flush pesticide residues from the machine.

Figure 11.5 There are many ways in which a business can reduce the risks posed by pesticides.

Other Tips for Cleaning Pesticide-Contaminated Clothing

- Starching cotton or cotton blend fabrics before drying may help pesticide residue removal after future wearings. Pesticides deposited on the starch fabric may link to the starch and be washed away with it during the next laundering. Starch must be reapplied after each washing to maintain effectiveness. If the pesticide user is uncomfortable in starched clothing, consider starching specific areas of the garment. For instance, heavy starching of lower pant legs of coveralls and jeans should not create an uncomfortable fit but would enhance residue removal from that area.
- Ammonia, chlorine bleach, and fabric softeners are not effective in reducing pesticide residues.
- Do not wash contaminated chemical-resistant gloves with clothing. Gloves tend to be more heavily contaminated than other garments, and washing them with the rest of the laundry could result in movement of additional pesticide residue to other items.
- Pesticide residue levels in clothing can be reduced with multiple washings; but consider the toxicity of the pesticide product, the product's water solubility, your time, and associated costs (such as water bills) compared to garment replacement costs.

SUMMARY

Personal protective equipment is your responsibility. Whether your business is farming, custom application, nursery management, greenhouse cultivation, or pest management—and whether you have no employees or 100—if you use pesticides, you need to learn all you can about PPE (Fig. 11.5). Every day, new types of PPE become available. Choosing among the many alternatives is serious business because PPE is effective only when it is carefully selected, properly fitted, routinely used, and fastidiously maintained. It may not provide protection if any of these points are missed.

APPENDIX 11.1 WRITTEN RESPIRATOR PROGRAM FOR OSHA 29 CFR PART 1910.134

Purpose

The purpose of this document is to outline the procedures which employees are required to follow when using a respirator.

Objective

The primary objective is to prevent exposure to airborne contaminants through good management practices. The use of appropriate respiratory protection equipment will be enforced as required by pesticide labels or specific workplace conditions.

Responsibility

1. ______________________ is responsible for implementing and supervising the res-
   (Name)
   piratory program.
2. Employees, in accordance with the instructions and training received, are responsible for using the respiratory protection provided. Employees shall immediately report to their supervisor any malfunction of respiratory equipment.

Program Elements

1. Respirators shall be provided by the employer when such equipment is necessary to protect the health of the employee.
2. The supervisor shall designate a respirator for each job on the basis of specific hazards. Only NIOSH/MSHA-approved respirators shall be used.
3. The user shall be instructed and trained in the proper use, limitations, and maintenance of respirators. Every employee who wears a respirator shall receive fitting instructions and witness demonstrations on fitting, adjusting, and verifying the seal; each user also shall be required to practice these elements.
4. A qualitative respirator fit test shall be performed to determine the ability of each individual respirator wearer to obtain a satisfactory fit when using a negative-pressure respirator.
5. Employees shall use respirators according to the training instructions.
6. Respirators shall be cleaned and disinfected regularly.
7. Respirators used routinely shall be inspected during cleaning. Worn or deteriorated parts shall be replaced. Respirators for emergency use (such as supplied air respirators) shall be thoroughly inspected after each use and monthly when not in use.
8. Respirators shall be stored in a convenient, sanitary location.
9. Written procedures shall be prepared for safe use of respirators in dangerous atmospheres that might be encountered in normal operations or in emergencies.
10. Refresher training shall be completed on an annual basis.

Medical Evaluations

1. No person shall be furnished a respirator nor assigned to tasks requiring the use of a respirator unless it has been determined that they are physically able to perform the work and use the equipment. A physician shall determine what health and physical conditions are pertinent based on a series of questions required by OSHA and on examination when required.
2. Respirator users' medical status shall be reviewed periodically. Physical examinations shall be repeated on a regular basis for all employees who are required to use respirators. The frequency of examination shall be determined by the attending physician.

Program Maintenance and Effectiveness

Regular evaluation of the program—and adjustments to policy, as necessary— shall ensure its continued effectiveness. Routine, in-house inspections shall be conducted to ensure compliance.

1. Appropriate surveillance of the work area shall be ongoing and records of employee exposure maintained.
2. There shall be continual inspection and evaluation to ensure the effectiveness of the program.
3. Observations of and discussions with new employees shall be conducted to confirm that they have been given proper and sufficient training.
4. All employees shall be consulted periodically concerning their work habits relative to respirator use.
5. The overall effectiveness of the respirator program will be evaluated on an annual basis by the program administrator; action will be taken to correct defects, as necessary.

Program Availability

Copies of this respiratory program are available in the main office and as a separate section to the company's OSHA Hazard Communication Standard plan.

_______________________	_______________________
(Name)	(Company Position)
_______________________	_______________________
(Signature)	(Date)

CHAPTER 12

THE EMPLOYEE BULLETIN BOARD: WHERE EMPLOYERS COMMUNICATE POLICIES, PROCEDURES, AND PRACTICES TO EMPLOYEES

The employee bulletin board should be an official site where the employer communicates policies and procedures to employees. Most employers intend to maintain their employee bulletin boards, posting new information and dutifully displaying labor and employment law posters. As businesses grow, however, there is increased pressure on employers to attract new customers, become more competitive, deal with daily crises, and find quality employees in a tight labor market. Unfortunately, the bulletin board often is relegated to the back burner and good intentions turn to neglect. Posters become outdated, employee policies turn yellow with age, information is passed on verbally instead of in writing, and employees begin using the board as a place to post items for sale and party notices! (Fig. 12.1.)

It is no wonder that employees do not look to bulletin boards for the latest information. If postings have not changed in years, or if the board is so disorganized that important information is lost in the clutter, why bother? Many employers feel that their employees would not read a bulletin board faithfully anyway, so what is the big deal? Well, it IS a big deal.

It is relatively easy to change an outdated, ineffective employee bulletin board into one that improves workplace performance, increases job safety awareness, instills employee respect, and builds a better relationship between management and the work force. This chapter provides guidance for employers looking to revitalize their employee bulletin board to ensure that

- it is up to date,
- it provides job-related information for employees,
- it displays mandatory (regulatory) postings, and
- it contains information that could insulate your company against certain legal liabilities.

Is This Your Bulletin Board?

Figure 12.1 An employee bulletin board where good intentions have turned to neglect.

LONG-LASTING BENEFITS

There are numerous benefits to be realized by devoting time and staff to the development and maintenance of an employee bulletin board.

Benefit 1: Effective Line of Communication

Posting information on the bulletin board is an efficient way to communicate information to all employees. It effectively reinforces company memos distributed to each employee, notices enclosed with employees' paychecks, and word of mouth down the chain of command.

Benefit 2: Informed Employees

Your bulletin board should keep employees informed about the company, their jobs, and technological advances. It can be used as a teaching tool, reinforcing employees' awareness of the following:

- Company's expectations
- State and federal employment regulations and employee responsibility
- Opportunities for advancement, available continuing education, and company events
- Whom to contact with questions, comments, and complaints
- The proper procedures to deal with grievances
- Company policy on day-to-day issues
- What to do in case of an emergency
- Changes in employee handbooks, manuals, and safety policies

Benefit 3: Employee Safety, Compliance, and Morale

There is no doubt that safety in the workplace, legal compliance, and a friendly working environment are key elements that employers must continuously work to improve. An attractive, well-organized bulletin board can help promote safety in the workplace. Safety notices and information reminding employees to put safety first are more effective than you may realize, and posters and notices that instill the company's position on infractions such as discrimination and sexual harassment in the work place do impact employee behavior. The bulletin board also provides a forum for recognizing employee accomplishments—a deed that is often neglected but which boosts employees' morale and undergirds their sense of worth to the company.

Benefit 4: Retention of Productive Employees

Many employers are familiar with the nearly impossible task of finding a qualified substitute for an injured employee or a replacement for an experienced employee who leaves the company. You simply cannot afford to lose talented, experienced employees to competitors due to lack of information or communication.

Keeping safety at the forefront reduces the likelihood of injury on the job, and a bulletin board where safety information and policies and procedures are posted provides a continuous point of reference that employees come to depend on.

Posting job listings on the bulletin board, before they are advertised publicly, affords employees first chance at openings and new positions and reinforces the company's policy to promote from within. Some companies also post the minutes of management meetings to keep employees informed on what is going on with the company, financially and otherwise. Such communication demonstrates the employer's intent to be open and direct.

Benefit 5: Limited Regulatory Liability

The employee bulletin board is not just for large companies with human resource specialists, health and safety professionals, and hundreds of employees. Small retail and service outlets are also obligated to display employment, right-to-know, and other compliance posters.

Many state and federal agencies conduct periodic inspections to ensure that your company displays required posters in a conspicuous area accessible to all employees. As far as the inspector is concerned, you either do or you don't. If you do, you are in compliance. But if you do not, fines and penalties may be assessed, even if failure to post is simply an oversight.

Benefit 6: Consistent Decision Making

Well-organized bulletin boards offer protection from vengeful employees, such as those who sue for wrongful termination or those who file complaints with state or federal agencies alleging that they were forced to do dangerous work without adequate training or appropriate safety equipment.

An employer may be able to use bulletin board postings, employee handbooks, company manuals, safety policies, and training documentation to demonstrate that

- policies were in effect prior to disciplinary action,
- the employee received the appropriate training,
- applicable steps were taken to inform/train all employees,
- company policies and procedures were consistent with state and federal laws during the time frame in question,
- the employee's grievance was appropriately and thoroughly investigated, and
- the course of action taken to address the complaint was fair, consistent, and warranted.

Benefit 7: Policy Forum for Employees

In many companies, new employees are given an employee handbook, told to read it cover to cover, then asked to sign a document stating that they have read it. In other situations, employees are informed of new policies via enclosures with their paychecks and asked to sign a form stating that they understand the content. For many, this is the first and only time that the employee handbook or policy will be read, if indeed it is read at all. Policies and procedures are often relegated to long-term memory; however, the out-of-sight, out-of-mind attitude does little for employee relations and good communication.

The employee bulletin board (and an adjoining shelf or other surface) should serve as the focal point where policies, handbooks, and safety plans are posted conspicuously for the benefit of all employees. Documents that are important enough to develop and update are likewise important enough to be made accessible to employees on a continuous basis.

DESIGNING AN EFFECTIVE EMPLOYEE BULLETIN BOARD

Size and Location

Your employee bulletin board should measure approximately 4 by 8 ft. The actual size may vary, depending on the space available and the emphasis you place on the board as a primary forum for communicating with employees. Compute the area needed to post required information, then add enough space to allow for miscellaneous additional postings.

The location of the employee bulletin board is critical. Do not install it in a narrow hallway where the tendency is to walk past it without seeing it. Position your bulletin board where it is most likely to catch employees' attention, that is, in a location where employees are likely to stop and visit. The employee break room and cafeteria are good spots. Remember that government agencies require that posters be displayed in a location where they can be easily seen by all employees.

The employee bulletin board should be constructed with a sliding glass or Plexiglass door that can be locked. Access to the employee bulletin board should be limited to specified company personnel to ensure that materials are not removed or displaced.

Allow space below the employee bulletin board for a shelf or other surface approximately 2- to 3-ft long to accommodate policy manuals, health and safety plans, etc. An emergency medical kit and a fire extinguisher should be located nearby.

Appearance

The employee bulletin board needs to capture employees' attention. Make it attractive and interesting. Be creative! Consider breaking it into sections, each with a large heading differentiating it from the others. Use illustrations and colored backgrounds, which typically draw more attention than standard text in black and white. Allow open or "white" space between sections so that each one stands out.

Dates

There are four dates that apply to bulletin board postings:

- Document date: the date on which the posted item was written
- Effective date: the date on which the policy, procedure, etc., goes into effect
- Posting date: the date on which the article is posted on the bulletin board
- Removal date: the date on which the posting is removed from the bulletin board

It is important that every item on the bulletin board bear a *document date;* and the *posting date* should be written in red ink in the upper-right-hand corner. Highlight the *effective date* of the provision if it is not otherwise noted conspicuously on the document. When removing or replacing an item on the bulletin board, record the *removal date* and note the authorized person responsible for conveying the information it contains. Both could be helpful if the content later becomes the subject of litigation. File the items removed from the bulletin board under "Previous Postings".

Number of Items Posted

Limit the number of postings on the employee bulletin board so that the eye is drawn to the most important documents. Make a concentrated effort to emphasize postings that are critical to the safe and efficient operation of the company.

Items such as federal and state employment posters must be displayed in their entirety. But lengthy materials such as work safety rules and company policy may be summarized: state the main points and direct employees where to access the complete document.

Readability

Bulletin board documents must be written clearly and concisely. Vague postings and those too lengthy to read quickly, simply defeat the purpose of the employee bulletin board.

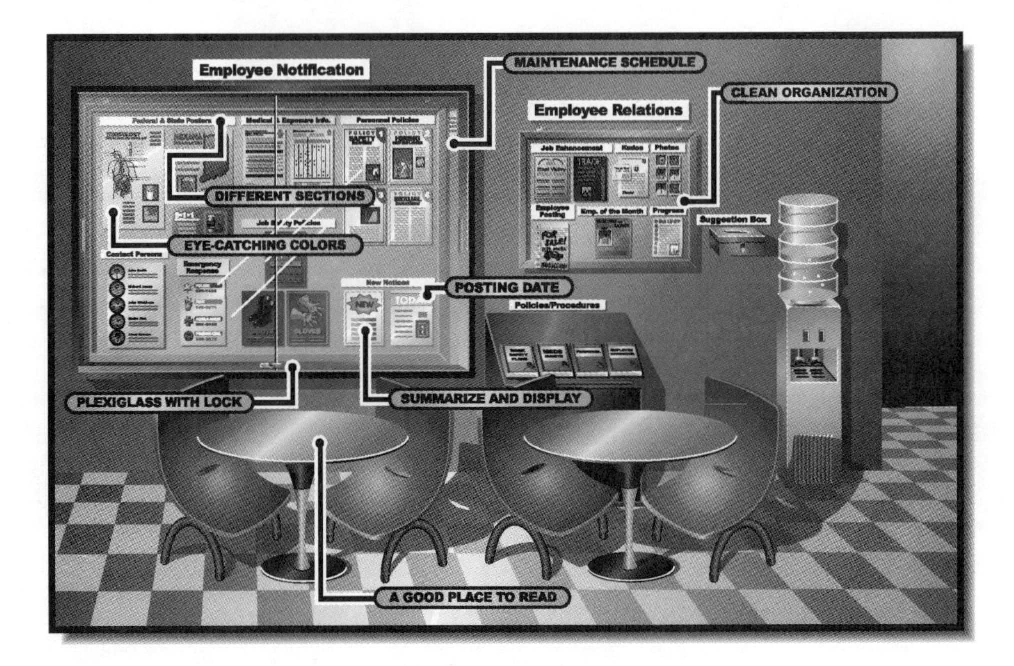

Designing an Effective Bulletin Board

Figure 12.2 Bulletin boards should be designed to encourage employees to read policies and new information.

Maintenance

Setting up an employee bulletin board is easy enough, but keeping it updated can be a challenge. Management personnel should be assigned to maintain the integrity of company postings, revising and replacing them as necessary. Some items may need replacing due to missing, torn, ripped, or illegible pages, even though they are still up to date. Federal and state posters are dated, and *the burden is on the employer to ensure that those posted are the most recent* (Fig. 12.2).

EMPLOYEE BULLETIN BOARD CONTENT

Labeling Sections

Distinguish your bulletin boards, one from the other, by labeling them in large letters across the top. A heading such as "Employee Policies, Procedures, and Practices" could be used for the employee bulletin board.

Also label each section to make it easy to find. Here are some examples of headings you might use on the employee bulletin board:

- Federal and State Posters
- Emergency Response
- Medical and Exposure Information
- Job Safety Policies

- Personnel Policies
- Contact Persons
- New Notices
- If You Don't Understand . . .

Building Each Section

These guidelines are intended to help you in formatting your employee bulletin board. Postings *required* by law are so noted, as are *optional* items.

Federal and State Posters (Required) Using an employee bulletin board to display federal and state posters brings continuity to the posting requirement; that is, employees learn that it is *the* place to look for information on policies and procedures.

Purchase posters only from those companies that have attorneys who review state and federal requirements. It is wise to work with a company that will notify you when new posters are developed and when there are updates available for those you have. Search the Internet for "federal posters."

Some posters can be downloaded from corresponding agencies' Web sites, but be certain that they meet each agency's legal posting requirements. Most do, but some may not download in appropriate posting format. Check them out, just to be sure. Use information on your current poster to search the Internet.

Remember that it is your responsibility to obtain and display all required posters, but you should not have to provide your name and business information to a government agency whose job it is to investigate or regulate your business. If your source asks for that information, you might want to consider obtaining your posters from a professional consultant or trade association.

Emergency Response When a facility is evacuated due to an emergency, it is critical that employees leave the facility as quickly as possible and proceed to a designated meeting place away from the building. This allows management to determine if there is anyone missing, which in turn dictates what course of action emergency personnel should take. See Chapter 13 for more information on emergency planning.

An *evacuation map* is a legal requirement for businesses with more than 10 employees. Therefore it is important to post evacuation information, including the map, on the employee bulletin board. The evacuation map should identify the location of the following items:

- Emergency exits
- Evacuation routes (each route in a different color)
- Primary and alternate meeting places
- Fire extinguishers
- Fire alarms
- Sprinkler controls
- First aid kits

- Bloodborne pathogen kits
- Biohazard containers
- Eyewash areas
- Emergency showers
- Circuit breaker boxes
- Gas and water shutoffs
- Spill control equipment
- Tornado shelters

Emergency Telephone Numbers *(legally required for most businesses).* Examples of phone numbers that may be posted are found in Appendix 10.3.

Emergency Response *(may be Optional or Required).* Following are examples of items to be posted under this section:

- Alarm system is by voice.
- Unless otherwise instructed, evacuate the premises in the event of an accident.
- Do not use a fire extinguisher unless you have received training.
- Remember the acronym PASS when using a fire extinguisher:
  - P stands for *Pull* the pin.
  - A stands for *Aim* the extinguisher nozzle at the base of the fire.
  - S stands for *Squeeze* the handle.
  - S stands for *Sweep* the hose from side-to-side at the base of the fire.
- Do not administer first aid if you have not been trained in the proper procedures. In the event of an accident, contact the individuals whose names are listed on the emergency medical kit near this bulletin board.
- When responding to an accident,
  - Call 911.
  - Have someone else call designated company personnel.
  - If there are personal injuries, care for the victims until trained company personnel or emergency responders arrive.
- When working with victims, remember A-B-C:
  - A, make sure the *Airway* is clear.
  - B, make sure the person is *Breathing*.
  - C, make sure the blood is *Circulating* (check pulse).

Medical, Exposure, and Illness/Injury Information (Required) Post the following right-to-know information: *You have the right to access your medical and exposure records on file with this company. Contact (name of person) for more information.*

Post your OSHA No. 300 log from February 1 through April 30 of each year. This is an injury and illness log that employers must complete based on the types and number of accidents that occurred within the company during the preceding year. It is a legal requirement for most companies and must be posted even when there are no accidents or injuries to report.

Job Safety Policies Posting bulletin board information on general safety, although voluntary in most cases, is vitally important for most companies. List work practices that are critical to running a safe business, that is, summarize the major points from written policy. Postings may vary from company to company, depending on what employees must know to operate safely in their respective work environments.

Following are examples of wording that may be used under the Job Safety Policies section of your employee bulletin board:

- These are only brief excerpts from company policies and employee handbooks. They are not meant to replace complete documents. Please review all written policies and procedures in full.
- Management's number one priority is safety. Be aware that your commitment to safety is a condition of employment, and that your compliance with safety requirements is evaluated along with job performance. Employees who violate safety standards, cause dangerous situations, or fail to correct or report unsafe conditions are subject to disciplinary action up to and including termination.
- Employees are to report to management any occupational safety and health risk—in fact, *any* unsafe condition—that they observe. There shall be no reprisal for disclosing the information, but employees may report anonymously if they so choose.
- If you need additional equipment or instructions to perform your job safely, notify your supervisor immediately.
- The certified pesticide applicators at this facility are (fill in the names). If you are a pesticide applicator, direct questions relating to your job to these individuals.
- Never use application equipment or personal protective equipment that you have not been trained to use.
- Never use a respirator unless you have been properly trained and fitted and have the approval of management.
- Nonfunctional or damaged equipment must be marked "Broken—Do Not Use" or "Out of Service" and turned in for replacement.
- Never use machinery or equipment that is broken or damaged, and never operate equipment that has damaged (or missing) guards or shields.
- Material Safety Data Sheets for chemicals that you use on the job are always available for review at (list the location).

Personnel Policies Civil lawsuits and government actions against employers have reached record highs, emphasizing the importance of posting key personnel issues that inform employees of the company's expectations in various areas: drugs and alcohol in the workplace, employee relationships, discrimination, and sexual and other forms of harassment. This information also should be included in the employee handbook.

Consider the following language for employee bulletin board posting:

- Following are some brief excerpts from company policies. For the complete text, see personnel notebook (fill in location).
- *Hiring* We comply with immigration regulations by only hiring individuals who are legally authorized to work in the United States.
- *Employment-at-Will* This company has an employment-at-will relationship with all employees. What this means is that any employee can terminate his or her employment at any time, with or without cause. Likewise, the company has the right to terminate the employment of any employee at any time, with or without cause. Only (fill in the name) has the authority to implement an employment agreement contrary to the employment-at-will policy. (The authorized person usually is the CEO or company president.)
- *Harassment* Sexual harassment, workplace violence, and any other forms of harassment (e.g., foul language, the display of sexually suggestive materials, dirty jokes) will not be tolerated. (Read the complete harassment policy in your employee handbook.)
- *Equal Opportunity and Antidiscrimination* This company is an equal opportunity employer. As such, we are committed to providing equal opportunity in all employment practices and to maintaining a workplace free of discrimination against applicants and employees based on race, color, religion, gender, age, disability, national origin, marital status, and other protected classes according to local, state, and federal requirements.

This policy extends to all terms, conditions, and privileges of employment, including but not limited to recruiting and hiring, working conditions, training programs, and use of company facilities. If you feel you have been discriminated against, report the incident promptly to (name of person to call) at (phone number). All allegations will be thoroughly investigated. We strictly prohibit retaliation of any type for filing such a complaint.

- *Traffic Violations* All employees who drive company vehicles on the job must notify their supervisor of any moving traffic violation they have been assessed, before the start of the next working day. This requirement extends to violations received while driving personal vehicles on the job as well.
- *Weapons Policy* All employees and visitors on company premises are prohibited from carrying a concealed weapon of any sort. Employees who are personally licensed to carry a concealed weapon may do so in their own vehicle but not in a company vehicle. Any employee found in violation of this policy shall be subject to immediate disciplinary action up to and including termination of employment. This company reserves the right to grant complete or partial exemption from this policy if warranted.
- *Drugs, Alcohol, and Tobacco* This facility and all company vehicles are considered part of our drug-, alcohol-, and tobacco-free workplace. We are committed to providing a safe work environment and fostering good health among our employees, and this commitment is jeopardized when any employee uses drugs illegally on the job; comes to work under the influence of drugs or alcohol; or possesses, distributes, or sells drugs in the workplace.

Policies, Plans, and Procedures Companies spend thousands of dollars developing written health and safety plans and countless hours training employees accordingly. Examples include the following: Hazard Communication Standard; Respirator Plan; Bloodborne Pathogen Plan; Emergency Response Plan; Lockout/Tagout Plan; and Confined Space Plan.

It is important to display all of these materials on or near the employee bulletin board (i.e., on a nearby shelf or table). You should maintain four notebooks: personnel policies; employee handbook; product labels and Material Safety Data Sheets; and work safety plans. Clearly label each notebook and secure it next to the employee bulletin board so that it can be read but not removed.

Include in this area of the employee bulletin board a section for training programs, outlines or minutes of meetings for employees who did not attend.

Contact Persons Whom to list as a contact depends largely on the size of the company and the structure of the organization. For example, the contact may be a department head, a supervisor, a safety manager, an office manager, a financial director, a general manager, the president, etc., but in any case the individual's name and phone number should be posted.

Contacts must have the authority to handle complaints, suggestions, and questions. List at least one alternate next to the primary contact, and provide alternate avenues for reporting concerns, ideas, or complaints. This becomes very important when the primary contact is the person who is creating the problem. It is also a good idea to list both male and female contacts to afford employees the choice; it is important that they feel at ease discussing sensitive matters with the person. Employees should be encouraged to approach the president or CEO if they do not feel comfortable addressing their concerns with the designated contact person.

Other Suggestions

New Notices When new or revised employee policies are being introduced, post them conspicuously on the employee bulletin board so that all employees are clearly aware of the changes. Leave them on the board long enough to counter absences due to vacations, maternity leave, etc. When a posting announcing a change is removed, replace it with instructions where to locate the revised document in its entirety.

Vacation Planner You may find it helpful to post an annual employee vacation planner to assist employees in scheduling time off.

For Those Who Don't Understand (Optional) In any given work force there may be employees with poor reading skills or whose primary language is something other than English; either way, their understanding of the material presented may be compromised. If that is the case with your company, consider posting the following message in English and in other languages represented among your employees:

If you have questions about company policies, training, safety issues, or other information posted on this bulletin board, please contact ___(name of person)___.

You could further accommodate employees for whom English is a second language by translating all documents and providing a second employee bulletin board. Perhaps your bilingual employees would be willing to assist, and getting them involved might serve a second purpose: generating self-esteem and a sense of worth to the company. If legal or otherwise technical language is used in your policies, consider hiring a certified translator.

EMPLOYEE RELATIONS BULLETIN BOARD: WHERE EMPLOYEES COMMUNICATE WITH THE EMPLOYER AND OTHER EMPLOYEES

The employee relations bulletin board provides a forum where employees may communicate with the company and with other employees. Information on special events, items for sale, interesting photographs, and noteworthy occasions are typical examples of items posted.

Set aside a portion of the employee relations bulletin board for company postings such as kudos to the work force or specific individuals, employee of the month recognition, and job-enhancement opportunities. Use employee photos, when possible, to accentuate positive reinforcement.

EMPLOYEE RELATIONS BOARD
SUGGESTION BOX

Suggestion Box

Place a suggestion box near your employee relations bulletin board and encourage employees to offer suggestions and constructive criticism, emphasizing that the company will not allow reprimands of any nature in response to employee input. Provide comment cards that do not require contributors to identify themselves unless they choose to do so, and keep the suggestion box locked at all times.

Employees often comment on how their jobs might be made easier or safer, or accomplished quicker or cheaper; but about 10% of all comments are complaints, and complaints are very important. They afford the employer an opportunity to solve little problems before they become big ones, and before a third party such as a union gets involved.

Comment Cards Employees should be given the option to sign their comment cards or to remain anonymous. However, employees must understand that management may not be able to react appropriately to their comments without knowing the source. Encourage employees to sign their names, and reassure them that every comment will be handled as confidentially as possible. Advise them to provide as many details as they can, particularly if they choose to remain anonymous. See the sample comment card.

COMMENT CARD

Name (optional): ______________________ **Date:** ______________________

Department (optional): ____

Type of work: ____

Comments: Please share any concerns or suggestions on how we can make our company the best it can be. Please be very specific to enable management to effectively address your question, comment, or concern.

____

____

____

May we post your comments and management's response on the employee relations bulletin board? Yes _______ No _______

Thank you for taking the time to share your opinions. Please provide your name only if you want to be contacted.

Company Response

Date Reviewed: ________________

Action Taken: ____

Manner in Which Employee is Notified: ______________________________

Follow-Up Action Making the suggestion box accessible to all employees is only part one of a good idea: follow-up is critical to your credibility and to the integrity of your company. Employees need to know that management will read their comments, take them seriously, and act on them as appropriate.

Posting an unsigned comment and management's response on the employee relations bulletin board is an effective way to communicate your actions to all employees. But only those comments preapproved by the employees who submitted them should be posted; note the space included on the sample comment card for employee consent.

Employee Posting

Outline and enforce specific guidelines for employees to follow in posting items on the employee relations bulletin board. For instance, inform them that attacks on the rights of others and posters that offend specific groups are illegal and therefore not allowed. Monitor bulletin board content by conducting periodic inspections.

Posting rules may include (but are not necessarily limited to) the following:

- The posting date must appear in the upper-right-hand corner.
- The name of the person posting the item must be indicated directly beneath the date.

- Items must be removed within 30 days of posting.
- The company has the right to remove any item that is perceived to be in poor taste, offensive, illegal, or otherwise detrimental. Examples may include information related to religion, race, and sexual matters; material that is or may be perceived to be discriminatory or otherwise inflammatory; solicitation materials; and political documents.

Management Posting

Kudos It is important that you post correspondence from customers expressing their appreciation for an employee's hard work and dedication and for delivering quality service. it gives the employee a sense of personal accomplishment and pride.

Job Enhancement Opportunities for outside training, announcements of classes to improve language skills, and notices of upcoming company-related events should be posted in the management section of the employee relations bulletin board.

Employee of the Month An employee of the month award can boost job performance and employee morale, and seeking nominations from the workers, themselves, fosters cooperation and teamwork. Consider using the following form and offering incentives such as gift certificates or time off with pay.

You did a great job, and we noticed!

Who: ______________________________

Department: ______________________________

Date: ______________________________

Details of good deed or excellent service:

____

____

____

____

Keep Up The Good Work

____

(Name and Title of the Person Expressing Appreciation)

Progress Management's "progress" section on the employee relations bulletin board is used to inform employees of projects, special jobs, new clientele, budget details, etc.

Employee Photos Depending on the size of your organization and the space you have, you might post photos of all employees, of those who have been in the news, or of recipients of merit or other in-house awards. Also consider a section solely for welcoming new employees.

USING THE EMPLOYEE BULLETIN BOARD FOR TRAINING

The employee bulletin board provides an organized way to disseminate company policies and information required by local, state and federal agencies. However, you should train employees on the information in addition to posting it. It takes only minutes at a time to review key concepts on a continual basis. The person designated by management to conduct the training should stress that it is the responsibility of each employee to read the postings, review policies and handbooks as advised, and direct any questions to the appropriate contact person. Employees should be asked to sign off on training received and on their reading of complete documents as directed by the trainer.

It may also prove useful to videotape your training programs. The tapes can be useful in training new employees, although you must ensure that they address required regulatory poster content as well as company policies and procedures.

Remember that as the employer, you are obligated to communicate regulatory information to your entire work force. If you employ individuals who cannot read well or cannot speak English, you may have the information read aloud for them or translated (verbally and/or in writing), respectively. If you have employees with learning disabilities, you must also accommodate their needs.

BEWARE: BULLETIN BOARDS CAN BECOME A LIABILITY

In addition to all the good they do, bulletin boards also can work against you. Never consider your obligation to enforce regulatory requirements met by merely posting the information on a bulletin board. You are obligated to ensure that employees understand what is required of them; likewise, you are obligated to ensure their compliance. Even the best bulletin board cannot do that for you: The responsibility for enforcement lies squarely with management (Fig. 12.3).

Once the company has stipulated that certain practices are disallowed, tolerance poses significant legal liability and compromises the company's credibility. For example, the company may have a written policy requiring employees to wear chemical-resistant gloves when using pesticides, but if supervisors knowingly allow employees to work without gloves, the company is liable. Management must enforce policies and regulations consistently to assure that their postings do not become "Exhibit A" in a lawsuit against the

Your Bulletin Board Can Testify Against You

Figure 12.3 Bulletin boards can be a liability to the company when the procedures outlined in company policies are not followed by management.

company. Obviously, management cannot monitor every employee, every moment on the job, but appropriate training, written materials, and enforcement consistency can deter liability.

GETTING ADDITIONAL HELP

When it comes to state and federal employment regulations, the burden to comply lies with the employer, and ignorance of the law does not excuse noncompliance.

Many companies do not have human resource departments or health and safety specialists to help them interpret personnel and labor issues; however, there are numerous outside resources available to assist even the smallest company. Consider the following when looking for advice on dealing with issues addressed in this publication (Fig. 12.4).

Human Resource Consultant

Consultants who specialize in human resources, employment, or labor relations can provide accurate and sound management advice on a wide variety of topics. They can provide timely assistance by fax, telephone, or on-site consultation or via the Internet. They can assist overworked human resource departments in large companies and even in the smallest ones.

Figure 12.4 Employers can turn to many professionals for assistance.

Contact the Forum of the Society for Human Resource Management (http://www.shrm.org) for a listing of consultants in your area. Always check references before hiring a consultant.

Labor and Employment Attorney

When hiring a labor or employment law attorney to handle a problem that has occurred in your company, consider the following:

- Does the attorney have expertise in the area in which you are having problems and in your specific industry? If you have a potential American with Disabilities Act lawsuit looming, has the attorney handled this type of case in the past? Was he or she successful? If so, why? If not, why not?
- Has the attorney actually represented other companies or competitors in your industry? Contact your local attorney referral service for the name of a reputable employment law attorney.
- What are the attorney's rates? Are they comparable to other attorneys' rates in your area?
- Many employment/labor law attorneys have free checklists and seminar materials to assist employers in complying with applicable laws. Legal seminars offered by these professionals lend valuable tutelage at minimal cost. Call local labor and employment attorneys to inquire about the seminars they offer.

Chamber of Commerce

Many chambers of commerce have general labor law check sheets to assist employers in understanding what is required of them. But you also must be familiar with specific requirements that may affect only your industry.

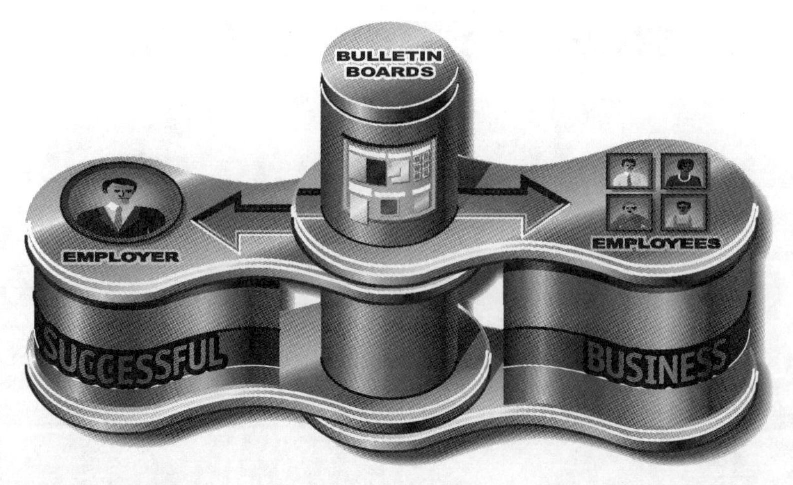

Figure 12.5 An effective way to communicate between employer and employees is by posting information on bulletin boards.

Employer Advisory Council

Employer advisory councils typically offer fee-based or membership-based assistance; their staff typically is composed of legal, human resources, and/or management professionals. Consult your local Yellow Pages or the Internet for employer advisory councils in your area.

Additional Resources

Other resources that can provide valuable assistance:

- Industry trade associations
- Printed and/or Web-based materials on regulatory/compliance/safety issues
- Insurance or workers' compensation loss control representatives on safety-related issues
- Government Web sites
- OSHA consultation services
- Land Grant University Cooperative Extension Service

SUMMARY

Make your bulletin board the central communications focal point of your company: an efficient and readily available source for regulatory information and company policy. Properly designed, a well-organized bulletin board fosters communication between employers and employees (Fig. 12.5). It keeps employees informed and offers an avenue for asking questions and offering suggestions in a constructive manner. Communication between management and the work force is the key to running a successful business, and the bulletin board is a very efficient and convenient method to that end.

CHAPTER 13

PLANNING FOR EMERGENCIES: PREVENTING AND REACTING TO EMERGENCIES IN THE WORKPLACE

Customer loyalty can mean the difference between prosperity and bankruptcy in today's highly competitive pest management markets, both urban and rural. Managers must continually adapt to changing markets by establishing sales goals, implementing business plans, taking advantage of business opportunities, and coupling effective, competitively priced products with quality service.

Equally important—but often overlooked—are potential financial threats from injuries, fires, spills, explosions, and other accidents. A fire or an accidental spill can be as catastrophic as losing your entire customer base. You must identify the potential for these events and, more important, develop a proactive plan to be prepared for such events. Many prudent business owners and managers recognize and acknowledge their responsibility for human and environmental protection and purposely develop a plan for handling emergencies (Fig. 13.1). But too many don't.

Postponing critical judgment until an unfortunate situation presents itself—be it a fire, a spill, or what have you—is an irresponsible stance. Decisions of this magnitude should not be made under stress fueled by an emergency situation: Snap decisions are seldom the best decisions. Wrong decisions may leave an individual or a business vulnerable to the ravages of fines (local, state, federal) and legal action as well as potentially astronomical cleanup costs. Once an emergency occurs, it's too late to plan (Fig. 13.2).

Business owners cite various reasons for failing to plan for emergencies:

- A belief that accidents will not happen to them
- A misconception that preplanning is expensive
- Ignorance of potential costs an emergency can impose
- Lack of understanding of the emergency potential of stored chemicals
- Oversight of the fact that environmental damages may require reimbursement or fines to a state agency
- Failure to recognize that on-the-job injuries are covered by workmen's compensation and that claims raise the cost to the business owner

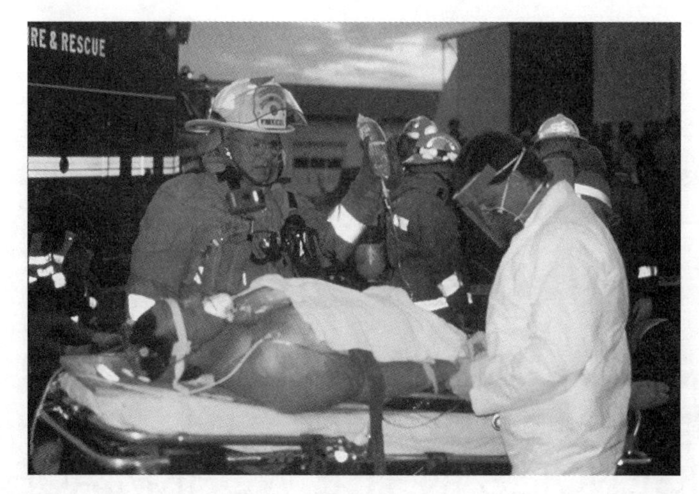

Figure 13.1 Emergency medical technicians respond to a person feigning pesticide injury during a mock training session.

- Failure to recognize that emergency-related claims for damage increase liability insurance premiums
- Failure to recognize that some emergency-related claims are not covered by insurance
- A belief that time dedicated to preplanning is wasted

Figure 13.2 A pesticide application truck involved in a traffic accident.

Figure 13.3 A fire that destroyed a farm building (foreground) nearly melted the plastic tanks that were near the building.

- Failure to comprehend that a serious emergency can quickly tarnish the spotless reputation that a business has spent decades building
- Failure to recognize that some laws require businesses to prepare contingency plans

CHARACTERISTICS OF EMERGENCIES

An Emergency Is a Situation Out of Control

An emergency is defined as a serious situation which is unanticipated and which demands immediate action. Once the situation is under control, it is no longer regarded as an emergency; but follow-up and cleanup may take days or weeks, or even longer (Figs. 13.3 and 13.4).

Figure 13.4 Diesel fuel spilled from a locomotive at a grain storage facility.

Figure 13.5 Emergencies can be major or minor.

Emergencies Can Be Large or Small

Some emergencies require professional assistance (police, firefighters, paramedics, environmental contractors), whereas others may be handled by properly trained company employees. Personal injuries may range from minor cuts, treatable with a first aid kit, to major injuries from exposure to toxic chemicals, which may require hospitalization (Fig. 13.5).

Small fires often can be extinguished with a portable fire extinguisher, whereas larger ones require trained firefighters and possibly emergency medical assistance.

Some spills can be controlled and contained and the area cleaned, using spill kits kept on-site. It is important that all employees know exactly where spill kits, fire extinguishers, and first aid kits are stored.

A large, uncontained spill from a ruptured thousand-gallon pesticide tank would likely require a trained hazardous materials response team to control the release, evacuate the area, coordinate remedial measures, contain the spill, clean and decontaminate the site, and dispose of contaminated waste.

Example of a Preventable Emergency A driver who is behind schedule decides to inform the next customer while filling an application rig with pesticide and water. The hose is left running inside the tank while the driver walks to the office to use the phone. While the driver is gone, the tank overflows, creating an emergency. Upon discovering the situation, the driver immediately shuts off the hose and contains the spill. With the situation under control, the driver's company must address its responsibility for the proper handling/removal of any pesticide-contaminated soil at the spill site and for appropriate actions to minimize or prevent ground water contamination resulting from the spill.

Responsible preplanning might prevent or minimize this type of emergency; that is, it should be an ingrained company policy that drivers must remain with their rigs when filling the tank. If drivers need to make any type of phone call, they must know to do so either before or after filling the tank—never during.

Emergencies Can Be Anticipated

You can never completely eliminate the risk of accident or injury. However, preventive planning can reduce the likelihood of emergencies through

- the implementation of appropriate, everyday operational procedures;
- the identification of potential hazards; and
- the development of contingency plans.

Contingency plans can minimize the severity of an emergency and the extent to which business is disrupted. A trained work force can minimize the immediate and long-term impact of fires, spills, and pesticide exposures. Good advance training equips employees to respond appropriately; and if they know exactly what to do, they are less likely to panic during an emergency (Fig. 13.6).

Stated simply, the objective of preplanning is to be ready for emergencies. Management personnel and employees should be trained on how to react in an emergency situation. Community response personnel should be brought on-site and familiarized with the various

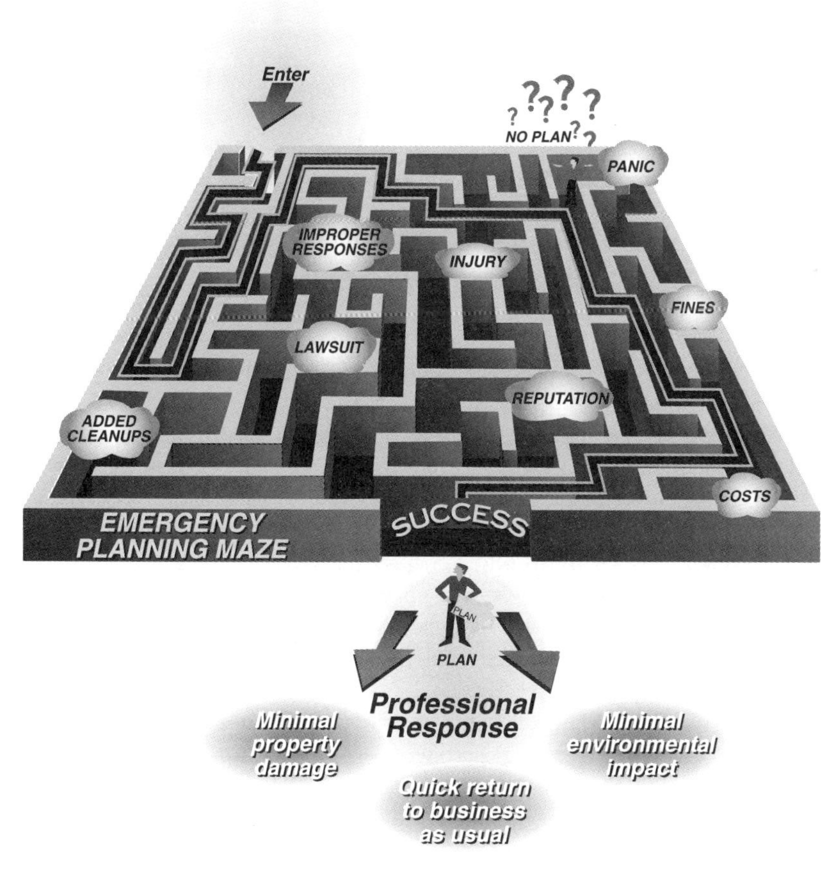

Figure 13.6 Emergency preparedness often reduces the severity of an accident.

chemicals stored there, their location on the property, and the actions recommended in case of a fire or spill, etc. Both company personnel and outside responders must understand what they need to do during an emergency to minimize injuries and adverse effects on public health, the environment, and the business itself.

Preplanning includes taking the time to resolve small problems as they occur. Don't let a problem snowball into something serious before addressing it. A sound problem-solving policy contributes greatly to the overall professional success of the company.

The following describes preventive strategies to assist companies that handle, store, and use pesticides in meeting health, safety, and environmental regulations. Implementing these strategies also helps

- protect company assets,
- safeguard employee health,
- reduce the potential for an emergency to impact surrounding communities and the environment,
- assure regulatory compliance,
- promote professionalism and goodwill, and
- potentially reduce insurance costs.

PREPLANNING FOR EMERGENCIES

Emergency planning should not be viewed as an insurmountable task, nor should it be viewed as just more paperwork required by bureaucrats who have limited familiarity with your business. An emergency can result in injury, environmental damage, financial loss, fines, and a damaged reputation for the business involved. But time invested in preplanning for emergencies can pay large dividends.

If you are skeptical, just ask businesspersons who have experienced a warehouse fire, a chemical spill, or the serious injury or death of an employee. Ask them what they wish they had done differently. Those who were unprepared will wish they could turn back the clock and develop a contingency plan; they will wish they had verified specific insurance coverage and conducted in-depth emergency response employee training. Those who were prepared likely will mention that they intend to implement changes to fine-tune what they had in place, such as upgrading an insurance policy or educating employees on a specific point.

The recommendations set forth in this chapter represent the professional opinions of the authors and should guide you in setting up a functional emergency contingency plan for your business. Simply read each section and complete the assigned tasks, and you will be well on your way to compliance with regulations pertinent to chemical emergency situations.

Each section heading in this chapter is followed by a citation, in parentheses, to facilitate the reader in locating pertinent regulations for review. The acronym OSHA stands for the Occupational Safety and Health Administration; CFR stands for Code of Federal Regulations. Various local, state, and federal agencies, as well as state and national associations, can be contacted for copies of specific regulations in their entirety.

WORKPLACE HAZARD ASSESSMENT (OSHA, 29 CFR PART 1910.132)

Employers are required to identify potentially hazardous conditions in their workplace. This is important because each facility is unique in its operations, the types of equipment it uses, and the chemicals which it stores and applies.

The term workplace can include more than just the main base of operation. It includes off-site locations where any phase of the business is conducted: places such as restaurants, hospitals, construction sites, farm fields, golf courses, and city parks. Workplace hazards can exist at any of these locations.

Assessment procedures to identify problem areas should involve both management and employees, and particular attention should be devoted to details relating to materials handled and work procedures followed.

A hazardous condition may exist due to such things as

- handling a particular pesticide,
- performing a particular task such as welding, or
- entering a particular area such as a hospital where there is the possibility of coming into contact with blood and other potentially infectious substances.

Hazardous situations might include such activities as

- mixing concentrated pesticides,
- handling anhydrous ammonia,
- cutting weeds with a string weed trimmer, or
- working in a grain elevator.

Workplace hazards (Fig. 13.7) could also include working in proximity to

- falling objects,
- projectiles,
- electrical or motorized equipment,
- confined space,
- harmful dust, or
- extreme temperatures.

How Do I Comply with This Regulation?

OSHA, 29 CFR Part 1910.132 states that "the employer should assess the workplace to determine if hazards are present, or are likely to be present, which necessitate the use of personal protective equipment." Thus, all businesses are subject to this regulation.

A survey form for use as a guide in assessing workplace hazards can be quite useful. When developing such a form for your operation, interview employees and identify the types of hazards they might encounter on the job: things such as contact with chemicals, sharp objects, excessive heat, etc. Do not, however, delegate hazard assessment solely to employees; a good mix of management and employee personnel is more likely to yield recognition of all potential hazards.

Figure 13.7 Employees should always be aware of their surroundings and the tasks being performed.

You must focus on the big picture when conducting a hazard assessment. Although a tendency to focus primarily on pesticide applicators might seem reasonable due to the complexity of their jobs, it would be shortsighted to limit the hazard assessment process to applicators only.

Consider hazards present in maintenance shop and office atmospheres. Give some thought to hazards that might be present in the mechanics' shop. Electrical, mechanical, and physical hazards may exist in any of these locations. Consider hiring a risk manager or train someone to serve as the business risk management specialist.

You should also refer to documentation (OSHA injury log and/or medical records) of previous workplace injuries. Is there a repetition of injuries? If so, identify and correct specific conditions leading to the recurring problem.

Although conducting an all-inclusive hazard assessment takes time, completion of the task will place you in a much better position to anticipate and control specific hazards in the workplace—and to prevent injuries (Fig. 13.8).

The person who actually conducts the hazard assessment can see firsthand the relativity of various workplace activities to hazardous situations. This observance, supplemented by employee interviews, can facilitate the identification of work practices that could lead to serious injury.

Hazard assessments are needed also to evaluate a business's level of compliance with current health and safety policies and regulations (i.e., how well those already in place are

Figure 13.8 A dangerous situation where the electrical outlet is not properly installed.

actually being followed). Spot checks and regularly scheduled safety meetings are important in enforcing compliance.

Once hazards are identified, they should be engineered out of the work process or otherwise controlled (e.g., by changing job procedures to exclude them). The use of personal protective equipment always should be the last resort. Consideration first should be given to engineering controls and alternative job design; exceptions would be pesticide label requirements and the use of specific tools or devices specified by the owner's manual for equipment in use.

Documenting the Assessment Your hazard assessment must be documented in writing (handwritten notes at a minimum). It must be documented as the Certification of Hazard Assessment, detailing all areas inspected: pesticide storage room, mechanics' shop, etc. The certification of hazard assessment also should bear the date of inspection and the printed name, signature, and title of the inspector. An example of a certification of hazard assessment appears in Appendix 13.1.

Personal Protective Equipment Selection Personal protective equipment (PPE) is essential for anyone who handles potentially hazardous chemicals or conducts hazardous tasks. Identify when, where, and under what circumstances employees and management personnel might be exposed, then implement—and enforce—a PPE program which includes proper training in their use. Make sure anyone who handles chemicals wears appropriate protective gear as required by the pesticide label and company policy. One or more of these items may be necessary: earplugs, gloves, splash aprons, coveralls, chemical-resistant or steel-toed boots, face shields, hard hats, respirators, back supports, nonslip shoes, goggles.

Companies are required to make available to all workers whatever protective equipment is required for handling the chemicals they use. An effective program details the PPE required (by the pesticide label and company policy) for each task and includes regimented company training on its use. Employees must be taught to always wear protective gear and to wear it properly, and management personnel should monitor them to make sure that they do.

Recommendations

Recommendation 1 Seek input from your employees; they may shed light on physical and chemical job hazards unique to their everyday routine—things that you might otherwise overlook. Their involvement in the development of the PPE program will make it their own, in a sense, and will enhance their appreciation for its importance.

Recommendation 2 Assess workplace hazards at least on an annual basis—more frequently, if possible. At a minimum, reassessment is appropriate as new chemicals are added to inventory and as new hazards are recognized or introduced into the workplace. Sometimes chemical hazards can be diminished or eliminated by selecting a different formulation or product or by implementing new handling procedures.

Recommendation 3 Identify and execute practices that can be modified to eliminate hazards noted during the assessment process, and document all changes.

Recommendation 4 Retain all documents for as long as you own or manage the business—written notes as well as audiotapes and videotapes recorded during the hazard assessment process—as an official record of each assessment performed. Be sure to label each record with the date and the name of the person logging the documentation. Keep copies both on- and off-site.

Recommendation 5 The type of contaminant and the anticipated level of exposure directly influence the selection of PPE, so seek professional advice about which equipment best meets your specific needs. Such information is available in PPE catalogs or through product manufacturers.

Recommendation 6 Train employees explicitly on how to use PPE and how to maintain it properly.

Recommendation 7 Issue employees their own set of PPE, and have them sign for it. Keep a dated record of exactly what was issued, and stress that they alone are to use the equipment. Experience has shown that employees take better care of equipment issued to them personally. Check all safety equipment regularly, and replace it as necessary.

PORTABLE FIRE EXTINGUISHERS (OSHA, 29 CFR PART 1910.157)

The effects of even a small fire can extend well beyond simple destruction of property. For instance, a fire that spreads to a pesticide storage room may produce deadly toxic vapors, creating the need to evacuate the community. Therefore, it is important to preplan:

- Train all employees to use portable fire extinguishers.
- Invite local fire department personnel to survey your facility to acquaint themselves with what is stored there and the precise location of chemical inventory. This may be required by your local emergency planning committee.
- Conduct a mock on-site emergency.

Water used to fight a chemical fire should be contained; afterward, you will likely have to deal with environmental contamination and regulatory oversight of remediation as well as insurance claims. But remember, such problems often can be avoided simply by training employees to operate portable fire extinguishers! Provide your employees the training they need, and make them aware that extinguishers can be effective in putting out initial-stage fires.

Protecting your business against fires involves more than simply mounting fire extinguishers throughout the building. Fire extinguishers contain only enough fire suppressant material to deal with a small fire; the contents last only for a matter of seconds. Improper discharge can waste the contents, diminishing any hope of putting out a small fire. Therefore, it is crucial to train your employees and to post visible instructions on how to use a fire extinguisher. Check with your local fire department or an extinguisher company to help with or conduct such training.

How Do I Comply with This Regulation?

OSHA does not require extinguishers to be placed in a building. Employers who preplan to evacuate their facility in case of fire need not train employees on the use of extinguishers as long as there are none on-site. Instead, obviously, employees should be trained to vacate the building at the first indication of fire.

If you preplan to evacuate and prefer not to install fire extinguishers, first check your insurance company's requirements and the local fire code; either of these might mandate extinguishers despite your plan to evacuate. And be aware that, even if you preplan to evacuate, if there are fire extinguishers on the premises, OSHA regulates their inspection and maintenance and requires that employees be trained to use them properly. In essence, their presence triggers the need for compliance under OSHA.

If fire extinguishers are to be installed, choose models recommended for the specific type of fire most likely to occur on the premises. Every fire extinguisher has a class designation that communicates the type of fire that it is designated to suppress. It is critical that you purchase the right class. Using the wrong extinguishers on a fire may worsen the problem.

OSHA regulations state that fire extinguishers must be located within a certain distance from employees:

- No more than 50 ft for Class B extinguishers
- No more than 75 ft for Class A, C, or D extinguishers

Also, extinguishers must be installed at certain heights and certain distances from exits. Check local fire codes or call your local fire department for those criteria.

Employers are required to visually inspect each extinguisher, monthly, to ensure that it is fully charged and sealed, properly mounted, and accessible without obstruction.

More sophisticated, annual inspections are required for refillable and rechargeable extinguishers. These inspections should be performed by a commercial fire extinguisher company representative. Refer to 29 CFR Part 1910.157, Table L-1, or consult with a fire extinguisher service for more information.

Employers are responsible for providing employee training on basic firefighting techniques. All employees who might need to use a fire extinguisher must be trained upon hiring or upon initial installation of extinguishers, and annually thereafter. Employees battling a small fire must be trained to recognize when it is beyond their control. Local fire departments and commercial fire safety companies often provide excellent training. In some communities, local universities or trade schools also may offer fire safety training.

FIRE EXTINGUISHER RECOMMENDATIONS

Class Designation	Use Specifications	Approved Markings on Fire Extinguisher
A	Ordinary combustion (paper, wood, cloth)	Green triangle with an A inside
B	Flammable and combustible liquids	Red square with a B inside
C	Electrical	Blue circle with a C inside
D	Combustible metals	Yellow five-pointed star with a D inside
A-B-C	Multipurpose	A-B-C markings, as above

Recommendations

Recommendation 1 Hire a company that specializes in fire extinguisher sales, service, and training. Make sure training is tailored to chemicals that might be present.

Recommendation 2 Conduct annual, hands-on training where each employee has the opportunity to handle and use a fire extinguisher. Advise your local fire department, when planning training, to see if a burn permit is required. They will know who you should call to get the burn permit. There are many excellent fire extinguisher training videos currently on the market which can be shown to complement—not replace—face-to-face training.

One teaching method often used by professional trainers is to have the employees remember the word PASS:

- P stands for "Pull the pin."
- A stands for "Aim the extinguisher nozzle at the base of the fire."
- S stands for "Squeeze the handle."
- S stands for "Sweep the hose from side-to-side at the base of the fire."

Answer the Following in Regard to Your Business

- Do you have the correct type of fire extinguisher for the materials that you store or transport?
- Do you have an appropriate number of fire extinguishers for the size of your facility?
- Are extinguishers placed according to OSHA distance requirements?

Figure 13.9 Fire extinguishers need to be readily accessible—not blocked by equipment.

- Do you verify the following information during your monthly inspections?

  Extinguishers are in their designated areas.
  Each extinguisher is easily identifiable.
  There is a clear path to each extinguisher (Fig. 13.9).
  Each extinguisher is properly mounted (e.g., not sitting on floor; Fig. 13.10).
  Each extinguisher gauge indicates a full charge.
  The plastic seal is intact on each extinguisher.
  All extinguisher tags and pins are in place.
  All extinguisher hoses are intact (not cracked, dry rotted, or plugged).
  All extinguisher pieces are present and functional.

- Are all previous monthly inspections documented?
- Are all extinguishers professionally evaluated once a year?

Figure 13.10 Fire extinguishers should always be mounted—never placed on the floor.

- Are monthly inspection and annual maintenance records maintained for a minimum of one year?
- Have all employees who might use a fire extinguisher been verifiably trained or instructed?
- Are all employees who might use a fire extinguisher retrained annually?

Recommendation 3 Verify all training with employees' signatures.

Recommendation 4 Although current OSHA regulations do not specify the height at which fire extinguishers are to be mounted, general guidelines prescribe placement of extinguishers in locations that are easy to see and easy to get to. Fire extinguishers weighing less than 40 lb should be placed no more than 5 ft off the floor; and those that weigh more than 40 lb should not be placed higher than 3.5 ft off the ground.

Recommendation 5 If extinguishers are not readily visible, post FIRE EXTINGUISHER signs that clearly identify the location of each unit so that employees can find them easily.

Recommendation 6 Many companies select A-B-C multipurpose extinguishers. But you should consult a professional for guidance in choosing the most appropriate size and type of extinguisher for your business.

Recommendation 7 Instruct employees to call for help and verify that it is on the way before attempting to fight any fire; otherwise, they may become trapped by the fire, with no help on the way. Employees should be taught not to attempt to fight a fire larger than what might be controlled by discharging one or (at worst) two fire extinguishers.

Recommendation 8 Always use the buddy system. Never respond to a fire alone because smoke and vapors can quickly overcome a responder.

Recommendation 9 Ask your local fire department to conduct an inspection and consider their input on fire potential at your facility. Such inspections also allow fire departments the opportunity to do some preplanning of their own.

Recommendation 10 Cover extinguishers located in dusty and/or corrosive environments, but make sure they remain visible. Consider using bright orange covers that are marked and have Velcro for easy removal.

Recommendation 11 Consider using the check sheet in Appendix 13.2.

Recommendation 12 Install 10 lb, B-C fire extinguishers in all vehicles that meet the definition of "commercial motor vehicle" as defined by the state or federal department of transportation (whichever is stricter).

Recommendation 13 Contact your local fire department or insurance company for help with selection and distribution of extinguishers.

Recommendation 14 Consider using the Let-It-Burn Policy in Appendix 13.3. Make sure to first check with your insurance provider before instituting such a policy.

FIRST AID KITS (OSHA, 29 CFR PART 1910.151)

First aid is the initial action taken to stabilize an injury or illness (i.e., to prevent it from becoming worse) until the victim can be treated by a qualified medical professional. But, unfortunately, some injuries can be made worse by inappropriate first aid administered by well-meaning but unqualified individuals.

You should not expect an untrained employee to aid an injured co-worker. First aid training, including information on bloodborne pathogens, is essential. OSHA requires the availability of first aid kits suitable for hazards most likely to occur in a given workplace, but it requires certified first aid personnel only if there are no local medical services within 3–4 min of the site. First aid training should be conducted at least every two years; CPR training should be repeated annually.

How Do I Comply with This Regulation?

First aid kits are required. The person administering first aid must be educated on proper techniques and must have supplies to deal with cuts, bruises, and other injuries that are likely to occur. A certified first aid person is required only when emergency responders are more than 3–4 min away.

Consider the following:

- Do you have at least one qualified employee on each shift available to administer first aid and CPR?
- Can you verify the training of employees assigned to respond to an injury?
- Are first aid kits inspected at least monthly and contents replenished (or replaced) as needed?
- Do first aid kits contain bloodborne pathogen safety equipment?
- If your business handles corrosive materials, do you have a specific area where the eyes can be flushed or the body showered for a minimum of 15 min?
- Are the names of designated, company-trained first responders posted on all first aid kits?

Recommendations

Recommendation 1 Purchase medical kits that contain the appropriate supplies in sufficient quantities to deal with the type of emergency that might occur in your business; make sure there are ample supplies to treat multiple victims if necessary. Before purchasing the kits, seek the advice of a doctor who specializes in occupational medicine; if buying the kits from a safety supply house, ask what they recommend for your specific situation.

Recommendation 2 If your business is not in close proximity to a hospital, contact the American Red Cross or the National Safety Council for first aid and CPR training. Being trained to administer first aid to an ill or injured worker or customer might make a major difference in that person's prognosis. A double benefit of CPR/first aid training is that it can be used for home and family life.

Recommendation 3 Train all full-time employees to administer basic first aid; if possible, train temporary and part-time employees as well.

Recommendation 4 Invite local emergency medical services (EMS) personnel to visit your facility. Building a relationship and educating these front-line responders will ease their fears and help to dispel any misunderstandings they may have regarding your business and the pesticide application industry in general. Go over in detail the steps that you take to prevent contamination and to make sure that products are handled and applied according to label directions. Discuss appropriate victim decontamination procedures that emergency personnel should follow in case of human injury.

Recommendation 5 Document that first aid kits are inspected monthly. Consider using the check sheet in Appendix 13.2, or develop your own. A good approach is to inspect medical kits at the time that the fire extinguishers are being checked.

Recommendation 6 Affix to all first aid kits a sticker outlining procedures to follow when dealing with an injury where the victim is bleeding.

Recommendation 7 Whenever a first aid kit is used, immediately replace the items used to treat the injury. And, while you're at it, check to make sure that product expiration dates are current on all contents.

Recommendation 8 Remove all oral medication—even aspirin—from first aid kits; any decision to use such products should be left up to the physician.

Recommendation 9 Inform your local hospital or emergency training center if you deal with pesticides that would require special handling of accident victims to minimize medical staff exposure.

Recommendation 10 If you have a company policy that no first aid is to be administered by one employee to another, make sure it is understood that the injured person should at least be provided a first aid kit. If the injured individual is unable to deal with the injury personally, arrange for medical assistance.

Recommendation 11 Mount medical kits properly and in a clean environment; this applies to vehicles as well as offices, warehouses, workshops, etc. Consider covering first aid kits with a clear plastic bag to protect them from dust and dirt.

Recommendation 12 Give employees who work with chemicals a small, unbreakable bottle of water to carry with them to flush their eyes in the event of exposure.

BLOODBORNE PATHOGENS (OSHA, 29 CFR PART 1910.1030)

Diseases spread by bloodborne pathogens (e.g., AIDS and hepatitis viruses) can cause lifelong debilitation and sometimes can be life threatening. Transmission occurs when an individual comes into contact with body fluids of a person harboring a bloodborne disease. Although transmission of the disease between two people can occur during a split second exposure, the manifestation of symptoms in the newly infected person may not occur for decades.

Transmission of these diseases in the workplace can occur when a worker aids a co-worker who is bleeding from an injury on the job. Although it is obviously important to provide first aid, it is equally important that those administering it be aware of the risk of infection by bloodborne pathogens and that they take steps to protect themselves. And it should be noted that, despite their name, bloodborne pathogens are present in other human fluids in addition to blood: saliva, tears, etc.

In addition to implementing a medical surveillance program to track levels of harmful substances in employees' blood (e.g., cholinesterase testing of employees who use organophosphates) employers must follow universal precautions and requirements under bloodborne pathogen regulations.

Consider this hypothetical situation: A truck driver suffers a severe heart attack while unloading bulk pesticides at your facility. A nearby employee determines that the driver is not breathing and immediately begins mouth-to-mouth resuscitation, bringing himself into direct contact with the driver's saliva. His quick actions may save the driver's life, but at what risk to himself? What if the truck driver is harboring a bloodborne pathogen? The heroic act that saves a life, today, may yield catastrophic consequences to the hero, tomorrow; symptoms of bloodborne diseases might not become apparent for decades after initial contact.

So the critical question is this: How can we encourage our employees to respond to injured co-workers and at the same time protect themselves from bloodborne pathogens?

How Can I Comply with This Regulation?

The rule says that "each employer having an employee(s) with occupational exposure as defined by paragraph (b) of this section establish a written Exposure Control Plan designed to eliminate or minimize employee exposure." Occupational exposure is defined in paragraph (b) as "reasonably anticipated skin, eye, mucous membrane, or parenteral contact with blood or other potentially infectious materials that may result from the performance of an employee's duties." Using this definition, we believe that all pest management businesses must comply with the bloodborne pathogen regulation.

It is the intent of the OSHA bloodborne pathogen regulation to compel employers to evaluate each job to determine whether the employee performing it could potentially come into contact with another person's blood or other body fluids in an emergency situation. If so, employers are required to develop a bloodborne pathogen plan: a detailed outline stating the company's policies on safeguarding employees who may come to the aid of others who become injured on the job.

Recommendations

Recommendation 1 Conduct bloodborne pathogen training for all employees who are expected to administer first aid and CPR. Conduct awareness training for all others. Current OSHA guidelines require training to include interaction with an individual who is knowledgeable on bloodborne pathogens and competent to answer pertinent questions.

Recommendation 2 Train all employees—custodial, field, office, part-time, and managerial staff—on bloodborne pathogens because most would react to a work-related accident by assisting the injured.

Recommendation 3 Use employees trained in first aid, CPR, and bloodborne pathogens as trainers for other employees only if they are qualified to train. Additional training normally is required to become a certified instructor.

Recommendation 4 Post on first aid kits the name, the phone number and extension, and the department of employees designated and trained to administer medical assistance.

Recommendation 5 Keep the original written plan as an official business record; it is important that it state the implementation date. If the written plan mandates training, itemize the training provided and include a list of names and signatures of employees who have completed the training. There can be an extended delay between the date of exposure/infection and the date that symptoms manifest themselves. Therefore, the list might become a very important record at some future date (e.g., if an employee were to sue a company for negligence concerning its training policies). The plan should be reviewed and the training repeated annually. Employees should sign updated lists annually as well.

Recommendation 6 Keep the most current written plan where it is readily accessible to every employee, and verify that all employees are knowledgeable of its content.

Recommendation 7 Have employees sign a form stating that they have read and understand the plan.

Recommendation 8 Document every incident where an employee performs first aid or CPR; indicate the name of the victim as well as the name of the person offering assistance. Complete an exposure record for the individual offering first aid, describing the incident in detail. If medical records result, OSHA requires that they be kept for 30 years.

Recommendation 9 Make sure that the first aid kit contains rubber gloves, splash shields, goggles, and CPR mouthpieces suitable for dealing with injuries involving blood or other body fluids.

Recommendation 10 Instruct employees not to eat, drink, or smoke during a bloodborne pathogen incident.

Recommendation 11 Consider using the model plan (see Appendix 13.4) when writing your own bloodborne pathogen plan.

Recommendation 12 If your company has a policy whereby employees are not to administer first aid to injured co-workers, make sure they at least know to provide first aid kits to injured individuals. If the injured cannot administer their own first aid, they should go (or be taken) to a medical facility for treatment.

EMERGENCY PHONE NUMBERS [OSHA, 29 CFR PART 1910.38(A)(2)(V) AND PART 1910.165(B)(4)]

Post emergency telephone numbers throughout your facility and inside company vehicles. This simple procedure will save valuable time in the event of an emergency—time that may make the difference between a controlled situation and a disaster.

How Can I Comply with This Regulation?

This regulation states that employers must make employees understand "the preferred means of reporting fires and other emergencies." Most companies emphasize that calling for help should always be the first action taken; emergency telephone numbers must be posted at or near every phone (Fig. 13.11).

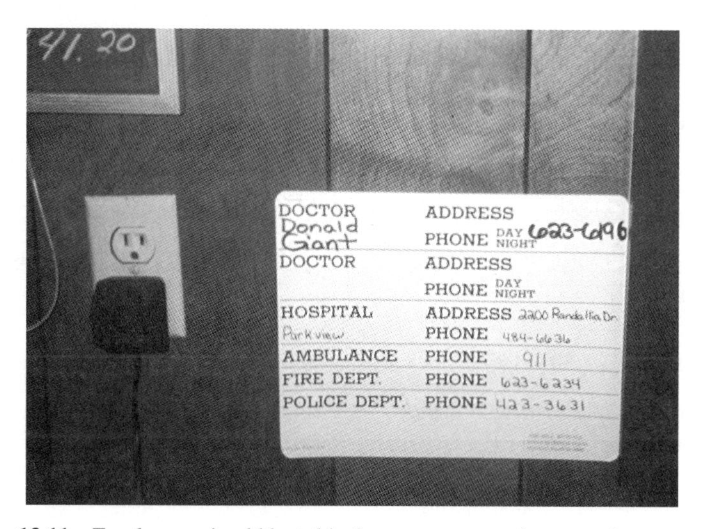

Figure 13.11 Employees should be told where emergency phone numbers are posted.

Recommendations

Recommendation 1 Develop a phone list with only those numbers that are extremely important. Other, less important numbers may be listed in the emergency response plan.

Recommendation 2 Verify phone numbers annually; mark your calendar and form a habit of checking them the same month every year. If there are changes, retype the list; handwritten notations can be difficult to read, especially in an emergency situation. Make sure the updated list replaces every old list posted.

Recommendation 3 Post emergency telephone numbers on all bulletin boards and near all phones.

Recommendation 4 Post signs listing emergency contact personnel conspicuously on the outside of all buildings. The signs should include the names of emergency contact personnel (company representatives), their phone numbers, and street address.

Recommendation 5 Provide company contact numbers to local law enforcement agencies, fire departments, and medical responders. Update them on all personnel or telephone number changes.

Recommendation 6 Post company emergency numbers on a clipboard or by other means in all company vehicles to facilitate notification efforts during a roadway emergency.

HAZARDOUS WASTE OPERATIONS AND EMERGENCY RESPONSE (OSHA, 29 CFR PART 1910.120)

This standard, better known as HAZWOPER, requires emergency response training for all employees who might have to react to a hazardous situation. Training is the key to competent, effective response, and the more employers expect employees to participate in emergency response, the more education and training they must provide.

There are five levels of first responder training: awareness, operations, technician, specialist, and incident commander. Employers should designate and train individuals at each level, for each site. This is determined by the rules and responsibilities of employees. This requirement can be fulfilled by training multiple individuals or by training one individual at all five levels.

First Responder Levels

First Responder Awareness Level All employees with the potential for contact with hazardous substances during an emergency must complete awareness level training. Employees who are likely to discover an accidental release of a hazardous substance should be instructed to evacuate the area and notify management personnel; they are to take no other action. This level of training is usually fulfilled within the context of the company's emergency response training.

First Responder Operations Level All employees with responsibility to protect other individuals, the surrounding property, or the environment during a hazardous substance release must complete operations level training. They must be trained to control the spill without actually trying to stop the release or coming into contact with the substance. OSHA mandates 8 hr of training at the operations level; the requirement usually can be fulfilled within the context of emergency response training.

First Responder Technician Level Employees responsible for stopping an accidental hazardous substance release must complete technician level training. Twenty-four hours of initial training are required, plus an additional 8 hr of field experience and 8 hr of refresher training annually. These individuals are designated on the basis of their training to plug, patch, or otherwise stop the release in an emergency situation.

First Responder Specialist Level Specialist training requirements are similar to those of the technician level, but specialists are required to know more about the specific substances used at the facility.

Incident Commander Level An employee trained as an incident commander may actually participate in the emergency response decision-making process. All first responders trained above the awareness level who could potentially assume control of an incident must complete commander level training. As the title implies, these responders are trained to take command of an incident and coordinate on-the-scene emergency operations. Initial training requirements consist of 24 hr of formal training and an additional 8 hr of field experience. In addition, commanders must complete 8 hr of refresher courses annually.

The five levels of competency are illustrated in the following example: *Imagine that a pesticide is leaking from a tank and flowing toward a nearby creek.*

1. An employee trained at the awareness level should evacuate the area and contact company management personnel.
2. An operations level employee should construct a dam ahead of the flow or otherwise contain the spill, thereby preventing the pesticide from reaching the creek.
3. An employee trained at the technician or specialist level should attempt to stop the leak (e.g., repair a malfunctioning valve) while wearing sufficient PPE and taking appropriate safety precautions.
4. A specialist should provide other emergency responders with detailed environmental and safety information on the specific product involved.
5. An incident commander or a specialist should work with community emergency responders to make decisions on defusing the situation and should coordinate and direct trained personnel involved in remedial action.

How Can I Comply with This Regulation?

It is recommended that all employees be trained to respond to emergencies. However, minor spills that can be handled by employees, such as releases incidental to their job, do not constitute an emergency situation.

The level of training required depends on company policy, that is, what management personnel expect of their employees. However, OSHA mandates that any business working with pesticides and/or fertilizers must at least train their employees (both full- and part-time) to the awareness level. Training to higher levels is required only if management expects employees to respond to an emergency.

Basic awareness training may be conducted in-house and should convey the following:

- The definition of a hazardous substance and associated risks
- The need to identify hazardous substances during an emergency
- Potential repercussions of a hazardous substance emergency
- The role of first responder awareness level employees as set forth in the employer's emergency response plan

Recommendations

Recommendation 1 Document all training with itemized agendas, tests, certificates, and signed attendance sheets. If materials are passed out, either list the exact contents or keep the packet of information on file for subsequent referral.

Recommendation 2 Provide a trained person to answer questions that employees may have after viewing videos or reading publications purchased or rented to help with awareness level training.

Recommendation 3 If awareness level training applies to new employees, provide it immediately. They should be instructed to call 911 in the event of an accident when company representatives are not present.

Recommendation 4 Train all full-time employees to the operations level, covering the following:

- Basic hazard and risk assessment techniques
- Implementation of the emergency response plan
- Proper selection and use of personal protective equipment
- Basic control and containment operation techniques
- Basic decontamination procedures

Recommendation 5 Train at least one person to the technician and incident commander levels; in smaller companies, this person likely would be the company owner or manager. The designated person must have thorough knowledge of the facility and what is stored there and be able to communicate effectively with fire department, ambulance, and regulatory personnel. The individual must have authority to make on-the-spot monetary decisions regarding emergency remedial action. The technician/incident commander should notify regulatory agencies of the mishap.

Recommendation 6 Invite local fire department, police, and emergency personnel, and possibly local and state regulators, to your facility. Making them familiar with your facility and the chemicals you handle could prove valuable in the event of an emergency.

Recommendation 7 Invite local emergency responders to speak at your training programs. Their training and experience will add credibility to your effort in teaching your employees how to deal with emergencies properly.

EXIT SIGNS (OSHA, 29 CFR PART 1910.35)

It is often necessary to evacuate a facility during an emergency, and a quick decision to do so can prevent injuries and save lives.

Employers often ask why they should post EXIT signs and escape routes when their employees already know how to exit the building. But we must recognize that excitement is a powerful emotion during an emergency, and people may panic and lose their way; new employees may not remember exactly where the exits are. Others such as salespeople, suppliers, customers, and the general public are often allowed to enter facilities without restriction; if an emergency were to occur while they are present, they might or might not remember how to get out.

How Can I Comply with This Regulation?

Place EXIT signs conspicuously and make sure doors serving as exits are side-hinged and accessible from both directions. Leave exit doors unlocked during business hours (Fig. 13.12).

Place an EXIT sign above each exit door. OSHA mandates illuminated signs only if the area is dimly lit. Check with the fire department to see if local/state regulations require illuminated signs.

Recommendations

Recommendation 1 With exits such as see-through doors that are not readily apparent, post EXIT signs with accompanying arrows indicating the route of escape.

Recommendation 2 Indicate all exits on the site evacuation map.

Recommendation 3 If EXIT signs are lighted, perform monthly inspections to verify their operation.

Recommendation 4 Perform periodic inspections to make sure all exits are adequately marked. This can be accomplished simultaneously with fire extinguisher and medical kit evaluations; documentation should be filed as a company record.

Recommendation 5 Post NOT AN EXIT signs on or over doors that are not exits, and those that could be confused as an exit (Fig. 13.13).

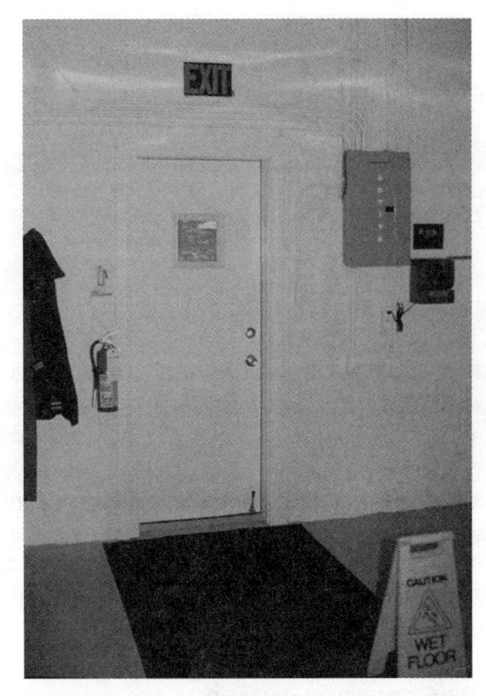

Figure 13.12 Exit doors should be clearly marked.

Recommendation 6 Keep emergency lighting available and inspect it regularly to make sure batteries are fully charged, etc.

Recommendation 7 Familiarize employees with fire alarm locations and conduct drills on how to signal other employees that there is a fire.

Figure 13.13 Doors that might be confused as an exit door should be marked accordingly.

Figure 13.14 Exit doors should never be locked when employees are present, and the paths leading to the doors should never be blocked.

Recommendation 8 Keep all exit routes open and accessible. There must be at least 28 inches of clearance around each exit (Fig. 13.14).

Recommendation 9 Post DO NOT BLOCK EXIT signs on the outside of the building, near each exit.

ELECTRICAL PANEL MARKINGS (OSHA, 29 CFR PART 1910.303)

Most employees seldom have reason to access the breaker panel at the facility where they work. In some emergencies, however, it may be necessary to turn off power to one room or one piece of equipment. For example, during a spill or fire the power to a specific area might need to be turned off to ensure that emergency responders can work without threat of electrocution. Meanwhile, it might be essential to maintain power to other rooms or buildings so that emergency teams can operate their equipment. Label each circuit in the breaker panel with what it controls, its voltage, and its amperage. And make sure all employees know where the panel is located as well as how to use it.

How Can I Comply with This Regulation?

The two most important items to be addressed under this regulation are as follows: Make sure the outside of the breaker box is marked MAIN ELECTRICAL PANEL, and mark each circuit with specifically what it controls (e.g., machine 3, air conditioning).

Keep covers in place over all breakers and the breaker box to prevent electrocution. All cover latches must be operational (Figs. 13.15 and 13.16).

Figure 13.15 Electric panel boxes should have each fuse marked according to what the specific switch controls.

Figure 13.16 An example where an electric panel box poses a serious threat to employees.

Recommendations

Recommendation 1 Make sure breaker box locations are noted on the evacuation map.

Recommendation 2 Include in your emergency response plan a list of what each circuit controls.

Recommendation 3 As stated previously, label each circuit in the breaker panel for what it controls, its voltage, and its amperage. Check breaker boxes regularly to confirm that each circuit is marked for specifically what it controls, that each such identification is current, and that labels remain legible. All markings must be in permanent ink, not in pencil.

Recommendation 4 If electrical panel work is done, be sure to update circuit identification markings.

Recommendation 5 Make sure the circuit box remains accessible at all times. Keep an area clear at least 28 inches in front of the box.

Recommendation 6 Plug any open knockout holes in the breaker box.

EVACUATION MAP (OSHA, 29 CFR PART 1910.38)

When a facility must be evacuated, it is important that it be done quickly and uniformly and that employees proceed immediately to a predesignated rendezvous point. A facility evacuation map should be posted conspicuously and employees made aware of the exit route to take in the event of an emergency (Fig. 13.17).

The importance of meeting at a designated site can never be overemphasized. Having everyone rendezvous immediately after evacuation facilitates accounting for all employees. This is critical since the emergency responders' strategy for controlling the situation may hinge on whether or not the area has been totally evacuated. For example, without confirmation that everyone is out, an incident commander might send firefighters into a burning building to look for persons trapped inside; verification that everyone is out of the building would prevent placing those firefighters in jeopardy. From another standpoint, as long as firefighters are inside a burning building, other emergency management strategies might have to be delayed. After everyone is accounted for, the rendezvous point allows the person in charge to delegate responsibilities if needed.

How Can I Comply with This Regulation?

This regulation requires that employees be trained on evacuation routes and rendezvous points. Employers with 10 or fewer employees can communicate verbally to employees about what they need to do in an emergency.

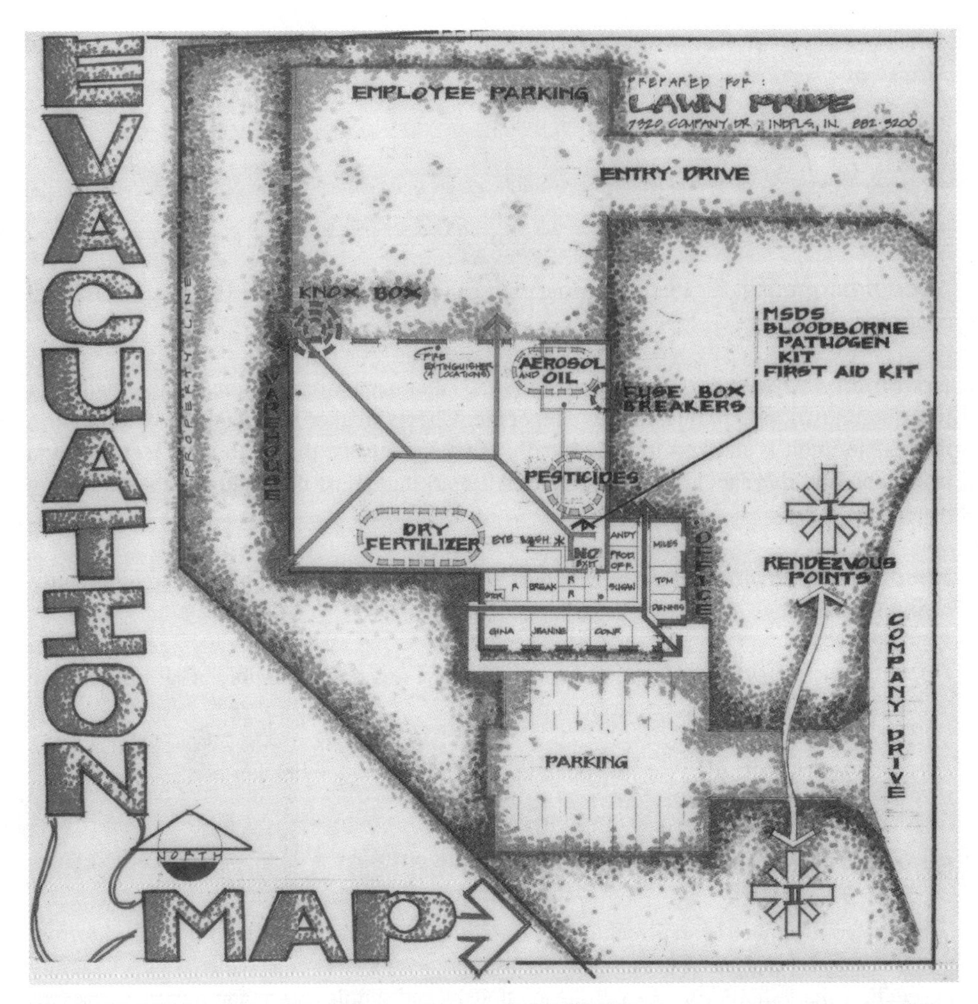

Figure 13.17 An evacuation map provides essential information on escape routes and meeting sites.

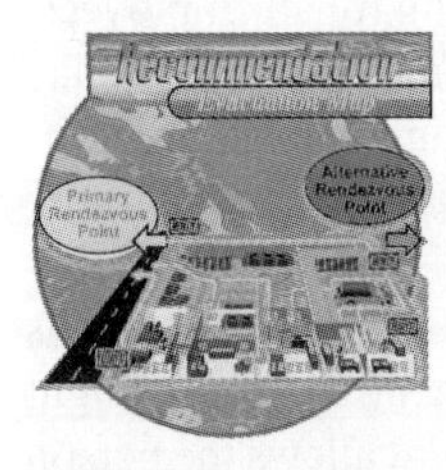

Recommendations

Recommendation 1 Create and post in employee break rooms, the main office, bulletin boards, and work areas an evacuation map for each building; copies should be filed with the emergency response plan and in company policy and training manuals.

Recommendation 2 Verify the evacuation map is up to date and make changes as necessary; always update it when the layout of the facility changes due to construction, the installation of new equipment, etc. Follow with immediate notification of employees. Schedule evacuation map revisions along with those for fire extinguishers, medical kits, and EXIT signs.

Recommendation 3 Designate a first- and second-choice rendezvous point where employees are to proceed following evacuation. The second-choice site should be used if the first is made undesirable by the emergency itself. For instance, if the first-choice site turns out to be downwind of a burning building, employees should proceed to the second site.

Recommendation 4 Mark locations of the following on the evacuation map:

- Emergency exits
- Evacuation routes (each in a different color)
- Primary and alternative rendezvous points
- Fire extinguishers
- Fire alarms
- Sprinkler controls
- First aid kits
- Bloodborne pathogen kits
- Biohazard containers
- Eyewash areas
- Emergency showers
- Circuit breaker boxes
- Gas and water shutoffs
- Spill control equipment
- Tornado shelters

Recommendation 5 Assign an individual to account for all employees and visitors at the rendezvous points. Decide on a bright-colored vest or cap which that person is to wear in the event of evacuation, and train all employees to check in with that person upon reaching the rendezvous point.

Recommendation 6 Institute a sign-in and sign-out ledger for guests. Managers and supervisors should be made aware (each day) of all visitors on the premises for accountability purposes in the event of an emergency.

Recommendation 7 Conduct annual, unannounced drills to determine the effectiveness of training provided on emergency evacuation procedures. Analyze the results and compliment employees on the aspects that were performed well; call to their attention the things that need correcting, then schedule training to emphasize those points. Document the drill and all training conducted as a follow-up. Critique is important.

SITE MAP (OSHA, 29 CFR PARTS 1910.38 AND 1910.120)

The site map is a requirement of OSHA's regulations dealing with emergency plans. Its preparation is one of the most important activities that any business can undertake. Essentially, it is a blueprint of the site which identifies all buildings and their exact locations; it also indicates exact locations within each building where chemicals are stored.

A detailed site map can convey vital information to emergency personnel in a matter of minutes. It can be a critical resource for first responders in assessing a situation and determining a plan of action.

How Do I Comply with This Regulation?

Site maps fall under OSHA's regulations on emergency planning. If an employer has fewer than 10 employees, a written emergency plan is not required, so neither is a site map. Nevertheless, it is strongly recommended that business owners and managers develop site maps for use in emergency situations and in coordination with emergency responders.

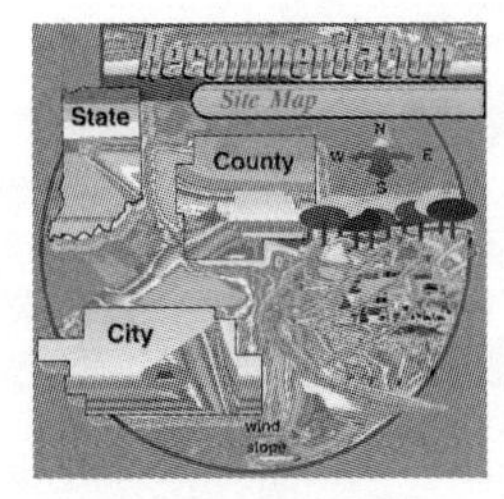

Recommendations

Recommendation 1 Post the site map in employee break rooms and the main office and on bulletin boards; include it in company policy and training manuals and the company emergency response plan. Also post it in a weatherproof exterior location where it is easily accessible to emergency responders. Review the map (and where to find it) with employees annually.

Recommendation 2 Draw the site map to scale.

Recommendation 3 Show the following on your site map:

- Direction legend: north, south, east, west
- Road numbers
- Ditches, rivers, and lakes on or near the property
- Points of access to the property
- The directional slope of the land and storm water flow
- Types and location of fencing on or around the property
- Name and exact location of each building
- Indoor and outdoor chemical storage areas
- Outdoor chemical mixing and loading areas
- Fuel storage areas
- Liquid propane tanks
- Anhydrous ammonia tanks
- Underground sewer, electrical, water, and gas lines
- Emergency disconnect sites for gas, water, and electricity
- Aboveground electrical service drop
- Tile drains (outdoors)
- Septic tanks
- Wells
- Fire hydrants
- Special equipment for cleanup of spills

Indicate the following with respect to the interior of each building:

- Building age
- Building dimensions

- Type of building construction
- Type of roof, window, and floor construction
- Location of MSDSs, first aid kits, fire extinguishers, and biohazard kits
- Chemical storage areas
- Chemical mixing and loading areas
- Emergency eyewash and shower areas
- Drains and where they lead
- Mustering sites

Keep a county map handy, showing the following:

- The business site
- The directional slope of the land
- Schools, hospitals, nursing homes, subdivisions, towns, etc., adjacent to the property
- Flow patterns of surface water
- Proximity of ditches, rivers, and lakes to the property
- Direction of prevailing winds

Recommendation 4 Consider hiring a local high school or trade school student with experience in drafting techniques to help with the layout of the site map, or consider using one of the many computer programs available.

Recommendation 5 Confirm storm sewer outfall locations, using tracer dye. This information is useful in the event of a spill in that responders can identify the point at which the spill must be contained.

Recommendation 6 Color code all drains to differentiate storm sewers from sanitary sewers.

Recommendation 7 Videotape the site, with narration, and make the tape available to local emergency plan committees, fire departments, and other emergency responders. Their review of the tape will help prepare them to address an emergency at the site, should one occur.

Recommendation 8 Update site maps annually or as changes occur.

Recommendation 9 Include worst-case scenarios in the emergency response plan.

EMERGENCY RESPONSE PLAN (FEDERAL REGISTER, VOL. 61, NO. 109, PP. 28642–28664)

The objective of contingency planning is to prevent emergencies; but if they do occur, the objective becomes a matter of reacting appropriately to minimize detrimental effects. Both aspects—prevention and reaction—require a well-organized effort on the part of business owners and management personnel.

Figure 13.18 A company temporarily places a herbicide in a tank while it constructs a containment system.

When government took the stance that businesses must adopt contingency plans to deal with emergencies, many state and federal agencies wrote regulations requiring the development of emergency response plans. These regulations, despite originating in different agencies, were often very similar. But despite the similarities, each agency had its unique requirements. And pest management companies, in attempting to comply, often duplicated their own efforts. That is, they began anew to satisfy each requirement instead of adding to a plan already in place (Figs. 13.18 and 13.19).

In 1996, the U.S. Environmental Protection Agency, the Department of Transportation, the Department of the Interior, and the Department of Labor agreed to a single consolidated plan: the National Response Team's Integrated Contingency Plan. Since that time, businesses have been able to develop and implement a single plan to satisfy all requirements. The resulting, more comprehensive plans are easier for training purposes and more efficient

Figure 13.19 A close-up view of the temporary tank where the herbicide corrodes the plastic, which allows most of the contents to drain to the ground.

Figure 13.20 The herbicide moved through an underground drain that lead to a city pond.

to maintain. There is no repetition. The focus on a single plan yields a well-informed employee base; workers are better prepared to react appropriately in emergency situations—a benefit to the business and the community.

A contingency plan is only as good as the information it conveys to employees and emergency responders. It is useless if the only people who comprehend its intent and how to execute it are those who wrote it. Employees must be educated to understand the purpose of the plan, and they must be trained to perform their assigned duties in an emergency situation.

It is essential that every employee and all emergency responders in the community be familiar with the plan. And it is equally important that the plan be updated on a regular basis to incorporate changes: phone numbers, new employees, new (company) emergency responders, new or reassigned position responsibilities, etc. A thorough review should be done at least annually, as should employee review and retraining (Figs. 13.20–13.23).

How Do I Comply with This Regulation?

A written emergency response plan is required for most emergency response businesses, and in developing a plan the approach should be to fit it to your industry, your facilities, and your geographic site. The goal is not to impress with bulk, nor is it only to meet the letter of the law; and it is more than just putting a plan in writing. The goal is to prescribe procedures and policies to minimize injury, death, and environmental damage that could result from an emergency. A vital component of preplanning is to educate and train employees on how to carry out the plan in an actual emergency.

Recommendations

Recommendation 1 Post a 24-hr number on the outside of all buildings so that emergency responders will know where to call if an emergency occurs when the business is closed and the premises vacant (Fig. 13.24).

Figure 13.21 Small fish began dying soon after the herbicide contaminated the water in this pond.

Recommendation 2 Train all employees and document all training on:

- the location of the written emergency response plan,
- the purpose and objectives of the plan,
- implementing the plan,

Figure 13.22 The herbicide was found in a previously unknown underground drain.

Figure 13.23 The herbicide also found its way into the city's sewer system, killing many of the biological organisms used to break down the waste.

- who to contact in an emergency,
- where to rendezvous following evacuation, and
- who should address the media.

Recommendation 3 Personally deliver copies of the emergency response plan to local responders: fire departments, law enforcement agencies, emergency medical services, and emergency planning committees. Review the plan with them, and document their receipt of the plan. Periodically contact them.

Such one-on-one contact is much more meaningful than their receiving the plan by mail. By showing this type of interest and initiative, you captivate the emergency responders

Figure 13.24 Emergency phone numbers placed on the outside of buildings also can be used by those responding to an emergency when the company employees are not present.

interest in your business and impress upon them that you have everyone's best interests in mind. Point out the following:

- Who to contact in an emergency
- Where employees are to rendezvous if evacuated
- Types of chemicals stored on-site
- Precise locations of chemical storage areas
- Important listings on the site map
- What you and your staff should and should not do in an emergency
- What expectations you have (e.g., a let-it-burn policy) when outside assistance is requested
- The on-site location of the written plan (i.e., where emergency responders can access it)
- Off-site locations where the plan is accessible (Figs. 13.25 and 13.26)

Recommendation 4 Add value to the plan by inviting emergency responders to walk through your facility. Consider incorporating videos and photographs with the written plan and ask for suggestions to make the plan better; follow up on any recommendations they offer.

Figure 13.25 Evacuation maps, site plans, a chemical inventory, and Material Safety Data Sheets can be stored in plastic tubes that can be used by emergency responders.

Figure 13.26 Mail boxes can be used to store emergency information.

Recommendation 5 Update the emergency response plan annually and as changes occur. This practice is often overlooked, but the consequences of having emergency responders act on an out-of-date plan can be critical.

Recommendation 6 Stage a mock accident or spill, annually, and critique the generated response. This can meet the requirements of annual refresher training for HAZWOPER.

Recommendation 7 Contact hospitals to see if they can treat patients for exposure to the chemicals you handle, and ask if they have decontamination capabilities. Hospitals are often overlooked when notifying local responders.

Recommendation 8 Use the emergency response plan in Appendix 13.5 as a model for developing your own.

Recommendation 9 Spill recovery contractors play an essential role in follow-up operations related to a spill emergency. Select a reputable firm that you are comfortable with. Always obtain references—and check them—before signing a contract. Hire someone who understands your business and is knowledgeable on prevailing state and federal requirements; it is important that they accept the obligation to comply with local, state, and federal paperwork and permit requirements.

Preplanning in cooperation with the company is important. Discuss your operations, describe a worst-case scenario, and ask for a firm estimate before signing. Determine their availability and the amount of time it would take them to respond to an emergency at your facility. Insist on a number where you can reach them 24 hr a day. Select a company that either employs or can easily and quickly subcontract industrial hygienists, engineers, heavy equipment, specialized equipment such as laboratory and monitoring devices, and supplies needed during an emergency.

SUMMARY

Prevention of pesticide accidents is the first line of defense, but of equal importance is being prepared for emergencies; this can never be overemphasized. Preparation can make the difference between a controlled emergency and total chaos. Preplanning can minimize damage, injury, and the cost of cleanup (Fig. 13.27).

Figure 13.27 Planning for emergencies protects the interest of the business and the health of the community.

But even with the best of training, accidents can and will occur. No company is immune to accidents around the shop, on the road, or on a customer's property.

Given the fact that fires and spills will occur, it is important that business owners and managers thoroughly understand their insurance coverage: what is and is not covered and what limitations or exclusions apply.

It is the sad truth that many who believe they are insured for fire, theft, and spills often are negatively surprised when they seek restitution for a loss. They may be told that a particular type of accident is not covered, that the amount of coverage is quite low, or that a steep deductible applies.

Emergencies are aggravating and stressful, and they often take longer than anticipated to resolve. It is impossible to predict how, when, and where an emergency will occur or how responders will deal with it. No two emergencies are exactly the same, but not knowing exactly how an emergency might unfold should not be used as an excuse to avoid preplanning.

Responders must be trained to assess each situation and make a decision on what action to take; it might mean proceeding as outlined in the emergency response plan or making split-second decisions to modify the plan.

It is imperative that emergency response plans be reviewed with employees on a regular basis. Information must be updated regularly so that management personnel and employees know precisely what procedures to follow if an emergency occurs. Stage a mock emergency annually to confirm that your work force is adequately trained, and follow up by repeating training in weak areas.

We all hope never to face an actual emergency, but we must prepare for the possibility. Only a real-life emergency can put an emergency response plan to the test. Only in that experience can the effectiveness of our preventive measures, safety practices, and employee acceptance be measured, for it is only then that we truly implement the plan. And it is only then that our skills and knowledge and those of our employees—must prevail. It is only then that we will be called upon to make quick decisions and to rely on the judgment of others trained to respond. Tip the scale in your own favor: Always be ready for the unexpected.

APPENDIX 13.1 CERTIFICATION OF HAZARD ASSESSMENT (OSHA, 29 CFR PART 1910.132)

This is to certify that an inspection of company premises, vehicles, and application equipment has been conducted to assess hazards associated with this firm: ______________________ The inspection addressed potential job site hazards as well.
Name of Company

____________________ ____
Date of assessment Facility assessed

____
Department(s) assessed

____
Location/address

PERSON(S) INTERVIEWED:

________________________________ ________________________________
Name Title

________________________________ ________________________________
Name Title

________________________________ ________________________________
Name Title

________________________________ ________________________________
Name Title

ASSESSOR:

________________________________ ________________________________
Name (Please Print) Title

____
Signature

TYPE OF INSPECTION: Initial __________ Follow-up __________

NUMBER OF PAGES IN THIS REPORT: __________

INCLUSIONS IN THIS REPORT:

Written notes ______ Audio ______ Video ______ Still photographs ______ None ______

LOCATION OF PREVIOUS ASSESSMENT REPORTS: On-site ______ Corporate headquarters ______

HAZARD EVALUATIONS

Consideration has been given to each of the basic hazard categories outlined in Appendix B, 29 CFR Part 1910.132. Listed here, for each category, are specifics on hazards observed and a description of the location of each.

1. Impact sources (e.g., vehicles, moving machinery, grinders, forklifts, hammers, movement of personnel that could result in collision with stationary objects)
2. Penetration sources (e.g., sharp objects which might pierce feet or cut hands, lawn debris, lawn mower blades, sharp edges or corners on equipment)
3. Compression sources (e.g., high-pressure compressed air, rolling or pinching objects which could crush, pneumatic or hydraulic mechanisms/equipment)

4. Chemical sources (e.g., pesticides, paints, fuels, solvents, hydraulic fluids)
5. High-temperature sources (e.g., welding torches, vehicle cooling systems, exhaust mufflers)
6. Low-temperature sources (e.g., high wind chill, propane, anhydrous ammonia)
7. Dust sources (e.g., grain bins, crawl spaces, basements, attics, fertilizer dust, dust from granular trace elements, silica sand from sand blasting, dry flowable and dry powder pesticides)
8. Light-radiation sources (e.g., welders)
9. Steam (e.g., steam cleaners, vehicle and machinery cooling systems)
10. Noise (e.g., grinders, saws, small engines, loaders and other heavy machinery, highway noise)
11. Electrical sources (e.g., exposed wiring, ungrounded wiring, damaged cords)
12. Other sources (e.g., atmospheric conditions, confined space, bloodborne pathogens)

WORKPLACE CHANGES TO REMOVE HAZARDS

The company, through internal review and discussion with employees, has agreed to introduce into the workplace the following listed procedures.

Pesticides

- Treatment options that do not require the use of pesticides will be given first consideration. If a pesticide application is necessary, general-use products will be given preference over those labeled for restricted use, thereby reducing risk potential.

Whenever possible, we will select products

- that have the signal word CAUTION instead of WARNING or DANGER,
- that are packaged to reduce the risk of exposure or spillage,
- for which Material Safety Data Sheets do not indicate long-term health effects, and
- that do not require users to wear personal protective equipment (PPE). (But if products requiring PPE are selected, we will train all designated pesticide users on PPE requirements as stated on the label.)

We will

- designate certain individuals to handle pesticides;
- train designated pesticide handlers on the specifics of handling, mixing, and applying the products they use;
- supplement training by providing pertinent literature (e.g., product labels, MSDSs, hazard communication plan);
- maintain proper storage of chemicals in inventory;
- ensure good ventilation in pesticide mixing, loading, and storage areas;
- provide proper secondary containment in mixing, loading, and storage areas;

- train personnel on proper personal hygiene and decontamination procedures (washing hands, laundering clothing, etc.); and
- enforce good housekeeping practices to reduce workplace exposure and accidents.

Maintenance Shop

- Access to the maintenance shop shall be restricted to authorized personnel only.
- All equipment shall have safety guards, as appropriate.
- OSHA's lockout/tagout program will be enforced.
- Store oils, solvents, paints, and other flammable materials away from welding, cutting, and grinding areas and away from pilot lights.
- Good lighting, ventilation, and ultraviolet screens will be provided in welding and cutting areas.
- Good ventilation and proper PPE will be provided in the battery-charging area.
- Control of drains in shop floors will be verified.
- Mechanics will be made aware of pesticides used and the PPE and/or cleaning practices required when working on pesticide application equipment. Mechanics also will be informed of products that are volatile or flammable and warned that all pesticides are potentially poisonous.

Vehicles

- All company vehicles will be properly maintained and serviced.
- Employees will receive training on safe driving techniques.
- All employees will wear seat belts when operating company vehicles.
- Vehicles will transport chemicals only in small quantities whenever possible.

General Employees

- All full- and part-time employees will be made aware of potential hazards in the workplace.
- Employees will be trained on how to lift heavy objects to prevent back injuries.
- Training in forklift (required by OSHA) and loader operations will include instructions to lower loads whenever possible, to use caution in proceeding up and down ramps, to use slow to moderate speeds, and to stay alert when turning corners and when backing up.
- Training and proper fall-protection equipment will be provided for employees working off the ground on such apparatuses as storage tanks, roofs, rail cars, and tanker trucks.
- Handling and storage areas for liquids will be kept clear of standing water and product contamination to prevent personal injury.

Personal Protective Equipment Evaluations

In recognition that not all potential hazards can be eliminated through engineering, the use of PPE will be enforced to safeguard employees. All workers with exposure potential—not only to chemicals, but to all workplace hazards—are required to wear appropriate personal safety equipment. Each employee will be instructed to wear safety equipment relative to their job assignment as determined by the type, level of risk, and injury potential from hazards that they might encounter in the workplace. We will enforce the use of personal safety equipment as indicated in the following list.

HAZARD CATEGORY	SPECIFIC HAZARD	PPE/PRACTICES REQUIRED
Impact	Drill press	Safety glasses or full-face shield
Chemical	Pesticides	Gloves and goggles for all pesticides, plus any additional requirements as stated on the label
Penetration	Grass trimmers	Safety goggles and work shoes
Other	Bloodborne pathogens	Gloves and mouthpiece for CPR

Employee Education

All employees will be trained on the proper use and maintenance of safety equipment. Training programs will be documented by a Certificate of Training.

OSHA, 29 CFR PART 1910.132 CERTIFICATE OF TRAINING

Name of company

Street address

City, State, Zip code

Training location | Date

Program agenda	Speaker	Major points covered
What jobs require PPE?		
What PPE is necessary?		
What are PPE limitations?		
How to fit and wear PPE		
How to dispose of PPE		
Company policies on PPE		

Employees trained

Printed name	Signature

APPENDIX 13.2 CHECK SHEETS FOR FIRE EXTINGUISHERS, EXIT SIGNS, AND FIRST AID KITS

Fire Extinguishers

__________ Date of inspection

__________ Name of person conducting the inspection

	Extinguishers located at the facility				Extinguishers located in vehicles			
Extinguisher number	**1**	**2**	**3**	**4**	**5**	**6**	**7**	**8**
Size of each extinguisher	___	___	___	___	___	___	___	___
Type of each extinguisher	___	___	___	___	___	___	___	___
Is each extinguisher . . .								
. . . on-site indicated on evacuation/site maps?	___	___	___	___	___	___	___	___
. . . in its assigned location?	___	___	___	___	___	___	___	___
. . . locatable with signs?	___	___	___	___	___	___	___	___
. . . easily accessible?	___	___	___	___	___	___	___	___
. . . properly mounted?	___	___	___	___	___	___	___	___
. . . fully charged?	___	___	___	___	___	___	___	___
. . . in good condition?	___	___	___	___	___	___	___	___
Are plastic seals intact?	___	___	___	___	___	___	___	___
Are all pieces present?	___	___	___	___	___	___	___	___

Exit Signs

__________ Date of inspection

__________ Name of person conducting the inspection

EXIT sign number	**1**	**2**	**3**	**4**	**5**	**6**	**7**	**8**
Is each EXIT sign present?	___	___	___	___	___	___	___	___
Are letters at least 6 inches high?	___	___	___	___	___	___	___	___
Are the bulbs in lighted signs functional?	___	___	___	___	___	___	___	___
Are exit doors free of obstructions?	___	___	___	___	___	___	___	___
Are doors that are not exits marked NOT AN EXIT?	___	___	___	___	___	___	___	___

First Aid

____________________ Date of inspection

____________________ Name of person conducting the inspection

	First aid kits located at the facility				First aid kits located in vehicles			
First aid Kit number	**1**	**2**	**3**	**4**	**5**	**6**	**7**	**8**
Is each first aid kit . . .								
. . . on-site indicated on the evacuation and site maps?								
. . . in its assigned location?								
. . . easily identifiable?								
. . . easily accessible?								
. . . mounted on the wall indoors?								
. . . clean and in good condition?								
. . . labeled with the name of an employee assigned to administer first aid?								
Are the contents complete?								
Is there an ample supply of all contents?								
Are all expiration dates current?								
If a bloodborne hazard kit, is sticker attached that indicates it is a bloodborne hazard kit?								
Does the kit contain protective gloves and eyewear?								
Is there a CPR mask and check valve in the kit or nearby?								

Topic	Fire extinguishers	EXIT signs	First aid kits
Date of correction			
Name of person who made the correction			

APPENDIX 13.3 LET-IT-BURN POLICY

Fire in a pest management business is always a concern to management, employees, and the community. The company has adopted an emergency response plan that will be activated in the event of a fire or other emergency situations, and we have taken steps to train our employees in the safe handling of chemicals, both on and off company property.

Heat is one of the best ways of breaking down hazardous materials; therefore, we ask that you support our LET-IT-BURN POLICY under the following circumstances:

- When the situation is deemed out of control and the use of water could create additional hazards
- When there is no additional danger to people by letting it burn
- When the use of water could increase airborne contamination
- When the use of water could adversely affect off-site properties

In recognition that environmental impairment is detrimental to society and must be held to a minimum, we request that the FIRE DEPARTMENT acknowledge certain conditions which favor letting a facility burn.

____________________________________ prefers that the fire department ap-
(Name of Company Representative)
ply only the smallest quantity of water necessary to prevent additional hazards.

______________________________ resolves to exclude the emergency agency from
(Name of Company Representative)
liability for situations that might arise from permitting the facility to incinerate. Our respective insurance companies are aware of this policy.

____________________	____________
Company name (printed)	Date
____________________	____________
Company official	Title
____________________	
Name of fire department	
____________________	____________
Fire chief	Date

APPENDIX 13.4 BLOODBORNE PATHOGEN POLICY

____
Name of company

Purpose and Compliance

The purpose of our Bloodborne Pathogen Exposure Control Plan is to educate our employees on recognizing and implementing safeguards to prevent the transmission of bloodborne pathogens. The plan, written in compliance with the Occupational Safety and Health Administration's Bloodborne Pathogen Standard (OSHA, 29 CFR Part 1910.1030), is an important addition to other company policies that deal with occupational and safety issues.

Determination of Exposure

OSHA requires that employers identify jobs that may bring employees into contact with blood or other body fluids. Management personnel, in consultation with employees, have concluded that the job classifications listed below could potentially include occupational exposure to bloodborne pathogens. The determination was made without regard to the use of personal protective equipment or to the likelihood of exposure being great or small. Employees holding these jobs are required to attend bloodborne pathogen training as a condition of employment.

Job classification	**Person to be trained**

Company Policy

We do not mandate testing for bloodborne pathogens as a condition of employment, nor are employees randomly screened when hired or thereafter. Instead, as a precautionary measure and universal precaution, the company operates under the hypothetical assumption that everyone is infected. This precautionary stance assures maximum protection of all employees through policies and guidelines enforced by management. Compliance with the following company policies is mandatory.

- Only persons trained in first aid response are to administer care to an injured person. However, in unique emergency situations where there is no trained employee available, untrained personnel may offer minimal assistance pending the arrival of emergency responders.
- Employees are to consider themselves at risk when providing assistance to an injured person. The presence of blood or other body fluids on victims, on their clothing, or in the surrounding area is cause for heightened concern on the part of the responder.
- Each employee must wear appropriate personal protective equipment when dealing with an emergency situation. Company first aid kits contain protective gloves, eyewear, and a shield for mouth-to-mouth resuscitation.
- Responding employees who come into contact with blood or body fluids other than their own must report the situation to their supervisor. A mandatory physical examination will follow.
- Clothing, towels, etc., that have come into contact with body fluids must be stored in a clearly marked biohazard container, pending proper disposal.
- Employees are not to perform cleanup procedures involving blood or other body fluids without specific approval from management; company-prescribed procedures must be followed.

Components of Implementation

Designated Personnel At this facility, only the following individuals (name) are trained to administer first aid. Their names are posted on all first aid kits. As a preventive measure, they have the option of receiving the Hepatitis B vaccine at no cost.

General Contact with someone else's blood or other body fluids carries the risk of pathogen transmission, so it is imperative that all employees follow company protocol, precisely, when dealing with an injured person or contaminated objects.

- Wear a mask with solid side shields or a chin-length face shield whenever there is potential for eye, nasal, or oral contact with someone else's blood or other body fluids.
- Never perform mouth-to-mouth CPR without using a protective mask.
- Wash hands, arms, and face with soap and water immediately after assisting injured parties.
- Store cosmetics and contact lenses away from areas where there is potential for bloodborne pathogen exposure.

- Do not share food and beverages with other employees, and store these items separately to prevent cross-contamination. Disposable paper cups are provided at drinking fountains and in service vehicles.
- Wear single-use rubber gloves when handling objects that are contaminated—or possibly contaminated—with blood or other body fluids.
- Wear single-use, disposable gloves when handling contaminated clothing. All contaminated clothing, as well as contaminated objects such as disposable gloves and paper towels, must be placed in the container marked biohazard. The biohazard container is located in the main office. Tape is available there as well for taping the lid to prevent easy access; the container is to be opened by authorized personnel only. The taped biohazard container must be given directly to the manager or supervisor for proper disposal.
- Stay away from or vacate the scene of any emergency unless you are a first aid responder. The manager or designated company representative will limit access by individuals other than first aid responders, and they will post employees around the perimeter to redirect traffic, etc.
- Do not clean areas/objects contaminated with blood or other body fluids unless you are appropriately trained to do so. The company assigns trained employees to decontaminate machinery, floors, walls, etc., or hires an outside decontamination contractor, depending on the situation.
- Do not eat, drink, or smoke during cleanup of a bloodborne pathogen incident.

Engineering and Work Practices Safe job performance and work environments minimize the likelihood of accidents and injuries. We train our employees and maintain our equipment in an effort to reduce accident potential, and supervisors must ensure that work areas remain uncluttered.

Personal Protective Equipment All personal protective equipment (PPE) selected for protection against bloodborne pathogens blocks body fluids from reaching the wearer's skin, eyes, or mouth. Gloves and safety goggles are distributed to trained responders throughout the company and are also available as follows: in the main office; in bloodborne pathogen emergency kits located in the break room; and in all first aid kits, both in the building and in fleet vehicles. The site evacuation map located in the main office and in the break room shows where first aid and bloodborne pathogen kits may be accessed. All personal protective items used to deal with bloodborne pathogen emergencies are inventoried regularly, and used or outdated items are replaced as necessary. All personal protective equipment used at this facility is provided to employees free of charge.

Contaminated Objects and Surfaces Objects and surfaces such as clothing, countertops, equipment, etc., are to be quarantined immediately upon contamination with blood or other body fluids. A designated company representative will determine if and when such items/areas can be put back into use following effective decontamination procedures.

Small contaminated objects are to be discarded in biohazard containers; plastic bags may be substituted if additional containers are needed. It is the responsibility of management personnel to consult with local solid-waste disposal authorities to determine appropriate disposal procedures.

Laundry Procedures No attempt is to be made to decontaminate gloves and clothing contaminated with blood or other body fluids; such items must not be washed or reused. All contaminated clothing, regardless of the extent of contamination, must be left at the facility for placement in a biohazard container, pending disposal. Employees are to be reimbursed for replacement costs of discarded clothing. Temporary clothing (e.g., a Tyvek suit) is provided by the company for interim use.

Postexposure Evaluation and Follow-Up Any employee exposed to blood or body fluids must receive a medical evaluation at the company's expense. The victim (the employee whose body fluids were the source of contamination) will be asked to submit to testing for hepatitis and AIDS. If the victim consents to testing, the results will be revealed only to those employees who might possibly have been contaminated. Victims who refuse testing must sign refusal forms. (*Note*: Before adopting this policy, clear it with your attorney since some states do not allow it.)

Education, Communication, and Documentation It is our policy to train all employees on bloodborne pathogens. The training must occur before any employee—newly hired or newly assigned—begins a job where exposure to blood or other body fluids might occur. This training is conducted as part of new employee orientation. All employees receive additional training on an annual basis. Training covers the following information:

- The OSHA Bloodborne Pathogen Standard and the company's written plan (includes information on where to access these documents)
- Epidemiology and symptoms of diseases transmitted by bloodborne pathogens
- How bloodborne pathogen transmission occurs
- Company policy as outlined in the written plan
- Identification and explanation of jobs for which some possibility of exposure to bloodborne pathogens exists
- Methods used to detect the potential for contact with bloodborne pathogens, including engineering controls, work practices, and PPE
- Instruction on how to deal with an injured person who is bleeding
- Actions that must be taken in the event of human contact with blood or other body fluids

Each person trained is required to sign a roster as evidence of attendance. This documentation is maintained in the company's file and includes the following:

- Date of training
- Content or summary of training conducted (agenda format)
- Names and qualifications of training instructors
- Names, job titles, and employee signatures of those in attendance

Recordkeeping All records required by OSHA, 29 CFR Part 1910.1030 are maintained by the following designated company representatives:

(printed name)____

(printed name)____

Availability of the Plan The written Bloodborne Pathogen Exposure Control Plan is on file in the main office and is available upon request to all employees and OSHA inspectors.

Inception and Revision of the Written Bloodborne Pathogen Exposure Control Plan

Inception date ____________________	Revision date ____________________
Prepared by ____________________	Revised by ____________________
Title ____________________	Title ____________________
Signature ____________________	Signature ____________________

APPENDIX 13.5 EMERGENCY RESPONSE PLAN

______________________________ ______________________________

Geographic address Telephone

____

Developed by

______________________________ ______________________________

Name Title

Revision date	**Employee training date**

INTRODUCTORY ELEMENTS

Purpose and Scope

The purpose of our company's emergency response plan is to prepare employees for emergency situations. Safeguarding the environment as well as the health of our own personnel and people in surrounding communities is paramount. Strict compliance with the plan will lessen the potential for accidents and equip us to respond to emergencies in an organized and professional manner.

A primary objective of this plan is to educate employees to react appropriately to emergencies, pending the arrival of emergency responders, or to evacuate if their well-being is threatened.

Prevention Philosophy

The company strives continually to develop and implement practices and procedures designed to minimize accident and emergency potential. All employees—full- or part-time, salaried or paid hourly—must comply with all company policies and all local, state, and federal regulations.

Facility Profile

The (branch or site) of (name of company) is a commercial facility for (services provided—e.g., sales, distribution, custom application) of (type of product—e.g., pesticides, fertilizers, fuels). Business numbers and a detailed site map are provided herein. The site map shows the exact location of the facility (with streets, roads, and highways clearly marked) and every structure on-site. It also includes directions on how to get there; recognizable icons such as large signs, chemical storage tanks, and structures that are visible from long distances are indicated, as are geographic landmarks that can be used by emergency response personnel in finding the facility. All entrances to the property are clearly marked on the map.

Chemical Profile

Our chemical inventory varies throughout the year. The months with the highest inventories are __________. See Chemical Inventory section for detailed information on chemicals that we store.

Employee Profile

We normally retain _____ full-time employees. We hire an additional _____ seasonal and part-time workers to meet production schedules during these months: ______________. We have ____ full-time clerical employees and ____ mechanics who work on-site. The facility is staffed between the hours of __________ A.M. and __________ P.M. (daily, M–F, etc.)

FACILITY INFORMATION

NAME OF FACILITY/BRANCH

____

Telephone ____________________ ____________________ Fax ________________

(main number) (24-hr number)

GEOGRAPHIC SITE ADDRESS

Street/Road/Highway ______________________________________

County ______________ Latitude ______________ Longitude ______________

Global positioning system coordinates ______________________________

MAILING ADDRESS

Street or P.O. box ______________________________________

City/State/Zip code ______________________________________

MANAGEMENT PERSONNEL

Name ____

Title ____

Telephone ______________________________ ______________________________

(work) (home)

______________________________ ______________________________

(cell phone) (pager)

Street or P.O. box ______________________________________

City/State/Zip code ______________________________________

County Map

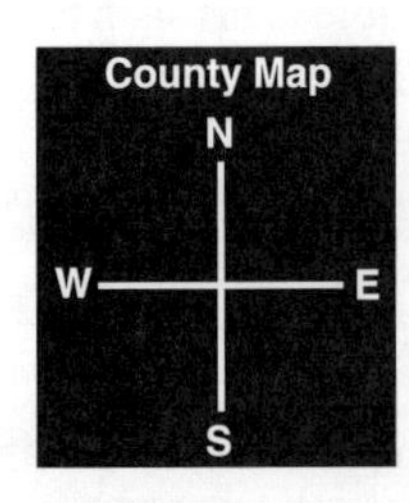

(Need box with directions.)
Geographic location of facility = X
Yellow highlight over solid black line ____ = most common route of access
Solid blue line ________ = alternate traffic route
Solid red line ________ = route for transporting hazardous materials

Verbal Directions for Most Common and Alternate Routes of Access
During an emergency situation it may become difficult to think clearly. These directions to the facility can be read easily when calling for assistance.

Most common route

____
____
____

Alternate route

____
____
____

Site and Evacuation Map

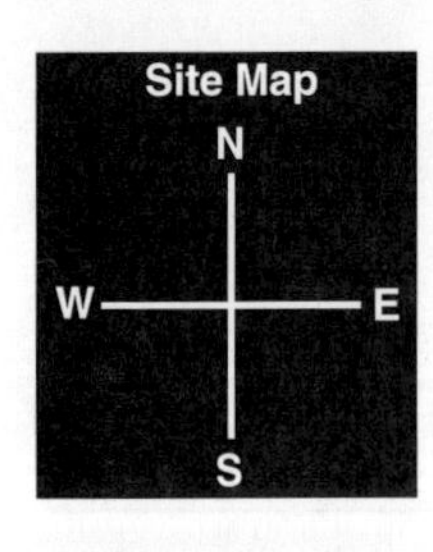

Code	Building, Storage Tank, or Other Structure
A	
B	
C	
D	
E	

Building or other structure
Site map code (from County Map)
Utilities are marked with a red star (*).
Evacuation routes = Exits = →
First-choice rendezvous point = √
Second-choice rendezvous point = χ

Chemical Inventory

Pesticides (separate sheet for fertilizer and fuel)

Company name ________________ Date of original inventory ________________
Geographic address ________________ Dates of revision ________________

Note: The amount of each pesticide in inventory may fluctuate throughout the year, but the maximum amount of each pesticide stored will not exceed the maximum stated here. The

storage location for each pesticide is identical in wording to the corresponding location on the enclosed site map.
A checked box χ indicates a chemical reported under Community Right-to-Know.

Pesticide Trade name	CAS Number	Storage location Building name and code	Maximum Amount Stored	Month(s) of Highest inventory	Size of Container
☐					
☐					
☐					
☐					
☐					
☐					
☐					

Revision History and Distribution of Emergency Response Plan

The company's emergency response plan is reviewed annually, at least, and more frequently as necessary. Revisions, employee training updates, and distribution records appear herein.

Regulatory Compliance

This emergency response plan was written using the template provided in the National Response Team's Integrated Contingency Plan Guidance (Federal Register, Vol. 61, No. 109, pp. 28642–28664). This plan, as written, is intended to comply with the EPA's Oil Pollution Prevention Regulation [40 CFR Parts 112.7(d) and 112.20], OSHA's Emergency Action Plan Regulation [29 CFR Part 1910.38(a)], and OSHA's HAZWOPER Regulation (29 CFR Part 1910.120). The sequence and titles of topics suggested in the integrated plan have been modified to meet our specific needs.

CORE PLAN ELEMENTS

Emergency Response Policy

The most important thing to keep in mind in any emergency is to avoid putting yourself or anyone else at risk. Emergency responders, both trained company staff and professionals, should give highest priority to the prevention of personal injury. Responders should also make every effort to minimize environmental damage to the site and to surrounding properties. Protection of structures, chemical inventory, and supplies should be of concern only after any threat to human health and the environment has been satisfactorily addressed. Emergency responders can meet all three goals—protection of people, the environment, and the facility—by implementing an integrated, preplanned, well organized emergency response plan.

Location of the Plan

The most current revision of this written plan, along with other emergency information, is located as stated below (fill in the blanks). If your personal copy bears a different date, use the more current version. Company management personnel will check this location periodically to ensure that all contents are intact.

A copy of the plan is also kept at the main office of this facility, at corporate headquarters, and at the homes of company-trained emergency responders.

A copy is on permanent display on the safety bulletin board located: ______________
____

Employee Information

Date original information was recorded ______________________ Updated __________
____
A box checked χ indicates an employee who has keys to all buildings. __________
____

Name	Title Position	Telephone numbers			HAZWOPER training?	
		Home	Cell	Pager	Yes/No	Level
☐						
☐						

Revision History of Emergency Response Plan

Date of revision ______________________

Revised by

Name ______________________ Title ______________________

Telephone ______________________ (work) ______________________ (home)

Revision reviewed and approved by

Name ______________________ Title ______________________

Telephone ______________________ (work) ______________________ (home)

Changes made ____
____
____
____
____
____
____
____

Employees trained on changes ______________________ (date)

Trainer ______________________ (name)

Location where training agenda and
record of employees trained may be accessed ______________________
____

Message to Community Responders

The company trains certain employees as emergency responders; therefore, someone is quickly accessible at all times to respond to emergency situations. Knowledgeable, authorized company employees are assigned to coordinate their efforts with those of professional emergency responders and to address questions posed by local fire, medical, and environmental emergency personnel.

See the list herein of our primary and alternate emergency coordinators. These individuals are authorized to make emergency decisions. They are familiar with the written emergency response plan, with the operations and layout of this facility, and with the on-site location of hazardous chemicals. Each person listed is authorized to commit company resources, manpower, equipment, etc., to manage emergency situations.

Company employees and emergency responders have been instructed to take no action without using the personal protective equipment necessary to deal safely with the specific emergency.

Employees are instructed not to participate in emergency response if they feel that doing so would pose excessive personal risk. In such situations, emergency response is to be handled by professionals from the fire department, the rescue squad, or other teams appropriately trained and duly assigned those duties by local, state, and federal authorities.

The company requests that local fire departments consider how actions taken during their response to an emergency might impact the facility, the environment, and the surrounding property. Runoff is a serious issue relative to soil and water contamination, and the decision to apply large volumes of water should be made as a last resort. First consideration should be to human safety and to the amount of product in inventory, adjacent buildings, prevailing atmospheric conditions, and alternative control tactics.

In certain situations, it might be more desirable to let a structure burn than to extinguish it with water (let-it-burn policy, Appendix 13.3).

It is acknowledged that representatives from local, state, and federal agencies have authority to investigate emergencies. However, while on-site, said individuals are expected to follow the specific instructions of a company-employed emergency coordinator. It is our policy that one of our employees shall be assigned to accompany all local, state, or federal authorities conducting investigations on-site and that all safety protocols be strictly followed.

In cases where an emergency is in progress, local, state, and federal authorities may contact the company employees on the following list. Agencies should direct questions to the primary or alternate emergency coordinator. It is our policy that employees with relevant and useful information be accompanied by the designated emergency coordinator when addressing questions during an emergency.

The responsibility to notify local, state, and federal authorities of certain emergency situations is delegated to emergency coordinators specified by company management. Under no circumstances should any employee who is not so-designated be ordered to initiate notification procedures.

Primary and Alternate Emergency Coordinators

Primary on-site emergency coordinator

Name ______________________________

Telephone ______________ ______________
(home) (work)

______________ ______________
(cell) (pager)

Alternate on-site emergency coordinator

Name ______________________________

Telephone ______________ ______________
(home) (work)

______________ ______________
(cell) (pager)

Primary corporate emergency coordinator

Name ____

Telephone ____________________ ____________________
(home) (work)

____________________ ____________________
(cell) (pager)

On-the-Scene Jurisdiction

It is our expectation that our on-site emergency coordinators will participate in decision-making processes during an emergency; They have received HAZWOPER training. They have been instructed to inform outside emergency responders when they believe that actions being considered would pose excessive danger to public health or the environment. Our emergency coordinators and other employees at the scene are to extract themselves from dangerous conditions when, in their opinion, actions being taken by the incident commander would jeopardize their personal well-being.

Company Response Management System

This response plan addresses a wide range of emergencies, including severe storms, floods, fires, explosions, injuries, deaths, and toxic gas releases. The company's employees are trained to quickly evaluate the circumstances surrounding an emergency before making decisions on remedial action. Guidelines set forth in this plan shall govern the decision-making process.

The company's emergency coordinators are instructed to take charge at the scene of an emergency. They are to manage the emergency according to the following guidelines.

Remarks to Employees Never attempt to deal with any emergency situation alone. Immediately upon recognition of an emergency, employees must alert the company's designated emergency coordinators, who will quickly interpret the cause of the emergency situation and implement steps necessary to stabilize it. But under no circumstances should employees attempt response measures clearly beyond their training. Employees unfamiliar with emergency response procedures should follow the instructions of designated emergency coordinators.

Remarks to Company Emergency Coordinators

A. Notify all company personnel and visitors that an emergency exists.

B. Address the situation in a controlled sequence.

  1. Assess the situation to determine the seriousness of the problem.
  2. Evacuate employees and visitors, if necessary.
  3. Take measures to prevent personal injuries.
  4. Control the source of the emergency, if applicable, to lessen the potential for adverse environmental impact.
  5. Notify authorities.
  6. Contact corporate headquarters.

C. Document the following:

  1. Details of the incident

2. Injuries and/or deaths
3. Impact of the emergency on surrounding areas
4. Actions taken
5. Contacts made with service and regulatory agencies

D. Material Safety Data Sheets (MSDS), the North American Emergency Response Guidebook, and professionals linked to MSDS emergency phone numbers are excellent sources for health and environmental guidelines when an emergency involves a hazardous chemical.

E. Determine if additional help is needed: fire department, rescue crews, ambulances, company response team. Phone numbers are posted near each phone and on the outside of the office. When calling for assistance, be sure to do the following:
1. Provide your name.
2. Tell the responder that your emergency plan is on file.
3. Provide your address and specific directions to the facility.
4. Describe the type of emergency.
5. Provide information on injuries or deaths resulting from the emergency.
6. Advise the responder if the situation poses a human health threat to the surrounding community.
7. Explain any potential impact on surrounding waterways or wellheads.
8. Describe prevailing weather conditions at the site.

F. Be sure to notify the Local Emergency Planning Committee in your county.

G. Secure the emergency site until help arrives:
1. Seal off the area to keep unauthorized persons at a safe distance.
2. Keep people upwind of the site.
3. Post company employees along the main road to direct emergency vehicles and to prevent unauthorized people from entering the area.
4. Document the name and agency affiliation of each person admitted on-site during the emergency.

H. Initiate emergency diking procedures to keep contaminated runoff on-site. Direct contaminated runoff away from wells, storm sewers, ditches, and creeks.

I. Turn off utilities within the emergency area. Call utility companies if assistance is warranted.

J. Do not provide transportation to local hospitals for slightly injured employees or bystanders; if injuries are not life threatening, direct injured parties into an isolated area to await evaluation by emergency medical personnel. This lends continuity to the effort of accounting for everyone.

K. Continue to coordinate on-site activities until help arrives, but relinquish control to the outside emergency coordinator if no one from the company has been trained as an emergency coordinator. This is termed transfer of command.

L. Begin writing detailed notes as soon as possible, or assign an employee do so; someone should write down the chronology of all details of (and responses to) the emergency, as they occur. Assign someone the responsibility for taking still photographs and videotapes to supplement written documentation. If your camera has the option to print a date on each negative, use it; and make sure that the time recorder on the video camera is engaged. If your video camera is equipped with a voice recorder, be sure to speak very clearly. As a general rule for shooting videotapes, do not zoom in and out frequently; and do not walk while shooting, if you can avoid it.

M. Notify local, state, and federal authorities.

N. Continue to document every detail in the days following the emergency. The more complete the file, the better the company will be able to address legal, safety, and environmental issues that may emerge days, weeks, months, or years later.

O. Make arrangements for bulldozers, backhoes, and general labor, as required. This plan contains a list of sources for such equipment. Management personnel of the firms listed have been informed of, and have agreed to, services that they might be called upon to provide during an emergency or during cleanup procedures following an emergency. Safety meetings held with representatives of these firms, following the incident, should be documented. Typically, safety meetings include topics such as a briefing on the situation at hand, the objectives of procedures to be performed, the risk potential involved, precautions that participants will need to take, and the types of personal protective equipment that should be worn. All participants brought on-site must have been trained for hazardous material operations, as required by OSHA.

P. Know your deadlines. But file written reports with appropriate agencies only after the Company's legal counsel has reviewed them for accuracy and language. If you cannot make the deadline, request an extension; and follow up by confirming the extension via certified mail. Log all conversations pertinent to the incident: date, time, parties involved, and the details of each discussion.

Q. Upon completion of all remedial actions following an emergency, debrief employees, critique the company's handling of the emergency, and investigate polices and reactions that failed. Make changes to the plan and train employees accordingly.

R. Maintain a complete incident file of original documents, on-site, and send a copy of the file to company headquarters.

S. (Name of company spokesperson) has been designated as the company's media spokesperson.

EVACUATION

A. The on-site emergency coordinator will promptly notify all employees and visitors that an emergency exists and order immediate evacuation of the premises. Notification will be by voice or telephone intercom, by audible alarm, or by flashing lights as predetermined in this plan.

B. Visitors will be instructed to follow the lead of company employees.

C. When asked to evacuate, employees are to proceed immediately to the nearest exit. All exits are clearly marked on the evacuation map.

D. Once outside the building, and without delay, everyone should proceed to the first-choice rendezvous point as indicated by a check mark (√) on the enclosed evacuation map. If that site is inaccessible due to the incident in progress, everyone should proceed to the second-choice rendezvous point indicated on the evacuation map with a χ.

E. Once everyone is safely outside, employees should promptly inform their supervisors of any prevailing condition, inside, that might impact the situation: equipment left running, for example.

F. Make sure to check in with the person designated to account for all employees.

G. Anyone safely outside who realizes that another employee is unaccounted for should immediately tell the person taking count. That person is responsible for reconciling the employee count with whatever attendance verification exists for the day in question. If all records of attendance are inaccessible due to the emergency, the person taking count should canvass those at the rendezvous point and ask for names of employees who were known to be absent that day as well as names of those known to have reported for work but who are not present at the rendezvous point.

H. Remain at the evacuation site until dismissed by the company's emergency coordinator.

EMERGENCY CALL LIST

Internal Emergency Coordinators

Employees are to contact the person at the top of the company's emergency coordinator list as soon as possible after an emergency situation is detected. In the event that this person is unavailable, try the second person on the list, then the third, etc., until contact is made. Assume responsibility until that person arrives on-site.

Emergency Call List: Information That May Be Requested

- Directions to the facility (County map, herein)
- Exact on-site location of the incident (Site map, herein)
- Date and time incident occurred or was discovered
- Chemicals involved *Notes*:
  - — Type
  - — Quantity
  - — Source
- Death
- Human injuries
  - — Number of persons injured
  - — Type and extent of injuries
    - Respiratory arrest
    - Loss of consciousness
    - Burns
    - Bleeding
    - Broken bones
  - — Accessibility of injured persons
- Potential threat to humans
  - — Chemical burns
  - — Inhalation hazard
  - — Acute toxicity
  - — Explosion
- Potential environmental hazards

- Movement of hazardous materials off-site
  - — Direction of flow
  - — Areas likely to be contaminated
    - Wells
    - Ground water
    - Streams, creeks, rivers, lakes, etc.
    - Sewers and storm drains
    - Soil
    - Air
- Prevailing weather conditions
- Identity of responders already on-site
- Potential for explosion or fire
- Company representative serving as emergency coordinator on-site
- Equipment necessary to deal with the emergency

All necessary information can be found in this emergency response plan.

Emergency Call List: Internal Emergency Coordinators

Call the person at the top of the company's emergency coordinator list as soon as possible after an emergency situation is detected. If that person is unavailable, put an χ in the box and try the second person on the list, then the third, etc.; put a check mark (√) in the box beside the name of the person with whom contact is established.

☐ Emergency coordinator ______________________________
(name)
Telephone ____________________ ____________________
(work) (home)
____________________ ____________________
(pager) (cell)
Time notified (voice contact) ______ A.M./P.M. ______________________
(signature of caller)

☐ Alternate coordinator ______________________________
(name)
Telephone ____________________ ____________________
(work) (home)
____________________ ____________________
(pager) (cell)
Time notified (voice contact) ______ A.M./P.M. ______________________
(signature of caller)

☐ Alternate coordinator ______________________________
(name)
Telephone ____________________ ____________________
(work) (home)
____________________ ____________________
(pager) (cell)
Time notified (voice contact) ______ A.M./P.M. ______________________
(signature of caller)

☐ Alternate coordinator ______________________________
(name)
Telephone ____________________ ____________________
(work) (home)
____________________ ____________________
(pager) (cell)
Time notified (voice contact) ______ A.M./P.M. ______________________
(signature of caller)

EMERGENCY CALL LIST

Fill in names and numbers as part of the emergency plan. During an incident, check each box as notification is accomplished.

- **Fire Department**
- **City Police**
- **State Police**
- **County Sheriff**
- **Ambulance**

DIAL 911

Local

❑ Hospital ____________________ (name) ____________________ (phone number)

❑ Local Emergency Planning Committee (LEPC)

❑ Chairman ____________________ (work) ____________________ (home)

❑ Other ____________________ (work) ____________________ (home)

❑ Local Emergency Management Agency ____________________ (phone number)

State

❑ State Emergency Response Commission .... ____________________ (phone number)

❑ State Fire Marshal (24-hr. number)..

❑ State Department of Agriculture...

❑ State Department of Environmental Management ....

Federal

❑ National Response Center.. (888) 233-7745

❑ EPA Region 5 CERCLA.. (888) 233-7745

Signature of Caller____________________ ____________ (date)

Emergency Call List: Adjacent Landowners and Tenants

Fill in names and numbers as part of the emergency plan. During an incident, check each box as notification is accomplished.

☐ *North of Emergency Site* ____________________
(name of occupant or contact person) (phone number)

____
(name of occupant or contact person) (phone number)

____
(name of occupant or contact person) (phone number)

☐ *South of Emergency Site* ____________________
(name of occupant or contact person) (phone number)

____
(name of occupant or contact person) (phone number)

____
(name of occupant or contact person) (phone number)

☐ *East of Emergency Site* ____________________
(name of occupant or contact person) (phone number)

____
(name of occupant or contact person) (phone number)

____
(name of occupant or contact person) (phone number)

☐ *West of Emergency Site* ____________________
(name of occupant or contact person) (phone number)

____
(name of occupant or contact person) (phone number)

____
(name of occupant or contact person) (phone number)

Signature of caller ____________________ ____________
(date)

Emergency Call List: Pesticide Manufacturers

Fill in names and numbers as part of the emergency plan. During an incident, check each box as notification is accomplished.

Notes:

☐ Aventis ____________________
(phone number)

☐ BASF ____________________
(phone number)

☐ Bayer ____________________
(phone number)

☐ Dow AgroSciences ____________________
(phone number)

☐ DuPont ____________________
(phone number)

☐ FMC ____________________
(phone number)

☐ Griffin ____________________
(phone number)

☐ Monsanto ____________________
(phone number)

☐ Novartis ____________________
(phone number)

☐ PBI Gordon ____________________
(phone number)

☐ Rhone-Poulenc ____________________
(phone number)

☐ Zeneca ____________________
(phone number)

Others:

☐ ____
(company name) (phone number)

☐ ____
(company name) (phone number)

☐ ____
(company name) (phone number)

☐ ____
(company name) (phone number)

☐ ____
(company name) (phone number)

☐ ____
(company name) (phone number)

☐ ____
(company name) (phone number)

☐ ____
(company name) (phone number)

Signature of caller ______________________ ____________
(date)

Emergency Call List: Miscellaneous Contacts

☐ Insurance Company ______________________________
(name of company or contact person) (phone number)

☐ Heavy Equipment Provider ______________________________
(name of company or contact person) (phone number)

☐ Department of Health ______________________________
(name of company or contact person) (phone number)

Spill Recovery Contractor

☐ (1) ______________________________
(name of company or contact person) (24-hr phone number)

☐ (2) ______________________________
(name of company or contact person) (24-hr phone number)

☐ ChemTrec (subscription required) . . . . (800) 424-9300

☐ InfoTrac (subscription required) . . . . (352) 323-3500

☐ Poison Control Center . . . . (800) 382-9097

☐ National Pesticide Information Center . . . . (800) 858-7378

☐ Physician ______________________________
(name) (phone number)

Utilities

Electricity ______________ Water ______________
(phone number) (phone number)

Natural gas ______________ Other ______________
(phone number) (phone number)

Contractors for

☐ Fill dirt or sand ______________________________
(name of company) (phone number)

☐ Hazardous-waste disposal ______________________________
(name of company) (phone number)

Other:

☐ ____
(name of company) (phone number)

Signature of caller ______________________ ____________
(date)

Emergency Supply Sources On-Site

Item	Location at the Facility List the building and the exact interior location where the article can be found.
Chemical-resistant gloves	
Respirators	
Splash suits	
Barricade tape	
Portable pumps	
Absorbent	
Push brooms	
Shovels	
Fire extinguishers	
First aid kits	
Recovery containers	
Diking materials	
Spill kits	

Emergency Supply Sources Off-Site

Item	Name of Company	24-hr phone number
Medical supplies		
Pumps		
Safety equipment		
Explosion-proof flashlights		
Portable toilets		
Sand		
Bottled water		
Self-contained breathing apparatus		
Food (e.g., Red Cross)		
Decontamination equipment		
Air cylinders		
Sandbags		

Adjacent Areas

The evacuation of public areas should be handled by the fire and police departments, but it is important for company personnel to cooperate in the effort. Preplan so that everyone will know what is expected of them, and review the plan at least annually.

Direct residents to a safe location—perhaps a school or a church—for accountability purposes, and instruct them to stay there pending notification that they may go. It is a good idea to designate multiple sites in the emergency response plan, just in case the primary location is inaccessible for any reason.

Sometimes it is impossible to get people to leave their homes, but do the best you can. Record the names of people who enter the shelter as well as those who refuse to evacuate.

A shelter should be staffed with law enforcement and emergency management personnel, emergency medical staff, and company representatives to keep evacuees up to date on the emergency situation. They should be kept informed about when they might be allowed to return to their homes.

Procedures for Handling an Uncontained Liquid Chemical Spill

- Send someone or call to inform the emergency coordinator that a chemical has been spilled.
- Call appropriate local agencies: fire department, police, local emergency planning committee.
- Consult the Material Safety Data Sheet and emergency response guidelines for the specific hazard(s), personal protective equipment, cleanup guidelines, and evacuation distances.
- Never physically contact an unknown material. Stay upwind when identifying a spilled substance.
- Inform the product manufacturer of the spill, and solicit advice in dealing with the accident and for cleanup suggestions. Keep the manufacturer on the line for easy access as the emergency unfolds.
- Control (stop) the spill at its source by shutting off leaking valves, etc. If the leaking substance is hazardous, only trained individuals should assume this task.
- Eliminate all ignition sources, including pilot lights and electrical lights.
- Evacuate all nonessential and unprotected employees to a predesignated site.
- Make certain that everyone who enters the spill area wears safety equipment as specified by the MSDS. If the chemical is unknown, emergency personnel must wear a respirator, chemical-resistant gloves and boots, goggles, and a Tyvek suit. Under no circumstances are employees to assist in the area of the emergency if they have not received formal instruction (employee training) on how to wear a respirator properly and unless they have been trained in the appropriate HAZWOPER category.
- Do not allow smoking, eating, or drinking in the emergency area.
- Do not allow nonessential personnel to walk or drive through the affected area.
- Persons trained in the proper HAZWOPER category can work outside the spill area to prevent the spill from spreading (e.g., by making a dike to contain it).
- Utilize all available spill control materials to contain the spill. Large spills may require the mobilization of bulldozers and backhoes to build larger berms.
- Be prepared to assist fire departments and police with equipment, MSDSs, extra personnel, and technical support.

- Initiate cleanup of a small spill according to directions provided by state and federal agencies, in-house specialists, or product manufacturers. Chemicals and contaminated absorbent materials may be placed in secure drums. Mark each drum with the date and the name of the product involved.
- Use remediation consultants where large spills are involved.
- Store debris from each spill separately. Combining chemicals can trigger adverse chemical reactions. Some waste may be considered hazardous and require special disposal. Check MSDSs for incompatibilities.
- Decontaminate all equipment and place the generated waste in labeled containers. These containers should then also be considered hazardous, so mark them "HAZARDOUS WASTE" and label them with the date and contents.
- Replace all equipment and supplies used during cleanup.
- Remember the three C's:

  Control the source.

  Contain the flow.

  Clean up the spill site.

Procedures for Handling a Contained Liquid Chemical Spill

- Cease all loading operations.
- Control the source (e.g., shut off valves). Only HAZWOPER-trained employees may take this action within the spill area; outside the area, anyone may operate valves.
- Turn off pilot lights on equipment if spill is flammable.
- Disconnect pumps and electricity if an explosion hazard exists.
- Turn on electricity and reassemble sump pumps after ensuring that explosion or fire is unlikely.
- Wash spill material into sumps.
- Place recovered materials and contaminated water into containers, and label each container with the date, time, and type of material stored.
- Analyze the stored materials for identification and concentration, or consider the contents to be 100% concentrate.
- Consult with your **department of agriculture** if a pesticide is spilled, before making an application of the stored materials to an appropriate site of application. Never apply at rates that exceed label recommendations.
- Use absorbent products properly and also understand that their use may create additional disposal problems. Consult the **department of agriculture** for advice on absorbent disposal. Whenever possible, use absorbent products that can be recycled or field applied, such as oil dry, ground corn cobs, peat moss, and fly ash. Pillows and fiber booms must be processed or hauled to a special landfill.
- Remember the three C's:

  Control the source.

  Contain the flow.

  Clean up the spill site.
- Dry chemical releases from normal handling operations should be cleaned up immediately to prevent accumulation, especially prior to rain.

- Return recovered dry material to the appropriate bins, or add it to a load of the same material being applied that day.
- Any waters collected which came into contact with these materials should be handled as product rinse water and applied to a site listed on the label of the product.

Procedures for Handling a Fire or Explosion

- Know the capabilities of the local fire department.
- Evacuate all employees and visitors to designated areas upwind of smoke.
- Only employees trained on the proper use of fire extinguishers shall attempt to contain a small fire.
- Evacuation maps show exact locations of fire extinguishers.
- Fires larger than a wastepaper basket should be left to professional firefighters.
- Immediately report any fire to the on-site emergency coordinator. In the event you are unable to notify the emergency coordinator, it will be necessary for you to contact the local authorities by dialing 911 or the local emergency number.
- Inform responders that limited supplies are on hand to assist emergency coordinators or professional firefighters in their efforts.
- Do not allow any person to walk or drive through the fire area. This may require the posting of guards around the perimeter of the fire area.
- Shut down all operations within the structure that is on fire. This should be done prior to everyone leaving.
- Turn off electricity and all other utilities associated with the building. Check the facility map for the location of the turnoff connections. Turn off electrical power to LP gas tanks.
- Do not use water on chemical fires, except to protect human health. However, the final decision is left to either the fire department or the incident commander at the scene.
- Be prepared to assist firefighters, but do not enter or get close to a burning building.
- Notify fire department of the available water supply on-site.
- Be prepared to dike around burning buildings if water is used as the extinguishing medium.
- Do your best to keep any contaminated water out of nearby ditches, streams, or drains.
- Implement your emergency response plan and spill notification to the environmental contractor.
- Do not enter the area until the incident commander gives permission.
- The only personnel allowed to clean up the debris from a contaminated area are trained in the appropriate HAZWOPER category.

Vehicle Emergency

- Have MSDSs, pesticide labels, emergency phone numbers, and extra fuses in the vehicle.
- Stop immediately if a chemical leak is detected or if the vehicle is involved in an accident.
- Park the vehicle in a safe location.

- Turn off the ignition and set the parking brake.
- Turn on emergency flashers.
- Put out safety triangles.
- If the accident involves human injury, do the following:
  a. Make sure that the person is breathing.
  b. Do not move the person unless their position is life threatening (e.g., if the vehicle is on fire).
  c. Call 911, then your supervisor or the emergency coordinator.
  d. Be prepared to describe the location of the accident and to provide pertinent information.
  e. Keep everyone except emergency personnel out of spill area.
  f. Repair the leak, if possible, but only if you have been trained to do so.
  g. Fill out your company's Incident/Accident Report form; it will contain information that your insurance company and/or company safety committee may need.
- If the accident involves an environmental release, follow these guidelines:
  a. Wear safety equipment.
  b. Repair the leak, if possible, but only if you have been trained to do so.
  c. Use shovels and spill material to build berms to prevent the material from entering creeks, waterways, or drains.
  d. Call 911, then your immediate supervisor or the emergency coordinator.
  e. Call the emergency number on the MSDS if the chemical enters a waterway, to determine any potential impact on water consumption and aquatic wildlife.
  f. Fill out your company's Incident/Accident Report form; it will contain information that your insurance company and/or company safety committee may need.
- Provide police with the following driver information:
  a. Your name and home address
  b. Company name and business address
  c. Your license number
  d. Vehicle license number
  e. Name of your immediate supervisor
  f. In the presence of the police, and/or other officials involved at the scene, remember the following:
    i. Be cooperative, but answer questions thoughtfully. Do not admit guilt!
    ii. Exchange pertinent vehicle, insurance, and driver information with any other drivers involved.

Petroleum Emergency

- Evacuate employees and visitors to prearranged areas.
- Check with supervisor for dealing with small spills; contact your local fire department for larger spills.
- Turn off all power and pilot lights. Remember that electrical switches create small arcs when being turned on or off, thereby producing a potential ignition source.
- Identify the product involved.

- Check the MSDS for hazard potential: pollution, fire, explosion, etc.
- Wear personal protective equipment as prescribed on the label.
- Use nonsparking tools.
- Contain the spill with absorbent materials or by whatever means possible.

Liquid Propane or Natural Gas Emergency

- Check with supervisor for dealing with small spills; contact your local fire department for larger spills.
- Evacuate all employees and visitors to the designated location.
- Eliminate all ignition sources within a 500-ft radius of the release.
- Turn off main valve if it can be done safely.
- Contact the propane or natural gas company for assistance.

Anhydrous Ammonia Emergency

- Anhydrous ammonia is a liquid stored under pressure that becomes a gas when exposed to the atmosphere.
- Consult the Material Safety Data Sheet
  — For small leaks, stop the flow of gas.
  — For large leaks, call the fire department.
- Skin contact with anhydrous ammonia can cause frostbite; inhalation can cause severe respiratory symptoms or death. Use appropriate safety equipment when dealing with any anhydrous ammonia release:
  — nonvented goggles
  — PVC gloves
  — boots
  — respiratory protection
  — do not wear contact lenses
- Evacuate an area of 1500 ft in all directions from around the anhydrous ammonia release. Keep all employees and others from entering an area with a 30-ft radius around the site. The local fire department will make a determination whether more or less distance is required.
- Evacuate to a location upwind of any anhydrous ammonia release.
- Do not reenter the evacuated area until given permission by authorities.

NATURAL DISASTERS

Source: Adapted from U.S. Department of Commerce National Weather Service Publications.

Tornadoes

Tornadoes can occur any time of the year. They are most likely to occur between 3 and 9 P.M. but may strike at any hour of the day or night. The average tornado moves from southwest to northeast.

Before the Storm

- Develop a plan.
- Conduct frequent drills.
- Keep a highway map nearby to follow storm movement reported in weather bulletins.
- Purchase an NOAA weather radio, with a warning alarm and battery backup, for receiving warnings.
- Listen to radio and television for information.

Listen for the Following

Tornado Watch: Tornadoes are possible in the area. Remain alert for approaching storms.
Tornado Warning: A tornado has been sighted or indicated by weather radar.

If a Tornado Warning Is Issued

- Shut down all nonessential energy sources.
- Account for all employees and visitors.
- Move to your predesignated place of safety.
- If an underground shelter is not available, move to an interior room or hallway on the lowest floor and get under a sturdy piece of furniture.
- Stay away from windows.
- Do not try to outrun a tornado in your car: Pull over, get out of the vehicle, and lie flat in a ditch or depression.
- If caught outdoors, lie flat in a ditch or depression.

Floods

Nearly half of all flash flood fatalities are auto related. When driving during a storm, or while a flood watch is in effect, look for flooding at highway dips, bridges, and low areas. Remember that 2 ft of water will carry away most automobiles.

Before the Flood

- Determine your elevation above flood stage to establish your flood risk.
- Keep your vehicle fueled; if electric power is cut off, gas stations may not be able to operate pumps for several days.
- Keep first aid supplies on hand.
- Keep an NOAA weather radio, a battery-powered portable radio, and flashlights in working order.
- Install check valves in sewer traps to prevent flood water from backing up into the drains.

Listen for the Following

Flash Flood or Flood Watch: Flash flooding or flooding is possible within the designated watch area—be alert.

Flash Flood or Flood Warning: Flash flooding or flooding has been reported or is imminent—take necessary precautions at once.

Urban and Small Stream Advisory: Flooding of small streams, streets, and low-lying areas, such as railroad underpasses and urban storm drains, is occurring.

Flash Flood or Flood Statement: Follow-up information regarding a flash flood or flood event.

Flash Flood Warning

- If advised to evacuate, do so immediately.
- Move to a safe area before access is cut off by flood water.
- Continue monitoring the NOAA weather radio, television, or an emergency broadcast station for information.

During the Flood

- Avoid areas subject to sudden flooding.
- If you come upon a flowing stream where water is ankle-deep, STOP! Turn around and go another way.
- Do not attempt to drive over a flooded road because the depth of water is not always obvious and the roadbed may be washed out. You could become stranded.

After the Flood

- If food has come into contact with flood water, throw it out.
- Boil all water for drinking and cooking. Wells should be pumped out and the water tested for purity before drinking. If in doubt, call your local public health authority.
- Do not visit disaster areas. Your presence might hamper rescue and other emergency operations.
- Check and dry electrical equipment before returning it to service.
- Use flashlights—not lanterns or matches—to examine buildings; this is a safeguard against igniting any flammables that may be present.
- Report broken utility lines to appropriate authorities.

Thunderstorms and Lightning

A typical thunderstorm is 15 miles in diameter and lasts for 30 min. All thunderstorms are dangerous. They produce lightning that kills more people each year than do tornadoes, and heavy rain from thunderstorms can cause flash flooding. Danger from strong winds, hail, and tornadoes also is associated with thunderstorms.

Who Is Most at Risk from Thunderstorms?

Lightning: People who are outdoors, especially near or under tall trees, in or on water, and on or near hilltops.

Flooding: People in vehicles.

Tornadoes: People in vehicles.

Listen for the Following

Severe Thunderstorm Watch: A thunderstorm watch indicates that conditions are right for severe thunderstorms to occur. Watch the sky and tune to a radio or television channel to receive a warning if one is issued. A watch is intended to heighten public awareness and should not be confused with a warning.

Severe Thunderstorm Warning: A thunderstorm warning is issued when severe weather has been reported by spotters or indicated by radar. A warning indicates imminent danger to life and property in the path of the storm.

Before the Storm

- Know the county in which you live and the names of nearby major cities.
- Check the weather forecast before leaving for an extended period outdoors.
- Postpone outdoor activities if thunderstorms are imminent.
- Watch for signs of approaching storms.
- Listen for weather updates on an NOAA weather radio, an AM/FM radio, or television.

When Thunderstorms Approach

- If you can hear thunder, you are close enough to the storm to be struck by lightning. Seek shelter immediately.
- Move to a sturdy building. Do not take shelter in small sheds or beneath isolated trees.
- If lightning is occurring and there is no sturdy shelter available, get inside a hardtop vehicle and keep windows up.
- Get away from water.
- Telephone lines and metal poles can conduct electricity. Unplug appliances not necessary for obtaining weather information. Avoid using the telephone and electrical appliances. Use phones only in an emergency.
- Turn off air conditioners. Power surges from lightning can overload the compressors.
- Get to higher ground if flash flooding or flooding is possible. Once flooding begins, abandon vehicles and climb to higher ground. Do not attempt to drive to safety. *Note:* Most flash flood deaths occur in automobiles.

If Caught Outdoors without Shelter

- Find a low spot away from trees, fences, and poles. Make sure the place you pick is not subject to flooding.
- If you are in the woods, take shelter under the shorter trees.

- If you feel your skin tingle or your hair stand on end, squat low to the ground on the balls of your feet to minimize your contact with the ground. Place your hands on your knees and your head between them—make yourself the smallest target possible.

Winter Storms

Sometimes winter storms are accompanied by strong winds creating blizzard conditions with blinding wind-driven snow, severe drifting, and dangerous windchill. Strong winds with these intense storms and cold fronts can knock down trees, utility poles, and power lines.

Everyone is potentially at risk during winter storms. The actual threat to you depends on your specific situation. Observations indicate the following are contributors to winter deaths.

Deaths Related to Ice and Snow

- About 70% occur in automobiles
- About 25% are people caught out in the storm
- Majority are males over 40 years old

Deaths Related to Exposure

- 50% are people over 60 years old
- Over 75% are males
- About 20% occur in the home

Listen for the Following

Winter Storm Watch: Severe winter conditions, such as heavy snow and ice, are possible. Prepare now!

Winter Storm Warning: Severe winter conditions have begun or are about to begin. Stay indoors!

Blizzard Warning: Snow and strong winds will combine to produce a blinding snow (near-zero visibility), deep drifts, and life-threatening windchill. Seek refuge immediately.

Winter Weather Advisory: Winter weather conditions are expected to cause significant inconvenience and may be hazardous. If caution is exercised, conditions should not become life threatening. The greatest hazard often is to motorists.

Frost/Freeze Warning: Below freezing temperatures are expected and may cause significant damage to plants, crops, and fruit trees. In areas unaccustomed to freezing temperatures, people without heated homes need to take precautions.

What to Have Ready at Work

- Flashlight and extra batteries
- Battery-powered NOAA weather radio and a battery-powered portable radio to receive emergency information

- Extra food and water. High energy food, such as dried fruit or candy, and food requiring no cooking or refrigeration is best.
- First aid supplies
- Ample heating fuel. Fuel carriers may not reach you for days after a severe winter storm.

Vehicles

- Fully check and winterize all vehicles before winter.
- Keep gas tanks near full to avoid ice in the tank and fuel lines.
- Try not to travel alone.
- Let someone know your timetable and primary and alternate routes when traveling in inclement weather.
- Carry a winter storm survival kit which includes extra clothes, snacks, flashlights, blankets, shovels, cell phones, etc.

Earthquakes

- Stay indoors. Take cover under furniture or brace yourself in a doorway.
- Stay near the center of the building and away from glass and windows.
- If outdoors, stay away from structures and electric lines.
- Keep employees together after earthquake is over.
- Keep all persons out of damaged buildings.
- Evacuate the area if necessary.
- Do not drink water or flush toilets until it is determined that lines are undamaged.
- Use binoculars to assess damage to tall structures.
- Notify fire departments when conditions warrant their assistance.
- Prepare a damage report, paying attention to downed power lines, especially, and potential fires, spills, and damaged buildings.
- Aftershocks can occur for several months after the main quake. Each occurrence may trigger additional damage to previously weakened structures.

BOMB THREATS AND THREATENING CALLS

- Take all bomb threats seriously.
- In the event of a bomb threat, keep the caller on the line as long as possible, and record the call if you have that capability. If you cannot tape the call, listen for background noise. Write down as much of the conversation as possible (see following checklist).
- Pass a note to a co-worker, informing them of the threat and telling them to call the police and begin evacuation procedures.
- Try to make the caller understand that if the bomb is detonated many people could be seriously injured or killed. Ask the caller to tell you where the bomb is located so that the area can be evacuated.

- Wait until the state police have inspected the area before resuming normal activities.
- Complete the following checklist.

Bomb Threats and Threatening Calls Checklist

Your name ____

Date __________________ Time __________________ A.M. P.M.

Information on the caller: ☐ male ☐ female ☐ adult ☐ child

Was the call ☐ local? ☐ long distance? ☐ from within the building or facility?

Caller's voice characteristics: ☐ loud ☐ soft ☐ deep
☐ high-pitched ☐ clear ☐ raspy

Caller's speech characteristics: ☐ fast ☐ slow ☐ distinct
☐ clear ☐ slurred ☐ distorted

Language skills: ☐ excellent ☐ good ☐ poor

Accent characteristics: ☐ local? ☐ foreign? ☐ Possible race ____________

Was the caller: ☐ calm? ☐ angry? ☐ rational? ☐ irrational?
☐ cursing? ☐ deliberate? ☐ emotional?

Background noise: ☐ factory? ☐ office machines? ☐ music? ☐ traffic?
☐ airplanes? ☐ trains? ☐ animals? ☐ crowd noise? ☐ silence?

Ask the caller these questions:

- When will the bomb go off?
- Where is it located (e.g., in which building, where in the building, where on the grounds)?
- What kind of bomb is it?
- How do you know it will go off?
- How do you know so much about this facility?
- What is your namc and addrcss?

MEDICAL EMERGENCY

General Guidelines

- Stop all activities immediately.
- Have someone call 911.
- Verify the safety of responders.
- Check on victims.
- Do not move victims unless their location puts their lives in danger.
- Calm the victims and stay with them until professional help arrives, but do not administer first aid unless you are trained to do so. First aid is to be administered only by trained personnel.
- Call the emergency number indicated on the product's MSDS. The manufacturer's medical personnel may offer valuable input on the immediate care of the injured; they also may advise what actions to take to guard against additional injuries as the emergency unfolds.

- Have nearby company employees assist in evacuating the area, directing emergency responders from the highway to the scene of the accident, keeping bystanders a safe distance away, etc.
- Establish the identity of all victims and notify their families.
- Try to find out exactly what took place and how the person was injured.

Plan of Action for Acute Pesticide Poisonings

A pesticide user should establish a plan of action to follow in case of a pesticide-related accident. Advanced planning and preparation should be routine. Make sure all employees are familiar with appropriate emergency procedures.

Contact Medical Personnel Step one in any poisoning emergency is to prevent further exposure and to make sure the victim is breathing; then call emergency medical personnel.

Maintain Vital Signs Administer first aid while help is on the way. Maintenance of vital signs is imperative, and cardiopulmonary resuscitation techniques may be required. The cause of death of most pesticide poisoning victims is respiratory failure. Many victims will recover if the supply of oxygen to the body can be maintained. Only a doctor will have the medication and equipment necessary to treat a poisoning victim properly. Always provide attending medical personnel with a copy of the pesticide label.

Eliminate Further Contamination

Ingested Pesticides If an individual swallows a pesticide, act immediately: Do not wait for symptoms to appear.

The pesticide label will indicate whether or not vomiting should be induced. Never induce vomiting if the victim is unconscious or convulsive. In cases where vomiting *can* be induced safely, fast action can mean the difference between life and death for the poisoning victim. Syrup of ipecac is useful for inducing vomiting; make sure the victim assumes a forward kneeling position or remains on his right side, if lying down, to prevent vomit from aspirating into the lungs. Gastric lavage—performed by a physician—is another method for removing stomach contents. The latter must be performed as soon as possible after ingestion of the pesticide—and no longer than 2 hr afterward. After 2 hr, the pesticide will have passed into the intestine, thus requiring a different approach to effect removal of the poison; physicians can administer absorptive charcoals to prevent the absorption of the pesticide from the intestine and to promote its elimination in the feces. Select a physician knowledgeable about pesticide poisonings early on.

It is important to remember to consult the pesticide label before proceeding with first aid. There are certain situations where inducing vomiting might only cause *additional* damage. Vomiting should not be induced if the pesticide formulation contains organic solvents or corrosives such as strong acids and bases since these materials can cause serious, permanent damage to sensitive tissues of the esophagus—or the lungs, if aspiration occurs.

Pesticides on the Skin Wash the pesticide off the victim as soon as possible to prevent continued exposure and injury.

- Remove clothing and drench the skin with water (shower, hose, faucet, pond, etc.).
- Cleanse skin and hair thoroughly with soap and water. (Don't abrade or injure the skin while washing.)
- Dry the person and wrap him or her in a blanket.

Chemical Burns of the Skin Taking immediate action is extremely important.

- Remove contaminated clothing.
- Wash skin with large quantities of cold running water.
- Immediately cover the affected area loosely with a clean, soft cloth.
- Do *not* use ointments, greases, powders, or other drugs recommended as first aid treatments for chemical burns.

Pesticides in the Eye It is very important to wash out the affected eye as quickly but as gently as possible.

- Hold eyelids open; wash eyes with a gentle stream of clean running water, at body temperature if possible.
- Continue washing for *15 min. or more*.
- Do not use chemicals or drugs in wash water; they may increase the potential for injury.

Inhaled Pesticides If the victim is in an enclosed area, wear an appropriate respirator when removing the person from the contaminated area.

- Immediately carry the victim to fresh air.
- Loosen all tight clothing.
- Apply artificial respiration if breathing has stopped or if it is irregular.
- Keep the victim as quiet as possible.
- If the victim is convulsing, watch breathing and protect the person from falling and striking his head. Pull the chin forward so that the tongue does not block the air passage.
- Prevent chilling. Wrap patient in blankets but do not overheat.

EMPLOYEE TRAINING GUIDELINES

Consider the following ideas when providing emergency response training for employees.

- The emergency response plan is located in the main office and is posted on bulletin boards for quick access and reference.
- Employees will be trained annually on the contents of the plan and what is expected of them.
- New employees will be trained on the contents of the plan as part of their new employee orientation.

- All employees will be informed immediately of new safety procedures that impact them.
- It is important to act quickly, but safely, during an emergency.
- Emergency phone numbers are posted throughout the facility.
- Employees who have not been trained to deal with emergencies are not to get involved.
- Employees who have not received training on respirators are not to get involved during an emergency that requires one.
- Employees who have not been trained to use fire extinguishers are not to use them.
- Employees are expected to report to their supervisors any unsafe practices which could lead to employee injury or chemical release.
- All employees of other companies who unload fertilizers, pesticides, and fuels must show proof that they have been trained to safely load and unload the products.
- Employees will be trained on proper evacuation procedures at least on an annual basis.
- All new employees will receive in-house training on safety in handling, mixing, using, storing, and disposing of hazardous chemicals; employees will not be assigned such duties until training is completed.
- Employees will be taught the three C's in dealing with a spill: Control the source. Contain the flow. Clean up the spill site.
- Employees will receive awareness training and will be instructed on how to recognize a problem, identify the cause, and secure the scene of an emergency; they also will be taught how to communicate the hazard (duty to warn).

MEDIA MANAGEMENT

Owner or Manager

Up-to-the-minute awareness of what is happening is crucial, and it is the responsibility of the emergency response coordinator to keep the media relations representative apprised of an emergency situation as it unfolds.

The more significant an incident, the more media coverage you can expect, so make sure that your spokesperson is equipped with enough details to satisfy media curiosity.

- Cooperate by making your spokesperson available throughout the event to coordinate media response activities.
- Be prepared: Investigate all elements of the crisis and organize your facts to ensure that all important points are addressed with the media relations representative.

Media Relations Representative

Management must make sure that all employees know who is the company's media relations representative (i.e., the designated company spokesperson) and that they know to defer all media questions to that individual.

If you are the spokesperson, don't wait for a crisis; make yourself known. It is important to develop a healthy relationship with the media, establishing yourself as a dependable and credible source of information. Positive communication heightens understanding and lessens confusion and mistrust in times of crisis. Good communication skills are essential.

When dealing with the media during a crisis, set a time and location for a news release, stick to it, and by all means tell the truth.

- Assume that microphones are on and cameras rolling.
- Respond quickly. Deal with an issue early to avoid misinterpretation and fear.
- Communicate by providing the media enough information to file their story.
- Let the media know the situation is under control. Anticipate questions and plan concise answers.
- Stay calm. Avoid confrontation. Never argue or lose your composure. If a question contains words you dislike, don't repeat them. Politely correct hostile or inaccurate remarks in your answer and avoid assigning blame.
- Be clear, concise, and consistent. Make sure your answers are easily understood. Don't use technical terminology or jargon unless you are prepared to explain it. If you must leave some information out, be sure to tell reporters what they need to know and use qualifiers so they don't feel misled by new information. State your policy and stick to it.
- Be factual. Avoid extreme positions. Don't be led into unfamiliar territory. Keep the interview on track by emphasizing the points you want to make.
- Never say "no comment." It invites speculation. Offer to update the media when more information is available.
- Don't say anything "off the record." There is no legal obligation for a reporter to keep anything off the record. If you have a comment that you do not want publicized, don't say it.
- Never speculate. If asked a hypothetical question, state what you know at the time and let the reporter know you'll be available, later, for follow-up questions.

LOCAL TELEVISION

Contact person ____

Phone ________________ Fax ________________ E-Mail ________________

LOCAL RADIO

Contact person ____

Phone ________________ Fax ________________ E-Mail ________________

LOCAL NEWSPAPER

Contact person ____

Phone ________________ Fax ________________ E-Mail ________________

CHAPTER 14

THE INSURANCE POLICY: PROTECTING YOURSELF AGAINST THE UNEXPECTED

Pesticide application business owners devote considerable time, energy, and money to position their companies favorably in a highly competitive market. Product availability, pricing structure, quality of service, employee hiring practices, and employee education must be continually reevaluated in the face of stiff competition and an ever-changing, demanding public. If your business is to prosper, you must deliver what the public believes to be an indispensable service, priced right and backed by a strong commitment to deliver. Businesses that do not adjust to public demands are seldom afforded a second chance. Customers move on to do business with companies more attuned to their needs, and your chances of winning them back are slim. The finality of business opportunities missed and customers lost and the inability to build a new clientele are the ultimate results of failure to deliver quality service. Yet, as critical as this is, it doesn't even begin to touch upon the role of a sound business insurance program.

Insurance? Failure to plan and manage your business insurance program can put at risk all that you have worked for: your business, your home, your savings—everything! You could end up owing thousands of dollars. It deserves your attention.

PLANNING FOR UNFORESEEABLE RISKS

Pesticide application business owners must plan for risks that have nothing to do with marketplace pressure: unforeseeable risks that may include partial or total destruction of their facilities by fire, flood, windstorm, or earthquake; employee injury or death on the job; and lawsuits by customers or other third parties demanding compensation for real or alleged personal injury or property damage.

Planning your business future means taking the time to examine all aspects of your operations:

- Identify the risks that could threaten your business.
- Implement a plan to minimize or eliminate those risks.

Figure 14.1 Insurance can help reduce the negative impact from accidents.

This is called risk management, and a sound insurance program is the nucleus. Insurance is an integral component of good business planning, but business owners need to be directly involved in its procurement. Those who are not may be subjecting themselves unnecessarily to the ravages of claims not covered by their insurance plan. Uncovered claims may result in cash outlay for remuneration and legal expenses. Factor in the possibility that your insurance agent and/or insurer might not be experienced or knowledgeable enough to handle pesticide-related claims and you have a nightmare that can be financially disastrous. Leaving insurance decisions solely to the discretion of others is not good business. Don't assume that all policies are the same (Fig. 14.1).

The unfortunate consequence of leaving insurance decisions to someone else may not strike home until a claim is turned down or disputed by the insurance company. For instance, a fire burns the main office to the ground. You quickly file a claim so that the process of rebuilding can begin and the business can resume operations. But, unfortunately, the policy was written years ago to cover the cash value of the building, not replacement cost; and the cost of rebuilding the structure according to current standards far surpasses what the insurance will pay. This is when—much too late—the importance of an annual evaluation of your insurance needs is recognized. This type of situation can be avoided simply by purchasing replacement-cost coverage.

Owners who understand what they are purchasing and who participate actively in the risk assessment and insurance procurement processes place themselves in a good position to protect their businesses from financial loss.

Figure 14.2 First-party insurance protects business owners against damage to their property.

INSURANCE COVERAGE: EXPECTING THE UNEXPECTED

Insurance is designed to make the policyholder "whole" by replacing what is lost. As a business owner, you have worked for years to increase earnings and build assets to support your lifestyle and impending retirement. But if you find yourself the defendant in a business-related lawsuit, a jury can award the plaintiff a sum that exceeds your ability to pay. Even if you win a business lawsuit, your attorney fees alone can be financially devastating. The safety net that insurance provides is deserving of your full participation in selecting and purchasing the right coverage.

The parties to an insurance contract are the policyholder and the insurance company. It is important to understand that the insurance company, in exchange for a premium, promises to defend and indemnify the policyholder to the extent set forth in the insurance policy.

There are two basic types of business insurance contracts: first-party and third-party. First-party coverage protects business owners from the cost of damage to their own property or other assets (Fig. 14.2). Third-party coverage protects the policyholder/business from liability claims arising from alleged damage or injury to others or their property.

Perspective of the Policyholder

The prudent policyholder plans for the unexpected and purchases insurance coverage for that contingency. It is preferable to budget reasonable insurance premiums rather than risk sustaining large, unexpected, and uninsured losses. The policyholder expects the insurance

Figure 14.3 A fire destroys a building and causes pesticide containers to leak.

company to respond quickly and fairly to an insurance claim and to award payment for the loss.

Perspective of the Insurance Company

Insurance companies are risk takers; they defend and indemnify (compensate) pesticide application companies for covered losses, for the cost of a premium. Depending on the policy, covered losses may include bodily injury, property damage, business income, environmental damage, and legal fees.

Insurance companies are unique in that they charge for a hypothetical product; that is, they cannot predict with certainty whether a claim will be made against them or what expenses they will incur. Therefore, they predict expenses based on statistical information from past claims. They set premiums to cover contingent expenses and to generate a reasonable profit (Fig. 14.3).

Insurance companies use actuarial formulas to predict loss; they also use statistical data to estimate the cost of associated litigation and other expenses. Insurance companies that experience low levels of loss may set lower premiums than those whose claims history shows higher losses.

Businesses that perform high-risk operations typically are charged higher premiums than those businesses whose operations are low-risk. Insurance companies review individual pesticide application businesses' loss experience and business practices to determine appropriate fees. Business-specific information on loss history and risk management determines whether the business is charged higher or lower premiums than are other businesses that perform similar operations (Fig. 14.4).

COVERAGE ALONE IS NOT ENOUGH

Purchasing insurance is easy. In fact, it's so easy that all you have to do for some policies is answer a few questions over the phone and fax or mail in an application. People believe that they can buy an insurance policy that covers just about everything; but although such policies make it easy to think you are fully protected, it is very likely that you are not.

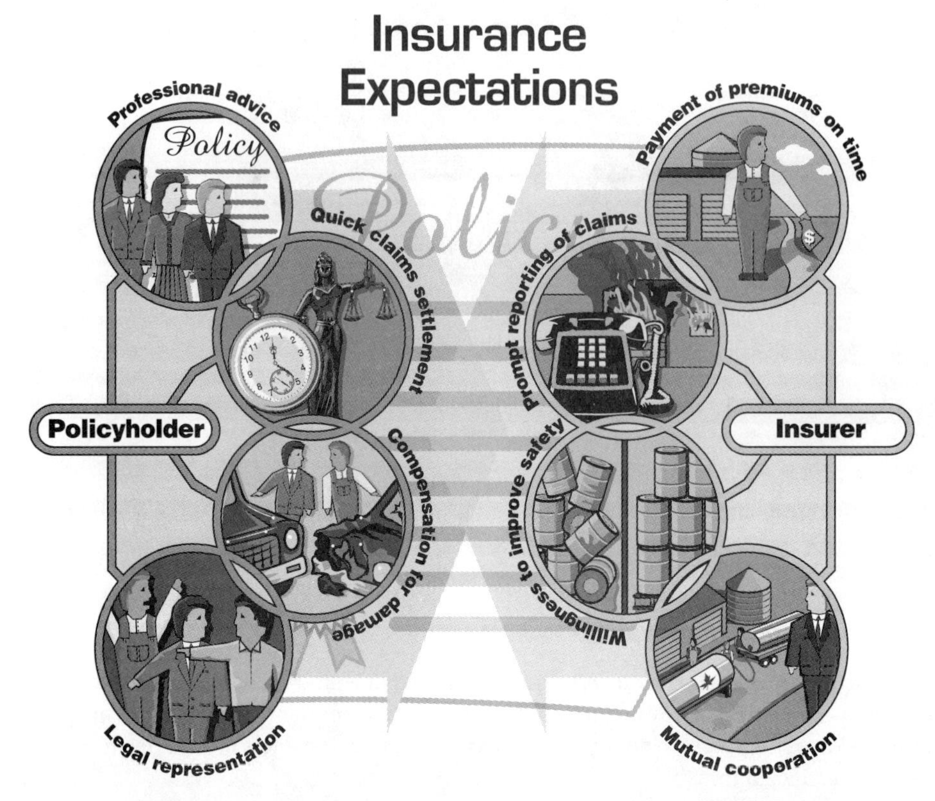

Figure 14.4 Insurance is a partnership between the policyholder and the insurer.

Never assume that owning insurance means total protection. This can be a costly mistake because every policy has exclusions. Do not assume coverage if you have not read your policy. The very claim you file might be refused by exclusion.

The Lowest Price Might Not Be the Best Deal

There's no doubt about it: Just shop around and you can find the company that quotes the lowest Vehicle coverage and another that offers the best deal on Property insurance. But although shopping on the basis of price alone may save you money up front, what happens when you file a claim?

It is generally recommended that pesticide application businesses purchase their insurance from a single insurance company—and from one agent who makes an effort to stay attuned to their insurance needs—instead of contracting coverage through multiple carriers.

Insurance companies use the policyholder's loss-to-premium ratio to determine account profitability; and when a business purchases its Vehicle, General Liability, Worker's Compensation, and Property Damage coverage from a single underwriter, more of the premium dollar is available to offset losses. In other words, paid claims have a smaller loss-to-premium ratio in proportion to the higher premium.

Other advantages include the following:

- Gaps and overlaps in coverage are avoided when one insurance company is used for all types of coverage. When multiple insurers are involved, the likelihood of claim disputes increases.
- Some insurance companies provide periodic physical inspections of the policyholder's business facilities. These safety and loss control inspections can pinpoint areas that raise liability concerns for both the business and the insurance company. The more insurance that is purchased and the more knowledgeable an insurer is about your business, the more comprehensive the review.
- Insurance companies generally offer rate discounts for purchasing multiple coverages.
- Insurance agents who provide all lines of coverage for your business develop a broader awareness of your needs, which leads to better advice and better service.

THE INSURANCE POLICY IS A LEGAL CONTRACT

Having a million dollars' worth of insurance sounds impressive; but if your loss is excluded, what good is it? It is critical that you read every word of your insurance policy because the old adage "The devil is in the details" was never more true.

Many insurance policies—be they Automobile, General Liability, or whatever—contain clauses that obligate the insurance company to cover all risks not otherwise excluded. For example; a liability policy might state that "We will pay all sums that you are legally obligated to pay resulting from. . . ." But it is the language following that statement that determines who, where, what, and when a policy will pay, so set aside an afternoon to read your policy in-depth. Call your agent for a full explanation of segments you don't understand. Know what you have purchased (Fig. 14.5).

Figure 14.5 It is important to understand what your policy does and does not cover.

Reading the Policy

What makes an insurance policy a legal contract? There are three key elements:

- An offer by the insurance company to provide coverage
- Payment by the insured of a premium for the coverage
- Acceptance of a premium by the insurance company

Read all insurance policies carefully! A policy may provide basic coverage, take it away wholly or partially by exclusion, and then add back part of the coverage by way of endorsement. Because of this give-and-take approach to writing insurance, careful review of policy language is warranted.

Standardized Insurance Forms

Insurance companies use standardized forms for most insurance policies. The forms often are copyrighted by an independent but industry-supported group that comprises the Insurance Services Office (ISO). Use of ISO forms provides a means of ensuring that the insurance company is providing coverage that meets industry standards. What this means to the policyholder is that the dos, don'ts, whys, and hows are uniform across the industry. Identical policies may be issued by various insurance companies that use the same ISO forms; the differences in coverage are determined by the endorsements added.

The copyright statement of the Insurance Services Office is located at the bottom of each page of the insurance policy; its absence indicates a nonstandard form. The variation might benefit the policyholder by providing better coverage, or it might offer less coverage than the comparable ISO provision. Ask your agent why the ISO form was not used and, more specifically, what differences in coverage result.

Key Elements of the Insurance Policy

An insurance policy describes the conditions under which the insurance company agrees to insure in exchange for the premium that the policyholder agrees to pay. The Declarations page and the Common Policy Conditions are the two sections that bind together the entire insurance policy.

Declarations: Overview of Coverage The Declarations page, normally the first page of the insurance policy, provides a quick reference to

- the type of coverage,
- the name and address of the policyholder,
- the policy period (beginning and ending dates of coverage),
- the policy number,
- the agent's name,
- the type of coverage provided,
- the covered amount,

- the policy limits (per occurrence as well as aggregate),
- the premium, and
- the method of payment.

It also lists all forms attached to the policy as part of the contract—and this is important.

Policy language determines the amount that is paid on a claim. For instance, consider a property damage policy on a building. Let's say that the building is insured for $100,000 and is completely destroyed by fire. The insurance company may legally pay less than $100,000 if

- the insured amount is less than that required by the coinsurance clause,
- the building has been vacant for more than 60 days,
- the building can be replaced for less than $100,000 with "like kind and quality" construction, or
- hazardous operations of which the insurer was unaware took place in the building.

The point here is that policy language can limit or nullify your coverage.

Multiple Locations and Corporations Some pesticide application companies operate out of branch offices scattered across the state or country. Others operate multiple corporations (e.g., lawn care, irrigation, snow removal, landscaping) out of a single location. In either case, the name of each and every business entity must appear separately on the insurance policy. They may be listed on the Declarations page or by an endorsement naming additional insureds.

Legal Statements There are a number of important legal statements included in insurance policies. One example states that "in return for the payment of the premium, and subject to all the terms of this policy, we (the insurance company) agree with you (the policyholder) to provide the insurance as stated in the policy." The interpretation is that the insurance policy is a legal contract, the terms of which determine how the policy will be interpreted by the insurance company and (in litigation) by the courts. More examples follow:

- "These declarations together with the common policy conditions, coverage parts, and forms and endorsements, if any, issued to form a part thereof, complete the above numbered policy." In other words, interpretation of the insurance policy must include consideration of its entire contents.
- "All words and phrases printed in boldface or in quotes are defined in the policy." Or, in other words, the interpretation of coverage will be based on how certain words are defined in the policy. The policy's Definitions section should be examined carefully.

Common Policy Conditions: Your Rights and Responsibilities The Common Policy Conditions comprise an important section outlining the rights of both the insurer and the policyholder. Examples of Common Policy Conditions include statements on cancellation, policy changes, right to examinations and inspections, transfer of rights, and payment of premiums.

Key Clauses Found in Common Policy Conditions

- The policyholder may cancel the policy for any reason, at any time, by simply mailing a notice of cancellation to the insurance company. The right of the insurance company to cancel an existing policy is more difficult, but common justifications for cancellation are nonpayment of the premium and discovery of hazards not identified to the insurance company at the time of policy issue. Even then, the insurance company is required to give at least a 10-day notice. All other reasons for cancellation require the insurance company to provide a 30-, 45-, or 60-day written notice of cancellation (time varies by state).
- The insurance company has the right to examine financial records of the insured company at any time during the policy period, and usually for up to 3 years afterward. Insurance companies reserve this right because premiums for some types of coverage (e.g., Worker's Compensation and General Liability) are based on payroll and sales. Inspection of the insured's records helps to verify that the policyholder has disclosed payroll and revenue figures accurately.
- The insurance company is allowed to conduct periodic inspections of the policyholder's property to verify that operations meet state and federal safety and environmental regulations and, more important, the insurer's own minimum underwriting standards and guidelines. The purpose of such inspections is to reduce or eliminate situations or practices which might lead to future losses.
- The policyholder should contact the insurance company promptly when something that could possibly trigger a loss occurs. If not, the insurance company may deny coverage based on late notice. The insurance company may be required to demonstrate prejudice when late notice is used as a basis for denial. The term prejudice, in this context, means that the insured has in some way (e.g., by not notifying the insurance company promptly) hindered the ability of the insurance company to investigate and settle the claim satisfactorily.

Carefully review the section of your policy that states what actions to take in the event of an occurrence, claim, or suit. Pay particular attention to who should be contacted and what form of notice is required (Fig. 14.6).

Subpolicy Components

Some insurance policies contain subpolicies that provide different types of coverage. Every subpolicy contains the following components that help to explain how the policy is to be interpreted, how the insurance company will respond to claims, and the amount that the insurance company will pay: Insuring Agreement, Conditions, Exclusions, Endorsements, and Definitions.

Insuring Agreement: Risks Covered The Insuring Agreement describes the risk that the insurer has agreed to assume (i.e., the coverage that it will provide). For instance, the Insuring Agreement in liability policies commonly begins with "We will pay those sums that the policyholder becomes legally obligated to pay as damages. . . ." The final wording depends on the type of coverage involved. This initial broad promise of coverage may be modified in sections dealing with Conditions, Exclusions, and Endorsements.

Figure 14.6 Firefighters preparing to deal with a pesticide spill.

Conditions: Claim Payment Limitations Statements in the Conditions section narrow the coverage provided by the policy. For instance, property policies usually contain a condition stating that the insurance company will not insure a vacant property for vandalism, sprinkler damage, glass breakage, water damage, theft, and attempted theft.

Key areas addressed in the Conditions section include the following:

- The maximum that the insurance company will pay
- How deductibles will be applied to a claim
- How losses will be valued (e.g., depreciated value vs. replacement cost, repair vs. replacement)
- Information on the use of an independent appraiser when claims are disputed
- The policyholder's responsibilities (e.g., reporting crimes to local police, providing prompt notice of claims to the insurance agent or insurance company)

Exclusions: Risks Not Covered Risks that the insurance company has not agreed to assume are listed in the Exclusions section. Common risks excluded from liability policies are pollution and damage to the property of others, real or personal, that is in the care, custody, or control of the policyholder.

There usually is an exclusion for contractual liability, including warranties or guarantees, whether verbal or in writing, that policyholders might enter into with their customers. It is important to remember that general liability policies are intended to cover claims arising from unexpected events or from the policyholder's negligence. A General Liability policy usually does not cover (and therefore is not priced to cover) claims arising from contractual obligations, except for specific types of contracts listed in the policy.

Exclusions limit, modify, or restrict how the Insuring Agreement applies. A risk may be excluded either because it is more appropriately covered under a different type of policy or because the insurance company does not want to assume the risk at the premium quoted. Examples are arson by the insured, poor workmanship, and intentional acts.

Figure 14.7 Reacting to a pesticide spill requires quick action.

Sometimes the subject of exclusion can be covered at a higher premium. Coverage for floods or pollution are good examples. In cases where there are particular risks for which you want coverage, but which your base policy does not cover, ask your insurance agent how you might be able to obtain that coverage. Although it may be excluded at the base rate for the policy, it may be available for an increased premium, in which case the policyholder would have the choice.

One typical exclusion from standard General Liability coverage deals with property under the care, custody, or control of the business. For example, a lawn or field is damaged when an applicator accidentally applies the wrong herbicide: In either case, the damaged property is directly under your care, custody, or control; therefore, the exclusion applies and the claim for reimbursement of the homeowner or the farmer may be denied (Fig. 14.7).

In both cases, the damage could have been covered by endorsement at the time of purchase. However, if those same applications drift onto a neighboring lawn or farm, the resulting damage is covered because those areas are not under the policyholder's care, custody, or control. It can get confusing, so always consult your insurance agent and your insurance company underwriter. The "Care, Custody, or Control" exclusion is one of the most commonly overlooked exclusions in a General Liability policy.

Endorsements: Additions, Deletions, Modifications of Coverage Endorsements can add, remove, modify, or clarify coverage specified in the Insuring Agreement, Conditions, Exclusions, and Definitions. Frequently it is these endorsements that actually differentiate policies among companies.

Endorsements give warning to the policyholder via the wording: "This endorsement changes the policy. Please read it carefully." Following this warning is the name of the endorsement (e.g., Pollution Liability coverage), followed by a statement that identifies the specific modifications provided by the endorsement.

An endorsement may refer the reader to a specific part of the policy: "This endorsement modifies the Pollution exclusion under section F2a." In these cases, it is wise to photocopy the original policy and mark the changes, then reread the policy. It is commonly difficult to make sense of the endorsement, itself, out of context.

Definitions: Meanings of Key Words The Definitions section defines key terms found in the policy. These definitions can be critical when claim disputes arise: It's all in the wording! When a term is not defined by the policy, most states require that it be interpreted from the perspective of an ordinary person of average intelligence. Many states require that ambiguous terms or phrases in insurance policies be interpreted in favor of coverage for the policy holder.

COMMERCIAL PROPERTY INSURANCE

Protect your real and personal property with good Commercial Property insurance. Ask your insurance agent as many pertinent questions as possible, early in the insurance buying process. Some examples are listed here:

- What property is covered by the policy? Property insurance should cover direct physical loss or damage to buildings and structures, fixtures, permanently installed fixtures, machinery, and equipment. It should provide coverage for personal property such as furniture, fixtures, computers, etc. Make a list of those items that are definitely worth protecting, and make sure that they are covered by the policy. Carefully review exclusions that remove coverage for certain types of property. Keep an itemized inventory list with values of furniture, equipment, etc., and provide a copy to your insurance agent.
- What perils are covered by the policy? Perils are the cause of a loss. Some property insurance policies cover certain perils (e.g., fire, lightning, explosion, water damage, wind) but exclude others (e.g., earthquakes, floods). If your business might face risk from an excluded peril, ask your insurance agent how much it would cost to purchase an endorsement for the coverage, and weigh that information against the likelihood of the event occurring in your location. Always thoroughly review the Exclusions section of your insurance policy.

Figure 14.8 Policies provide different coverages.

- Are the insurance limits adequate? The maximum amount of insurance recovery is set by the limits of the policy. Any loss over those limits is uninsured. Be careful when buying a low-priced policy since the lower premium may reflect unacceptably restricted coverage.
- Does my policy satisfy the coinsurance requirement? Property insurance policies may include a coinsurance clause applied separately to your buildings and contents. The coinsurance clause requires a policyholder to carry a minimum amount of property insurance, such as 80, 90, or 100% of the replacement cost of the insured property. Insuring property for less than the required coinsurance amount can result in a penalty at the time of loss, regardless of the size of the claim.
- What type is my property policy form? There are three general types of forms: Special (all risk), Broad, and Basic. The Special form offers the widest coverage available, in that it covers all perils not specifically excluded. The Broad form covers only those perils that are specifically listed, such as fire, wind, theft, and collapse. The Basic form is very restricted. It covers only a few perils, usually just fire and lightning (Fig. 14.8).

Endorsements

Endorsements are provisions that modify the standard policy. Although they contain standardized language developed by the insurance industry, subtle differences in wording can affect the scope of coverage. Examples of property insurance endorsements include the following:

- Flood and Earthquake—Consider adding this endorsement if your business is located in an area that is susceptible to earthquakes, floods (e.g., high creeks), and sewer backups.
- Business Income—Otherwise known as business interruption coverage, the Business Income endorsement covers indirect losses that result from a fire or other direct damage. When insured damage causes an interruption of business, revenues are lost but expenses may continue. Additional expenses may be incurred as well, such as rent or lease of a temporary facility, overtime pay, and phone lines and other utilities. Extra expenses incurred in maintaining business operations after an incident may even surpass normal operation expenses. Statistics show that most small businesses fail to reopen after a catastrophic experience, due to insufficient capital.
- Off-Premises Utility Interruption—Utility interruption that occurs away from your insured premises is excluded under property coverage if your building or its contents are not directly damaged. The Off-Premises Utility Interruption endorsement covers loss of business due to utility interruption (power, water, gas, telephone) resulting from an insured occurrence away from your premises.
- Replacement Cost—Property insurance can be purchased to cover either replacement cost or actual cash value. Replacement cost is the amount it would take to replace the building or other property at current prices. Actual cash value is adjusted for depreciation and can be significantly less than the replacement cost for an older building.
- Building Ordinance or Law—Some communities specify by ordinance that any undamaged portion of a damaged building must be demolished prior to reconstruction. But demolition costs and the cost of replacing the undamaged portion may not be covered by property insurance. Changes in building codes over a period of time may increase the replacement cost of a building beyond its insured value. Building Ordinance or Law endorsements provide coverage for these situations.
- Agreed Value or Agreed Amount—The policyholder and the insurance company agree on the value of the property before the policy is written. With this endorsement, the possibility of a coinsurance penalty is removed, even if the property is underinsured.
- Inflation Guard—This is an endorsement that increases coverage limits for your property, annually and automatically, to keep pace with inflation.

COMMERCIAL CRIME INSURANCE

One has only to read the local newspaper to realize that businesses are regularly confronted with crime—and many find themselves the victim of their own employees! A variety of Commercial Crime insurance is available for protection against theft, robbery, burglary, forgery, embezzlement, etc. (Fig. 14.9). Ask your insurance agent for help in answering these questions:

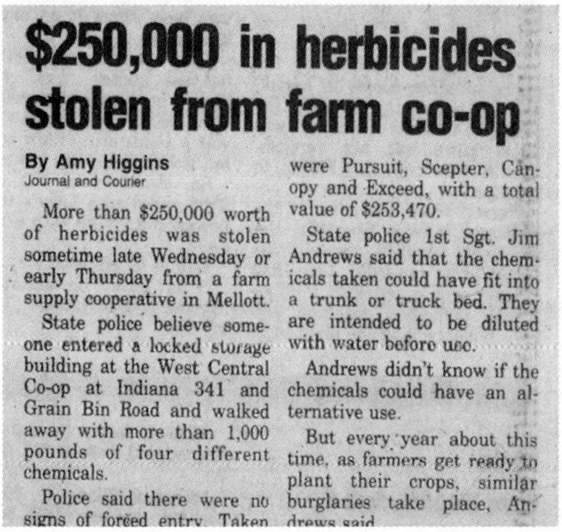

$250,000 in herbicides stolen from farm co-op

By Amy Higgins
Journal and Courier

More than $250,000 worth of herbicides was stolen sometime late Wednesday or early Thursday from a farm supply cooperative in Mellott.

State police believe someone entered a locked storage building at the West Central Co-op at Indiana 341 and Grain Bin Road and walked away with more than 1,000 pounds of four different chemicals.

Police said there were no signs of forced entry. Taken were Pursuit, Scepter, Canopy and Exceed, with a total value of $253,470.

State police 1st Sgt. Jim Andrews said that the chemicals taken could have fit into a trunk or truck bed. They are intended to be diluted with water before use.

Andrews didn't know if the chemicals could have an alternative use.

But every year about this time, as farmers get ready to plant their crops, similar burglaries take place, Andrews said.

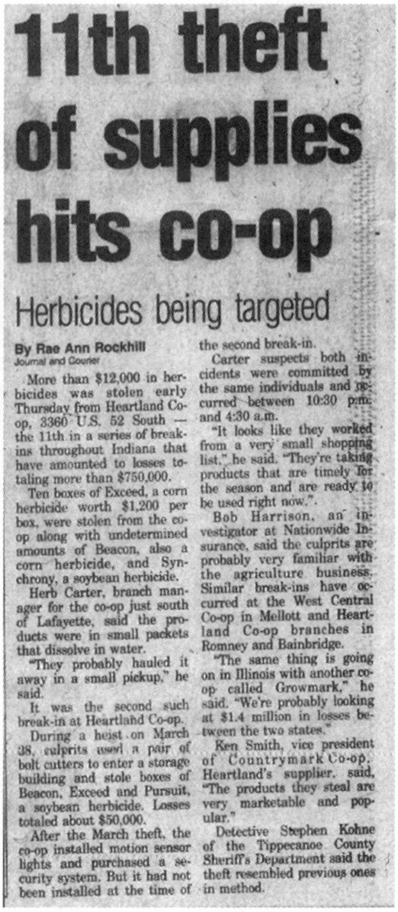

11th theft of supplies hits co-op

Herbicides being targeted

By Rae Ann Rockhill
Journal and Courier

More than $12,000 in herbicides was stolen early Thursday from Heartland Co-op, 3360 U.S. 52 South — the 11th in a series of break-ins throughout Indiana that have amounted to losses totaling more than $750,000.

Ten boxes of Exceed, a corn herbicide worth $1,200 per box, were stolen from the co-op along with undetermined amounts of Beacon, also a corn herbicide, and Synchrony, a soybean herbicide.

Herb Carter, branch manager for the co-op just south of Lafayette, said the products were in small packets that dissolve in water.

"They probably hauled it away in a small pickup," he said.

It was the second such break-in at Heartland Co-op.

During a heist on March 28, culprits used a pair of bolt cutters to enter a storage building and stole boxes of Beacon, Exceed and Pursuit, a soybean herbicide. Losses totaled about $50,000.

After the March theft, the co-op installed motion sensor lights and purchased a security system. But it had not been installed at the time of the second break-in.

Carter suspects both incidents were committed by the same individuals and occurred between 10:30 p.m. and 4:30 a.m.

"It looks like they worked from a very small shopping list," he said. "They're taking products that are timely for the season and are ready to be used right now."

Bob Harrison, an investigator at Nationwide Insurance, said the culprits are probably very familiar with the agriculture business. Similar break-ins have occurred at the West Central Co-op in Mellott and Heartland Co-op branches in Romney and Bainbridge.

"The same thing is going on in Illinois with another co-op called Growmark," he said. "We're probably looking at $1.4 million in losses between the two states."

Ken Smith, vice president of Countrymark Co-op, Heartland's supplier, said, "The products they steal are very marketable and popular."

Detective Stephen Kohne of the Tippecanoe County Sheriff's Department said the theft resembled previous ones in method.

Figure 14.9 Pesticides should be locked and inaccessible to unauthorized persons.

- Do you need to insure against burglary of cash, checks, securities, or valuable papers, or against employee theft?
- If you have employees who handle money, should you insure against embezzlement?
- Do you need to be insured for the possibility that an employee may steal property from the premises of a customer?
- Do you need coverage for yourself or employees who might be robbed at the facility or en route to the bank to make a business deposit?

Endorsements

- Employee Dishonesty—This coverage protects the employer from theft of money, securities, or property by an employee. The insurance company can step in, investigate aggressively, and prosecute. If the employer provides a qualified retirement plan, another form of Employee Dishonesty coverage is required under the Employee Retirement Income Security Act: An employer must carry an Employee Dishonesty limit of at least 10% of the total assets of the retirement plan. But what if an employee steals from a customer? Employee Dishonesty coverage does not cover such a loss because the employer is not the target of the theft. However, certain types of surety bonds might apply. Coverage can be very expensive, so if you feel that you cannot rely on your own internal procedures and controls, ask your insurance agent for advice on how to proceed.

- Depositor's Forgery or Alteration—Forgery is the generation of a document or signature that is not genuine. Conversely, alteration is changing a document in a manner that is neither authorized nor intended by the insured. The loss to the policyholder is covered when forged or altered documents (checks, drafts, promissory notes, bills of exchange, etc.) are used to withdraw money from the insured's account.
- Theft, Disappearance, and Destruction—Theft is any act of stealing; disappearance is an unknown cause of loss; and destruction is the loss of certain property. Disappearance lacks the element of knowing whether the disappearance resulted from theft, burglary, or robbery, or if the missing property is simply misplaced. Disappearance insurance is the broadest form of coverage available for money and securities.
- Robbery and Safe Burglary— Robbery includes an element of threat of personal harm in taking away someone's property. Safe burglary is the taking of property from a safe, where there are visible signs of forced entry and where no other persons are present.

COMMERCIAL VEHICLE INSURANCE

The consequences of vehicular accidents are among the most frequent and potentially severe situations that business owners must insure themselves against. And, whether you own one truck or a whole fleet, your business is at risk. Each accident—whether it's a fender bender with no injuries, whether someone is permanently disabled, or whether someone is killed—carries the potential for serious consequences. Vehicle insurance may cover vehicle repairs or replacement, bodily injury to persons in your vehicles or others' involved, and representation in the event of a lawsuit. But it is essential that you ascertain

- exactly which vehicles are covered,
- the type and amount of coverage,
- who may drive the insured vehicles, and
- the beginning and ending dates of coverage.

Offer safe driving programs for employees. Teach drivers to get as much information as possible when involved in a vehicular accident, and teach them to spot fraud. Be especially watchful for "swoop and squat" insurance fraud where a driver intentionally cuts in front of your vehicle and slams on the brakes, causing your vehicle to hit his in the rear (Fig. 14.10).

Physical Damage

This part of your Vehicle insurance policy explains the conditions under which your damaged vehicle will be repaired or replaced. There are two types of coverage for physical damage: collision and comprehensive.

Figure 14.10 Emergency responders deal with a pesticide spill on a highway.

- Collision coverage ensures that your vehicles will be repaired or replaced if involved in an accident. It pays no matter who is at fault, but only up to the actual cash value of the vehicle at the time of the accident. In some cases, if your vehicle is nearly new and is financed, insurers will extend coverage to include the difference between the actual cash value and the balance on your loan. Collision coverage is subject to a prescribed deductible.
- Comprehensive insurance provides coverage for vehicle damage from situations other than collision (fire, theft, vandalism, wind, floods, etc.). Comprehensive coverage is subject to a prescribed deductible and may include reimbursement for transportation expenses (e.g., rental cars) incurred while a vehicle is being repaired or replaced; it also may cover towing and labor expenses.

Liability Coverage

How does Vehicle insurance protect your business when injured persons file claims against it? Under Auto Liability coverage, the insurer agrees to defend the policyholder and to pay all legal defense costs and resulting judgments arising from the ownership, maintenance, or use of a vehicle covered by the policy subject to the policy's limits. Attorney and other defense fees are payable in addition to the policy limits.

Ask your insurance agent the following questions:

- If I am driving and have an accident involving another vehicle, whose policy pays my medical bills and/or those of my passenger?
- If my vehicle is stolen while my friend's computer is in the trunk, whose insurance covers the loss of the computer? What if it is my wife's computer?
- If I am hauling a trailer, and the trailer comes loose and damages other vehicles, whose insurance pays for damage to those vehicles and damage to the trailer? What if my vehicle is insured but my trailer isn't? What if the trailer belongs to a friend?
- What if I fail to list a titled or leased business vehicle on my policy? If the unlisted vehicle incurs or causes damage, will my policy pay? (Fig. 14.11)

Figure 14.11 Make sure all vehicles are listed on the insurance policy.

The answers to the preceding questions are not necessarily straightforward, but depend on the circumstances in each situation. The discussion between you and your agent, however, will help uncover your needs and eliminate potentially unpleasant surprises.

Expect to be asked the following questions by your insurance agent:

- Do your employees use their own vehicles during the course of their work? If so, a Hired and Non-Owned Auto Liability endorsement must be included on your General Liability policy. Always. No exceptions—even if your business does not own any vehicles. An employee's private Vehicle policy may not protect your business if it is named in a lawsuit, so make sure you carry this coverage.
- Do you transport chemicals in or on your vehicle? If so, be aware that your Vehicle policy may not cover a pollution claim nor provide cleanup coverage. These incidents can be covered through special endorsements or Pollution coverage under special General Liability policies.

Endorsements

- Rental Reimbursement—Rental reimbursement pays the cost of renting a substitute vehicle while the damaged vehicle is being repaired or replaced or after a vehicle is stolen (Fig. 14.12).
- Broad Form Drive-Other-Car Coverage—This endorsement provides personal Vehicle liability coverage for business owners or specified employees who do not own or drive a personal vehicle other than the company-provided vehicle. Under this endorsement, the commercial Vehicle policy provides personal Vehicle liability coverage in situations where the specified driver is involved in a non-work-related accident while driving any vehicle other than a company-provided vehicle.
- Towing and Labor—Coverage for towing and labor pays towing charges up to the limit selected, each time a covered vehicle needs such services.

Figure 14.12 Rental reimbursement coverage can help pay for a rental unit until the damaged vehicle can be repaired or replaced.

- Personal Injury Protection (PIP) or Medical Payments—PIP also is known as "no-fault insurance." It is available at various optional limits to pay for the medical care of persons injured in a covered vehicle, regardless of whether the policyholder is liable.

In addition to medical coverage, PIP provides reimbursement for loss of earnings, for additional living expenses, for funeral costs for pedestrians, and for occupants of the policyholder's vehicle other than those covered under other policies. No-fault vehicle insurance is designed to compensate victims of vehicular accidents without the necessity of proving negligence on the part of those involved.

- Uninsured/Underinsured Motorist—This provides protection for the policyholder and passengers against bodily injury caused by the negligence of an uninsured or underinsured motorist. Instances where such protection becomes important include a hit-and-run accident, an insolvent insurer, and an insurer that denies coverage to such other motorists.

Schedule of Coverage

The business Vehicle policy contains a schedule (list) of the aforementioned types of coverage and specifies the vehicles to which they apply. These specifications appear as numeric symbols that can affect coverage significantly; therefore they should be reviewed carefully. Symbol "1" refers to "any vehicle." If it appears next to a particular coverage, it means that any vehicle you drive is covered, regardless of who owns it or whether it is listed on the schedule. Symbol "7" refers to "scheduled vehicles," limiting coverage to only those vehicles that appear on the schedule; a vehicle inadvertently left off the schedule, or a new vehicle that has not yet been added, may not be covered.

INLAND MARINE FLOATER

Someone breaks into your locked toolbox inside your truck and walks away with all of your tools and equipment. You ask your insurance agent if the stolen property is covered, and he indicates that it is covered by a policy known as an Inland Marine Floater which is used to protect tools and equipment (in this case) but only if there are signs of forced entry on the vehicle.

The Inland Marine Floater typically covers property that is mobile. Equipment that is moved from site to site is one example. Other Inland Marine examples are valuable papers and specialized personal property such as computer equipment. Ask your insurance agent for help in answering the following questions:

- Should I insure all of my equipment against theft?
- Do I have to schedule (list) every piece of equipment that I want covered by the insurance policy?
- Does an Inland Marine policy cover the replacement cost of stolen equipment, or only the depreciated value at the time it is stolen?
- Do I need to cover accounts receivable in case records are stolen or damaged and cannot be reconstructed?

- What deductible should I carry per piece of equipment, and per aggregate loss?
- If several pieces of property are stolen at one time, will I have to pay a deductible on each item?

Other Inland Marine Endorsements

- Equipment Floater—This coverage protects your equipment or tools on the business premises, in transit, at job sites, and even at your employees' homes. These tools and equipment may be scheduled, or they may be covered on a blanket basis. It generally is less expensive to schedule them, although additional effort is required to maintain a current schedule. The Equipment Floater can be endorsed to automatically cover items that you rent or lease temporarily, or you may choose Rental Reimbursement coverage to assume the expense of temporary rentals for use until a damaged or lost item is replaced.
- Installation Floater—This coverage protects materials and supplies that are in transit or at job sites: things such as seed, chemicals, trees, shrubs, landscaping materials, etc.
- Electronic Data Processing (EDP)—This coverage provides broad protection for computer equipment, data, and media, and it can include coverage for mechanical breakdown and electronic erasure. Coverage applies both on and off the business premises, and in transit. EDP coverage is generally superior to traditional property insurance when insuring computer equipment and software.
- Accounts Receivable—This covers the policyholder's inability to collect accounts receivable balances when records are destroyed by a covered peril such as fire. It is difficult to determine the actual dollar amount of losses where records are destroyed, so the insurance adjuster will compute the loss based on prior years' income. An accounts receivable policy is written with a limit sufficient to compensate the insured for catastrophic loss, usually covering the following:
  - Accounts receivable which are uncollectible if records cannot be reestablished
  - Expenses incurred by the policyholder in reestablishing records of accounts receivable following loss or damage
  - Collection expenses in excess of customary expenses, made necessary because of a loss
  - Interest charges on interim loans (from time of loss until reimbursement is received)
- Valuable Papers and Records—This covers direct physical loss of valuable papers and records owned by the policyholder. The definition of valuable papers is broad. It includes documents and records such as books, maps, films, drawings, abstracts, deeds, mortgages, and manuscripts. The term valuable papers does not include money and securities.

COMPREHENSIVE GENERAL LIABILITY INSURANCE

Although being sued is a possibility that all businesses face, pesticide application companies are particularly vulnerable. Dealing effectively with lawsuits resulting from real or alleged negligence requires experience, expert handling of claims, and expert legal advice. Suits may be filed for various reasons: property damage; bodily injury; failure to manage pests;

(a)

(b)

Figure 14.13 (a) A lawsuit brought before a jury often requires discussion of complex issues. (b) The General Liability insurance policy will provide legal counsel.

pesticide exposure from applications, spills, drift, etc. That's not to say that all lawsuits have merit, but going to court—win, lose, or draw—costs you in terms of time, attorney fees, and a multitude of related expenses. Liability resulting from vehicular accidents is not covered by General Liability insurance; only a vehicle policy covers liability arising from situations involving "covered vehicles."

If you are held legally responsible by a court of law for bodily injury or property damage due to negligence, it is your General Liability policy that pays. The insurance company must defend any claim (or portion thereof) asserted against the policyholder that falls within the policy's scope of coverage. Claims often are settled out of court to avoid expensive litigation (Fig. 14.13).

With so much at stake, it is not surprising that most owners of pesticide application companies are willing to purchase General Liability insurance. In fact, they would likely purchase unlimited liability coverage, if available, but insurance companies offer only limited protection. This inherent conflict between the policyholder and the insurer necessitates careful scrutiny of policy language.

Most General Liability policies are written on an Occurrence form, which means that the act or condition giving rise to liability must take place within the policy year; but many times a business owner is unaware of a claim until long after his policy has expired. Under the Occurrence form, however, a situation may be covered even if not discovered and reported until after the applicable policy has expired. Therefore it is important to keep expired policies in a safe place, permanently; they can be recalled to pay claims presented well after their expiration date.

The other General Liability form available is a Claims Made form. Unlike the Occurrence form, it requires that claims must occur and be reported during the active policy period. The Claims Made form is commonly used for Specialty Liability coverage such as Errors and Omissions, but pest management companies should avoid it. Significant coverage or claims reporting gaps may occur when switching from an Occurrence to a Claims Made policy.

The following are key areas that should be reviewed in General Liability insurance:

- Policy limits
- Deductibles
- Rating basis
- Exclusions
- Endorsements

Policy Limits

General Aggregate The General Aggregate (total) is the most the insurance company will pay for all combined claims within the policy year—except products and completed operations claims, which have separate annual aggregate limits.

Products and Completed Operations Aggregate The Products and Completed Operations Aggregate is a separate, annual aggregate which applies to products and completed operations claims. Products generally refers to something that the policyholder sells or manufactures—something used by others, either in whole or as part of something else used by others. Completed Operations is the insurance term that most accurately applies to pesticide applications. Applications, inspections, and other integrated pest management activities or services, once completed, and where the insured has left the serviced premises of the customer, fall under this Completed Operations Aggregate.

Personal and Advertising Injury Personal and Advertising Injury is a separate limit of coverage for certain types of claims such as libel, slander, false arrest, invasion of privacy, advertising, etc.

Each Occurrence The dollar amount specified under Each Occurrence is the limit of coverage for each single claim or occurrence of loss.

Fire Damage Limit If the portion of the premises that you lease is damaged by fire, and if you are liable, the Fire Damage Limit applies; the limit applies to each fire, and the general aggregate limit applies.

Medical Payments The Medical Payments limit applies, per person, and pays regardless of legal liability on the part of the policyholder. The general aggregate limit applies.

Deductibles

Deductibles are used in General Liability policies to curb the filing of small claims (< $1000). They may apply only to bodily injury, only to property damage, or to both types of loss combined. Deductibles may apply per claim or per occurrence. Insurance companies generally adjust and pay all claims, after which the policyholder is obligated to reimburse the insurance company up to the deductible limit.

Rating Basis

The premium for General Liability policies is based on the policyholder's total sales receipts or payroll for a given activity. Specific rates—usually per $100 or $1000—may apply to revenue or payroll for professional activities such as landscaping, lawn care, snow removal, and tree pruning. Payroll and sales figures are estimated annually for the upcoming policy year. The insurance company conducts an audit of actual claims at the end of the policy year to determine the premium due from or the refund due to the policyholder.

Exclusions

- Pollution—Most liability policies, and many Property insurance policies, contain a broadly worded Pollution exclusion (See section Pollution Insurance later in this chapter).
- Care, Custody, or Control—This exclusion can be particularly troublesome because most policyholders are unaware of its existence. It may not cover damage to the real or personal property of others which is under your care, custody, or control. In other words, unintentional damage that you may cause to others' property during your course of work may not be covered. An example is the application of the wrong chemical to a customer's lawn. The lawn may be considered under your care, custody, or control and excluded from coverage. However, damage to adjacent property which is not under your care, custody, or control usually is covered. The Care, Custody, or Control exclusion is subject to interpretation. Some insurance companies offer coverage for property under your care, custody, or control at an additional premium—and usually with a reduced limit. Be sure to check with your insurance agent about the availability of Care, Custody, or Control coverage.
- Professional Services—This exclusion normally applies when engineering, architectural, financial, inspection, and other professional services are offered by the policyholder. This is a key exclusion for operators who conduct termite inspections for real estate transactions, act as consultants, perform radon testing, etc.

Figure 14.14 The application of pesticides requires special insurance known as "Herbicide and Pesticide Applicator" coverage.

Endorsements

- Herbicide and Pesticide Applicator Coverage—This endorsement modifies the Pollution exclusion under the General Liability policy by deleting a portion of the Pollution exclusion to provide coverage at a job site (Fig. 14.14).
- Product Misdelivery—This endorsement provides coverage when a product or chemical has been applied to the wrong premise.
- Per Project/Location General Aggregate—The General Aggregate limit in the General Liability policy specifies the limit available for all claims for an entire policy year. Any General Liability claim during the policy year (other than product and completed operations claims) reduces the general aggregate limit available for additional claims within the policy year. This endorsement reinstates the full aggregate limit after each claim (or each insured "location") so that additional claims within the policy year are covered for the full amount. "Per occurrence" limits do apply.

WORKER'S COMPENSATION INSURANCE

Hundreds of workers are injured on the job each day. Many sustain disabling injuries that prevent them from ever returning to work. States mandate that employers carry a reasonable amount of insurance to partially compensate workers injured on the job. We refer to it as "Worker's Comp," short for Worker's Compensation.

Worker's Compensation premiums are based on rates per $100 of covered business payroll. The National Council for Compensation Insurance publishes an Experience Modification Factor (EMF) for most businesses that reach a certain premium threshold. Each business earns its own EMF based on its Worker's Compensation loss experience during the applicable 3-year rating period. The EMF may change each year and is applied directly to your premium as either a debit or a credit. Excellent loss experience results in an EMF that can dramatically reduce your Worker's Compensation premium; conversely, poor loss experience results in an increased premium. Safety in the workplace protects employees

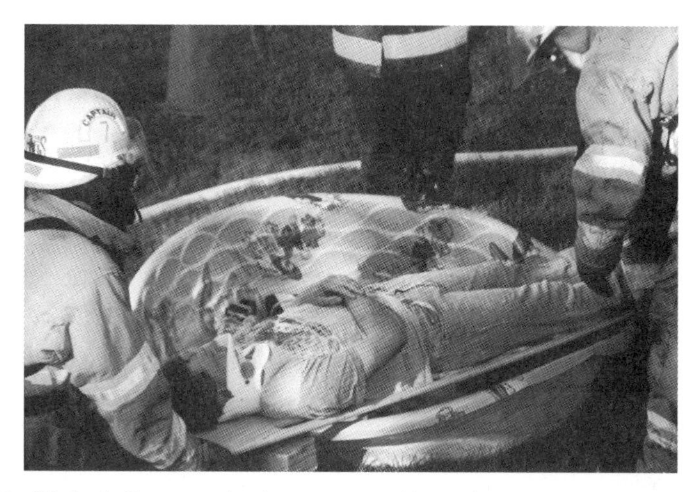

Figure 14.15 Worker's Compensation insurance provides assistance to employees who are injured on the job.

and reduces Worker's Compensation premiums for the employer (Figs. 14.15 and 14.16). It should be promoted diligently.

Worker's Compensation laws vary by state, depending on the types of employees covered, the types of injuries covered, benefit limits, and who is required to carry the insurance. All states include some form of "exclusive remedy" provision to prevent employees from suing their employer when Worker's Compensation coverage is applicable.

Employers who do not provide Worker's Compensation coverage are in violation of the law and may be subject to lawsuits, penalties, and fines. Your insurance agent can describe Worker's Compensation provisions in your state.

Because Worker's Compensation laws differ by state, premiums also vary. All states approve rates for each occupation (salesperson, clerical employee, applicator, landscape

Figure 14.16 Emergency responders practice dealing with a mock pesticide injury on a farm.

gardener, etc.), and insurance companies apply those rates to estimated payroll. Premium discounts apply to policies with premiums of $5000 or more per year.

Worker's Compensation policies pay prescribed benefits for the cost of medical care and a percentage of lost wages during the period of disability.

- Medical bills are paid, up to state prescribed limits.
- An injury may result in a worker's becoming disabled and missing work for an extended period of time. Worker's Compensation provides an insurance benefit to the injured employee, generally up to two-thirds of their average weekly wage, subject to a state mandated benefit limit. Ask your insurance agent for assistance in answering the following questions:
  - — Is there coverage when one employee injures another employee?
  - — What is covered by the employer's liability section of the Worker's Compensation policy?
  - — Which classification is used for payroll determination if an employee is doing more than one job, such as sales and service work?
  - — If you hire nonemployees as independent contractors and they are injured while doing work for you, are they covered by your Worker's Compensation policy? What if they are injured while doing work for themselves or others, and no other Worker's Compensation policy exists? Will your policy cover them?
  - — If your company does work in more than one state, are your employees and subcontractors covered under your Worker's Compensation policy for injuries suffered out-of-state if the state is not listed on your policy?
  - — Are the benefits for employee injuries the same in another state?
  - — How does the insurance company or a court of law determine whether a person is your employee or an independent contractor? Are there any legal regulations that apply in determining who is an employee?
  - — Does your Worker's Compensation policy cover stress on the job? Does it cover long-term carpal tunnel syndrome? Does it cover claims other than for physical injury?

Following Form Excess or Umbrella

The Umbrella policy provides supplemental liability insurance for a nominal premium increase. Its limits apply after primary insurance limits have been reached. The Umbrella policy usually provides limits in excess of the underlying General Liability, Vehicle Liability, and Employer Liability policies (under Worker's Compensation). Umbrella policies also may offer coverage for hazards not addressed in the policyholder's underlying liability policies (Fig. 14.17).

Excess liability policies are issued as either Following Form Excess or Umbrella. Policies issued as Following Form Excess provide the same coverage and conditions as those contained in applicable underlying policies. An Umbrella policy may provide broader or narrower coverage than those provided by applicable underlying policies. Ask your insurance agent for assistance in answering the following questions:

Figure 14.17 Pollution insurance can help cover the costs associated with cleaning up large spills such as this fertilizer release.

- What primary coverage is the Excess or Umbrella policy designed to cover?
- What minimum limits must be stipulated in the underlying liability policies to secure Excess or Umbrella coverage?
- Must a deductible or self-insured retention amount be satisfied on the underlying policies before the Excess or Umbrella coverage kicks in?
- How do I know whether I have purchased a Following Form Excess policy or an Umbrella policy?
- Should the effective dates of the underlying policies and the Excess or Umbrella policy be the same?
- Is there an advantage to buying my Excess or Umbrella policy from the same insurance company that provides my underlying liability policies?
- What is the primary reason for purchasing additional liability insurance provided by Excess or Umbrella policies?
- What kind of loss is most likely expected to exceed the primary liability limits?

POLLUTION INSURANCE

Virtually all commercial property and liability policies contain a Pollution exclusion. Prior to 1986, the standard General Liability form covered pollution claims for "sudden and accidental" incidents.

Some state supreme courts have held that "sudden and accidental" means "unexpected and unintended." Under this interpretation gradual contamination such as that caused by pesticide leakage over time would be covered unless it was somehow expected or intended. Other state courts have ruled that the term "sudden" describes a single incident (Fig. 14.18).

Figure 14.18 Selection of pollution coverage is critical to those who store, transport, and mix pesticides and fertilizers.

In the mid-1980s, the insurance industry largely replaced the sudden and accidental exclusion with the broadly worded "absolute" Pollution exclusion. The term "pollutant" in the absolute exclusion is defined as "any solid, liquid, gaseous, or thermal irritant or contaminant, including smoke, vapor, soot, fumes, acids, alkalis, chemicals, and waste." The exclusion eliminates coverage for any "loss or damage caused by or resulting from the discharge, dispersal, seepage, migration, release, or escape of pollutants." This broad wording has prompted many courts to conclude that the literal interpretation excludes coverage for virtually any routine business mishap that an ordinary person would not have called "pollution" even though it may involve chemicals. Some states allow this literal interpretation whereas others don't. This means that since pesticide application businesses routinely store, transport, and use pesticides, fertilizers, and fuels, they could be essentially uninsured for pollution, as defined: They lie vulnerable to uninsured risk!

Insurance company denials of chemical-related insurance claims, based on Pollution exclusions, have spawned a tremendous amount of litigation. Some claims involve damage to the environment, such as soil or ground water contamination. Others involve alleged bodily injury or property damage arising from exposure to noxious fumes or chemicals. Policyholders argue that the policy language is ambiguous and must be construed in favor of the insured. Insurance companies argue that the exclusion must be interpreted literally. The rules of insurance policy interpretation vary considerably from state to state, hence the outcome of these lawsuits varies. Many courts have ordered insurance companies to provide coverage for chemical-related claims despite the Pollution exclusion; other courts have denied coverage. Pesticide application businesses should stay informed of legal developments on this issue in states where their offices are located and/or where they do business.

Controversy over the scope of the Pollution exclusion subjects policyholders to considerable uncertainty and delay. And even when they prevail against their insurers in court, their legal costs alone can be substantial.

Figure 14.19 Pollution coverage generally does not pay for mismanagement of chemicals.

Some insurance companies now offer specialized Pollution coverage. But, as with any policy provision or endorsement, policyholders must read the fine print to make sure that the coverage they intended to purchase is not removed by exclusion. The fine print also may impose requirements such as immediate reporting of a loss to the insurance company as a prerequisite to coverage. It is advisable to hire an attorney who represents policyholders in coverage disputes to review the intricate wording of your Pollution coverage provisions (Figs. 14.19 and 14.20).

Pesticide contamination claims can originate

- at the site of application,
- in transit,
- in storage,

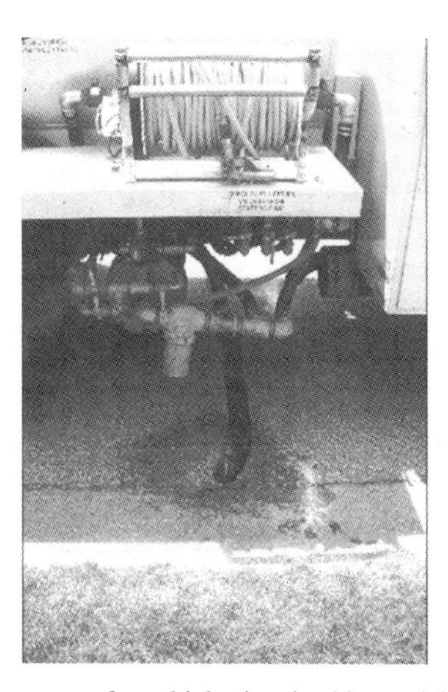

Figure 14.20 Pollution coverage for vehicles involved in pesticide applications is a must.

- from pesticide spills (remediation or cleanup costs),
- at the site of cleanup (bodily injury or property damage),
- at pesticide and container disposal sites, and
- during product manufacturing and formulation.

It is critical for business owners to consult with their insurance company and private counsel for definitions of exactly what their general and specialized policies cover as well as the coverage limits. For instance, Pollution coverage is of very limited help if it covers only a fraction of the cleanup cost following an incident. It also is important to determine the conditions under which the insurance company will pay for contamination cleanup.

Consider putting the following questions in writing to your insurance agent to start the review process. More than likely, your inquiry will be forwarded to and answered by your insurance company's legal department. In your letter to the agent, request that all responses be put in writing, then take the written responses and the policy to your attorney for an interpretation of what the policy covers—and what it does not.

Will my insurance policy pay for investigation, assessment, remedial action, litigation, and fines under the following scenarios? If so, how much and under conditions?

• Example: A fire breaks out in the building where I store pesticides and fertilizers. The fire department extinguishes the fire with water, resulting in soil contamination to my property and to neighboring properties.

• Example: My truck overturns, releasing pesticides onto the roadway. Runoff from the spill reaches a nearby creek, resulting in a fish kill.

Will my insurance pay to replace lost equipment, supplies, contents, and the building under the following scenario?

• Example: A fire breaks out in the warehouse where I store pesticides, fertilizers, trucks, and other equipment used in my business. I call the fire department and they quickly arrive at the scene. Their first reaction is to extinguish the fire, but in my emergency plan I have informed them not to put water on the fire unless it is absolutely necessary. The fire chief decides to let the building burn since the smoke does not pose a risk to nearby neighbors.

Does the Pollution exclusion on my policy disallow coverage when a worker or bystander inhales or otherwise contacts and is injured by chemicals in association with my business operations?

Are the coverage limits adequate to address the financial risks that my business faces?

EMPLOYMENT PRACTICES LIABILITY INSURANCE

Many insurance carriers have recognized the demand for Employment Practices Liability and have begun to offer the coverage. Employment Practices Liability policies are designed to protect the employer when an employee files a complaint or sues for alleged acts of sexual harassment, failure to hire or promote, wrongful termination, discrimination, etc.

Thousands of such cases are backlogged in federal agencies and courts, each awaiting an appropriate response or investigation. New rulings continually redefine what constitutes sexual harassment and job discrimination, both of which pose real liability problems for employers. Be sure to ask your insurance agent about programs and resources that you can implement to reduce premiums (Fig. 14.21).

Figure 14.21 Employers must be constantly alert to posters in the workplace that can be offensive to employees.

DIRECTORS AND OFFICERS LIABILITY INSURANCE

Directors and Officers Liability coverage protects the pest management company's directors and officers in lawsuits alleging errors or omissions related to poor management decisions that result in financial loss to the business. Lawsuits against directors and officers can be filed by employees, competitors, government entities, investors, and others.

ERRORS AND OMISSIONS LIABILITY INSURANCE

Remember, the typical General Liability policy only covers liability resulting in property damage or bodily injury to others; it does not cover claims resulting from a monetary loss to third parties if, as an example, improper advice is offered. The General Liability policy also contains a Professional Services exclusion.

These two conditions create the need for coverage that responds to professional services other than bodily injury and property damage. Such coverage is known as Errors and Omissions Liability, and it is essential for pesticide application operations that provide inspection or consulting services. Without it, claims resulting from inaccurate assessment on the existence of termites, for example, are not covered. Errors and Omissions is specialized coverage that may not be available from a standard insurance company. Discuss your need for Errors and Omissions Liability coverage with your insurance agent.

HOLD HARMLESS AGREEMENT

Some contracts require the contractor to hold the customer harmless for any claims arising from work at the customer's premises. Before signing such a contract, check with your agent, because the insurance company may require copies of any such agreements between the policyholder and its customers. Those agreements may claim to commit the insurance company to cover risks they will not cover and which are not covered by the policy.

Keep in mind that insurance companies are not automatically bound by agreements that you make with others; so never assume liability that your policy will not cover. Speak with your insurance agent before entering into any agreements. Modifications to your insurance coverage on behalf of customers or other third parties may result in additional premiums. Ask your insurance agent the following questions:

- Will my insurer charge to add "additional insureds"?
- Will my insurer be willing to alter the policy required by my customers?
- Should I speak with you to confirm that I have adequate coverage before signing any agreement?
- What is the difference between a customer's (or prospective customer's) asking for a Certificate of Insurance and his or her asking to be named as an additional insured on my policy?
- Do you need to review the actual contract under which I am asked to provide a Certificate of Insurance?

THE INSURANCE AGENT: A VALUABLE ADVISOR

The insurance agent may wear one of two hats. Depending on whether he is talking to a prospect or to a customer, he may go from salesperson (able to close a sale) to serviceperson (able to provide customers with objective advice on insurance companies and coverage). Some agents are better at sales than at service, and vice versa. The policyholder's goal is to find the best insurance professional—and it's not easy. Ideally, your agent will assume a risk management role, helping you to identify risks, eliminate some, and choose the best way to address those remaining (Fig. 14.22).

Figure 14.22 Insurance agents are professionals who develop an insurance portfolio designed to meet your specific needs.

Hopefully, your insurance agent will guide you in making decisions regarding the types of coverage that you need; certainly, he or she should be working in your best interest. You should choose to insure your pesticide application business with special coverage, due to the nature of the risks you face. An informed agent will discuss the risks common to your business, advise you on the coverage available to protect against those risks, and possess the expertise to assist you in settling claims.

It is important to understand that insurance agents can be influenced by considerations that differ from—even conflict with—your long-term best interest: their commission rates, for example. Therefore it is important to evaluate the insurance products that your agent recommends and, sometimes, to get a second opinion. A good agent will ask numerous questions about your business to align your needs with the best products available. His or her analysis on what risks to insure and at what cost should guide your decision-making process.

Develop a List of Insurance Agents

Being a professional insurance agent entails much more than passing a state insurance examination and hanging a shingle. For certain, it's more than handing out business cards with catchy slogans; running slick advertisements on television, radio and in print; and quoting lower prices than the competition.

Ask owners of similar businesses in your area where they purchase their insurance; if a couple of names surface repeatedly, chances are those individuals are good at what they do. Make appointments to speak with them, individually, and get a feel for their desire to handle your account.

Local, state, and national trade associations may generate a list of insurance agents and/or companies that cater to pesticide application businesses. When you attend association meetings, sit down with vendors who exhibit their insurance products and services. If you find them knowledgeable and easy to talk with, make an appointment to discuss your specific needs.

Ask each agent the following questions:

- How many years have you been writing commercial insurance for businesses like mine?
- How long has your agency been in business?
- Would you provide a telephone list of pesticide application businesses that you currently insure? May I call them?
- What is the claims history of the insurance companies you represent with businesses like mine (i.e., what are the types of claims that I might file)?
- Would you furnish the name of someone who has recently filed a claim with you or your insurance company? May I call them to see how the claims were handled?
- How do you advise businesses whose insurance claims have been denied?
- Have you exhibited at any of our state, regional, or national workshops and conferences in the past two years?
- May I see your current state insurance license?
- Have you attained any other professional designations (e.g., Chartered Property and Casualty Underwriter or Certified Insurance Counselor) beyond the state license?

Gather Key Underwriting Information

Select one agent to set up the completed insurance application by type of coverage: General Liability, Vehicle, Worker's Compensation, and Property. Ask him or her to establish a proposal outline or bid specification sheet that other agents may use when submitting proposals. Some agents charge for this service, but the cost is justified by the simplicity of proposal comparison among various insurance companies. Once a format is developed, annual updates can be accomplished with minimal effort.

Call the agent who will develop the outline and ask for a written list of information that he or she will need. Based on the list, establish a format to use with all companies from which you solicit proposals. Following are some items that most companies request.

- Replacement value of buildings and contents at each location
- A description of each building: construction type, occupancy, year built, and square footage. A description of what each building is used for, the town protection class of each building (sprinkler system, distance from hydrants, response time based on proximity to fire department), and the type and proximity of neighboring businesses
- Equipment inventory: description, serial number, model number, and current replacement cost
- A schedule of all licensed vehicles: makes and models, what they cost new, current motor vehicle reports (MVR) if available (insurance companies generally can supply them free of charge), uses, and radius of operation
- A list of authorized drivers: full name, date of birth, and driver's license number and state of issue
- Payroll and sales estimates for the upcoming policy year for each operation (e.g., lawn care, mowing and maintenance, mower repairs, landscaping, snow removal, termite work)
- Brochures or other documentation that describes your company
- Copies of written safety programs and a description of safety devices in place
- A detailed claims report for each insurance policy held over the past five years, prepared by your current and past insurance carriers

Use the Outline to Evaluate Insurance Products

Send the outline to the agents on your list at least 60 days prior to the expiration date of existing coverage. Although agents may be generally knowledgeable of the pest management industry, there is no way they can know the specifics of your operations. Discuss your insurance needs with them and use your proposal outline or bid specifications to provide the details they need. The more the agents know about your situation, the better they can serve your needs (Fig. 14.23).

Conversations between you and the insurance agents you have selected should not be one-sided. You should provide as much information as possible on your business operations, and the agents should make you aware of all pertinent insurance coverage available in the marketplace. Ask the agents to evaluate your current coverage and offer insight into how you might decrease premiums (e.g., by reducing policy overlap or by increasing deductibles). Ask for an evaluation of your total insurance portfolio to make sure that there are no gaps in coverage; if gaps exist, additional coverage may be required. The agents

Figure 14.23 A tractor-mounted sprayer releases pesticide into a ditch bank following an accident.

should provide specific recommendations for improving your insurance program—not just duplicate current coverage.

Presentation of a Proposal

Agents are unable to provide formal premium quotes at your first meeting. They usually do not have underwriting authority and must submit paperwork to their company for risk evaluation. Their underwriters review the information and determine whether the insurance company is willing to insure your business and at what cost.

The underwriter reviews your documentation as submitted to him by the insurance agent. The underwriter will consider prior insurance claims, quality of management, type of operation performed, and business safeguards in place—safety plans, vehicle and physical inspections, sprinkler systems, fire extinguishers, training programs, etc.—in his final proposal to the agent.

The quotation process may take 30 days. Most proposals are typed outlines detailing the types of coverage offered: limits, deductibles, coverage, rates, and premiums. The agents should discuss with you the details of their proposals. For instance, he or she may point out that a property quote is based on a limit of \$200,000 on the building and \$50,000 on its contents, each with a \$1000 deductible.

The agents should reveal each insurer's financial rating. A.M. Best is one firm that rates the financial strength of insurance companies; try to purchase coverage with an A.M. Best rating of A− or better, and confirm that they are licensed in the states in which your business operates.

Ask the agents to deliver along with their proposals a specimen copy of the proposed General Liability policy, including any limitations or endorsements that will be attached to the policy at issuance. Insurance agents should explain all exclusions that are written into each policy and should answer any questions that you have. They should identify all coverage gaps that exist in the proposed insurance program and offer you secondary coverage, if available; whether or not to cover the gaps is up to you.

Require—not request —your agent to provide a proposal, including premium quotes, no later than 14 days prior to the expiration date of your existing policies. Some agents will agree to do so, then fail to deliver; they may ultimately provide a quote only one or two days prior to your policy expiration date so that you don't have adequate time to compare their coverage and premiums with those of other companies. Be fair with agents and respect the time that they devote to your insurance needs, but impress upon them that you require time to contemplate various companies' proposals and ask questions that they provoke. Make it clear that you will not accept bids submitted after the 14-day deadline. Reputable insurance agents will be happy to oblige.

You must compare quotes, but it is impossible to do so unless you provide each company with identical information. It is especially important to verify that all quotes are computed on the same premium base. Coverage such as General Liability and Worker's Compensation usually is based on estimated sales/receipts and payroll figures, respectively, that you provide. Obviously, agents using different payroll estimates will quote different premiums; so make sure that all quotes are based on the quote specifications. Review policies carefully and ask your agent if differences exist among the various companies' forms.

CHOOSING THE RIGHT INSURANCE POLICY

As a pesticide application business owner, you have a lot riding on the insurance coverage that you select. The difference between continuing in business and closing the doors after a fire, chemical spill, or vehicular accident may hinge on whether or not you have selected the right types and amounts of coverage. And to do that, you must identify your short- and long-term risks. Assessing your needs may take time that you don't feel you can spare. But businesses big and small must be protected against risk, and the first step is identification.

Taking the time to assess insurance needs may be easier for large businesses with supervisors, managers, and technicians who generate sales and maintain operations while the owner handles insurance and other corporate matters. Small companies often do not enjoy such an elaborate infrastructure, and the owners themselves may assume all roles: supervisor, manager, applicator, and secretary. They may be spread too thin to recognize the importance of setting aside time to research and select the right kinds of insurance. But with the business at stake, the time is well spent.

Consider the following points when deciding to purchase or update a commercial business insurance policy:

- Ask the agents to tailor coverage to your needs.
- Keep an open mind. Listen to your insurance agent's advice about the coverage you need, even if you feel that you don't need it or can't afford it.
- Review all of your options and costs.
- Ask for written proposals.
- Determine that coverage limits are adequate.

You may be surprised to learn that certain additional coverage may not increase your insurance premium substantially. So leave your options open: Ask to review and discuss all types of coverage, then pick and choose based on your needs. Ask if there are measures that you can take to decrease your premium (e.g., installing safety equipment).

Your Decision to Purchase

Don't let price or coverage alone influence your decision. Give some thought to the services offered by insurers: customer education, claims processing, and risk management. Factor into your decision the financial stability and reputation of the insurance companies whose quotes you are considering. See Appendix 14.1 for more information on what should be reviewed.

The Written Binder

The insurance agent confirms with the insured the types and amounts of coverage purchased; any coverage declined is marked on the original written proposal and initialed by the insured. The initialed proposal, then, indicates the entire realm of coverage offered and the exact coverage declined. It documents that the agent recommended certain coverage that the insured decided not to purchase. This is especially important to the agent in the event of errors and omissions issues.

Policyholders should always require a written binder, effective and delivered prior to the expiration date of existing coverage. The initialed proposal forms the basis for the written binder—a single-page document, usually—stating the types and amounts of insurance purchased, the inception date, the policy period, etc. The binder is considered a legal document; both the insurance agent and the policyholder should have a copy on file.

The agent signs the binder and presents it to the policyholder (in exchange for a premium deposit) as proof of insurance until the actual detailed policy can be written; the policyholder signs it, as well, binding himself to the contract. Always require the agent to deliver the issued policy within 30 days of inception.

Instruct your agent to send certificates of insurance directly to state agencies that require them; provide a list of other entities that require certificates of insurance and ask your agent to forward those as well. Request copies of the agent's correspondence with all recipients. Maintain an organized file of all insurance transactions.

Reviewing the Insurance Policy

Ask your agent to review your policy with you when he or she delivers it. Check it for accuracy. Look and listen for discrepancies between the limits, deductibles, types of coverage, premiums, rates, sales and payroll estimates, etc., and those originally proposed. Keep a copy of the policy on file at the office so that it is handy for reviewing coverage as your situation changes (e.g., if you purchase additional equipment or if your Worker's Compensation rates and classes are updated). Maintain the original policy off-site so that it cannot be destroyed if there is a fire or other mishap on the business premises.

The Legal Perspective—Claims

Most insurance claims are handled amicably, that is, the insurance company, the agent, and the policyholder work together to verify loss and determine the appropriate claim and payment, and the insurance company pays for the loss. However, some insurance claims do not progress so easily. The insurance company may deny or contest a claim based on their interpretation of the policy provision or exclusion (Fig. 14.24).

Policy provisions can be ambiguous, leaving room for debate on the scope of coverage. It is common for insurance companies and policyholders to take opposing views on the

Figure 14.24 Sometimes courts are used to settle disputes between insurance companies and their clients.

definition of a term or phrase. For example, consider an intentional conduct exclusion. The insurance company may take the position that a claim resulted from the unexpected consequences of intentional conduct and therefore is not covered; the policyholder may view the consequences accidental, thereby expecting payment of the claim. Insurance policy interpretation usually is a matter of state law, and it varies radically from state to state.

Ask your agent for assistance in contacting the insurance company, directly, for written interpretation of your coverage; or consider having your policy reviewed by an attorney or an outside consultant who represents policyholders in insurance coverage disputes.

Lawyers view insurance policies from a different perspective than do insurance companies. Let your attorney know, up front, how much time and money you can allocate to the review. Stipulate questions of critical importance for which you need answers in writing. Following are some examples:

- How do the courts in this state construe standard policy provisions with regard to the specific risks my business faces?
- How might exclusions affect my claims?
- Will my policy cover chemical pollution as a result of a warehouse fire? If so, are the limits sufficient to cover judgments and remediation expenses?
- Will my vehicle policy cover claims arising from an accident in which chemicals are released into storm drains or creeks? If so, are the limits sufficient to cover judgments and the cost of remediation?
- Will I be covered if a customer claims exposure to pesticides?

Annual Update and Review

A good agent and a sharp business owner begin reviewing existing coverage 60–90 days before its expiration. Update the insurance specification list to correspond to changes involving vehicles, gross receipts, payroll, sales, building and equipment values, etc. Make the agent aware of any new services or operations that you now perform, such as snow removal, aquatic pest management, fumigation, etc. Discuss any type of exposure that is new since the inception of, or excluded from, your existing coverage.

Reevaluate annually to determine what your business has to lose and what risks and liabilities should be covered. For example, if the building in which you operate cost $100,000 to build in 1985, today's replacement cost might be $150,000 or more. Working at least 14 days prior to the expiration of the existing policy allows your agent to solicit quotes on the updated information and to write a proposal for discussion.

Retaining Insurance Policies

Liability policies should be kept permanently—even if the business is sold. Coverage that you had years ago might apply to a claim based on something that occurred during the active policy period even if the claim arises years after the policy has expired. Environmental contamination and the long-term effects of an employee's exposure to a chemical, as an example, may not be discovered for years—even decades. An "occurrence" policy still applies if the claim arises from an occurrence covered during the policy period. Documentation of past coverage can be difficult if the policy has been discarded.

Different types of coverage can be included in a single business policy and can include policy definitions or other information which can be useful in determining occurrence-based coverage. Therefore, the retention of all insurance policies, liability and otherwise, is a prudent business practice. Dedicate a space in your office for filing copies of all insurance information; maintain the originals off-site.

Staying with the Insurance Company

As the policyholder, consider staying with the same insurance company for as long as it can meet your needs. As with most business relationships, the longer the affiliation, the stronger the loyalty. Insurance companies are more likely to pay questionable claims to a loyal client than to a newcomer. And it is less likely that long-term, profitable clients' coverage will be canceled due to a large claim.

THE CLAIM: WHO DOES WHAT, WHEN, AND WHERE

You buy insurance for the claims coverage that it provides, so the coverage—not the price—should be most important to you at the time of purchase. Determine the expertise of your agent and investigate the claims experience of the insurer for the types of loss your business might suffer.

Addressing complaints does not guarantee that everyone will walk away happy, but a quick response often prevents the problem from getting worse. Ignoring complaints or failing to address claims against you or your business can make

- small problems large,
- minor claims major, and
- out-of-court settlements nearly impossible.

Allowing a complaint to reach the courts can lead to lengthy hearings, increased litigation expenses, and contentious bickering. The history lesson is simple: Missed opportunities to resolve claims promptly often lead to more expensive resolutions.

The Policyholder: Reporting and Participating

Common policy conditions place the responsibility on the policyholder to immediately report claims to the insurance company or its agent. All insurance policies include a requirement for prompt notification; and, in some cases, a delay can cause an otherwise-covered claim to be denied. Sometimes there is litigation between the policyholder and the insurance company on the issue of late notice. But even if the insurance company and the policyholder settle differences, or if a court determines that coverage exists, there may be a delay of months or years before the insurance company pays the policyholder or claimant. Carefully review the notice provision in your policy to be sure that notice is given properly and promptly.

In addition to reporting and notification requirements, the policyholder also is required to

- notify the police when a crime is committed;
- provide a detailed explanation of the accident, injury, or loss;
- furnish the names and addresses of injured persons;
- produce a list of witnesses and their addresses;
- send to the insurance company copies of requests, demands, or summonses concerning the claim;
- cooperate with the insurance company in its investigation, settlement, and defense of the claim;
- give signed statements under oath;
- authorize the release of medical records to the insurance carrier; and
- submit injured parties to medical examination by physicians selected by the insurance carrier.

Agreeing to Fault The policyholder should never agree to or make an offer for settlement of a claim. Most policies specifically exclude coverage for voluntary payments made by the insured. A common rule of thumb is for the policyholder to inform the claimant that all pertinent information will be turned over to the insurance company for review. Visit the site of contention and write a detailed report, substantiated with pictures, to provide the insurance adjuster. You may tell the claimant to call you if the insurance company has not contacted him within three days; this will allow you time to speak to or meet with your agent to discuss details of the claim. Once you have done so, it is the responsibility of the insurance company to handle, settle, and pay the claim.

Many policyholders believe that paying their own claims helps to keep premiums low; the truth is, a few minor claims do not have an adverse impact on your premium. Insurance companies expect to be notified of all claims: Policy language requires it. Even if the claim is equal to or less than your deductible, you are required to report it promptly. Failing to do so may jeopardize coverage if the small claim turns into a big one. Let the insurance adjuster do his job. Timely notification allows him to respond quickly to contain the loss. It is his responsibility to satisfy the claim in such a manner as to guard against future demands from the claimant.

The Agent: Collector of Information

Most insurance claims are submitted directly to the agent who services the policy. The agent's role is to get as much pertinent information as possible, organize the facts into a written claim report, and send the material to the insurance carrier. Agents may wish to view the scene of serious or substantial accidents; firsthand observation helps in communicating the claim details to the insurance company. An agent should ask the following questions about a vehicle claim:

- When and where did the accident occur?
- What vehicle was involved?
- Who was driving?
- Why was that person driving?
- What is the extent of bodily injuries, if any? Who was injured?
- Is the vehicle damaged? If so, what portions of it?
- What is the estimated cost of repair?
- What are the details surrounding the accident?
- Is there a police report? If so, what is the name of the police officer, the jurisdiction, and where a copy of the report can be obtained?
- Were there any witnesses? If so, can you provide names, addresses, and telephone numbers?

The agent is under the same constraints as the policyholder as far as making commitments to settle a claim. Although he or she may investigate and document a claim, he or she usually is not in a position to judge who is at fault or the dollar amount of damages.

Independent agents represent the insurance company but are not company employees. Even agents directly employed by the insurance company are not authorized to pay or settle claims; that is the job of the company's claims department.

A good agent should help you in the handling of a claim even if it doesn't involve his policy. In cases where you are not at fault and your own policies are not involved, your agent should help you deal with the other party's carrier to arrive at a fair settlement.

The Insurance Company: Investigate and Reimburse

Claims are forwarded to an adjuster who decides whether the policy covers the claim and, if so, how much will be paid.

If coverage applies, the claims adjuster may request additional records (e.g., service agreements, service slips, chemical application logs, customer contracts) or direct more detailed questions to the policyholder. Most of the initial work by the adjuster can be conducted over the telephone. One common method of documenting questions and answers is to record conversations between the adjuster and the policyholder and other persons linked to the case.

Adjusters in many cases can offer a settlement within a few days on first-party claims involving only the insurance company and its policyholder. The resolution of a claim on third-party legal liability where there is a question whether coverage applies, or where a substantial amount of money is involved, takes longer. In such cases, adjusters or their representatives personally visit the affected person or property to examine and document the seriousness of the claim; they may conduct interviews with involved parties and document injuries and damages via photographs and videotape.

Claims adjusters also may ask independent, professional appraisers for estimates of loss. For instance, if an orchard is damaged by spray drift, how much is the loss worth? If a fire destroys a building, how much will it cost to repair or replace it? The appraiser's estimate, policy provisions, and the claims file are used by the adjuster to estimate a settlement.

Disagreement between the Policyholder and the Insurance Carrier

In most cases, the insurance company, the agent, and the policyholder work together to verify a loss and come to agreement on the settlement, and the insurance company then pays the claim. But some claims are not settled so easily.

An insurance company's initial denial may be incorrect, so do not hesitate to press a claim. You may disagree with the adjuster's interpretation of your policy or with the amount of insured loss he assigns. The disagreement may prompt you to hire your own insurance attorney and/or appraiser to negotiate and/or litigate differences. Unresolved minor issues may be decided in small claims court or through arbitration; to the extreme, a bad-faith denial of coverage can subject the insurance company to the payment of punitive damages.

Some aspects of business insurance are more likely than others to result in controversy and litigation. Denials of environmental claims, for example, are litigated routinely nationwide. Courts in some jurisdictions find coverage for environmental claims, pursuant to the terms of standard comprehensive General Liability policies, even when the policy contains a Pollution exclusion. Some courts find coverage for environmental or chemical-related losses under first-party property insurance policies.

Insurance law varies from state to state, so it is impossible to generalize statements on policy provisions. Policy language is fairly standard, even in policies issued by different insurance companies; but, even so, most states review general statements and request changes based on state laws.

Third-Party Litigation and Lawsuits

If the insurance company has assumed the policyholder's defense, the claim may be settled—even if the policyholder wants to fight it—because it is cheaper than hiring an attorney and going to court. Consider, for instance, a case where a person asks for $1000 for damages to her garden and trees, allegedly caused by pesticide drift from an adjacent property. The pesticide applicator wants the insurance company to fight the claim, tooth and nail, because he believes his chemicals did not drift. An attorney will take the case, estimating that litigation will cost $7500. So the insurance company must decide whether to spend $1000 to close the case—even if its client is innocent—or spend $7500 in attorney fees, court costs, etc. It's clearly cheaper to pay the $1000 and settle!

Any admission of liability may be harmful to the policyholder's business reputation and may invite subsequent claims. Therefore, it is important that your opinions regarding claims be clearly communicated to the insurer. The decision, however, generally rests with the carrier. Adjusters routinely require a claimant's signature after settlement to show that "full and final" payment has been made.

There are numerous examples where both sides—the insurance company and the third party—just cannot reach a settlement, where negotiations reach an impasse and all discussions seem counterproductive. Litigation may be the only remedy; yet, even as the process begins, each party tries to find middle ground on which to resolve their differences. If all efforts fail, it's left to the judicial system.

AUDITS AND SURVEYS: PAYROLL, SALES, AND SAFETY

The Insurance Audit

Appropriate pricing of insurance coverage is a challenge to the insurance carrier. Unlike other businesses, insurance companies set prices for their products without knowing what claims they ultimately might have to cover. Actuarial studies help anticipate losses due to various exposures, thereby serving as a starting point in determining rates. Rates are established for the various unknown exposures and adjusted based on limits purchased, deductibles, perils covered, etc.

Rates for property coverage are based on characteristics specific to the property item. Pricing for General Liability and Worker's Compensation insurance is more difficult because exposures cannot be exactly predicted. Exposures, or premium bases, usually are computed on sales or payroll projected for the upcoming policy year. Insurance companies price these policies on the basis of assuming more or less risk when sales increase or decrease; more sales might mean more customers, therefore more concern for General Liability claims. Employers may have to hire additional staff to accommodate the increased workload, fueling concern for Worker's Compensation claims and forcing premiums up (Fig. 14.25).

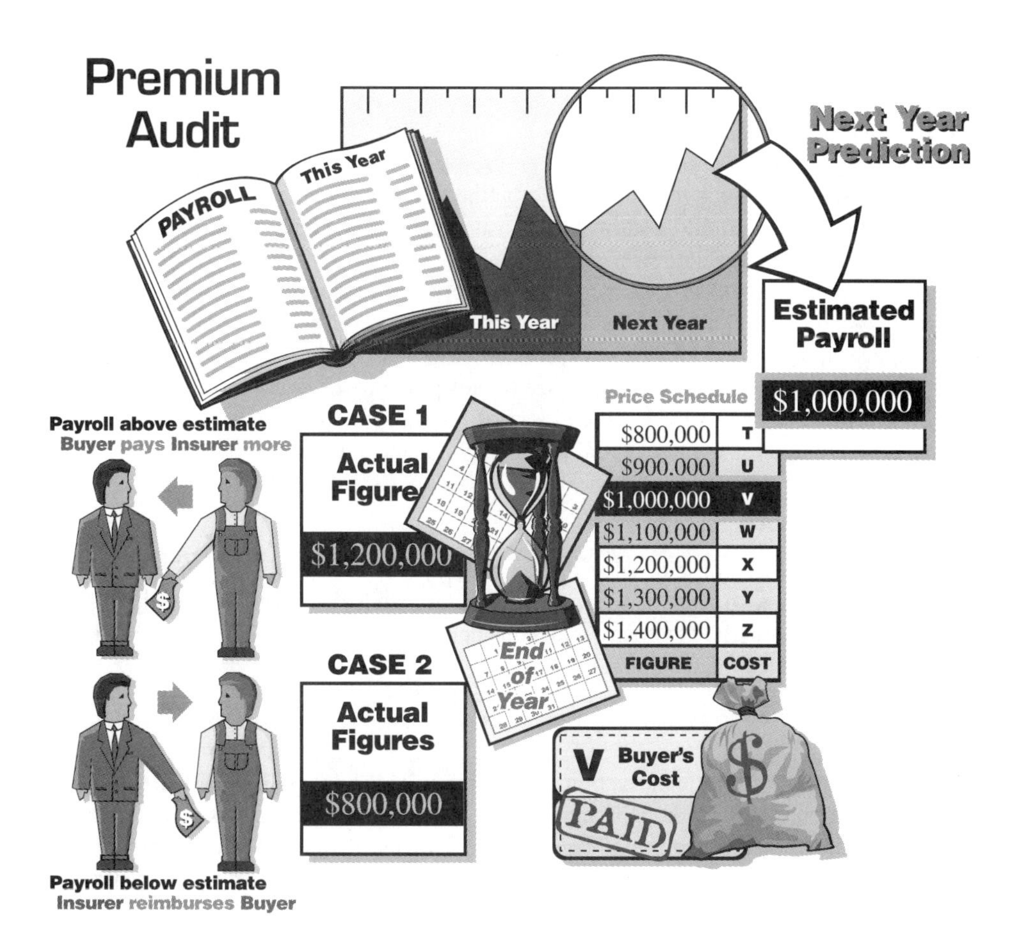

Figure 14.25 The premiums of many policies are based on sales and/or payrolls.

The insurance audit adjusts the differences between projected and actual payroll and sales. It ensures that the final earned policy premium is truly reflective of the intangibles insured, that is, the coverage provided. Insurance audits can be conducted voluntarily by the insured or on-site by the insurer.

A voluntary audit is less expensive but riskier to the insurance company. The policyholder is asked, at the end of the policy period, to complete and return an audit form detailing truthfully and accurately the final sales and payroll.

The on-site audit—more expensive to the insurance company, but less risky—is conducted when the insurance auditor makes an on-site visit to personally inspect the policyholder's actual payroll and sales records.

Loss Control Surveys

Some insurance companies send safety specialists to evaluate the day-to-day activities of the businesses they insure. They observe how work is performed on the premises and at job sites and complete a loss control survey.

A good survey evaluates all aspects of the business operations (e.g., how chemicals are stored, how internal training is conducted, how work and office areas are maintained). Loss control surveys are designed to offer the policyholder suggestions to help prevent claims. Businesses which fail to implement reasonable and standard procedures suggested by loss control specialists may face nonrenewal, restricted coverage, or higher premiums.

RISK MANAGEMENT

The purchase of insurance may not always be the best way to mitigate hazards. There are alternatives that may modify your need for insurance or reduce the risk of business assets being impaired. This is called risk management.

The risk management process begins with the identification of risks inherent to your business. As an example, consider a hypothetical lawn care business and its associated risks: physical damage to the business property, lawsuits, and loss of income. Explore the process necessary to eliminate or minimize threats posed by its operations.

Once ABC Company has identified its areas of risk, prudent economic solutions must be exercised to minimize the potential for negative financial impact.

Risk Assumption

Risk assumption is the conscious decision to assume some or all of the risk. In other words, it may be determined that the benefits of a particular activity or situation outweigh its risk potential; therefore, ABC Company may consider the risk worth taking and be willing to absorb any associated financial loss.

Risk Avoidance

The act of avoiding a particular activity or operation that poses risk is called risk avoidance. For example, ABC Company may believe that lawn mowing poses significant risk: Employees are vulnerable to injury, equipment is expensive and difficult to maintain, and the cost of labor is high; also, there is potential for third-party injury or property damage. Consequently, insurance premiums are high. ABC can avoid these risks—and therefore the cost of insurance—by not offering lawn mowing services.

Loss Prevention

A business owner may assume a risk but implement loss prevention measures, both physical and procedural. In our example, ABC Company might incorporate loss prevention procedures in an attempt to make mowing a less hazardous operation. Employees would be properly trained to use the equipment (fewer injuries); a regular equipment maintenance schedule would be implemented (longer serviceable life); and equipment would be secured, inside, at night—or locked, if left at the job site (fewer incidents of theft). Loss prevention measures afford the business better control over sustainable loss.

Loss Reduction

Loss reduction means simply to minimize loss. ABC Company may choose to transport chemicals in a 200-gallon tank instead of a 600-gallon tank. The smaller load would be easier to contain than the larger load and possibly less costly to clean up, should a spill occur. Such measures also improve ABC Company's defense against negligence claims if an accident occurs.

Risk Transfer

Shifting the consequences of loss from one entity to another is called risk transfer. The classic example is insurance, where ABC pays a premium to relegate its risk to the insurance company. Other methods of risk transfer may include the use of "additional insured" endorsements, hold harmless agreements, and other contractual methods to place responsibility with other parties.

A combination of risk management solutions may be employed. Let's say that ABC Company decides not to conduct lawn mowing operations; instead, they will subcontract the work. In doing so, ABC transfers to the subcontractor the risk of employee injury and the cost of mowing equipment, its maintenance, and insurance against theft. Although third-party liability might still apply, the subcontract shifts primary responsibility to the company that conducts the mowing operation. The transfer is validated by the subcontractor's certificate of insurance in evidence of its General Liability, on which ABC Company is named as an additional insured, and its Worker's Compensation coverage.

Through the use of multiple risk management techniques, ABC adequately addresses the hazards associated with lawn mowing, while increasing revenue and maintaining its image as a full-service operation. Risk management is an ongoing process of recognizing loss potential and implementing solutions to minimize or eliminate risk.

SUMMARY

Your first and best line of defense against claims is careful attention to minimize risk. Insurance should never be viewed as a substitute for risk management: It does not reduce the occurrence or dollar amount of claims against you; it merely pays for them. Risk management addresses all facets of an operation, whether it involves writing a better customer contract; providing safer work areas; conducting employee education programs; inspecting service vehicles; posting signs, placards, and posters; conducting preemployment physicals; or requiring drug or alcohol testing. Companies that train and expect their employees to perform their work in the safest way possible take major strides in reducing claims (Fig. 14.26).

Figure 14.26 Getting insurance coverage designed for your needs is critical in protecting a business against accidents.

Even safety-conscious business owners incur claims. And because even the best-run businesses are susceptible to situations that result in claims, it is prudent to transfer a portion of the risk to an insurance company. Insurance decisions never should be based solely on price. As owner of a company that conducts pesticide application services, you know very well that the lowest bid does not necessarily mean the best value. Price is just one component of insurance; others are agent and company services, financial stability of the insurance company, and details of the proposed coverage.

When selecting an insurance agent, choose a person that you can work with effectively: someone you can trust; someone with the expertise and experience to advise you and to handle any claims or audits that might arise. Select someone who exudes confidence. If the agent is successful in gaining your trust, chances are that he or she will have a similar impact on insurers, claims adjusters, and auditors—on your behalf.

Select agents, insurers, and attorneys with experience in your industry. They should be familiar with the kinds of risk involved and the challenges that you face.

Never assume that the insurance protection that you have purchased conforms exactly to your needs. Take time to read the policy and to compile questions to ask your agent. Ask questions about claim scenarios (Would I be covered if. . .?), and insist on answers in writing. Stay informed of legal controversies regarding the meaning and scope of standard insurance provisions and exclusions.

You have a lot at stake in deciding what insurance to purchase, and from whom, so participate in the process: It's your business! Take time to listen to your agent, and be prepared to make important coverage decisions up front. Once a claim is filed against you, it's too late to make sure it's covered!

APPENDIX 14.1 AN INSURANCE POLICY CHECKSHEET

When your insurance agent delivers your policy, make sure that its specifications match the application you filled out as well as those in the agent's proposal. The following checklist is helpful in checking key points.

Declarations

Yes No Is the policy term (beginning date/ending date) correct?
Yes No Is the name of the policyholder spelled correctly?
Yes No Is the name on the policy exactly the same as the legal name of the business?
Yes No If you own, rent, or lease at more than one location, are the names and addresses of all branches and locations spelled correctly on the policy?
Yes No Is the type of legal entity (e.g., sole proprietor, partnership, or corporation) stated correctly?
Yes No Are the types of coverage agreed to in the agent's proposal indicated on the policy?
Yes No Are the premiums exactly those quoted by your agent during policy procurement?
Yes No Do you feel comfortable that the limits per occurrence will cover your worst-case scenario?
Yes No Are all of the endorsements listed on the Declarations page attached to the policy?
Yes No Do the types and amounts of coverage meet the requirements of your customers and state agencies?

Property

Yes No Do you have "special form" perils covered?
Yes No Is your property insured to its current value? At what percentage?
Yes No Does your policy cover replacement cost instead of actual cash value?

Inland Marine

Yes No Do you have an Inland Marine Floater for equipment leaving the business premises?
Yes No Do you have an Inland Marine Floater for computer equipment?
Yes No Is every piece of equipment (that you want to insure) scheduled on the Equipment Floater?
Yes No Are equipment values accurate and up to date?
Yes No Have you purchased replacement-cost coverage for your equipment and tools?

Crime Coverage

Yes No Do you have Employee Dishonesty coverage?
Yes No Do you have insurance to cover the theft of money and securities, both on and off your property?
Yes No Do you have a qualified pension plan such as a 401K? If so, is your Employee Dishonesty coverage adequate to meet the legal requirements?

General Liability

Yes No If you damage the property that you are working on, will your insurance cover it? If so, how much will it pay? If not, is there coverage available that will compensate for the Care, Custody, or Control exclusion?

Yes No Are your gross receipts and payroll stated correctly?

Yes No Is yours an Occurrence Form policy?

Yes No Are your deductibles per occurrence rather than per claim? For instance, if 25 people in an office complex say they were sickened by your application and each went to see his or her physician, you would face 25 claims; in this situation, a per claim deductible would apply for each of those 25 claims. In the same situation, a per occurrence deductible would apply only once.

Yes No Are your gross receipts within 10% of what is reasonably expected? Check this carefully because an audit that reveals higher gross receipts would indicate that you had underpaid the previous policy. You would be required to pay the premium audit bill along with higher premiums to extend the coverage for a year.

Yes No Is an Umbrella policy available? At what price?

Vehicle

Yes No Do the serial numbers for each vehicle appear exactly correct on the vehicle schedule page?

Yes No Do you have a policy that automatically provides coverage for new vehicles as they are added to your business?

Yes No Do you really want comprehensive and collision coverage for all vehicles owned by the business (symbol 2 in the vehicle policy)? Older vehicles may not be worth insuring for collision.

Yes No Are the gross vehicle weights listed correctly? The heavier the truck, the more damage it creates when it is involved in an accident.

Yes No Are the use designations—commercial, service, personal—correct?

Yes No Is the work area radius correct for each vehicle? These are designated as local (within 50 miles), intermediate (50–200 miles), and long haul (greater than 200 miles).

Yes No Have you included the value of permanently attached equipment as part of the vehicle value?

Worker's Compensation

Yes No Are the payroll estimates within 10% of the projections?

Yes No Are the job classifications correct?

Yes No Is your experience modification factor correct?

Exclusions

Yes No Does an exclusion take away coverage for a substantial risk that your business faces? If so, contact the insurance agent to discuss whether an endorsement is available, affordable, and/or appropriate.

Endorsements

Yes No Do you understand endorsement changes? If not, contact your insurance agent for clarification.

Insurance Services Office (ISO) Forms

Yes No Are any forms used in your policies not copyrighted by ISO? If yes, ask the agent for specifics on why the insurer used their own forms in lieu of the industry-standard ISO forms.

CHAPTER 15

EDUCATING THE COMMUNITY AND THE WORKFORCE ABOUT HAZARDOUS CHEMICALS

Imagine a fire in a pesticide warehouse on the edge of town. The first emergency units to arrive at the scene of a fire—police, medical professionals, and firefighters—must react quickly. A portion of the structure is engulfed in flames. Billowing smoke drifts into surrounding neighborhoods. But because the firefighters are aware of the types, amounts, and exact locations of pesticides stored in the warehouse, they are able to quickly and safely extinguish the fire. Their prompt and effective response eliminates the need to evacuate the surrounding area, limits damage, and reduces environmental pollution. This is obviously the desired result in emergency situations associated with fires, spills, or releases involving hazardous chemicals (Figs. 15.1 and 15.2).

Unfortunately, emergency personnel often are expected to battle fires and react to spills and other chemical emergencies without knowledge of the substances involved, their quantities, their potential impact on health and the environment, or their exact locations. Responders may be forced to devote valuable time to asking questions, gathering facts, and making decisions on how to react to the emergency. Given limited time to determine the level of danger, often the course of action is to evacuate the area, ensure the safety of those responding to the emergency, and minimize adverse environmental effects by simply containing the fire and allowing it to burn itself out. Communities seldom expect individuals battling blazes, spills, or other accidents to perform heroic acts when the hazards and potential dangers are unknown. Not knowing what chemicals are involved in these situations can lead to loss of life, serious injury, damage to the environment, and the destruction of valuable records, equipment, vehicles, and structures.

CHEMICALS IN THE COMMUNITY

The federal Emergency Planning and Community Right-to-Know Act (EPCRA) enacted in 1986 requires businesses (including farms) that store certain quantities of hazardous substances to report to local and state agencies. In turn, these agencies coordinate emergency

Figure 15.1 Emergency responders can best deal with emergencies when they are aware of what chemicals are stored inside a building.

response plans for implementation in the event of fires, spills, accidents, and acts of vandalism. They also distribute information about stored hazardous substances, available upon request by individuals or public organizations.

Linking Federal and State Right-to-Know Programs

The Environmental Protection Agency (EPA) is responsible for administering the Emergency Planning and Community Right-to-Know Act at the federal level through oversight of and coordination with state EPCRA programs. The federal oversight task has been assigned to the EPA's Office of Chemical Emergency Preparedness and Prevention. The exchange

Figure 15.2 Pesticide distributors inform their local fire departments of the large quantities of pesticides stored in their warehouses, which allows the local responders to preplan for an emergency at the facility.

Figure 15.3 Rural volunteer fire departments play an important role in protecting property.

of Community Right-to-Know information is greatly enhanced by an EPA 24-hr, toll-free hotline (800-535-0202) that provides answers to technical and regulatory questions. The EPA also provides technical assistance by offering training to states, local communities, and affected industries.

Planning for chemical emergencies is one of the central themes of EPCRA. The governor of each state is responsible for appointing a State Emergency Response Commission (SERC). Representatives serving on the state commission typically represent a diverse cross section of persons involved with issues such as environmental protection, emergency response, health care, transportation, commerce, emergency training and planning, and education. SERC's primary functions are to assist local communities' development of emergency plans, to review those plans annually, and to coordinate Community Right-to-Know and data collection activities at the local level.

Risk Management at the Local Level

The success of the Right to Know program hinges on the effectiveness of the Local Emergency Planning Committee (LEPC). Each LEPC is assigned an emergency planning district by SERC. The local committee draws on the expertise and experience of individuals with diverse backgrounds: state and local elected officials; law enforcement, emergency management, and firefighting personnel; first aid, health, and hospital workers; transportation experts; journalists; educators, community and environmental groups; and representatives of EPCRA-regulated facilities. Typically, each committee consists of 15–20 individuals appointed by the state commission. The administration of local programs generally is assigned to existing local units of government which, in turn, receive assistance from committee members who volunteer their time (EPCRA Section 301) (Fig. 15.3).

LEPC is assigned responsibility for understanding and managing potential risks posed by chemicals in the community. This responsibility includes the development of an emergency and chemical risk management plan (EPCRA Section 303). The local emergency response plan is prepared using the chemical information submitted to LEPC by local industries, businesses, and farms. Compilation of the submitted information enables the committee to conduct a community-wide analysis of the location and types of chemicals stored, as well as vulnerable areas and populations at risk. These analyses form the basis for short- and long-term emergency response plans. Successful management of hazardous chemical release is achieved by LEPC in harmony with hazardous materials teams, local firefighters, emergency management personnel, medical professionals, and the regulated community; together, they prepare for and respond to chemical emergencies.

Elements of a Local Emergency Plan

- Utilizes information provided by industry and farms to identify facilities where hazardous substances are present and to identify transportation routes leading to and from such sites.
- Establishes evacuation plans and alternate traffic routes.
- Establishes emergency response procedures, including evacuation plans, for dealing with accidental chemical releases.
- Designates a Community Emergency Coordinator and sets up notification procedures for those who will respond in an emergency.
- Establishes methods for determining the occurrence and severity of a chemical release and the areas with populations likely to be affected.
- Establishes ways to notify the public of a chemical release.
- Identifies emergency equipment available in the community.
- Establishes a program and schedule for training local emergency response and medical workers to respond to chemical emergencies.
- Establishes methods and schedules for conducting simulation exercises to test the emergency response plan.

The LEPC is responsible for disseminating information regarding the identity, quantity, location, and hazardous properties of chemicals stored in or transported through the community. Residents may request such information from their Local Emergency Planning Committee, exercising their right to know; thus, valuable information on the types of chemicals stored and used within their emergency planning district is readily available. This provides greater assurance that the community can be protected in the event of a chemical emergency.

Community Awareness

Identifying and communicating the existence of hazardous substances within the community is vitally important and lays the groundwork for emergency planning and preparedness. Community planners who can anticipate and are prepared to respond to emergencies provide their citizenry maximum health and property protection.

Extremely Hazardous Substances The potential hazards of chemicals—relative to toxicity, reactivity, volatility, combustibility, and flammability—have been evaluated by the EPA. Chemicals which pose the most serious hazard potential during release are found on the list of Extremely Hazardous Substances (EPCRA, Section 302, 40 CFR Part 355, Appendixes A and B). The Extremely Hazardous Substances (EHS) list consists of 360 substances, including pesticides. The current listing, Title III List of Lists, is available

TABLE 15.1 Threshold Planning Quantities for Some Chemicals

CAS Number	Chemical Name	Threshold Planning Quantity (TPQ) Listed in Title III List of Lists
116-06-3	Aldicarb	100
94-75-7	2,4-D acid	—
13194-48-4	Ethoprophos	1000
1910-42-5	Paraquat	10
7664-41-7	Ammonia (anhydrous)	500
71-43-2	Benzene	—

through your LEPC or SERC or through the EPA hotline in Washington, D.C. (800-535-0202). It is also found at http://www.epa.gov/cepppo/pubs/title3.pdf.

Threshold Planning Quantities The EPCRA list of Extremely Hazardous Substances does more than identify specific hazardous chemicals: it includes information on the Threshold Planning Quantity (TPQ), which is a reportable storage quantity for each listed chemical.

The easiest method to determine the TPQ for a specific chemical is to locate the Chemical Abstract Service (CAS) number either on its Material Safety Data Sheet or under the alphabetical listing of the chemical name in the Title III List of Lists.

The TPQ for each EHS chemical is 500 lb or the amount listed, whichever is lower. An EPA administrative ruling mandates that any TPQ above 500 must be reduced to the 500-lb TPQ. For example, the following table lists the TPQ for ethoprophos as 1000 (Table 15.1), but due to the administrative ruling it would be changed to 500. If a threshold number is not assigned to a pesticide on the EHS list, a TPQ of 500 lb may be assumed. For example, a TPQ is not listed for 2,4-D acid; therefore, the reportable quantity is 500 lb. In addition, all hazardous substances not on the EHS list, but for which Material Safety Data Sheets (MSDSs) are required under OSHA's Hazard Communication Standard, must be reported under EPCRA Sections 311 and 312 if they are present at the facility. All such hazardous substances have a TPQ of 10,000 lb.

Facilities that manufacture, store, or use extremely hazardous substances must be aware of the assigned TPQ (EPCRA, Section 302). If the threshold planning quantity of a listed substance is present at a facility at any time, it must be reported to the State Emergency Response Commission, the Local Emergency Planning Committee, and the fire department. The report is due within 60 days of the first day of storage of the TPQ at the site.

Determining the Amounts of EHS On-Site It is important to maintain accurate records of all pesticides, fertilizers, and other chemicals brought on-site. The resulting list of products can be checked against the Extremely Hazardous Substances list to identify those present at any time in quantities which must be reported. Chemicals present in quantities less than the corresponding TPQs need not be reported. To make this determination, review the records or your current chemical and fertilizer inventory and develop a list of all of the materials you have had in stock at any time during the previous calendar year. Compare what has been stored or used on your property with those chemicals listed on the EHS list.

Even though reporting may not be required under EPCRA, local emergency units such as fire departments should be informed of chemical inventories and storage locations. It is advisable to include a map or diagram showing where chemicals are stored. A reporting form and a descriptive list of the chemicals are advised, as is information on *when* they are on-site. Good communication and advance planning are the best defense against emergencies involving hazardous chemicals.

Pesticides Containing EHSs If a pesticide used or stored at any time appears on the EHS list, it must be determined if the amount (by weight) meets or exceeds the TPQ assigned to that product. If the pesticide product is on the EHS list, the entire product is reportable. If only an ingredient of the pesticide product is listed, only the quantity of that specific EHS ingredient is reportable. Remember that reporting requirements are based on the total amount (above the TPQ) of a given chemical that is stored at the facility, regardless of location, number of containers, method of storage, or how it is packaged—including differences in formulation.

The following worksheet can be used to calculate the total amount of active ingredient in EHS products at a given facility.

Step 1

Amount of total formulation: Amount of the ingredient in formulation:

total dry weight of product ____ × the percentage of the ingredient ____.

or

total gallons of product ____ × the total lb of the ingredient per gal. ____.

Step 2

Multiply figures in Step 1 to establish the amount of Extremely Hazardous Substance: ____ (lb).

Step 3

Repeat Steps 1 and 2 for other formulations with the same active ingredient.

Step 4

Add the weight of the ingredient in each formulation.

____ + ____ + ____ = ____ (lb)

Step 5

For each active ingredient calculated in Steps 1–4, fill in the Threshold Planning Quantity

(Table 15.1) ____ (Threshold Planning Quantity)

Step 6

If Step 4 is equal to or greater than Step 5, you must report that you have an Extremely Hazardous Substance on your premises. If not, you are not obligated to report.

Example: Solid Formulation The active ingredient in Counter 15G insecticide is terbufos. Terbufos makes up 15% (by weight) of the product and is found on the EPA list of Extremely Hazardous Substances. If a storage facility contains thirty, 50-lb bags of this insecticide, the total weight of the stored Counter 15G is 30 × 50 lb, or 1500 lb. Fifteen percent of the 1500 lb equals 225 lb of terbufos. The 225 lb of terbufos is above the 100-lb Threshold Planning Quantity and therefore must be reported.

Example: Liquid Formulation The active ingredient in Dyfonate 4-EC is fonofos, which is on the EPA list of Extremely Hazardous Substances. There are twenty-five, 5-gallon containers at the storage site, for a total volume of 125 gallons. There are 4 lb of fonofos per gallon of Dyfonate 4-EC, so the total weight of fonofos contained in the stored product is

4 lb × 125 gallons, or 500 lb. The TPQ for fonofos is 500 lb, so the on-site quantity must be reported.

Example: Fertilizer Anhydrous ammonia is listed as an Extremely Hazardous Substance, with a TPQ of 500 lb. Therefore, even a single, 850-gallon nurse tank must be reported. At 60°F, anhydrous ammonia weighs 5.2 lb/gallon. The calculation is 850 gallons (total stored) times 5.2 lb of product, for 4420 lb of anhydrous. The TPQ is exceeded; therefore reporting is required.

Notification and Community Planning for Emergencies An initial one-time letter of notification is required of all facilities (e.g., commercial dealers, golf courses, lawn care companies, greenhouses, or farms) whose storage or possession of an EHS meets or exceeds its Threshold Planning Quantity. Although it is only a one-time requirement, a facility's EHS inventories should be reviewed frequently to account for product changes and fluctuations in quantities stored. Owners and operators are required to report that their facilities are subject to the emergency planning requirements of EPCRA Section 302; they must notify the SERC and the LEPC within 60 days of first meeting the TPQ for an EHS. The notification requirement does not require that the actual names and amounts of pesticides be reported, but the LEPC can legally request this information to develop or implement the local emergency plan. Someone at the facility will need to be assigned the responsibility of serving as Facility Emergency Coordinator to work with the LEPC. When a new entry is added to the EPA's EHS list, facilities that store the new EHS at or above its TPQ are required to provide the SERC and the LEPC with a follow-up notification letter.

Date____________________

Address of State Emergency Response Commission:	Address of Local Emergency Response Committee:
________________	________________
________________	________________
________________	________________
________________	________________

This letter is to notify the State Emergency Response Commission and the Local Emergency Planning Committee that I have a facility that falls under the requirements of the Emergency Planning and Community Right-to-Know Act, SARA Title III, Subtitle A, Section 302.

1. Name of the facility or farm ____________________
2. Address of the facility or farm ____________________
   ____
3. County of location ____________________
4. Name of emergency contact person ____________________
5. Phone number of facility representative ____________________

6. List and quantities of EHS chemicals (optional)

   Chemical(s) ______________________ Quantity(ies) ______________________

   ______________________ ______________________

   ______________________ ______________________

   ______________________ ______________________

   ______________________ ______________________

Signature ______________________

Printed name ______________________

Mailing address ______________________

Copy to company or farm file (attach returned certified mail card to letter).

Providing Details Completes the Circle of Awareness

Material Safety Data Sheets Assist Communities in Evaluating Hazards (EPCRA Section 311) Any facility that manufactures chemicals and prepares Material Safety Data Sheets or is required to maintain files of MSDSs under OS0A's Hazard Communication Standard may have to comply with the additional requirements of Section 311 of EPCRA.

Reports are necessary if a pesticide or another chemical listed as an Extremely Hazardous Substance is present in inventory at the Threshold Planning Quantity. The TPQ for an EHS is 1, 10, 100, or 500 lb, depending on the substance. Any substance requiring an MSDS on file under OSHA's Hazard Communication Standard and stored in excess of its respective threshold quantity (or 10,000 lb if no threshold is listed) also must satisfy requirements under Section 311. Any facility whose individual chemicals exceed Threshold Planning Quantities for EHSs or for non-EHSs must provide to its LEPC, SERC, and local fire department copies of each chemical's Material Safety Data Sheet or a list of those chemicals and their specific hazards: flammability, reactivity, sudden release of pressure, immediate health hazards, and/or delayed health hazards. Information on the hazards can be obtained from the MSDS or directly from the chemical manufacturer.

Five Exemptions Apply to Section 311 The following are exemptions from the MSDS reporting requirements under EPCRA Section 311.

1. Any substance used in routine agricultural operations at the point where it is being used or stored by the end user; any fertilizer stored for sale to the end user. (Although farms generally are not covered, one possible exception is that a farm which is part of a larger business operation may be required to comply with Section 311.)
2. Any food, food additive, color additive, drug, or cosmetic regulated by the Food and Drug Administration.
3. Any chemical present as a solid in any manufactured item where exposure to the substance does not occur under normal use conditions.
4. Any substance to the extent that it is used for personal, family, or household purposes, or to the extent that it is present in the same form or concentration as a product packaged for distribution to and use by the general public.
5. Any substance to the extent that it is used in a research laboratory or in a hospital or other medical facility under the direct supervision of a technically qualified individual.

A list of substances identifies hazards, generally, but does not provide information on the amounts in inventory or their seasonal occurrence at the facility. If the list is chosen as the method of reporting, be aware that the LEPC may request that the owner make available all or specific MSDSs. Either the MSDSs or the list is a one-time submission requirement of the SERC, LEPC, and the local fire department. Facilities have 90 days to update the original list or to provide new MSDSs to the three agencies when new chemicals falling under the MSDS provisions are added to the inventory or when stock is increased above a threshold limit.

EPCRA Section 311 Report Form

Name of facility ____________________

Physical address (Street or P.O. box) ____________________

City, State, Zip code ____________________

County where facility is located ____________________

Mailing address (Street or P.O. box) ____________________

City, State, Zip code ____________________

Printed name of facility contact person ____________________

Facility phone number ____________________

Phone number of contact person (if different from facility phone number) ____________________

Printed name of person submitting information ____________________

Signature of person submitting information ____________________

Certified mail number (**optional**) ____________________

Material Safety Data Sheets enclosed? Yes ____ No ____

If yes, is this an initial submittal or an update? Initial Update

If no, will the submittal of the list comply with Section 311 of EPCRA?

Yes ____ No ____

Has either the list of chemicals or the MSDSs been submitted to the:

State Emergency Response Commission? Yes ____ No ____

Local Emergency Planning Committee? Yes ____ No ____

Fire department? Yes ____ No ____

Common name	Chemical name	Acute health	Chronic health	Sudden release	Fire	Reactivity

Copy to facility file (attach returned certified mail notice to form).

Inventory Information Provides Valuable Clues to Protecting Communities (EPCRA Section 312) If a facility is required to comply with EPCRA Section 311, compliance with Section 312 also is required. Section 312 of the Emergency Planning and Community Right-to-Know Act requires that either a Tier One or Tier Two inventory form be completed on or before March 1 of each year; check with your own state authorities regarding which form to use. State and federal contacts can be found at http://www.epa.gov/ceppo/staloc.htm. The inventory forms present the maximum amounts of pesticides and other substances stored on-site at any one time during the preceding calendar year. The inventory forms are submitted to the State Emergency Response Commission, the Local Emergency Planning Committee, and the fire department that has jurisdiction over the facility. Always keep a copy of the completed forms and the certified mail receipt in your files for future reference and review. The cost for reporting varies, depending on state regulations. Not all states have filing fees.

Tier One Inventory Forms This reporting form assigns to the reportable chemicals specific "qualitative hazards." The basic feature of the Tier One inventory form is not to identify specific pesticides, fertilizers, or other chemicals but, rather, to categorize all of the chemicals according to physical hazards (fire, sudden release of pressure, reactivity) and health hazards (acute or chronic toxicity). The inclusion of information regarding the maximum amounts stored, average daily amount, number of days on-site, and the storage location for each of the hazard categories makes the Tier One inventory form different from the reporting required by EPCRA Section 311. Tier One inventory forms can be obtained by contacting your LEPC, SERC, or the EPA hotline in Washington, D.C. (800-535-0202).

Tier Two Inventory Forms Specific chemical and location information is provided by the Tier Two inventory form. In general, most SERCs, require that the Tier Two inventory form be completed because it is more useful for emergency planning purposes than is the Tier One form. In the event of a fire or an explosion, a Tier Two inventory form provides a good indication of chemicals' identities, locations, physical characteristics, health hazards, inventory quantities, and conditions of storage. EPCRA Section 312 also allows an LEPC and/or a SERC to request information from a facility even when no threshold planning quantities are on-site. These details provide emergency responders with necessary information for planning and responding to an emergency. The forms and accompanying information can be obtained by contacting your LEPC or SERC, or the EPA hotline in Washington, D.C. (800-535-0202).

Accidental Releases and the Reporting of Emergencies

In dealing with the accidental release of pesticides and other substances which threaten people and the environment, it is very important that those providing information fully understand the spill or reporting provisions. The reporting requirements are based on the Reportable Quantity (RQ) found alongside the Threshold Planning Quantities. The RQ sets the amount of an active ingredient which is reportable in the event of release.

The chemicals and their RQ values that need to be reported are provided in the EPA publication, Title III List of Lists. Be aware that RQ numbers are listed under the Section 304 EHS and CERCLA columns. For example, 2,4-D acid has a CERCLA Reportable Quantity of 100 lb in Table 15.2.

TABLE 15.2 Reportable Quantity (RQ) Listed in Title III List of Lists Section 304

CAS Number	Chemical Name	EHS	CERCLA
116-06-3	Aldicarb	—	1
94-75-7	2,4-D acid	—	100
13194-48-4	Ethoprophos	1	—
1910-42-5	Paraquat	1	—
7664-41-7	Ammonia (anhydrous)	—	100
71-43-2	Benzene	—	10

If an EHS has no listed RQ, the CERCLA Reportable Quantity is to be used. For example, the EHS Reportable Quantity for anhydrous ammonia is 100 lb because that is the CERCLA Reportable Quantity for anhydrous ammonia. If no RQ is listed for a CERCLA chemical, 1 lb is to be used as the RQ. For example, ethoprophos would have a CERCLA Reportable Quantity of 1 lb.

Section 304 of EPCRA classifies a reportable release as one which meets the following criteria:

- Listed as either an EHS or a CERCLA chemical
- Meets or exceeds the RQ of such substance
- Escapes the facility into the environment to potentially impact persons off-site

If a facility has a release of a CERCLA hazardous substance at or above the RQ, the facility must provide immediate notification to the National Response Center (800-424-8802), LEPC, and SERC. In the case of EHS reportable releases which escape the facility, only LEPC and SERC have to be notified; however, contacting the National Response Center is advisable. Each LEPC must designate a Community Emergency Coordinator to receive notification in the event of a release of Section 304 designated substances. If a release occurs from a facility where the chemical is produced, used, or stored, the Community Emergency Coordinator for the LEPC must be notified immediately by telephone or radio or in person. It was the intent of Congress that reporting should take place within 15 min of the release. The following information should be reported:

- Chemical name or identity of any substance involved in the release
- Location of the release
- An indication of whether the substance is on the list of Extremely Hazardous Substances
- An estimate of the quantity of the substance released into the environment
- The time and duration of the release
- Medium (i.e., soil, water, and/or air) to which the release occurred
- Any known or anticipated acute or chronic health effects posed by the emergency and, where appropriate, advice regarding medical attention necessary for exposed individuals
- Proper precautions to be taken as a result of the release, including evacuation (unless such information is readily available to the Community Emergency Coordinator pursuant to the emergency plan)
- Name and telephone number of the person(s) to contact for further information

As soon as practical after a release which requires notice, the owner or operator of the facility must provide a written follow-up emergency notice detailing and updating the initial information. The report should indicate actions taken to respond to and contain the release; any known or anticipated acute or chronic health effects associated with the release; and, where necessary, advice regarding medical attention for exposed individuals. The written follow-up report must be submitted to the LEPC and SERC. Depending on the nature of the release, the facility may be obligated under other laws, such as a state spill reporting rule, to provide additional notification. An attorney should be consulted and a copy of pertinent information filed for further reference. Sending the report via certified mail is recommended (Fig. 15.4).

Public Access to Right-to-Know Information

The Local Emergency Planning Committee is required by EPCRA to publish an annual notice in newspapers, stating where submitted emergency response plans, Material Safety Data Sheets, and inventory forms are available for public review. This provides the public an opportunity to become better aware of chemicals stored and used in their community.

Section 313 of EPCRA requires manufacturers (defined as having Standard Industrial Classification Codes of 2000–3999) who have the equivalent of 10 or more full-time employees and annually manufacture or process 25,000 lb, or otherwise use 10,000 lb of a listed toxic chemical to report releases of any chemical listed on the Toxic Chemical Release Inventory (TRI). The need to report is not based on the size of releases but on the amount of chemical used at the facility.

Both routine and accidental releases must be reported. Reporting is done on Form R, which is filed with the EPA in Washington, D.C., and with the SERC of the state in which the facility is located. Reports are due by July 1 for the preceding calendar year. The EPA assembles all information from Form R submissions into a single database which is accessible to the public.

Toxic Release Inventory data provide communities with specific information regarding the kinds and quantities of toxic chemicals being released in their area. Although EPCRA does not require LEPCs to receive and make available these data, it does require each SERC and the EPA to do so.

CHEMICALS IN THE WORKPLACE

During the past several decades, there has been a growing awareness of the importance of safety in American industry. This awareness has been brought into sharper focus by the news media, the government, and the public, triggered in part by major chemical accidents such as the Bhopal, India, chemical explosion in 1984.

Safety is no less an issue for the pest management industry. Government regulations and industry programs place emphasis on the safe manufacture, distribution, use, and disposal of pesticides. Examples of government action include: the Occupational Safety and Health Administration (OSHA) Hazard Communication Standard for workers, and EPA regulations under the Emergency Planning and Community Right-to-Know Act for community emergency response. On the industry side, there is the Responsible Care Program of the Chemical Manufacturers Association.

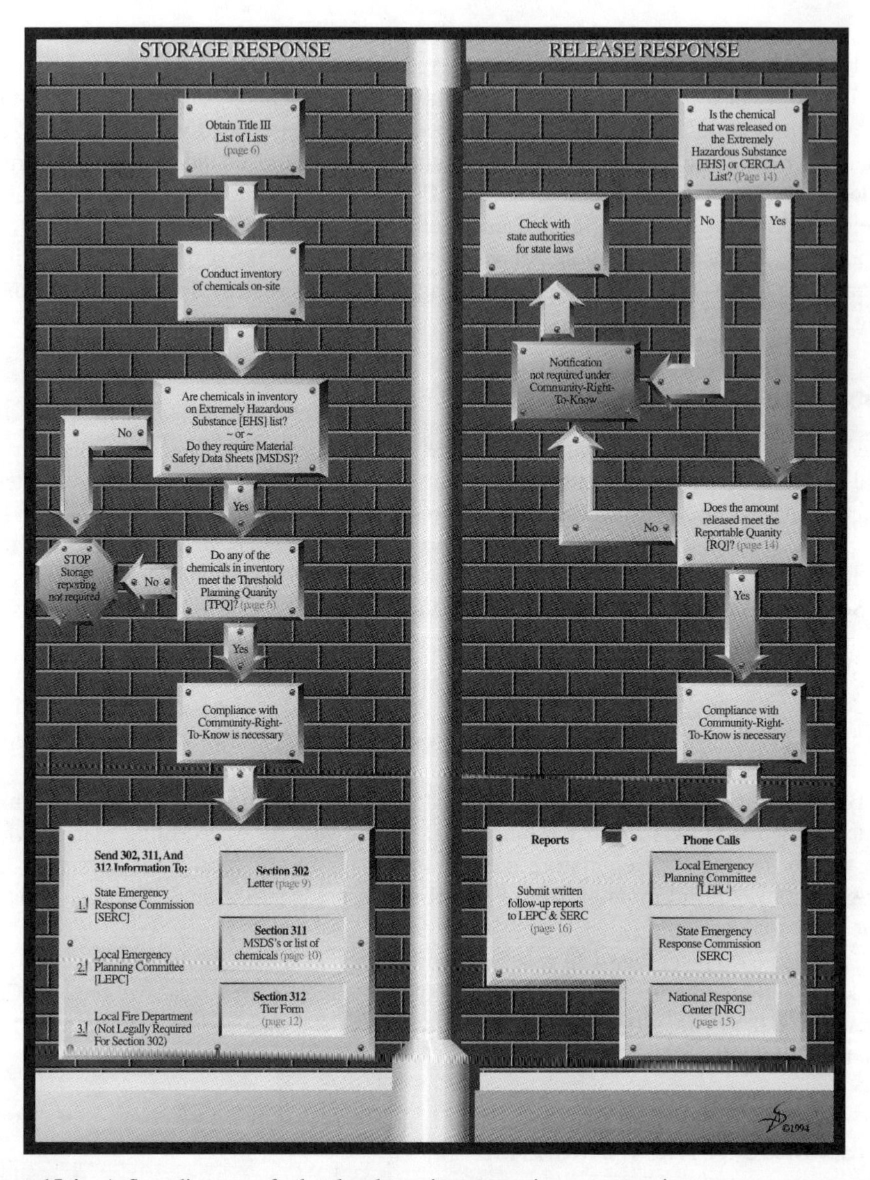

Figure 15.4 A flow diagram of what legal requirements trigger companies to report storage and releases of pesticides.

The Hazard Communication Standard

Over time, state and federal government has assumed more and more responsibility for ensuring that workers are provided a safe working environment. The Occupational Safety and Health *Act* of 1970 led to the creation of the Occupational Safety and Health Administration (OSHA) within the U.S. Department of Labor. Many preexisting federal worker safety laws were consolidated under OSHA. This law also authorized the establishment of regulations to ensure that employees are provided information on the hazards to which they may be exposed and that precautions are taken to prevent or reduce exposure to hazardous chemicals (Fig. 15.5).

Figure 15.5 The availability and accessibility of Material Safety Data Sheets is an integral right-to-know component of informing workers on the health risks that the products can pose and on what steps can be taken to reduce such risks.

The Occupational Safety and Health Act also provides OSHA with a mechanism for entering into cooperative agreement with state agencies to enforce worker protection rules within the states' boundaries. Federal and state programs mandate safety in the workplace, through regulations, and promote such safety through educational and technical assistance.

OSHA enacted the federal Hazard Communication Standard (HCS) (29 CFR Part 1910.1200) in 1983. Only employees in manufacturing industries were covered by the original version of HCS; but in 1989 it was expanded to cover all employees who (potentially) may be exposed to hazardous chemicals in their work areas—regardless of the place of employment or nature of the industry.

The HCS is the cornerstone of each employee's right to know. Its strategy is to ensure that hazardous chemicals are fully evaluated and their hazards communicated to workers by means of labels, Material Safety Data Sheets, and training programs.

Many OSHA regulations have specific goals, whereas HCS sets only general performance standards which employers have the flexibility to adapt to their specific workplace, work force, work practices, and work situations. Information on each chemical is customized to ensure that all potential hazards are properly addressed.

The HCS clearly requires each employer using hazardous chemicals to develop, implement, and maintain at the workplace a written hazard communication program for employees. The employer also must provide training to inform employees (1) about the nature of hazardous chemicals in each work area, (2) on how to detect a release of hazardous chemicals in the work area, and (3) on what measures to take to protect themselves from hazardous chemicals. The employer also must ensure that all containers of hazardous chemicals in the work area are properly labeled and stored. The worker safety program must be organized into a formal written plan and made available along with MSDSs to employees upon request.

Creation and Distribution of the MSDS

The Hazard Communication Standard requires that chemical manufacturers and importers thoroughly evaluate chemicals that they produce and import, respectively, to determine their hazard potential.

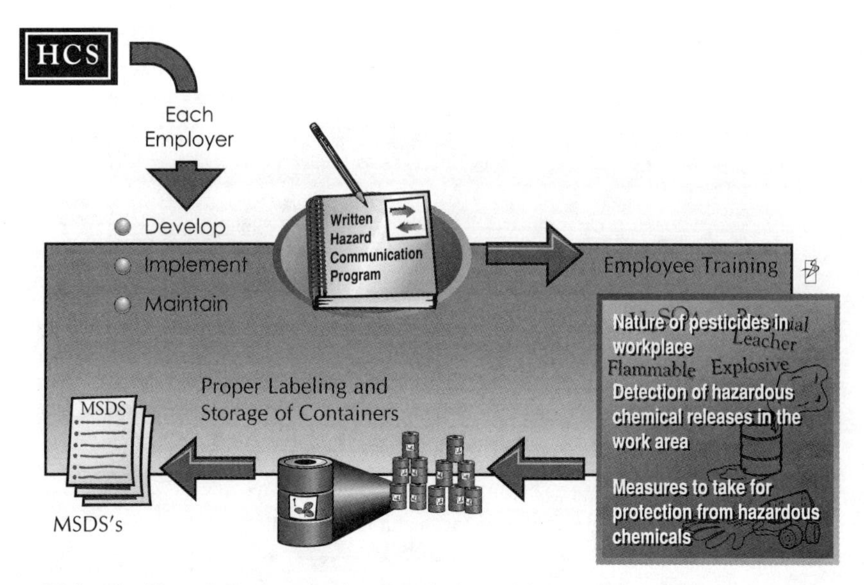

Figure 15.6 The Hazard Communication Standard provides product information and employee training as part of a comprehensive right-to-know law.

If a chemical presents a hazard, a Material Safety Data Sheet must be developed to communicate its hazard potential to users. The first step in preparing an MSDS for a hazardous chemical is to identify its composition. The product may be pure (consisting of just one component; namely, the pesticide active ingredient) or it may be a formulation of two or more chemical ingredients. Once the composition has been established, information on hazards can be collected (Fig. 15.6).

Companies that prepare MSDSs may perform their own in-house tests to collect product hazard information, or they may rely on data from testing others have conducted (e.g., as when one company allows another to use its active ingredient in their product).

A large amount of pesticide hazard information is generated in the course of fulfilling regulatory requirements for product registration. The EPA requires that toxicological, environmental effect, and physical property data be generated, much of which can be used in preparing an MSDS.

Once the composition of the product is determined and the necessary data collected, an MSDS can be prepared. The steps actually used to create an MSDS vary from company to company. The process varies from (1) simply collecting the required product information in written form and typing it on a standardized MSDS form to (2) using electronic databases. Regardless of which procedure is used, a draft of the MSDS is reviewed by key personnel within the company. Upon approval of the draft, the MSDS is printed.

The Hazard Communication Standard places specific responsibility for the flow of product information on manufacturers, importers, distributors, and employers. Chemical manufacturers are required by the standard to provide an MSDS to the purchaser of the product at the time of the first order and, thereafter, anytime the MSDS is significantly revised. The MSDS may be included with the pallet, submitted electronically, or delivered by mail. As the pesticides are further distributed to satellite suppliers, dealers, or users, a copy of the MSDS must accompany their original orders. This assures that the MSDS follows the same distribution route as the product itself so that employees of the manufacturer, distributor, and end users will have access to the information. Thus, MSDSs are disseminated along the

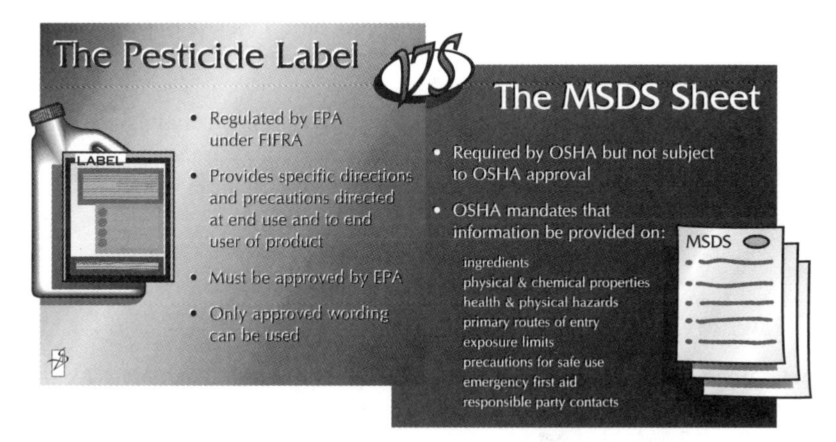

Figure 15.7 The pesticide label and the Material Safety Data Sheet are the key documents used to inform pesticide applicators.

distribution chain until they eventually reach businesses whose workers actually use the products. Some states also require that pesticide applicators provide MSDSs to their customers.

Each MSDS must be updated within 90 days of significant new information becoming known or available to the chemical manufacturer. The addition or alteration of significant MSDS information requires a new effective date for the document; as with the original MSDS, the revision must be passed along the distribution chain.

MSDS: Key to Communication

The Material Safety Data Sheet is used to communicate information vital to the safe use and handling of each chemical or product. Although both the MSDS and the pesticide label contain information intended to address potential hazards, they differ significantly.

Pesticide labels are regulated by the U.S. EPA under FIFRA and provide specific directions and precautions for the end user. The EPA requires specific wording on pesticide labels and must approve each one.

MSDSs are required by OSHA; however, they are not subject to OSHA approval. OSHA does mandate that each MSDS provide information on hazardous ingredients, physical and chemical properties, accompanying hazards, primary routes of entry, exposure limits, precautions for safe handling and use, emergency first aid procedures, and responsible party contacts. Most pesticide manufacturers also elect to include on the MSDS additional information which may be useful to those handling the material along the supply chain; for example, regulatory sections dealing with the reporting of chemical releases, Right-to-Know regulations, National Fire Protection Association ratings, and Department of Transportation product classification (Fig. 15.7).

Organization of the Material Safety Data Sheet The MSDS is a basic part of the OSHA Hazard Communication Standard, but no specific format is prescribed for the presentation of information. Therefore, MSDSs from various manufacturers may differ dramatically in order and format yet still present the required data. To help bring some order to the MSDS format, the American National Standards Institute has published the American National Standard for Hazardous Industrial Chemicals—Material Safety Data

Sheets—Preparation (ANSI Z400.1-1993). This voluntary standard prescribes the division of MSDSs into 16 sections, the mandatory order and titles of which are intended to create consistency from manufacturer to manufacturer.

ANSI Z400.1-1993 is used to illustrate key concepts for understanding and interpreting MSDSs. Although the 16 section titles and their order of appearance are the same from manufacturer to manufacturer, the order and amount of information within a given section is left to the discretion of the manufacturer.

SECTION 1 CHEMICAL PRODUCT AND COMPANY IDENTIFICATION

Product name: Zapo Aminophos	Specific brand name of product.
Manufacturer: Acme Chemical Company 123 Main Street Hometown USA 12345-1234	Where to write for information.
General phone: (800) 555-0000	Where to call for answers to nonemergency questions.
Company emergency phone: (517) 123-4567	Manufacturer's number to call for emergency assistance.
CHEMTREC: (800) 424-9300	Chemical Transportation Emergency Center phone number. Note that this number is for a transportation emergency.
EPA Registration number: 99999-999	Unique number assigned by the EPA to a registered pesticide product.
Date prepared: January 28, 2002	The date the MSDS was prepared.
MSDS number: 000708	The number assigned to the MSDS by the manufacturer.
Product code: 45386	Specific product identification number assigned by the manufacturer.

SECTION 2 COMPOSITION, INFORMATION ON INGREDIENTS

Active ingredient (AI): Gratol (1,1′ Dimethyl) = 25%	The ingredient that controls the target pest. The AI can be identified by a common name or by a chemical name. In this example, the common name is Gratol and the chemical name is 1,1′ Dimethyl.
CAS #999999-99-9	Active and inert ingredients also may be specifically identified by their Chemical Abstract Service (CAS) number.
Inert ingredients = 75% Kaolinite clay CAS #001332-58-7	Inert ingredients help deliver the AI. Inert ingredients known to contribute to the product's hazard potential must be listed by name unless they are trade secrets.

SECTION 3 HAZARD IDENTIFICATION

Emergency overview: Orange crystalline solid. Slight acrid odor. Causes eye irritation. Harmful if swallowed or inhaled. Heating to temperatures above 158°F can lead to rapid pressure buildup. Toxic fumes are released in fire situations.	A quick overview of potential hazards and description of the product's appearance. This information is intended for emergency response personnel.
Potential Health Effects	
Primary routes of entry: Eyes, lungs, skin.	Exposure sites through which the chemical might enter the body.
Eye contact: May cause irritation, redness, and tearing.	Effects resulting from contact of the chemical with the eye. Eye contact is a form of dermal exposure.
Skin contact: May cause skin irritation with redness, pain, and allergic reaction based on toxicity studies and human experience.	Effects resulting from contact of the chemical with the skin. Skin contact is a form of dermal exposure.
Skin absorption: Not known to be absorbed through the skin. Exposure is not likely to result in the material being absorbed.	Effects resulting from the chemical being absorbed through the skin. Absorption through the skin is a form of dermal exposure.
Ingestion: Small amounts swallowed incidental to normal handling operations are not likely to cause injury.	Effects resulting from swallowing chemicals. Ingestion is called oral exposure.
Inhalation: Inhalation of dusts may cause respiratory tract irritation.	Consequences of breathing the chemical into the lungs.
Cancer information: Not listed as a carcinogen or potential carcinogen. Not considered to be carcinogenic in lifetime feeding studies.	Information on any cancer-causing capabilities indicated during testing of the product ingredients (both active and inert).
Teratology: Birth defects are unlikely.	Information on any birth defects that occurred in laboratory animals during testing of the chemical.
Reproductive effects: In laboratory animal studies, this product has not been shown to interfere with reproduction.	Description of impacts of the chemical on reproductive processes.
Mutagenicity: Not mutagenic in either bacterial or mammalian cells.	Indicates whether or not the chemical may damage genetic material (as evidenced in laboratory animals).

SECTION 4 FIRST AID MEASURES

Eyes: Hold eyelids apart and flush eyes with a gentle stream of water for 15 min. See an eye doctor immediately.	What to do if the product gets into the eyes.
Skin: Wash with plenty of soap and water. Get medical attention if irritation persists.	What to do if the product gets on the skin.
Ingestion: Drink 1 or 2 glasses of water and do not induce vomiting. Call a physician or poison control center.	What to do if the product is swallowed.
Inhalation: Remove victim to fresh air. Get medical attention immediately.	What to do if the product is breathed into the lungs.
Note to physician: Gratol is a mild cholinesterase inhibitor. Treat symptomatically. In case of exposure, plasma and red blood cell cholinesterase tests may indicate significance of exposure (baseline data are useful). Atropine, by injection, is the preferable antidote.	Specific instructions to the physician. Users of pesticides should be familiar with where this is found on MSDSs so that in an emergency this information only can be given to the physician quickly.

SECTION 5 FIRE-FIGHTING MEASURES

Flashpoint and method: >200°F	The minimum temperature at which a liquid gives off vapor in sufficient concentration to ignite near the surface of the liquid or in the test vessel used.
Flammable limits: UEL: 12.0% @ 150°F LEL: 6.5% @ 150°F	The upper explosive limit (UEL) and the lower explosive limit (LEL) concentrations in air that will produce a flash of fire when an ignition source is present.
Extinguishing media: Use water, fog, foam, or CO_2; foam preferred. Water, if used, must not enter sewers.	Specific instructions to firefighters on how to extinguish a fire involving the chemical.
Fire and explosion hazards: Toxic, irritating gases may be formed above 320°F (160°C).	Important instructions for emergency responders dealing with fire or explosion.
Fire-fighting equipment: Use positive pressure, a self-contained breathing apparatus, and full protective clothing.	Description of safety equipment and clothing that firefighters should use in case of fire involving the chemical.

SECTION 6 ACCIDENTAL RELEASE MEASURES

Action to take for spills: Isolate and post the spill area. Keep animals and unprotected persons out of the spill area. Sweep up small spills with material such as Hazorb, Zorball, or soil. Thoroughly wash body areas that come into contact with the product. Contain spills to keep out of sewers or streams. For larger spills, consult Acme Chemical Company or CHEMTREC.	Actions to take when dealing with a spill. Intended to be used by emergency responders.

SECTION 7 HANDLING AND STORAGE	
Handling: Mechanical handling can cause formation of dusts. To reduce the potential for dust explosion, do not permit dust to accumulate outside of equipment designed to handle potentially explosive dusts. Wash thoroughly with soap and water after handling.	Procedures for handling to minimize the risks of accidental exposure or release of the product.
Storage: Do not contaminate water, food, or feed storage or disposal areas. Store in cool, dry place. Under normal handling and storage conditions, avoid heating above 158°F (70°C). Store in original containers.	Procedures and conditions for product storage that will minimize hazard potential.

SECTION 8 EXPOSURE CONTROLS, PERSONAL PROTECTION	
Engineering controls: Provide general and/or local exhaust ventilation to control airborne levels below exposure guidelines.	Procedures used to maintain airborne levels below TLV (Threshold Limit Value) or PEL (Permissible Exposure Limit).
Exposure guidelines: Gratol: American Conference of Governmental Industrial Hygienists TLV and OSHA PEL are 0.2 mg/m^3 respirable. PELs are in accord with those recommended by OSHA, as in the 1989 revision of PELs.	TLV and PEL identify the amount of a chemical in the air, below which workers would not be expected to experience health problems from exposure during a 40-hr workweek.
Eye/face protection: Use safety glasses.	Describes protective measures to reduce the likelihood of the pesticide getting into the eyes.
Skin protection: Mixers, loaders, applicators and other handlers must wear a long-sleeved shirt, long pants, chemical-resistant gloves, and shoes and socks. Wash the outside of gloves before removing. Users should remove clothing immediately if pesticide gets inside. Keep and wash personal protective clothing (and any other clothing worn while handling the chemical) separate from other laundry. Personal protective equipment required for early entry permitted under the Worker Protection Standard is coveralls, waterproof gloves, and shoes and socks.	Describes protective measures to reduce the possibility of the pesticide getting on the skin (dermal exposure).
Respiratory protection: Atmospheric levels of the chemical (either dust or vapor) should be maintained below the exposure guidelines. When respiratory protection is required for certain operations, wear a National Institute for Occupational Safety and Health (NIOSH) approved air purifying respirator. Mixers and loaders must wear a dust/mist filtering respirator, NIOSH-approval number prefix TC-21C.	Describes conditions under which workers must wear specific, NIOSH-approved respiratory protection to minimize the potential for breathing the chemical (inhalation exposure).

SECTION 9 PHYSICAL AND CHEMICAL PROPERTIES	
Vapor pressure: Approx. 46 mmHg @ 20°C	Relates to volatility of the material.
Specific gravity (SG): 1.18	The weight of a material compared to the weight of an equal volume of water. Insoluble materials with an SG less than 1 will float on water, whereas those greater than 1 will sink.
Solubility in water: 1 g/100 ml	A term expressing the amount of a material that will dissolve in water.
pH: 7	pH 7, neutral; above 7, alkaline; less than 7, acidic; pH values from 0 to 2 and from 12 to 14 are usually corrosive to skin and eyes.
Boiling point: 165°F (74°C)	Temperature at which a liquid becomes a vapor.
Vapor density: 2.00 (air = 1)	Weight of a vapor or gas as compared to air. Weight of air is 1. Vapors with weight values less than 1 will rise. Heavy vapors—those with weight values greater than one—sink and concentrate.
Freezing point: 41°F (5°C)	Temperature at which material freezes.
Odor: Weakly pungent odor	Describes the product odor for detection purposes.
Appearance: Clear	Describes the physical appearance of the chemical.

SECTION 10 STABILITY AND REACTIVITY	
Chemical stability: Stable under recommended storage conditions. Unstable at temperatures above 266°F (130°C).	Describes the stability of the material, usually in general terms.
Conditions to avoid: Accelerated rate calorimetry (ARC) data indicate that an exotherm occurs at temperatures as low as 288°F (142°C); a significant rise in pressure is associated with this exotherm. Certain conditions may facilitate occurrence of exothermic events at temperatures significantly below 288°F. Therefore, it is important to maintain product processing temperatures by whatever means necessary below 158°F (70°C).	Describes conditions under which the product may react.
Incompatibility with other materials: Gratol may be inactivated by certain sulfur-containing fertilizers.	Describes other materials that may react with the product.
Hazardous decomposition products: Under fire conditions, hydrogen chloride and ethyl sulfide can be formed by the breakdown of the diethyl sulfide, and nitrogen oxides can be formed.	Provides information on what by-products are formed when the product burns or under other conditions.
Hazardous polymerization: Will not occur.	Tells if product will react dangerously with itself to form a new product.

SECTION 11 TOXICOLOGICAL INFORMATION

Acute Studies

Eyes: Mild irritant to rabbit eye.	Consequences of short-term exposure to eyes (irritancy or blindness).
Skin contact: Short, single exposures may cause skin irritation.	Consequences of short-term exposure to the skin.
Skin absorption: The LD_{50} for skin absorption in rabbits is > 2000 mg/kg. A single prolonged exposure is not likely to result in the material being absorbed through the skin in harmful amounts.	Toxicity by absorption through the skin. LD_{50} is the dose level that is expected to cause the death of 50% of the test animals.
Ingestion: The oral LD_{50} for female rats is 272 mg/kg. Product is considered moderately toxic by ingestion.	Toxicity of short-term exposure from ingestion.
Inhalation: The LD_{50} for rats is greater than 1.5 mg/l for 4 hr. Considered toxic by inhalation.	Consequences of short-term exposure from breathing. LC_{50} is the concentration of dust, fume, or mist (vapor) expected to kill 50% of the test animals.
Sensitization: Not a contact sensitizer when tested on guinea pig skin.	An allergic reaction on tissue after repeated exposure to a chemical.

Chronic Studies

Chronic toxicity: At high doses in one of the species studied (rats) there was growth retardation and a decrease in red blood cell counts. Additional effects included increases in liver, kidney, and thyroid weights. Administration of 0, 100, 500, and 2500 ppm Gratol fed to male and female young adult dogs for 6 months produced no observable effects.	Adverse health effects resulting from long-term exposure to chemical(s), or long-term effects from short exposure.
Teratology: Abortion, fetal death, resorption, and developmental abnormalities occur only at maternally toxic doses.	Effects of exposure to materials that may have the capacity to cause birth defects.
Reproduction: A three-generation study in rats (conducted with dietary concentrations of 35, 100, and 300 ppm) showed no influence on reproduction.	Effects of exposure that may affect the ability to reproduce viable offspring.
Mutagenicity: Not mutagenic in either bacteria or mammalian cells.	Effects of exposure to a substance that may change the genetic material in a living cell.
Carcinogenicity: Not listed by national and international organizations as a carcinogen or potential carcinogen. Not judged to be carcinogenic in lifetime feeding studies.	The ability of a substance to cause cancer.

SECTION 12 ECOLOGICAL INFORMATION

Ecotoxicological Information

96–hr LC_{50} , bluegill sunfish: > 1000 ppm 96-hr LC_{50} , rainbow trout: > 1000 ppm 48-hr LC_{50} , daphnia magna: > 1000 ppm Oral LC_{50} , bobwhite quail: > 2250 ppm Dietary LC_{50} , bobwhite quail: > 5620 ppm Dietary LC_{50} , mallard duck: > 5620 ppm	LC_{50} values state the concentration of the product (in the water) at which 50% of test animals die.

Chemical Fate Information

Hydrolysis Study (28 days): half-life of 2 hr at pH 11.	The amount of material that is broken down through reaction with water. Half-life describes the length of time it takes for half of the material to disappear.

SECTION 13 DISPOSAL CONSIDERATIONS

Disposal method: Open burning or dumping of this material or its packaging is prohibited. Pesticide wastes are toxic. Improper disposal of excess pesticide, spray mixture, rinsate, or other pesticide waste is a violation of federal law. If these wastes cannot be disposed of by use according to label directions, contact your state pesticide or environmental control agency.	Directions and limitations for disposal of the material.

SECTION 14 TRANSPORT INFORMATION

DOT UN Identification No: UN2783	Number assigned by U.S. Department of Transportation (DOT) to a specific hazardous material, as defined in the Hazardous Materials Table 49 CFR Part 172.101.
DOT hazard class: Class 6.1	Department of Transportation recognizes 9 classes of hazardous materials, including flammables, poisons, and corrosives.
Proper shipping name: Organophosphorus—pesticide, solid, toxic.	The name that should appear on DOT shipping papers.

SECTION 15 REGULATORY INFORMATION

Superfund Amendment and Reauthorization Act (SARA) Hazard Category: This product has been reviewed according to EPA "Hazard Categories" promulgated under Sections 311 and 312 of SARA 1986 (SARA Title III), and it is categorized as an immediate health hazard, a delayed health hazard, and a fire hazard.

Toxic Substances Control Act (TSCA): This product is a pesticide and is exempt from TSCA regulation.

OSHA Hazard Communication Standard: This product is a "hazardous chemical" as defined by the OSHA Hazard Communication Standard 29 CFR Part 1910.1200.

National Fire Protection Association (NFPA) ratings: 0 = least; 1 = slight; 2 = moderate; 3 = high; 4 = extreme

Category	*Rating*
Health	2
Flammability	1
Reactivity	0

SECTION 16 OTHER INFORMATION	
Product registration: This product is registered under EPA/FIFRA regulations, EPA Reg. No. 99999-999.	All pesticides must be registered by the EPA before they can be offered for sale or distribution.
Reason for issue: Revised Sections 2 and 14.	Explains where the revised MSDS differs from the one it supersedes.
Supersede date: 06/30/93	Provides the date the previous MSDS was issued.
Responsibility for MSDS: Acme Chemical Company Registration and Regulatory Affairs 123 Main Street, Hometown USA 12345-1234	Name of company responsible for the MSDS.

Components of the Hazard Communication Program

The latest version of the Hazard Communication Standard was published in the Federal Register (59 FR 6126) dated February 9, 1994; contact your state or federal OSHA for a copy, or obtain one from your trade association. The Hazard Communication Standard should be included in its entirety in the company procedures manual.

The Hazard Communication Standard requires each employer to develop a written hazard communication program when hazardous materials are used by one or more workers. The plan should not be viewed as a complicated document that serves solely to comply with OSHA requirements; a well-designed plan serves as a blueprint to increase employee awareness of chemical hazards in the workplace.

Appendix 15.1 provides an example that employers might use to develop and implement their own written hazard communication program. Employers must determine the language and construct programs which best serve their own needs within the context of the Hazard Communication Standard. The following is a guide for writing a hazard communication program for your company.

Responsibility for the Hazard Communication Program Under the Hazard Communication Standard, employers are responsible for developing in writing, implementing, and maintaining a hazard communication program. For instance, they must write a company-specific plan and revise it as necessary, obtain and maintain MSDSs, provide initial and follow-up training on hazardous chemicals, and ensure that labels are provided on hazardous chemicals.

Employers should distribute to each employee an annual reminder of the company's hazard communication program. It should reiterate the physical location of the plan and the purpose for having it in writing. Such annual updates should be entered as appendixes in the company procedures manual.

Hazardous Chemicals List Employers should assume that chemicals accompanied by an MSDS are hazardous as defined by the OSHA Hazard Communication Standard unless the MSDS states that they are not. The standard requires that all hazardous chemicals in a workplace be identified on a hazardous chemicals list. Headings for the list may include date of entry, product brand name, product manufacturer, first date in inventory, MSDS on file (yes or no), and training requirements met (yes or no); the person making the entry should sign beneath the list. See following sections for more information.

Company personnel in charge of purchasing chemicals must be aware of the hazardous chemicals list. Purchases of chemicals not previously listed need to be brought to the attention of those responsible for implementing the Hazard Communication Standard. Introduction of a new hazardous chemical mandates acquisition of the corresponding MSDS and triggers additional employee training. Internal audits should be conducted periodically to ensure that newly introduced hazardous chemicals get added to the list. Such inspections could be documented as an appendix to the company procedures manual.

MSDSs for Chemicals on Hazardous Chemicals List Suppliers are required to provide an MSDS for each product sold. All MSDSs should be easily accessible, on-site; for instance, they might be placed in a binder and/or incorporated into the company procedures manual. They may be organized alphabetically by product name or by the type of hazard they present (e.g., flammable, explosive, toxic). Employees should be informed that they can access all MSDSs on file.

When it is determined that MSDSs for products in inventory are not on file, they should be requested in writing from the manufacturer or by downloading the MSDS from the Internet; and it is recommended that manufacturers be consulted concerning any product for which the hazardous chemical status is unknown. It is advisable to save copies of all letters written to manufacturers along with their responses; they should be entered in the company procedures manual.

Whenever MSDSs are received from manufacturers and suppliers, they should be checked automatically against the effective date of those already on file for the same products. Those determined to be updates should replace older versions in the company procedures manual, and those representing new acquisitions should be entered. The hazardous chemicals list should be checked periodically to confirm its accuracy and to confirm that MSDSs for all hazardous chemicals used by the company are on file; discard those for products no longer used.

Labels and Other Forms of Warning The Hazard Communication Standard requires that workers be informed of the potential dangers of all hazardous chemicals in the workplace, and labels are required on all containers. Employers must ensure that minimum requirements are met. The identification of all products and their corresponding hazards must be clearly identified on the product container and communicated to workers in writing and/or in pictures, with symbols, or via any combination thereof. If both of these conditions are met on the label attached to the container, employers are not required to attach additional information. Pesticide products are labeled according to EPA mandates and are exempt from OSHA labeling requirements.

Supervisors should conduct periodic (e.g., monthly) inspections to ensure that all containers are properly labeled. Records of these formal inspections should be entered in the company procedures manual.

Nonroutine Tasks Nonroutine tasks are those assigned to workers who do not normally perform them as part of their job assignment. When nonroutine tasks involve hazardous chemicals, workers must be warned of potential hazards and must utilize the corresponding safety equipment and clothing necessary to protect themselves. All records of training conducted should be logged as appendixes to the company procedures manual.

Training of Company Employees Employers are required to provide worker training (either product- or hazard-specific) on the Hazard Communication Standard and the safe use of hazardous chemicals. Worker training must provide

- information on how and where to access the written hazard communication program;
- instruction on how to use the hazard communication program;
- locations where hazardous chemicals are present;
- detection methods to determine the presence or release of a hazardous chemical in the work area;
- an overview of potential hazards of chemicals present in the work area;
- employee safety instructions for protection from potential hazards;
- tips on using and understanding pesticide labels; and
- instructions on reading and understanding an MSDS.

Those responsible for the training portion of the Hazard Communication Standard must be continually alert to new or additional hazardous chemicals introduced into the workplace. Any chemical added to the hazardous chemical list should be addressed in training prior to use.

Periodically, employees should be asked to retrieve specific MSDSs to make sure they can find them. As a part of the training program, you might ask a series of questions about an MSDS to see if employees can locate and understand the information. The results will reveal the strengths and weaknesses of your hazard communication program, and you may wish to modify your training accordingly.

Document Training Activities Keep a log on all training programs, including in-house staff training, correspondence courses, certification workshops, and continuing education meetings that employees attend. Agendas should be developed for in-house training. A copy of the agenda should be initialed by employees in attendance at the conclusion of each program. The agendas may be logged as appendixes to the company procedures manual.

Outside Contract Employees Workers not directly employed by a company, but who will be working there (e.g., electricians, truck drivers, subcontractors), must be notified about chemical hazards in the workplace. A work environment where such situations exist is termed a "multiemployer workplace." The hazard communication program must contain information on how these unique employees can access MSDSs, the labeling system, and instructions on essential precautionary measures to protect against hazards under normal operating conditions and in emergencies. These requirements apply equally to all employers who operate a multiemployer workplace where hazardous chemicals are present. Individuals should sign a form verifying that they have been informed of the preceding issues, and these forms should be maintained as appendixes to the company procedures manual.

Access to Workers and OSHA Inspectors The company's written plan, MSDSs, and chemical information lists must be immediately available to workers and OSHA inspectors. Make sure the plan is accessible to all employees—in the main office or general work area—during normal working hours. The availability of this plan and all of the supporting documentation satisfies all Right-to-Know and Right-to-Access obligations.

Employers should have all employees who use information contained in the written plan record their names and the products or types of information they are using in a log attached to the plan. This documentation should be maintained as an appendix to the company procedures manual as documentation of accessibility to those seeking Right-to-Know information. It also may serve to highlight chemicals of concern for which specific training should be implemented. It is important to maintain the integrity of the system; therefore, a specific individual should be delegated the responsibility for ensuring that any information "borrowed" is retrieved and refiled appropriately. Workers themselves should not be allowed or required to refile information.

Modifications to the Hazard Communication Standard

The Hazard Communication Standard provides specific requirements for laboratories handling hazardous chemicals and businesses that store sealed containers (e.g., warehouses).

The HCS applies to laboratories only as follows:

- Labels on containers of hazardous chemicals brought into the laboratory must not be removed or defaced.
- Employers need to maintain MSDSs on all hazardous chemicals received into the workplace; MSDSs must be accessible to workers.
- Although training is required, laboratories are exempt from developing a written hazard communication plan.

The HCS applies, as follows, to businesses that handle only sealed containers:

- Marine cargo handling, warehousing, and retail sales are examples of activities that may involve sealed containers.
- Labels on all incoming containers of hazardous chemicals must not to be removed or defaced.
- Employers must maintain MSDSs accompanying incoming containers of hazardous chemicals; all MSDSs must be accessible to workers.
- Employers must obtain MSDSs for all sealed containers of hazardous chemicals received without MSDSs.
- Employees must be trained on what measures to take to protect themselves in the event of a spill or container leak involving a hazardous chemical.

SUMMARY

Safety in the workplace is everyone's responsibility. The Hazard Communication Standard provides a mechanism for assuring that every employee who handles pesticides directly, or who may be exposed to them, is made aware of the potential hazards associated with those products. Once trained, employees are better equipped to do their part in maintaining a safe working environment for themselves and their co-workers. Employers and employees should view the Hazard Communication Standard not as a burden but as a tool to promote safety.

The federal Emergency Planning and Community Right-to-Know Act of 1986 requires the development of community management plans for emergencies involving hazardous substances, and it serves as a source of information on routine toxic emissions. The law is flexible, allowing state and local governments to devise plans specific to the needs of individual locales.

Who benefits from hazardous materials planning and reporting? The benefits of compliance with EPCRA extend from regulated businesses to the surrounding community. EPCRA planning and the reporting on which it is based reflect good business sense while protecting public welfare. Accurate reporting of hazardous substances used, produced, or stored facilitates the protection of employees, property, and chemical investment. Emergency personnel who lack information on the presence of hazardous substances during a fire may elect to simply let the fire burn, perhaps resulting in unnecessary damage to inventory and property. Compliance serves to improve business relations within the community, whereas evasiveness serves only to accentuate people's fears concerning the potential impact on human health.

EPCRA benefits emergency personnel in a chemical spill or fire situation. Prior knowledge of chemical types and storage locations equips emergency responders to reduce their own exposure and that of others. Concern for the welfare of neighboring hazardous materials storage areas also is demonstrated by compliance with EPCRA. Adjacent property owners have a large stake in hazardous materials reporting, and owners of facilities where hazardous materials are stored need to respect that interest. The reporting of pesticides in storage and the occurrence of spills, along with educating the workforce on hazardous chemicals, is a key component in protecting the community.

APPENDIX 15.1 A HAZARD COMMUNICATION PROGRAM

A. Policy on Compliance with OSHA Hazard Communication Standard

This is to inform employees that (name of company) is in compliance with the OSHA Hazard Communication Standard (Title 29 Code of Federal Regulations Part 1910.1200). The OSHA Hazard Communication Standard (herein referred to as "the standard") applies to all company operations where there is potential for exposure to hazardous chemicals during normal conditions of use (routine job assignments) or in emergency situations. In compliance with the standard, employee Right-to-Know information is provided via the following mechanisms: a list of hazardous chemicals, accessible Material Safety Data Sheets, labeled containers, and instruction on proper handling procedures for hazardous chemicals. Workers will be informed of hazards associated with nonroutine tasks. The responsibility for implementing the program outlined in this document has been assigned to (name of person). Copies of the program and attachments may be reviewed in the main office.

B. Policy on Compliance with OSHA Hazard Communication Standard

(Name of company) has available to all employees a hazardous chemical list identifying, by brand name, all hazardous chemicals used by the company. The list is maintained by (name of person) and may be accessed in the main office.

C. Policy on Material Safety Data Sheets for Hazardous Chemicals

Material Safety Data Sheets for all hazardous chemicals used by (name of company) are on file in the main office. Employees must read and understand those pertinent to their job and are encouraged to reread them periodically to stay current. Also located in the main office is a binder containing a hazardous chemicals list. The responsibility for acquiring MSDSs on all hazardous chemicals used by this company has been assigned to (name of person). MSDS acquisitions are accomplished through manufacturers and vendors from whom the products are purchased.

D. Policy on Labels and Other Forms of Warning

All hazardous chemicals at this facility must be properly labeled. Labels shall list at least the chemical identity, appropriate hazard warnings, and the name and address of the manufacturer. If multiple stationary containers in the work area contain similar chemicals with similar hazard potential, signs must be posted to convey their common hazard information. If chemicals are transferred from labeled containers to portable ones intended for immediate use, no labels are required on the portable containers.

E. Policy on Nonroutine Tasks

When employees are required to perform nonroutine hazardous tasks (e.g., cleaning tanks, stripping floors, washing out 55-gallon drums with cleansers, entering confined spaces), special training sessions will be conducted to inform them of hazardous chemicals to which they may be exposed. The training also will address the implementation of proper precautions to reduce or avoid exposure.

F. Policy on Employee Training

Everyone who works with hazardous chemicals or potentially could be exposed to them must receive initial training on the Hazard Communication Standard. Classroom instruction incorporating verbal presentations, written publications, and audiovisual materials will be the primary method of delivery. The initial training program must be conducted around an agenda covering, at a minimum, the following topics:

- Summary of the Hazard Communication Standard and the company's written program
- Chemical and physical properties of hazardous materials (e.g., flash point, reactivity) and methods that can be used to detect the presence or release of chemicals
- Physical hazards of chemicals (e.g., potential for fire, explosion)
- Health hazards, including signs and symptoms of exposure and any medical condition known to be aggravated by exposure to the chemical
- Procedures to protect against hazards (e.g., the proper use and maintenance of personal protective equipment; work practices or methods to assure proper handling of chemicals; and procedures for emergency response)
- Work procedures to follow during emergencies
- The physical location of MSDSs, how to read and interpret information on pesticide labels and MSDSs, and how employees may obtain additional hazard information
- Availability of training when new hazardous chemicals are introduced into the workplace, or when a new hazard relative to chemicals already in use is recognized. Regular safety meetings also may be used to review and update the information presented in initial training. Workers may be required to attend external programs to receive training on the safe use of hazardous chemicals.

G. Policy on Outside Contract Employees

The provisions of the standard dealing with multiemployer workplaces includes contractors. As multiemployer workers, outside contractors must be notified of any chemical hazards that they may encounter in their normal course of work on the premises. Notification includes information on the labeling system in use, protective measures to be taken, and safe handling procedures. Individuals must be notified of the location and availability of MSDSs. Each contractor bringing chemicals on-site must provide the corresponding hazard information, including labels and precautionary measures.

H. Policy on Access to Information

All employees and OSHA inspectors can obtain further information on this written program, the Hazard Communication Standard, applicable MSDSs, and chemical information lists in the main office.

CHAPTER 16

EDUCATING YOUR CUSTOMER CLIENTELE: A HOLISTIC APPROACH TO PEST MANAGEMENT

A family on a camping vacation applies mosquito repellent. Suburban homeowners spend Saturday morning spreading "weed-n-feed" products on their lawns. An apartment dweller cleans the bathroom with bleach and disinfectant. Parents sprinkle a few mothballs into boxes while packing their children's winter clothes for storage.

What activity do these people have in common? They are all using pesticides. Although, they may not realize these products are pesticides. Literally, pesticides are chemical compounds used to manage pests. Examples include insecticides, which manage insects; rodenticides, which manage rodents; fungicides, which manage the spread of fungal diseases; and herbicides, which manage weeds and other plants.

But the legal definition of a pesticide is much broader. A pesticide is any product that makes a claim to kill or repel pests. Pests can include plants or animals that carry disease, damage our landscapes and gardens, or become a nuisance and detract from the quality of life (Fig. 16.1).

PEST MANAGEMENT OPTIONS FOR REDUCING PESTICIDE USE AT HOME

The first step in pest management is to identify the pest that is causing the problem. It then becomes essential to learn about the life cycle and behavior of the pest to facilitate the development of a plan to manage it. The goal should be to reduce pest populations and damage to economically and aesthetically tolerable levels. Complete eradication may not be possible, practical, or desirable. The following methods should be considered in developing a pest management strategy.

Exclusion

One of the safest and most effective ways to manage pests in the home, lawn, and garden is to deny them access. This is called exclusion, or pest-proofing, and may be accomplished via one of the following options.

Figure 16.1 Dandelions cover the lawn of this mansion.

Barriers and devices such as fences, traps, lights, row covers, and noisemakers are examples of mechanical exclusion methods used to keep pests away from garden plants and out of homes. For example, nuisance wildlife such as rabbits can be excluded from gardens and landscape plantings with fencing. Some insects can be kept away from vegetables by covering them with netting. Birds can be banished from fruit crops by covering trees, bushes, or vines with plastic netting.

Rodents and other mammals such as bats can be excluded from homes by permanently closing entrance holes with caulking, steel wool, or structural repairs. Many insects can be kept out by caulking holes and cracks and ensuring that doors and windows are tightly sealed and screened. The use of yellow light bulbs (instead of the more conventional white bulbs) outside entrances will attract fewer insects. Storing food products and pet food in tightly sealed containers will guard against insect and rodent infestations.

Glue boards for cockroaches and traps for wildlife are other examples of devices that can be used to keep pests away from homes and plants. High-pressure water can be used to dislodge insects such as aphids and spider mites from host plants because such slow-moving insects often simply die before they have a chance to crawl back onto the host.

Cultural

Most plants resist pests when they are in good health. Therefore, keeping a potential host healthy can help prevent pest damage.

Plant Selection It is important to select species and cultivars of crops and ornamental plants recommended for the locale. Professionals in the community should be consulted to determine which kinds of plants grow best and without significant pest problems. Avoid those plants that are known to have a questionable history and those recognized as marginally hardy. Cold temperatures can predispose tender woody and perennial plants to pest damage.

Some plant cultivars resist or tolerate pest damage. Examples include tomato cultivars that are resistant to wilt diseases, apple cultivars resistant to Japanese beetles and apple scab diseases, and plants bred to produce more surface hairs that will discourage insect feeding. Cultivar selection should be based on the plants' known resistance to common pest problems, thus limiting loss potential and reducing the likelihood that a pesticide application might be needed.

Planting Dates There are recommended planting intervals for most crops, and it is wise to recognize their importance. Careful selection of planting dates enhances crops' defenses against disease and insect infestations. Planting too early in the spring can result in plants weakened by cold, wet soil conditions. Late spring frosts can damage or kill crops planted too early. Root and seed rots usually can be avoided by choosing later planting dates with soil conditions more favorable for seed germination and plant growth. Certain insect pest problems also can be avoided by choosing appropriate planting dates. Growing a combination of early, mid and late season crops may decrease the potential losses due to pests, because the risk of damage by any one pest is spread across the season.

Crop Rotation If space permits, crops should be planted in different areas of the garden each year to prevent buildup of pests in the soil. Removing the host for a season often reduces the pest of this host.

Sanitation Sanitation is perhaps the most important cultural practice that can be used to help manage pests. It consists of removing plants or plant parts suspected of harboring insects, disease, or a source of weed seed. For example, leaves, twigs, and branches of dogwoods infected with anthracnose should be removed and destroyed to help prevent the disease from spreading. Another example is the removal and destruction of certain plant parts that may be diseased, such as spotted tomato leaves that are infected with a fungal disease. Removal of diseased plant parts reduces the disease pressure to the plant. It is important to always buy from a reputable source good quality seeds and plants known to be free of insects and disease. Examine "gifts" from neighbors and tactfully decline those that obviously display pest symptoms. Remove garden weeds before they mature and produce seeds.

Sanitation is also important at the end of the growing season. Plant residue from annual crops, as well as the tops of herbaceous perennials, should be removed from the garden in the fall or thoroughly tilled or plowed into the soil. Those not infested with insects or infected with disease can be added to a compost pile.

Other examples of cultural management through sanitation include removal of dead or diseased limbs from trees and shrubs; garbage management to discourage flies and rodents; careful attention to pet food areas; scrupulous cleanup of food crumbs in the home; burning, chipping, debarking, or burying of trees lost to borers to prevent continued emergence of adults from the logs; and elimination of paper bags, newspapers, and other materials that provide food and shelter for pests such as cockroaches and rodents.

Other Cultural Methods Good cultural practices include providing plants the best possible growing situation: proper spacing, watering, and fertilization. Weed management and the timely harvesting of produce also help to maintain healthy plants. The management of weeds with mulch is a good cultural practice. Mulch also contributes to plant health by moderating soil temperatures and conserving moisture and by reducing the splashing of soil onto foliage or fruit during rain or irrigation.

Manipulation of a pest's environment can also be an effective method of cultural management. For example, venting the crawl space beneath a house will allow the space to dry, rendering the area unfavorable for the development of allergy-causing mildews and wood destroying organisms such as termites and decay fungi.

Biological

Biological control is the use of living organisms such as pathogens, predators, or parasites to kill pests. These organisms are commonly called the natural enemies of pests. Although some natural enemies can be purchased and released around the home, many natural enemies are already there. Homeowners can use biological management more successfully if they start by making their yards and gardens places where natural enemies can thrive. This can be accomplished by reducing the use of pesticides that kill natural enemies and providing food and shelter to encourage them to stay around the home. Planting a mix of flowers that provide season-long sources of pollen and nectar will attract insects such as predatory lady beetles that feed on pests. Mulching can provide a cool place for ground beetles to hide in the day so they can feed on foliar insect pests at night.

Chemical

Chemicals have been used for hundreds of years to manage certain pests. Pesticides are chemical compounds such as roach sprays, weed killers, and rat and mouse baits that are formulated to manage pests. Pesticides not only include typical insect and weed sprays but also bleach, toilet bowl cleaners, disinfectants, humidifier tablets, insect repellents, indoor air foggers, flea collars and shampoos, and many cleaning products normally used in the home. To determine whether or not a product is a pesticide, look for an EPA registration number (e.g., EPA Reg. No. 3120-280) on the container. If a number is present, it automatically identifies the product as a pesticide. The first set of numbers (3120) identifies the specific registrant (manufacturer), whereas the second set of numbers (280) identifies the specific product; this information might become crucial in the case of a problem with the product. The EPA registration number assigned to a product signifies to the user that all federal requirements for testing have been met. It also means that all of the instructions, directions, and precautions associated with the pesticide collectively comprise the pesticide label and become legal requirements with which the user must fully comply.

Figure 16.2 Pesticides in a home include more than just insecticides and herbicides.

Pesticides may consist of one or more active ingredients, and the active ingredients can be either organic or inorganic. Organic compounds are based on carbon chemistry and are formulated from molecules that contain carbon, hydrogen, and oxygen. Some organic compounds used in pesticides occur naturally and are obtained from plants or bacteria; these are called natural, biological, or botanical pesticides. For example, the organic active ingredient in pyrethrin insecticides is obtained from a certain chrysanthemum flower. Organic pyrethrins and their synthetic counterpart pyrethroids are both often used to manage flying and crawling insects in the home, as well as on pets and garden plants.

Inorganic compounds do not contain carbon but are derived from mineral sources. Some inorganic compounds used in pesticides also occur naturally. An example is a copper-based product used for algae management in ponds.

Pesticides containing synthetic (manufactured) compounds comprise the largest number of products used to manage pests. Most synthetic pesticides consist of organic compounds. These represent the group that most people consider when contemplating the use of pesticides.

Many synthetic pesticides have a mode of action similar to that of natural pesticides. Both natural and synthetic pesticides may vary in their toxicity to people and pets. Do not assume that just because a pesticide is "natural" it is not toxic. Some natural pesticides can be quite toxic. Always read the product label and follow the precautions stated regardless of the source of the active ingredient. Follow all instructions carefully.

Some pesticides manage pest problems without killing them. The following are examples of pesticides that provide pest management alternatives (Fig. 16.2).

Repellents Topical products which lessen human attractiveness to ticks, chiggers, mosquitoes, biting flies, etc., are called repellents. These products help make outdoor activities more enjoyable and contribute to the prevention of diseases transmitted by certain insects. Treating lumber with wood preservatives extends the lifetime of outdoor structures and furniture by repelling wood-destroying organisms. Repellents can be used to discourage deer or rabbits from feeding on the bark of valuable trees.

Attractants Chemicals used to lure pests into a trap are called attractants. For instance, an insect sexual attractant (called a pheromone) may be used inside a plastic bag, luring Japanese beetles inside. When the bag becomes filled with beetles, it can be tied off and discarded. However, Japanese beetle traps should be used with caution. These traps work by attracting the beetles, and in many parts of the country an individual will end up with more beetles instead of fewer. Other examples of attractant devices are glue boards for cockroaches and traps used to monitor the presence of other insects.

Growth Regulators Pesticides developed to adversely affect the development of specific insects are categorized as growth regulators. They work by preventing the immature stages of certain insects from maturing into adults or by rendering adults sterile or killing them. Some flea and roach products are growth regulators.

Desiccants and Smothering Agents Desiccants such as diatomaceous earth are abrasive to the outer covering of some insects, dehydrating the pest and resulting in its death. Soaps and oils smother insects, causing them to suffocate or to dehydrate. Some chemicals, such as stylet inhibitors, paralyze the muscles of the sucking insect and render it unable to feed; the insect simply starves to death.

Home Remedies Some people try to formulate their own pesticides using household products. Although they may be effective, this can be a dangerous practice. Many household chemicals are toxic, and mixing several of them together can result in combinations that are injurious to people, pets, and plants. The likelihood of injury is especially great when the concoctions are applied to food crops. Remember that home remedies have not been tested and registered by the Environmental Protection Agency. Many home remedy applications can also lead to plant injury even when common household products such as soaps, oils, carbonated beverages, gasoline, mouthwash, and urine are used.

PEST MANAGEMENT AT HOME: PATIENCE, PERSISTENCE, AND PRACTICE

When various pest management strategies—exclusion, cultural, biological, and/or chemical—are used in conjunction with regular inspection of the home environment to monitor and manipulate pest activity, the process is recognized as integrated pest management (IPM). In most cases, an IPM approach is considered sensible and environmentally sound and will keep pest levels and damage below economically or aesthetically injurious levels.

Although it is extremely important to remember that total eradication of a pest population is not the goal in most cases, it is equally important to recognize that sometimes it is. Eradication is by all means desirable when termites are damaging a structure or when pests present the possibility of disease transmission to people or pets (e.g., fleas).

A preferred strategy in most pest management situations is to think in terms of reducing pest activity to a level which poses only minimal potential for damage to or annoyance of

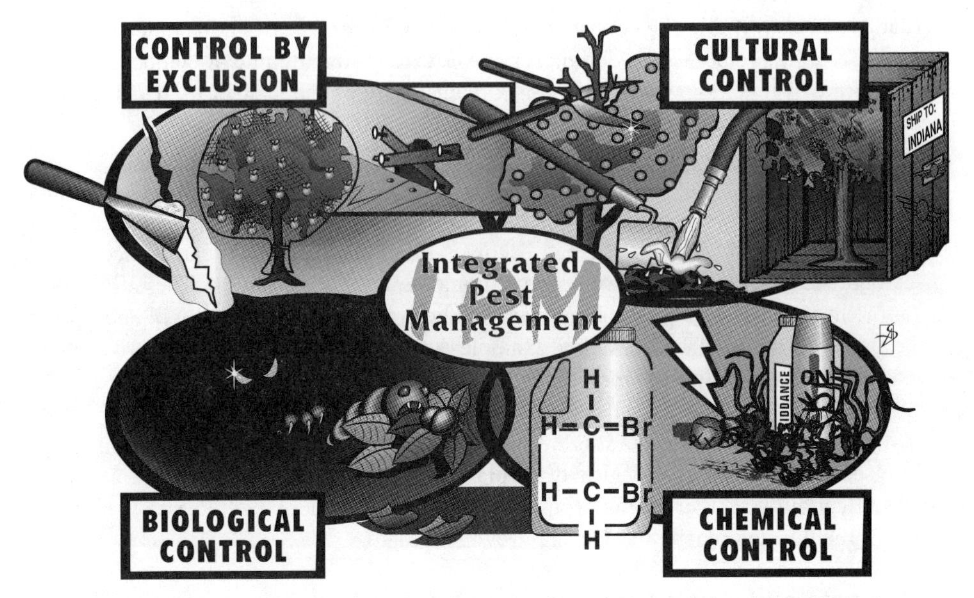

Figure 16.3 There are many proven methods and approaches that can be used around the home to manage pests.

the host, be it plant, animal, or structure. For instance, it is not necessarily desirable to kill all spiders in the home environment, because most of us are willing to tolerate a spider here and there, but we're not willing to share our kitchens with even a single German cockroach or mouse.

Selection of the most appropriate pest management method in a particular situation should be preceded by accurate pest identification and a survey of the site to determine the exact location and extent of the pest population. In some cases, successful pest management strategies require patience, persistence, and a long-term commitment (Fig. 16.3).

LAWN PEST MANAGEMENT

Healthy turf can compete successfully with weeds, survive insect attacks, and fend off disease. The key is to create an optimum environment where grass plants have every advantage for vigorous growth. The following suggestions can assist the homeowner in establishing and maintaining healthy turf with no (or minimal) use of pesticides.

Choose an Adapted Grass with Pest Resistance

Select grass species and cultivars that are insect and disease resistant. Choose those that are best adapted to grow under the sun or shade conditions of the home landscape. The amount of maintenance which will be required and the intended use of the turf also should be considered in the selection process. A blend of two or three varieties of a species or a mixture of two or more compatible species provides better disease resistance and adaptation to the site. Planting a single cultivar invites problems.

Homeowners with many large trees in their landscape often become frustrated because it is difficult to grow a beautiful lawn in heavy shade. The grass plants gradually disappear and weeds, moss, and algae take over. Designing a woodland garden for such problem areas would be a better choice than repeated annual attempts to manage pests with chemical inputs. Use ground covers (even mosses), perennial flowers, and shrubs that are adapted to heavy shade and root competition. Combine these with compatible mulch, decks, outdoor furniture, and other accessories to create a pleasant retreat.

Seed the Lawn When Conditions Are Favorable

Cool season grasses are best seeded in August and September. Seeding turf areas in winter and spring may result in increased weed, disease, and insect pressure. Sodding or sprigging warm season grasses should occur in early summer. Some warm season grasses are available by seed and this should be planted in early summer.

Determine the Lawn's Needs with a Soil Test

Lawn grasses require ample nutrients for healthy growth and resistance to pest damage. Test the soil about every 3 years to identify nutrient deficiencies and changes in soil acidity. A routine soil test will determine the soil's pH (acidity or alkalinity), its lime index, and its levels of phosphorus and potassium. If requested, the results will be accompanied by a recommendation for fertilizing the lawn. Turf grass is tolerant of a broad range of pH and lime should never be applied to a lawn unless a soil test indicates it is needed. County Cooperative Extension Service offices or garden store personnel can identify soil testing laboratories in the area.

Fertilize the Lawn for Sustained Growth

It is important to understand how nutrients are packaged in fertilizer products. The three numbers on the label are required by law to inform the consumer of the percentage (by weight) of nitrogen, phosphorus, and potassium in the package. For example, the numbers 24-6-12 on a fertilizer label indicate that the product contains 24% nitrogen, 6% phosphorus (phosphate, P_2O_5), and 12% potassium (potash, K_2O). A good general-use fertilizer for turf contains a nutrient ratio of about 4 parts nitrogen, 1 part phosphorus, and 2 parts potassium. It is normally recommended that the fertilizer product used for routine maintenance of a lawn provide 2–4 lb of nitrogen/1000 ft^2 of lawn per year. For example, to apply 1 lb of nitrogen/1000 ft^2 with a 25% N fertilizer (25-6-12) one would need to apply 4 lb of the fertilizer; with a 10% fertilizer, 10 lb would be needed.

Cool season grasses respond best to fall applications of fertilizers. Two-thirds of the annual nitrogen requirement for a lawn should be applied in the fall, followed by the remainder in mid to late spring. Early spring applications of fertilizer will promote lush turf that is more susceptible to disease and will lead to more mowing. Warm season grasses like Zoysia or bermuda should be fertilized only in late spring or summer.

Mow Properly for a Healthy Lawn and Fewer Weeds

Improper mowing practices result in more damage to lawns than any other cultural factor. Many lawns are mowed too short, infrequently, and with a dull blade. These practices

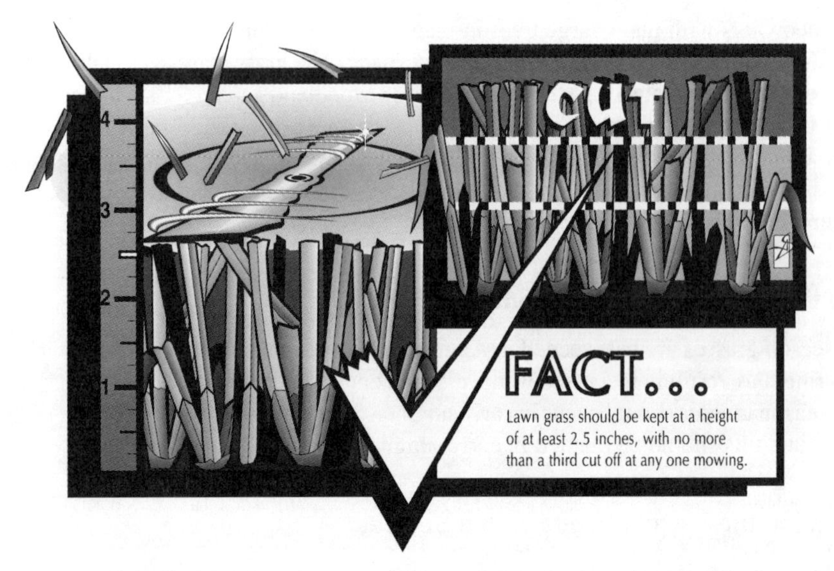

Figure 16.4 Grass height is important in keeping the plants healthy.

reduce the grass plant's ability to feed itself, which restricts root growth and increases insect, disease, and drought damage—ultimately resulting in more weeds.

The grass species in a lawn should determine the mowing height. Kentucky bluegrass, perennial ryegrass, and fine fescues should be mowed to a height of 2.5 inches whereas a mowing height of at least 3 inches is best for turf-type tall fescues. Warm season grasses can be mowed slightly lower.

It is recommended to remove no more than 1/3 of each leaf blade each time the grass is cut. This may mean mowing several times each week in the spring and fall, but usually only weekly or biweekly in the summer. Keeping mower blades sharp allows for a clean cut and better mower performance. Leave the clippings on the lawn unless they are needed for mulch or compost. This important practice will return nutrients to the soil to be taken up by the grass plants; it will not increase the buildup of thatch. When too many clippings remain on the surface of grass plants after mowing, spread the clippings uniformly with a rake or recut the lawn. Frequent mowing when the lawn is dry helps disperse the clippings properly and facilitate clipping breakdown and the return of the nutrients to the soil. Removing clippings is like removing free fertilizer from the lawn. The widely held belief that clippings contribute to a thatch layer is only a myth (Fig. 16.4).

Reduce Diseases and Weeds with Proper Irrigation

Improper watering is the second largest cause of lawn pest problems. Irrigation for established lawns should be thorough, with each watering wetting the soil to a depth of about 6 inches. Watering should not be repeated until the turf begins to show signs of drought stress, such as a bluish-gray color or footprints that do not disappear as someone walks across the lawn. Such signs are not cause to worry; the grass can withstand mild stress and will recover when watered thoroughly.

It is best to water between 4 and 8 A.M. because the evaporation rate is low early in the morning, allowing most of the water to soak into the soil, and in urban areas this is when

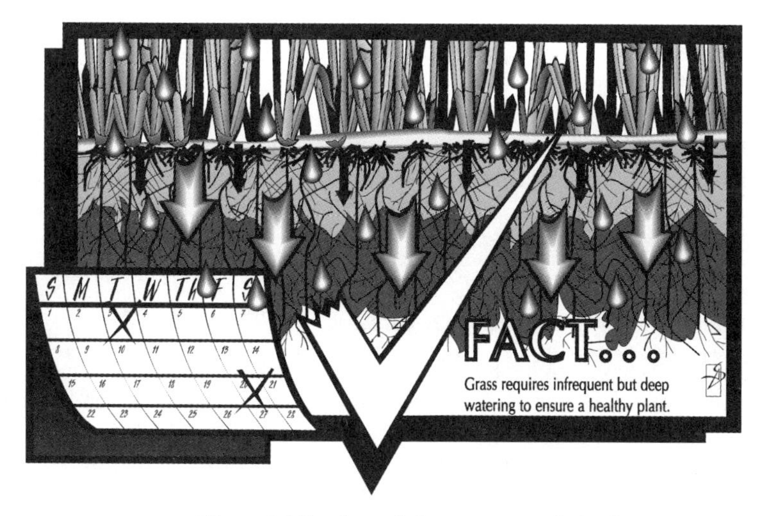

Figure 16.5 Grass thrives on proper irrigation.

municipal water pressure is highest. Midday watering is not advisable due to the likelihood of rapid evaporation. Watering early in the evening is not advisable because grasses, flowers and other landscape plantings are more likely to stay wet, making them more vulnerable to disease. Irrigation systems operated on a timer make sense only if the soil needs water at the time the system waters them (Fig. 16.5).

Reduce Pests by Controlling Thatch

Microorganisms and earthworms help decompose dead organic materials in and on the soil. This activity releases nutrients into the soil to be taken up by grass roots. Grass clippings decompose easily in this manner, and they do not contribute to thatch.

The thatch layer in lawns is composed of dead and living shoots, stems, and roots of grass plants. These parts of grass plants resist decay and accumulate on the soil surface, forming thatch. A small amount of thatch is desirable. However, the accumulation of more than 1/2 inch of thatch limits water and air movement, reduces the effects of fertilizer and pesticide applications, promotes shallow rooting, and increases disease and insect damage. Excessive thatch buildup can be managed with proper applications of moderate amounts of fertilizer and water.

Removal of excessive thatch is difficult and expensive. It may require the use of aerification equipment or, at worst, the physical removal of the sod (including the thatch layer) and the reestablishment of a new lawn. Power rakes (dethatching machines) are effective in minimizing thatch, but they are not effective in removing excessive layers of thatch.

Renovate the Lawn to Correct Major Problems

Homeowners often desire a simple solution to a major problem. A lawn that has been seriously damaged by insects or disease cannot be repaired with pesticides. These chemical compounds are formulated to prevent pests; but once serious damage occurs, pesticides can't bring dead grass back to life.

Pesticides are useless on neglected or poorly managed lawns or where soil and other environmental conditions prevent healthy grass plant growth. When the majority of a lawn consists of weeds and dead grass, complete renovation maybe necessary. The homeowner may choose to do the renovation with the aid of information available from the local Cooperative Extension Service; in some cases, however, it may be better to employ a professional firm to renovate the lawn. In either case, lawn renovation is a major undertaking that has to be done properly and completely.

Use Pesticides Only for Major Pest Outbreaks

Good cultural practices result in healthy grass that can withstand some damage from insects, weeds, and diseases. Regular inspections will help detect early insect infestations, the presence of weeds, and symptoms of disease. Pest populations and the damage they cause should be monitored closely. If it is determined that the situation does warrant a pesticide application, it is essential that the person making the application know what they are doing.

First, the pest must be positively identified, followed by the selection of a pesticide product suitable for both the pest and the site to be treated. The pesticide label must be read carefully and followed explicitly. If any part of the label is unclear to the applicator, it is important that a professional be consulted for clarification. Children and pets must be kept well away from the area during treatment; and in the case of lawn spray applications they must be kept away until the pesticide has dried completely. When granules are applied to lawns, they should be watered thoroughly into the soil and the grass allowed to dry. Some pesticide labels state specific periods of time during which people must stay off a treated lawn. Consideration of neighbors should be exhibited by posting "keep off" signs which indicate that a pesticide application has been made to the area.

Most lawn diseases go essentially unnoticed in the early stages. Once they have advanced to the point of recognition it is very unlikely that good control can be accomplished, even with fungicides. Most disease damage in lawns can be minimized by good maintenance practices such as fertilization and watering.

The best management for lawn weeds is to mow and fertilize appropriately. A dense lawn, mowed as needed and at the proper height, will prevent many annual weeds from becoming a problem. When a weed does create a problem, a positive identification becomes the first step in achieving control. If chemical control is determined to be the best approach, then comes product selection, followed by application according to label directions. It should be noted that some perennial weeds are difficult or impossible to manage; in those cases, a licensed professional lawn care company may need to be selected.

HOME GARDENS

Home vegetable and fruit gardening is a popular outdoor recreational activity in the United States. Among a variety of benefits to the gardener is the satisfaction of growing food crops at home. However, insects, weeds, and diseases can become a problem. The best pest management plan for home gardens is based on prevention strategies. The following pest prevention methods are recommended.

Plant Disease-Resistant Cultivars

Controlling infectious diseases once their symptoms appear is difficult; therefore, emphasis should be placed on disease prevention. The first step in preventing disease problems is to determine what diseases have the greatest potential to cause damage to the planting. The second step is to choose cultivars known to be resistant to those potential diseases. If disease-resistant cultivars are unavailable (or not chosen), preventive pesticide applications might be warranted, depending on the specifics of the particular disease, the type of planting, and other environmental conditions.

If a fungal disease is identified on a planting, a fungicide registered for management of that particular disease on the specific host plant must be applied on a regular basis (usually every 7–14 days). To be effective, thorough coverage of the healthy leaf and stem surfaces of the entire planting is necessary since most fungicides available to homeowners are not curative, but preventive; that is, they must be applied to healthy leaves to prevent invasion by the fungus.

Remove Plant Debris

Since many pests can overwinter in dead plant debris, it is important to remove vegetation as soon as its produce is harvested, thereby eliminating possible harborage of pests. Removal of expended plant material will render the garden less attractive to insects overall, thus lessening the potential for infestation of plants not yet harvested. Weeds should be removed before they have a chance to produce seed; if the plant tissue is healthy, it may be composted. In orchard situations, fallen fruit and leaves should be collected and removed or thoroughly mowed/mulched to ensure that they do not host pest populations destined to become a problem the following year.

Purchase Healthy Plants

Plants should be inspected carefully before purchase to be sure they are free of insects and disease. Discoloration and stunting may indicate pest damage. Plants purchased on sale are not always a bargain if they are stressed or come complete with diseases or insects.

Improve the Soil

Organic matter, such as compost, should be mixed into the soil at the beginning of each gardening season to improve fertility, aeration, and water penetration. Garden soil should be tested every 3–5 years and amended by fertilizing as required. A loose, fertile soil promotes healthy plant growth. Healthy, vigorous plants can usually out-compete pests and yield more high quality produce than those plants weakened by poor soil conditions. Uncomposited manures should probably not be applied to gardens during the growing season.

Use Proper Gardening Practices

Optimum growing conditions provide plants with fewer pest problems. Planting, spacing, watering, fertilizing, and managing weeds according to prescribed regional guidelines will result in gardening success. Support plants with stakes and trellises to foster air circulation, reduce plant-to-soil contact, and facilitate foliage exposure and drying.

Figure 16.6 Mulching provides many benefits to landscape plants.

Mulch to Prevent Weeds

A thin layer of grass clippings (not recently treated with herbicides), shredded leaves, straw, sawdust, wood chips, or compost can be spread around plants to manage weeds. As these organic materials decay, they add nutrients to the soil. Mulch helps to retain soil moisture and to moderate soil temperatures in the summer, reducing plant stress while reducing foliage and fruit contact with soil that can harbor pathogens (Fig. 16.6).

Black plastic can be used as a ground cover to conserve moisture and manage weeds. It will warm the soil in the spring but may retain too much heat for some plants in the summer. Plastic mulches should be used only with plantings of warm season annual vegetables and flowers, especially those that will quickly produce enough foliage to cover the plastic. Conversely, in wet clay soils, plastic mulches used around perennial plantings may keep soil continually wet, causing anaerobic conditions to develop. It also may encourage shallow rooting, leaving the plants more susceptible to cold injury.

Rotate Crops and Diversify Plantings

Vegetable crops should be rotated to different locations in the garden each year in an effort to reduce the potential for buildup of pest problems in the soil. If a serious problem occurs, the crop involved should not be replanted in the questionable area for at least 5 years. Care should be taken to avoid following like crops in sequential years—e.g., tomatoes following peppers, potatoes, or eggplants, or cucumbers following muskmelons, pumpkins, or squash.

Encourage Beneficial Insects

Less than 1% of all known insect species are considered pests. It is important to know the difference between beneficial and harmful insects. Beneficial insects include ladybugs, bees, green lacewings, praying mantids, dragonflies, and wasps. Spiders are examples of beneficial arthropods. Make your garden more hospitable to natural enemies by planting pollen and nectar flowers to feed them.

Scout for Pests and Symptoms of Pest Damage

Plants should be inspected for pests several times each week. Insects feeding on plants present the gardener with two choices: tolerate the damage, or try to manage it. Management methods that do not require pesticides include hand removal of the insects, early harvest, or tolerance of cosmetic or minor amounts of damage. For example, corn earworm damage to the tips of sweet corn ears can be cut off; outer leaves of cabbage damaged by insects can be removed; and superficial blemishes on the skin of an apple can be peeled off.

An insecticide application may be necessary when plants are in danger of being severely damaged or destroyed. Before using a pesticide the home gardener should understand the following.

- The pesticide label is a legal document.
- It is the user's responsibility to read and follow the label explicitly.
- Pesticides must be applied only at rates specified on the label (more is not better).
- Pesticide labels may stipulate waiting periods—called preharvest intervals—which must elapse between application and harvest of the crop for human consumption; fruits and vegetables always should be washed thoroughly before eating.

LANDSCAPE PLANTS

The diversity of plant material in the residential landscape translates to a wide range of potential pest problems. As mentioned previously, keeping plants healthy can prevent many of the potential problems. Plants that are stressed from undesirable cultural or site conditions grow poorly, and these less vigorous plants become prime targets for pests. For example, a landscape plant is more likely to have spider mite problems when it is planted in a hot, dry area. Likewise, plants that have been damaged by a lawn mower or string trimmer are more likely to have borer problems than undamaged, healthy ones. The best way to discourage pest problems is to fulfill the individual plants cultural requirements.

Select Recommended Plants

It is wise to choose proven performers—plants known to do well in the area intended for planting. Those known to have a history of pest problems should be avoided. Resistant plant species and cultivars should be used when available. When a pest becomes a recurring problem, consider replacing the plant with a nonsusceptible cultivar. Resistant varieties are well documented in Cooperative Extension Service publications.

Design a Diverse Landscape

Increasing plant diversity in the landscape makes it more difficult for pests to spread between plants. For example, there are three pests—the mimosa webworm, the honey locust spider mite, the honey locust plant bug—that can cause serious damage to honey locust trees. Merely planting other tree species among the honey locusts will reduce the damage potential.

Inspect Plant Materials before You Buy

Plants should always be inspected for pests before they are purchased. New (infested) plants can introduce pest problems to the landscape.

Provide a Properly Prepared Site

Site selection is critical: The site must be compatible with plant requirements. Things to consider are exposure to sunlight, drainage, soil pH, and nutrition. The soil must be prepared carefully and well beyond the current root zone, using soil amendments only as required for healthy growth. Lime should not be added unless a reliable soil test has indicated the need. "Heavy" clay-based soils require special care and preparation to allow and encourage roots to grow beyond the planting hole.

Plant at the Proper Depth

Planting at the proper depth will prevent damage that could make plants more attractive to pests. The hole for planting should be dug just deep enough to accommodate the root ball of the plant. Loosening soil under the root ball will result in settling and subsequent injury to the roots. Be sure the planting hole is wide enough to comfortably accommodate the existing roots.

Provide Proper Fertilization and Irrigation

Plants need proper and adequate nourishment for healthy growth, and deciding what fertilizer to use and how much to apply is crucial to successful landscaping. Soil tests are helpful in determining the needs of the soil itself, but it is equally important to address the nutritional needs of each plant as well. Transplanted trees, shrubs, and other landscape plants may benefit from a small application of fertilizer at planting, but care should be taken not to overdo it. Established plants usually receive adequate nutrients when the lawn is fertilized. Fertilizer should be applied not near the base of the plants, but rather where the feeder roots are which is often at or even beyond the drip line of the tree or shrub (Fig. 16.7).

Apply Mulch around Landscape Plants

Hardwood mulch will conserve moisture, protect plant roots from extreme temperatures, and prevent bark injury from lawn mowers and string trimmers. It should be applied and maintained at a depth of 2–4 inches and kept at least 2 inches away from tree trunks and the bases of shrubs. Black plastic mulch is sometimes used to conserve moisture and manage weeds in annual plantings. However, in landscape beds, plastic can prevent water and air

Figure 16.7 Care of transplanted plants is critical in maintaining their long-term health.

from reaching the roots of trees and shrubs. An alternative may be to use a "breathable" landscape fabric or just organic mulch.

Know When to Prune

There are good and bad times to prune. Caution should be taken to ensure that the pruning process does not leave trees and shrubs more vulnerable to damage from other factors. For example, oak wilt, a serious disease of red and white oaks, is spread by a fungus that beetles carry from infected trees to fresh wounds on other oaks; therefore, oaks should not be pruned in spring and early summer when insect activity is high.

Most plants can be pruned in late winter or early spring as long as the wood is not frozen, but trees and plants that flower in the spring should be pruned after the blooms fade if a floral display is important in the following season. Plant parts that pose a hazard to human health or person and property may need to be pruned at times that are less than optimal. However, never apply pruning tar to the cut.

Manage Noninfectious Problems Promptly

Problems that result from the transmission of living organisms (such as fungi and bacteria) from one plant to another are called infectious diseases. Many landscape plant problems are also caused by noninfectious factors.

Plant problems resulting from soil compaction or mechanical damage to the bark are caused by people, not pests. Environmental factors such as extreme cold or hot temperatures, drought, flooding, and lightning also can cause problems, many of which can be diminished by implementing proper cultural practices. Watering during periods of drought, mulching, improving soil drainage, and fertilizing appropriately can minimize injury from noninfectious factors. Pesticides cannot prevent damage caused by noninfectious factors.

Inspect for and Monitor Insects and Diseases

Regular inspections help prevent serious pest damage. Early season infestations can be more damaging than those that show up later.

Physically Remove the Pest

Physical removal and destruction is one effective way to manage many pest problems. For example, the Eastern tent caterpillar starts in a small group of eggs attached to a twig in a tree. After the eggs hatch, the caterpillars feed on leaves at night. During the day they hide from birds in webbed tents built in the forks of tree limbs. Much of their damage can be avoided by removing and destroying these tents when the caterpillars are still small and inside. Pruning out pest-infested branches can be an effective way to avoid further damage on vigorously growing plants. Euonymus scale on euonymus shrubs can also be controlled in this manner.

Identify the Location of Injury

When pest damage is identified, it is important to determine which part of the plant is being affected. This will help in deciding whether or not a pesticide is needed; pest damage to tree leaves usually is less serious than damage to trunks, stems, or roots, so pesticides might be ruled out if only the leaves are involved, especially late in the season.

Use Biological Management Methods

Biological management involves the use of natural enemies to reduce or prevent pest damage. Not all insects and mites are harmful to plants; many, in fact, are predators and parasites. It is important to distinguish between pests and their natural enemies. When the natural enemies are present, they should be given time to become established and manage the pest (Fig. 16.8).

Predators attack, kill, and eat large numbers of pests. Parasites lay an egg in or on a pest; and when the egg hatches, the new parasite consumes and usually kills the pest as it matures. Pathogens are free-living microscopic organisms (bacteria and fungi) that invade the pest and cause a disease that weakens or kills it. Some bacteria and fungal pathogens are also used as biological management agents. Predators, pathogens, and parasites are generally very specific in what they feed on.

Use Pesticides as a Last Resort

Pesticides should be used only when the pest threatens to cause serious damage to the host. When more than one pesticide is available to manage the pest on the specified site, the one that is least harmful to natural enemies should be selected.

PEST MANAGEMENT IN THE HOME

Practical methods are available for managing pest infestations in the home. The homeowner should be aware, however, that a one-time pesticide application will not usually provide

Figure 16.8 Natural enemies help reduce pest numbers.

long-term management except in some cases where roaches and rodents have been managed with a thorough, one-time applications of baits. The following methodology applies to many situations and should be considered in developing pest management strategies for use in the home environment.

Identify the Pest

Proper and complete identification of the pest is critical. For example, identifying an insect pest as a cockroach is not sufficient. German, American, Oriental, wood, and brown-banded cockroaches have distinct behavior patterns. A thorough understanding of those characteristics is essential in the development of a sound management strategy. It is from such information that one can determine how, when, and where to inspect for the pest. Cooperative Extension personnel are trained professionals available to assist homeowners with pest identification. They can provide information on the pest and how to manage it—and they also can advise the homeowner on whether or not a pest management professional should be consulted (Fig.16.9).

Inspect the Home for Pests

Routine and detailed visual inspections are critical to long-term pest management. The inspection should identify conditions favorable for pest infestations. Some insects (e.g., cockroaches, silverfish, carpenter ants) and rodents are active at night; therefore, nocturnal inspections conducted perhaps an hour or two after dark are important in determining where they are nesting, feeding, traveling, etc. The search (or inspection) should occur quietly and with the aid of a powerful flashlight, covering all areas that might possibly provide the pest with food, water, warmth, or shelter. During warm weather, the inspection should be extended to the outdoor perimeter of the structure (Fig. 16.10).

Figure 16.9 The proper identification of the insect, weed, or plant disease is the first step in solving a pest problem.

Use Good Sanitation Practices

Elimination of a pest's source of food, water, and shelter will almost certainly reduce the infestation and may even eliminate it altogether. Good sanitation practices for the home include proper management of household garbage; the avoidance of long-term storage of food products and clutter in garages, attics, and basements; and regular cleaning around appliances and in areas frequented by pets (Fig. 16.11).

Figure 16.10 Once the identity of the pest is known, nonchemical and chemical approaches can be used to reduce, remove, or eliminate the pest.

Figure 16.11 Good sanitation reduces pests.

Pest-Proof the Home

The best method of managing pests in the home is to exclude them—in effect, to "pest-proof" the house. Caulking cracks and crevices, repairing a leaky roof, adding or repairing screens, and sealing thresholds and chimneys all are ways of pest-proofing a home. All openings larger than 1/4 inch must be closed to exclude mice, whereas smaller openings and cracks must be sealed to exclude insects. Many "how to" home repair and improvement books offer tips on pest-proofing the home.

Use Traps for Minor Problems

Although traps are not effective in managing major, established pest infestations, there are many types of traps that can be used for small infestations of certain pests. Examples include snap traps for mice, sticky (glue) boards for cockroaches and mice, yellow jacket sugar traps, and sticky fly strips. The key to success with traps is placement. Since pests will not travel out of their way to locate traps, it is important to determine where the pest is most active or where the pest population is concentrated. A sufficient number of traps—for mice, 3 or 4—should be placed in those areas. It is always advisable to use more than one trap, and check them routinely several times per day.

Use Pesticides Properly

Understanding pest biology and behavior can reduce unnecessary or excessive use of pesticides in the home. Pesticides should be applied only as spot treatments in pest-populated areas identified during an inspection. Only pesticides labeled for indoor use should be used inside the home. It is important to realize that one application of a pesticide inside the house may not provide long-term control. Multipurpose pesticides labeled for outdoor use should not be applied indoors unless the label specifically permits it. Routine application of pesticides on baseboards and shelves and in attics is not necessary or effective for most pests.

Pesticide treatments around the perimeter of a building can be effective for some pest problems. In many cases, pest management professionals are better trained and equipped than the homeowner to make perimeter treatments. Perimeter treatments should be used only when needed, not as a cure-all for keeping insects out of the home. It is more important to identify and correct food, water, and shelter conditions that favor pests, both inside and out.

CHOOSING AN OVER-THE-COUNTER PRODUCT

The pesticide selection process begins once all other avenues of management have been exhausted. Start by asking yourself these questions:

What Is the Pest?

Accurate identification of the pest is essential to selecting a pesticide because pesticides are categorized according to the pest or problem that they manage. For instance, an insecticide will manage insects, yet will not manage fungi.

Where Will the Pesticide Be Applied?

Pesticide products are also labeled according to the site of application; for example, one flea management product might be labeled for outdoor use on turf, another for indoor use on carpets, and yet another for use on pets. Therefore, it is essential to identify the site to which the chemical will be applied.

Which Formulation Is Most Appropriate to Use?

Pesticides are packaged in a variety of dry and liquid formulations. Determine what formulation best suits your situation by considering the location and any related human, wildlife, or environmental elements.

Also consider application and safety equipment required. Ready-to-use products are handier than concentrates that require dilution and mixing. They often come in containers designed to double as the application device and usually require only basic safety equipment. Although they may be more expensive, their convenience and safety often justifies the cost.

How Much Do I Need?

Choose the quantity of product based on anticipated need, not just the best value. Excess pesticides in storage may lose their effectiveness over time and eventually require disposal—which often is not easily accomplished. Plan ahead and purchase only in quantities that can be used efficiently within the same year.

PESTICIDE LABELS FOR THE HOMEOWNER

Pesticide data submitted by a manufacturer undergoes intensive review by the EPA, which is similar to the federal Food and Drug Administration review of human medicines. EPA registration numbers are assigned to labels only after each product's active ingredient has been thoroughly investigated and only after scientific tests have been reviewed. The pesticide label is extremely important because this is where users find specific instructions for using the product effectively and safely. Pesticides are developed by manufacturers, registered with the EPA, and marketed and sold to the public with the assumption that users will comply

Figure 16.12 Following label instructions will help manage the pest in a manner that is responsible to family, pets, and the environment.

with label directions. A point that can never be overemphasized: reading, understanding, and strictly adhering to label instructions will allow for effective pest management and reduce the risk of negative health or environmental consequences (Fig. 16.12).

Pesticide products generally are recognized by their advertised brand names, such as Ortho Rose and Floral Dust, Fertilome Sevin, D-Con Kills Rats, and Hi Yield Diazinon. Every pesticide label must list the total amount of active ingredient (the actual compound that kills or repels the target pest) in the product. The ingredient statement is clearly marked and easily identifiable on the front of the pesticide label.

Labels provide the information necessary to answer questions on pesticide application, safety, storage, and disposal. Becoming familiar with the types of information on a typical pesticide label will help improve your understanding of the product. Once you understand how a pesticide label is organized, it takes only a few minutes to read all of the information found on most pesticide products. When you have doubts about the interpretation of information contained on the label, contact your county Extension Educator or other professionals who can assist you. They can help with pest identification; suggest appropriate management tactics; and make recommendations about products, rates, and application methods (Fig. 16.13).

Figure 16.13 Many garden centers are using signs to encourage their customers to follow label directions.

PRODUCT INFORMATION

Terminology	Sunniland Rose Dust	D-Con Kills Rats
Manufacturer	Sunniland	D-Con
Active ingredient	Chlorothalonil	Brodifacoum
Active ingredient (%)	20	0.005
Inert ingredient (%)	80	99.995
EPA registration number	9404-12-16	3382-66
EPA establishment number	16-VA-1	3282-OH-1
Formulation	Dust	Pellets
Classification	Insecticide	Rodenticide

SAFETY INFORMATION

Terminology	Ortho Weed-B-Gon Weed Killer	Lysol Disinfectant
Signal words	CAUTION	WARNING
"Keep out of reach of children."	Present	Present
Route of entry	Skin	Eyes, skin, mouth
Protective clothing	Goggles, long pants,	Rubber gloves shoes, gloves
Practical treatment	Wash skin with soap and water.	If swallowed, drink a large quantity of milk.
Emergency phone	800-457-2022	Not indicated
Environmental toxicity statement	Toxic to aquatic invertebrates	Not indicated
Specific action	Do not apply directly to water.	Do not get into eyes or reuse containers.
Reentry statement	Do not permit children or pets to enter treated areas until spray has dried.	Not indicated

Definitions

Active ingredient Component of a pesticide formulation that is toxic to the pest.

Brand name The name under which a pesticide is marketed.

Environmental toxicity statement Precautions for protecting the environment.

EPA establishment number Identification number for the specific manufacturing location.

EPA registration number Number that is assigned to a particular registrant's product.

Inert ingredients Carriers that are not toxic to the target pest.

Practical treatment Procedures for responding to a human exposure emergency.

Protective clothing The *minimum* safety equipment that must be worn.

Reentry statement Precautions about reentering treated areas.

Route of entry Site where pesticides might enter the human body: mouth, skin, eyes, lungs.

Signal words Indicators of acute toxicity to humans: DANGER means highly toxic; WARNING means moderately toxic; CAUTION means slightly toxic.

Specific action *Do not* and *avoid* are directives to look for.

Use inconsistent with the label It is a violation of federal law to use any pesticide product in a manner inconsistent with its label. Use includes everything from purchase to the disposal of the container.

The label is the law!

USE DIRECTIONS		
Terminology	Off Insect Repellent	Natural Guard Pyrethrum Powder
"Use inconsistent with the label is a violation of federal law."	Present	Present
Pests controlled	Mosquitoes	Fleas, ticks, gnats
Sites of application	Human skin	Cats and dogs and clothing
Application rate	Use enough to cover skin.	Dust thoroughly.
Application method	Hold container 6–8 inches from skin or clothing.	Not indicated
Storage	Store away from heat.	Not indicated
Disposal	Wrap container, put in trash.	Approved waste disposal facility
Spills	Not indicated	Stop this spill by repositioning or repairing.

PURCHASING AND USING GRANULAR SPREADERS

Granular formulations are dry materials for distribution via a piece of equipment commonly called a spreader. There are two types: the drop spreader and the rotary spreader.

Drop Spreaders

Drop spreaders are popular with do-it-yourself applicators for applying dry fertilizers, weed and feed products, and granular insecticides to turf. The product is poured into a hopper mounted between two wheels. At the bottom of the hopper is a rotating agitator bar, driven by the movement of the wheels, which stirs the product and keeps it flowing freely and uniformly.

Drop spreaders are aptly named because the granules drop straight down between the wheels through a set of adjustable openings, collectively called the gate, in the bottom of the hopper. The amount of product that falls through the gate is governed by adjusting the openings according to the settings or numbers on the hopper's scale. Higher numerical settings correspond to larger openings and will result in more of the product being dropped.

The size of individual granules affects the rate at which they can fall through the gate openings. Switching to a new product that contains smaller or larger granules, without changing gate openings, set previously for another product, can result in significantly more or less of the new product being applied (Fig. 16.14).

(a)

(b)

Figure 16.14 (a) A drop spreader. (b) Numerical markings used to set the application rate for a drop spreader.

Combination products consisting of large fertilizer granules mixed with small pesticide granules affect the delivery rate of a drop spreader significantly. Always consult the product label to determine approximate settings for your equipment.

The Good News about Drop Spreaders

- Drop spreaders are accurate and easy to use. The applicator usually can see the wheel tracks from the previous pass across the area and use them as a guide, overlapping slightly to ensure uniform coverage. Sometimes it is also possible to see the product on the ground after it is dropped.
- Precise placement is easy to achieve. The product drops down between the wheels, preventing granules from moving off-target or falling onto an area where it isn't intended such as a sidewalk, driveway, dog kennel, goldfish pond, flower bed, street, storm sewer, pool, sandbox, etc.
- Drop spreaders are efficient in small areas. They are easy to operate on small lawns and narrow strips of grass in a side yard or between the street and sidewalk.

The Bad News about Drop Spreaders

- Drop spreaders leave little room for error. There is no forgiveness when passes are not overlapped correctly. Failure to ensure that the wheel tracks overlap slightly means that some turf will remain untreated, and pests living in untreated strips may not be controlled. Turf may appear striped when fertilizer is not applied uniformly (Fig. 16.15).

  Too much overlap is even worse. Overlapping too far can burn or kill grass plants because twice the required amount of product is applied. Overapplication of a product wastes the product, money, and even time. Running out of product before treating the entire area will require yet another trip to the store to make an additional purchase.
- A different setting is required for each product used. Settings must be changed for each product, because granules differ in size, weight, and shape. Carrier materials in granules also can affect their flow through the spreader.

Figure 16.15 The improper use of a drop spreader can result in a "striped" lawn when the applicator fails to overlap each pass.

- Corrosion can occur. Drop spreaders must be thoroughly cleaned with water and dried properly after each use, especially when a fertilizer is used. Fertilizer products can rust metal hoppers and may render the adjustable gate openings or wheel inoperative. Lubricate all moving parts with a light oil after cleaning.
- There is no automatic shutoff. Granules continue to drop after the spreader stops moving. The gate openings must be closed manually—and quickly—each time the spreader is stopped.
- Clogging can occur. The gate openings can become clogged with the product when the bottom of the hopper brushes against tall wet grass. Some pesticides must be applied to wet turf so that the granules will stick to the grass blades (in the case of fungicides) or to the weeds (in the case of herbicides).

Rotary Spreaders

Push-type rotary spreaders are gaining popularity among homeowners who treat larger yards or irregular or hilly landscapes. Gravity causes granules to drop through an adjustable gate opening (or openings) in the bottom of the hopper onto a spinning plate activated by the turning wheel axle as the spreader is pushed across the treatment area. The spinning plate slings the granules 8–12 ft out in front of the equipment in a semicircular arc. Smaller, hand-held rotary spreaders are available for use in smaller areas; they have a crank that is turned by hand to throw the granules outward (Fig. 16.16).

As with the drop spreader, the openings on rotary equipment can be set by selecting the appropriate number (or letter). Read the product label or your equipment manual to determine the initial settings to use in calibrating the spreader.

Rotary spreaders do not distribute granules uniformly because of the tapered, fan-shaped pattern created by the spinning plate. More of the product is distributed in the center of the arc than on either side. This uneven distribution requires that you overlap each pass by 30–50% as you move back and forth across the site, thus throwing additional granules onto the outer edge of the previous pass.

It is important to read the product label or the spreader instruction manual for instructions on how to determine the amount of overlap required. If that information is not provided, space the passes far enough apart so that the outside edge of the distribution arc just touches the wheel tracks made during the previous pass. Overlapping each pass in this manner will ensure uniform, complete coverage.

The Good News about Rotary Spreaders

- Rotary spreaders save time. They dispense the product over a wider swath, per pass, than do drop spreaders, allowing quicker treatment of the total area.
- Low maintenance is required. Most rotary spreaders are manufactured primarily from plastic parts, which eliminates rust problems. They are easy to clean with a hose after each use. Be sure to lubricate any metal parts with a light oil after cleaning.
- Rotary spreaders are compatible with many products. They permit the use of many different products, including some that have large granules.
- The hoppers of rotary spreaders are higher than the hoppers of drop spreaders. Wet, tall grass will not clog the gates.

(a)

(b)

(c)

Figure 16.16 (a) A rotary spreader. (b) Fertilizer pellets fall onto the spinning plate. (c) Settings for a rotary spreader, which determine how much fertilizer falls on the spinning plate.

The Bad News about Rotary Spreaders

- Uneven distribution is a problem with combination products. Small, light particles tend to fall to the ground close to the spreader, whereas larger, heavier ones are thrown farther out. Products that contain a mixture of large fertilizer granules and small pesticide granules, for instance, may be difficult to disperse evenly.
- Uniform coverage may be more difficult to achieve than with drop spreaders. Proper overlap is essential. It is important to overlap the right amount on each pass and to do so consistently.
- Off-target placement can be a problem. When operating a rotary spreader, you must prevent the product from being thrown onto areas where it should not be applied: sidewalks, driveways, patios, play areas, sensitive landscape plants, gardens, etc. In addition to managing the swath pattern, you must be aware of cross winds that could possibly blow granules off-target.

PURCHASING AND USING LIQUID SPRAYERS

Liquid pesticides are needed when the aboveground parts of plants require treatment to manage pests. Some liquid formulations come packaged in their own sprayer, ready to use; no dilution is required. Liquid concentrates, however, must be diluted with water in the spray tank, as do wettable powders, which are also applied as liquids. There are some points to consider when using liquid pesticide application equipment.

- **Agitation** A few products will dissolve in water and remain permanently in solution, but most form suspensions. Over time, the suspended particles separate from the water and settle to the bottom of the sprayer. Continual agitation during application will keep most liquid products suspended in the water inside the sprayer; a good rule of thumb is to agitate every 5 min.
- **Spray tank pressure** Tank sprayers are equipped with a plunger in the lid for pumping/forcing air into the tank to increase the pressure inside. An application wand and nozzle are attached to the tank with a hose; and when the trigger on the wand is squeezed, the air pressure inside the tank forces liquid out through the nozzle.

  As the sprayer is used, the pressure inside drops; if you do not repump the tank, eventually the pressure will drop too low to force the liquid out of the tank. Therefore, it is important to pump the tank periodically during an application to maintain a consistent flow and to ensure uniform coverage of the area being treated. When the tank is nearly full, it takes only a few seconds to pressurize it; but as the product is sprayed out and the pressure level drops, pumping it back to full pressure takes a little longer.

  The amount of pressure inside the tank determines the force with which the spray exits the nozzle. When the pressure is high, the liquid exits rapidly as a fine mist; low pressure results in a slower exit of larger spray droplets. Controlling the pressure, and therefore the sprayer output, can facilitate applications where drift is a concern. Nozzle adjustments may be necessary as well. For example, when spot spraying (treating individual weeds), low pressure and large nozzle openings help produce a thick stream of liquid that can be directed at the target. Conversely, when full coverage is needed, high pressure and small nozzle settings help produce a mist that covers the entire plant. A shielded sprayer that can reduce unwanted contact of herbicides to desirable plant materials can be constructed with a gallon milk jug or bleach bottle. Simply cut the bottom out of the gallon bottle, insert the nozzle through the cap hole and adjust to a height and spray pattern.

- **Spray nozzles and tips** As described previously, pressure inside the spray tank forces liquid out through a nozzle on the end of the wand when the trigger is squeezed. The opening in the nozzle is called an orifice, and the diameter (width) of the orifice regulates how much liquid can pass through.

  Some sprayers have adjustable nozzles that enable users to adjust the size of the orifice. When the nozzle is tightened clockwise, the size of the opening is reduced and tiny spray droplets exit the nozzle as a fine mist. If the nozzle is loosened counterclockwise, the diameter of the orifice increases, producing larger spray droplets that form a coarse spray pattern or a solid stream.

  Some specially designed nozzles have interchangeable tips shaped to deliver the spray in a specific pattern. For example, a fan tip delivers the spray in a lens-shaped or rectangular pattern and is useful for broadcast spraying lawns. Hollow- and solid-cone tips deliver the spray in a circular pattern, which is good for spraying plants. Check with your equipment dealer for tips that can be used on the nozzle you have. In some instances, nozzle tip kits containing several types of nozzles are available.

- **Walking speed and pattern** Walk at a consistent, comfortable speed when applying liquid products. The walking speed affects the amount of product that falls onto the treatment area, because the faster you walk, the less product reaches the target. Conversely, as you walk slower, more product is applied. In addition to maintaining a consistent walking speed when applying a liquid product, it is important to walk back and forth across the lawn in a precise pattern.

  Keep an eye on the amount of mixed product that is being applied as the job progresses: especially, note how much has been used at the midway point. If more or less than half remains, slow down or speed up the rate of application, respectively.

- **Hand and arm movement** Some liquid applications require the applicator to swing the spray wand from side-to-side while walking in a straight line. Moving your arm quickly while spraying a broad swath gets the job done quickly, but the greater your speed, the less product is applied. A narrow swath (and less material applied per unit area) and slow arm movement will result in more product being applied.

- **Nozzle height above the ground** When applying a liquid product, the height that the nozzle is held above the ground determines the swath (width) covered and the amount of product applied. Raising the nozzle farther from the ground creates a wider swath with less product applied per area, and lowering the nozzle creates a narrower swath with more product applied per area.

Hose-End Sprayers

Hose-end sprayers, which attach directly to the end of a garden hose, are easy to use. They are constructed with three main parts: a plastic container to hold the product, a lid that attaches the hose to the container, and a siphoning tube.

Water moving through the hose creates a suction that pulls the product from inside the container up through the siphoning tube and distributes it uniformly into the water flow. Always obtain quality equipment that has an antisiphoning device to prevent backflow into your water system. Backflow can result if the water is turned off at the faucet or if a sudden drop in water pressure occurs. There are three main types of hose-end sprayers: fine spray (mixing required); coarse spray (mixing required); and dial-in (no mixing required).

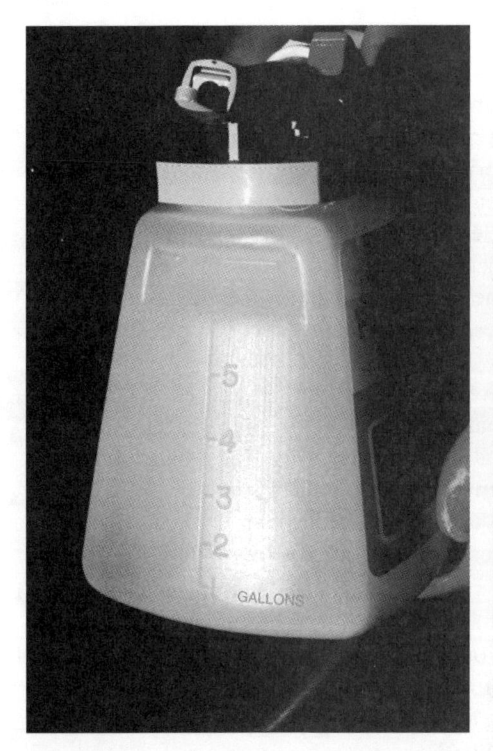

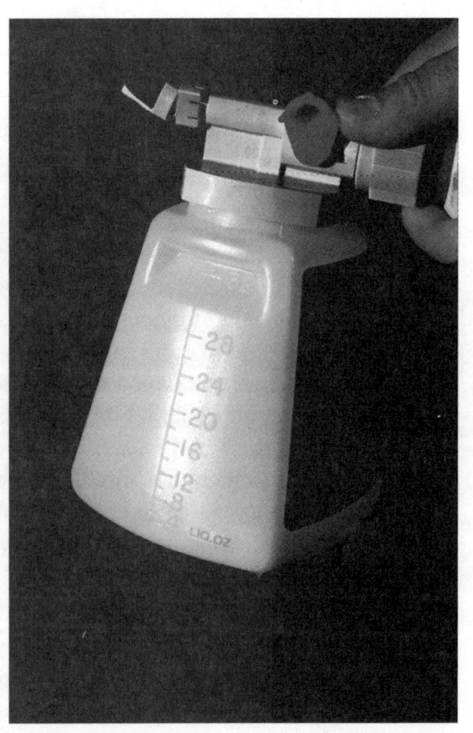

Figure 16.17 Some hose-end sprayers require water and pesticide to be mixed together in the container.

Fine-Spray Hose-End Sprayers *(Mixing Required)* These sprayers deliver a fine spray of small droplets to coat the outer leaf surfaces of flowers, small trees, and shrubs. They are excellent for applying fungicides for disease management and for applying insecticides to insects that feed on plant leaves. Fine-spray hose-end sprayers cannot be used to manage soil-dwelling insects such as white grubs because the mist of pesticide dries quickly on the blades of grass and the pesticide does not make it down into the soil to contact the pest.

The pesticide product is diluted with water inside the hose-end container. The plastic container is marked conspicuously with two scales: ounces for pesticide and gallons of water. Read the product label to determine how many ounces are needed for the job. Pour the required amount into the container and add water to the gallon level needed for the application (Fig. 16.17).

Coarse-Spray Hose-End Sprayers *(Mixing Required)* This type of equipment works in a manner similar to the fine-spray type but delivers larger particles as a heavier spray. This delivery method is desirable when the spray material must reach past the grass plants, into the soil. Coarse-spray hose-end sprayers have limited use in treating flowers, shrubs, and trees because the large droplets do not cover leaf surfaces uniformly. The large droplets also do not stick to leaf surfaces as well as fine-spray particles, and, in fact, bounce off. The instructions for filling coarse-spray hose-end sprayers are the same as those given previously for fine-spray hose-end sprayers.

Dial-in Hose-End Sprayers *(No Mixing Required)* These sprayers usually cost a little more than the fine- and coarse-spray hose-end sprayers, but the extra cost is justified by

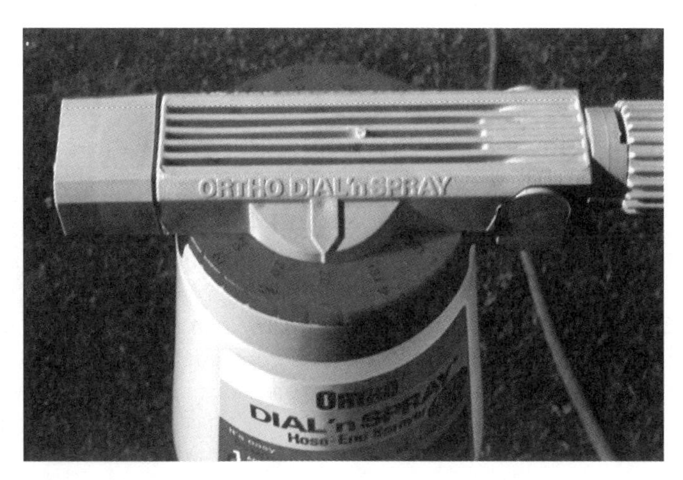

Figure 16.18 Dial-in hose-end sprayers allow the user to dial in the desired rate.

one distinct advantage: The need to measure and mix the product with water within the container is eliminated. Only the concentrated product is poured into the container; then the user simply dials in the desired rate on the sprayer as specified by the product label. (Fig. 16.18). The dial mechanism meters the correct amount of concentrate for each gallon of water that passes through the sprayer. There are no markings on the sides of the container since no mixing is required. If unused concentrate is left in the sprayer after the job, it should be returned to its original container.

The Good News about Hose-End Sprayers

- They are easy to use. Hose-end sprayers are uncomplicated, lightweight, and portable.
- They require very little maintenance. Rinsing after finishing the job is the only requirement.
- They are versatile. Hose-end sprayers can be used to apply products to the lawn, landscape plants, and vegetable plants.
- They are inexpensive. Hose-end sprayers cost between $8 and $20.

The Bad News about Hose-End Sprayers

- Leftover dilution creates a storage/disposal problem. Leftover diluted material cannot be returned to the original product container so it should be used up on the plants being treated.
- Application accuracy requires measuring walking speed (when treating large areas) or determining the amount of time required dispersing the right amount of spray (when treating individual plants). Failure to maintain a consistent walking speed for the proper length of time when treating lawns, bushes, or trees can result in too much or too little product being applied.
- The outside surface of your garden hose can become contaminated with the product as it is pulled across treated areas of the lawn.
- Extra garden hose may be required. The cost of additional garden hose needed to reach the farthest point of the treatment area may offset the low initial cost of a hose-end sprayer.

- Most hose-end sprayers are not flexible. The simplicity of their design does not permit adjustment of nozzle tips or spray patterns.

Compressed Air Sprayers

Two common types of compressed air sprayers for home use are the backpack and hand-held tank sprayers. Both are extremely useful when only a few gallons of material are needed. Generally, these sprayers hold 1–3 gallons of water.

Backpack Sprayers Backpack sprayers are carried on the back of the user and held by adjustable shoulder straps that allow the weight of the tank to be distributed evenly. A hose extends from the bottom of the tank around one side of the user. Attached to the hose is a spray wand with a trigger mechanism that the user squeezes to expel the tank mixture. Many backpack sprayers are equipped with a pump lever that extends from the bottom of the equipment around one side of the user. This permits the applicator to pump the sprayer regularly, with one hand, while moving the spray wand with the other hand. This maintains the pressure inside the tank and keeps the flow rate uniform (Fig. 16.19).

Figure 16.19 A backpack sprayer.

Figure 16.20 A hand-held tank sprayer.

The nozzle tip on backpack sprayers usually can be adjusted to create various spray patterns. This important feature allows the user to select the pattern most appropriate for uniform coverage of the target weed, plant, etc. Backpack sprayers generally are sold with a hollow cone tip which expels spray in a circular pattern; it is often the best choice for spraying shrubs or other plants. A fan tip, which creates a rectangular spray pattern, is best for spraying lawns. Check with the dealer where you purchased your equipment to obtain this special tip.

Hand-held Tank Sprayers The only real differences between a hand-held tank sprayer and a backpack sprayer are the size of the tank and how it is carried by the user. A hand-held tank sprayer typically holds 1–3 gallons of liquid and is carried in one hand while operating the spray wand with the other. When the pressure inside the sprayer begins to decrease, the user must set the sprayer down and pump it. As with backpack sprayers, hand-held sprayers come with adjustable nozzles (Fig. 16.20).

Minitank Bottle Sprayers When only a small amount of pesticide is needed, such as for spot treating a few weeds in a lawn, you might prefer a 1-quart bottle sprayer similar

Figure 16.21 A minitank bottle sprayer.

to those in which window cleaning products are sold (Fig. 16.21). Empty spray bottles are available at discount and department stores as well as at your neighborhood garden center. Purchase one that has an adjustable nozzle that will allow you to select the setting most appropriate for the job at hand. Using a permanent marker, label the bottle sprayer in large block letters with the name of the product to be used, and always store the empty bottle sprayer in a locked cabinet with other pesticide products and equipment. Do not use a bottle that has previously held, or is labeled for, any other product!

When dealing with very small quantities of pesticide, there is no room for error. It is important to carefully measure the amount of pesticide that you put into a bottle sprayer. Too little or too much can render your application useless—or dangerous.

Since it is usually not practical to calibrate a bottle sprayer, you must carefully estimate the total amount of product needed for a job, since it is important to use the entire quantity mixed. Never save and store pesticide products that have been mixed with water, because they lose their potency over time.

Pressure is created in a bottle sprayer by squeezing a trigger on the lid. Each time the trigger is squeezed, pressure forces a small amount of liquid through the nozzle.

The Good News about Compressed Air Sprayers

- Compressed air sprayers are portable. Backpack and hand-held compressed air sprayers allow the user to move about freely.
- They are efficient. Compressed air sprayers allow more precision than larger equipment in treating individual or small groups of plants.
- Their flexibility is a plus. Adjustable nozzles and interchangeable nozzle tips allow the applicator to select the spray pattern most appropriate for the job.
- They are easy to use and affordable. Most tank sprayers are moderately priced. Factors that affect price are the material they're made of (plastic, brass, or stainless steel) and features such as adjustable nozzle tips and pressure gauges, size, and overall quality.
- Compressed air sprayers allow precise direction of the spray. A predictable spray pattern permits the user to accurately place the product where it is needed, such as along sidewalks or around flower beds, and helps to prevent off-target applications.
- They are durable. When maintained according to the manufacturer's instructions, compressed air sprayers will last for many years. Individual parts usually can be obtained, permitting repair rather than replacement when the sprayers malfunction.

The Bad News about Compressed Air Sprayers

- Usefulness of compressed air sprayers is limited. Due to their small tank capacity, they are inefficient for treating large areas or large groups of plants.
- Experience governs effectiveness. The user has to develop a feel for the amount of pressure that must be maintained to ensure uniform and complete coverage.
- Proper maintenance is essential. Backpack and hand-held compressed air sprayers generally require more maintenance than hose-end or bucket sprayers.
- Compressed air sprayers are cumbersome. Lifting and carrying a full compressed air sprayer—especially a backpack sprayer—can result in muscle strain and fatigue.

Bucket Sprayers

Bucket sprayers are devices that use a bucket for mixing and supplying the pesticide product.

Trombone Bucket Sprayers Trombone sprayers are useful in treating tall trees that cannot be covered by small compressed air sprayers and hose-end equipment. They are particularly useful for spraying fruit trees.

Trombone bucket sprayers have a sprayer wand made up of two pieces. The front piece slides back and forth on the other, much like the slide on a trombone. This movement creates a suction that pulls the mixture out of the bucket through a small hose clipped to the side. The opposite end of the hose is attached to the sprayer (Fig. 16.22).

Figure 16.22 A trombone sprayer.

The bucket is not stationary and is carried to the job. There is no attachment to a faucet, nor is a garden hose used to deliver the product.

The Good News about Trombone Sprayers

- Trombone sprayers can be used to treat tall shrubs and small trees. Fully primed, a trombone sprayer can reach trees as tall as 40 ft.
- Their range of treatment is unlimited. The sprayer is not tethered to a garden hose and does not have to be used in proximity to a water faucet.
- Contamination possibilities are minimal. A garden hose is not used to deliver the product.
- Trombone sprayers are easy to operate. Equipment is self-contained, portable, and relatively easy to maintain; and it requires minimal storage space.
- They are easy to monitor. The level of the bucket mixture is observable while spraying.

The Bad News about Trombone Sprayers

- Trombone sprayers are relatively expensive.
- They are somewhat inconvenient. Larger jobs require the user to stop and refill the bucket several times.
- They are awkward. The pesticide applicator has to lift and carry the bucket as the job progresses, and it is very heavy when full.
- Buckets are easy to turn over.
- Drift can be a factor. Movement of the spray off-target, due to a breeze or wind, can be a problem when spraying trees at high pressures.

OWNER'S MANUAL AND INSTRUCTIONS

New pesticide application equipment generally is sold with the manufacturer's instructions on how to use it properly. Instructions may be found in an owner's manual or come as loose-leaf papers; sometimes they are printed on the application equipment itself. Most equipment manuals provide

- a product registration card,
- the model number of the equipment,
- the phone number of the manufacturer's customer relations department,
- an address where a new set of instructions or an owner's manual can be obtained,
- a diagram and corresponding numbered parts list (and sometimes prices),
- instructions on how to set up and use the equipment,
- routine maintenance procedures needed to extend the life of the equipment, and
- the manufacturer's warranty and related disclaimers.

Operating instructions for equipment used only occasionally are sometimes misplaced or inadvertently thrown away. Guard against these possibilities by designating a safe place where instructions can be accessed quickly and easily. A file box works well and can also accommodate other household appliance and tool manuals. Instructions printed directly on equipment can wear off, or they can become scratched or soiled beyond recognition. Copy the instructions onto a piece of paper when the equipment is new and place it in the file box for safekeeping.

WEARING PERSONAL PROTECTIVE EQUIPMENT

Most pesticides that homeowners use contain instructions for reducing exposure (e.g., avoid contact with eyes or skin). Most labels state the types of clothing that must be worn during the handling, mixing, and application processes. Protective equipment requirements differ from product to product. For instance, whether the product comes ready to use or as a concentrate that requires mixing influences the protective clothing requirement assigned to the product. The potential health hazards and the precautions necessary to prevent injuries differ dramatically between those two types of products. For example, ready-to-use sprays and baits are accurately mixed and packaged by the manufacturer to ensure the proper

Figure 16.23 Chemical-resistant gloves may be required by a pesticide label.

diluted concentration. As such, these diluted products pose little risk to the user, provided the products are used according to label directions (Fig. 16.23).

Some products are packaged so that the user handles and mixes a concentrated form of the pesticide. Improper mixing, storage, and disposal of a concentrated pesticide constitutes misuse which can pose serious risks to people, pets, or wildlife. A long-sleeved shirt, long pants, shoes, and chemical-resistant gloves are minimal requirements for reducing exposure to pesticide concentrates or sprays. By covering hands, forearms, and legs, the dermal exposure potential can be reduced by 95%. Additionally, it is a good idea to wear safety glasses when handling pesticides because of the corrosive nature of some of these materials. With your next pesticide purchase, remember to ask the salesperson for chemical-resistant gloves and glasses if required by the label. Only by reading the label will you know whether additional protective clothing is needed to adequately safeguard yourself against exposure. It's better to be safe than sorry!

Handling Pesticide-Contaminated Clothing

Always assume that clothing worn while working with pesticides has been contaminated. It should be laundered separately after each use: Never launder pesticide-contaminated clothing with the family wash. The longer pesticide-contaminated clothing remains unwashed, the more difficult it is to remove the pesticide. See Chapter 11 for full details on laundering pesticide-contaminated clothing.

REDUCE PESTICIDE DISPOSAL THROUGH SMART BUYING

The amount of time committed to pest management in and around the home often governs the kinds and amounts of pesticides homeowners should purchase. Buying more than is needed often results in storage and disposal problems associated with leftover pesticides. For instance, yard enthusiasts get much satisfaction from the aesthetic beauty of a well-maintained landscape around their property. These individuals will devote considerable time looking for early signs of pests in their lawns, vegetable gardens, shrubs, and flowers. Such enthusiasts seldom accumulate unused pesticides because of their experience in knowing which pests to expect in a typical year, which products have worked in the past, and how much of a given product they will need. Those who are less experienced often allow surplus pesticides to accumulate around the home. For the following reasons, pesticides often are relegated to storage shelves: they are difficult to mix and apply; they are not suitable for the task at hand; or too much product was purchased. When pesticide use is a necessity, purchase ready-to-use products or concentrates that can be used up within a short period of time.

Most homeowners purchase pesticides with the intent of using them within a reasonable time period. However, surveys have shown that 25% of all home-stored pesticides remain unused either on the shelf or under the kitchen sink, for over a year. This ultimately can cause problems. Exposure of products to hot and cold temperatures is a problem associated with leftover pesticides. Most labels stipulate that the product be protected from freezing temperatures—a goal seldom achieved if products are stored in the garage. When pesticide products are exposed to extreme temperature fluctuations, their effectiveness is greatly reduced. Therefore, it is imperative that you not store concentrated or ready-to-use pesticides for prolonged periods.

Unfortunately, some households dispose of leftover pesticide products by dumping them in the household trash or by pouring them into the sink, toilet, street, gutter, sewer, or onto the ground. Such disposal "sites" are unacceptable. Household pesticide product labels generally will indicate that partially filled containers may be wrapped in several layers of newspaper and discarded in the outdoor trash. For many people this disposal option is neither acceptable nor environmentally sound. Unused pesticides are best disposed of by using the products on the sites indicated on the label. Additionally, discard empty containers in the household trash so that they are not reused.

PREVENTING ACCIDENTAL POISONINGS IN THE HOME

It is scary to think that your children could be poisoned. It can be overwhelming to look around your home and see all of the products that could potentially harm your child. Don't be discouraged and don't assume it's an overwhelming task to deal with these products and to make a safe place in your home to store them. Many times moving a few products to a higher shelf or locking the products up in a cabinet can eliminate most problems (Fig. 16.24).

Children are quick learners too. Encourage older children to avoid hazardous materials and reward them when they do so. Tell them what the safety rules are, and most children will quickly learn what they must and must not do. A few minutes spent talking to your children and putting items out of reach will go a long way toward avoiding having to call 911 or the poison control center.

Even with our best intentions, poisoning accidents can still occur around the home. Preparing yourself and the other adults in your home to handle poisoning emergencies, if they occur, is also wise. That's why you need to plan what you would do in the event you are presented with an emergency situation. Spending a few minutes taking these fundamental safety precautions and planning what to do in the event of an emergency could make the difference between a happy ending and a tragic one. Seconds do matter!

Figure 16.24 Spend a few minutes implementing these fundamental safety precautions.

How Poisonings Happen

Most childhood poisonings occur to children under the age of 5 years:

- Children this age are in a period of learning and discovery. Whatever they see, they try to pick up and examine, and often the object ends up in their mouth.
- During these years, children's mobility increases and areas that were previously inaccessible are no longer. Parents are sometimes surprised to find that their toddler can devise a way to reach something he wants, such as pushing a chair or table into position to reach things on the kitchen counter or in the medicine cabinet (Fig. 16.25).
- Young children are attracted to bright colors, pretty packaging, and appealing fragrances. And since they cannot read labels, they don't know that the lemony-smelling furniture polish is not something to drink.
- Children often imitate adult behavior. If they see their parents swallowing medication, they may assume that the tablets or capsules are edible.
- Children who are hungry or thirsty are more likely to ingest medicine or other household products than children who are not. The very fact that they want something to eat or drink increases the likelihood that they will ingest something that smells good to them.

Figure 16.25 Toddlers can find many ways of reaching areas that you don't want them reaching.

- Childhood poisonings often occur when parents or caretakers are busy or distracted. A good example is when meals are being prepared. It is common for children to have free run of the house as adults focus attention on preparing the meal.

Primary Sources of Poisonings

Pharmaceuticals (prescription and over-the-counter medications, vitamins, etc.) account for approximately half of all poisonings reported to poison centers. Cleaning supplies and personal care products represent approximately 20% and poisonous plants, insect bites and stings, and pesticides each account for approximately 5% of poisoning cases. Most children are poisoned by ingestion rather than by inhalation or dermal contact.

Understanding Emergency Response Teams

Ask adults whom they would you call if they had a poisoning emergency at home and most will answer 911 or poison control. But do most people know what each of these agencies does or what information would be needed to help during an emergency?

Figure 16.26 911 can respond immediately to an emergency.

911—We're On Our Way! Calling 911, first, when someone is exposed to toxic materials, may not be necessary; in fact, most poisonings are nonemergency events. Over 96% of all children exposed to toxic substances suffer few, if any, effects. However, if a poisoning victim of any age is unconscious, having difficulty breathing, or having seizures, call 911 immediately (Fig. 16.26).

"Hello. 911. What is the emergency?" It is a phrase that we hear time and time again in the news and on television. The 911 system is a countywide emergency program that coordinates hospital, ambulance, and fire department response teams in emergency situations.

Most 911 systems are *enhanced,* which means that the dispatcher answering the incoming call is instantly provided the name and address that correspond to the phone number from which the call is placed. This computerized information system is crucial when the caller is either unable to speak or so frantic that they cannot be understood (e.g., a child too scared to talk, or an individual screaming incoherently in the heat of the moment).

The dispatcher must ask the caller for the address, even if it is on the computer screen, and he may ask for directions. He will continue to collect pertinent information about the event while dispatching emergency personnel to the scene. He may ask how many people are present and the names of those involved in the poisoning incident, and he will keep the caller on the phone until help arrives.

Poison Information Centers Poison control centers operate 24 hr a day, 7 days a week. Trained professionals answer questions, help determine the seriousness of a poisoning, and give specific advice on how to deal with the incident. Poison information professionals also interact with and give advice to health care professionals, including emergency room

Figure 16.27 Poison centers can offer advice on how to treat someone who has been accidentally exposed to a pesticide.

physicians, intensive care nurses, school nurses, workplace safety personnel, dentists, veterinarians, 911 dispatchers, and paramedics.

Poison control centers are staffed by specially trained nurses, pharmacists, physicians, and toxicologists. These specialists usually can determine over the phone whether or not a poisoning incident constitutes a true emergency; if it does not, they reassure the caller that the situation is not as serious as they had feared. And if the situation does have emergency potential, they offer step-by-step advice to help the caller deal with it. If complications are likely, they explain what to expect and how soon; and they generally continue to check with the caller over a period of days until the victim is both feeling better and out of danger (Fig. 16.27).

Poison center staff will quickly ask the nature of the emergency and the name of the product involved, as well as the following:

- The age and weight of the child, since toxicity usually is based on these factors
- When the child was exposed
- How much the child ate, drank, or spilled
- The child's general health status
- Whether the child has any signs or symptoms
- What actions have already been taken

Details are critical in delivering proper medical advice over the phone, and each poison exposure must be assessed and managed individually. Poison center personnel have access to emergency information that equips them to manage most poisonings, but they need *the exact name of the product involved* (brand name, common name, or active ingredient). Be prepared to communicate product information directly from the pesticide container; if you cannot access the container from where you are calling, have someone bring it to you.

The poison specialist will use product information in prescribing what you should do; that is, he will be able to determine what treatment is appropriate; whether or not the victim should seek medical attention; and what symptoms, if any, may be expected. If necessary, he will coach the caller on stabilizing the victim.

How to Prevent Poisonings

Prevention of poisoning through proper selection, storage, use, and disposal of poisonous products is very important. Poisonous products include household cleaners, medications, vitamins and herbal preparations, and pest management products.

Despite our efforts to the contrary, poisonings do happen; and it is particularly important that parents recognize this potential and exercise steps to minimize the risk. All adults and older children need to know and understand that all cleaners and chemical products are potentially hazardous; they should know basic first aid for poisonings and how to access the poison control center. The following checklist will assist you and your family—including older children—in preventing, preparing for, and handling emergencies.

- *Store toxic products out of reach of children.* Keep prescriptions and over-the-counter medications, pesticides, and other toxic substances on shelves higher than 6 ft. Better still, store them in locked cabinets or containers. You can purchase child-resistant devices for lower cabinets or drawers.
- *Purchase products in child-resistant packaging.* Choose products marketed in containers with safety caps when purchasing chemicals. But remember that no container is child *proof.*
- *Keep products in their original containers.* Never store medications, cleaning products, paints, pesticides, or other toxic materials in food or beverage containers. And never transfer chemicals or medicines into a container bearing a label from another product.
- *Retain the identity of the product.* Make sure all product labels remain intact and legible. If a label is lost or damaged, discard the product by taking it to a household hazardous-waste disposal day; or throw it in the trash—if you know *positively* that doing so is appropriate. If you do not know label instructions for disposal, check with the retailer where the product was purchased or with the manufacturer.
- *Never combine products.* Never mix household cleaning products together because certain combinations can be explosive or produce toxic fumes.
- *Never give a child medication meant for an adult.* Never give a child any type of medication meant for an adult, and never administer a prescription drug to anyone but the person for whom it was prescribed.
- *Never refer to medicines as candy.* Do not tell children that a medicine is candy, that it tastes like candy, or that it is delicious. This applies to both prescription and over-the-counter medications.
- *Read and follow product labels.* Be sure that the product is appropriate for your intended use, that you are willing to use the product only as directed, and that you have the required safety equipment (such as gloves) specified on the label.
- *Select the least toxic product.* Read the label before you buy; and if you are uncomfortable with label information on a product, make another selection. If you are purchasing a pesticide, choose one with the signal word CAUTION, meaning slightly toxic, instead

of a product labeled WARNING, which indicates moderate toxicity. If there are small children in the household, avoid chemicals with the signal word DANGER, meaning highly toxic or corrosive. *Note*: Signal words are based on human toxicity.

- *Purchase ready-to-use chemicals rather than concentrates.* Any chemical in its concentrated form is more hazardous than its diluted counterpart; that is, there is less chance of serious injury from exposure to a dilution than from exposure to a concentrate. The potential for applicator exposure is greatest during mixing and loading procedures—when the product is being handled in its most concentrated form.

  Avoid purchasing concentrates; but if you must use one, buy only the quantity that you will need to complete the application and dispose of the container—according to label directions—immediately upon finishing the job. If you do have to store concentrates, secure them in a locked compartment—both out of sight and out of reach of children.
- *Purchase products in quantities that you will use within a short period of time.* Storing "leftovers" increases the likelihood of a spill or poisoning incident.
- *Reduce inventory.* Write the purchase date on each product, when you buy it, and take inventory of stored quantities regularly. Make a concerted effort to use whatever you have on hand before purchasing new product. If you have had something for a long time and have no intention of using it, dispose of it according to label directions or call your local extension office, county health department, or solid-waste management district for information on upcoming household product collection and disposal days. If you have outdated medications on hand, flush them.
- *Eliminate poisonous plants from your home and garden.* Do not cultivate poisonous plants, indoors or out, if there are small children in the household. A list of poisonous plants can be obtained free of charge from poison centers, libraries, or the Cooperative Extension Service.

How to Prepare for an Emergency

- Teach your children their address as soon as they recognize letters and numbers. If they are too young to memorize it, or if they have difficulty remembering, call the 911 *nonemergency* number (check your local directory) and verify your address in their database. If your house is in a rural area or otherwise difficult to locate, call the 911 *nonemergency* line *before* an emergency occurs to provide specific directions to your house (Fig. 16.28).
- Be aware of potentially complicating factors. Language barriers, disabilities, special medications, allergies to medications (such as codeine) or products (such as latex) could be dangerous in the event of an emergency. Call the 911 nonemergency number and have these things added to the database. Post a card beside your telephone, listing your name, phone number, address, and directions to your home so that if a nonresident adult (e.g., a babysitter) needs to call 911 due to an emergency, the information will be readily available.
- Have pertinent emergency numbers handy at all times. Post the telephone number for your local poison center near your telephone, or affix a sticker that includes the number. Also post the number on your refrigerator or some other prominent place, and tell everyone in your family, as well as caregivers for children or elderly adults, where to find it. If you program the poison center number into your phone for quick dial, you still should post it near the phone and at least at one other site. It is a good idea to affix

Figure 16.28 Teaching children what to do when they call 911 can be fun and informative for the child.

the poison emergency sticker to your wallet or purse so the phone number is always available (Fig. 16.29).

- Post emergency information on the refrigerator because responders are trained to look there for emergency information; it may be posted in additional locations as well. You may want to mark a red border around the information to make it stand out. Make sure

Figure 16.29 Being prepared for an emergency results in a quicker response in the event that one occurs.

all family members and caregivers who visit your home routinely know exactly where the emergency information is posted.

Post the following information:

- Poison center telephone number
- Long-term medications for each member of the household
- Medications and other substances to which household inhabitants are allergic (indicate who is allergic to each)
- Names and phone numbers of persons to contact in the event of an emergency
- Directions to your home

Handling a Poisoning Emergency

- Reduce the exposure: Take away any hazardous substance that a person is swallowing. Remove people from the area of toxic fumes. If a poison has been spilled on a person, remove their clothes and flood the skin thoroughly with water. Get the victim into a shower if at all possible, being careful to avoid getting the poison on yourself in the process (Fig. 16.30).

Figure 16.30 Handling a poisoning emergency in a responsible manner can mean the difference between life and death.

- Never try to give anything by mouth to an unconscious person. Doing so could cause the victim to choke.
- Call the right emergency service. Every emergency situation is different, but the rule of thumb for a poisoning incident is this:
  - — If the person is conscious, call the poison center.
  - — If the person is unconscious, call 911.
- Identify the poison. Poison centers can offer specific medical advice only if they know what product or substance the victim has been exposed to. When calling the poison center, have the product container at hand, if at all possible, for quick access to information that the responder might request.

Teaching Children about Poisonings and Emergencies

- All products can be poisonous. Teach your children that all substances have the potential to be harmful and that there are certain things they should never attempt to use without adult approval and supervision. Even medicines, vitamins, and other products routinely taken by mouth or applied to the skin can be seriously harmful if misused or abused (Fig. 16.31).
- Dangerous products can be attractive. Teach children that even if something is packaged in a pretty wrapper—and even if it smells or looks really good, or if it is their favorite color, or if it tastes good—it could still make them sick. Chewable vitamins and flavored liquid antibiotics prescribed for children are good examples. Keep toxic substances out of reach of children, and out of sight, regardless of whether or not you think they understand this concept.
- Spills are to be handled by adults. Make sure your children know that they must not touch or go near any chemical spill, even if they have caused it. Assure them that they will not get into trouble for telling you about it. Stress that their safety is your first priority and that you will clean up the spill.

Figure 16.31 These are a few basic messages that parents should teach their children about pesticides and other chemicals.

Figure 16.32 Many educational programs allow children to practice what to do in an emergency situation that could happen at home.

- Neighbors can help. Teach children to go to a neighbor for help at the first sign of a fire (smoke or heat), a chemical spill, or any unusual smell, or if an exposure occurs. Establish a safe meeting place where everyone in the family is to go in the event of an emergency. Tell your neighbors that you have instructed your children to go to them in an emergency if you are not at home. Never leave small children home alone.
- Emergency responders will need the name and address of the caller (or the emergency site). Have your children practice reciting their names and address. Make sure they know where emergency information is posted (e.g., on the refrigerator, by the phone).
- Children should be told to call 911, first, during an emergency. Assure them that they will not be punished for calling 911, even if it turns out not to be an emergency. Unplug the phone and have young children practice dialing 911. Teach your kids that *they must not hang up* after dialing 911 and that they need to stay on the phone and follow directions given by the dispatcher. Make sure they understand that dialing 911 is for *real emergencies* and that they *must not* dial it while playing.
- Earlier in this chapter we mentioned programming the number for the poison center into your phone. Although older children may be able to distinguish between a poisoning and another emergency, younger children may not. Instruct young children to always dial 911.
- Learning about poisons and what to do in an emergency is important—and it can be fun! Make a game of the learning process. Ask your children questions and give positive rewards for correct answers. Use the EPA's "interactive home" Web site (http://www.epa.gov/oppt/kids/hometour/), that shows hazards in and around a household. See if the kids can locate similar things in their own home. The poison control center Web site (http://www.aapcc.org/games.htm) has learning games for older children and a variety of information for parents on making the household poison-proof (Fig. 16.32).

SEEKING ADVICE ON NONEMERGENCY QUESTIONS

There are many sources that can answer nonemergency questions. When calling *poison* control or 911 with general questions, it is important for the caller to indicate to the specialist in poison information at the beginning of the call that this call is not an emergency but simply a request for information so the specialist can appropriately prioritize the calls relative to other calls.

If you have general questions about poisonous substances, you may contact poison control or other sources for answers. For general information about medications, over-the-counter drugs, vitamins, and dietary supplements, ask your pharmacist first. For information about pesticides or poisonous plants, contact your county Cooperative Extension Service office, listed in the phone book. The National Pesticide Information Center at (800) 858-7378 is another excellent source of pesticide information. Most pesticides, disinfectants, sanitizers, and other cleaning products should also list the telephone number of the product manufacturer. The product manufacturer can be called during regular business hours for general information regarding the product, its use, and safe disposal.

CHOOSING A COMMERCIAL SERVICE

If you are interested in improving the appearance of your landscape but feel that you cannot make the necessary time commitment, consider hiring a professional to maintain it until you are free to handle it yourself. Others may lack either the confidence or the time to read and execute label directions properly, making the choice of using a commercial business to manage lawns, shrubs, and home pests a good one. Remember, you are hiring a service because of its professional knowledge relative to managing pests—which may involve some applications of pesticides. Selecting the best and most professional pest management service requires more than asking about price. Implement the following suggestions before you select a pest management company.

- Make sure the service has met all legal and educational requirements that give them the privilege to service your home. Ask to see the following credentials: a valid state pesticide certification number and a state license.
- Check with your local Chamber of Commerce or Better Business Bureau regarding the firm's past work performance.
- Ask plenty of questions! Ask for local referrals and affiliations with professional trade associations. Your pest management service company should be willing to explain company policies and give detailed explanations of its methods and solutions for managing pests. If you don't understand the answers, ask for them to be repeated until you do. Answers to your questions should give you a good indication of the present and future reliability, expertise, and commitment of the company.
- Expect your pest management professional to provide you with answers to a wide assortment of pesticide-related topics, such as personal safety and environmental impact, as well as standard and alternative insect, weed, and disease management tactics.

- Ask for pesticide labels when interviewing company representatives. Many answers to your questions will be found in the label instructions. Beware of companies whose representatives answer questions contrary to the instructions, directions, or precautions contained on the label.
- Selection of the company should not be made hastily. Take a few days to reflect on what you heard and to read the documentation supplied by the companies. Once a company has been selected, your loyalty should be based on a strong professional commitment toward effective, safe, and friendly pest management service.

SUMMARY

Homeowners take great pride in beautiful and productive landscapes that complement their homes and increase their property's value. A lush green lawn, large shade trees, healthy shrubs, bountiful fruit trees and vegetable gardens, and beautiful flowers blend to create the perfect picture.

Landscape plants must be maintained properly to protect the homeowner's investment, and those who assume the task themselves must become knowledgeable of the needs of landscape plants and the appropriate and timely cultural practices required to keep them healthy (Fig. 16.33).

Remember that many landscape pest problems can be avoided by following recommended cultural practices:

- Select plants that are hardy in your climate.
- Plant them in the proper location.
- Provide timely and accurate fertilization, as needed.
- Prune, water, weed, and mulch properly.

These practices will help ensure that your lawn and landscape plants are as healthy as they can be, and, in turn, healthy plants have fewer pest problems minimizing the need to use chemicals.

A homeowner must constantly nurture his or her green thumb! Some homeowners have an inherent desire to interact directly with their landscape plants and do so with an ongoing, intensive study of horticultural practices. They derive great pleasure and satisfaction from gardening and the challenges it presents. They exert an effort to learn about landscape maintenance and actually take charge, performing whatever activities are required.

Other homeowners are less interested in gardening but share gardening enthusiasts' love of a beautiful landscape and appreciation for the value it adds to their property. Their choice is to contract with professional lawn services to maintain their yards and gardens.

Even though an ounce of prevention is worth a pound of cure—sometimes pests get the upper hand. Therefore, despite our use of good cultural practices, even well-cared-for plants occasionally experience weed, insect, or disease problems. Some pests can kill plants or stifle growth and flowering. Others may reduce a plant's vigor, leaving it susceptible to a variety of pests. Many render the landscape unsightly. Therefore, pesticide applications are sometimes necessary.

Figure 16.33 Consider nonchemical approaches to managing pests, but when pesticides are needed, be sure to follow all label instructions.

If you decide to apply pesticides remember these steps:

1. Make sure the pest is correctly identified and in numbers worthy of management.
2. Select the appropriate product to manage the pest.
3. Buy only the amount you will need to avoid having to store leftover product even if the larger quantity is on sale.
4. Read the label carefully and understand how to mix and use the product safely and effectively (Fig. 16.34).
5. Notify others of your intention to apply the pesticide.

Figure 16.34 Handling pesticides safely around the home is everyone's responsibility.

6. Always wear long pants, a long-sleeved shirt, gloves, socks, and shoes when mixing or applying a pesticide. Read the label for additional advice.
7. Select the proper application equipment for the job and calibrate it accurately.
8. Mix just enough pesticide to complete the job or to complete each increment if the job requires making a series of applications over time.
9. Know and practice all safety measures referenced on the product label.
10. Never place rodent baits where children, pets, or wildlife may find them.
11. Remove or cover food, dishes, utensils, pet food, water dishes, fish tanks, and children's toys before spraying pesticides.
12. Always keep children and pets away from mixing and application areas.
13. Wash your hands thoroughly with soap and water after each application.
14. Wash the clothing worn during the application process separately from other laundry.
15. Clean, store, and maintain the application equipment according to the manufacturer's instructions.
16. Know what to do in case of an emergency.

These steps may appear complicated to a first-time user of pesticides and pesticide application equipment, but they are very straightforward and can be mastered with study and experience.

CHAPTER 17

RISK COMMUNICATION: INTERACTIONS AND DIALOGUES WITH THE PUBLIC

Communicating the complex and controversial topics surrounding pesticides is challenging, and there is no shortage of issues, information, or opinion. There are no magic words or secret recipes for communicating a fair, informed message, and there are no prescriptive rules for educators to follow.

Although no communication rules exist, the battle to win the hearts and minds of consumers goes something like this: Group A calls a press conference to announce that, based on their findings, pesticides in our children's diet pose serious health concerns. After many painstaking hours of searching through government files, pesticide testing results, and public policy documents, Group A determined that the U.S. government and pesticide manufacturers are subjecting children to unnecessary risk. Their press conference concludes with the rhetorical question, Can *any* risk to our children be justified when we do not know all of the (potential) adverse effects that pesticides may have on them? (Fig. 17.1.)

Group B's experts, who say that the risk to children is very low and that government standards are set to exceed all likelihood of adverse effects, quickly refute Group A's arguments. They say that Group A uses scare tactics in citing an invisible, imaginary, and theoretical risk. Group B calls Group A's report junk science, claiming that their conclusions are based on faulty assumptions; that the data are misunderstood and misused; and that their report was neither peer-reviewed nor published in a reputable scientific journal. They counter Group A's additional argument—that there are effective alternatives to pesticides available for managing weeds, insects, and diseases that threaten farmers' crops—with the realization that pesticide use remains a necessary component in assuring an abundant, affordable, and reliable U.S. food source.

The public draws from these sound bites and headlines in forming opinions on the benefits and risks that pesticides pose. They are left to judge which group of dueling experts is more credible and which has the consumer's best interest at heart. The critical role of the educator in this plight is to help consumers cut through volumes of technical, contradictory, and sometimes frightening information in concluding where to place their confidence (Fig. 17.2).

Figure 17.1 People have many views when it comes to pesticides, even when reviewing the same set of facts.

The issue of consumer (mis)understanding is a major one. Consumers are asked to make increasingly difficult risk decisions for themselves and their families based on complex scientific and technical information. This publication outlines the history of commercial pesticide development and regulation and describes the educator's role in helping the consumer form an educated opinion.

THE HISTORY OF PUBLIC DEBATE ON PESTICIDES

The Early Years: A Public Focused on Benefits

A review of the early years reminds us how productive the debate has been and piques our concern for the issues we face today. When commercial pesticides were first used in agriculture in the 1930s and 1940s, the public in general welcomed and applauded them. DDT, the first widely used and recognized synthetic pesticide, was of such obvious benefit that it spurred the development of new pesticides for use in the home, on the farm, and

Figure 17.2 The media is the main conduit for getting information out to the public.

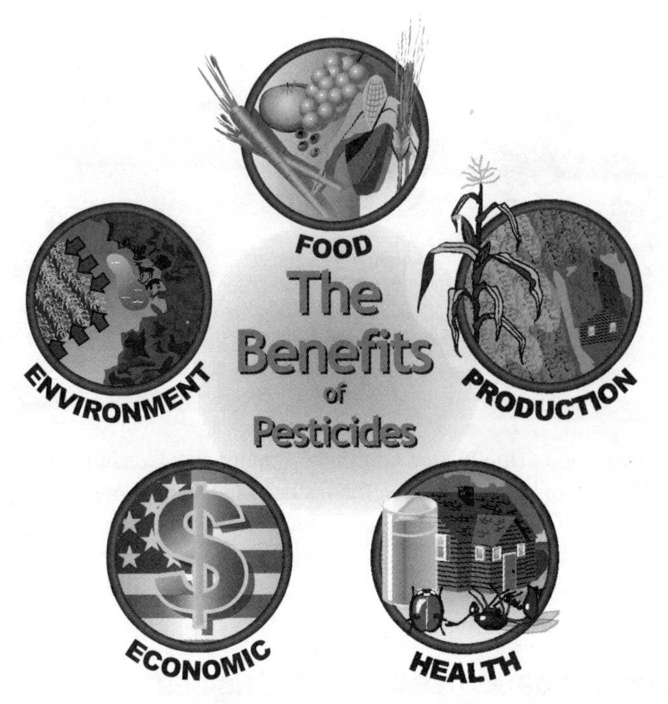

Figure 17.3 There are many benefits associated with the use of pesticides.

in the workplace. The benefits of pesticides were easily demonstrated and observed—and very convincing. There prevailed a strong public conception that technology could solve all problems (Fig. 17.3).

Food Pesticides reduce the negative impact of pests on crop production and facilitate sustainable yields on fewer acres of farmland. They protect our crops, our stored grain, and our processed food; they contribute significantly to our abundant, high quality, economical food supply (Fig. 17.4).

Figure 17.4 The use of herbicides to manage weeds in a soybean field.

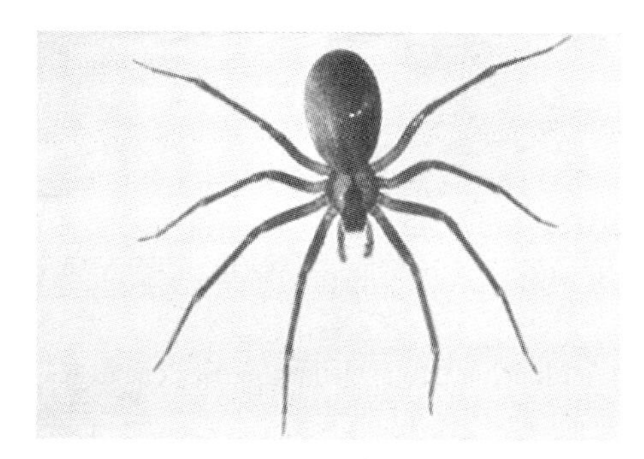

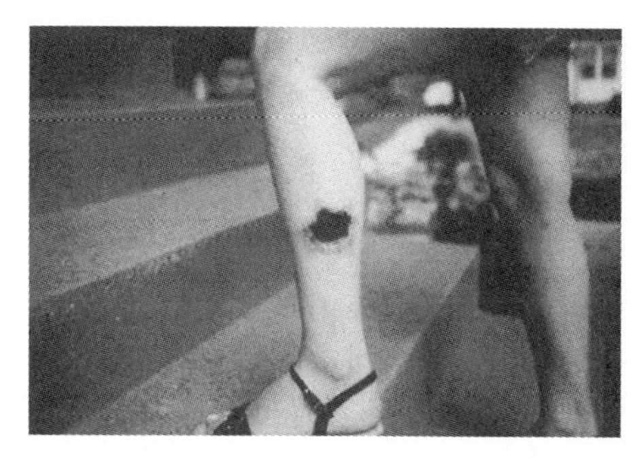

Figure 17.5 Brown recluse spiders are a serious health threat.

Pesticides are an integral part of the crop production equation that enables one American farmer or rancher to produce enough food to feed more than 100 people per year. It takes less than 2% of the American workforce to produce enough grain, meat, and fiber to feed the nation, freeing the remaining 98% to pursue other vocations.

Health Pesticides manage pests at home, at school, and at work. They reduce the incidence of waterborne and insect-transmitted diseases such as malaria and West Nile virus, protect consumers against potentially lethal toxins in molds, protect pets from fleas and ticks, and facilitate vegetation management on rights-of-way, which contributes to safer transportation and electricity transmission (Fig. 17.5).

Wildlife and Environment Pesticide use on fertile farmland increases production, facilitating the return of marginal farm acreage to wildlife habitat. Pesticides protect the diversity and quality of natural habitat by managing invasive, nonnative species. Pesticides also contribute to improved water quality and aquatic habitat by reducing soil erosion: they manage weeds in no-till farming systems, where the soil is not disturbed (to erode) by disking.

Economics Pesticide manufacturers, users, industries, and associated businesses contribute positively to the balance of trade, providing good jobs, and a tax base to support local, state, and federal governments.

As we contemplate pro-pesticide arguments of today, we recognize their similarity to those of the past. Perhaps the most significant *difference* is the audience: today's consumers are more suspicious—even pessimistic. The public is less willing to accept the premise that pesticides are beneficial, overall. They are wary of scientific authority and less willing to accept and rely on the positive without knowledge of the negative. They want substantiated proof that the benefits outweigh the risks.

USDA Regulations and Product Registration

In 1947, shortly after World War II, the U.S. Department of Agriculture (USDA) was required under the Federal Insecticide, Fungicide, and Rodenticide Act (FIFRA) to register all pesticides and establish standards for label content. Within the USDA, the Pesticide Regulation Division (PRD) was assigned the responsibility of registering pesticide products and was divided into the Registration Branch and the Enforcement Branch.

The Registration Branch was responsible for registering all products before they entered the market. Manufacturers wishing to register a product had to

- provide efficacy data demonstrating product performance;
- substantiate their claim that the product met USDA safety and health criteria;
- document truth in labeling: that the contents of the product were exactly as stated on the label; and
- include use directions (on the label) that were clearly stated so that the user could gain maximum benefit from the product, and describe use precautions necessary to ensure human and environmental safety.

Government involvement continued to increase. In 1958, the Food and Drug Administration (FDA), through the Miller amendment, worked with the USDA to set tolerances for residues in food. Thus, manufacturers were required to submit data to show that pesticide residues in food from treated crops did not exceed established tolerances. During this period, most tolerances were around 7 *parts per million,* which corresponded to the sensitivity limit of analytical methods available at that time. Pesticides *without* detectable residues were assigned a "no residue" registration, and those *with* detectable residues were assigned a "residue" registration.

The residue/no residue registrations were challenged as improvements in technology allowed scientific detection of residues in *parts per billion.* This increased capability raised concerns because pesticides previously registered through the USDA as "no-residue" products could now be shown to contain residues in parts per billion. The result was elimination of the "no residue" registration.

FIFRA was amended in 1964 to require the USDA to refuse registration of pesticides determined unsafe or ineffective and to revoke registration of and remove such existing products from the market. It also requires that all pesticide labels bear a USDA registration number; that the front label of all pesticides display a signal word—CAUTION, WARNING, or DANGER—and the phrase "Keep Out of Reach of Children"; and that all claims made about the "safety" of a product be removed from the label.

Silent Spring Introduced the Public to Risks

In 1962, Rachel Carson published *Silent Spring,* a book that refocused and energized public debate on pesticides. Carson was a scientist with the U.S. Fish and Wildlife Service, and in her book she described the devastating effects of DDT on the environment. *Silent Spring* influenced the public, the farming community, scientists, and government officials to quit thinking of pesticides as miracle chemicals and to acknowledge the danger they posed to wildlife. From that time on, the terms *risk* and *environmental pollutant* were linked to pesticides; and, for many, the preconceived notion that pesticides were "good" was replaced with serious doubt (Fig. 17.6).

Reactions to Carson's book set into motion a wave of public participation in the political debate on pesticides. Environmental advocacy groups drew public attention to associated risks, whereas industry and trade associations continued to extol the benefits of pesticide use. Congress became the middleman, and lines were sharply drawn for a debate that continues today.

Figure 17.6 *Silent Spring* introduced the public to the risks that pesticides pose to the environment.

The impact of *Silent Spring* went far beyond the eventual banning of DDT in 1971. The book legitimized public concern and public participation in decisions on pesticides. *Silent Spring* made people realize that their government was not necessarily telling them the whole story. Suspicion of government was heightened as advocates made themselves heard, and there was a major shift in how society perceived science and scientists: technology was viewed skeptically and more critically (Fig. 17.7).

Integrated Regulation: Creation of the Environmental Protection Agency

In 1972, public outcry on environmental pollution in general—and pesticides in particular—gave rise to a new federal agency: the U.S. Environmental Protection Agency (EPA). Responsibility for enforcement of FIFRA was transferred from the USDA to the EPA, and the focus of federal pesticide policy shifted from controlling the quality of pesticides used in agriculture to *the reduction of unreasonable risk to human health and the environment.* In addition, the authority to establish pesticide tolerances for food was transferred from the Food and Drug Administration to the EPA, placing the agency in full control of the pesticide registration process (Fig. 17.8).

Figure 17.7 The public's perception shifted.

Figure 17.8 The EPA addressed the policy shift over time from the benefits of pesticides to a focus on the risks that pesticides may pose to people, wildlife, and the environment.

Public Policy Sets the Stage for Rules Governing Pesticides It is interesting that the development of policy to correct one potential problem—unreasonable risk to human health and the environment—in fact fueled additional pesticide concerns. This cause-and-effect scenario shaped key historical decisions from which pesticide rules and regulations have emerged (Fig. 17.9).

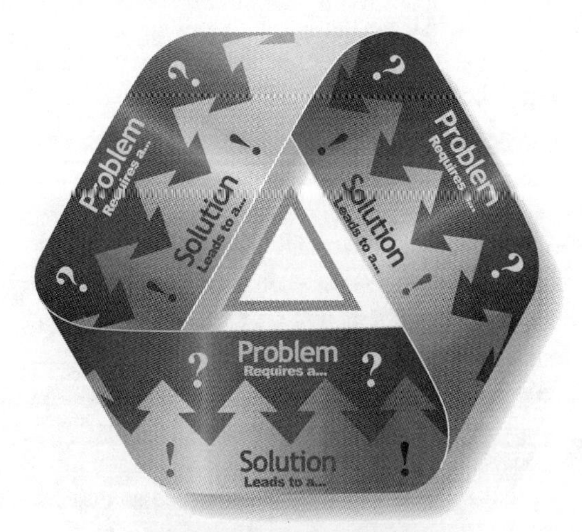

Figure 17.9 Each policy decision and each solution to a problem raises more questions that need answering.

- The furor surrounding pesticides marked them for extensive governmental scrutiny. FIFRA was amended by regulation to protect human health and the environment; and with the EPA responsible for the regulation of pesticides from registration to final disposal, critics then had but one agency to target.
- The party in power appoints the EPA's upper management personnel, and their decisions in turn reflect political influence. All decisions are public.
- The EPA must decide how to implement public policies that meet the imprecise definition of *safe*. As defined by Congress, *safe* means that when the product is used according to its label it will not cause "unreasonable adverse effects on human health or the environment"; and that there is "reasonable certainty that no harm will result from aggregate exposure to the pesticide chemical residue, including all anticipated dietary exposures and all other exposures for which there is reliable information."

What on the surface seems to be a clear and reasonable definition of *safe* is actually quite murky. What is reasonable certainty? What effects are unreasonable? One person's definition of *safe* can be another's definition of *dangerous*. For example, a homeowner who uses pesticides in her garden may view them as harmless—that is, *safe*—except to targeted pests; but the neighbor who watched her pet die after ingesting a pesticide last year might consider them *dangerous*. Regulatory interpretation lies with each generation of policymakers who command their own historical, political, and scientific perspectives. What was deemed unacceptable in the past may be viewed as acceptable or even desirable now; and what is acceptable today may not be tomorrow.

- The EPA is responsible for both environmental research and environmental regulation, but it is often faulted for failing to fulfill its obligations.
- The EPA requires pesticide manufacturers to perform numerous tests and to submit data in support of their product for registration. The data are evaluated to determine whether use of the pesticide will adversely affect human health or the environment. Even so, a degree of risk remains because testing cannot identify absolutely all risks nor answer every question that may arise. Registration data are scrutinized by consumers, manufacturers, and scientists looking for evidence to challenge the EPA's interpretations.
- Decisions on pesticides are based on the premise that human health and the environment are not adversely affected by low-level exposure. But what level is *safe?* The level of minimal risk has to be determined for each and every product, and the responsibility lies with the EPA to interpret scientific data and to assign *safe* exposure levels.
- An EPA registration is not a recommendation for, nor an endorsement of, the product registered. It simply indicates that, based on its evaluation of available data, the EPA considers the product *safe* for use according to label instructions. However, there remains the possibility that others might interpret the exact same data quite differently; so it is easy to see that registration can be perceived as endorsement, particularly by groups that oppose the registration of a given product.
- Restricted-use pesticides may be legally purchased and used only by certified applicators and those who work under their direct supervision. Certification signifies that an individual has demonstrated (by testing and/or training) the competency to handle pesticides safely and judiciously.

- The EPA is shaping the pesticide marketplace by encouraging the development, registration, and use of reduced-risk pesticides and by accelerating the registration process for biological control agents and other products that meet *low risk* criteria: low toxicity in test animals and nontarget organisms; short-term persistence in the environment; low potential to contaminate surface and ground water; low risk of human and environmental exposure; and compatibility with integrated pest management strategies. It is noteworthy that the implementation of this reduced-risk policy places the EPA in a position to promote one pesticide over another.
- Pesticide regulations in any given state may differ from federal regulations. That is, federal law constitutes *minimum* requirements, but states are given the latitude to impose more stringent pesticide regulations as they see fit; that is, some states may elect to be more "protective" than the federal government.
- The product label is the main avenue of communication between the pesticide manufacturer and the user—and the best source of safety information. The EPA bases its registration decisions on the premise that users follow label instructions: there is no alternative. The pesticide label is a legal document, and any use of a product inconsistent with its labeling constitutes *mis*use. The label may be cited as the basis for enforcement actions such as fines, probation, and license revocation; and it is often questioned why the EPA and its state counterparts do not police pesticide applications more diligently to deter or identify misuse incidents.
- Pesticide manufacturers are required by law to report (to the EPA) any problems associated with their products. This requirement facilitates follow-up on products that adversely impact human health and the environment. But although the law *requires* manufacturers to report suspected problems, they are trusted to do so *voluntarily.*

THE PUBLIC'S VIEW OF PESTICIDES

Risk assessment is based on a complex mix of perceptions, social considerations, and science. Consider the true example of an applicator who was observed pouring a herbicide from a 55-gallon drum into his sprayer without wearing gloves and safety glasses as required by the label. When asked why he was not wearing the safety equipment, he responded, "This stuff's not so bad. My children are all right. It didn't hurt them, so it won't hurt me." When asked whether he'd allow photographs to be taken while doing his work, his response was, "Yes, but let me put on my safety gear!" The point is that people's perceptions of a purely physical phenomenon—the hazard—are not solely a function of the hazard but, instead, the product of hazard, experience, and risk tolerance. In this example, the applicator was more worried about what others might think than about any danger presented by the hazard.

Personal Judgments Reflect More Than the Facts

Deciding whether or not to make (or contract for) a pesticide application may be likened to deciding whether or not to have surgery! You must evaluate the risk factor—high, low, or moderate—and weigh it against the projected benefits.

The physician can easily describe the surgical procedure and the invasiveness of the surgical technique, inform you of the potential complications and side effects, project the odds on partial or full recovery, and estimate the long-term prognosis. But his view is subjective: he will not be undergoing the procedure *himself.* As the patient, you have to consider elements such as cost, quality of life, comfort level before and after the operation, risk potential, and your confidence in both the physician and the diagnosis. You (the customer) and the physician (the professional) view the surgery from two different vantage points, each in the context of your own knowledge and experience. In deciding what to do, you must weigh his professional opinion against your own perception of the consequences.

Judgments about pesticides are more complex than simply understanding government risk assessments, reports, charts, and figures. It is one thing to read that the risk is being managed, but it is quite another to realize that your neighbor has allowed a pesticide to drift onto your property.

A representative study of Indiana residents (nonfarmers) and farmers clearly illustrates that our vantage point affects how we view pesticide risk. Farmers have a vested interest: pesticides make farming easier and increase crop productivity/profit. And indeed the study (Table 17.1) illustrates that farmers are less concerned about pesticide risk than are citizens whose need for and use of pesticides is limited—and whose income is not directly affected. Overall, Indiana residents are more fearful of pesticides than are farmers. So, who is right? It all depends on the vantage point: Who is faced with risk? Who manages the risk? and Who benefits from pesticide use?

The Real Issue

We make judgments based on our values and experiences, on the information available, and on the credibility of our source. Ideally, we should gather all the facts before passing judgment: to use or not to use; to allow or to ban. But everyday situations often provoke spontaneous decisions, even without all the facts and even when an immediate response is unnecessary. Risk communicators—anyone in a position to influence others for or against pesticides—must provide enough of the right information for the audience to make informed decisions.

For instance, a high school freshman doing a science project asked 50 people if they would sign a petition demanding strict control or total elimination of the chemical dihydrogen monoxide—and for plenty of good reasons (Fig. 17.10):

- It can cause excessive sweating and vomiting.
- It is a major component in acid rain.
- It can cause severe burns in its gaseous state.
- It can kill if aspirated.
- It contributes to erosion.
- It decreases effectiveness of automobile brakes.
- It has been found in tumors of terminal cancer patients.

Forty-three of the people surveyed said yes, six were undecided, and one said no; yet, if the student had called dihydrogen monoxide by its common name—water—the results would have been a unanimous no. Perception and context are critical to good judgment.

TABLE 17.1 Perceptions of Risks from Pesticides[a]

Study Question	Residents	Farmers
Are the risks of pesticides understood by the public? (1 = Risk known precisely; 7 = Risks not known)	5.9	5.5
Is the risk of death from pesticides immediate or is death likely to occur at some later time? (1 = Effects immediate; 7 = Effects delayed)	5.9	5.5
Do pesticides pose risks for future generations? (1 = Very little threat; 7 = Very great threat)	5.2	3.7
Is the risk from pesticides new and novel or old and familiar? (1 = New; 7 = Old)	4.9	4.6
Do people face the risks of pesticides voluntarily? (1 = Voluntarily; 7 = Involuntarily)	4.8	3.9
Are the risks from pesticides increasing or decreasing? (1 = Decreasing greatly; 7 = Increasing greatly)	4.8	3.7
Pesticide risk affects how many people in your community? (1 = Few; 7 = Many)	4.6	3.5
To what extent do the benefits from using pesticides make up for any risk of using pesticides? (1 = Benefits make risk okay; 7 = Risks make benefits unacceptable)	4.3	2.7
How easily can the risk from pesticides be reduced? (1 = Easily reduced; 7 = Not easily reduced)	4.3	3.8
Can pesticides cause large-scale death and destruction across the whole world? (1 = Low potential; 7 = High potential)	4.2	2.8
Are you at risk from pesticides? (1 = No risk; 7 = Great risk)	4.1	4.0
Can you control your risks from pesticides? (1 = Cannot control; 7 = Can control)	3.9	5.1
When there is a mishap or illness from pesticides, how likely is it that the consequences will be fatal: (1 = Certainly fatal; 7 = Not fatal)	3.9	4.6
Is the risk from pesticides a risk that people have learned to live with and can think about reasonably, or is it one that people are afraid of? (1 = Live with; 7 = Afraid of)	3.7	3.9
Is the risk from pesticides one that kills people one at a time or a risk that kills a large number of people at once? (1 = One at a time; 7 = Large numbers)	3.5	2.6
Are the risks of pesticides known to science? (1 = Known; 7 = Unknown)	3.5	3.6
Can the risk of pesticides be controlled by preventing accidents or by reducing what happens after an accident occurs? (1 = Prevent before; 7 = Control after)	3.0	2.6
Are the harmful effects of pesticides easily seen by the public? (1 = Not easily seen; 7 = Easily seen)	2.2	2.6

[a] *Source:* R. A. Feinberg, F. Whitford, and S. Rathod. Perceived risks and benefits from pesticide use: The results of a statewide survey of Indiana consumers, pesticide professionals, and extension agents.

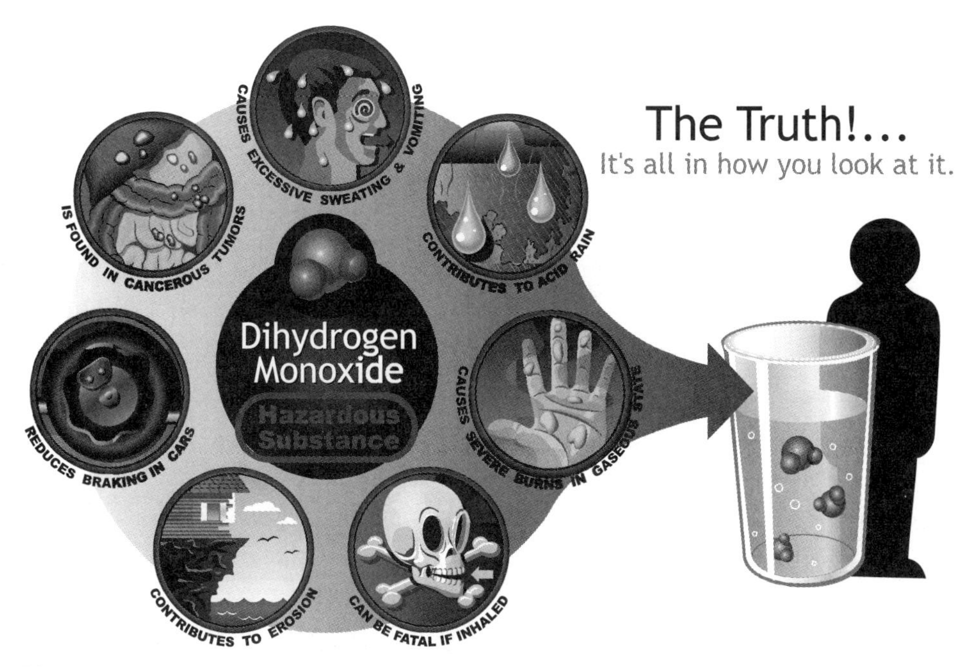

Figure 17.10 Citizens who are not provided with balanced and complete information cannot be expected to make good decisions.

THE REAL CONFLICT: PESTICIDES ARE SAFE AND UNSAFE

Pesticides are used to kill or alter the behavior of certain organisms. They are beneficial, yet they pose risks. So, what are the dangers from any particular pesticide? How many organisms are at risk? Are we willing to accept the risks in pursuit of the benefits? These questions have to be addressed and, ultimately, the EPA must decide what constitutes acceptable risk. It's difficult to do because there is a fine line between *safe* and *dangerous.* The issue is not only whether pesticides are dangerous, but to whom or what they are dangerous and to what degree.

Where Is the Line between Safe and Unsafe?

Risk *assessment* and risk *management* are vital in determining the *level* of risk posed by a given pesticide. The EPA's current policy is that the risk assessment process should identify methods and criteria for estimating the level of risk. The policy also mandates release to the public of all scientific information on which the EPA bases its conclusion.

Risk assessment is the science-based process of quantifying and characterizing risk, that is, estimating the likelihood of occurrence and the nature and magnitude of potential adverse effects. (See Chapters 2 and 5 for more information.)

Risk management is the process by which judgments and decisions are made on the acceptability of the level of risk identified during risk assessment. Risk managers must integrate the results of risk assessment with social, economic, and political factors. They may classify a product for restricted use, lower the application rates, restrict the number of

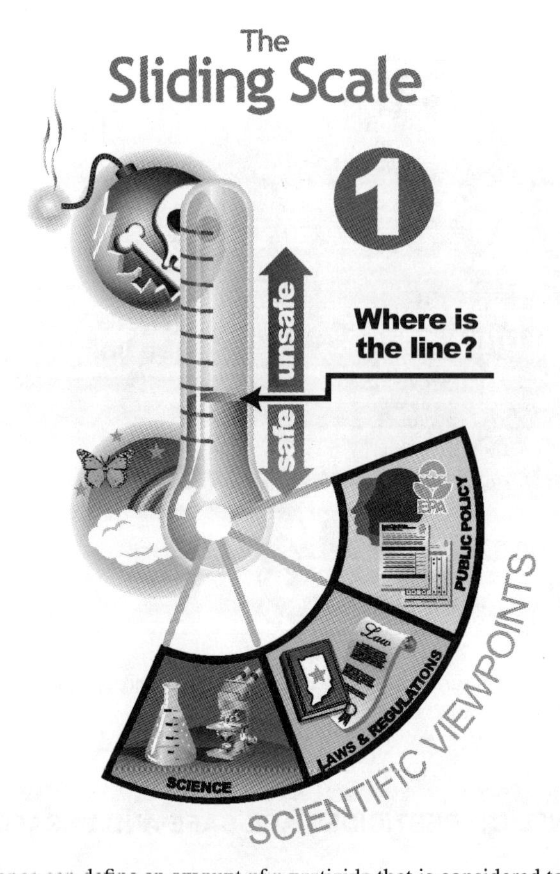

Figure 17.11 Science can define an amount of a pesticide that is considered to pose minimal risk.

applications, increase application intervals, stipulate longer intervals between application and harvest (in agriculture), or prescribe alternative application methods. These measures often take the form of label changes designed to reduce the amount of pesticide used or to lower human exposure potential. They may even decide not to register the product.

Uncertainty about Where the Safety Line Should Be Drawn

Safety determinations are based on scientific information and public opinion on what constitutes acceptable risk. But science is not exact: there are uncertainties in evaluating the safety of any substance, including pesticides. The EPA must incorporate scientific information, policy guidelines, and professional judgment in estimating whether a pesticide can be used beneficially within the limits of acceptable risk (Fig. 17.11).

A product is assumed safe from a scientific point of view if associated risks are minimal. However, the following four points must prevail to substantiate that assumption:

- Conditions must not change to the extent that the assumptions and methods used in the supportive risk assessment may be rendered invalid.
- The user must follow label directions explicitly.

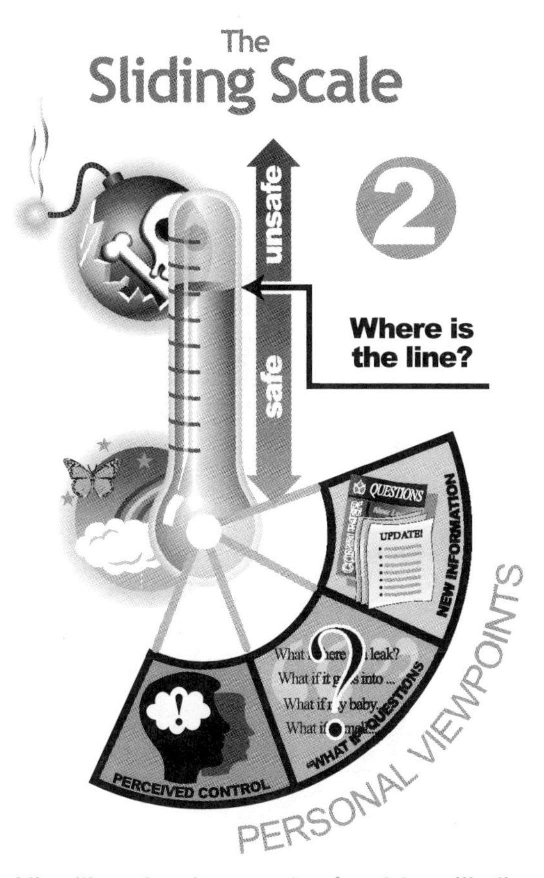

Figure 17.12 The public will use the science as a benchmark but will adjust the safety line based on their own personal viewpoints.

- The product must perform as anticipated, once it is released into the environment.
- Use of the product must not create adverse effects previously undetected in lab and field test data used for risk assessment.

Can the Line Be Drawn at All?

When asking where to draw the line, start with the question, Is it safe? In reality, we will never know with complete certainty that a pesticide is or is not safe: the line between safe and dangerous is never as defined in real life as it is in science. Pesticides are developed to work with reasonable certainty and minimal risk. But they exist in a world of *what-ifs* that loom outside the realm of verifiable scientific information; and often it is the what-ifs that alert policymakers to data gaps (Fig. 17.12).

Based on evaluation of the best data available on a pesticide at a particular point in time, scientists can state in all honesty that no significant problems exist with it. But in reality there are many reasons why we may never know whether it is safe under all circumstances, nor can we predict with certainty its performance in hypothetical or future situations. Scientific investigation is bound by the tools and techniques available, and new developments continually redefine our capabilities.

Scientific and Personal Viewpoints Determine the Line

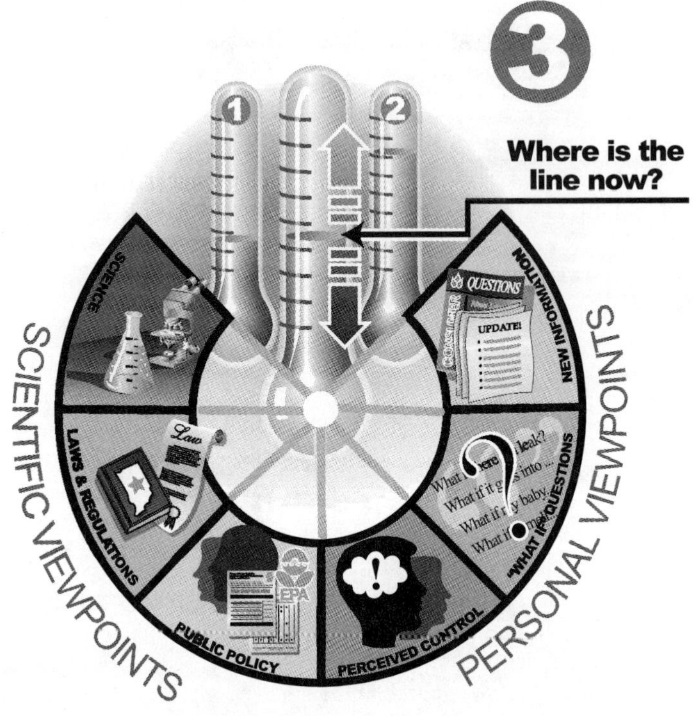

Figure 17.13 Ultimately, government standards and public policy positions are based on scientific facts and public opinion.

Pesticide manufacturers are required by the EPA to present extensive scientific data in support of products submitted for registration. But science can go only so far in addressing pesticide issues. The more data we have, the more questions we ask—and science often stops short of definitive answers. Problems can span many disciplines (e.g., medicine, chemistry, and biology), which makes solutions evasive.

We will never know if pesticides are safe in the absolute sense of the word. Science may never define safety, nor prove it. But the *what-ifs* will continue to drive regulatory agencies, manufacturing, marketing, public interest groups, application industries, judicial processes—and science. And there is an interesting, unintended side effect.

The fact that data analyses are disputed among scientific, government, and industrial interests cultivates a public mind-set of distrust and disbelief. On one hand, we extol the power of science; on the other hand, we caution that science cannot answer the *what-ifs*. We school the public to rely on experts, but we caution them that experts disagree! (Fig. 17.13.)

It is questions such as these that challenge risk communicators:

- Do we know all the right questions to ask to establish data requirements for all pesticide registrations?

- Why are products registered when science cannot answer my questions?
- Why should someone else decide whether I am exposed or not?
- If the EPA says that product registration is not an absolute guarantee of safety, then what does registration mean to me?
- Why can't the EPA guarantee that its decisions will ensure no harm?
- Why does the EPA rely on the manufacturer's data on their own products in making registration decisions?

So, can the line be drawn? The answer is yes, but it may have gaps. From a distance, the line may look solid; but, up close, you may see spaces: spaces that represent the information which we *do not* and *may never have.* The line between *safe* and *dangerous* is drawn only as definitively as our knowledge allows, and it is the uncertainties that challenge researchers and educators alike.

THE ART AND SCIENCE OF RISK COMMUNICATION

Professionals who make their living managing risks and hazards think principally about the physical characteristics of substances. If the hazard they manage is a pesticide, they think about its toxicity and volatility, its effectiveness in managing the target pest, the likelihood of its contaminating ground water, its persistency in the environment, etc.

In contrast, when people have concerns regarding hazards managed by *others,* they tend to think less about the substance itself and more about personal impact, fairness, and control. If the hazard is a pesticide, people may worry about whether there are pesticide residues in or on the foods they buy or whether their children are exposed to pesticides at school. Those who fear exposure to a hazard beyond their control may be justifiably concerned: Why am I or my children being exposed?

Table 17.2 summarizes factors shown to influence risk perception. These factors help to explain why farmers or commercial applicators who have used pesticides for years might perceive their risk as lower than urbanites' who view pesticides as unnatural, nonbeneficial, and beyond their control.

We Say Risk, They Hear Danger

We make presentations on risk every day that are right on the money, but sometimes we are dismayed by questions from the audience. For example, perhaps you lead a discussion on the level of risk posed by pesticide residues in food and drinking water. You discuss how laboratory animals are exposed to pesticides and how the toxic response in the mother and the fetus is evaluated. You state that, under most circumstances, detectable pesticide residues in food do not cause birth defects.

But upon your conclusion a pregnant woman asks, "What is the bottom line? Can these levels cause birth defects in my unborn baby? You say the odds are low, but what about my baby?" *So how did she miss that? Why didn't she hear what you said?*

Actually, she did hear you, but she does not feel in control. What she "heard" was that someone else has control over her personal level of exposure. So the risk to her and her baby looms larger than if she herself were in control. She wants more assurance that she will not be the one in a million who is affected.

TABLE 17.2 Factors Influencing Risk Perception[a]

Questions about the Product	Positive Perception (Viewed as Less Risky)	Negative Perception (Viewed as More Risky)
Who makes it?	Occurs in nature	Man-made
Who benefits from it?	I do	Others do
How important are the benefits?	Compelling	Vague
Is the risk familiar?	Yes	No
Are effects immediate?	Yes	No (delayed)
How serious are the effects?	Not very	Dramatic
Is exposure controllable?	Yes	No
Who controls my exposure to it?	I do	Others do
Is there visible risk?	Yes	No
Is its use a moral issue?	No	Yes
Has it ever received memorable media attention?	No	Yes
Is my exposure voluntary?	Yes	No
Is there a fairness issue?	No	Yes
Are there scientific answers?	Yes	No
Is the risk old or new?	Old	New
Who does it affect?	Not me	Me
Where is it used?	Not in my backyard	In my backyard
Is the risk controllable?	Yes	No

[a] Information contained in this table was obtained from many published papers by Peter M. Sandman.

Another person asks, How can it be safe to eat fruits and vegetables if they have pesticides on them? Don't pesticides kill or injure living organisms? Do you really know the effects of consuming food that contains pesticide residues?

You point out that the scientific consensus is that the benefits of eating a diet rich in fruits and vegetables far outweigh the risk of ingesting pesticide residues. The person listens politely, but you can tell by the look in his eyes that he doesn't believe a word you are saying, and then comes his rebuttal: It just doesn't make sense. How could eating pesticides be good for you?

Then someone in the front row says he read in a magazine or saw on television that even small amounts of pesticides can impact a child's hormone system and that there is scientific proof that some children have developed learning disorders and behavior problems as a result of pesticide exposure. He related the story of young Mexican children who had been exposed to pesticides and could not draw pictures like unexposed kids their age. He said that his retired neighbor, a pediatrician, said that pesticides also have been proven to cause leukemia.

He wants to know how you can claim that pesticides detected in or on food are insignificant when the chemical has been scientifically proven harmful. He questions how you can be certain that trace levels pose no risk to his grandchildren and to that woman's unborn baby. And *you* wonder why he does not acknowledge the complexities of science and how difficult it is to answer his questions with a definitive yes or no!

You have the facts, and you have already explained how we know that any risk posed by registered pesticides on food is low and that the odds of someone reacting adversely are minuscule. So why hadn't these people listened to what you said? Why are they questioning your information and expertise? Why are they looking for a guarantee when *nothing* we do is risk-free?

People smoke and think nothing of being overweight. They skip their annual physicals and seek medical attention as infrequently as they can. They fly. They drive—even *drink and drive!* They invest long, stressful hours in their careers. They ingest tons of chemicals known as prescription drugs. Many have multiple sex partners. Yet they worry about trace levels of pesticides in their food? Why does science have so little impact?

The simple answer is that people's experiences, perceptions, beliefs, and values contribute more to their life choices than does scientific fact. This is not to say that their conviction is right or wrong: we all see things from our own point of view. But people tend to want total assurance against negative consequences of pesticide use, and that is simply impossible.

A representative survey of Indiana residents shows the magnitude of danger that citizens associate with pesticides (and there is no reason to believe that Hoosiers are any more or less fearful than citizens in other parts of the country).

Citizens were asked to rate on a scale of 1–7 how fearful they are of various potential risks: 1 = I am not very afraid; 7 = I am very afraid. Pesticides used on farms ranked in the most-feared category along with nuclear accidents, pollution, smoking, handguns, nerve gas, auto accidents, and chain saw accidents. It can be derived from this response that pesticides spell catastrophe in the minds of many; and it is easy to understand why weak explanations and exclamations of pesticide safety have minimal impact on the public mind-set.

DO YOU FEAR THE FOLLOWING?

Respondents were asked to rank their fears on a scale of 1–7: 1 = I am not very afraid; 7 = I am very afraid.

Activity/Technology	Resident Responses
Nuclear accident	5.6
Pollution	5.4
Smoking	5.3
Hand guns	5.1
Nerve gas accident	5.0
Auto accident	5.0
Food tampering	4.8
Pesticides used on farms	4.8
Chain saws	4.6
Auto exhaust	4.5
Pesticide use in homes	4.1
Fireworks	4.1
Pesticide use in the garden	4.0
Biotechnology	3.5
X-rays	3.0
Caffeine	2.4
Microwave ovens	2.4
Water fluoridation	2.4
Antibiotics	2.3
Bicycles	1.7

Figure 17.14 Science talks about the risks to populations, whereas the public personalizes the potential danger.

There is a second aspect of this that makes communication difficult: Scientists view pesticide risk as a professional issue, but consumers look at it as a danger imposed on them by others. For example, the probability of getting injured or killed by a handgun is remote. But you cannot convince a person with a handgun pointed at his temple that the risk is low. In much the same way, it is difficult to convince some consumers that the risk posed by pesticide residues on food is low. Similarly, it is difficult to convince those who feel comfortable using pesticides around home or on the job to take extra precautions.

Public Policy Decides Societal Risk *Risk* is the chance of injury, damage, or loss; the degree or probability of loss; the act of exposing oneself to a risk or taking a chance. And scientists and government officials address risk in terms of probability for *populations,* not for individuals (Fig. 17.14).

Individuals Personalize Risk Risk, by definition, implies that someone or something is at risk, even if the risk is minimal. To the scientist, risk is a continuum from low to high—not an absolute. Individuals hear the word *risk* and think *danger.* The word *danger* is defined as *a thing that may cause injury, pain, etc.*

Each of us personalizes *danger* relative to any phenomenon that has the possibility of injuring us, our family, our community, or our environment. Our interpretation is based on our experience, lifestyle, and expectations. The critical question is whether the risk is important or dangerous to us or to people or things we care about.

Figure 17.15 To be an effective communicator, you must guide your audience through various levels of understanding before they can comprehend the message.

The Foundation of Communication

Communicating about pesticides and their risks often is an uncomfortable process. As William Dury once said, "When your views on the world and your intellect are being challenged, and you begin to feel uncomfortable because of contradictions, and you have detected what is threatening your current model of the world or some aspect of it, pay attention—you are about to learn something." If everyone followed Dury's advice, risk communication would be simpler!

Generally, before people are receptive to risk information, they must believe that the source of that information is trustworthy, credible, and fair. So when designing their message to convey pesticide risk, speakers must understand the communication difficulties they may face. In risk communication, there are five common barriers listed in the mnemonic CAUSE (Fig.17.15):

- Lack of Confidence
- Lack of Awareness
- Lack of Understanding
- Lack of Satisfaction
- Lack of Enactment

First, risk communicators often confront suspicion, so they need to employ strategies that earn the *confidence* of their audience. Second, risk communication often is impeded by unfamiliarity with the subject, so the communicator must create *awareness* of the scope of information available. Third, because risk communication involves concepts that may be difficult to grasp, communicators need to ensure that their audience is *understanding* the message: information should be presented at the audience's level, from the audience's vantage point. Fourth, *satisfaction* with solutions is critical: Risk communicators must offer plausible precautionary approaches to risk management. Fifth, the audience must be stimulated toward *enactment;* that is, the risk communicator must deliver the message so that the audience will embrace and implement his recommendations.

Good communicators depend on research and experience for overcoming obstacles described by the mnemonic CAUSE. These obstacles must be addressed in order: C-A-U-S-E. For instance, *Understanding* cannot be addressed until the risk communicator has earned the audience's *Confidence.*

Earning Confidence: Working with the Audience People are often skeptical of large companies, big government, or any organization that wields power. So when representatives of academia, government, or industry offer assurance relative to pesticide safety, their effectiveness may hinge on trust—or lack of it.

Trust pertinent to risk management is earned by acknowledging people's fears and by providing information on which they can base their own informed choices. It is molded by competence and character: *Competence* is a matter of relevant expertise, whereas *character* is reflected by the integrity and fairness of the individual.

Trust in an individual—a risk communicator—begins with a good first impression. If your audience perceives that you care about their well-being, they will trust what you tell them. But trust is best earned by matching words with actions.

Reflecting the Audience's Perception The best way to reach an audience is to address their concerns. When you commit to making a presentation, ask the program sponsor to identify any uncertainties the audience might have about the subject for discussion; then develop your message to fit their need.

On the day of the program, arrive early and mingle with the group; talk with people and determine what is on their minds. Acknowledge their concerns at the beginning of your presentation to capture their attention, and follow through by addressing them.

If people are mostly frightened or upset about pesticides, offer whatever information you can to ease their minds. Talk about ways they can exercise some degree of control over pesticide applications. Explain how they can contact government officials to express their concerns, identify groups that share and can assist in projecting their opinions, and offer tips on monitoring pesticide use. On the other hand, if your audience is generally supportive of pesticides, you can focus more on risk education than on empathy.

Sorry, But I Need to Run Every program has the speaker that "runs": here one minute, gone the next. He arrives with only a minute to spare; talks for an hour without taking questions; and then says, Sorry, but I must leave for important business. Never be that person: set a better example.

As mentioned previously, it is important and effective to spend some time talking with people and listening to their concerns before your presentation. But it is equally important to field questions from the audience during your wrap-up and to stay through the next break to talk with folks who didn't want to ask questions in front of the group. Pass out business cards and encourage people to call you later if they think of additional questions. Going out of your way to make yourself available adds credibility to the message just delivered. *Never* tell an audience that you must leave immediately after your presentation—it sends the message that something else is more important.

Don't Be Held Hostage by the Lectern The lectern separates you from the audience, so break the barrier by stepping away from it. Whenever possible, walk into the group so that you are *among* them. Ask the program sponsor (well ahead of time) to provide a hand-held microphone with an extended cord or a wireless mike that will allow you to escape the podium prison!

Experience Means More Than Degrees After being introduced by the moderator, greet your audience and let them know, briefly, what qualifies you to speak on the topic. But don't talk about your background in terms of educational degrees. Academic credentials may imply credibility, but it is your own involvement—your own personal experience in the field—that adds credence to your message.

Be Confident Speaking with confidence builds trust, but cockiness has just the opposite effect. This is one reason why it is important to practice your presentation in front of people whose opinions you value and whom you can trust to offer honest feedback.

$5 Words Do Not Impress Clarity instills trust, so use common language. If the audience doesn't understand your message, you will lose them; so keep your message simple and to the point. Don't risk sounding evasive or arrogant, and don't be a know-it-all!

Be Courteous Courtesy demonstrates respect and can earn you the trust of your audience. As you draw people into your presentation, be aware of cultural differences and be careful not to embarrass anyone. A word or gesture meant to embellish your talk may offend someone taken out of context, so be very cautious; a remark gone wrong is a high price to pay to make a point or to generate a laugh. And remember: The genuine, old-fashioned handshake is still an effective method of recognition that tells people you are pleased to see them and interested in what they have to say.

Address People by Name Addressing people by name is an effective way to connect with the audience. At many programs, each person wears a name tag. Call people you know by name; but if the person to whom you are speaking is neither a friend nor an acquaintance, address them by their last name preceded by Mr. or Ms. When someone asks a question, ask his name and use it in response.

Listen to Others A true expert is always interested in learning about others' experiences relative to his field. So be a good listener. Wherever you go, ask people to share their personal knowledge on pesticides and pesticide management. They may add a local twist that you have not encountered elsewhere—and you might learn something! People love to tell their own tales, and your interest in what they have to say will establish good rapport and expand your expertise.

Answer Questions Some speakers instill audience participation by taking questions throughout their presentation. But this approach works for some speakers and not others. It serves people who cannot focus on whatever else you have to say unless and until they get their question answered, but it can be distracting for you and others. Also, questions breed questions; and it is easy to get caught up answering questions and deplete your allotted time without addressing everything you had planned.

If you cannot maintain your momentum when fielding questions, simply tell your audience, up front, that you will leave plenty of time for questions and answers following your presentation. Then do so. Never tell them you will and then not follow through.

Listen Do not interrupt a question: It's rude. Many times, the core of the question is expressed last, so don't anticipate that you already know what it is. Let the asker finish. Then, make sure you understand the question, think about it briefly, and repeat it for the whole audience before you address it. Give an honest reply. If you need a few seconds to organize your thoughts, throw in something like "That's a very good question," or "Good point!" or "The point of your question is actually quite important." It will buy you a little time and make the asker feel good; it might even encourage others to ask questions.

What They Asked Was . . . When someone asks a question, make sure you understand what is being asked. If it is lengthy, quickly summarize it and confirm that you understand exactly what the question is. Then, repeat it succinctly into the microphone to ensure that everyone in the audience has heard it. This approach serves not only to clarify the question; it also gives you a short window of time to compose your response.

Acknowledge Their Feelings When people start a question with *I feel,* your first comment should be to acknowledge their feelings. Begin your response with a personal statement such as *I understand how you feel.* There are not always good answers to emotional questions, but at least acknowledge the person's concerns. Try to draw out their underlying fears; often they can be calmed, at least in part, by fact. Provide the facts of the situation they describe, and explain issues associated with their concern. Try to balance their apprehension with extenuating circumstances typical of the real world trade-offs we face.

Admit When You Don't Know People appreciate honesty, so don't try to fool them when you don't know the answer. If you offer a best-guess response, qualify it by *admitting* that it is your best guess. If you just do not know the answer, simply admit it. Get the person's name and write the question down; follow up, later, by researching the question and getting back to the person with a response.

What Do You Think about That? Draw on the experiences of others in the audience. Ask how they would answer the question or solve the problem posed. This demonstrates to the asker that situations can be approached from different points of view. It helps them gain a better understanding of the complexity of the issue. Once you and others (if solicited) have answered, ask the person if their question has been addressed satisfactorily. If not, get their name and phone number and follow up after the program.

I'm Here to Help Let people know that you can be reached whenever they have questions. Announce that you will leave a few business cards in the front of the room for those who might like to contact you. Express not only your availability, but also your sincere interest.

Respect People's Time People watch the clock, so always finish up slightly early to allow for questions and thoughtful responses. Do not take advantage of your audience and program sponsors by speaking longer than your allocated time.

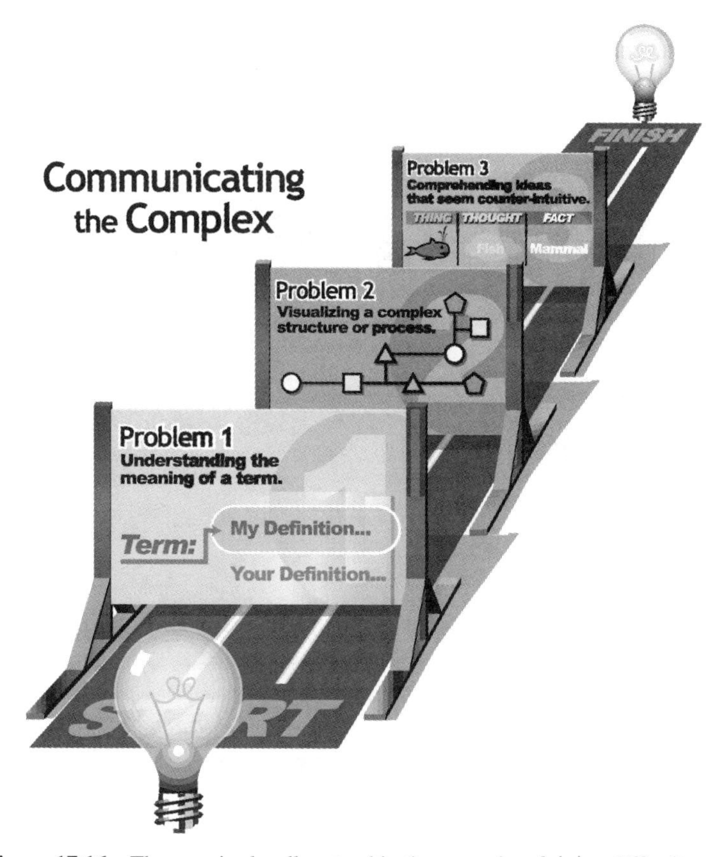

Figure 17.16 Three major hurdles stand in the way of explaining difficult concepts.

The More You See Them, the More They Trust You Trust is developed when people perform a task reliably over a period of time. The more times you address the same group and satisfy their concerns, the more comfortable they become with you.

Explaining the Difficult

Three major facilitators in grasping difficult concepts have been identified (Fig. 17.16):

- Understanding the meaning of common terms
- Visualizing a complex structure or process
- Comprehending ideas that seem counterintuitive (not instinctive)

Explaining Essential Meanings of Commonly Used Words Research in instructional design shows that people master a word or concept by learning to distinguish its meaning *as intended by the communicator* from other associated but unintended meanings. According to many studies, this learning process is most likely to occur when a person's attention is called to the distinction between the intended meaning and the unintended meaning. For example, a communicator might remind an audience that the term *chemical,* within

Figure 17.17 Explaining how genes from bacteria are inserted into corn.

the context of his or her presentation, means *chemical compounds comprising everything in the universe.* In advertising, on the other hand, the word *chemical* sometimes refers only to manufactured substances injected into food. The distinction is critical to the essence of the communicator's message (Fig. 17.17).

There are many, seemingly familiar words that may be misinterpreted in discussions on pesticides. When people hear the word *risk,* for example, they may think *danger*. But Purdue University ethics professor Paul Thompson advises that risk communicators should tell audiences that the word *risk* means *probability*—not danger—relative to pesticides. Risk communicators use the words *pesticide, chemical,* and *science* routinely and may assume that their audiences recognize them in context, but in actuality they may not. As an example, the term *pesticide* often refers to farm chemicals; but disinfectants, although technically considered pesticides, are never thought of as such. Risk communicators should define key words to assure understanding.

Explaining Complex Structures or Processes Complexities that are difficult to envision also can be obstacles to comprehension. For example, difficulty arises when people try to understand why a tumor develops under some circumstances and not others; how a company is going to create a safer pesticide; how genes that code for pesticides can be inserted from a bacterium into a corn plant; or what a risk of one part per million means. Given the latitude for misunderstanding, always assist your audience in comprehending key terms and concepts referenced in your presentation (Fig. 17.18).

Studies in educational psychology have identified many techniques effective in helping people build mental images: pictures, diagrams, and analogies are particularly useful. Other effective strategies are preview statements ("The four key steps in protecting yourself when using pesticides are . . .") and scenarios (how hormones function . . . why DDT persists in fatty tissue).

Analogies and comparisons that simply make an abstract notion more concrete, familiar, or visual can be very helpful. For instance, you might help an audience understand the importance of dosage in determining pesticide risk, with this example: You take a bottle of aspirin and it kills you. You take two aspirins, you feel better. You take just a little flake

Figure 17.18 The use of real examples can help the audience understand complex issues and processes.

of one and you feel nothing, even though you have detectable levels in your body. This analogy can be used when trying to explain why low pesticide exposures encountered in the workplace, in food, and in drinking water do not necessarily cause harm.

Explaining Implausible and Counterintuitive Ideas A third source of confusion lies with ideas that are hard to understand because they are counterintuitive. These scientific ideas frequently conflict with deeply held lay theories. For instance, some people have difficulty believing that exposure to any pesticide level is *safe* or that pesticides can be used safely in schools. The conflict between lay and scientific accounts can lead people to reject, ignore, or misunderstand fundamental aspects of science which can affect their health and safety.

Like scientists, people in general do not give up their theories easily. Consequently, explanations that help people question lay notions must

- state the common view,
- acknowledge its apparent legitimacy,
- present an alternative viewpoint, and
- explain why the scientific view has merit.

For example, how does one explain why otherwise healthful foods contain natural toxins? Following is an account of the hard-to-believe notion that many foods contain natural pesticides.

State the Common View It seems reasonable to believe that natural foods are composed of entirely healthful substances.

Acknowledge the Apparent Legitimacy of the Common Theory There are many good reasons for believing natural foods are good for us. Clearly, eating healthful food is associated with well-being, and many long-lived people are known for their healthful eating. Personal experience tells us we feel better when we eat a balanced diet.

Present the Alternate Viewpoint However, it is not the case that all things *natural* are healthy. We know that some mushrooms are poisonous, that forests can contain poison ivy, and that simply eating too much food is bad for us. So perhaps we should not be too surprised to learn that healthful foods such as fresh baked bread, shrimp, potatoes, and peanuts often contain natural toxins and human carcinogens.

Explain Why the Scientific View Has Merit Why would natural toxins exist in foods? Plants produce these toxins to protect themselves from their natural enemies: diseases, insects, and predators. For example, aflatoxin is a known human carcinogen. It is a natural toxin produced by fungi that form on (and contaminate) stored products: wheat, corn, nuts, and carbohydrate foods such as peanut butter. Aflatoxin also is found in the milk of cows that eat moldy grain. In essence, the idea that natural foods contain *only* healthful substances is unsound.

As this example shows, the key to good explanation of counterintuitive ideas is recognizing that when people have deeply held theories they do not reject them easily. These theories exist because they seem to work. So, good communicators need to acknowledge that such theories are apparently reasonable.

However, we cannot simply reject a common theory by way of a good explanation. Instead, as in the example, the explanation must remind an audience that their theory does not account for certain phenomena. In this example, people are reminded that some natural entities such as poison ivy are harmful.

In summary, good risk communicators diagnose situations before communicating. They identify the principal difficulties in risk communication situations—the obstacles—and consider steps to overcome them. Most important, they know that trust must be earned before education can occur.

The Myths of Risk Communication

The following myths of risk communication are adapted from the 1987 publication of C. Chess, B. Hance, and P. Sandman, "Improving dialogue with communities: A short guide for government risk communication"; untitled work by Thomas J. Hoban, North Carolina State University.

Myth 1: There Is Not Enough Time nor Resources for Communicating about Pesticides Risk communication does take time and staff. Nevertheless, if you do not make an effort to interact with the public, you may be forced to deal with communication disasters that typically take even more time and resources to fix.

Myth 2: Communicating with the Public about Risk Is Likely to Alarm People Risk communication itself can be risky, but not giving people a chance to express their concerns is more likely to *increase* alarm than to decrease it. Balanced communication of pesticide benefits and risks is more likely to *decrease* public concern.

Myth 3: If We Could Only Explain Risks Clearly Enough, People Would Accept Them Although explaining risk is important, there are many factors beyond our control that influence individual perceptions of risk.

Myth 4: We Should Not Go Public Until We Have Solutions to the Problems
There may be some logic in the notion that problems are better accepted when coupled with solutions; but, when you get right down to it, the public wants a say in their own destiny. They want to be apprised of the negatives as well as the positives. They want the chance to voice their own opinions. And sometimes they propose solutions that the experts have not considered!

Myth 5: These Issues Are Too Difficult for the Public to Understand Issues can be complex. Nevertheless, citizen groups throughout the country have demonstrated that lay people are quite capable of grasping difficult concepts associated with complex scientific issues. We cannot communicate successfully by talking down to the public: they become justifiably angry.

Myth 6: Technical Decisions Should Be Left with Technical People Technical personnel may be well versed, scientifically; but policy is determined not only on the basis of science but also on public values. And an informed public is more likely to reach a sound decision than one that is not.

Myth 7: Risk Communication Is Not My Job True, you probably were hired on the basis of other credentials, but you still have a responsibility to deal with people. Failure to communicate may result in policy that damages good science.

Myth 8: Interest Groups Are Responsible for Stirring Up Public Concerns
Activists work to bring about change. They do not create the concerns; they merely arouse and channel attention to those that already exist.

Risk Communication in Practice

Our radio and television programs are interrupted with the following weather update: A tornado watch is in effect. Conditions are right for a tornado to occur in our area within the next 15 min. Prepare to take shelter immediately.

So, why don't we? Why do 50% of us ignore the alert? Why do we normal, perfectly sane people—scientists included—dismiss or totally ignore this kind of information? Why don't we take shelter?

Why do we ignore safety advisories on the use of seatbelts, helmets, and chemical-resistant gloves? Why do we ignore our doctors' advice to lose weight for the sake of our health? The list is endless.

The mystery about why risk messages have little or no impact is one of the many puzzles facing social scientists today. And it is even more puzzling with regard to *pesticide risk* because the science is complex and unclear: there are no simple answers.

Messages that communicate pesticide risk often run counter to prevailing or logical beliefs. For whatever reason, most people have a preconceived belief that pesticides are either safe or risky. And any message that claims otherwise is likely to be dismissed without consideration. The emotions attached to pesticides—My children are in danger! or, I won't be able to maintain the family farm if pesticides get taken off the market!—set up mental roadblocks to logical judgment. Moreover, there are many well-spoken, well-respected people who reinforce the inherent tendency to believe that pesticides are either safe or dangerous.

It is complicated and difficult to effectively communicate the trade-offs associated with pesticide use. Unlike directions to someone's house, they just cannot be drawn on a map: numerous communication and psychological difficulties are attached.

To illustrate, give an audience the following information on a hypothetical product and ask for their reactions. The product

- contains a chemical that causes cancer in laboratory animals,
- causes serious injury to millions of people,
- kills 40,000 people per year,
- kills millions of animals per year,
- causes fires when ignited,
- requires tremendous resources for production,
- causes major air pollution problems,
- produces toxic gases,
- causes billions of dollars per year in property damage, and
- destroys millions of acres of land for roads to facilitate it.

When presented with this information, the public unanimously agrees that the product should be banned immediately. But divulge to the same audience that these facts are actually linked to the automobile, and their reaction is just the opposite (Fig. 17.19). The risks are accepted,

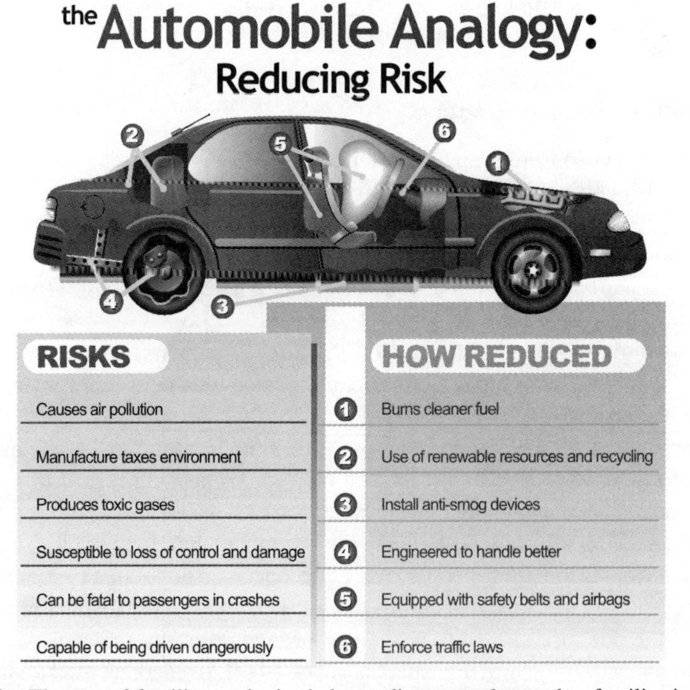

Figure 17.19 The use of familiar analogies helps audiences understand unfamiliar issues. In this example, the automobile has benefits and risks. Pesticides, too, have benefits and risks, and, as with the automobile many work on reducing those risks.

then, because the car is part of the American way of life, because individuals believe they have control over the risks, and because there is no good alternative to the automobile!

The public is quick to discuss how government, industry, and advocacy groups have worked to reduce risks associated with the automobile. The public demands, via pressure for legislation, that gasoline burn cleaner, that cars be equipped with antismog devices, and that alternative renewable resources (e.g., electric batteries) be explored. They have prescribed that we utilize mass transit systems, that government set air quality standards, and that cars be engineered for increased mileage. Manufacturers incorporate safety features such as seat belts and air bags, public service announcements warn us not to drink and drive, and law enforcement officials arrest us for driving while under the influence of alcohol. Clearly, the public sees both sides to this story and is able to arrive at decisions on the dangers of driving an automobile by weighing both the benefits and the risks. In contrast, unfamiliarity with pesticides makes it more difficult to encourage the careful management of associated risk.

There are no formulas, magic bullets, or workshops on how to master the art of communicating and educating people on the trade-offs that pesticides pose. Communicators who want to deliver an effective message to the public must be prepared with supporting facts, must develop trust with the audience, and must convey a balanced message.

These simple components of persuasion are required whether the message is presented in writing or in person and whether the intent of the message is to teach, inform, inspire, persuade, or entertain. Risk communicators can succeed by using common sense in presenting risk information to an individual, to a small group, or to a national audience.

Their message must be (Fig. 17.20)

- fairly portrayed;
- balanced;
- accurate, clear, and concise;
- easily understood; and
- respectful of the audience's values, beliefs, and perceptions.

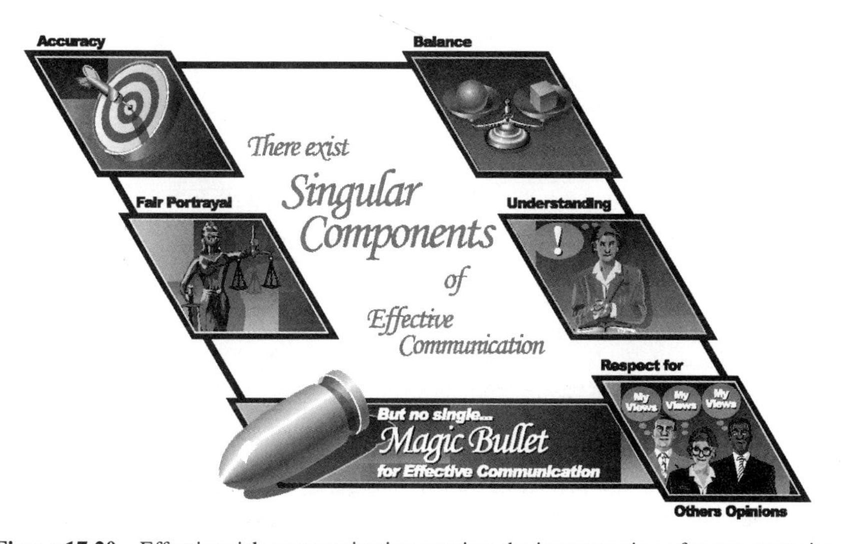

Figure 17.20 Effective risk communication requires the incorporation of many strategies.

The Old Way Doesn't Work Anymore Most risk communicators have backgrounds in either science or public administration. However, seldom does the person writing or speaking on pesticide issues have professional training in the art of communication. Most risk communicators learn their skills through trial and error, by observing others, or from on-the-job experience. But whatever you do, don't make the following mistakes!

Don't Just Talk Can you remember the last time you witnessed a highly respected scientific expert addressing a group, talking nonstop, then running out of time? How many presentations lack a concise summary? How often do speakers fill the time slot without time for questions and answers? Very bright people—experts—often give speeches as though they were in school: They present the facts, using graphs and figures. They discuss their findings but offer no speculation. They limit their conclusion to the facts and often leave no time for questions.

Don't Ignore Your Audience Scientific presentations that work at professional meetings usually fail when presented to the general public. Your qualifications as an expert are not the last word on the subject, and speaking before a group and quoting scientific assurances does not necessarily resolve an issue. Sometimes there are no absolute answers.

The public has views and opinions that may or may not oppose those of scientists, government policymakers, and elected officials. And although their objections and arguments may have nothing to do with the facts as presented, they do reflect public sentiment on the issue.

Today's communication experts acknowledge that it is critical to develop a relationship between themselves and their audience and to recognize and respect various interests among the group; otherwise, what they say will not be heard, nor will what they write be read.

Personal Preparation: Know What You Are Talking About Preparedness is the backbone of effective communication. Do your homework and know your subject—or don't even bother!

I Didn't Study That in School Knowing more about pesticides than your audience knows does not make you "the expert." Pesticides represent a complicated area of study, and even the most seasoned professionals in the field may not be experts; typically, their expertise is in only a small subpart of the issue. So know your stuff!

He Says, She Says We often imply that scientists agree on the facts in their field, that is, that science speaks with one voice. But nothing could be further from the truth. In fact, it is common for reputable, well-known scientists to *dis*agree!

Do not be selective in assembling evidence to support your stance. An important aspect of risk communication is to understand and appreciate the spectrum of scientific opinion on the subject, particularly when fronted by ingrained attitudes. Research shows that fixed attitudes can be modified only through balanced persuasion. If you do not present a reasoned discussion to an audience that has an adamant attitude *counter* to your message, that audience will react against your argument and intensify their belief to the contrary.

I've Summarized the Research Be sure to read and review articles written for scientific journals by leading experts in the field. These works are essential reading because they merge published literature (past and present) and often project future needs. They also provide citations that will lead you to additional scientific literature on the issue. Authors do not always refer to the same literature base; so, for a balanced perspective, read all you can find on the subject.

That Research Sure Set the World on Fire Keeping up to date is vital. Be ever mindful of what is hot in the media. Read published articles and other scientific literature, and research the issue on your own. Learn as many details as you can. Generally, people focus on the headlines, so go beyond that. Equip yourself with additional information and be prepared to speak intelligently on the issue when asked.

I'm Not That Public, I'm Me Individuals form a group because they share common values and objectives, but that does not mean that each person thinks like the others on all issues. Each individual has their own unique interests, needs, concerns, and level of knowledge on the subject; so avoid categorizing extreme positions on either side of the issue. The majority in most audiences will take a wait-and-see attitude. They will listen to facts and conclusions and weigh the merits of your risk analysis before forming their opinion.

Be Smarter Than the One There is almost always a person in the audience who understands the fine details of the pesticide issue—possibly better than you do! So stay at the top of your game; fumbling with the facts in front of an audience means losing face. Be well read, attend specialized workshops, and ask the experts plenty of questions to fend against problems with the complexities of communicating risk management.

Been There, Done That There are facts you get from reading, and there are facts you get by seeing and doing. Work with the groups to whom you normally speak. Observe what they do and how they do it. Try to understand how they think. Take plenty of photographs and use them in your presentations to show the audience that, although you might not be one of them, you are making an effort to view the issue from their perspective. However, be careful not to assume by virtue of having visited a few times and taken pictures that you know *exactly* what they are experiencing. Get to know your audience. Get a feel for the particular group, their issues, and their concerns. You may need to fluctuate communication strategies, depending on your audience. Try to find at least one common thing that you can personally identify with to help bridge the gap between yourself and your audience.

People Don't Want to Know All You Know Don't let details get in the way of education. Every issue has crucial baseline facts, but it is critically important to separate the *have-to-know certainties* from the *nice-to-know details.* The more information you provide, the more you must explain—and the greater your chance of losing the audience to a sea of information: Your message may not get through!

Cull the most important points from your subject matter and present them as simply as you can. Give your audience enough information to form their own opinions, but do not fuss with "but it depends" qualifications. Save the nonessential, nice-to-know details for the question and answer period, and interject them skillfully into your responses.

Know That Science Cannot Answer All Questions If you were writing a book on any pesticide issue, some chapters would be blank due to lack of data or to conflicting scientific interpretation of available data. Be honest about uncertainties, be very careful

Figure 17.21 Allowing people to make their views known is an important element in risk communication.

about promoting your own interpretations, and always let the audience know when you are speculating or when available data are inconclusive.

What's Your Point of View Audiences want unbiased information, but they also expect to hear your view on the risks that pesticides pose. If asked directly, be honest; and base your response on the facts. Be ready to answer the question, Is it safe or is it dangerous? because that is what people really want to know (Fig. 17.21).

You Can't Educate Them All Going into a presentation believing you are going to change everyone is unrealistic. No matter how hard you try to balance your point of view with those of others, there will be instances when even your best efforts fail.

The fact is, some people think pesticides are risky—or completely safe—and no amount of scientific argument will persuade them otherwise. Some people's goals are best achieved by maintaining a controversy and by designating you, or the group you represent, as the opposition. Quite frankly, a risk communicator can easily become the enemy to both pro-pesticide and anti-pesticide groups!

The Delivery: Making an Audience Receptive to Your Message You can help people become more comfortable in an uncomfortable situation by being approachable yourself. Begin your presentation by communicating your interests: If you care about people's health, make that clear. If you care about good science and a safe food supply, make that clear. *Never* begin by focusing on an argument that needs to be settled.

Frequently, you will share more interests with your audience than either you or they realize, and finding common ground gets the speaker and the audience on the same side. If you truly care about your audience, make yourself available and responsive when they have questions. Matching your words with action is critical in earning their confidence.

Don't Just Tell Them What They Want to Hear You can be a hero to your audience by telling them what they want to hear. It's simple: Tell pesticide users that pesticide issues have been blown out of proportion; and tell health and environmental advocates that pesticides are dangerous! But risk communicators who pander to the audience will eventually lose credibility.

Work within Your Comfort Zone, but Learn from the Best Capitalize on your skills and expertise by doing what comes naturally. Enhance your natural abilities by watching closely and incorporating positive aspects of others' speaking techniques that suit your style; however, stay within your comfort zone at all times. Just as important, take note of any annoying characteristics in other speakers and consciously exclude them from your own delivery.

People Expect to Be Entertained Learning is easier when you are having fun. And meaningful communication between you and your audience is most easily established when you enjoy speaking and can make them glad to be there. Education and entertainment often go hand in hand, and comedy is conducive to a good time for both speaker and audience—but don't make a joke of pesticide issues. Stay appreciative and respectful of people's fears and concerns even when approaching your audience lightheartedly.

Stop Talking before They Quit Listening If you know that your audience views a situation as unfair, you need to listen to and show empathy for their concerns. You might spend your visit with them most productively by setting up a context for listening. Be flexible. Don't get locked into the notion that you have to speak every minute of your time frame. For example, limit your presentation to perhaps half the allotted time and spend the remainder listening to your audience and addressing their specific anxieties. Draw them into the subject. Listen to them. Carefully. And discuss their concerns candidly. They will remember who you are, and they will seek you out the next time they have concerns. You will depart with new friends and new contacts. And they will, too.

You've Got 20 Min to Set the Hook The human brain works in cycles of attentiveness and inattentiveness, and the attention span of the learner generally is about 20 min. After 20 min, attention often is diverted elsewhere, and a period of inattentiveness may prevail for several minutes before the brain refocuses on the speaker. This cycle can repeat several times during a 60-min presentation, but adding a twist to your delivery every 15–20 min helps rejuvenate your listeners (Fig. 17.22).

State Your Purpose Continually remind people what you are trying to accomplish and where you are in the program.

I Can't Hear You! Always use a microphone in a large room. It is important that everyone be able to hear your presentation without difficulty.

Connect the Dots to the Bigger Picture Think of the issue in stages or steps, and present them in order—but don't be boring! Explain and demonstrate the logical steps in drawing valid conclusions. Impress upon the audience the basics of reasonable, accurate risk assessment; that is, replace the mystery component with knowledge and technique.

What I Know about Pesticides Comes from DDT and Agent Orange Offer complete information on the benefits and risks of the pesticide being discussed. Most issues involve

Figure 17.22 Add the unexpected to your programs to keep the audience interested.

trade-offs, so present both ends of the spectrum and encourage discussion. Conclude by summarizing what has been addressed and offer your own slant, if appropriate; sometimes it is better left to the audience to form their own conclusions.

It's Show and Tell Time *A picture is worth a thousand words.* This statement is true even when speaking about pesticide benefits and risk. Pictures and show-and-tell items facilitate the communication of concepts and key scientific principles much more effectively—and enjoyably!—than mundane facts and figures (Fig. 17.23).

It's Really Very Similar to . . . People understand concepts better when they can relate to something familiar. Use examples or analogies that people can identify with to

Figure 17.23 Show-and-tell items may be one of the most effective teaching tools available to risk communicators.

illustrate points or to convey facts, but be very careful to avoid comparing risks that are not similar.

What Did You Say? The language of science—abbreviations, acronyms, and jargon—is unfamiliar to most and difficult to teach, so don't go there! Use short sentences and familiar words to make understanding as easy as possible. Everyone prefers simplicity to complexity, summary to detail, and certainty to uncertainty. Use common, everyday language and deliver your presentation as if you are talking to your own family and friends.

That Just Doesn't Make Sense Government agencies such as the EPA, state departments of agriculture, and state environmental divisions set safety standards that guide pesticide manufacturers and pesticide users; and public policy, science, and research help define and determine acceptable risk. But concepts often do not seem to make sense, and people ask questions:

- How can eating pesticides on food be safe?
- What is the relevance of pesticide research on rats to human exposure?
- What do you mean that exposure does not necessarily imply harm?
- Do plants produce cancer-causing compounds?

These questions are best answered in perspective to the concepts of toxicity, exposure, risk assessment, and risk management.

That's Them. What about Me? Science often addresses effects on populations, not on individuals. But individuals are concerned about themselves and their families. They want to know if they personally will be affected.

That's Not What They Said There often is conflicting evidence on an issue, which is inevitably brought up by someone in the audience. If you have done your homework, you are aware of the conflict and prepared to address it. Briefly state both points of view, then demonstrate your justification for the stance you have taken.

Key Points Begin and end your talk by stating your key points. Plan your presentation around a few major points and tell the audience what they are, as you begin. Communicate your take-home message and emphasize its importance. At the end of your presentation, ask the audience to restate your key points; if they leave some out, fill in the blanks as you conclude.

It Doesn't Have to Be My Way Every Time Risks loom large when they are involuntary or even *feel* imposed. For instance, farmers are upset to learn that the EPA may take a popular, inexpensive, effective pesticide off the market, thus increasing the cost of crop production. But explaining what farmers can do to keep the product on the market affords them an opportunity to impact the EPA's final decision. People who do not want pesticides applied to or near their property have a right to influence the decision as well.

Your job as a risk communicator is to inform your audience of their choices and the probable, corresponding consequences. If there are several solutions to a complex problem and each is technically equal, the ultimate decision likely will reflect the values of the majority affected. And when people are thus allowed to exercise some control, they feel more confident in the decision.

No Way Do I Believe It's Not Risky Confrontational remarks such as there's no way that I believe it's not risky bring everyone to the edge of their seats, awaiting your response. Ask the person who commented what it is that they don't believe, and respond with a calm, rational discussion. In many cases, it's simply best to say that it is all right to disagree, that you recognize and respect their position, and that you hope they can appreciate yours.

I've Never Been Hurt! *How can something be risky if it's never hurt me?* Convincing people to change their behavior to avoid risk is difficult. Try telling stories that you have heard where someone, in fact, *has* been injured or hurt as a result of misuse of a pesticide product.

I Just Don't Care! Some issues that you view worthy of discussion may not be important to some members of your audience. It might be that other problems are more pressing, that you didn't connect with them, or that they have taken a fatalistic approach: I can't control the risk anyway! This is their personal decision. Sometimes, all you can do is provide the information: it is up to the listener to decide if an issue is important to him- or herself.

If Only I Could Do It Over It is best to reflect on the program as soon as it is over. Ask yourself what you would change if you could do it over. Were there points that you had difficulty explaining? Were there questions you couldn't answer, or for which you felt your responses were inadequate? Did you learn something from the audience that you will be able to use in another program? What did you do well?

Make it a habit to write down key points and ideas immediately after your presentation; waiting even a short time fades your recollection. Incorporate changes so that you are continually improving your communication skills. As risk communication expert Peter Sandman has indicated, explaining risk information is difficult but not impossible—if the motivation is there.

THE ROLE OF THE INTERNET

The ability of anyone, anywhere, to post their view on anything and everything—for the whole world to see—is both the promise and the horror of the Internet. Evidence supporting any *view* and any *thing* can be found on the Net. Unfortunately, the *source* of information is not always clear and evident. In general, people treat information on the Internet as factual; there is a general assumption that everything on the Net comes from a reliable source, even when no source is listed. This raises the stakes—and opportunities—in the risk communication business.

Not only must we, as risk communicators, be up front and forthright with our one-on-one and group interactions, but as advocates of a balanced approach to truth we also must project our message on the Internet. There is no reason why people seeking pesticide information on the Internet should be able to access only the positive or only the negative. We must jump on the Internet bandwagon and post our message, and we must connect with every search engine available. We must **stand out** and earn our own recognition.

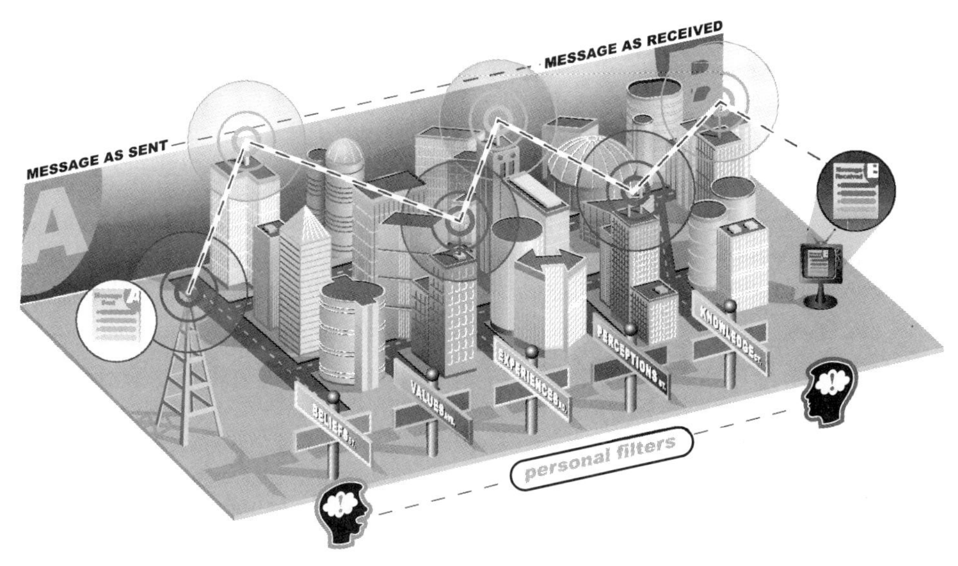

Figure 17.24 The message delivered and the message heard can be very different depending on the personal filters that the message passes through.

SUMMARY

Unquestionably, the use of pesticides has been and always will be controversial in our society. It involves very real and important trade-offs that concern people. Can one reasonably expect to educate the public amid so much background noise? And what about all the complications surrounding pesticide risk? Well, sometimes it is the noise that creates the *interest* . . . and people listen!

Is it possible to educate the public on pesticide risk when the issues are couched as *good news versus bad,* or *them versus us?* Do people have the patience—or the interest—to listen to more of the facts? Will they tolerate descriptions of risk–benefit trade-offs instead of the either–or scenarios of advocates? Do people have the ability to understand science-based, reasoned explanations of the need for pesticides and the consequences of use, both good and bad? Will they listen with an open mind before making their own decisions? The answer is yes. The public will listen to a credible communicator who earns their respect. And the backbone of respect is knowledge and effective communication.

Interest in risk communication is at an all-time high as government officials, industry representatives, scientists, and health and environmental safety advocates strive to communicate why the public should or should not worry about pesticide risk. Delivery of a clear and effective message through teaching, conversation, writing, or speech is a difficult proposition, even in the best of situations.

It is difficult to get people to understand and accept risk. It is also difficult to get those who ignore risk to acknowledge and respect it. As individuals, we base our beliefs on what we know; and what we know depends largely on our source of information. A person's knowledge on pesticides, coupled with their own personal values, forms the basis for their stance on the issue (Fig. 17.24).

There are myriad views on pesticide risk. But people tend to key into concepts that complement their own agenda, that is, concepts that validate their own preconceptions.

As risk communicators, our success in *educating* the public hinges largely on our skills in *public relations.* Instead of talking *to* an audience, we need to talk *with* them, to engage them in healthy dialogue. We must acknowledge and respect the audience's point of view, even if it is unfounded or inappropriately skewed. We must afford the audience an opportunity to validate their concerns, and we must share our points of view and identify areas of agreement.

Technical information alone will not address public concerns effectively, nor will it necessarily reduce regulatory restrictions. The key is interaction and dialogue.

We must stay abreast of scientific developments and changes in pesticide policy if we are to achieve effective risk communication—and we must understand our audience. We must present concepts that are clear, understandable, and nonthreatening. We must embrace questions and treat discussions, responses, and what may seem to be unreasonable concerns (or lack of concern) with equity.

Skillful risk communication techniques and useful examples that audiences can relate to are paramount to our dissemination of understandable, useful pesticide information. Our goal is to reduce unreasonable fears, heighten awareness, foster support, and steer good public policy.

CHAPTER 18

TODAY'S DISCUSSIONS, TOMORROW'S ISSUES

We understand more of the nature and scope of pesticides today than ever before. Science and technology continually open new windows of learning; and the more we know about pesticides, the more complex our questions—and answers—become. For example, enhanced detection methodologies allow us to identify minute pesticide residues in or on food, and our understanding of the interactive complexities of ecosystems generates questions on the short- and long-term effects of pesticides. More and more, people want and need information on such issues, and laws have been developed to fulfill government's responsibility to provide it.

"WHAT-IF" ISSUES CHALLENGE EXISTING SCIENCE

The what-if questions associated with pesticides continue to drive regulatory agencies, manufacturing businesses, application industries, and public interest groups. Often, the what-ifs reveal to public policymakers that data or credibility gaps do exist; and sometimes this leads to debate on what the policy of government—specifically, the EPA—should be.

Following are examples that demonstrate the complexity of pesticide issues up for debate in the United States.

HUMAN HEALTH ISSUES

Endocrine Disruptors

What If Pesticides Masquerade as Hormones? The endocrine system consists of glands—e.g., pituitary, pineal, thyroid, adrenal, ovaries, testes—that secrete hormones into the blood. Hormones bind to various receptor cells found in the mammary glands, reproductive organs, prostate, liver, and kidneys. Once attached, hormones signal these cells to turn on or turn off processes involved with metabolism, development, stress, and reproduction.

Excesses or deficiencies of hormones make the difference between normal and abnormal processes.

Some chemicals believed to be endocrine disrupters can mimic hormones and interrupt normal processes by activating receptor sites or by blocking hormones from binding to the sites. In recent years, scientists and public interest groups have spoken out over the use of suspected endocrine-disrupting chemicals (EDCs: also known as environmental estrogens, environmental endocrine modulators, and estrogen mimics).

Some argue that exposure to EDCs can cause irreversible damage to the developing fetus, increase the incidence of breast cancer in women, and decrease sperm concentrations in males. A specific example of an endocrine-disrupting chemical that has been well documented and frequently cited in scientific literature is the drug diethylstilbestrol (DES). DES was a very potent synthetic estrogen used by women from 1948 to 1971 to prevent spontaneous abortions. Follow-up studies on daughters whose mothers took DES revealed reproductive organ dysfunction, abnormal pregnancies, reduced fertility, immune system disorders, and periods of depression.

Some endocrine-disrupting chemicals are persistent and mobile in the environment and have been detected in lakes, oceans, and food products. High levels of EDCs have been associated with abnormal thyroid function and decreased hatching success in birds and fish, decreased fertility in birds and shellfish, and feminization of male fish.

Proponents for more EDC oversight point out that current EPA pesticide testing requirements—which include studies on acute and chronic toxicity, mutagenicity, reproduction, residues, and environmental fate—fail to account for EDCs. They argue that EDCs are not acute toxicants, mutagens, or oncogens, and that therefore the current system does not provide enough information to judge their *safety* relevant to endocrine disruption.

There is general agreement that *high-dose* ingestion of the drug diethylstilbestrol and significant exposure to environmental pollution (e.g., from spills) may be detrimental to humans and wildlife. The controversy concerns the impact of *low-dose* exposure. Some scientists challenge claims that pesticide residues below established tolerance levels in food, water, and air are significant. They argue that concentrations of natural and synthetic EDCs are thousands to millions of times less potent than DES. Others argue that the mere presence of weak estrogenic substances in the environment or in the human body is insufficient cause to conclude that DESs are capable of producing adverse effects.

This debate also has promoted discussion on what to do with a pesticide that is suspected to be an endocrine-disrupting chemical: Should EDCs be regulated under the current risk assessment process? What are the measurements that would be used to evaluate an EDC?

These discussions took on new meaning with changes in government policies. The *Food Quality Protection Act* and the *Safe Drinking Water Act* require the EPA to develop, validate, and implement a screening program to determine whether pesticides are likely to cause disruption of the endocrine system. The EPA assembled an expert committee called the Endocrine Disruption Screening and Testing Advisory Committee (EDSTAC) to advise them on various approaches. Based on input from EDSTAC, the EPA released a conceptual framework that focuses attention on estrogen, androgen, and thyroid hormones in mammals, wildlife, and fish, since these hormones are the most studied and understood. The screening process places pesticides and other chemicals into one of four categories:

Category 1: Hold There is sufficient scientific information on pesticides in this category to conclude that they are not likely to interact with hormones. No testing is required.

Category 2: Priority Setting There is insufficient information on pesticides in this category to evaluate hormonal impact. The EPA likely will require information in support of lowering the classification for any Category 2 pesticide to Category 1—or raising it to Category 3.

Category 3: Tier 2 Analysis There is sufficient information to suggest that pesticides in this category have the potential to mimic hormones. Based on analysis data, Category 3 pesticides will be reassigned to Category 1 or moved into Category 4. Each pesticide will be required to undergo assays that could include a two-generation mammalian reproductive toxicity assay, an avian reproductive toxicity assay, a fish life cycle toxicity assay, a shrimp toxicity assay, and an amphibian developmental and reproductive assay. The mammalian, avian, and fish studies are commonly conducted already.

Category 4: Hazard Assessment The EPA will characterize the level of risk that pesticides in this category pose to people and wildlife. Amended label directions and/or the advisability of continued registration will then be considered.

Immune System

What If Pesticides Interfere with the Immune System? The immune system defends the body against infectious organisms: bacteria, fungi, and viruses. When antigens (foreign agents) gain access into the body, the immune system reacts by sending white blood cells to attack them. The white cells secrete antibodies that attach to and attempt to destroy the invaders. Antibodies may neutralize the antigens immediately.

The immune system includes the thymus, spleen, lymph nodes, lymph tissues, stem cells, white blood cells, antibodies, and lymphokines. The lymph system, known as the body's drainage system, is a network of vessels that collect excess lymph (body fluid) for bathing body tissues. The lymph carries white blood cells and is filtered through lymph nodes located in the neck, armpits, abdomen, and groin. Any bacteria or viruses not initially destroyed are continually attacked by white blood cells and, ultimately, the lymph fluid (with its white blood cells) is discarded into the circulatory system.

Scientists speculate that pesticides may have the potential to interfere with the immune system. They offer limited clinical evidence, primarily from experimental animals, that pesticides can suppress the immune system, produce a hyperactive reaction (allergy), or cause an autoimmune response (rheumatoid arthritis).

Pesticides that impair white blood cells and damage organs might lead to immune system suppression. Epidemiological evidence suggests that a suppressed immune system can render a person more susceptible to infection. Thus, individuals exposed to pesticides might exhibit a higher susceptibility to infectious illness than persons not exposed.

Multiple Chemical Sensitivity

What If Current Health and Safety Standards Do Not Protect Individuals Who Are Sensitive to Low Levels of Pesticides? Some individuals suffer from a medical condition commonly called *multiple chemical sensitivity.* Chemical ecologists believe that low levels of environmental chemicals produce significant and sometimes debilitating reactions in certain people. In the past, these reactions were classified as *mysterious,* or, as with other illnesses with similar symptoms, simply not classified or identified at all.

Today, multiple chemical sensitivity may be labeled chemical hypersensitivity, environmental illness, multiple chemical syndrome, chemically induced immune dysregulation, total allergy syndrome, or ecologic illness. Recognition of this syndrome as a legitimate illness would allow sufferers to seek the same health insurance benefits, vocational rehabilitation, and disability benefits afforded those with other illnesses.

Multiple chemical sensitivity is thought to result from high-level chemical exposure that triggers the onset of low-level exposure sensitivity. Specific environmental chemicals and other irritants implicated include insecticides; detergents; fresh newspapers; new automobile interiors; and chemicals found in perfumes, cigarette smoke, diesel particulates, dyes, and hair sprays.

Treatment of patients seeming to suffer from multiple chemical sensitivity is preceded by asking a battery of questions to establish a link between the individuals' symptoms and substances found in food, water, and the environment. The usual process of diagnosis and treatment involves removing sensitized individuals from exposure to the chemicals believed to trigger the symptoms. Patients may be housed in a controlled environment in which chemical exposure is reduced to the lowest feasible level. Acclimating patents to the controlled environment allows physicians to reintroduce suspected chemicals and monitor patients for symptoms of exposure.

Clinical ecologists speculate that multiple chemical sensitivity may involve damage to the nervous system or to the immune system. Patients diagnosed as "chemically hypersensitive" are described as having flu-like symptoms: headaches, irritability, depression, memory difficulties, gastrointestinal problems, joint and muscle aches, and shortness of breath. However, the symptoms generally are manifested in the absence of physical findings or pathologic abnormalities. One complication of diagnosis is that there are no diagnostic markers for laboratory tests.

Medical associations such as the American Academy of Allergy, Asthma, and Immunology; the California Medical Association; the American College of Physicians; and the American College of Occupational and Environmental Medicine argue that the causes identified and therapies practiced by chemical ecologists are not supported by rigorous scientific and medical testing. Those opposed to classifying multiple chemical sensitivity as a "real" disease argue that similar symptoms associated with allergies, infections, and toxicities can be diagnosed on the basis of physical and pathological symptoms and supported by laboratory testing. They reject the notion of classifying multiple chemical sensitivity symptoms as a disease in the absence of quantifiable, repeatable scientific testing.

Proponents counter that the lack of physiological and pathological evidence does not detract from the fact that several groups—industrial workers, occupants of tight buildings, and communities whose residents are subject to a contaminated environment—do show symptoms that can be described as multiple chemical sensitivity. These groups differ in professional and educational achievement, age, gender, and exposure to the types of chemicals that trigger sensitivity. Physicians who recognize the existence of multiple chemical sensitivity as a disease remind the skeptics that similar arguments were made about multiple sclerosis and lupus prior to their being determined specific diseases.

We are left with real people who have debilitating physical and/or psychological problems. Their suffering is complicated by debate among medical professionals: some claim they do have a disease; others claim they don't. And in the meantime, these people suffer. The science is unclear, and all points of view are illustrated and supported by thousands of Web sites easily accessible to all. The desire of sufferers for a definitive name for their condition—and a cause—is quite understandable; and they will push science to ultimately clarify the uncertainties.

Sperm Counts

What If Sperm Counts Are Affected by Pesticides? Studies that deal with the effects of pesticides on human reproduction have focused primarily on pregnant mothers and unborn babies. But the *British Medical Journal* published an article in 1992 that concluded that males had experienced a 50% decline in sperm count since the 1930s. In 1997, however, a second group of scientists analyzed the same data and concluded that the decline in sperm count was even more dramatic: 1.5% per year.

Research indicates five variables that influence sperm-count data:

- *Source* Sperm counts vary greatly among volunteers recruited from sperm banks, fertility hospitals, and vasectomy clinics.
- *Age* Older men have lower sperm counts.
- *Abstinence* Prolonged sexual abstinence results in higher counts.
- *Season* Men have higher sperm counts in winter.
- *Geography* Significant regional differences exist within the United States and between continents.

Some scientists argue that the 1992 and 1997 analyses are flawed because they did not control for all variables in their meta analysis of approximately 60 studies. They also argue that not accounting for or eliminating "background noise" left open the question of whether sperm counts were lower during their study than in previous decades.

But why are pesticides discussed as influential if the scientific community has yet to resolve their differences on how to conduct these studies and analyze the data? All issues seem to have a pivotal event that everyone uses to legitimize their concerns, and such an event occurred in 1977 when it was found that occupational exposure to the pesticide dibromochloropropane (DBC) resulted in male infertility.

So where does this leave the sperm-count issue? It seems that those who speculate on the role of pesticides in declining sperm counts have outpaced the ability of science to deliver the facts. But there is no question that the issues of data collection and analysis will be resolved as more studies incorporate better investigative techniques. If resulting data indicate that sperm counts are in fact dropping, pesticides will be one of many environmental factors that scientists will examine closely.

Cognitive Skills of Adults and Children

What If Pesticides Impede the Mental Growth of Children? An anthropologist from the University of Arizona asked whether physical growth and mental development are affected when children are exposed to pesticide; as a result, Elisabeth Guillette and her colleagues from Mexico began searching for a population that differed only in its exposure to pesticides. Their search culminated with the identification of a population of indigenous people living in northwestern Mexico.

Researchers found that the population had divided itself into two groups. One group farmed by traditional methods in the foothills. The other group took up commercial farming in the valley. Observational studies found that poverty level, genetic background, diet, cultural patterns, social behavior, education, and availability of health services were similar between the two groups. One major difference was that the valley farmers relied more on pesticides than did those farming in the foothills.

Having found potential study groups, researchers designed a testing protocol to measure developmental progression and physical growth. They developed a quick screening method—the Rapid Assessment Tool for Preschool Children—that involved interviewing parents, playing games with the children, and taking physical measurements of each child. Examples of methods used are as follows:

- Immediate memory was measured by having the children repeat numbers.
- Intermediate memory was evaluated by having the interviewers say "Hello. My name is (name of interviewer). I will ask a few questions and we will play some games. When I am finished I will give you a red balloon. Do you like balloons?" Since some of the children did not know colors by name, an object was pointed out as being red. The child was expected to ask for a red balloon at the conclusion of the interview.
- Gross eye–hand coordination was determined by having children catch balls thrown to them from various distances.
- Fine eye–hand coordination was measured by having them drop raisins into a bottle.
- Stamina was evaluated by having the kids participate in jumping games.
- Mental development was measured by having each child draw a person.
- Physical development was evaluated by taking measurements of the head, chest, height, and weight.

Preschoolers enrolled in the study were evaluated in the home for approximately 30 min each. The result indicated that children from the two groups were similar in their physical development and instant recall. However, children living in the foothills had better intermediate memory, could draw a better representation of a person, had more stamina, and had better eye–hand coordination. Looking at the entire study, the researchers found that children from the foothills were more advanced psychologically and physiologically than those from the valley. The authors concluded that exposure of the valley children to pesticides may have played a key role in the differences found between the two groups.

The study has generated much debate among scientists. Some consider the findings alarming, whereas others challenge the study due to its lack of pesticide exposure measurements. Although controversial research is always hotly debated, the fact remains that this is one of the most thought-provoking studies published on the subject.

What If Pesticides Impact the Physical and Mental Health of Growers? Scientists today are opening new lines of inquiry into whether low-level exposure to pesticides impacts the mental and physical health of farmers and commercial pesticide applicators. These studies are not based on traditional animal toxicology (dose and measure); instead, the trend is to question what impact low-level pesticide exposure has on human memory, concentration, coordination, mood, and personality.

A group of physicians looked specifically at low-level applicator exposure. A group of New Jersey tree-fruit growers—pesticide applicators—who had never sought medical attention nor been hospitalized for overexposure to pesticides was selected. The physicians selected a class of pesticides—organophosphates—that exhibit easy-to-document neurophysiological symptoms when humans are overexposed. With these selections, the authors studied whether low-level exposure of tree-fruit growers to organophosphate pesticides used in their work influenced their physical and mental health.

Volunteers were male, self-employed, licensed pesticide applicators. Each grower completed a detailed questionnaire about himself and his use of organophosphates. Each underwent a series of evaluations to measure concentration, language, memory, and visual-motor skills, as well as a battery of neurological tests that examined cranial nerves; motor function; sensation to light touch, pin prick, and vibration; cerebellar function; deep tendon reflexes; and gait. Psychiatric assessments measured each volunteer for hypochondria, depression, hysteria, paranoia, and schizophrenia. The tree-fruit grower volunteers exposed to organophosphates were compared to unexposed blueberry/cranberry growers and hardware store owners.

The tree-fruit grower volunteers averaged 48 years old and had farmed for an average of 27 years. Their longevity in the industry implied continual exposure since early adulthood, making them prime candidates for measuring symptoms from chronic low-level exposure to pesticides. The results were encouraging: Tree-fruit growers who had experienced long-term occupational exposure to organophosphates showed no major deficits in neurobehavorial performance, personality, or emotion.

The results showed that farmers who use organophosphate pesticides as directed on the label are no more prone to problems than is the general public. However, this is just one study. Until further studies address a larger cross section of pesticide users and pesticide products, the issue will remain open.

WILDLIFE AND ENVIRONMENTAL ISSUES

Wildlife Habitat

What If We Can Safeguard Wildlife by Intensively Cultivating Productive Farmland? The world's population is expected to reach eight billion by the year 2025. How much additional land will be needed to support 10 billion people? Where will "new lands" be found? And will lands be spared for nature? Increasingly, arguments are being proposed that link human populations, biodiversity, and wildlife habitat with high-yield, technology-based agriculture. The assumption that pesticides can be good for wildlife has become a controversial issue for those who believe that yields per acre must be dramatically increased if people are to be fed, and if wetlands, forests, and grasslands are to be conserved for plant and animal species. The feasibility of maintaining land set aside for nature will depend on maximizing production on cultivated lands; feeding people and saving wildlife cannot become an either/or proposition (Fig. 18.1).

Today's farmers are cultivating about one-tenth of the world's arable land to feed nearly six billion people: twice the number of people on the same amount of land that was farmed in 1955. Soil conservation and the use of technological inputs—pesticides, fertilizers, biotechnology, and pest-resistant plants—are cited for these remarkable increases in output per acre that are outpacing population growth.

Grain surpluses resulting from the use of chemicals, new technologies, and proven agronomic practices have afforded the United States the luxury of idling one-fifth of its cropland. Around the world, 10 million square miles of wildlife habitat have been spared the plow, due in part to chemically supported and technology-based farming; other conservation efforts have contributed as well.

Figure 18.1 Trees and native land interspersed with farms.

Without increasing productivity, it is estimated that current practices can supply enough food to provide a vegetarian diet to 10 billion people. However, maintaining crop yields at current levels would not support the nonvegetarian diets that affluent nations demand, nor the diets that developing countries are striving to provide their populations. More food—not less—will be needed. Because the most productive farmlands are already under production, these "new," less productive and less sustainable lands would come at the expense of forests, marshes, and grasslands: lands necessary for biodiversity.

Those who support high-yield farming that relies on the use of multiple technological input point to increasing yields as the solution to feeding people and sparing land for nature. Conversely, they argue that simply maintaining or reducing current yields will exert greater pressure to bring marginal lands into production as governments attempt to feed ever-growing populations.

Detractors point out that high-yield farming will result in more soil and water contamination from chemicals, increased soil erosion, less wildlife habitat on acres being intensively managed, and a system that requires tremendous input and energy. They point out that this intense management approach will not be sustainable over the long run.

The questions being discussed warrant great consideration. Whether or not lands will be spared for conservation may depend on whether we pursue a technology-based approach that leads to increased production per acre, or an approach where reduced input may require cultivation of additional "new" lands. Although scientists argue about the merits of conventional, biotechnology, high-yield, sustainable, and organic farming, this fact remains: Farmers will occupy a pivotal position for feeding the world's population and conserving land for nature.

Amphibian Deformities

What If Pesticides Are Involved in Frog Deformities? A school project aimed at understanding the ecology of a local wetland captured national attention when students found frogs with paralyzed, deformed, missing, or extra legs. The initial detections were investigated quickly by the Minnesota Pollution Control Agency (MPCA) and the University

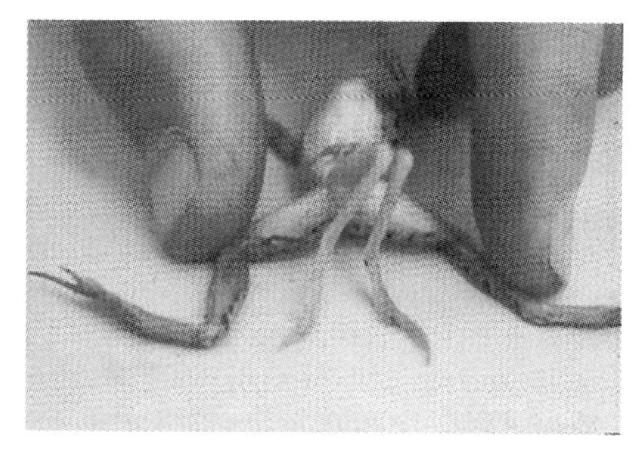

Figure 18.2 Deformed frogs found in Minnesota ponds.

of Minnesota, and it was concluded that one-third of the frogs collected from a single wetland were deformed. Surveillance of an additional 54 Minnesota counties turned up deformed frogs from urban, agricultural, and natural areas, as well as state parks (Fig. 18.2).

The culprit has thus far eluded researchers. Speculation is that the deformities resulted from the interaction of something in the environment with the eggs and developing tadpoles in frog populations. Researchers currently are pursuing a number of potential possibilities by testing mud sediments, water samples, and frog tissues for abnormal levels of metals, PCBs, pesticides, and parasites.

The story is even more complicated because amphibian populations are declining quickly around the world. There has been speculation that the causes of declining frog populations are habitat destruction and fragmentation, disease, ultraviolet radiation, air pollution, and pesticides.

Monarch Butterfly and Bt Corn

What If Bt Corn Impacts the Monarch Butterfly? Biotechnology companies are successfully introducing into the marketplace new medicinal and agricultural products. Bt corn is one such commercially successful biotech product that has captured a sizable portion of corn acreage in the Midwest.

Bt is short for *Bacillus thuringiensis,* a bacterium long known to produce a toxin (delta-endotoxin) that has "natural" insecticidal effects on many species of caterpillars. Plant geneticists insert the Bt into corn, enabling it to produce the endotoxin (a protein). Pest caterpillars such as the European corn borer die when they consume plant tissues containing the endotoxin.

These new genetically modified corn hybrids produce endotoxins known as plant pesticides. Although all plants produce natural pesticides, these endotoxins are different in that they are derived from incompatible sources; for example, genes from bacteria are engineered into corn. The EPA policy is not to register the corn hybrid (e.g., Bt corn) but, rather, the active ingredient endotoxin—the plant pesticide. In Bt corn, the delta-endotoxin is the active ingredient. There are currently plant pesticides registered for use in potatoes, cotton, popcorn, and field corn.

Bt corn became embroiled in a controversy concerning the monarch butterfly. The monarch's life history has been the subject of many books, television specials, and magazine articles. It overwinters in Mexico, migrates northward in early spring, and reaches peak populations in the Midwest during summer. The monarch larva is one of the few insects that utilize the milkweed plant as a food source; it is poisonous to most predators of butterflies. Monarch larvae and adults accumulate milkweed toxins in their tissues, making them taste bad to predators.

But milkweed plants grow near cornfields; and as pollen falls from corn it may land on the milkweed surface, where both the milkweed plant and corn pollen can be consumed by monarch larvae. A university entomologist published a laboratory study showing that monarch larvae feeding on milkweed plants intentionally dusted with Bt corn pollen were smaller and more likely to die than larvae feeding on milkweed dusted with traditional corn pollen. Thus, the finding that Bt pollen increases mortality raises questions whether or not large acreage planted to Bt corn will seriously reduce the monarch population.

The entomologist's findings that the most recognized butterfly in the United States was affected by a bioengineered product appeared on television, in newspapers, and in newsletters. Within days of the news reports, EPA policymakers, industry specialists, and environmental groups were embroiled in yet another controversy over the use of bioengineered products. Environmental groups quickly filed petitions with the EPA, seeking protection of monarch butterflies and asking the EPA to review whether endangered species whose habitats are in close proximity to Bt corn might also be in harm's way.

The biotechnology industry argued that the study published in *Nature* was a preliminary laboratory study, that it was technically flawed, and that it bore little relevance to field conditions. The author and critics agree that the report merely identified a hazard: it did not evaluate risk and exposure.

The EPA has made a preliminary decision that the monarch is not threatened by Bt pollen. Even so, a variety of studies are being conducted to substantiate—or invalidate—the EPA's regulatory position.

One aspect of these studies is to note the natural occurrence of milkweed plants in proximity to cornfields; the distance pollen moves outside of cornfields; when, in relation to pollen shed, monarch larvae feed on milkweeds; and quantification of Bt pollen to plants, relative to levels known to harm monarch larvae. The EPA will place restrictions on the use of Bt corn if adverse effects on monarchs are substantiated by field studies.

Pesticide Transport by the Atmosphere

What If Pesticides Move Farther Than Where They Are Applied? Scientists are just beginning to document that atmospheric water (rain, snow, fog) contains measurable concentrations of pesticides. Environmental monitoring of the atmosphere is gaining momentum as scientists cite evidence that pesticides can be moved great distances from the site of application.

A review of 132 studies conducted mainly in California and the Great Lakes region, indicates that atrazine, alachlor, metolacholor, DDT, dieldrin, simazine, chlordane, cyanazine, and toxaphene were detected at measurable concentrations in at least 50% of the areas sampled. Higher concentrations in precipitation are linked to areas with higher use. For example, the highest levels of triazine and acetanilide herbicides in precipitation occur in the Midwest region during the spring and summer. It is estimated that the amount of pesticide returning in precipitation is less than 1% of the total applied.

Evidence indicates that pesticides in the atmosphere can move hundreds or thousands of miles from the site of application. A study published in *Science* describes the regional and global transport of organochlorine pesticides. Tree bark from 90 sites around the world was used to track the movement of 22 organochlorine compounds. Tree bark was selected because it is an excellent scavenger of even minute traces of organochlorine pesticides. The evidence collected from tree bark indicates that many of the organochlorine pesticides are present around thc world.

Organochlorine concentrations were found in tree bark samples from very remote regions of the world, such as the Orinoco Rain Forest in Venezuela, Guanacastle National Park in Costa Rica, the rain forest of Ecuador, and the Marshall Islands in the Pacific Ocean. The authors of the study claim that compounds can enter the atmosphere and be carried by global atmospheric currents from the warmer latitudes into the colder areas of higher altitudes. Once they have entered these cooler areas, pesticides can condense onto plants and soil and into water.

The presence of pesticides in the atmosphere and their ability to be transported and distributed worldwide are just beginning to generate questions on their potential impact on drinking water supplies, stream water quality, aquatic organisms, and food chain bioaccumulation. The interaction of pesticides with the atmosphere and the movement of pesticides over long distances will continue to attract scientific investigation.

REGULATORY ISSUES

Crop Quality Standards

What If Government Standards Increase the Use of Pesticides? The federal government sets standards that characterize fruits and vegetables based on external appearance, and critics maintain that this compels farmers to use pesticides to maximize the aesthetics of their produce. But does it, actually?

Critics of quality standards say yes. They argue that the external appearance of fruits and vegetables has no bearing on nutritional content, storage life, or flavor, and that farmers therefore choose to apply pesticides to enhance physical appearance in pursuit of top dollar in the marketplace. They argue that the public is not well served because "cosmetic" standards motivate growers to apply otherwise unnecessary pesticides, and that this denies the public the advantage of price differentials between blemished and unblemished produce. The bottom line is this: Aesthetic quality drives the price up—for farmers, wholesalers, and grocery stores alike. It is also agreed that these standards result in unnecessary pesticide application and residue in and on produce.

Proponents of government standards cite the difficulty in separating cosmetic factors from those that affect yield, taste, and nutrition. For example, product color, shape, or size may indicate maturity, which in turn may affect flavor or texture. Proponents argue that if quality standards based on appearance were eliminated, pesticides still would be necessary to protect the internal quality of fruits and vegetables from bacterial rots and other decay organisms that enter at points of exterior insect damage.

Plants generate their own natural pesticides to protect themselves from rot, pests, and other injuries, that is, for the *health* of the plant, not for the *appearance*. So unblemished plants may be *naturally* more healthful as well as more attractive and tasty.

Points for debate on quality standards:

- Some consumers are willing to pay premium prices for produce with cosmetic defects in exchange for reduced pesticide residues (i.e., reduced risk).
- The role of industry is as important as state and federal standards—or perhaps more important—in determining the aesthetic quality of produce ultimately available to consumers.

- It is difficult to identify a clear-cut case where a pesticide was used to manage a pest that causes cosmetic damage only.
- Appearance quality standards have little effect on some crops because pesticides applied for other reasons facilitate aesthetics as well.
- Existing federal grade standards allow for lower grade (cosmetically blemished) produce; that is, if the consumer is willing to forgo "pretty" fruits and vegetables for reduced pesticide residues and/or lower retail prices, the current grade standards will suffice.

Biotechnology and Pesticide Usage

What If Herbicide-Tolerant Crops Increase Pesticide Use? Agronomists improve crop species by recognizing and enhancing the genetic diversity found within plants. They have developed varieties that have revolutionized crop production systems by combining positive traits (e.g., higher yields) and eliminating negative traits (e.g., disease susceptibility). Today, growers can select plants that are tailor-made for their local environment.

Collaborative research between seed companies and pesticide manufacturers began during the 1980s. These investigations were aimed at finding plants that could tolerate specific herbicides and therefore minimize inadvertent crop damage caused by pesticide applications. Plant physiologists in the laboratory not only uncovered a number of physiological mechanisms that could convey herbicide tolerance to crops, they worked with agronomists to develop varieties for field testing. Today, there are crop varieties in the marketplace with reduced herbicide sensitivity at the molecular site of action. For example, some plants degrade herbicides through metabolic pathways, whereas other plants avoid herbicides through lack of uptake or by sequestering the herbicide within the plant.

Biotechnology, too, has contributed to scientific understanding and development of herbicide-tolerant crops. Scientists have isolated microbial genes that code for an enzyme that degrades herbicides, and they have inserted the genes into the crop's genome (inheritable traits). This method is different from traditional plant breeding programs because, in this case, the actual genes used for resistance may be foreign to the crop species. These transgenic (genetically altered) crops can then pass on the herbicide-tolerant traits to their progeny.

Some public interest groups and scientists believe that herbicide-tolerant crops promote continued reliance on chemical weed management. They strongly argue that the real beneficiaries of herbicide-tolerant crops are the companies that have a vested interest in expanding their herbicide markets. It also is argued that crops may pass on the resistant genes to closely related weed species, incidentally producing "super weeds" that will be difficult to manage. Weed scientists have expressed concern that the expanded use of herbicides (e.g., labels on products for weed management in corn extended to include soybeans) may result in a dramatic increase in weed resistance.

Controversy over herbicide-tolerant crops may escalate as more become available to growers and as they are processed into food and feed. Proponents for the development and use of transgenic plants indicate growers will gain the flexibility of selecting "environmentally friendly" herbicides in favor of those that pose excessive risk to humans, wildlife, or the environment.

Environmental Justice

What If Certain Segments of the Population Shoulder a Disproportionate Share of the Risk? Environmental justice involves the legal, scientific, and public policy inquiry into the distribution of environmental health hazards. The EPA definition of environmental justice is the fair treatment of all races, cultures, and income levels with respect to the development, implementation, and enforcement of environmental laws, regulations, and policies. But of particular relevance is whether environmental *racism* exists: Do low-income, racial, and ethnic minorities shoulder a disproportionate share of the risk?

Churches, advocacy groups, and legal professionals are asking state and federal decision makers to address why low-income and ethnic minorities bear more of the brunt of environmental health hazards than do affluent and nonminority communities. The debate has grown cynical, with suggestions that environmental decisions (e.g., citing of hazardous facilities) are racially motivated.

Those concerned with environmental racism argue that hazardous facilities, such as hazardous-waste incinerators that contribute to air pollution and agrichemical operations that store pesticides, usually are located near minority populations and low-income communities.

The first step in assessing whether or not environmental justice prevails is to pinpoint all hazardous facilities on a map; then draw concentric circles around each facility, representing a radius of one-half mile, one mile, and two miles. Plot the radii against a series of demographic maps (e.g., age, income, gender, race) to define the populace of each circled area. Data from these analyses disclose the equity or disparity of distribution, thereby facilitating assessment of potential acute and chronic environmental health risks assumed by those in close proximity to hazardous facilities.

Federal response to this issue has been remarkably swift. President Clinton signed Executive Order 12898 that calls on federal agencies to develop strategies to deal with environmental justice issues. For example, information on emissions, violations, and permit types from EPA programs—superfund, hazardous and municipal waste, air, and water—are merged with demographic data to provide insight into the burden of environmental health risk distribution.

Advocates for low-income and minority rights believe that environmental injustice will remain an issue. The lack of political power and scarcity of minorities in key, decision-making, government positions are cited in support of that premise. Many believe the only route to environmental injustice is through litigation.

The legal profession is actively studying environmental justice, not as a single legal theory but as an amalgamation of environmental and civil rights laws. Environmental justice has been designated as a priority area of study and litigation by many legal associations (e.g., the National Bar Association, the National Lawyers Guild, and the Minority Environmental Lawyers Association). The New York State Bar Association, for example, has created a committee on environmental justice to address issues of race and economic status in relation to environmental burdens and benefits.

The EPA is reexamining the basis for its earlier decisions to cite hazardous facilities in Mississippi and Louisiana. Interestingly enough, the EPA's reexaminations on its decisions were instigated not under environmental laws but under the 1964 Civil Rights Act, which applies to all federal actions.

The legal community and advocacy groups are following the current EPA studies closely. The findings may reveal major avenues of litigation by which low-income and minority

Figure 18.3 A Right-to-Know sign posted at a golf course.

communities (that believe their civil rights have been disregarded) may pursue environmental justice through legal redress.

RIGHT-TO-KNOW ISSUES

The public is entitled to pesticide information under many state and federal right-to-know laws. The issue often is simply one of opening government records or making industry reports more accessible to the public (Fig. 18.3).

Inert Ingredients

Why Doesn't the EPA Require the Pesticide Registrant to Identify Inerts on Each Label? Pesticide products generally contain both active and inert ingredients. Active ingredients are the chemicals that manage the target pest, whereas inert ingredients (e.g., carriers and adjuvants) are not toxic to the *target pest* but are added to the pesticide formulation for other purposes. (*Note:* Some inert ingredients may be toxic to organisms *other than* the target pest.)

Federal law requires that the active ingredient(s) of every pesticide product be clearly stated on the label attached to the container and on the Material Safety Data Sheet. Additional information on active ingredients is accessible in public documents, books, and journals; and from these sources it is possible to construct a health and environmental profile on the active ingredient(s). However, since federal law does not mandate that *inert* ingredients be identified on pesticide labeling, it is impossible to research them; and without knowing what they are, it is impossible to determine the full toxic potential of the product.

It is this lack of full public disclosure that has been a point of contention—for decades—among pesticide manufacturers, the EPA, and public interest groups.

Why don't manufacturers end this debate by listing the inert ingredients on the label? Every pesticide formulation (active plus inert ingredients) is evaluated by the EPA during the registration process, so why are public interest groups so adamant about full disclosure and public protection?

Pesticide manufacturers' primary arguments for protecting trade secrets are that

- full disclosure enables others to "steal" what manufacturers have taken years to perfect, and that
- consumers surveyed by the EPA's Consumer Labeling Initiative Project indicated that information on inerts is of little importance.

Claims made by public interest groups arguing for product-specific disclosure include the following:

- The public is not protected because inerts are not evaluated for long-term effects.
- The general public has the right to know what inerts are contained in the products they use.
- Pesticide users are entitled to all information on the hazardous products they use. Without testing and disclosure of inert ingredients, workers are not fully informed.
- Total contents of household items such as toothpaste, shampoo, and cereal are fully disclosed, so why should pesticide manufacturers be exempt from listing inert ingredients?
- Inerts are not trade secrets because manufacturers can identify the ingredients in their competitor's products through reverse engineering.

Federal law mandates that manufacturers disclose all ingredients—active *and* inert—to the EPA during the product registration process. This information is obtained when the pesticide registrant completes a document known as a Confidential Statement of Formula, which fully discloses to the EPA the chemical name, trade name, Chemical Abstract Service number, and the amount and percentage by weight for each active and inert component formulated into the product. The Confidential Statement of Formula stipulates that federal law prohibits the EPA from making accessible to the public any information that the EPA administrator judges to contain trade secrets or confidential information.

The EPA's long-standing policy is to "reduce the potential for adverse effects from the use of pesticide products containing toxic inert ingredients," since some inert ingredients used in pesticide products may pose long-term risk to human health or the environment. The EPA implements its policy by developing a toxicological profile for each inert ingredient, which is then assigned to one of four lists.

List 1. Inerts of Toxicological Concern These include eight inerts that have been linked to chronic health effects or harm to the environment. These substances must be disclosed on the pesticide label or their use discontinued.

List 2. Potentially Toxic Inerts/High Priority for Testing Fifty-two inerts have been identified as requiring animal toxicity studies to determine the actual level of concern. The animal test results can then be used to move the inert from List 2 to List 1 or List 4.

List 3. Inerts of Unknown Toxicity List 3 inerts include approximately 1700 chemicals of unknown toxicity.

List 4. Inerts of Minimal Concern These 430 inerts are assigned to either a 4A List (inerts commonly consumed in foods) or a 4B List (sufficient information to indicate no adverse public health or environmental impact).

Lastly, the EPA now requires that inert ingredients that have not been used previously in an approved pesticide formulation undergo a series of tests to assess residue chemistry, toxicology, ecotoxicology, and environmental fate.

The EPA's procedures are designed to walk a fine line between disclosing confidential information and protecting the public. Nevertheless, these guidelines still do not get to the heart of the argument. That is, the public still does not know what inerts are in the product. In 1996, public interest groups challenged the EPA's interpretation of inerts as confidential information, and the U.S. District Court for the District of Columbia ruled that common names and chemical abstract numbers of inert ingredients are not automatically exempt from full public disclosure. One of the results of the case was the EPA's development of a searchable Web site that provides the chemical abstract number, chemical name, and inert category for approximately 2500 inerts used in pesticide formulations. Although this can be construed as full disclosure, some are not impressed because it doesn't tag specific inerts to specific products.

There appears to be a compromise in the offing. Some members of industry have said that consensus could be reached by listing the most commonly used inert ingredients on the label or by providing a way of describing them without revealing their exact identity (e.g., by disclosing EPA List 1, 2, 3, 4 on which the inert appears). It remains to be seen whether the ideas expressed by these few will be acceptable to the industry overall in protecting their investments and to public interest groups in assuring full disclosure.

Human Testing

Does the Use of Human Adult Volunteers for Exposure Studies Raise Ethical Questions? The 1996 Food Quality Protection Act (FQPA) established new standards for the EPA to follow in assessing pesticide risk. Prior to FQPA, the relative risk posed by an active ingredient was determined by separate consideration of dietary, residential, and occupational risk; so three risk determinations were possible. FQPA requires that dietary and residential risk be combined for one determination, so now only two risk assessments are possible.

FQPA also requires that exposure data on pesticides that share a common mechanism of toxicity be combined to derive a single exposure estimate. In other words, exposure to similar active ingredients (e.g., organophosphates) might be lumped together to determine potential exposure to multiple compounds that produce a common toxic effect.

An additional tenfold safety factor may be included to assure extra protection of children. However, FQPA may have contributed unintentionally to human subject participation in risk assessment. Several news organizations have reported that pesticide manufacturers have developed risk assessment information for FQPA through clinical trials using human volunteers. Reportedly, volunteers were paid \$780 in a four-day study and \$1500 dollars in a seventeen-day study. Pesticide doses were administered to human volunteers in a gelatin capsule or in orange juice.

Volunteers were observed by physicians before, during, and after the study to detect effects. Some were monitored to ascertain how the human body absorbs, metabolizes, or excretes the pesticide. It is hoped that data derived from human trials can be compared effectively with those from animal studies. Perhaps they might offer insight about whether the standard safety factor of 1000 (10 × for differences between animals and man, 10 × for differences in sensitivity among human adults, and 10 × for sensitivity in infants and children vs. adults) is warranted.

Arguments in Favor of Using Human Volunteers

- The use of multiple, conservative, worst-case assumptions has so exaggerated pesticide risk that the only recourse is to test pesticides on human volunteers.
- The dose given to humans is based on levels determined to be safe in animal studies.
- Humans will not replace laboratory animals in developing pesticide toxicity profiles.
- Human studies are conducted according to stringent ethical and medical guidelines.
- Subjects volunteer after full disclosure of the possibility of being given a pesticide and of potential associated risks.

Arguments against Using Human Subjects

- There are no standardized protocols for human studies.
- Adult volunteers focus on monetary compensation rather than on personal risk.
- Some companies have misled volunteers by referring to pesticides as drugs.
- Information derived from adult volunteer studies is irrelevant in determining risk to children.
- Children are not little adults and the extra margin of safety may still be needed.

This debate surely will continue as organizations whose interests include the relevance of pesticides to food safety, consumer protection, public policy, and child safety issues are drawn into the discussion. There is little doubt that paying people for their participation in clinical studies will raise ethical, scientific, and medical questions just as it does in testing experimental drugs or new surgical procedures on paid volunteers.

Notification by Utilities

Should Utility Companies Give Notice and Options about the Use of Pesticides to Maintain Rights-of-Way? Public utilities are expected to provide constant, reliable, and affordable sources of electric power. Electric utilities have long been given a right-of-access to run their distribution lines across private property. Being granted this privilege, they also assume the management and upkeep of their easement properties by employing multiple methods to manage obstructing vegetation through herbicide application, mowing, and cutting (Fig. 18.4).

Tall vegetation (trees, shrubs, etc.) impedes utility workers in repairing and replacing utility poles, lines, and transformers. Trees and shrubs allowed to grow into power lines can interrupt customer service, especially during storms. They also can serve as attractive nuisances, luring children into close proximity to electric lines. Utility companies use herbicides to limit the growth of these rights-of-way obstacles.

Herbicides, mowers, and chain saws are used to clear trees and other obstacles from utility rights-of-way. Utility companies generally favor assumption of the minimal risk that herbicide application presents over the hazards imposed by mowing and cutting. That is, their employees are less likely to become injured during herbicide application than when mowing or cutting.

However, some disagree that herbicide application is the best method of clearing rights-of-way, citing potential effects on people, wildlife, and the environment. Questions are being asked: Should property owners have a say in whether or not herbicides are used on

Figure 18.4 Right-of-way areas can encompass diverse terrain and the property of many different owners.

the portions of their property that serve as easements for utilities? Do they have a right to know, ahead of time, when utility companies plan to apply herbicides?

The right to know and the opportunity to participate in vegetation management decisions have led to an unusual agreement among consumers, environmental groups, and electric utility companies in North Carolina. Four investor-owned utilities have agreed to include easement property owners and leaseholders in decisions on how to manage easement vegetation. The five-year agreement includes the following:

Customer Notification Utility companies have agreed to inform all of their customers of the methods that will be used to manage vegetation on easement properties. They most likely will notify consumers via inserts with their monthly billing. If herbicides are to be used, the information will include when and how they will be applied, the product name and EPA registration number for each chemical that will be used, and the telephone number of a utility company contact person.

Each notice also will include this statement: "We have found that many of our customers are pleased with this herbicide program. However, for organic farmers and others with concerns, we can provide other options. To discuss these options, contact (name of contact person)."

Internet Accessibility Each utility company will develop an Internet home page that will display the same information as their monthly billing insert. In addition, the Web site will be linked to pesticide labels and Material Safety Data Sheets for products they use.

Right of Herbicide Refusal It is recognized that some landowners and tenants prefer that utilities not use herbicides on their easement properties. Although it is the customer's responsibility to request that utilities not use herbicides, the utilities have agreed to honor such requests by mowing and cutting in lieu of spraying. Those choosing the no-spay option are required to post "No Spray" signs on their easements.

SUMMARY

Public interest and involvement is critical to the decision-making processes; but, unfortunately, the modern ease of communication saturates us with diverse information from varied, sometimes uncertain, sources.

Advances in science and the inability to provide definitive, black and white answers to risk questions ensure that the resolution of pesticide issues will become more complex. There are always questions to answer, and future resolution of today's issues will generate new questions.

The challenges of answering new risk questions, making appropriate regulatory decisions, and communicating effectively with the public can be overwhelming. However, when regulators, educators, and environmental and industry groups work together constructively to resolve issues, these challenges can be transformed into exciting opportunities—to protect public and environmental health and to ensure maximum benefits with minimum risk from pesticide use.

CHAPTER 19

CONCLUSION

An effective pesticide policy requires public confidence in the regulatory, scientific, and business communities. Such trust can be gained only through open, balanced, fact-based knowledge and effective communication of pesticide safety evaluations. Research, education, and cooperation are the critical elements of pesticide risk management.

The public and scientists are constantly exposed to a steady barrage of pesticide safety issues: endocrine disruptors, multiple chemical sensitivity, bioengineered foods, reproductive problems, cancer, and others. They ask, "Are pesticides safe?" Experts and a diverse public agree that no pesticide is risk-free—and certainly no pesticide can be described as absolutely safe.

Citizens pose the question, Why are pesticides used? and they want assurance that pesticides are not contributing to health problems. The public clearly expects sound science and empirical data to support government decisions on pesticides.

Countless "what-if" questions on the impact of pesticides drive legislators, regulatory agencies, producers, pest management industries, the media, and public interest groups. The pesticide debate subsides and intensifies in cadence with controversial issues addressed in the public forum. Discourse on some issues spans years as opposing points of view shape public opinion and regulatory policy. The issues take knowledge to the fringes of contemporary science. Government agencies must adjust their registration requirements to keep pace with scientific capabilities. Pesticide manufacturers must adapt their research to reflect the latest, most sophisticated science available, thereby lending credence to the data collected supporting registration.

Good regulatory decisions depend on documented scientific research, an understanding of the strengths and weaknesses of the science, and sound professional judgment in drawing conclusions from compiled data. Risk assessments should clearly identify pertinent facts *and* any assumptions deemed necessary to accurately evaluate a given pesticide. If risk assessments are clear, concise, and thorough, they add a vitally important dimension to the EPA's decision-making process. Clarity and openness in the risk assessment process permits informed debate on pesticide use. Ultimately, the registration and use of a pesticide must withstand scientific inquiry, public scrutiny, and legal review.

Through three decades of public policy decisions, the EPA has directed pesticide manufacturers concerning the types of data required and the necessary steps for pesticide registration. The EPA's message to the pesticide manufacturer is clear: provide beneficial products with minimum risks. This policy exerts a strong influence on the kinds of pesticides being developed; that is, the shift is toward the development of effective, reduced-risk pesticides.

Society at large—manufacturers, governments, professionals, consumers, educators, the media, environmental advocates, and the public—share responsibility for understanding and reducing pesticide risks to human health and the environment. State and federal agencies must continue to require comprehensive testing of all pesticides and require new tests, when warranted, to answer emerging questions. Government must balance its oversight of pesticide risk with the benefits of modern pesticide technology. It is imperative to guard against unacceptable risk to the public and the environment while simultaneously recognizing the benefits of prudent pest management. All pesticide users—homeowners as well as private and commercial applicators—shoulder some of the responsibility for safety, and precautions stated on pesticide labels must be heeded. Industry and academia must take the lead in enhancing educational methods for delivering pesticide-related information. The media should promote accurate, balanced reporting of pesticide issues.

At the beginning of this book, we asked the questions, Are pesticides safe? and Do they have benefits? We have answered yes to both questions, but only in the context of a pesticide paradigm which balances risks with benefits and integrates science with societal needs. We are no closer to "absolute truth" because, as we have shown, the answers are not that definitive. Our challenge as regulators, manufacturers, educators, and users is to understand how the public may react politically, psychologically, socially, and economically. We all share a responsibility to effectively balance diverse and sometimes contradictory information in communicating understanding of pesticide benefits and risks. If we communicate these truths, the public will benefit.

Society should expect maximum benefit and minimum risk from the use of pesticides. As the preventive philosophy of current regulatory policy merges with new technologies, the development of low-risk, effective pest management strategies will enhance protection of human health and the environment. But society must take responsibility for basing decisions and opinions on sound *science,* rather than sound *bites*. The public needs to look at both sides of an issue before concluding that one side is right and the other is wrong.

We intend this book to promote a better understanding of pesticide issues and the science and policy behind pesticide management. We hope that it provides useful tools for effectively communicating to a diverse public the benefits and risks of pesticides and for forming opinions concerning pesticide use.

INDEX